Recent Advances in the Modeling of Hydrologic Systems

NATO ASI Series

Advanced Science Institutes Series

A Series presenting the results of activities sponsored by the NATO Science Committee, which aims at the dissemination of advanced scientific and technological knowledge, with a view to strengthening links between scientific communities.

The Series is published by an international board of publishers in conjunction with the NATO Scientific Affairs Division

A	**Life Sciences**	Plenum Publishing Corporation
B	**Physics**	London and New York
C	**Mathematical**	Kluwer Academic Publishers
	and Physical Sciences	Dordrecht, Boston and London
D	**Behavioural and Social Sciences**	
E	**Applied Sciences**	
F	**Computer and Systems Sciences**	Springer-Verlag
G	**Ecological Sciences**	Berlin, Heidelberg, New York, London,
H	**Cell Biology**	Paris and Tokyo
I	**Global Environmental Change**	

NATO-PCO-DATA BASE

The electronic index to the NATO ASI Series provides full bibliographical references (with keywords and/or abstracts) to more than 30000 contributions from international scientists published in all sections of the NATO ASI Series.
Access to the NATO-PCO-DATA BASE is possible in two ways:

– via online FILE 128 (NATO-PCO-DATA BASE) hosted by ESRIN,
Via Galileo Galilei, I-00044 Frascati, Italy.

– via CD-ROM "NATO-PCO-DATA BASE" with user-friendly retrieval software in English, French and German (© WTV GmbH and DATAWARE Technologies Inc. 1989).

The CD-ROM can be ordered through any member of the Board of Publishers or through NATO-PCO, Overijse, Belgium.

Series C: Mathematical and Physical Sciences - Vol. 345

Recent Advances in the Modeling of Hydrologic Systems

edited by

David S. Bowles
Utah Water Research Laboratory,
Utah State University,
Logan, Utah, U.S.A.

and

P. Enda O'Connell
Department of Civil Engineering,
University of Newcastle upon Tyne,
Newcastle upon Tyne, U.K.

Springer Science+Business Media, B.V.

Proceedings of the NATO Advanced Study Institute on
Recent Advances in the Modeling of Hydrologic Systems
Sintra, Portugal
July 10–23, 1988

ISBN 978-94-010-5538-3   ISBN 978-94-011-3480-4 (eBook)
DOI 10.1007/978-94-011-3480-4

Printed on acid-free paper

This book contains the proceedings of a NATO Advanced Research Workshop held within the programme of activities of the NATO Special Programme on Global Transport Mechanisms in the Geo-Sciences running from 1983 to 1988 as part of the activities of the NATO Science Committee.

Other books previously published as a result of the activities of the Special Programme are:

BUAT-MENARD, P. (Ed.) – *The Role of Air-Sea Exchange in Geochemical Cycling* (C185) 1986

CAZENAVE, A. (Ed.) – *Earth Rotation: Solved and Unsolved Problems* (C187) 1986

WILLEBRAND, J. and ANDERSON, D.L.T. (Eds.) – *Large-Scale Transport Processes in Oceans and Atmosphere* (C190) 1986

NICOLIS, C. and NICOLIS, G. (Eds.) – *Irreversible Phenomena and Dynamical Systems Analysis in Geosciences* (C192) 1986

PARSONS, I. (Ed.) – *Origins of Igneous Layering* (C196) 1987

LOPER, E. (Ed.) – *Structure and Dynamics of Partially Solidified Systems* (E125) 1987

VAUGHAN, R. A. (Ed.) – *Remote Sensing Applications in Meteorology and Climatology* (C201) 1987

BERGER, W. H. and LABEYRIE, L. D. (Eds.) – *Abrupt Climatic Change – Evidence and Implications* (C216) 1987

VISCONTI, G. and GARCIA, R. (Eds.) – *Transport Processes in the Middle Atmosphere* (C213) 1987

HELGESON, H. C. (Ed.) – *Chemical Transport in Metasomatic Processes* (C218) 1987

SIMMERS, I. (Ed.) – *Estimation of Natural Recharge of Groundwater* (C222) 1987

CUSTODIO, E., GURGUI, A. and LOBO FERREIRA, J. P. (Eds.) – *Groundwater Flow and Quality Modelling* (C224) 1987

ISAKSEN, I. S. A. (Ed.) – *Tropospheric Ozone* (C227) 1988

SCHLESINGER, M.E. (Ed.) – *Physically-Based Modelling and Simulation of Climate and Climatic Change* 2 vols. (C243) 1988

UNSWORTH, M. H. and FOWLER, D. (Eds.) – *Acid Deposition at High Elevation Sites* (C252) 1988

KISSEL, C. and LAY, C. (Eds.) – *Paleomagnetic Rotations and Continental Deformation* (C254) 1988

HART, S. R. and GULEN, L. (Eds.) – *Crust/Mantle Recycling at Subduction Zones* (C258) 1989

GREGERSEN, S. and BASHAM, P. (Eds.) – *Earthquakes at North-Atlantic Passive Margins: Neotectonics and Postglacial Rebound* (C266) 1989

MOREL-SEYTOUX, H. J. (Ed.) – *Unsaturated Flow in Hydrologic Modeling* (C275) 1989

BRIDGWATER, D. (Ed.) – *Fluid Movements – Element Transport and the Composition of the Crust* (C281) 1989

LEINEN, M. and SARNTHEIN, M. (Eds.) – *Paleoclimatology and Paleometeorology: Modern and Past Patterns of Global Atmospheric Transport* (C282) 1989

ANDERSON, D.L.T. and WILLEBRAND, J. (Eds.) – *Ocean Circulation Models: Combining Data and Dynamics* (C284) 1989

BERGER, A., SCHNEIDER, S. and DUPLESSY, J. Cl. (Eds.) – *Climate and Geo-Sciences* (C285) 1989

KNAP, A.H. (Ed.) – *The Long-Range Atmospheric Transport of Natural and Contaminant Substances from Continent to Ocean and Continent to Continent* (C297) 1990

BLEIL, U. and THIEDE, J. (Eds.) – *Geological History of the Polar Oceans – Arctic Versus Antarctic* (C308) 1990

SHEN, H.W. (Ed.) – *Movable Bed Physical Models* (C312) 1990

GIESE, P. (Ed.) – *Improvement of Joint Interpretation of Geophysical and Geological Data* (C338) 1991

Preface

Modeling of the rainfall–runoff process is of both scientific and practical significance. Many of the currently used mathematical models of hydrologic systems were developed a generation ago. Much of the effort since then has focused on refining these models rather than on developing new models based on improved scientific understanding. In the past few years, however, a renewed effort has been made to improve both our fundamental understanding of hydrologic processes and to exploit technological advances in computing and remote sensing. It is against this background that the NATO Advanced Study Institute on Recent Advances in the Modeling of Hydrologic Systems was organized.

The idea for holding a NATO ASI on this topic grew out of an informal discussion between one of the co–directors and Professor Francisco Nunes–Correia at a previous NATO ASI held at Tucson, Arizona in 1985. The Special Program Panel on Global Transport Mechanisms in the Geo–Sciences of the NATO Scientific Affairs Division agreed to sponsor the ASI and an organizing committee was formed. The committee comprised the co–directors, Professor David S. Bowles (U.S.A.) and Professor P. Enda O'Connell (U.K.), and Professor Francisco Nunes–Correia (Portugal), Dr. Donn G. DeCoursey (U.S.A.), and Professor Ezio Todini (Italy).

The overall objective of the ASI was to present a systematic review of the current state–of–the–art in comprehensive hydrologic modeling, current modeling issues and technological trends which are influencing advances in the field, and potential future directions. The ASI was held in Sintra, Portugal from July 10–23, 1988. It was attended by one hundred hydrologists from every NATO country, except Luxembourg, and from five non–NATO countries.

This book contains twenty–nine chapters based on invited lectures and several contributed papers which were presented at the ASI. An introductory part provides an historical overview and a perspective on the need for developing a more scientific basis for hydrology. The second and third parts present state–of–the–art reviews of modeling individual and linked components of the hydrologic cycle, respectively. Current modeling issues and technological trends which are influencing the development of a new generation of hydrologic models are addressed in Parts IV and V. In Part VI the synthesis of hydrologic models for different physical settings and various types of applications are discussed. The last part of the book contains two chapters which explore various aspects of the relationship between experimental studies and hydrologic modeling. An appendix contains abstracts of other presentations which were made at the ASI, but which could not be included in this book.

In addition to NATO funding, support for the ASI was provided by the Directorate General for Natural Resources in Portugal, the Utah Water Research Laboratory at Utah State University, the Department of Civil Engineering at the University of Newcastle–Upon–Tyne, and the National Science Foundation. Significant additional funding was provided by the Utah Water Research Laboratory for the preparation of this book.

We wish to thank each of the authors of chapters in this book for their contributions. Each is a recognized leader in his or her field. We are very grateful to them for taking the time from their very busy schedules to prepare their manuscripts and participate in the ASI. Also we express our appreciation to Colleen A. Riley, Jane Sorenson, Paul Jarvis, Jan Cochley, Chris Gunter, and Dr. Zhida Song at the Utah Water Research Laboratory for their work in preparing, or supporting the preparation of, this manuscript. Jane Sorenson's marathon efforts are worthy of particular mention—thanks Jane for a job well done! Thanks also to a very patient publisher and in particular, Mrs. Nel de Boer, of Kluwer Academic Publishers.

It is our sincere hope that this book will be as valuable to the reader as the ASI was to those of us who were able to attend it.

David S. Bowles
P. Enda O'Connell

List of Contributors

M. B. Abbott
International Institute for Hydraulic
  and Environmental Engineering
Oude Delft 95
2601 DA Delft, The Netherlands

James C. Bathurst
University of Newcastle upon Tyne
Department of Civil Engineering
Natural Environment Research Council
Water Resource Systems Research Unit
Newcastle upon Tyne, NE1 7RU, United Kingdom

Keith Beven
University of Lancaster
Centre for Research on Environmental Systems
Institute of Environmental and Biological Sciences
Lancaster, LA1 4YQ, United Kingdom

Wilfried Brutsaert
Cornell University
School of Civil and Environmental Engineering
Hollister Hall
Ithaca, New York 14853-3501 U.S.A.

David R. Dawdy
3055 23rd Avenue
San Francisco, California 94132 U.S.A.

Donn G. DeCoursey
USDA-ARS
Hydro-Ecosystem Research Unit
243 Federal Building
301 South Howes, P.O. Box E
Fort Collins, Colorado 80522 U.S.A.

Edwin T. Engman
NASA Goddard Space Flight Center
Hydrologic Science Branch
Code 974
Greenbelt, Maryland 20771 U.S.A.

Efi Foufoula–Georgiou
University of Minnesota
Department of Civil and Mineral Engineering
St. Anthony Falls Hydraulic Laboratory
Minneapolis, Minnesota 55414-2196 U.S.A.

Konstantine P. Georgakakos
The University of Iowa
Department of Civil and Environmental Engineering
Iowa Institute of Hydraulic Research
Iowa City, Iowa 52242 U.S.A.

L. Douglas James
Utah State University
Utah Water Research Laboratory
Logan, Utah 84322-8200 U.S.A.

James T. McCord
Sandia National Laboratory
Waste Management Systems Division
P.O. Box 5800
Albuquerque, New Mexico 87185-5800 U.S.A.

E. M. Morris
British Antarctic Survey
Ice and Climate Division
High Cross, Madingley Road
Cambridge, CB3 0ET, United Kingdom

Francisco Nunes Correia
National Laboratory of Civil Engineering
Av. Brasil, 101
1799 Lisboa Codex, Portugal

P. E. O'Connell
University of Newcastle upon Tyne
Department of Civil Engineering
Claremont Road
Newcastle upon Tyne, NE1 7RU, United Kingdom

Soroosh Sorooshian
University of Arizona
Department of Hydrology and Water Resources
Tucson, Arizona 85721 U.S.A.

Børge Storm
Danish Hydraulic Institute
Agern Allé 5, Dk-2970
Hørsholm, Denmark

Ezio Todini
E. Todini & Partners, S. r. 1.
Via Zanardi, 16
40131 Bologna, Italy

Bertel Vehviläinen
National Board of Waters and the Environment
Water and Environment Research Institute
P.O. Box 436
00101 Helsinki, Finland

Table of Contents

Part I – Introduction

Chapter 1

A Historical Perspective

P. E. O'Connell
University of Newcastle upon Tyne
Department of Civil Engineering
Claremont Road
Newcastle upon Tyne, NE1 7RU, United Kingdom

Abstract. The evolution of hydrological system modeling is reviewed under two headings: descriptive modeling which aims to enhance our understanding of catchment system behavior, and prescriptive modeling which is directed towards the solution of engineering problems. Progress in descriptive physically based modeling has been initially slow but some catchment–scale models and modeling systems are now emerging. Although a large amount of effort has been invested in prescriptive modeling, this research has not followed any well defined development path, and there has been an excessive proliferation of models without achieving any real advances in predictive capacity. There has also been an excessive preoccupation with the pursuit of systems methodology as an end in itself. However, "technique–driven" research has now started to give way to a desire to gain a more fundamental understanding of causal mechanisms within the rainfall–runoff process. This new fundamentalist trend in hydrological system modeling holds exciting prospects for the future.

1. Introduction

The relationship between rainfall and runoff has been one of the central themes of hydrological research for many years. With the growth of digital computing power in the fifties and sixties, a tremendous upsurge in hydrological modeling activity took place. Since then, the science, and sometimes art, of modeling has developed in response to the research community's perceived opportunities for advancing the state of knowledge in the subject, and the engineering community's need for predictive hydrological tools. The historical perspective presented here will be primarily a research one, but with due consideration of the impact of real world needs. The chapter will concentrate on identifying major trends, pathways followed, and notable milestones, rather than presenting a historical catalogue of the plethora of models which has been developed throughout the world over the past thirty years.

In assessing progress and achievements in mathematical modeling to date, two standpoints can be taken:

(1) What are the contributions which have been made to our scientific understanding of catchment response?

(2) How has our capacity to solve engineering problems been advanced through mathematical modeling?

The scientific standpoint (i) is essentially concerned with why catchments behave as they do (WHY – driven research); put very simply:

"What happens to the rain?" (Penman, 1961)

3

D. S. Bowles and P. E. O'Connell (eds.), Recent Advances in the Modeling of Hydrologic Systems, 3–30.
© 1991 *Kluwer Academic Publishers.*

For the purposes of this review, modeling which is essentially concerned with this question will be referred to as "descriptive modeling." The engineering standpoint is concerned with how to make engineering predictions (HOW – driven research). In an ideal world, engineering predictions should be underpinned by scientific understanding; in practice, expediency demands that predictions of an acceptable accuracy be generated by whatever means available. This need can in principle be fulfilled by drawing on modeling techniques from the domains of statistics and systems engineering of which there is always more than an adequate supply. This approach may be scientifically acceptable provided that the assumptions underlying the modeling technique are consistent with hydrological reality and that predictions are not made outside of the observed range of system behavior. However, it is not easy to guarantee that these conditions are fulfilled. Modeling which is directly concerned with how to make engineering predictions will be referred to as "prescriptive modeling."

2. Discussion

2.1. The Evolution of Prescriptive Modeling

2.1.1. The Systems Engineering Approach to Modeling

The origins of this particular branch of modeling are found in the desire to solve engineering problems, notably reservoir spillway design, land drainage system design, and urban sewer system design (Todini, 1988). The rational method developed by Mulvaney (1851) represented the first formal relationship to emerge for predicting design (peak) discharge from rainfall. This method, which was derived for small or mountainous catchments, was based on the concept of time of concentration: the peak discharge caused by a given rainfall intensity occurs when the rainfall duration equals or exceeds the time of concentration. The rational method has stood the test of time well; it was introduced into storm sewer design by Lloyd–Davies (1906) and is still in use, in a modified form, in storm sewer system design at the present time (Hydraulics Research Ltd., 1981).

In the early part of this century, efforts were made to modify the rational method to account for the nonuniform distribution, in space and time, of rainfall and catchment characteristics. This was achieved by introducing the concept of isochrones or lines of equal travel time, which led to the concept of the time–area diagram and, its mathematical derivative, the time area concentration curve. This modified rational method constituted the first model capable of predicting the time–dependant response of a catchment to rainfall; the shape and parameters of the response or transfer function could be derived from topographic maps and the use of Manning's formula to evaluate the different travel times. However, when originally formulated, a graphical formulation of the method was used to predict peak discharge only.

Sherman (1932) introduced the concept of the unit hydrograph based on the use of the superposition principle, later to be shown, in systems theory terminology, to be equivalent to the assumption of a time invariant linear system. The introduction of the unit hydrograph was to mark the beginning of a notable era in rainfall runoff modeling. A tool had now emerged which could be used to predict the runoff hydrograph, albeit with a number of difficulties:
  · The separation of surface runoff from base flow.
  · The calculation of effective rainfall.
  · The derivation of the unit hydrograph.

Dooge (1973) discusses some of the earlier efforts made to overcome these difficulties. He also discusses the further development of the time–area method, notably the contributions of Zoch (1934, 1936, 1937) and Clark (1945) who, in order to account for storage effects, suggested that the time area curve might be routed through an element of storage to obtain the response of the catchment. The parameters of these responses could then be related to catchment characteristics leading to the concept of the synthetic unit hydrograph which could be used to predict the response of an ungauged catchment. Of considerable practical interest was the contribution of O'Kelly (1955) and his co-workers who observed that no appreciable loss of accuracy occurred if the routed time–area diagram were replaced by a routed isosceles triangle.

The unit hydrograph had by this time an established position in the field of engineering hydrology, even if the propositions on which it was based (linearity and time invariance) were recognized by some to be dubious:

"All these prepositions are empirical. It is not possible to prove them mathematically. In fact, it is a rather simple matter to demonstrate by rational hydraulic analysis that not a single one of them is mathematically accurate. Fortunately, nature is not aware of this." (Johnstone and Cross, 1949)

Indeed, these prepositions opened the door to a surge of research activity in the late 1950's and the 1960's when the research hydrologists of the day recognized the relevance of systems engineering and statistical techniques to the study of the unit hydrograph. The impulse response function of a linear time invariant system was seen to equate with the concept of an instantaneous unit hydrograph (IUH) which in turn could be interpreted in terms of assemblages of linear reservoirs and linear channels. These assemblages came to be known as "conceptual models": the best known of these is the "Nash cascade" of n linear reservoirs with equal time constants K (Nash, 1958). An elegant and unifying treatment of unit hydrograph theory was presented by Dooge (1959).

From a practical standpoint, a conceptual model had the attraction that its IUH could be expressed in terms of few parameters which could, in principle, be related to catchment characteristics. Nash (1960), in an extensive study of British catchments, demonstrated how this could be done by first finding relationships between the statistical moments of the input data, the output data, and the parameters of the IUH, and then using a regression approach to relate the latter to catchment characteristics.

With the more widespread availability of digital computing power in the early 1960's, a new strand of unit hydrograph research developed which came to be known as "black–box" (or input–output) modeling of the IUH; this involved the direct inference of the unit hydrograph from observed input (effective rainfall) and output (surface runoff) data (the inverse problem in systems engineering terminology). In principle this gave greater freedom in determining the shape of the unit hydrograph since no a priori assumption was made about its shape, in contrast to conceptual models. However, all the problems associated with estimating the impulse response from noisy data were subsequently encountered. O'Donnell (1960) proposed the use of harmonic analysis as a solution to the problem, with the attendant difficulty of deciding on the level of truncation to be adopted. Snyder (1961) and Nash (1961) proposed a least–squares matrix inversion technique for estimating the T–hour unit hydrograph; however, physically unrealistic unit hydrographs with negative ordinates could result from noisy data.

Following these earlier seminal papers, research on what had now become generally known as the systems approach in hydrology intensified. However, a law of diminishing returns was associated with much of this effort, particularly that concerned with the development of new variations on existing conceptual models, since neither scientific understanding nor predictive capability were being advanced. From a practical standpoint, the inverse problem deserved some further attention and received it. Eagleson et al. (1966) formulated the inverse problem in terms of the Wiener-Hopf equations, and then adopted a linear programming approach to solving them in which non–negative constraints were imposed on the ordinates of the unit hydrograph. Dooge (1965) explored the use of Laguerre functions, and later their discrete time equivalents, Meixner functions, in unit hydrograph estimation (Dooge and Garvey, 1978). Natale and Todini's (1977) constrained linear system approach had both theoretical and practical appeal; non–negative constraints could be imposed on the unit hydrograph and the solution of the resulting equations involved a relatively straightforward computational approach based on quadratic programming. Moreover, Natale and Todini introduced a multiple–input, single–output system formulation in which both rainfall and upstream flow inputs could be incorporated, with the latter subject to continuity constraints. Further work on the inverse problem has been carried out by Bree (1985) and Bruen (1985).

A new dimension to input–output analysis had been added by Amorocho and Orlob (1961) through the introduction of nonlinear system theory; however, this was addressed to the non-linearity in the effective rainfall–storm runoff relationship only, not to the overall nonlinearity in the response of a catchment to rainfall. Later, Amorocho and Brandstetter (1971) developed a general nonlinear, black–box inversion technique; however, it was not apparent that, in the face of noisy data, the additional complexity would generate proportional returns in predictive ability.

With the emergence of several different techniques for solving the inverse problem, a considerable amount of effort has been devoted to identifying the "best" technique. Two types of experiments have been used to do this:

(1) Controlled Monte Carlo experiments where noise with specified properties is introduced into the input and/or output data and the effect on impulse response estimation explored.

(2) Experiments with real world data in which fitting is carried out on a proportion of the data and an independent test of predictive performance is carried out on the remaining data.

Both of the above approaches have been demonstrated by Natale and Todini (1977) and Bruen (1985), amongst others. If there is a general conclusion to be drawn from this work, it is that if a priori information (e.g. constraints) is introduced into the estimation problem, then improved predictive performance can be achieved at the expense of a slightly worse fit. This conclusion also extends to conceptual models, where physically realistic shapes for the impulse response function are imposed.

A new flurry of algebraic activity, primarily in the domain of conceptual modeling, resulted from the publication of Box and Jenkin's (1970) text on time series analysis, forecasting, and control. Much of this work was concerned with the equivalencing of the autoregressive moving average formulation of a linear system with the earlier more traditional formulations (Spolia and Chander, 1974). A rigorous mathematical treatment of this subject is given by O'Connor (1976). However, as Todini (1988) points out, these developments, although satisfactory from a mathematical and philosophical point of view, lost more and more of their connection with the hydrological problems of the real world, and became more and more concerned with the mathematical generality of their approaches.

Stimulated by technological developments in data sensing and transmission techniques, and in associated computer hardware and software, hydrologists turned their attention increasingly to the opportunities for employing rainfall runoff models as real–time forecasting tools. Again, the systems engineering literature appeared to have much relevant methodology to offer, and the "state–space" approach emerged as a further means of formulating linear models of hydrological systems. This and the Kalman filter, a tool for recursively deriving a minimum variance estimate of the "state," are discussed in Section 2.1.3, paragraph heading--Kalman Filtering.

2.1.2. The ESMA Approach To Modeling

Origins. The acronym Explicit Soil Moisture Accounting (ESMA) has been coined by Todini and Wallis (1977) to describe a generation of rainfall runoff models which emerged in the early 1960's. The term "conceptual model" is also frequently applied to such models; indeed, it may be argued that, to the extent that all models embody concepts of one kind or another, the term "conceptual" is not particularly informative. Linsley (1982), in an overview of rainfall runoff models, traces the origins of these models to various studies of the component processes of the hydrological cycle and of flood routing (e.g. Horton (1933) on infiltration; McCarthy (1938) on Muskingum routing). Linsley and Ackermann (1942) described trials of an elementary soil moisture accounting procedure which used measured evaporation and a simple infiltration process to calculate daily runoff values; this was to be the basic technique adopted in building ESMA models. Further development of this approach was not possible until the necessary digital computing power became available.

The Stanford Watershed Model (SWM) I was unveiled in 1960; this was a relatively simple model using daily rainfall, a simple infiltration function, and a unit hydrograph and recession function to produce the mean daily flow hydrograph. In line with the underlying philosophy of modeling total catchment response rather than storm runoff, the SWM I underwent further development (Crawford and Linsley, 1962, 1966) to emerge as the SWM 4 (Figure 1). In this model, efforts were made to represent the behavior of the various component processes within the hydrological cycle through storage and routing processes; quasi–physical functional relationships were used to describe the movement of moisture between these storages, which was later to earn this and similar models the somewhat irreverent title of "pots and pipes" models. An overall mass balance between inputs, outputs, and storage changes was maintained at all times. Fleming (1975) lists the data needed to simulate daily streamflow, together with the 34 "physically based" parameters involved if snowmelt is to be included; this reduces to 25 if snowmelt is omitted. Fleming (1975) indicated that only four parameters needed to be derived through a calibration procedure. Here the term "physically–based" cannot be interpreted in a strict sense since most of the parameters involved did not have a sound basis in physics; in the context of the SWM Model, the term was apparently used in relation to parameters which could be "evaluated from maps, surveys, or existing hydrometeorological records" (Fleming, 1975).

The advent of the SWM gave rise to a proliferation of similar ESMA models of varying degrees of complexity; Fleming (1975) gives details of 19 of these models. For a research hydrologist, there was the attraction of building a model with his/her "customized" representation of the manner in which rainfall was transformed into streamflow, thus earning it a separate identity from all previous models. However, since there are only a relatively small number of component processes within the hydrological cycle, and an even smaller number of hydrological laws with a basis in physics, each successive model did not represent a better way of representing these processes. Rather, the plurality of representations of different processes underlined the non–uniqueness of these representations in simulating streamflow.

The Advent of Optimization. The problem of matching model output to observed streamflow was seen by some research hydrologists as requiring special attention, particularly since model calibration was at that time a somewhat subjective process carried out through the interpretation of the results of successive model runs, a process best carried out by the developer of the model. Dawdy and O'Donnell (1965) proposed that this might be done through an optimization procedure whereby a function:

$$F^2 = \sum \left(Q_i - \hat{Q}_i \right)^2 \tag{1}$$

8

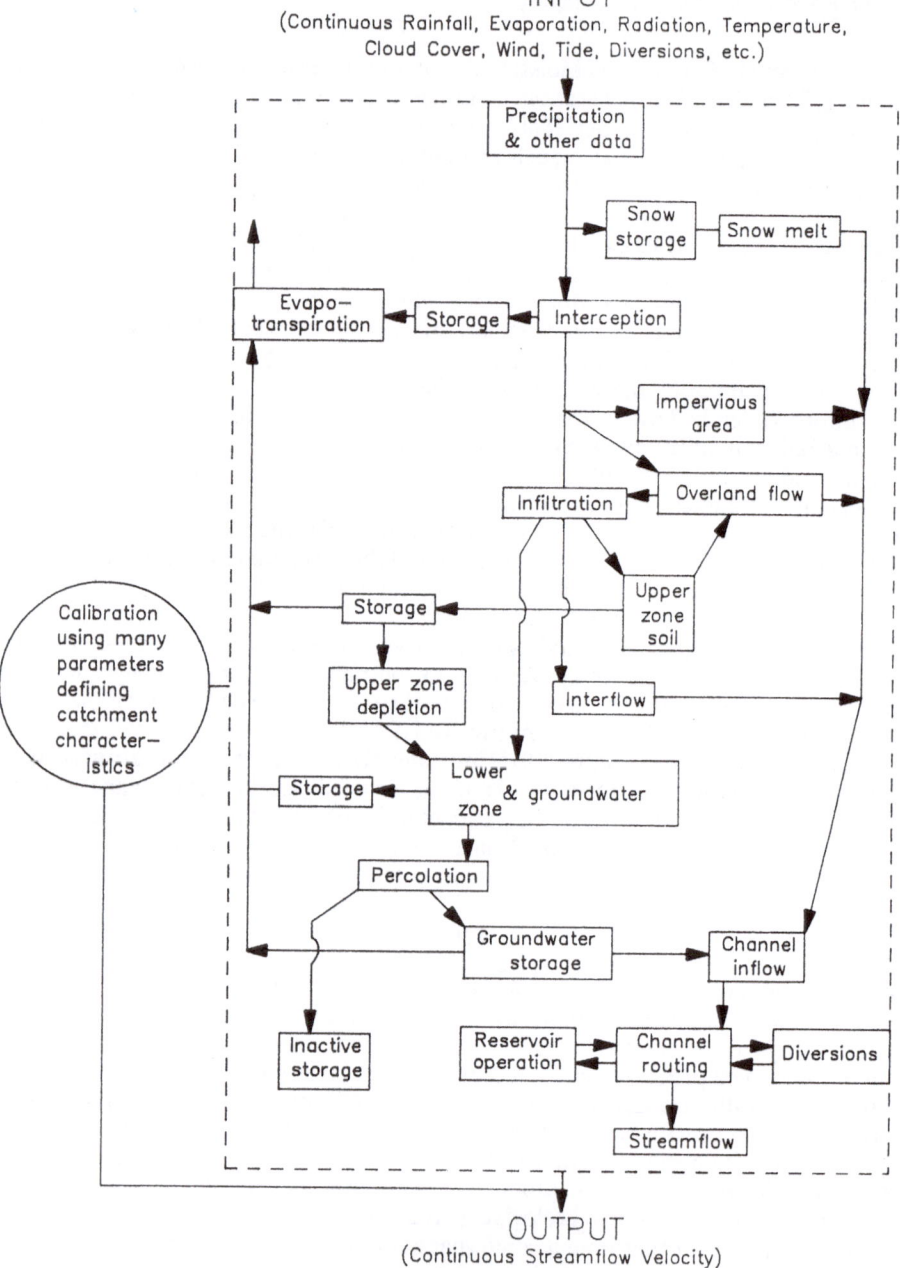

INPUT
(Continuous Rainfall, Evaporation, Radiation, Temperature,
Cloud Cover, Wind, Tide, Diversions, etc.)

Precipitation & other data

Snow storage — Snow melt

Evapo-transpiration — Storage — Interception

Impervious area

Infiltration — Overland flow

Upper zone soil

Calibration using many parameters defining catchment character- istics

Storage

Upper zone depletion

Interflow

Storage — Lower zone & groundwater

Percolation

Groundwater storage — Channel inflow

Inactive storage

Reservoir operation — Channel routing — Diversions

Streamflow

OUTPUT
(Continuous Streamflow Velocity)
Stage

Figure 1. Schematic of the Stanford Watershed Model IV (after Fleming, 1975).

is minimized through the successive automatic adjustment of the model parameters; here Q_i and $\hat{Q}_i$ represent observed and simulated streamflow, respectively. The optimization procedure employed by Dawdy and O'Donnell was the Rosenbrock (1960) method which is a "direct search" procedure that does not require the evaluation of the derivatives of F^2. They employed a relatively simple, nine parameter model (in comparison with the SWM model) to facilitate an assessment of the performance of the optimization approach. This was carried out by generating, for a specific record of model input, a "synthetic" output record corresponding to a particular set of model parameter values. By initiating the parameter search away from the known optimum, the ability of the optimization technique to recover the optimum could be assessed. However, the nature of the response surface was found to present problems, resulting in premature convergence away from the optimum; in particular, it was found that the error function F^2 was insensitive to changes in some parameter values. This implied that, for some areas of the parameter space, the associated model components were redundant, at least for the particular data record in question. Dawdy and O'Donnell (1965) suggested that parameter sensitivity might be assessed by perturbing each parameter in turn while holding the remaining parameters constant at the end of an optimization run.

An extensive study of the performance of optimization techniques was carried out by Ibbitt (1970) using the Dawdy and O'Donnell and the SWM models. He found that a modified version of the Rosenbrock technique performed best for the models considered; however, several problems were uncovered, notably the undesirable effects of threshold parameters on the response surface and the existence of multiple optima in the parameter space. Local search techniques, such as the Rosenbrock method, are not designed to find a global optimum, and so the ad-hoc expedient of using different starting points for the parameter search was suggested. Some of the more pathological effects of model structure (e.g. threshold parameters) could be alleviated by model restructuring (O'Donnell and Ibbitt, 1974). In all of this work a split sample testing procedure was advocated in which a part of the available flow record not used in model fitting was used to assess the predictive power of the model (Figure 2).

Nash and Sutcliffe (1970) defined some principles which might be employed in building ESMA models. Rather than assume that a complex model incorporating a large number of parameters was best a priori, they advocated that a simple model structure should be formulated initially, and new parameters/components added only if a significant reduction in the explained variance was achieved. This was to be measured by a coefficient of determination R^2 defined as:

$$R^2 = \frac{F_o^2 - F^2}{F_o^2} \tag{2}$$

where F^2 is defined by Equation (1) and F_o^2 is defined as:

$$F_o^2 = \sum (Q_i - \overline{Q})^2 \tag{3}$$

where $\overline{Q}$ is the mean discharge for the period of record.

This approach recognized that, where model parameters were to be determined on the basis of minimizing an objective function, there were certain similarities to a regression problem where a very good fit could be achieved by using a large number of parameters, but at the expense of large parameter variances and poor predictive power. However, a rigorous statistical framework did not exist to do this, but split sample testing could provide a good indication of predictive performance. These principles were subsequently applied in a number of modeling studies (O'Connell et al., 1970; Mandeville et al., 1970) in which some of the problems encountered in parameter optimization by Ibbitt (1970) were also experienced. Points to emerge from this work were:

SYSTEM SYNTHESIS

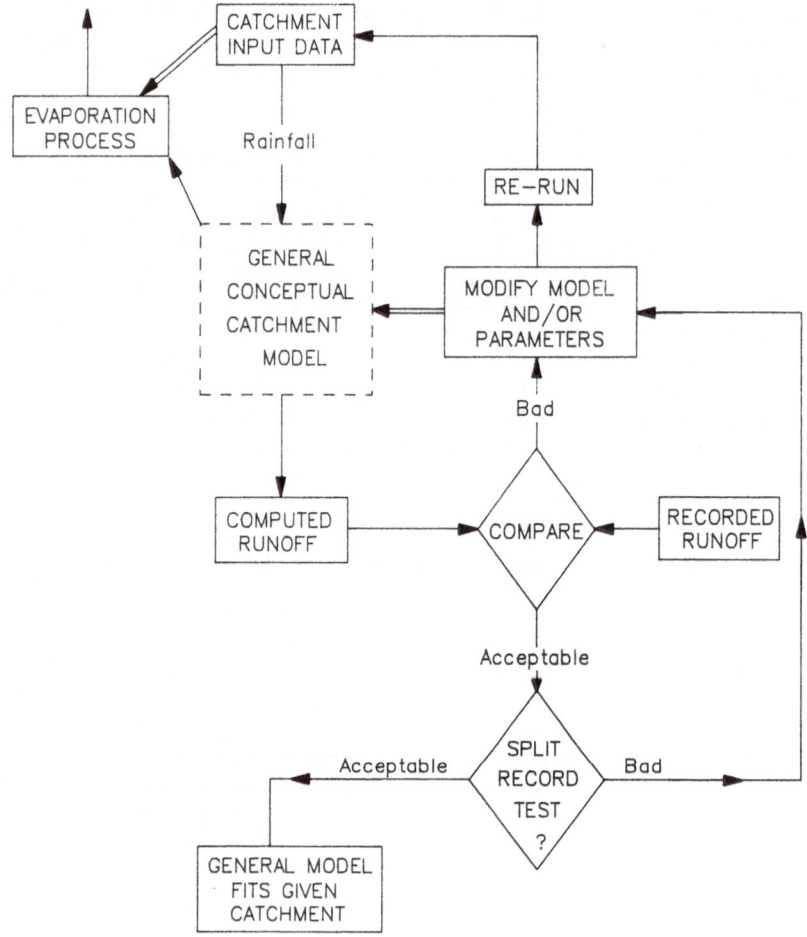

Figure 2. General catchment modeling strategy (after O'Donnell, 1986).

(1) A relatively simple model structure could be used to achieve an R^2 value of the order of 0.80.

(2) A high degree of interdependence between some model parameters was encountered whereby the effect of a change in one model parameter on R^2 could be compensated for by an adjustment in another model parameter.

One of the implications of (2) is that, while a sensitivity analysis of each parameter in turn would reveal a significant change in R^2, parameter interdependence would not be revealed by such an analysis, so interacting parameters would remain poorly determined (i.e. with large sampling variances).

Two schools of thought had now emerged on model calibration. One school advocated the use of subjective judgement and operator experience within a trial–and–error framework for parameter estimation while the second school pursued the optimization approach. With the former approach, parameter estimation was destined to become a subjective art of which the original model developer was probably the best exponent, while the optimization approach remained somewhat of an intractable approach for complex models of the SWM type.

The WMO Intercomparison Study. By the mid 1970's there was increasing interest in utilizing ESMA models for operational, real–time forecasting. However, choosing a model for operational use from the plethora of alternatives available at the time was not a straightforward task. To provide some guidance in this area, the World Meteorological Organization (WMO) organized a timely intercomparison study (1975) in which ten models were compared on the basis of results obtained from six data sets. The latter were chosen to cover a range of climatic characteristics, and consisted of two distinct periods : a calibration period (six years) followed by a validation period (two years). The model owners were supplied with both input and output data for the calibration period and input data only for the validation period. Model performance was then assessed by WMO by applying a set of agreed performance criteria to the observed and simulated data for the validation period.

The models which took part in the Intercomparison Study were classified into 3 groups:

(1) Explicit Soil Moisture Accounting (ESMA)
(2) Implicit Soil Moisture Accounting (ISMA)
(3) Systems approach with the use of indices.

Some of the ESMA models taking part were derivatives of the original SWM (e.g. the National Weather Service Model, the Sacramento River Forecast Center Model) which had been adapted for operational forecasting by the US National Weather Service. The models falling in category (2) had derived from the "tank–type" model first published by Sugawara in 1961 (Sugawara, 1961), and involved the construction of an assembly of linear storages in series and parallel to reflect the behavior of a particular catchment. The precise assembly of tanks to be used was a matter of judgement and experience and formed part of the calibration process. In the category of the systems approach (3), the CLS model discussed earlier in Section 2.1.1 took part; however, by using a threshold approach based on an antecedent precipitation index, the precipitation input was split into a number of separate inputs representing different states of catchment wetness (Figure 3).

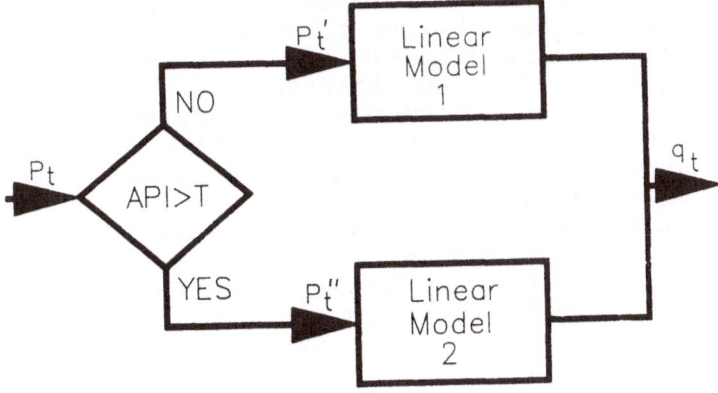

Figure 3. Schematic of CLS Model with threshold.

In assessing the results, it was concluded that there was little difference in model perform- ance for humid basins; however, for semi–arid regions where soil moisture exercises very strong nonlinear effects on catchment response, ESMA models would be expected to perform better. However, these conclusions were governed by the following overall conclusion that, in the presence of poor quality data, models based on the systems approach could be expected to give better results than ESMA models. ISMA (Tank) models demonstrated considerable adaptability and flexibility in dealing with different climatic and physiographic conditions, and performed consistently well overall; however, this performance was heavily dependent on the skill of the operator. On the issue of whether to use a manual or an optimization approach to model calibration, the conclusion drawn was that manual calibration, while expensive and time consuming, could produce a good calibration while optimization, which was cheap and fast, could produce unrealistic parameter values which in turn could lead to poor predictions. A combination of manual and optimization techniques was suggested as the best compro- mise.

The WMO Intercomparison Study also suggested that, where possible, a generally accepted set of criteria should be used in assessing model performance, and recommended a set of graphical and statistical criteria which had been used during the Intercomparison Study. Subsequently, Garrick et al. (1978) suggested that high values of the R^2 criterion did not necessarily imply good performance since the criterion was defined with reference to the relatively crude "no model" situation which was taken to equate with the mean flow $\overline{Q}$. For highly seasonal streamflow regimes, a greater degree of discrimination between models would be provided if R^2 was defined with reference to a seasonal mean flow Q_s i.e.:

$$R_s^2 = \frac{F_s^2 - F^2}{F_s^2} \quad , \tag{4}$$

where:

$$F_s^2 = \sum (Q_i - Q_s)^2. \tag{5}$$

In the case of a daily flow hydrograph, Q_s for each day could be taken as the mean discharge over the number of years of record for that day, or a smoothed version of the latter derived through harmonic analysis. When this criterion was applied to one of the ESMA models (the SSARR Model: Rockwood, 1958, 1968) which had taken part in the WMO Intercomparison Study, it was found that the seasonal mean daily discharge Q_s over the calibration period for the Sanaga River was a better predictor than the SSARR model; Garrick et al. (1978) pointed out that a similar result might be expected for other ESMA models. The term "Peasant's Model" was subsequently coined for the simple seasonal average; the arguments for using a complex model in this particular case seemed rather thin. Nash and Barsi (1984) proceeded to develop a more elaborate version of the Peasant's Model by relating the deviations from the seasonal mean discharge to the deviations from the seasonal mean of the daily rainfall input using a transfer function. However, this approach might only be expected to perform well on catchments with highly seasonal daily streamflow regimes.

New Insight into ESMA Models: Identification and Parameter Estimation. In the late 1970's, interest in using rainfall–runoff models for real–time forecasting intensified. In this context, the widespread implementation of ESMA models using manual, expert–based, model cali- bration procedures would prove exhaustive on resources, while automatic parameter optimi- zation, even for a limited number of parameters, did not present a reliable alternative to the manual approach. This led to a critical and rigorous reappraisal of model structure and pa- rameter estimation for ESMA models. The techniques of systems theory brought new insight to the problem, notably the concept of model identifiability which relates to the ability to estimate the parameters from the available data. Some of the work carried out on the problem related to "on–line" estimation and is considered later; work on "off–line" estimation (the traditional model calibration problem) is considered briefly here. In reviewing the structure of the components of the US National Weather Service River Forecast System Model (NWSRFS), Sorooshian and Gupta (1983) observed that better functional representations of the component model processes could be obtained without compromising physical integrity; this approach was demonstrated for the particular case of the percolation function in the NWSRFS model. However, the pathological influence of threshold parameters on the performance of optimization algorithms was also noted, but this problem remains essentially unresolved, as noted by Sorooshian, Chapter 20.

In a reassessment of the calibration and validation of ESMA models, Sorooshian and Dracup (1980) argued that a subjective choice of estimation criterion (i.e. the function to be minimized) which did not properly reflect the structure of the estimation errors would lead to biased and possibly physically unrealistic parameter estimates. This could lead to poor performance when used for forecasting (sometimes referred to as model divergence). The use of generalized least squares (Williams and Yeh, 1982) and maximum likelihood (Sorooshian and Dracup, 1980) objective functions where the errors are considered to be both auto–correlated and heteroscedastic have been shown to lead to improved response surface behavior. As a consequence, efficient gradient techniques can be employed, and better parameter estimates and better model performance obtained (Sorooshian and Gupta, 1983).

Further work on model identifiability (TASC, 1980) illustrated the interaction between parameter estimation and the information contained in the model inputs; unless the input data can "excite" the system sufficiently, the model parameters cannot be estimated. Research on the soil moisture accounting component of the NWSRFS model illustrated this point; increasing the length of record available may not necessarily resolve the problem.

Having reviewed the difficulties associated with the optimization approach to parameter estimation for ESMA models, Moore and Clark (1981) adopted a totally new approach. The catchment is considered to consist of a statistical population of soil moisture stores; runoff is generated through a soil moisture accounting procedure, and a probabilistic description of travel times is used for runoff routing. The resulting number of model parameters is small; moreover, the derivatives of the F^2 objective function are shown to be continuous in the parameters, allowing the use of efficient gradient procedures for optimization. Finally, the variances of the parameter estimates can be recovered. Although an elegant theoretical formulation, the practical potential of the model remains unproven, and it is not clear if its simple structure can account for the variation in catchment response associated with different climatic regimes without extending the model to the point that its original advantages are lost.

With the advent of interactive computer graphics in the early 1980's, prospects for making manual calibration procedures easier and more efficient have improved. Smith and Brazil (1980) reported the development of an Interactive Forecast Program (IFP) for the NWSRFS model while a similar study by Klatt and Schultz (1981) pointed to the possibility of using both manual and automatic parameter estimation within an interactive graphics framework.

2.1.3. Towards the Real World: Real–Time Flow Forecasting

Origins. Technological developments in the estimation of rainfall by radar, real–time data capture, digital computing, and mathematical modeling in the 1960's stimulated interest in the possibility of using these models for flow forecasting in real–time. In the UK, the Dee Weather Radar and Real–time Hydrological Forecasting Project was initiated in 1966 to explore the potential which these technological developments could generate for multipurpose reservoir management. The real–time flow forecasting system implemented on the River Dee in 1975 represented the culmination of this research program (Dee Steering Committee, 1977); considerations governing the choice of model for real–time flow forecasting on the River Dee are discussed by Lambert and Lowing (1980). Sub–catchment rainfall–runoff models were calibrated and linked to a main channel flow routing model; the former were simple in structure and based on an input–storage–output relationship (Lambert, 1972). One of the characteristics of this rainfall runoff model is an inbuilt capacity to update its forecasts recursively in real–time.

An alternative approach to forecast updating was proposed by Jamieson and Wilkinson (1972) within the framework of the River Dee Research Project. They suggested that a time series model should be used to forecast the discrepancy (noise) between observed and forecasted discharge; this had the advantage that a rainfall–runoff or flow routing model could be calibrated "off–line," and "simulation–mode" forecasts from the latter updated through the noise model, thus avoiding the more difficult problem of updating the forecasts from the rainfall–runoff or flow routing model directly.

Kalman Filtering. The problem of real–time forecasting and control has been a subject of major interest in the field of systems engineering for many years. By the late 1970's, a flux of papers had appeared based on the application of systems techniques to real–time hydrological forecasting (e.g. Chiu, 1978; Wood, 1980; O'Connell, 1980). Notable among the techniques employed was the state–space/Kalman filtering approach for linear dynamic systems; here, the system model is formulated within a state–space framework which accommodates both model and measurement error. The Kalman filter is then used to derive minimum variance estimates of the system "state" which, in the hydrological context, provides the necessary information to generate flow forecasts. Moreover, the error variances of the states are also provided.

While the state space/Kalman filter approach had considerable theoretical appeal, there were formidable difficulties associated with its implementation in hydrology. Firstly, state–space model formulations, which could accommodate the quantities to be filtered (e.g. rainfall, flow) and the model parameters, led to nonlinear estimation problems which then had to be solved using the extended Kalman filter. Here, the optimality properties of the linear Kalman filter are lost, and the properties of the resulting state estimates are difficult to establish. An alternative approach is to couple two filters interactively, one for estimating the states and the other for estimating the parameters (Todini, 1978; Todini et al., 1980). Yet another approach is to write the state–space model in equivalent ARMAX form (an ARMA model with an eXogeneous input) and to then use maximum likelihood estimation for the ARMAX model parameters (Cooper and Wood, 1982a,b).

A further problem with the application of the Kalman filter in hydrology and some other disciplines is that the properties of the system and measurement noises, particularly their variance–covariance matrices, are unknown and have to be estimated from the available data. Some recursive algorithms are available but their convergence properties are generally unknown.

While the state–space/Kalman filter approach has considerable theoretical appeal because of its ability to handle model and measurement error, and, suitably extended, parameter uncertainty, it is not clear that the practical results obtained have always fulfilled the user's expectations. An example is given in O'Connell and Clarke (1981) where the performance of some state/space Kalman filter models, based on one–step ahead forecast errors, was compared with the results from a conventional and more empirical estimation procedure; the one–step ahead forecast errors had virtually identical properties in all cases and were suboptimal in that they were not white and also displayed time variance in the neighborhood of flood events. These results suggest the assumptions which are conventional in applying the Kalman filter are not satisfied in this particular case; this is because most of the noise (model and measurement error) enters the system during flood events (errors in rainfall inputs; model error), which suggests that the original assumption of homogeneous and independent noise variances is unjustified. A careful appraisal of the underlying assumptions is thus warranted before a lot of effort is expended on application.

The estimation procedures described above for ESMA models have been carried out under the assumption of a deterministic model and input. Efforts have also been made to implement maximum likelihood estimation procedures with stochastic state–space representations of such models (Goldstein and Larimore, 1980; Restrepo–Posada and Bras, 1982). The development of such sophisticated techniques to represent model and measurement errors is desirable, provided that the conceptual structure of the model is not compromised too much to accommodate the use of a particular estimation algorithm. Much of this recent work on parameter identification and estimation is encouraging in that it recognizes the specific nature of the model estimation problem and is seeking to develop an appropriate solution rather than taking the easier route of using an "off the shelf" procedure.

Real World Implementation. The practical problems associated with implementing real–time forecasting models have been discussed in depth by Nemec (1985, 1986) who presents a valuable account of the experience gained from many WMO projects. While the forecasting model is of central attention to most hydrologists, this is but one component of a real–time forecasting system; considerable attention needs to be given to the interrelationships between this component and the other components of the system, notably the real–time data collection sub–system. Other important issues discussed by Nemec include the interaction between data quality and availability, the available manpower resources, the choice of model, and forecasting system performance. The role of the Hydrological Operational Multipurpose System (HOMS) in the transfer of forecasting technology to the real world is also discussed.

Both the theoretical and practical aspects of real–time flow forecasting are reviewed in an interesting collection of papers edited by Kraijenhoff and Moll (1986). Here Schultz (1986) discusses the link between theory and practice and notes the lack of appreciation, on the part of those involved with theoretical developments, of the practical problems associated with implementing real–time forecasting systems. The important role occupied by the forecaster and his base of subjective knowledge is also discussed. Several case studies are presented by Kraijenhoff and Moll (1986) that illustrate the diversity of forecasting problems encountered in practice and the diversity of the solutions adopted. The potential of the microcomputer in making forecasting more widely available at low cost is illustrated in a case study reported by O'Connell et al. (1986).

2.2. The Evolution of Descriptive Modeling

2.2.1. Rationale

As discussed in Section 1, the motivation behind descriptive modeling in hydrology is deemed to be the quest for a better understanding of the behavior of hydrological systems. The experimental approach must clearly occupy an important role in this, as must the discipline of physics in explaining what is going on. Hydrology has a very small inheritance of well accepted physical laws deriving from this approach; Darcy's law (1856) is clearly in this category while the St Venant equations of surface flow might also be regarded as part of this inheritance. However, studies of the component processes of the hydrological cycle have greatly enhanced the level of understanding of these processes, even if new hydrological laws have not been discovered. A review of relevant advances in hydrological process research is beyond the scope of this chapter; however, some selected developments are discussed which have had a bearing on the evolution of physically based distributed models of basin scale hydrology.

2.2.2. Overland Flow and Kinematic Wave Models

Interest in the dynamics of overland flow can be traced back to the experimental and theoretical work of Horton (1938) and Izzard (1946). The basic problem of interest was to determine the flow off the downstream end of a sloping plane for specified physical conditions and a given pattern of lateral inflow along the plane. By coupling a continuity equation with an empirical equation relating outflow to storage on the plane, a lumped parameter differential equation was obtained which was solved analytically for a number of cases. The motivation behind the work in Horton's case was to describe overland flow on natural catchments and, in Izzard's case, on paved surfaces. Theoretical work by Lighthill and Whitham (1955) laid the foundations for the kinematic wave model where the empirical storage–discharge–relationship in the Horton–Izzard approach was replaced by a simplified dynamic equation in which the friction slope S_f is assumed to be equal to the bed slope S_o. An early application of kinematic theory was reported by Henderson and Wooding (1964) and Wooding (1965) who formulated a kinematic wave model in which the catchment is represented as a tilted V–shaped, double–plane surface in which the valley forms the channel; both analytical and numerical solutions were obtained for this combined overland–channel flow system.

In the late 1960's and early 1970's, the Agricultural Research Service of the U.S. Department of Agriculture conducted a research program in which kinematic wave modeling was integrated with laboratory and field experiments. Liggett and Woolhiser (1967) summarized and compared numerical methods for the solution of the combined overland–channel flow equations. Woolhiser (1969) obtained a kinematic wave solution for overland flow on conic sections, a basic geometrical shape on small watersheds. Subsequently, a small–scale experimental rainfall–runoff facility was constructed at Colorado State University which incorporated a conic sector, thus allowing a comparison of the results of controlled experiments with theory (Woolhiser and Schultz, 1973). However, it was recognized that such small–scale experimental facilities may not reproduce certain flow phenomena that may be important in "real–world" hydrological systems. The mathematical properties of a kinematic cascade (a sequence of n discrete overland flow planes or channel segments in which the kinematic wave equations are used to describe the unsteady flow) were explored by Kibler and Woolhiser (1970), and the conditions under which the kinematic shocks could form in the flow field were defined. The kinematic cascade model was applied to the modeling of surface flow on twelve, small rangeland catchments characterized by different levels of grazing intensity (Woolhiser et al., 1970); roughness parameters were derived for one set of storms and then used to predict a second set of storms with good results.

The kinematic cascade model was subsequently coupled to a one–dimensional vertical unsaturated subsurface flow model of infiltration by matching boundary conditions at the soil surface. The model was tested by comparison with data from a laboratory soil flume and with data from a small experimental catchment (Smith and Woolhiser, 1971).

Work on kinematic wave modeling was also carried out elsewhere, notably by Brakensiek (1967), Huggins and Monke (1968), and Schaake (1970). Schaake's model for urban basins was later extended and refined into a comprehensive distributed model of surface water hydrology by Leclerc and Schaake (1973) and Wilson et al. (1979).

Since kinematic flow theory is based on the assumption that velocity is directly proportional to depth, this predicts that, as rainfall intensity increases, the time of response of a catchment decreases. Thus, unit hydrograph theory is immediately seen to be at variance with this simple physically based model of surface flow.

The limitations of kinematic wave modeling derive primarily from a unique single–valued relationship between stage and discharge. An order of magnitude examination of the terms in the full dynamic routing equation (e.g. Henderson, 1966) can help to establish the conditions under which kinematic flow models are appropriate.

2.2.3. Hillslope Hydrology and Runoff Production Mechanisms

Origins. The origin of the scientific debate on how runoff occurs can be traced to Horton's (1933) work on the role of infiltration in the hydrological cycle. Horton's theory assumes that runoff is generated where and when rainfall intensity exceeds the rate at which water enters the soil. Over the years, an extensive amount of field research has been carried out, mainly on hillslopes, which has demonstrated that Hortonian overland flow is a relatively rare occurrence in the field and that other physical mechanisms are needed to explain runoff production; this has led to the emergence of the sub–discipline of "hillslope hydrology."

As early as the 1940's, doubt was being cast on the infiltration excess runoff production mechanism (Hursh and Brater, 1944); nevertheless, while surface storm runoff was not observed during rainfall, characteristic flood hydrographs were still produced by heavy rain. Betson (1964) put forward the theory that infiltration–excess runoff occurs from a relatively small part of the catchment area (the partial area concept). Ragan (1968) analyzed a series of storms which showed that only a small portion of his catchment ever contributed to the storm hydrograph; the same conclusion was also reached by Betson and Marius (1969). Rawitz et al. (1970), in a two year investigation of a 16.2 hectare Pennsylvanian catchment, observed that summer storms produced practically no overland surface runoff, but the streamflow hydrographs produced by such storms possessed all the characteristics attributed to surface runoff.

By the beginning of the 1970's, two additional theories for runoff production had emerged. The first concept, that of subsurface storm flow, emerged from work carried out by Hewlett and his co–workers (Hewlett and Hibbert, 1967; Hewlett and Nutter, 1970). Early work by Hewlett and Hibbert (1963) showed the feasibility of such a flow mechanism experimentally, and Whipkey (1965) measured lateral inflows from subsurface sources in the field. The prime requirement for subsurface storm flow is a shallow surface soil horizon of high permeability, such as occurs in forested catchments.

The second mechanism, and that more fully supported by field evidence at the time, is that of overland flow generated by rainfall falling on small partial areas that quickly become saturated during rainfall events i.e. saturation occurs from below rather than above. Studies carried out by Ragan (1968) and Dunne (1970) suggested that the partial contributing areas were likely to be wetlands with location controlled by the topographic and hydrogeological configuration of the basin; they are likely to be located adjacent to stream channels or near seepage zones that fed small intermittent tributaries. The time lags for overland flow from these areas are likely to be very short, thus enabling them to contribute rapidly to the storm hydrograph. These contributing areas would vary dynamically in response to rainfall, thus earning them the title "variable source areas."

Modeling of Runoff Production Mechanisms. In a remarkable series of computer simulation experiments conducted using a three–dimensional fully integrated surface/subsurface flow model (Freeze, 1971; Freeze, 1972a,b), valuable insight into the competing subsurface runoff and the Dunne (1970) saturation excess runoff production mechanisms was obtained. This was an excellent example of how the power of the computer could be used to support scientific inquiry; Freeze embedded as much physical information as possible within his model from the experimental evidence available at the time. The simulations carried out with rainfall events on hypothetical hillslope areas showed that significant contributions to stormflow from the subsurface runoff could only occur under the restrictive conditions of a convex hillslope feeding a deeply incised channel, and then only if saturated soil conductivities are very large. On concave slopes, with lower permeabilities, and on all convex slopes, hydrographs are dominated by direct runoff through very short overland flow paths from saturated near-channel areas.

Freeze (1972a) also examined the mechanisms governing base flow contributions to channel flow, and found that there was no rationality behind baseflow separation. He also found that, for baseflow dominated streams, the contributing physical mechanisms create such wide variations in watershed response that simple black–box relationships are meaningless.

Stephenson and Freeze (1974) used a mathematical model of subsurface flow to complement a field study of snowmelt runoff for a hillslope area in the Reynolds Creek Experimental Catchment near Boise, Idaho. Here the transient saturated–unsaturated surface/subsurface flow model of Freeze (1971) was introduced "to the perils of the real world" (Stephenson and Freeze, 1974). The field data were derived from piezometer nests, soil moisture tubes, tensiometer nests, and a snowcourse. Additional observation wells were placed 100 feet away on each side of the piezometer nests for use in collection of more detailed subsurface information. Two V–notch weirs were used to monitor the flow contribution to the channel. The computed snowmelt input to the model covered a 100 day period; output from the model took the form of plots of the pressure head, total head, and moisture content fields for each time step. The calibration–validation procedure that was adopted relied heavily on physical reasoning and attempted to maintain internal consistency in the agreement between simulated and observed quantities. Commenting on the results, Freeze (1978) states:

"The matches are neither particularly good nor particularly bad. They are probably representative of the type of results that can be expected from the application of complex theoretical models to complex field situations under the constraints created by data limitations and the resulting uncertainties in the calibration procedures."

The results suggested that unsaturated/saturated subsurface stormflow was the main mechanism for generating runoff in this case. The authors concluded that the model proved a valuable aid to the unified interpretation of the field measurements; however, they did not feel that the approach was then ready for routine application on large watersheds. Deterrents included data requirements, computer limitations, and constraints on calibration procedures.

A variable source area simulation model for forested basins has been developed by Hewlett and Troendle (1975). The model, which was designed for application to rectangular, convergent or divergent segments of a first or second order basin, assumes that runoff occurs from three sources:

(1) The saturated area of the basin along perennial intermittent and ephemeral channels. Subsurface flow exfiltrates through this area to the channel.

(2) The horizontally projected area of the saturated surface of the basin on which rain or snowmelt occurs.

(3) Runoff from impervious areas which contribute Hortonian overland flow to perennial or expanding channels.

Within the model, the Richards equation is used to simulate both unsaturated and saturated flow through a number of soil elements comprising a basin segment. A number of soil horizons can be accommodated within the model, and segments can be subdivided according to various rules when saturation or desaturation are occurring. Lateral water fluxes on divergent and convergent segments are apparently accounted for in the selection of the segments and in the computation of elemental dimensions and volumes.

A first version of the variable source area model (VSAS1) was applied to a densely instrumented 38 hectare basin in the Appalachian Highlands. Twenty weirs were located within the area, and extensive information on the depths of soil layers and soil characteristics curves was obtained through extensive resistivity surveys and soil cores/boreholes. Antecedent soil moisture data were derived from 38 neutron probe access tubes. The model was applied to 12 storms without any calibration on observed outflows; comparisons were in terms of stormflow derived through baseflow separation since the model did not apparently have the capacity to represent an antecedent baseflow condition. This detracts somewhat from using a physically based model in the first place. However, the simulated responses all lay within – 55 percent to + 25 percent of the observed values, with two thirds of the responses lying within ± 20 percent. Agreement for the internally gauged points were also reasonable although errors from different segments tended to cancel out. An improved version of the simulation model (VSAS2) has been developed which has eliminated mass balance errors in VSAS1 (Bernier, 1982); however, further improvements are judged to be needed to the computational solution schemes employed. A review of variable source area models is provided by Troendle (1985).

In a review of field studies of hillslope flow processes, Dunne (1978) reviewed the field evidence for runoff production mechanisms and concluded that:

"Water can reach stream channels by several routes. The processes that deliver stormflow, and the volumes and timing of their contributions vary with topography, soil properties, and rainfall characteristics and indirectly with climate, vegetation, and land use. Even within a particular basin the dominant runoff process may vary with the characteristics of rainstorms. Even the highest infiltration capacities of forest soils will not accommodate the highest recorded rainfall intensities. The various models of storm runoff, therefore, are complementary rather than contradictory. Differences of emphasis between studies of runoff reflect the physical geography of the regions in which experiments were carried out."

Hence, as Beven (1983) observes, the field evidence suggests a complex pattern for runoff production, with all three mechanisms occurring in a catchment at the same or different times. Firstly, there is the problem of "at a point" runoff prediction, given adequate soil, vegetation, and meteorological data, and, secondly, at the hillslope and catchment scales, with the inherent lack of knowledge of the spatial variability of the above quantities. This has heightened interest in the subject of spatial variability and scale which are considered to hold the key to evolving a sound basis for modeling runoff production at the hillslope and catchment scales (Beven et al., 1988).

A catchment–scale variable source area model (TOPMODEL) has been proposed by Beven and Kirkby (1979) which attempts to combine the important distributed effects of channel network topology and dynamic contributing areas with the advantages of simple lumped–parameter basin models. Quick response flow is predicted by a storage/contributing area relationship derived analytically from the topographic structure of a unit within a basin. Average soil water response is represented by a constant leakage subsurface water store. A simple nonlinear routing related to the link frequency distribution of the channel network completes the model and allows distinct basin subunits, such as head water and sideslope areas, to be modeled separately. Beven and Kirkby (1979) claim that the model parameters are physically based and can be determined directly by measurement, and that the model can be used at ungauged sites. TOPMODEL has since been improved and employed in studies of the interaction between scale, catchment morphology, and catchment response as specified by Beven (Chapter 17), and Beven et al. (1988).

2.2.4. Physically–Based Distributed Modeling at the Catchment Scale

Origins. Freeze and Harlan (1969) outlined a blueprint for a physically based digitally-simulated hydrological model (Figure 4). They noted that the necessary numerical solutions to mathematical boundary value problems for groundwater flow, unsaturated porous media flow, overland flow, and channel flow were already available, and that developments in digital computer technology at the time could provide the impetus for a necessary redirection of

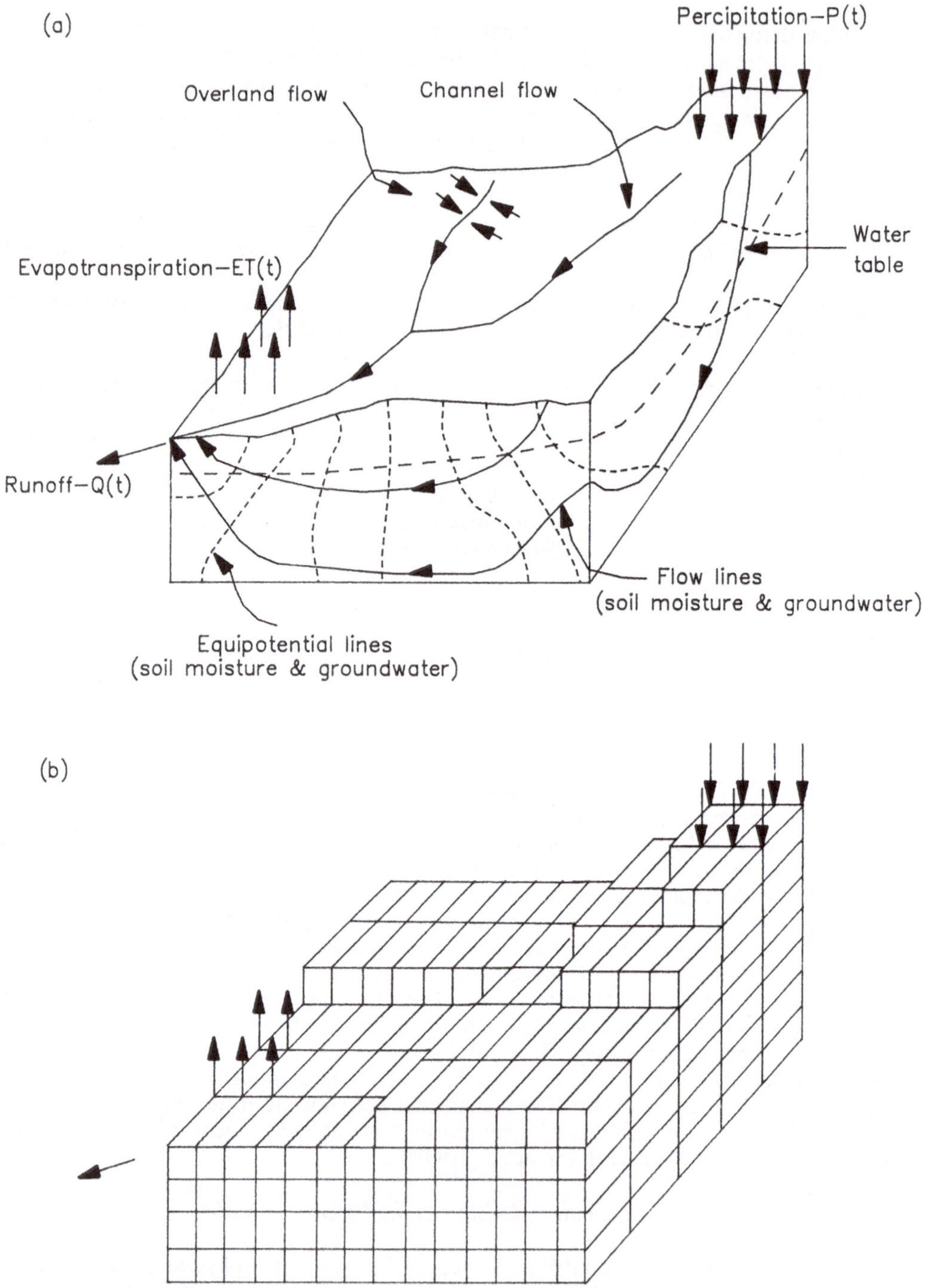

20

(a)

Percipitation—P(t)

Overland flow   Channel flow

Water
table

Evapotranspiration—ET(t)

Runoff—Q(t)

Flow lines
(soil moisture & groundwater)

Equipotential lines
(soil moisture & groundwater)

(b)

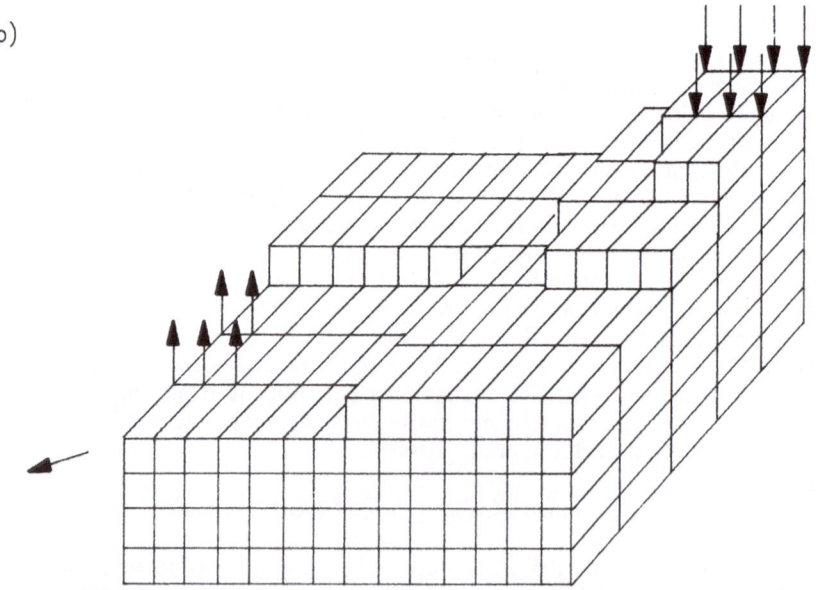

Figure 4. Schematic of a physically based distributed model (after Freeze and Harlan, 1969).

research in digital simulation. Some research requirements aimed at enhancing the prospects for implementing a physically based distributed model of basin hydrology were outlined.

Woolhiser (1973), in a state of the art review of watershed modeling, noted that the developing concern of society for environmental quality required consideration of water as a transporting medium for pollutants. Added to Penman's (1961) fundamental question quoted at the start of this chapter were the additional questions of "What happens to the fertilizer?" and "What happens to the pesticide?" Woolhiser (1973) noted that the partial differential equations for free surface flow and for flow in porous media constituted a rigorous mathematical description of the movement of water within a catchment. He also reviewed the then current generation of lumped–system models and noted that such models could probably prove useful in predicting the transport of substances naturally present in the environment, and which were currently being monitored. However, if the environmental impact of new substances had to be evaluated before they were released or if the transport system itself were changed, then physically–based distributed modeling appeared to be the only possible approach.

Beven (1985) notes that the development of distributed modeling has been a slow, faltering process. The modeling of individual component processes had been the subject of numerous papers, but relatively few papers have appeared on models involving interacting processes and on the application of catchment–scale models to real world problems. However, during the past decade, a number of organizations invested the resources needed to develop catchment–scale distributed models, notably the Système Hydrologique Européen (SHE) developed through a collaborative Danish–French–UK effort (Beven et al., 1980; Abbott et al., 1986a,b; Bathurst, 1986a,b), the USDA Agricultural Research Service Small Watershed Model (SWAM: Alonso and DeCoursey, 1985), and the UK Institute of Hydrology Distributed Model (IHDM: Morris, 1980). Brief summaries of the evolution of these models and modeling systems are given here.

The Système Hydrologique Européen (SHE). The motivation to develop SHE came from the perception that conventional rainfall runoff models were inappropriate to many pressing hydrological problems, notably those related to the impact of man's activities on the flow, water quality, and sediment transport regimes of river basins. A physically based distributed modeling system such as SHE appeared to provide the appropriate framework for predicting such impacts. The SHE is a comprehensive modeling system rather than a comprehensive model; SHE can generate models of different situations where one or more hydrological processes, and the interactions between them, must be represented. This approach is made possible through a flexible system architecture where new versions of the component models can be easily integrated into the system. Such a modeling approach provides a continuous upwards development path for the system as opposed to the "one–off" empirical modeling philosophy.

Within SHE, the hydrological processes of water movement are modeled either by finite difference representations of the partial differential equations of mass, momentum, and energy conservation, or by empirical equations derived from independent experimental research. Spatial distribution of catchment parameters, rainfall input, and hydrological response is achieved in the horizontal by an orthogonal grid network and in the vertical by a column of horizontal layers at each grid square (Figure 5). Each of the primary processes of the land–phase of the hydrological cycle is modeled in a separate component as follows: interception, by the Rutter accounting procedure; evapotranspiration, by the Penman–Monteith equation; overland and channel flow, by simplifications of the St Venant equations; unsaturated zone flow, by the one–dimensional Richards equation; saturated zone flow, by the two-dimensional Boussinesq equation; and snowmelt, by an energy budget method. Overall control of the parallel running of the components and the information exchanges between

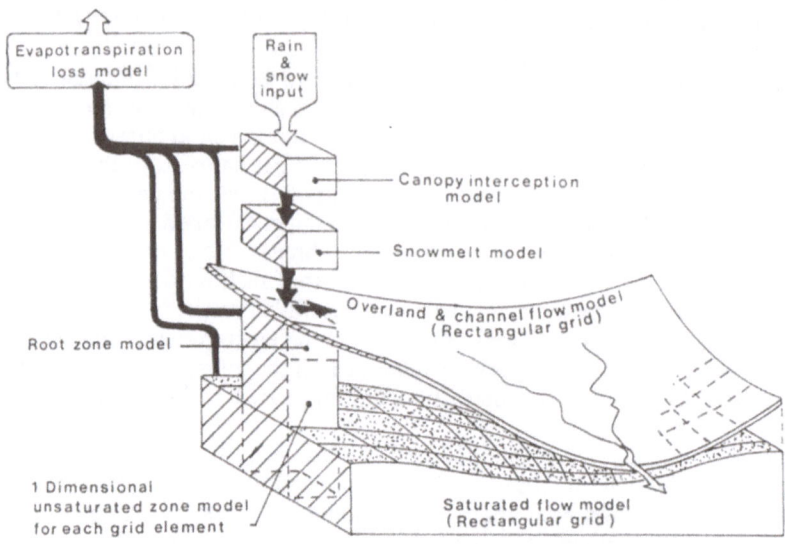

Figure 5. Schematic of the Système Hydrologique Européen (SHE).

them is managed by a FRAME component. Careful attention has been devoted to a modular construction so that improvements or additional components (e.g. water quality and sediment yield) can be added in the future. Considerable operating flexibility is provided through the ability to vary the level of sophistication of the calculation mode to match the availability or quality of the data.

An application of the SHE to an upland catchment (the Wye : area 10.55 km^2) in mid–Wales is described by Bathurst (1986a) in which an approximate calibration and validation of the model is achieved on the basis of physical reasoning and through consideration of the variation of runoff response from different parts of the catchment. Basic parameter values used in the application were derived from field measurements or published data. An approximate calibration was achieved for one hydrograph by varying only a few of the parameters and the model was then validated against four different hydrographs, largely on the basis of changes in the initial level of the phreatic surface. Some deficiencies in the simulations were noted, derived in part from the inability of the current version of the SHE to model subsurface flow in natural pipe networks and along the interface between soil layers. However, the generally good agreement between observed and simulated responses in all parts of the catchment, the use of parameters based firmly on field measurements, and the ability to improve the simulations by physical reasoning indicate the considerable power of this type of modeling system.

A sensitivity analysis of the SHE has been carried out by Bathurst (1986b) in which an assessment was made of the sensitivity of the system's response to changes in soil, roughness and vegetation parameters as well as such "nuisance" parameters as grid size and time step. The results suggested that those parameters to which the simulations were found to be most sensitive could be evaluated with sufficient accuracy at a few representative field sites, while the less sensitive parameters (e.g. vegetation coefficients) could be evaluated using data from the literature. The scope for achieving equally satisfactory calibrations based on different combinations of parameter values appeared limited as long as several different hydrographs were considered. The spatial distributions of rainfall and soil properties had a relatively minor effect on the simulations for the application considered.

Results from applications of the SHE to a range of catchments (areas 0.3 km^2 to 600 km^2) in a variety of climatic settings have now been obtained. Extensions of the SHE to model pollutant transport (Ammentorp et al., 1986) and erosion and sediment transport (Wicks and Bathurst, 1990) are in progress, while the validation of SHE for ungauged catchments is also receiving attention (Parkin and Mackay, 1990).

The Institute of Hydrology Distributed Model (IHDM). With the IHDM approach, the catchment is subdivided along the lines of greatest topographic slope using one–dimensional (downslope) overland flow elements, one–dimensional (downstream) channel reaches and two–dimensional (vertical slice) unsaturated surface/subsurface flow components (Figure 6). Each hillslope segment is simulated independently with outflows that cascade into the channel network. Flow down a slope element or a channel element is described by the St Venant equations for shallow water flow over a plane surface (Morris, 1979), while the Richards equation is used to describe unsaturated/saturated subsurface flow; finite difference solutions are obtained to the coupled surface and subsurface equations.

The IHDM was applied to the Wye (grassland) and Severn (forest upland catchments in mid–Wales) with a view to assessing the ability of the model to predict the impact of different land uses on flood flows. For two basins and their sub–basins, it was found that good predictions of flood flows could be made using a roughness parameter specified a priori from the vegetational characteristics and independent of geometrical and meteorological factors (Morris, 1980).

Further extensions of the IHDM model have since been carried out including the incorporation of a finite element hillslope model (Beven et al., 1987; Calver and Wood, 1989).

USDA–ARS Small Watershed Model (SWAM). SWAM (Alonso and De Coursey, 1985) has been designed to demonstrate the effect of land use or management changes on the hydrological and chemical response of a small catchment. SWAM uses an updated version of the Chemical Runoff and Erosion from Agricultural Management Systems (CREAMS) (USDA, 1980) model, CREAMS II, as a building block; the latter predicts the hydrological response of a field–size area. The CREAMS II model assumes that a field can be described by a series of overland flow areas and channels; a kinematic solution of the flow equations is adopted, thus

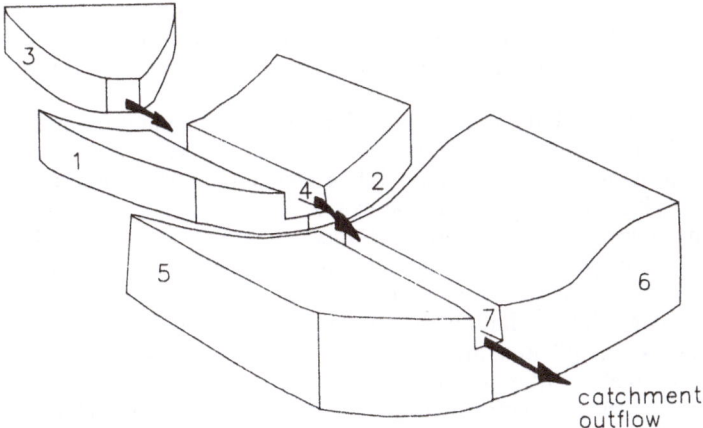

Figure 6. Schematic of the Institute of Hydrology Distributed Model (IHDM). Numbering indicates order of simulation of components.

drawing on the earlier work by Woolhiser and his collaborators. Water, sediment, and chemical outputs from each field in the catchment are then routed through the channel and any impoundment system to the downstream point; groundwater is added if appropriate. A reservoir model calculates profiles of temperature and sediment concentration as well as the effects of biological activity on nutrient levels. The SWAM model is still at a development stage; sensitivity analyses are being carried out to enable the model to be simplified and to find efficient ways of aggregating areas so that a basin–scale model can be developed.

3. Conclusions

Towards the end of the 1970's, a certain amount of unease existed about the state of hydrological research. It was felt that too much effort had been devoted to developing and refining ad–hoc empirical solutions to short term problems at the expense of more fundamental research which would provide a sound scientific basis for tackling several long–term water management problems (e.g. groundwater contamination, nonpoint pollution, acid rain, drought/desertification, etc.). An overall assessment of the research on descriptive and prescriptive modeling discussed above tends to support this view. Apart from the efforts of a few individuals (e.g. Freeze, Woolhiser), very little had been done to enhance our understanding of catchment response, and more and more effort was being devoted to prescriptive modeling activities which were subject to a law of diminishing returns or no returns at all. One of the main reasons for this was that ESMA modeling had not provided an upward development path for catchment modeling. Research effort had been dissipated into developing a vast array of these models, and no credible objective scientific method had emerged through which they could fulfil their intended role.

Linsley (1982) regrets this dissipation of effort and points out that the use of relatively few models on many catchments would have saved much effort and led to the evolution of a few very good multipurpose models. However, the nature of the SWM made this an unlikely scenario. Firstly, the descriptions of the component hydrological processes were not unique, widely accepted, or physically based; this invited the construction of alternative representations to see if they "worked" any better. Secondly, there was the problem of calibrating the model. While some of the many parameters could be evaluated in a reasonably objective manner, others could not, and their evaluation demanded considerable skill on the part of the user. Since the most effective user was likely to be the original developer of the model, the development of a new model had this attraction. The laudable objective of trying to make parameter estimation more objective and scientific through optimization had also been fraught with difficulty. ESMA modeling had evolved into an art rather than a scientific method.

A further problem associated with the ESMA modeling approach is that it has not added to our scientific base of knowledge of "what happens to the rain?"; rather a better understanding of the internal mechanisms of ESMA models has been obtained (by a limited fraternity) but not of the internal mechanisms of the real world.

While the systems engineering approach had performed a valuable role in providing the necessary framework for developing and applying the unit hydrograph, the introduction of "blackboard hydrology" in the 1960's led to a growth industry in the use of the latest systems engineering tools in hydrology. With a limited number of exceptions, most of the efforts in this area have shown a lack of imagination and critical insight; assumptions made elsewhere for processes exhibiting completely different behavior to the rainfall runoff process have been accepted without question when rudimentary hydrological considerations would have shown them to be inappropriate.

Perhaps one of the main reasons why hydrological modeling has developed in the manner it has is that a catchment can be treated as a closed system. While this was advantageous initially in developing the unit hydrograph as a predictive tool, the difficult route of treating the rainfall–runoff process as a boundary value problem had been avoided because of the possibility of calibrating a model on discharge at a single outflow point in space. In other fields where the assumption of a closed system has not been possible, (e.g. the hydrodynamic modeling of estuaries, dynamic atmospheric modeling, ocean modeling), physically based modeling has developed more rapidly, and the resources needed to support such research have been made available. In these fields, experimental programs and mathematical modeling have been more closely integrated.

Over the past several years, a philosophical debate has developed on the state of hydrological research, and of hydrological system modeling in particular. Good philosophical discussions of the issues involved are given Klemes (1986, 1988) and Beven (1987). A new research trend is now evident which reflects the desire to gain a more fundamental understanding of causal mechanisms within the rainfall runoff process. The emergence of the geomorphological IUH (Rodriguez–Iturbe and Valdes, 1979) was an early indication of this trend, and a quest for hydrological laws had begun (Dooge, 1986) which could explain the influence of factors such as basin topography, basin scale, and stream network geometry on runoff production. Allied to this work has been the development of models of rainfall fields in space and time, which could be coupled to rainfall runoff models to gain a better understanding of the interaction between spatial variability in rainfall and runoff production. Two distinct strands of modeling research had now emerged: the physically based distributed modeling approach based on solving the coupled surface and subsurface hydraulic flow equations which finds its origin largely in Freeze's work; and the simpler, geomorphologically based approach due to Rodriguez–Iturbe. Developments in remote sensing, GIS, and computing in recent years have considerably enhanced the viability of the former approach. A central issue in both approaches is how to deal with the problems of scale and spatial variability, particularly for physically based distributed models (Beven, 1989). Those involved with these new modeling developments are now turning their thoughts to the design of field experiments to tackle these issues and to test new emerging theories. This new fundamentalist trend in hydrological systems modeling holds exciting prospects for the future. However:

"An important scientific innovation rarely makes its way by gradually winning over and converting its opponents: it rarely happens that Saul becomes Paul. What does happen is that its opponents gradually die out, and that the growing generation is familiarized with the ideas from the beginning." (Max Planck)

4. References

Abbott, M. B., J. C. Bathurst, J. A. Cunge, P. E. O'Connell, and J. Rasmussen: 1986a, 'An Introduction to the European Hydrological System – Système Hydrologique Européen, "SHE", 1 : History and Philosophy of a Physically–Based, Distributed Modelling System,' *J. Hydrol.* **87**, 45–59.

Abbott, M. B., J. C. Bathurst, J. A. Cunge, P. E. O'Connell, and J. Rasmussen: 1986b, 'An Introduction to the European Hydrological System – Système Hydrologique Européen, "SHE", 2 : Structure of a Physically–Based, Distributed Modelling System,' *J.Hydrol.* **87**, 61–77.

Alonso, C. V., and D. G. De Coursey: 1985, 'Small Watershed Model,' in D. G. De Coursey (ed.), *Proc. of the Natural Resources Modelling Symposium*, Pingree Park, CO, USDA-ARS-30, pp. 40–46.

Ammentorp, H. C., K. H. Jensen, T. H. Christensen, and J. C. Refsgaard: 1986, 'Solute Transport and Chemical Processes: The Present State of the Unsaturated Zone Component of the SHE Modelling System,' *Proc. Int. Conf. on Water Quality Modelling in the Inland Natural Environment*, BHRA, Cranfield, pp. 187–197.

Amorocho, J., and G. T. Orlob: 1961, *'Nonlinear Analysis of Hydrologic Systems,'* Water Resources Center Contrib. 40, University of California, Los Angeles.

Amorocho, J., and A. Brandstetter: 1971, 'Determination of Nonlinear Functional Response Functions in Rainfall–Runoff Processes,' *Wat. Resour. Res.* **7**(5), 1087–1101.

Bathurst, J. C.: 1986a, 'Physically–Based Distributed Modelling of an Upland Catchment Using the Système Hydrologique Européen,' *J. Hydrol.* **87**, 79–102.

Bathurst, J. C.: 1986b, 'Sensitivity Analysis of the Système Hydrologique Européen for an Upland Catchment,' *J. Hydrol.* **87**, 103–123.

Betson, R. P.: 1964, 'What is Watershed Runoff?,' *J. Geophys. Res.* **69**(B), 1541–1552.

Betson, R. P., and J. B. Marius: 1969, 'Source Areas of Storm Runoff,' *Wat. Resour. Res.* **5**(3), 574–582.

Bernier, P. Y.: 1982, *VSAS2: A Revised Source Area Simulator for Small Forested Basins,* PhD Dissertation, University of Georgia, Athens, GA.

Beven, K. J.: 1983, 'Surface Water Hydrology – Runoff Generation and Basin Structure,' *Reviews of Geophysics and Space Physics* **21**(3), 721–730, U.S. National Report to International Union of Geodesy and Geophysics, 1979–1982, American Geophysical Union, Washington, DC.

Beven, K. J.: 1985, 'Distributed Models,' in M. G. Anderson, and T. P. Burt (eds.), *Hydrological Forecasting*, John Wiley & Sons, Chichester, pp. 405–435.

Beven, K. J.: 1987, 'Towards a New Paradigm in Hydrology,' in *Water for the Future: Hydrology in Perspective*, Intl. Assoc. Hydrol. Sci. Publ. No. 164, pp. 393–403.

Beven, K. J.: 1989, 'Changing Ideas in Hydrology – The Case of Physically-based Models,' *J. Hydrol.* **105**(1/2), 157–172.

Beven, K. J., and M. J. Kirkby: 1979, 'A Physically–Based Variable Contributing Area Model of Basin Hydrology,' *Hydrol. Sc. Bull.* **24**(1), 3, 43–69.

Beven, K. J., R. Warren, and J. Zaoui: 1980, 'SHE: Towards a Methodology for Physically–Based, Distributed Modelling in Hydrology,' in *Hydrological Forecasting*, Intl. Ass. Hydrol. Sci., Pub. No. 129, pp. 133–137.

Beven, K. J., A. Calver, and E. M. Morris: 1987, *'The Institute of Hydrology Distributed Model,'* Report No. 98, Inst. of Hydrology, Wallingford, U.K.

Beven, K. J., E. F. Wood, and M. Sivapalan: 1988, 'On Hydrological Heterogeneity: Catchment Morphology and Catchment Response,' *J. Hydrol.* **100**, 353–375.

Box, G. E. P., and G. M. Jenkins: 1970, *Time Series Analysis Forecasting and Control*, Holden Day, San Francisco, pp. 553.

Brakensiek, D. L.: 1967, 'A Simulated Watershed Flow System for Hydrograph Prediction: A Kinematic Application,' *Proc. International Hydrology Symposium*, Fort Collins, CO.

Bree, T.: 1985, *'Numerical Aspects of General Linear Analysis,'* Unpublished PhD Thesis, Civil Engineering Department, University College, Dublin.

Bruen, M.: 1985, *'Black Box Methods of Systems Analysis Applied to the Modelling of Catchment Behaviour,'* Unpublished PhD Thesis, Civil Engineering Department, University College, Dublin.

Calver, A., and W. L. Wood: 1989, 'On the Discretization and Cost-Effectiveness of a Finite Element Solution for Hillslope Subsurface Flow,' *J. Hydrol.* **110**(1/2), 165–179.

Chiu, C. L. (ed.): 1978, *'Applications of Kalman Filter to Hydrology, Hydraulics and Water Resources,'* University of Pittsburgh, Pittsburgh, PA, pp. 783.

Clark, C. O.: 1945, 'Storage and Unit Hydrograph,' *Trans. Am. Soc. Civ. Engrs.* **110**, 1416–1446.

Cooper, D. M., and E. F. Wood: 1982a, 'Identification of Multivariate Time Series and Multivariate Input-Output Models,' *Wat. Resour. Res.* **18**(4), 937–946.

Cooper, D. M., and E. F. Wood: 1982b, 'Parameter Estimation of Multiple Input-Output Time Series Models: Application to Rainfall-Runoff Processes,' *Wat. Resour. Res.* **18**(5), 1352–1364.

Crawford, N. H., and R. K. Linsley: 1962, *'The Synthesis of Continuous Streamflow Hydrographs on a Digital Computer,'* Tech. Report No. 12, Dept. of Civil Eng., Stanford University, CA.

Crawford, N. H., and R. K. Linsley: 1966, *'Digital Simulation in Hydrology: Stanford Watershed Model IV,'* Report No. 39, Dept. of Civil Eng., Stanford University.

Darcy, H.: 1856, *Les Fontaines Publiques de la Ville de Dijon*, Dalmost, Paris.

Dawdy, D. R., and T. O'Donnell: 1965, 'Mathematical Models of Catchment Behaviour,' *J. Hydraul. Div., Proc. Am. Soc. Civ. Engrs.* **91**(HY4), 123–127.

Dee Steering Committee: 1977, *'Dee Weather Radar and Real–time Hydrological Forecasting Project,'* Central Water Planning Unit, Reading, U.K.

Dooge, J. C. I.: 1959, 'A General Theory of the Unit Hydrograph,' *J. Geophys. Res.* **64**(2), 241–256.

Dooge, J. C. I.: 1965, 'Analysis of Linear Systems by Means of Laguerre Functions,' *J. SIAM Control, Ser. A* **2**(3).

Dooge, J. C. I.: 1973, *Linear Theory of Hydrologic Systems,* Tech. Bull. U.S. Dept. of Agriculture, No. 1468, U.S. Govt. Printing Office, Washington, DC.

Dooge, J. C. I.: 1986, 'Looking for Hydrologic Laws,' *Wat. Resour. Res.* **22**(9), 465–585.

Dooge, J. C. I., and B. J. Garvey: 1978, 'The Use of Meixner Functions in the Identification of Heavily Damped Systems,' *Proc. Roy. Irish Acad., Sect. A* **78**(18).

Dunne, T.: 1970, *Runoff Production in a Humid Area,* Rep. ARS 41-160, Agr. Res. Serv., U.S. Dept. of Agr., Washington, DC, pp. 108.

Dunne, T.: 1978, 'Field Studies of Hillslope Flow Processes,' in M. J. Kirby (ed.), *Hillslope Hydrology,* London, pp. 227–294.

Eagleson, P. S., R. Mejia, and F. March: 1966, 'Computation of Optimum Realizable Unit Hydrographs,' *Wat. Resour. Res.* **2**(4), 755–764.

Fleming, G.: 1975, *Computer Simulation Techniques in Hydrology,* Environmental Science Series, Elsevier, NY, pp. 333.

Freeze, R. A.: 1971, 'Three-Dimensional, Transient, Saturated–Unsaturated Flow in a Groundwater Basin,' *Wat. Resour. Res.* **7**(2), 347–366.

Freeze, R. A.: 1972a, 'Role of Subsurface Flow in Generating Surface Runoff. 1. Baseflow Contributions to Channel Flow,' *Wat. Resour. Res.* **8**(3), 609–623.

Freeze, R. A.: 1972b, 'Role of Subsurface Flow in Generating Surface Runoff. 2. Upstream Source Areas,' *Wat. Resour. Res.* **8**(5), 1272–1283.

Freeze, R. A.: 1978, 'Mathematical Models of Hillslope Hydrology,' in M. J. Kirkby (ed.), *Hillslope Hydrology,* Wiley, pp. 126–177.

Freeze, R. A., and R. L. Harlan: 1969, 'Blueprint for a Physically–Based, Digitally–Simulated Hydrologic Response Model,' *J. Hydrol.* **9**, 237–258.

Garrick, M., C. Cunnane, and J. E. Nash: 1978, 'A Criterion of Efficiency for Rainfall–Runoff Models,' *J. Hydrol.* **36**(3/4), 375–381.

Goldstein, J. D., and W. E. Larimore: 1980, *Application of Kalman Filtering and Maximum Likelihood Identification to Hydrologic Forecasting,* Tech. Report, Analytic Science Corporation, Boston, MA.

Henderson, F. M.: 1966, *Open Channel Flow,* Macmillan, pp. 522.

Henderson, F. M., and R. A. Wooding: 1964, 'Overland Flow and Groundwater Flow from a Steady Rainfall of Finite Duration,' *J. Geophys. Res.* **69**(8), 1531–1540.

Hewlett, J. D., and A. R. Hibbert: 1963, 'Moisture and Energy Conditions within a Sloping Soil Mass During Drainage,' *J. Geophys. Res.* **68**(4), 1081–1087.

Hewlett, J. D., and A. R. Hibbert: 1967, 'Factors Affecting the Response of Small Watersheds to Precipitation in Humid Areas,' in W. E. Sooper, and H. W. Lull (eds.), *Forest Hydrology,* Pergamon Press, Oxford, pp. 275–290.

Hewlett, J. D., and W. L. Nutter: 1970, 'The Varying Source Area of Streamflow from Upland Basins,' *Proc. of the Symp. on Interdisciplinary Aspects of Watershed Management,* Am. Soc. Civil Engrs., New York.

Hewlett, J. D., and C. A. Troendle: 1975, 'Non–Point and Diffused Water Source: A Variable Source Area Problem,' in *Watershed Management,* Logan, Utah, Am. Soc. Civil Engrs., pp. 21–45.

Horton, R. E.: 1933, 'The Role of Infiltration in the Hydrologic Cycle,' *Trans. Am. Geophys. Union* **14**, 446–460.

Horton, R. E.: 1938, 'The Interpretation and Application of Runoff Plot Experiments with Reference to Soil Erosion Problems,' *Proc. Soil Sci. Soc. Am.* **3**, 340–349.

Huggins, L. F., and E. J. Monke: 1968, 'A Mathematical Model for Simulating the Hydrologic Response of a Watershed,' *Wat. Resour. Res.* **4**(3), 529–539.

Hursh, C. R., and E. F. Brater: 1944, 'Separating Storm-Hydrographs from Small Drainage Areas into Surface and Subsurface Flow,' *Trans. Am. Geophy. Union* **22**, 863–871.

Hydraulics Research Ltd.: 1981, *Design and Analysis of Urban Storm Drainage: The Wallingford Procedure, Vol. 1, Principles, Methods and Practice,* pp. 173.

Ibbitt, R. P.: 1970, *Systematic Parameter Fitting for Conceptual Models of Catchment Hydrology,* Unpublished PhD Thesis, Dept. of Civil Eng., Imperial College of Science and Technology, London.

Izzard, C. F.: 1946, 'Hydraulics of Runoff from Developed Surfaces,' *Proc. Highway Research Board,* 26th Annual Meeting.

Jamieson, D. G., and J. C. Wilkinson: 1972, 'River Dee Research Programme 3. A Short–Term Control Strategy for Multipurpose Reservoir Systems,' *Wat. Resour. Res.* **8**(4), 911–920.

Johnstone, D., and W. P. Cross: 1949, *Elements of Applied Hydrology*, Ronald Press Co., New York, pp. 270.

Kibler, D. F., and D. A. Woolhiser: 1970, *'The Kinematic Cascade as a Hydrologic Model,'* Hydrology Paper No. 39, Colorado State University.

Klatt, P., and G. A. Schultz: 1981, 'Improvement of Flood Forecasts by Adaptive Parameter Estimation,' *Proc. Int. Symp. on Rainfall Runoff Modelling*, Water Resources Publications, Littleton, CO.

Klemes, V.: 1986, 'Dilettantism in Hydrology: Transition or Destiny,' *Wat. Resour. Res.* 22(9), 177s–188s.

Klemes, V.: 1988, 'A Hydrological Perspective,' *J. Hydrol.* 100, 3–28.

Kraijenhoff, D. A., and R. J. Moll (eds.): 1986, *River Flow Modelling and Forecasting*, Reidel Pub. Co., Dordrecht, pp. 372.

Lambert, A. O.: 1972, 'Catchment Models Based on ISO–Functions,' *Jour. Inst. Water Engineers* 26, 413.

Lambert, A. O., and M. J. Lowing: 1980, 'Flow Forecasting and Control on the River Dee,' in *Hydrological Forecasting*, Intl. Ass. Hydrol. Sci. Publ. No. 129, pp. 525–534.

Leclerc, G., and J. C. Schaake: 1973, *'Methodology for Assessing the Potential Impact of Urban Development on Urban Runoff and the Relative Efficiency of Runoff Control Alternatives,'* Report No. 167, R. M. Parsons Lab., Dept. of Civ. Eng., MIT, Cambridge, MA.

Liggett, J. A., and D. A. Woolhiser: 1967, 'Difference Solutions of the Shallow–Water Equations,' *Jour. Eng. Mech. Div., Proc. Am. Soc. Civ. Engrs.* EM 2, 39–48.

Lighthill, J., and G. B. Whitham: 1955, 'Kinematic Waves, I. Flood Movement in Long Rivers,' *Proc. Royal Soc. London* (A)229, 281–316.

Linsley, R. K.: 1982, 'Rainfall-Runoff Models – an Overview,' in V. P. Singh (ed.), *Proc. Int. Symp. on Rainfall–Runoff Modelling*, Water Resources Publications, Littleton, CO.

Linsley, R. K., and W. C. Ackerman: 1942, 'A Method of Predicting the Runoff from Rainfall,' *Trans. Am. Soc. Civ. Engrs.* 107, 825–835.

Lloyd–Davies, D. E.: 1906, 'The Elimination of Storm Water from Sewerage Systems,' *Proc. Inst. Civil Engrs.* 164, 41–67.

Mandeville, A. N., P. E. O'Connell, J. V. Sutcliffe, and J. E. Nash: 1970, 'River Flow Forecasting through Conceptual Models III. The Ray Catchment at Grendon Underwood,' *J. Hydrol.* 11, 109.

McCarthy, G. T.: 1938, 'The Unit Hydrograph and Flood Routing,' *Proc. Conf. North Atlantic Division, US Army Corps of Engineers*.

Moore, R. J., and R. T. Clarke: 1981, 'A Distribution Function Approach to Rainfall–Runoff Modelling,' *Wat. Resour. Res.* 17(5), 1367–1382.

Morris, E. M.: 1979, 'The Effect of the Small–Slope Approximation and Lower Boundary Conditions on Solution of the Saint–Venant Equations,' *J. Hydrol.* 40(1/2), 31–47.

Morris, E. M.: 1980, 'Forecasting Flood Flows in Grassy and Forested Basins Using a Deterministic Distributed Mathematical Model,' in *Hydrological Forecasting*, Intl. Assoc. Hydrol. Sci. Publ. 129, pp. 247–255.

Mulvaney, T. J.: 1851, 'On the Use of Self–Registering Rain and Flood Gauges in Making Observations of the Relations of Rainfall and Flood Discharges in a Given Catchment,' *Trans. Inst. Civil Engrs. Ireland* IV(II), 19–33.

Nash, J. E.: 1958, 'The Form of the Instantaneous Unit Hydrograph,' *Intl. Assoc. Sci. Hydrol.* 45, 114–121.

Nash, J. E.: 1960, 'A Unit Hydrograph Study, with Particular Reference to British Catchments,' *Proc. Inst. Civil Engrs.* 17, 249–282.

Nash, J. E.: 1961, 'A Linear Transformation of a Discharge Record,' *Proc. Ninth Convention, Intl. Ass. Hydr. Res.* 3, 13.1–2.

Nash, J. E., and J. V. Sutcliffe: 1970, 'River Flow Forecasting through Conceptual Models. I – A Discussion of Principles,' *J. Hydrol.* 10, 282–290.

Nash, J. E., and B. I. Barsi: 1984, 'Hybrid Model for Flow Forecasting on Large Catchments,' *J. Hydrol.* 65(1–3), 125–137.

Natale, L., and E. Todini: 1977, 'A Constrained Parameter Estimation Technique for Linear Models in Hydrology,' in T. A. Ciriani, U. Maione, and J. R. Wallis (eds.), *Mathematical Models for Surface Water Hydrology*, John Wiley & Sons, Chichester, UK, pp. 109–147.

Nemec, J.: 1985, *Design and Management of Hydrological Forecasting Systems*, D. Reidel, Dordrecht.

Nemec, J.: 1986, 'Design and Operation of Forecasting Operational Real–Time Hydrological Systems (FORTH),' in D. A. Kraijenhoff, and J. R. Moll (eds.), *River Flow Modelling and Forecasting*, D. Reidel, Dordrecht, pp. 299–327.

O'Connell, P. E. (ed.): 1980, *'Real–Time Hydrological Forecasting and Control,'* Institute of Hydrology, Wallingford, UK, pp. 264.

O'Connell, P. E., G. P. Brunsdon, D. W. Reed, and P. G. Whitehead: 1986, 'Case Studies in Real–Time Hydrological Forecasting from the UK,' in D. A. Kraijenhoff, and R. Moll (eds.), *River Flow Modelling and Forecasting*, Reidel, Dordrecht, pp. 195–240.

O'Connell, P. E., and R. T. Clarke: 1981, 'Adaptive Hydrological Forecasting – a Review,' *Bull. Intl. Ass. Hydrol. Sci.* **26**(2), 179–205.

O'Connell, P. E., J. E. Nash, and J. P. Farrell: 1970, 'River Flow Forecasting Through Conceptual Models II. The Brosna Catchment at Ferbane,' *J. Hydrol.* **10**, 317–329.

O'Connor, K. M.: 1976, 'A Discrete Linear Cascade Model for Hydrology,' *J. Hydrol.* **29**, 203–242.

O'Donnell, T.: 1960, 'Instantaneous Unit Hydrograph Derivation by Harmonic Analysis,' *Intl. Ass. Hydrol. Sci. Publ.* **51**, 546–557.

O'Donnell, T.: 1986, 'Deterministic Catchment Modelling,' in D. A. Kraijenhoff, and R. J. Moll (eds.), *River Flow Modelling and Forecasting*, Reidel Pub. Co., Dordrecht, pp. 11–37.

O'Donnell, T., and R. P. Ibbitt: 1974, 'Designing Conceptual Catchment Models for Automatic Fitting Methods,' *Intl. Ass. Hydrol. Sci. Publ.* **101**, 461–475.

O'Kelly, J. J.: 1955, 'The Employment of Unit Hydrographs to Determine the Flows of Irish Arterial Drainage Channels,' *Proc. Inst. Civ. Engrs. Ireland* **4**(3), 365–412.

Parkin, G. P., and R. Mackay: 1990, 'A Framework for Assessing the Predictive Capability of Physically–Based Hydrological Models,' Paper Presented at XV General Assembly, European Geophysical Society, Copenhagen, EGS Abstract *Annales Geophysicae*, Special Issue, p. 95.

Penman, H. L.: 1961, 'Weather, Plant and Soil Factors in Hydrology,' *Weather* **16**, 207–219.

Ragan, R. M.: 1968, 'An Experimental Investigation of Partial Area Contributions,' *Intl. Assoc. of Sci. Hydrol.* **76**, 241–251.

Rawitz, E., E. T. Engman, and G. D. Cline: 1970, 'Use of Mass Balance Method for Examining the Role of Soils in Controlling Watershed Performance,' *Wat. Resour. Res.* **6**(4), 1115–1123.

Restrepo–Posada, J., and R. L. Bras: 1982, *'Automatic Parameter Estimation of a Large Conceptual Rainfall–Runoff Model: A Maximum Likelihood Approach,'* Report No. 267, Dept. of Civil Eng., M.I.T., Cambridge, MA.

Rockwood, D. M.: 1958, 'Columbia Basin Stream Flow Routing Computer,' *Trans. Am. Soc. Civ. Engrs.* **126**(4), 32–56.

Rockwood, D. M.: 1968, 'Application of Streamflow Synthesis and Reservoir Regulation –SSARR–Program to Lower Mekong River,' *Intl. Assoc. Sci. Hydrol. Publ.* **80**, 329–344.

Rodriguez–Iturbe, I., and J. B. Valdes: 1979, 'The Geomorphological Structure of Hydrologic Response,' *Wat. Resour. Res.* **15**(6), 1409–1420.

Rosenbrock, H. H.: 1960, 'An Automatic Method of Finding the Greatest or Least Value of A Function,' *Comp. Jour.* **3**, 175–184.

Schaake, J. C., Jr.: 1970, *'Deterministic Urban Runoff Model,'* Burban Water Syst. Inst., Colorado State University, Fort Collins.

Schultz, G. A.: 1986, 'Relationship between Theory and Practice of Real–Time River Flow Forecasting,' in D. A. Kraijenhoff, and J. R. Moll (eds.), *River Flow Modelling and Forecasting*, Reidel Publ. Co., Dordrecht, pp. 181–193.

Sherman, L. K.: 1932, 'Streamflow from Rainfall by Unit–Graph Method,' *Eng. News Record* **108**, 501–505.

Smith, F. G., and L. E. Brazil: 1980, 'Real–Time Hydrologic Forecasting Using Estimation Theory and Computer Graphics,' *Proc. of the JACC Conf.*, San Francisco, CA, Aug.12–15, Vol. II, FA–6A.

Smith, R. E., and D. A. Woolhiser: 1971, 'Overland Flow on an Infiltrating Surface,' *Wat. Resour. Res.* **7**(4), 899–913.

Snyder, F. F.: 1961, *'Matrix Operation in Hydrograph Computations,'* TVA Res. Paper No. 1, Knoxville.

Sorooshian, S., and J. A. Dracup: 1980, 'Stochastic Parameter Estimation Procedures for Hydrologic Rainfall–Runoff Models: Correlated and Heteroscedastic Error Cases,' *Wat. Resour. Res.* **16**(2), 430–442.

Sorooshian, S., and V. K. Gupta: 1983, 'Automatic Calibration of Conceptual Rainfall–Runoff Models: The Question of Parameter Observability and Uniqueness,' *Wat. Resour. Res.* **19**(1), 260–268.

Spolia, S. K., and S. Chander: 1974, 'Modelling of Surface Runoff Systems by an ARMA Model,' *J. Hydrol.* **22**, 317–332.

Stephenson, G. R., and R. A. Freeze: 1974, 'Mathematical Simulation of Subsurface Flow Contributions to Snowmelt Runoff, Reynolds Creek Watershed, Idaho,' *Wat. Resour. Res.* **10**(2), 284–298.

Sugawara, M.: 1961, 'An Analysis of Runoff Structure About Several Japanese Rivers,' *Japanese J. Geophys.* **2**.

TASC: 1980, *'Application of Kalman Filtering and Maximum Likelihood Parameter Identification to Hydrologic Forecasting,'* The Analysis Science Corp., Reading, MA.

Todini, E.: 1978, 'Mutually Interactive State Parameter (MISP) Estimation,' in C. L. Chiu (ed.), *Applications of Kalman Filter to Hydrology, Hydraulics and Water Resources*, University of Pittsburgh, Pittsburgh, PA.

Todini, E.: 1988, 'Rainfall Runoff Modelling: Past, Present and Future,' *J. Hydrol.* **100**, 341–352.

Todini, E., P. E. O'Connell, and D. A. Jones: 1980, 'Basic Methodology: Kalman Filter Estimation Problems,' in P. E. O'Connell (ed.), *Real–time Hydrological Forecasting and Control*, Institute of Hydrology, Wallingford, pp. 66–98.

Todini, E., and J. R. Wallis: 1977, 'Using CLS for Daily or Longer Period Rainfall Runoff Modelling,' in T. A. Ciriani, U. Maione, and J. R. Wallis (eds.), *Mathematical Models for Surface Water Hydrology*, John Wiley and Sons, Chichester, pp. 149–168.

Troendle, C. A.: 1985, 'Variable Source Area Model,' in M. G. Anderson, and T. P. Burt (eds.), *Hydrological Forecasting*, John Wiley & Sons, Chichester, pp. 347–403.

USDA: 1980, *'CREAMS: A Field Scale Model for Chemicals Runoff and Erosion for Agricultural Management Systems,'* USDA Conservation Research Report No. 26, pp. 643.

Whipkey, R. Z.: 1965, 'Subsurface Stormflow from Forested Slopes,' *Bull. Int. Assoc Sci. Hydrol.* **10**(2), 74–85.

Wicks, J. M., and J. C. Bathurst: 1990, 'SHESED: A Physically–Based Distributed Sediment Yield Model for the SHE Hydrological Modelling System,' Submitted for Publication to *J. Hydrol.*

Williams, B. J., and W. W. G. Yeh: 1982, 'Parameter Estimation in Rainfall–Runoff Models,' *J. Hydrol.*

Wilson, C. B., J. B. Valdes, and I. Rodriguez–Iturbe: 1979, 'On the Influence of the Spatial Distribution of Rainfall on Storm Runoff,' *Wat. Resour. Res.* **15**, 321–328.

Wood, E. F. (ed.): 1980, *Recent Development in Real–Time Forecasting/Control of Water Resources Systems*, Pergamon, Oxford.

Wooding, R. A.: 1965 'A Hydraulic Model for the Catchment–Stream Problem, 1. Kinematic Wave Theory,' *J. Hydrol.* **3**(3/4), 254–267.

Woolhiser, D. A.: 1969, 'Overland Flow on a Converging Surface,' *Trans. Am. Soc. Ag. Engrs.* **12**, 460–462.

Woolhiser, D. A.: 1973, 'Hydrologic and Watershed Modelling – State–of–the–Art,' *Trans. Am. Soc. of Ag. Engrs.*, pp. 553–559.

Woolhiser, D. A., and E. F. Schultz: 1973, 'Large Material Models in Watershed Hydrology Research,' *Proc. Int. Symp. on River Mechanics*, Intl. Ass. Hyd. Res., pp. C6.1–C6.12.

Woolhiser, D. A., C. L. Hanson, and A. R. Kuhlman: 1970, 'Overland Flow on Rangeland Watersheds,' *J. Hydrol.* **9**(2), 336–356.

WMO: 1975, *'Intercomparison of Conceptual Models Used in Operational Hydrological Forecasting,'* Operational Hydrology Rep. 7, WMO, No. 429, Geneva.

Zoch, R. T.: 1934, 'On the Relation Between Rainfall and Streamflow,' U.S. Dept. of Commerce, *Monthly Weather Rev., Part I* **62**, 315–322.

Zoch, R. T.: 1936, 'On the Relation Between Rainfall and Streamflow,' U.S. Dept. of Commerce, *Monthly Weather Rev., Part II* **64**, 105–121.

Zoch, R. T.: 1937, 'On the Relation Between Rainfall and Streamflow,' U.S. Dept. of Commerce, *Monthly Weather Review, Part III* **65**, 135–147.

Chapter 2

Hydrology: Infusing Science into a Demand-Driven Art

L. Douglas James
Utah State University
Utah Water Research Laboratory
Logan, Utah 84322-8200 U.S.A.

Abstract. In content, hydrology is a science; in practice it is an art. Scientists study physical processes; practitioners seek numbers. Sciences are shaped by discovery, hydrology by user demands. Hydrologists have been too prone to seek the favor of institutions that seek an image of good performance and feel uncomfortable when forsaking simple methods with official blessings. Because the discipline has stagnated, conceptual breakthroughs are now essential to solve pressing water problems. However, a discipline dominated by scientists tends to understand practical problems poorly. Supply side advances in understanding should be coupled with demand–side advances in addressing problems under innovative leadership who are willing to take risks in the interest of better science and who come from professional societies where both academics and agencies participate. Universities must do better at infusing science into professional curricula. Employers must do better at rewarding excellence.

1. Introduction

1.1. The Art and Science of Hydrology

Hydrologists who prepare histories of their specialty (Biswas, 1970; Dooge, 1988) tell of the presentation, validation, and application of theories as they, using an analogy from economics, write with the viewpoint of suppliers of hydrologic understanding. A well–written, *supply–side* history presents the development of a science through a time series of innovative ideas (each supported by reasoning and credited to a contributor) and ends with a current *state of the science* for practitioners to use and for researchers to advance.

Users of hydrology write from a different perspective. Demand–side histories have chronologies of how hydrologic information has been used by engineers in the designs of public works generally (Armstrong, 1976) or of individual projects particularly (Barnes, 1987). A well–written *demand–side* history presents the development of a technology as a time series of innovative applications of science in landmark projects (each described in benefits to humanity and credited to an agency) and ends in a current *state of the art* that practitioners can apply and that researchers can refine.

This contrasting perspective between scientists and the users of science is found in all disciplines. However in hydrology, there is a difference. The traditional sciences of chemistry, physics, and biology have largely been shaped from the supply side whereas demand–side forces have shaped the content of hydrology.

1.2. The Exchange of Hydrologic Information

The road to this situation began as engineers responded to demand–side requests with judgments extrapolated from principles and experiences. Scientific validation was neglected

31

D. S. Bowles and P. E. O'Connell (eds.), Recent Advances in the Modeling of Hydrologic Systems, 31–43.
© 1991 *Kluwer Academic Publishers.*

whenever engineers would not risk losing a contract by objecting to using a scientifically weak or outdated method (Linsley, 1986). The problem began as users received multiple judgments, gained power to select among them, and eventually became able to require methodologies that gave the answers they wanted. Over time, the pursuit of user satisfaction has made hydrology a demand–driven art. The practice of reviewing answers for institutional acceptability contrasts with the supply–driven process of peer review.

The two contrasting processes have evolved out of differences in the arena where the supply and demand for information interact. The producers and users of most sciences exchange information in a forum in which both sides have relatively equal voice and no one can dominate the choices. In contrast, the delivery of hydrologic information is dominated by large governmental agencies driven by institutional goals that are able to transcend supply–side constraints. Using our analogy from economics, the result is much like what one would expect from the difference between a free market and monopoly control.

Government decisions are driven by popular vote not scientific facts. The ideal paradigm of water resources planning is that decision makers should secure sound scientific information to make objective decisions. In reality political and bureaucratic pressures often overwhelm objectivity. One of the greater pressures is described in Chow's (1972) discussion of the conflict between the completeness preferred by hydrologists and simplicity desired by popular institutions. Whether by simplification or by lack of understanding, hydrologic information is profoundly influenced by institutionally–oriented "satisficing" to use the term coined by Simon (1955). We may not like to see studies stop once people are satisfied, but we must consider this institutional tendency in our efforts to bring the art and the science together.

2. Discussion

2.1. The Present Situation

Given the tendency of institutions toward satisficing, governmental agencies are apt to overlook science as they respond to the strong public pressures that follow the chance occurrences of extreme hydrologic (water shortages and water pollution) and socio–political (financial difficulties, health or environmental disasters, elections) events (Grigg, 1985, p. 56). Judgments are made under the pressure to satisfy people in distress and become precedents that fix standards. Thereafter, bureaucracies impose barriers that keep hydrologists using the same tools despite changing needs for information on such topics as floods and droughts. In fact the practice of hydrology is dominated by methods that have changed little, except for computerization, since their adoption as much as fifty years ago. This stagnation in hydrology contrasts sharply with the continual updating of the methods used by agencies with clear scientific missions (space exploration, medicine, weaponry, etc.).

At the present time, updating is constrained because hydrologic science has taken hydrologic practice about as far as it can. As the art of hydrology becomes increasingly outdated, hydrologists express growing frustration with the archaic methods of their profession; Klemes (1986) wonders whether hydrology will join alchemy and astrology in the annals of dilettantism. Pilgrim (1986) implies that hydrology has been going downhill ever since the contributions of a few scholars of the first rank in the early 20th century.

The pressure for new methods will continue to mount. Fuller utilization of our water resources increases the demand for more exacting hydrologic information. The aging of the methodology used in hydrologic analysis is increasingly limiting water resources management and is contributing to a rising probability of "surprise" failures. Since new methodology to meet the long–term needs of water resources planning and management will require supply–driven advances in underlying hydrologic science, research is essential.

However, as one more manifestation of the problem, research funding in the United States has declined. Most of the remaining current hydrologic research is supported by agencies with development and regulatory missions; these agencies are pressured by circumstances into seeking "fixes" where the state of the art (demand side) works poorly. The few funds appropriated to advance the state of knowledge (supply side) support small, fragmented studies.

While the relative strengths of the supply–side and demand–side forces in hydrology vary from one country to another, the problem is worldwide; and one goal of the NATO Institute should be that of bringing the discipline together to address its need. Scientists and engineers from all nations must work together in advancing the science and the art of hydrology for the good of all people.

2.2. A Negative Feedback Hypothesis

Bemoaning accomplishes little. We should be analyzing the present stagnation to find its fundamental causes and to find ways to effect change. To begin this discussion, I offer a negative feedback hypothesis:

A discipline driven by institutionalized demands may advance in understanding problems but stagnates in developing ways to solve them.

In other words, research and application develops a discipline, and practitioners satisfy their funding sources. The negative feedback begins when funding agencies become satisfied with the methods that they have and cease searching for yet better ways.

The evidence is largely empirical. Agencies that began making hydrologic studies long ago follow older methods than those that began more recently. Old methods are kept by users who like their outcomes and never worry about the underlying assumptions. Established institutions shy away from critical inquiry and distrust new methods because of their complex representations, incomprehensible mathematics, and specialized vocabulary.

Interestingly enough, institutions outside the water agencies are beginning to sense a problem. Members of the U.S. Congress are making proposals that would allocate substantial funds to establish new research centers. Many see the self–satisfied mission agencies as spending significant sums for research with no documented progress.

2.3. Four Resisting Forces

To break out of this situation, we must understand the forces holding us within it and find ways to overcome them. For a start, the fact that universities do not recognize hydrology as a discipline and that agencies do not recognize it as a profession weakens the ability of hydrologists to resist demand–side pressures. Second, research is easily postponed because it lacks urgency. Agencies that fund according to "urgency ratings" are able to avoid research when conditions are normal and to demand rapid fixes during crises. They become satisfied with their emergency ends (usually in a time frame far shorter than that required to complete meaningful research) and thereafter give poor reviews on research proposals because they consider the problems solved.

Third, agencies continue outdated practices to avoid admitting mistakes and losing political support from parties who like their prior information. Conformance to recognized practice is often more important than scientific truth in resisting liability suits. Fourth, agencies that supply "collective goods" must satisfy the general public, whereas firms that produce goods for market sale need only satisfy their customers. Some people complain about change, and others want to try new things. Government protects the public interest by listening more to them, and business makes sales by providing opportunities for their customers. Consequently, agencies resist while firms thrive on new methods.

Hydrology has not been able to breakout of its negative feedback scenario because of its lack of recognition as a discipline, agency orientation to immediate crises, the primary use of hydrology in producing collective goods, and the advantage that institutions see in using proven methods to protect themselves against liability. These factors reinforce each other to make the situation progressively worse. Some way is needed to overcome this cycle.

2.4. A Companion Supply–Side Hypothesis

One response would be to seek a way to shift to a supply–driven mode. However, before those of you with a scientific bent become too excited, I must submit a second hypothesis:

A discipline driven by institutionalized supply may advance in the sophisticated representation of physical processes but stagnates in practice.

Here the empirical evidence is that the same water management problems seem to continue indefinitely. Agency people lack background to appreciate and time to apply complex ideas and have little incentive to make an extra effort in this direction. Agencies are never quite able to sue scientific principles to find hard solutions. Where the private sector has mechanisms for turning new ideas into profit; the public sector, served by hydrology, is more likely to view new ideas as a source of trouble. Returning to our analogy from economics, monopoly power over the exchange of ideas is harmful whether it comes from the supply or the demand side. Demand–side monopoly stifles the development of new concepts, and supply–side monopoly stifles their use.

2.5. Initiating Change

At this point in our analysis, we have concluded that science goes astray when research directions (approaching monopolistic control) are determined by either the demand side, users who make applications, or the supply side, scientists who pursue new understanding. Presently, hydrology is dominated from the demand side, and the demand–side institutions have four strong forces behind them. The logical way to address this problem is by building supply–side institutions to develop counter forces that can achieve a balance.

The leadership needed for this effort will come neither from the agencies nor the academic community. It must come from the discipline (Pilgrim, 1986). Professional hydrologists from both sides must take leadership responsibility. The expert leadership of a recognized discipline can interact with users in demonstrating the dangers of repeated rapid fixes while showing how abstract principles can be used in long–term problem solving. If we are to succeed, we must forcefully demonstrate that the day is coming when the water management institutions will no longer be able to satisfy the public with one more quick fix. We will then gain institutional support so that present and projected needs and present and projected capabilities can be joined in planning for the future.

2.6. Demands on Hydrology

This reasoning brings us to projecting the future demands on hydrology. The needs of water resources planning and management for hydrologic information are moving from point estimates to spatially–distributed mapping and from present estimates to projections of changes over time. Hydrologists will be asked to deliver greater precision and reliability in performing more complex analyses. Eight specific demands are listed below in order of increasing complexity. The time trend is for practicing hydrologists to move from long–accepted procedures for doing tasks at the beginning of the list to new procedures for doing the latter tasks. The eight demands are:

(1) *Conveyance Design*. At the simple end of our list of tasks, hydrologists are asked for a flow rate to use as a capacity for a channel or pipeline to supply water or to contain floods. Since agencies cannot afford to design for really extreme events, water planners dimension these facilities to convey more ordinary flows (10 to 100 year). Hydrologists can respond by estimating a single value with empirical relationships that have performed sufficiently well so local engineers see no reason for greater sophistication and believe that the occasional failures are unavoidable acts of God. However, the increasing utilization of land and water resources is causing a *trend* toward placing greater value on precision and reliability because small estimating errors can add greatly to the cost of either the facilities or the losses depending on whether the estimates are too high or too low. Changing water and land uses make estimation more difficult.

(2) *Storage Design*. At the next step of increasing complexity in estimating, the design of reservoir storage increases the information need from peaks to volumes so that the engineer can dimension a reservoir for holding water for later use or for reducing downstream flood peaks. Hydrologists are asked for entire flow hydrographs, and engineers use them to determine storage requirements, allocate the space among uses, and dimension outlet works and spillways. They initially responded by using infiltration indices for estimating runoff volumes and distributing the flows with unit hydrographs. As storage systems become more complex, the *trend* is toward needing sets of simultaneous inflow hydrographs that preserve the serial and cross correlations found in nature. The design engineers can then route multiple simultaneous hydrographs through systems of reservoirs, connecting channels, and select sizes for facilities throughout the system.

(3) *Economic Analysis*. The step from engineering design to economic justification and optimization adds a need for flow probability distributions. Many design dimensions are still selected by empirical standards. However, most agencies recognize the need to extend the hydrologic information from a single set of design hydrographs to a probability distribution of hydrograph sets by frequency so that net benefits can be computed, compared, and maximized. In the future one can expect *trends* toward increased costs to make the savings that can be achieved through system optimization more important. Hydrology will be asked for spatially–distributed sets of hydrographs for various frequencies.

(4) *Safety Design*. The step to safety analysis adds a need for greater precision in quantifying the flood and drought tails of the frequency distributions. The hydrologic study must extrapolate statistically, with guidance from the bounds of the physical laws, to estimate events much larger or smaller than are likely to be found during a period of record. The United States is witnessing a *trend* toward increasing institutional concern over failures of engineering designs during extreme events. The large amounts of money required to increase structural safety and service dependability will place increasing demands on precise hydrologic information. The present large estimating uncertainties can only be reduced through major methodological breakthroughs. Perhaps the specter of liability from poor estimates that lead to catastrophic failure will become a major incentive for bringing more science into the hydrologic art.

(5) *Operation and Maintenance*. The next step shifts from formulating facilities to developing and implementing rules for their operation and maintenance. The hydrologic study must expand to tracking flows and storages in real time to provide information for making operating decisions; this requires indices that can easily be measured while events are in progress. Some important demand *trends* are for advances in longer term weather prediction, projecting storm paths, using monitored indices to update estimates of flows in real time, and

automating control. These needs are further magnified as planners face pressures to operate systems of projects for greater water use efficiency and benefits at the river basin scale.

(6) *Extension to Nonstructural Measures.* The step from structural to nonstructural measures takes hydrologist away from the delivery of technically sophisticated information to professional engineers for project design and operation. For planning nonstructural measures, hydrologists must deliver descriptive information in popular format to managers who have limited hydrologic experience and who must explain the technical information to a public with even less hydrology experience. For example, hydrologists map flood plains for selected frequencies (principally the 100–year event in the United States), managers use the information to set regulatory zones, and property owners make the actual land use decisions. Most damages are caused by small, ungaged streams, and the limited availability of data, experts, and methodology to conduct floodplain information studies limit the reliability of such studies. As a *trend* for the long run, hydrology must deliver reliable, spatially varied information on such topics as floodplain hazard, drought susceptibility, and exposure to waste pollution. The best nonstructural practices vary by place and time. People need information pertaining to their place and time, need it when they have time windows for using it, and want opportunities to ask for clarification. The information systems of the future must maintain accessible data bases for user–friendly query on personal computers.

(7) *Extension to Environmental and Social Concerns.* Still additional hydrologic information is necessary for assessing environmental (such as fish habitat) and social (such as recreational opportunities) impacts. So far, hydrologists have largely supplied hydrologic information for impact assessments. This *trend* will surely lead to instances in which faulty estimates will lead to unanticipated disasters. Research will be needed to define the hydrologic processes causing favorable and adverse impacts, to develop methods for their estimation, and to apply the results in impact amelioration.

(8) *Extension to Sediment and Water Quality.* One last extension of hydrologic information is to estimate water transport of sediments, salts, organic chemicals, and other natural and manmade waste loads. Each transport impacts all seven preceding categories. The *trend* is toward greater concern over movements of transported substances through hydrologic systems. Most present transport functions are little more than gross empirical approximations. We need to quantify spatially distributed flow and transport processes to cover the entry (scour and dissolution), transport, and deposition (sedimentation and precipitation) of many physical and chemical substances (Golubev and Biswas, 1985).

2.7. Institutional Goals

In pursuing these goals, hydrologists must consider the needs seen by the agencies. Demand–side people often complain that hydrologic literature is too specialized to apply; this is often true because the increments of progress reported in most scientific articles are too small for individual applications. Most journals report intermediate findings that must later be synthesized for practical applications. Hydrologists should be doing a better job of this synthesization; our failure in this area may be contributing to the lack of support by water management institutions for scientific research in hydrology.

In the free–enterprise system, market demands express consumer preferences that are filtered through personal values and expressed as choices among products offered by suppliers (generally industry). Conversely, water resources planning and management use hydrologic information to serve economic and social needs that are filtered through the perceptions of governments and expressed in official authorizations and appropriations. This interpretive role of government makes the process more demand driven.

As projects and programs are selected and deliver new services, both the water users and the agencies "acquire" demands. Just as people acquire tastes for new goods, agencies gain satisfaction in accomplishments. These acquired institutional demands may differ from the interests of the people they serve, and these are driving the art of hydrology. In propagating counter forces, we must deal with cases where serving the public can increase the risk to the agency by, for example, giving the agencies new missions with other satisfactions.

Where individuals pursue profit, agencies seek funding, authority, and programs. Each of these is tied to reputation. New agencies are often willing to be innovative (try new methods of hydrologic analysis in this case) to build their reputation, but once they feel that they have arrived they want to protect their reputation by sustaining dependable service. Agency management resists ideas that have not been proven reliable in performance, and agencies become increasingly biased against change as they become more entrenched. Sometimes the aging of agency staff adds to the problem.

Agencies use hydrologic information to protect their reputation in three ways:

(1) *Performance Protection.* Agencies do not like surprises that may threaten project performance. They seek protection against sudden (dam collapse) and gradual (reservoir silting) physical failures as well as against increasing costs that erode their budgets. Advances in hydrologic science can appeal to this need by providing sound hydrologic information to increase safety and operating efficiency and reduce costs.

(2) *Legal Protection.* Agencies want to protect themselves against liability claims, costly environmental mitigation requirements, loss of water rights, and other restrictive legal actions. Lawyers advise that unjustified changes in technical procedures lead to trouble. However, sound hydrologic information can be used to win court cases.

(3) *Image Protection.* The erosion of public support is the greatest threat to a water agency. Public support for new water resources development projects is generally softening in North America. Irrigation is particularly affected. Water quality programs and municipal and industrial water supply fare more favorably. Sound hydrologic information can be used to detect pending problems and to document program benefits.

2.8. Institutionally Preferred Information Characteristics

The drawback in using these three contributions of hydrologic science is that they are all the incentive for the needed institutional change; however, change must begin by addressing the short–term institutional need for self protection. Here, we find that short–term forces have shaped the demands for hydrologic information for water resources planning and given it six characteristics:

(1) *Simple.* The Hydrologic information transferred through institutional channels is reduced to concepts that are clearly understood by everyone involved. Complex concepts are simplified because the individuals explaining them limit what they say to their comfort zone for answering questions. For this reason, an organization is often reluctant to accept methods that were not taught when the leadership of its technical staff were in school (Linsley, 1986). This process is not a problem in the private sector where people individually choose products that work for them. The difference is that in the public sector the information must be personally convincing to agency decision makers who select structural projects and who attempt to achieve consensus among individuals that implement nonstructural measures.

(2) *Deterministic.* Decision makers at both levels like fixed numbers as a way to avoid the doubts raised by probability distributions and unquantified uncertainties. Construction agencies are more risk adverse in their treatment of uncertainty than objective analysis would dictate. Regulatory agencies find that fixed numbers are more easily enforced, and they overlook the advantages of flexibility for greater program effectiveness. Both kinds of agencies consider explanations to the public that contain uncertainty to be a negative factor in image building. One way to begin to reverse this feeling is that people recognize the hydrologic variability separating wet years from dry years is readily recognized. This concept offers a

starting point for applying the concept of hydrologic risk for more cost effective program management.

(3) *Favorable.* Agencies, even more than individuals, like to appear in a good light. Institutions present data that make their programs appear more valuable.

(4) *Averaged.* Public agencies feel obligated to treat everyone equally. This preference for uniformity causes average flood, drought, or pollution risks to be used to portray conditions over large areas. Average data and common methods come to be applied inappropriately in quite different situations. Hydrologic spatial variability is widely overlooked in such non-structural programs as flood plain management and water conservation. For example, losses to properties subject to flooding by a 20–year event vary greatly with the severity of the floods that actually occur.

(5) *Consistent.* Facilities require a large investment in infrastructure. Effective regulation requires a training investment in human capital. A policy of consistency over time adds security for cost recovery that makes these investments more attractive.

(6) *Formal Blessing.* Since governmental agencies must defend their use of hydrologic information, agencies require that their procedures be formally documented. Thereafter, compliance with adopted policy becomes the single most important attribute of the information that is used. Policy changes are resisted, and the time lags that occur in making changes become a major obstacle in updating to the changing state–of–the–art.

In reviewing these six characteristics, we find once more that the better information a more scientific approach to hydrology could provide would help overcome the real institutional concerns. The information becomes more simple as it can be more clearly explained; it appears more certain because the uncertainties are better quantified; it is more favorable to agencies because they make fewer mistakes; it treats everyone equally in ways that can be shown quantitatively; it adds consistency by reducing needs for future change, and it provides better documentation for obtaining agency approval. All six points can be used to advantage.

2.9. Institutional Constraints

Thus far, we have examined the long–term institutional goals in using hydrologic information and the short–term institutional preferences on the form in which information is supplied. Our next task is to examine the constraints that these goals and preferences cause institutions to use in selecting information. As these constraints have a large part in inadvertently promoting the stagnation of the practice of hydrology (Linsley, 1986), they should be carefully examined.

The examination should begin by recognizing that agency standards have an important role in quality control. They are necessary to assure reliable measurements, guide sound professional practice, and prevent premature changes. However, they also allow institutions to filter out negative information when formalizing their preferences. Acceptance or rejection of a proposed standard ideally depends on its support in theory and data and practically depends on judgments based on institutional goals.

The bases for hydrologic standards change with time. Periodic reassessment is necessary to determine when a situation has altered sufficiently to make a change. When change is needed, a specific proposal must be processed through official channels where precedent has a stronger role and theoretical soundness a lesser role than scientists and engineers like to see. When changes are made, credible explanations must be given to the public.

Presently, new science must be infused into advancing the hydrologic state–of–the–art by changing past standards. The logical starting point is to investigate current standards and identify changes that would strengthen the scientific component of hydrology. Five types of standards to consider are:

(1) *Data Standards.* Data collection standards provide the quality control necessary to achieve measurement reliability by defining acceptable instrumentation, space–time gage networks, and data storage and retrieval systems. Existing standards were largely set to match the capabilities of the technology available when the present data collection system was inaugurated. They then became entrenched by the cost of changing the instrumentation and the consequences of losing the homogeneity of the data series. The instruments and networks are largely managed to serve institutional rather than scientific goals, and data from more recent tools, such as radar, have not been recorded for hydrological applications. Institutional needs have long been a driving force in data collection; Dooge (1988) notes that the first hydrological measurements on the Nile were used to assist taxation rather than hydraulic design.

(2) *Support Standards.* The standards for such services as personnel, purchasing, and computers have largely been established to procure the needed support for water resources planning by methods that treat suppliers fairly and minimize cost. These methods favor established procedures and simple information because they are less costly, and the methods set procurement and hiring at minimum acceptable levels rather than to achieve maximum contributions.

(3) *Design Standards.* Agency manuals on design practices seek to ensure that a bureaucracy acts consistently over time and treats all publics uniformly. Too often the manuals are written more to minimize liability than to maximize productivity. Later, a change may be interpreted as an admission of error; while adoption of change may expose the innovator to liability for discarding sound engineering practice. Criteria are needed to expedite responsible changes.

(4) *Professional Standards.* An agency tries to motivate people to do their best work as professionals; however, formal statements of expectation generate inertia. People trained and experienced in a way of doing things must make considerable effort to do something new. Change is even more difficult for institutions because of complex interpersonal relationships; some people resist change and have greater difficulty in developing new systems for information detection and transmission (Adams, 1986). Motivationally, innovative people who achieve needed changes should be given the recognition that makes their effort worthwhile.

(5) *Communication Standards.* Agencies divide responsibilities among their subunits and establish routes for information flow through the bureaucracy. Information items that should be assessed by experts may, in the routing, bypass hydrologists as policies are formed to support institutional rather than technical goals. Expert hydrologists have an important role in expediting the communication flow through channels for use in decision making.

2.10. Problems of the Commons

One of the major obstacles to change is the lack of motivation. The underlying problem is that hydrologic knowledge is a common resource that people can exploit and thus follows an incentive corollary to the Law of the Commons:

People exploit new knowledge before the opportunity for profit is lost.

However, in contrast to the grass in the village green or ground water, the use of knowledge by one party does not reduce its subsequent availability to others; in fact, the experiences gained by new applications is more likely to increase knowledge. Nevertheless, there is a major disincentive in what can be called the Reverse Law of the Commons:

People do not contribute to a resource pool when others get the credit or profit.

This law operates to some degree among scientists; institutions are probably more selfish in this regard than are individuals.

2.11. Comparing Surface and Ground Water Hydrology

Before going on, it is useful to reflect on the institutional differences between surface and groundwater hydrology. In the past, groundwater development could not provide institutions the glory of large projects, and ground water programs were thus seldom considered politically important. This may be changing as the growing threat from pollution to drinking water safety attracts political interest. Here we come to a paradox, greater institutional attention can both help and stifle science. In the recent past, the science of groundwater hydrology has been advancing more rapidly than that of surface water because, in part, it had enough attention to be funded but not so much attention that water management agencies established a vested interest in the specifics of the results. However, as the institutional setting changes, the expanded agencies and the profession itself will have to make a strong effort to prevent undue influence from the demand–side forces.

2.12. Impacts of Institutional Goals on Information Demands

The goal of protecting their institutional image affects (some would say dominates) the way that agencies use hydrology. When operating in their normal mode, agencies reduce their short–term risk by following established procedures for using hydrologic information. However, when these fail, new procedures must be hastily formulated. Haste increases mistakes, and institutional inertia continues them. In the end an agency exaggerates its failures (physical, environmental, economic, social, or political) by continued rejection of new ideas.

Conversely, an agency enhances its long–term image by finding and applying new methods to provide the continued fair and consistent service that the law requires. Sound hydrologic information can be used to show that people are being treated fairly when legal issues are raised in administrative review and court hearings. Such equity is impossible when management uses information with large random errors (Linsley, 1986).

Agencies use their public image when they justify their requests for additional programs and budgets. A suggestion that problems are being ignored within an agency and hidden from the public is damaging. For example, it is far better to use hydrologic information to quantify volumes of water delivered or values of damages prevented. Examples of problem solving by finding and using more reliable information can prevent subsequent problems.

2.13. A Positive Feedback Alternative

We now have both a pressing need and promising tools to advance the science of hydrology. Recent advances in field measurement, data processing, and systems analysis offer many opportunities. The remaining necessary ingredient is a leadership that can sell the long–term and short–term advantages of making hydrology a functioning science. For charting a way out, I offer hope in one more hypothesis, one that can be used to change the system from a negative to a positive feedback scenario:

A discipline that combines supply–side advances in understanding science with demand–side advances in understanding user problems can help both sides by adding to the body of knowledge and developing new practical applications.

To begin this positive feedback process, the supply (or scientific) side should make some innovative proposals. Even though the available funding is limited, some research can be redirected to more fundamental work. If this positive feedback hypothesis works, successes will show the agencies what they have been missing and attract more funding from them. Simultaneously, the accomplishments and funds will attract academics.

In keeping the process going, a balanced leadership would identify ideas and tools from science, test their practical and theoretical contributions, and make recommendations. The process would give hydrologic science a credence that would facilitate institutional change. In addition, the leadership could manage data banks as growing sources of useful information. Leadership could also train agency employees with new capabilities in instrumentation and computation that would make them feel that their jobs were enhanced rather than threatened.

Over time, the practice of hydrology would advance as the state of the science offered better representations of physical reality expressed by more precise mathematics. Water use efficiency and water project benefits would increase through more efficient water resources management and fewer "surprise" disasters from dambreaks to rapidly polluted aquifers. A track record would be established to prove that better science can significantly improve service efficiency and that research makes science better. The integration of the supply with the demand side would inspire ideas for using the state-of-the-science to enhance the state-of-the-art. The state-of-the-art could escape from the mire of familiar past techniques by adapting advances in mathematics, science, measurement, and computation. The resulting synergism could help turn hydrology into a new field-scale science.

2.14. Needs in Education

One finds no supply-side departments of hydrology in colleges of science. Courses in hydrology are orphaned without scientific prerequisites in colleges of engineering and of other applied professions. They do not follow the scientific method of applying basic principles to synthesize methods used in engineering design. Textbooks generally divide into halves. The first half surveys scientific principles and measurement methods (the supply side). The second half describes needs for hydrologic information (the demand side), selects supply-side principles that can be applied with available data, and suggests empirical design methods. Students are left to their imaginations in fitting the two halves together.

Unlike most professionals, hydrologists do not have a common core academic curriculum. They take a few courses in hydrology while working on degrees in engineering, geology, forestry, and agriculture. Most of the hydrology courses are taught from the perspective of the home discipline. The teaching commonly skips holistic characterization of field processes and critical evaluation of the physical-mathematical or the empirical constructs used to represent them (Bella, 1985). Consequently, the training commonly lacks the scientific depth required for the graduates to determine what to do when the applications that they have learned do not work. New principles from mathematics and physics should be taught. One of the greatest present needs is to infuse courses in electronic instrumentation and computerized data management into hydrologic curricula. Courses are also needed for the continuing education of prior graduates.

On the demand side, the ways that agencies fill hydrology positions give little incentive to obtain the needed training. Employees are hired from a broad variety of disciplines, called professional hydrologists, and expected to learn hydrology on the job. Afterwards, they work in agencies with varied missions in water resources development, environmental protection, agricultural productivity, land management, urban planning, etc. They spend their careers serving a few clients by working on large projects. From beginning to end, the process stresses agency practice and places scientific understanding in a secondary role. This "pressure to conform" may well be driving young people from hydrologic careers even though the job opportunities are increasing.

3. Conclusions

Institutional constraints are definitely hindering the expansion of hydrologic knowledge and threatening the long–run efficiency of water resources management. The discipline is caught in a web of demand–side domination in which the "delivering a collective good," bureaucratic management pursues goals that stifle scientific progress. The supply siders who work on specialized problems are handicapped by limitations in their access to data and other research support and by their separation from demand–side problem characterization. A negatively reinforcing feedback loop has become institutionalized in hydrology and water resources management.

In contrast, society sees continuing advances in the provision of computers, medicine, communications, and many other goods and services supplied through the private sector. Positive feedback loops seem to prevail in sciences serving industry. Somehow supply–side thinking must be integrated into demand–side management to find a way out.

Pilgrim (1986) suggests that professional societies should take the lead, and the societies that include both scientists and practitioners as members may be in a particularly strong position. Committees could be formed in which scientists from the supply–side would manage the research program and practitioners making applications would lead in refining applications. In a united management style, the scientific advances would be tested to the satisfaction of potential users so they can be more readily adopted. The testing should cover all three institutional goals (protection from poor performance, legal liability, and public blame) so that the agencies will truly feel comfortable from both scientific and popular viewpoints.

Funding and data support for research to build hydrology as a science must be backed by water resources management institutions. The major selling point is that scientific advances reduce uncertainty, enhance performance, and provide the information that vested interests need to achieve their long–run goals of legal and political security. Practices that violate scientific principles will eventually lead to conflicts and make those who employ them vulnerable to competing institutional interests.

Presently, we stand at the threshold of being able to make hydrology a field–scale science. The results would add benefits through more efficient water resources management and reduce losses from "surprise" dambreak and pollution incidents. If we fail to act, the inefficiencies and losses will grow more severe and lead to conflict. Neither legal nor political processes are effective at conflict resolution (Hart, 1974); scientific analysis supporting objective institutional review and market trading offers a way out.

Specifically, the leadership in hydrology should move the profession toward meeting future needs through:

(1) Universities that weave research results into professional curricula and continuing education.

(2) Employers in both public and private sectors that review the training in hydrology when making new hires and encourage their employees to keep up with the literature in the work they do in hydrology.

(3) Governmental agencies that take seriously their responsibility to the public to manage water resources from sound hydrologic principles.

(4) Professional societies that stimulate higher professional standards throughout the hydrologic community.

In conclusion, we need scientific hydrologists to add to our base of scientific understanding, and we need professional hydrologists to work within the agencies to move scientific advances into water resources management with greater efficiency. The proper incentives and support for scientific work can overcome the reverse law of the commons, make hydrology a science, and better serve the needs of long–run water resources management.

4. References

Adams, J. L.: 1986, *'The Care of Feeding of Ideas: A Guide to Encouraging Creativity.'*

Armstrong, E. L.: 1976, *'History of Public Works in the United States, 1776–1976,'* American Public Works Association.

Barnes, D. H.: 1987, *'The Greening of Paradise Valley,'* Modesto Irrigation District, Modesto, CA.

Bella, D. A.: 1985, 'The University: Eisenhower's Warning Reconsidered,' *Journal of Professional Issues Engineering American Soc. Civil Engineering* **111**, 12–21.

Biswas, A. K.: 1967, 'Hydrologic Engineering Prior to 600 B.C.,' *J. Hydraulics Division, ASCE* **93**(HY5), 115–135.

Biswas, A. K.: 1970, *'History of Hydrology,'* North Holland Publishing Co., Amsterdam.

Chow, V. T.: 1972, 'Hydrologic Modeling,' *Selected Works in Water Resources*, pp. 81–107, International Water Resources Association.

Dooge, J. C. I.: 1988, 'Hydrology Past and Present,' *Journal of Hydraulic Research* **26**(1), 5–26.

Golubev, G. N., and A. K. Biswas: 1985, *'Large Scale Water Transfers: Emerging Environmental and Social Experiences.'*

Grigg, N. S.: 1985, *Water Resources Planning*, McGraw–Hill Book Company.

Hart, H. C.: 1974, 'Toward a Political Science of Water Resources Decisions,' in L. D. James (ed.), *Man and Water*, University of Kentucky Press, Lexington.

Klemes, V.: 1986, 'Dilettantism in Hydrology: Transition or Destiny?,' *Water Resources Research* **22**(9), 177S–188S.

Linsley, R. K.: 1986, 'Flood Estimates: How Good Are They?,' *Water Resources Research* **22**(9), 159S–164S.

Pilgrim, D. H.: 1986, 'Bridging the Gap Between Flood Research and Design Practice,' *Water Resources Research* **22**(9), 165S–176S.

Simon, H. A.: 1955, 'Behavioral Method of Rational Choice,' *Q. J. Econ.* **69**, 99–118.

Part II – State–of–the–Art in Modeling Individual Components of the Hydrologic Cycle

Chapter 3

Hydrologic Advances in Space-Time Precipitation Modeling and Forecasting

Efi Foufoula–Georgiou
University of Minnesota
Department of Civil and Mineral Engineering
St. Anthony Falls Hydraulic Laboratory
Minneapolis, Minnesota 55414 U.S.A.

and

Konstantine P. Georgakakos
The University of Iowa
Department of Civil and Environmental Engineering
Iowa Institute of Hydraulic Research
Iowa City, Iowa 52242 U.S.A.

Abstract. The spatial and temporal rainfall characteristics influence the runoff hydrograph of a catchment, while accurate forecasting of precipitation is crucial to the prediction of floods and flash floods. This study presents a review of recent hydrologic advances in space–time rainfall modeling and forecasting. The strengths and weaknesses of the available models are discussed and directions for further research and improvements are given.

1. Introduction

1.1. Significance of Precipitation

Precipitation is the source component of the hydrologic cycle. As such, it regulates water availability and thus land use, agricultural and urban expansion, maintenance of environmental quality, and, in general, operation and management of water resources systems. Extreme rainfall events, and the floods they produce, determine the design and operation of large hydraulic structures, shape up the land by the drastic geomorphic changes they induce, and are often responsible for loss of property or sometimes lives due to flooding.

1.2. Spatial and Temporal Rainfall Characteristics

The importance of the spatial and temporal rainfall distribution on the runoff hydrograph of a basin has been demonstrated by many studies. For example, Wilson et al. (1979) concluded that the spatial distribution of rain and the accuracy of the precipitation input considerably influence the volume of storm runoff, time–to–peak, and the peak runoff of small catchments.

D. S. Bowles and P. E. O'Connell (eds.), Recent Advances in the Modeling of Hydrologic Systems, 47–65.
© 1991 *Kluwer Academic Publishers.*

Hamlin (1983) and Nicks (1982) reached similar conclusions for drainage basins of less than 3,000 km^2. Tabios et al. (1987) examined the effects of the dynamic characteristics of a storm, e.g., storm orientation and speed, on the streamflow hydrograph and concluded that these characteristics are important and should not be neglected in current design practices. The effect of storm size on the rainfall–runoff relationship was examined by Milly and Eagleson (1982) for relatively large catchments and Milly and Eagleson (1988) for cases where the storm size is of comparable size to the modeled area. Using simple approximations of the spatial rainfall distribution and duration, they concluded that the infiltration–excess runoff of a large area is extremely sensitive to the storm size and areal rainfall distribution and that, in general, spatial variability of precipitation produces increased surface runoff as compared to a uniform rainfall of the same volume. Eagleson (1984) and Eagleson and Wang (1985) stressed the importance of storm to catchment scales on the runoff production. They derived (for an idealized case of circular catchment and circular storm) the first two moments of the catchment area covered by a stationary storm; they found that the expected value of the covered area for most storm and catchment scales of hydrologic interest is much less than one, illustrating the danger of invoking the one–dimensionality assumption in hydrologic studies.

1.3. Precipitation Research

The study of rainfall has been an active area of research in the last decade. Under the lead of the American Geophysical Union, a Precipitation Committee was formed in 1982 composed of hydrologists, meteorologists, physicists, mathematicians, and statisticians. One of the main goals of the committee (AGU, 1984) was to bring together the isolated efforts of all these disciplines and bridge the gap between them. Precipitation studies were reported in several conferences and special sessions of the American Geophysical Union meetings. Collection of articles from two of these meetings were published in Water Resources Research (Vol. 21, No. 8, August, 1985) and Journal of Geophysical Research (Vol. 92, No. D8, August, 1987). The main theme of precipitation research in the last decade has been the understanding of the physical mechanisms producing rainfall and the incorporation of the precipitation dynamics and meteorological and physical evidence of rainfall organization patterns in the structure of stochastic rainfall models. Research in this area has rapidly evolved. One needs only to compare the review article of Court (1979) with that of Georgakakos and Kavvas (1987) to realize the tremendous volume of new developments and new ideas in the last decade and to sense the beginning of an emerging interdisciplinary effort. In the area of real–time hydrometeorology, the first physically–based models were developed that were compatible in scale and form with operational hydrologic models. Thus, a great potential exists in this area for the development of hydrologically useful precipitation models that integrate all available remote and on–site observations at the local hydrologic scale.

The areas of precipitation research can be broadly classified as: (a) precipitation measurement and estimation, (b) precipitation modeling in space and time, and (c) quantitative precipitation forecasting. Our review will concentrate on the areas of statistical precipitation modeling and quantitative precipitation forecasting at spatial and temporal scales of interest to hydrology, i.e., at the basin–scale. The reader is referred to the recent article by Georgakakos and Kavvas (1987) for a review of new developments on precipitation measurement via remote sensors, i.e., radar and satellite. This article is also complementary to our review in the other two areas of modeling and forecasting. The focus of our presentation is towards the operationally useful precipitation models for hydrologic simulation and forecasting and the critical assessment of the state–of–the–art models in terms of limitations and need for further developments.

2. Discussion

2.1. Precipitation Modeling

Rainfall is the result of complex atmospheric processes evolving continuously over space and time. Space–time rainfall modeling, based on the mathematical deterministic description of the underlying atmospheric processes, is extremely complicated and has little operational usefulness in hydrologic applications. The lack of complete understanding of the physical and dynamic processes involved in the formation of precipitation imposes limits on the ability of these models to make deterministic predictions beyond a limited temporal and spatial resolution (Cho, 1985; Cho and Chan,1987). The need for mathematically tractable descriptions of rainfall for operational purposes has motivated the treatment of rainfall as a stochastic process. In this section, we first review the stochastic rainfall models in time over a single site or over a number of sites. We then review the stochastic rainfall models which describe the rainfall field as a continuous random field in space and time.

2.1.1. Temporal Rainfall Models at a Single Site or at a Number of Sites

Due to the intermittency of the small time–increment rainfall (e.g., hourly or daily), two characteristic processes are of importance: the rainfall occurrence process and the process of the non–zero rainfall amounts. These two processes can be modeled either simultaneously (a compound process) or separately, and then be superimposed. Regarding the modeling approach of the rainfall occurrences, the temporal rainfall models proposed to date can be classified in three main categories: (a) the "wet–dry spell" approach; (b) the discrete time series approach; and (c) the point process models approach (in continuous and discrete time).

The "Wet–Dry Spell" Approach. In the "wet–dry spell" approach, any uninterrupted sequence of wet days or hours defines an "event." Such an occurrence process is completely specified by the probability laws of the length of the wet periods (storm duration) and the length of the dry periods (time between storms). This model structure, with exponential distributions for the lengths of the dry and wet periods, was used by Thom (1958) and Green (1964), among others. Grace and Eagleson (1966) used a Weibull distribution for the wet-period lengths and applied the model to short time increment (on the order of minutes and hours) rainfall occurrences. Other studies using this modeling approach include those of Todorovic and Yevjevich (1969) and Eagleson (1978).

In probabilistic terminology, the wet–dry spell model is an alternating renewal model. The term renewal stems from the implied independence between the dry and wet period lengths, and the term alternating is used to indicate that a wet (dry) period is always followed by a dry (wet) period, i.e., no transition to the same state is possible. In many early studies, such a model with exponential distributions for the dry period lengths was referred to as a Poisson model. This is inaccurate terminology resulting from the assumption that an event, which in this case corresponds to a wet period, occurs instantaneously at the middle or end of the wet period.

A recent study of an alternating renewal model for daily rainfall was reported by Galloy et al. (1981). They used discrete negative binomial distributions for the wet and dry period lengths and implemented the theory of point processes to derive the statistical properties of intervals (times between events) and counts (number of events in a time interval). Also, Roldan and Woolhiser (1982) compared a first–order Markov chain model with an alternating renewal model that had truncated geometric distribution for the wet periods and a truncated negative binomial distribution for the dry periods. They found that the Markov chain model was superior to the alternating renewal model for the four U.S. daily rainfall records studied.

Small and Morgan (1986) derived the relationships between a continuous–time renewal model and a Markov chain model and applied this analysis to daily rainfall occurrences.

There are certain disadvantages associated with the wet–dry spell approach. The first is the definition of independent events or storms (see, for example, Restrepo–Posada and Eagleson, 1982). This problem is even more pronounced in hourly data where wet sequences separated by one or several dry hours may still correspond to the same rainfall event. The second problem stems from the varying duration of the events which requires that the cumulative rainfall amounts corresponding to each event should be conditioned on the duration of the event. The identification and fitting of conditional probability distributions to the rainfall amounts may pose a problem especially with short records and for events of extreme duration. Finally, once the total storm rainfall has been modeled, it has to be redistributed within the wet period (internal storm characteristics); this requires additional statistical information to be extracted from the limited data available. Thus, the wet–dry spell approach seems more appropriate for the study of the external, rather than the fine–scale internal, storm characteristics.

The Discrete Time–Series Approach. The daily (or hourly) rainfall occurrence series can be viewed as a binary series of zeroes and ones, with zero corresponding to a dry day, and one to a wet day. The simplest probabilistic model for such a binary series is the Bernoulli process, which, due to its complete independent structure, is not adequate to describe the clustering present in the short–time–increment rainfall occurrences (see, for example, Smith and Schreiber, 1973). Markov chain models are the next simplest models with a dependence structure and have been extensively used for modeling daily rainfall. Gabriel and Neumann (1957, 1962) used a first–order homogeneous (i.e., constant parameters) Markov chain for the winter daily rainfall occurrences at Tel–Aviv, while Caskey (1963) and Weiss (1964) used a nonhomogeneous (i.e., time varying parameters) Markov chain for several stations in the Northern U.S. Hopkins and Robillard (1964) used a first–order Markov chain for the daily rainfall occurrences in Canada and found that it was not adequate to describe the months with few rainy days. Feyerherm and Bark (1967) showed the inadequacy of a first–order Markov chain in describing the higher–order dependence structure present in daily rainfall, and they proposed a second–order Markov chain for the daily rainfall occurrences in Indiana, Iowa, and Kansas. Wiser (1965) and Green (1965) also concluded that the geometric memory of the first–order Markov chain is not adequate to describe long droughts or long wet spells. Smith and Schreiber (1973) found a nonhomogeneous first–order Markov chain superior to an independent Bernoulli model for the seasonal thunderstorm rainfall in the Southwestern U.S. Woolhiser and Pegram (1979) studied Markov chain models with seasonally varying parameters using Fourier series.

In deciding the order of a Markov chain, Tong (1975) used the Akaike Information Criterion (AIC), while Hoel (1954) presented a likelihood ratio goodness of fit test. Chin (1977) identified the Markov chain orders of 25–year daily rainfall records in the U.S. and illustrated their dependence on the season and geographical location. A recent study by Stern and Coe (1984) contains a nice exposition of Markov chain models of rainfall sequences.

Although Markov chain models provide a simple mathematical representation of the daily rainfall occurrence process and may be adequate for some specific sites and seasons, their Markovian structure cannot describe the long–term persistence (i.e., long wet or dry spells) and the effect of clustering (i.e., higher likelihood of having an event due to an event at a previous time) present in the short time–increment rainfall occurrences. For example, long–term persistence in the daily rainfall may be caused by cyclonic activity persisting during certain seasons (Petterssen, 1969), and clustering may be the result of frontal thunderstorms with a relatively long life cycle (Kavvas and Delleur, 1981).

A more general class of binary discrete time–series models is the class of discrete autoregressive moving average (DARMA) models. A DARMA(p,q) model, where p is the order of the autoregressive and q the order of the moving average component, is a sequence $\{X_n\}$ formed by a probabilistic combination of elements of a sequence $\{Y_n\}$ which is independent and identically distributed (i.i.d.). For the binary DARMA models, Y_n is assumed to be i.i.d. with a Bernoulli distribution, i.e., $P(Y_n = 0) = \pi_0$, $P(Y_n = 1) = \pi_1$, and $\pi_0 + \pi_1 = 1$. For further details on these models and derivation of their statistical properties, see Jacobs and Lewis (1978a,b). DARMA models for daily rainfall were first used by Buishand (1978) to analyze wet–dry spells in the Netherlands. Chang et al. (1984) further developed and applied DARMA models to daily rainfall sequences in Indiana and reported satisfactory results.

Although DARMA models may be an improvement over Markov chains, in the sense that they can accommodate longer term persistence in a more parsimonious way than a higher-order Markov chain, in our opinion, their disadvantage is the lack of physical motivation for the model structure and also the discontinuous memory they exhibit (see, for example, Keenan, 1980). Also, their mathematical framework seems only to permit derivation of interval properties (i.e., probability distributions of run lengths) and not of counting properties (i.e., distributions of number of events in a time interval). Given that analysis of the second order properties of intervals and counts are in general unequivalent (see, for example, Cox and Lewis, 1978), it is advantageous to be able to use both for model identification and fitting; this can be done effectively in the point process mathematical framework.

Rainfall Amounts Models. Several marginal probability distributions have been proposed for the non–zero daily or hourly rainfall amounts. Todorovic and Woolhiser (1971) and Richardson (1981) used an exponential distribution, but Skees and Shenton (1974) and Mielke and Johnson (1974) suggested that the exponential distribution has a thinner tail than the one observed in daily rainfall amounts. The mixed exponential distribution was explored by Smith and Schreiber (1973), Woolhiser and Pegram (1979) and Roldan and Woolhiser (1982), among others. This last study compared several distributions (chain–dependent and independent exponential, gamma, and mixed exponential) and found, using the Akaike Information Criterion, that these distributions ranked from best to worst as mixed exponential, independent gamma, chain–dependent gamma, and exponential for five daily records in the continental U.S. The gamma distribution has been extensively used (see, for example, Ison et al., 1971; Buishand, 1978; Carey and Haan, 1978) with satisfactory results. Mielke (1973) and Mielke and Johnson (1974) proposed the use of the kappa or generalized beta distribution. Chain–dependent distributions, assuming that the rainfall amounts are independent but that the distribution function depends on whether the previous day was wet or dry, were studied by Katz (1977) and Buishand (1978). The chain–dependent distributions have the disadvantage of overparametrization. If dependence is present in the non–zero rainfall amounts, ARMA models with skewed marginal pdf's can be considered. Such models have been extensively used for streamflow modeling (e.g., Obeysekera and Salas, 1983) and for serially correlated rainfall durations (e.g., Raudkivi and Lawgun, 1972) but not very much for non–zero daily rainfall amounts because of their almost independent structure. Finally, several studies have considered the simultaneous modeling of daily rainfall occurrences and amounts via multiple–state Markov chain models (e.g., Khanal and Hamrick, 1974; Haan et al., 1976; and Carey and Haan, 1978).

The Point–Process Approach. A point process (PP) is a stochastic process which describes the occurrence (position) of "events" in the modeling space (e.g., the time axis, for the temporal rainfall occurrences). When a magnitude or intensity is attached to each occurrence, then the process is called "a marked point process." Waymire and Gupta (1981 a,b,c) present a nice review of the theory of point processes and their relevance and potential for the stochastic modeling of hydrologic systems. A continuous–time PP permits the events to occur anywhere on the time axis, whereas a discrete–time PP permits them to occur only on the marks specified by equally spaced increments, e.g., one day apart. Since rainfall is in reality a continuous–time process recorded over discrete–time intervals, both continuous and discrete

PP's can be explored and appropriately applied. The majority of the statistical literature (e.g., Cox and Lewis, 1978; Cox and Isham, 1980) has dealt with continuous–time PP's. This, together with the appealing idea of describing the rainfall process at the continuous time scale (and thus, by aggregation, at all other sampling scales) has motivated the application of continuous–time PP's in hydrologic modeling. Foufoula–Georgiou and Lettenmaier (1986) illustrate the problems associated with some early applications of the theory of continuous–time PP models to discrete–time rainfall data.

A continuous–time point process framework assumes the existence of an underlying continuous–time rainfall process with outcome that is only observed as the integral of the continuous process over a fixed sampling interval (e.g., hours or days). The problem in such a framework becomes that of inferring the properties of the underlying, unobserved process from the observed discrete data. The important question that naturally arises in such an approach is that of scale dependency. That is, are the parameters of the postulated unobserved continuous–time process consistent when estimated from data at different time scales, such as hourly and daily? Some of the models are clearly scale dependent, while some others seem to yield constant parameters over a range of time scales. Clearly, the models that are scale independent are preferable for two reasons. First, they may provide useful insights into the structure of the underlying rainfall process, especially when a physical meaning can be associated with the scale–independent parameters of the model. Second, these models provide general descriptions of practical value in that a model fitted to hourly rainfall data will guarantee the preservation of the daily rainfall characteristics. Below, we first present a review of the state–of–the–art continuous–time PP rainfall models, discuss their time–scale dependency, and draw comparisons between them. We then proceed with a discussion of the discrete–time PP rainfall models.

The simplest continuous–time point process is the Poisson process in which the events occur completely at random. In a Poisson process, the times between events are independent and exponentially distributed, and the number of events in a time interval is independent and Poisson distributed. The Poisson process has been extensively used to model the rainfall occurrences (see, for example, Todorovic and Yevjevich, 1969; and Gupta and Duckstein, 1975; for some of the early studies). More recently, Rodriguez–Iturbe et al. (1984) have studied two marked Poisson processes. The marks associated with each event are either instantaneous rainfall amounts (Poisson white noise, PWN, model) or rectangular pulses (Poisson rectangular pulses, PRP, model). The pulses are characterized by an intensity and a duration which are assumed to be i.i.d's and independent of each other. For these processes, Rodriguez–Iturbe et al. (1984) derived the moments of the aggregated rainfall amounts at any time scale, T, and used these parameters to fit the model to daily rainfall occurrences in Denver, Colorado and Agua Fria, Venezuela. They observed that both the PWN and PRP models are scale dependent in that the models yielded significantly different parameter estimates when fitted to hourly and daily data. Moreover, the Markovian dependence structure of the PRP model could not reproduce the empirical correlation functions of the hourly data. The inadequacy of the PWN model for short–time increment rainfall was also demonstrated (using the second order properties of a general marked point process) by Foufoula–Georgiou and Guttorp (1986, 1987).

Kavvas and Delleur (1981) observed that the daily rainfall occurrences in Indiana exhibit a clustering which might be satisfactorily modeled by the class of Poisson cluster models and, in particular, by the Neyman–Scott (N–S) models. A N–S process is a two–level process. At the primary level, the rainfall generating mechanisms (RGM) occur according to a Poisson process. Each RGM gives rise to a group, or cluster, of rainfall events. Within each cluster, the occurrence of events is completely specified by the distribution of the number of events and the distribution of their positions relative to the cluster center. Kavvas and Delleur (1981) assumed a geometric distribution for the number of rainfall events in a cluster and an exponential distribution for the distances of events from their cluster centers. Rodriguez–Iturbe et al. (1984) further studied the N–S white noise (NSWN) process, and derived the second–order properties of the accumulated rainfall amounts over different time scales. They found this model superior to the Poisson models for the analyzed hourly and daily rainfall data at Denver and Agua Fria. A more detailed analysis by Valdes et al. (1985) re-examined the time scale sensitivity of the NSWN, PWN, and PRP models by using temporal rainfall series generated from the space–time rainfall model of Waymire et al. (1984). They found that the N–S model performed the best at different time scales but that none of these models was able to preserve the statistics of extremes. Foufoula–Georgiou and Guttorp (1986) observed that since one of the parameters of the N–S model (the one related to the dispersion of events in a cluster) is uniquely determined by the rate of decay of the autocorrelation function of the cumulative rainfall amounts, the NSWN model cannot be time–scale invariant. In addition, these authors examined alternative fitting procedures by employing the properties of the discrete–time occurrence series derived by Guttorp (1986) and pointed out the insufficiency of the second order statistics in identifying the underlying continuous process, at least from daily data. An extensive simulation study illustrating other estimation problems and the properties of estimators for these models was reported by Obeysekera et al. (1987).

Motivated by the inadequacies of the NSWN model, Rodriguez–Iturbe et al. (1987a) introduced the N–S rectangular pulses (NSRP) model and the Bartlett–Lewis (B–L) rectangular pulses (BLRP) model. The rectangular pulse is characterized by a random intensity and duration. The B–L process differs from the N–S process only in the way in which the cells are positioned within a cluster. In the N–S process, the position of the cells is determined from the storm origin according to an exponential distribution. In the B–L process, the intervals between successive cells (and not the distances from the storm centers) are i.i.d. and exponentially distributed. Both the cases of a Poisson and a geometric distribution of the number of cells within a cluster were considered. Based on a detailed analysis of data from Denver, at different aggregation levels, Rodriguez–Iturbe et al. (1987b) concluded that these models are capable of preserving several rainfall characteristics at different levels of aggregation while the parameters of the model remain essentially constant. Also, these models seemed to preserve some extreme order statistics although the probability of the long dry periods was overestimated by both models (see, Rodriguez–Iturbe et al., 1987a). A modified version of the BLRP model was developed by Rodriguez–Iturbe et al. (1987c). This modification consisted of allowing for different characteristics among storms by randomizing several parameters of the distributions of the number of cells per storm, cell positions, and cell durations. These modifications seemed to improve the representation of the extreme events.

A point process model of a different structure was introduced by Smith and Karr (1983). This process, namely the doubly stochastic Poisson process (also known as the Cox process), has a rate of occurrence which alternates between two states, one zero and the other positive. During periods when the intensity is zero, no events can occur. Smith and Karr (1983) assumed that during periods with positive intensity, events occur according to a Poisson process and that the sequence of states visited forms a Markov chain. This model is a renewal model (i.e., interarrival times are independent) and was termed the Renewal Cox process with Markovian intensity (RCM). The authors applied this model to the daily rainfall occurrences of the summer season (July to October) rainfall in the Potomac River Basin. Smith and Karr (1985b) developed statistical inference procedures and maximum likelihood methods of estimation for Cox, N–S and renewal processes. These methods can be used for direct comparison of two classes of point processes to determine which one provides a better model for a given data set.

A different type of model was proposed by Kavvas and Herd (1985). Their model is a radar–based conceptualization of ground rainfall at a location and can describe rainfall at time scales smaller than or equal to one day. Under their conceptualization, the rainfall intensity distribution over the trajectory formed while the rainfall field that passes over the ground location is used to describe the rainfall at that ground location. This model has the structure of a filtered Poisson cluster process, and its parameters are estimated from the combined radar–raingage data. The advantage of this conceptualization is that it can explicitly use the almost continuous–time radar observations, although calibration with raingage data is essential. The authors applied this model to hourly and daily time scales of rainfall in Kentucky. They used 5–min increment radar microfilm and hourly raingage data for calibration.

The advantages of the continuous–time point process models will be fully realized if they are proved to be independent of the time–scale at which the fitting of the model is done. Note, that a time–scale dependent model is still a perfectly valid model, but only for a particular scale. Also, no physical interpretation should be attached to its parameters, even if the model has a physically–based structure. For example, no inferences about the rate of the passage of fronts should be made based on the time–scale dependent parameters of the NSWN process, as some earlier studies have suggested. At present, although some models, e.g., the NSRP and the BLRP models, hold promises, more simulation studies are needed to test the performance of these models and to test the sampling properties of the estimated parameters. In general, it is understood that there is a limit to the fine–structure information that can be retrieved from the aggregate data. For example, if only daily data are available the identification of the cluster structure may be impossible, in the same way that the fine structure of a raincell cannot be possibly identified from ground measurements at distances larger than the dimension of a cell ($1 - 10$ km^2). In those cases the description of the process at the aggregation level of the observations may be the only meaningful approach.

Foufoula–Georgiou and Lettenmaier (1987), motivated by the statistically flexible and physically attractive structure of the point process models and the scale dependency of the continuous–time PP's (at least the ones available at the time of the study), proposed the construction of discrete–time PP models. They introduced a Markov Renewal (MR) model for the description of daily rainfall occurrences. In the MR model the sequence of times between events (defined as any wet day or hour) is formed by sampling from two geometric distributions according to transition probabilities specified by a Markov chain. The motivation behind this structure is that daily rainfall occurrences may be the result of the interaction of several rainfall generating mechanisms. For example, the first rainy day in a wet period may be the result of a frontal storm passing over a region, whereas subsequent rainy days in the same wet period may be considered secondary events. The MR process is a clustered process and has as a special case the Markov chain models. It differs from a Markov chain in the sense that the probability of having a rainy day does not depend on the condition (rain, non-rain) of the previous day, rather on the number of days since the last rain. The authors derived the statistical properties of the MR model and developed maximum likelihood estimation procedures. Combining it with a mixed exponential distribution for the amounts, they applied the model to several daily rainfall sequences in the U.S. and found it adequate, in that it preserved both the short-term and long-term (e.g., monthly totals) structure. Following this approach, Smith (1987) introduced another discrete–time PP model, namely the Markov Bernoulli (MB) process which can be viewed as a sequence of Bernoulli trials with randomized success probabilities. He derived the key statistical properties of the process, developed likelihood–based inference procedures, and applied it to daily rainfall sequences in Washington, DC. The above two discrete-time point process models, namely the MR and MB processes, are a clear advancement over the long established Markov chains and should prove useful for operational purposes, especially when only daily data are available.

2.1.2. Space–Time Rainfall Models

Significant ideas on space–time rainfall modeling were presented in the early 1970's by J. Amorocho and co–workers (e.g., Amorocho and Slack, 1970; Amorocho and Morgan, 1974; and later Amorocho and Wu, 1977; and Amorocho, 1981). Their model had a physically-based structure but it was of a simulation nature with no analytical capabilities; this probably hindered its further implementation and development. A major contribution in space–time rainfall modeling was the theoretical work of Gupta and Waymire (1979) and Waymire et al. (1984). They employed empirical observations of extratropical cyclonic storms and formulated a physically–based kinematic, stochastic representation of the rainfall field in space and time. Their formulation was based on the empirical evidence (e.g., Austin and Houze, 1972; Hobbs, 1978; Hobbs and Locatelli, 1978; and Houze, 1981) that precipitation patterns have a high degree of hierarchical organization and definable characteristics and behavior. In particular, Austin and Houze (1972) categorized precipitation areas according to their areal extent and lifetimes as synoptic (meso–α scale), large mesoscale (meso–β scale), small mesoscale (meso–γ scale), and raincells. Synoptic rainfall fields cover areas of the order of 10^4 km^2 and have a lifetime of one to several days; large mesoscale areas (LMSA), also called rainbands due to their elongated shape, have an extent of 10^3 – 10^4 km^2 and a lifetime of several hours; small mesoscale areas (SMSA) have extent of the order of 10^2 – 10^3 km^2 and a lifetime of approximately one hour; raincells have areal extent of the order of 1–10 km^2 and lifetimes of a few minutes to at most 1/2 hour. A systematic analysis of radar observations and raingage records for several storms (see, for example, Austin and Houze, 1972) revealed a preferred hierarchical organization of the rainfall fields, in the sense that every precipitation area at any of the above scales contained one or several of each of the smaller sized precipitation areas. For instance, the large synoptic areas (which are present with cyclonic storms, but not with frontal passage or air mass thunderstorms) usually contain several (one to six) LMSA's which appear as elongated rainbands that build, move, and dissipate within the synoptic area. LMSA's (as those within the synoptic areas and the frontal bands which themselves are LMSA's) contained several SMSA's, whereas SMSA's were rarely observed outside the LMSA's. Cells were almost always located either within the SMSA's (air mass thunderstorms being themselves SMSA's) or in a cluster (within the LMSA) forming a similar size.

Waymire et al. (1984) described in mathematical terms the above organization of extratropical cyclonic storms. In their stochastic model, the location of rainbands is specified relative to a fixed spatial origin. Multiple rainbands can occur in time and their arrival time is specified (relative to a fixed time origin) according to a homogeneous Poisson process. Each rainband (LMSA) contains circular regions, called cluster potential regions (SMSA). Upon the arrival of a rainband, the centers of the cluster potential regions occur within the rainband according to a spatial Poisson process. Once a cluster potential center is located, cells are born within it according to a three–dimensional (space–time) process. The rainbands and the cells move with constant, but different, velocities. From the above morphological description of mesoscale cyclonic precipitation fields, Waymire et al. (1984) derived the theoretical mean and covariance function of the space–time rainfall intensity at the ground. An interesting feature of that model is that it is consistent with the empirical observations of Zawadzki (1973) concerning Taylor's hypothesis. Taylor's hypothesis of turbulence (Taylor, 1938) implies, in the context of precipitation fields, that the autocovariance at some time lag, Δt at a fixed but arbitrary spatial point, is the same as the spatial covariance at a fixed but arbitrary time between two points separated by a distance Δx when space is converted to time through the constant velocity of the storm, i.e., $\Delta x = U \bullet \Delta t$. Zawadzki (1973) found that Taylor's hypothesis was valid only for periods shorter than 40 minutes; this was reproduced by the model of Waymire et al. (1984) (see also Gupta and Waymire, 1987). It should be mentioned that some of the previous models, such as the model of Bras and Rodriguez–Iturbe (1976), had assumed the validity of Taylor's hypothesis at all times.

Following these developments, Sivapalan and Wood (1987) developed a similar stochastic model to describe the space–time rainfall characteristics within a single rainband. Valdes et al. (1985) used the kinematic–stochastic model of Waymire et al. (1984) to simulate two–year traces of rainfall at a number of stations. By aggregating these sequences at different time scales they compared the performance of several temporal rainfall models (see discussion of temporal rainfall models).

Kavvas et al. (1987a) studied the stochastic description of extratropical cyclonic rainfields at the synoptic and subsynoptic scales. Their model, although of the same general structure as that of Waymire et al. (1984), differs from it in several issues. Some of these issues are: (a) the synoptic scale description which accounts for the existence and movement of several cyclonic rainbands having preferred locations, orientations, and spatial characteristics relative to their cyclone center; (b) the relaxation of some assumptions related to the hierarchical organization structure of rainfields thus permitting the cores and cells to appear not only within but also outside the rainbands and cells to appear outside the cores, etc.; and (c) the relaxation of some assumptions related to the velocity of the SMSA's and cells relative to the rainbands. More details on the differences between these models can be found in Kavvas et al. (1987a); see also the work of Kavvas et al. (1987b). At present, this model has only been used in a simulation mode with parameters in agreement with radar–based observations reported in several studies. If the model is to be used for operational hydrologic studies, analytical derivations of some of its statistical properties are needed.

A simpler model for space–time rainfall was proposed by Smith and Karr (1985a,b). This model is composed of a description of the temporal occurrence of storms (assumed a homogeneous Poisson process), a description of the spatial distribution of the raincells within a storm (assumed a spatial Poisson process), and the description of the rainfall patterns within a cell. For this five–parameter model the authors developed maximum likelihood estimation methods but commented that this "would become intractable under formulations only slightly more general".

Rodriguez–Iturbe et al. (1986) studied the spatial structure of the total rainfall depths within a stationary storm. The models they developed consist of stationary cells distributed in space according to a Poisson or an N–S process. The rainfall depth at the center of the cells is an i.i.d random variable, exponentially distributed. Rainfall is distributed within each cell by a spread function of a specified form (exponential, quadratically exponential, and linear). The rainfall depth at any point is the superposition of the depths from all the contributing cells. They showed that for the case of exponentially distributed center depth and quadratically exponential spread function, the total storm depth at any point has a Gamma distribution. This may be a useful tool in partially justifying the structure of the raincell distribution. They applied their model to data from the Upper Rio Guaire Basin in Venezuela. The model was further applied and evaluated by Eagleson et al. (1987). This later study used eight years of summer storm data from 93 stations in the 154 km^2 Walnut Gulch catchment in Arizona and compared the relative performance of the spatial Poisson process models with different spread functions for the raincell depths. They concluded that the models with the exponential and quadratic exponential spread functions were able to describe the spatial rainfall distributions of the essentially stationary air mass thunderstorms in the Southwestern U.S.

An effort to incorporate radar and raingage observations in the model development and parameter estimation was reported by Smith and Krajewski (1987). These authors, recognizing the limitations of using only ground data for space–time rainfall estimation, combined the strength of the radar data in delineating wetted areas with the accuracy of ground measurements into a single estimation technique. In their model, the structure of rainfields within a wet period consists of circular cells organized within Poisson distributed rainbands. The cells have random but constant intensity. The temporal evolution of the wet–dry periods is governed by a Markov chain. The authors developed parameter estimation methods and applied the model to daily rainfall fields in the tropical Atlantic region covered by the GATE experiment. They note that a limitation of the model is its inability to deal with low–intensity rain hiding more apparent structure of a rainfall field.

An important area of precipitation research which is not reviewed in the present article is the scaling behavior of precipitation fields and the fractal models of rain. The reader is referred to the work of Waymire (1985), Lovejoy and Schertzer (1985), Lovejoy and Mandelbrot (1985), and Schertzer and Lovejoy (1987). Georgakakos and Kavvas (1987) review the fractal rainfall models of Lovejoy and co–workers and draw some comparisons between these models and the other classes of space–time rainfall models. As these authors comment, the fractal models of Lovejoy et al. are of a simulation nature with no analytical capabilities; at present, this hinders their application to operational hydrology.

2.2. Quantitative Precipitation Forecasting

Crucial to the prediction of floods in real–time is the accurate prediction of the precipitation rate. This is especially important for the improvement of forecasts for small watersheds where the lag time between rainfall occurrence and outflow from the basin is short. A recent (1982) Program Development Plan for improving Hydrologic Services in the U.S., illustrates that 50 percent of the forecast points for communities across the U.S. have potential forecast lead times (maximum possible lead times with uniform distribution of rainfall over the basin) of less than 10 hours while 25 percent have less than 4 hours of forecast lead times (Georgakakos and Hudlow, 1984). Clearly, accurate precipitation predictions for even a few hours into the future would result in valuable increases in effective lead time. Several models and procedures for the real–time quantitative precipitation prediction exist (see Georgakakos and Hudlow, 1984, for a review). However, it was only recently that dynamic precipitation models, suitable for use in hydrologic models of real time forecasting of floods and flash floods, were presented in the literature (Georgakakos and Bras, 1984a,b).

Distinctive characteristics of the latter models are their hydrologic spatial and temporal scale, and their state–space form which allows the use of a state estimator for the automatic, real–time update of the precipitation model state from surface precipitation observations (see analysis in Georgakakos, 1986). The state estimator also gives the capability for probabilistic forecasts that quantify uncertainties due to observation sensor errors, model structure and parameter errors, and input error. The term stochastic–dynamic model in this context denotes a dynamic model complemented by a state estimator. Previous reviews of this class of models were included in the works of Georgakakos and Kavvas (1987), Georgakakos (1987), and Georgakakos and Brazil (1987). A short review of the physical aspects of the formulation of hydrologic precipitation models follows along the lines of Georgakakos (1987).

The Georgakakos and Bras (1984a,b) station precipitation model was used to generate precipitation predictions based on surface air pressure, temperature, and dew–point temperature data. The model equations can be written in the following form:

$$dx_p(t)/dt = f[\mathbf{u}(t)] + h[\mathbf{u}(t)] \; x_p(t) \tag{1}$$

and

$$y_p(t_k) = \Delta t \; \phi[u(t_k)] \; x_p(t_k), \quad k = 1, 2, \ldots \tag{2}$$

where $x_p(t)$ is the condensed water–equivalent mass (or volume) at time t in a unit area column that extends from the bottom to the top of the clouds (in kilograms per square meter or millimeters per square meter); $u(t)$ is the vector of the precipitation model inputs (i.e., surface air temperature, pressure, and dew–point temperature) at time t; $\{h[u(t)] \, x_p(t)\}$ is the outflow mass (or volume) rate of condensed water equivalent from the cloud column at time t (in kilograms per square meter per second or in millimeters per square meter per second); and $f[u(t)]$ is the inflow mass (or volume) rate of condensed water equivalent into the cloud column at time t (in kilograms per square meter per second or in millimeters per square meter per second). Outflow is due to precipitation or local cloud–top anvil formation, while inflow is due to condensation and air mass ascent. Subcloud evaporation reduces the precipitation mass that reaches the ground to a value smaller than the one obtained at the cloud bottom level. The precipitation mass (or volume) reaching the ground is collected between time instants t_k and t_{k+1}, for $k = 1, 2, ...,$ $(\Delta t = t_{k+1} - t_k)$ and is represented by $y_p(t_k)$ (in kilograms per square meter per Δt or millimeters per square meter per Δt). The instantaneous precipitation rate at ground level is given by $\{\phi[u(t_k)] \, x_p(t_k)\}$ (in kilograms per square meter per second or millimeters per square meter per second).

The details of model derivation and expressions for $f(\ )$, $h(\ )$, and $\phi(\ )$ are given by Georgakakos and Bras (1984a,b). The model is based on a convective parameterization of the vertically averaged updraft velocity as a function of the surface–air meteorological model input variables:

$$v = \varepsilon_1 \sqrt{c_p}(T_m - T_s')$$ (3)

where ε_1 is a model parameter, proportional to the square root of the ratio of kinetic to maximum thermal energy per unit mass of ascending air; c_p is the specific heat of dry air under constant pressure; T_m is the cloud temperature at a certain pressure level p'; and T_s' is the corresponding potential ambient air temperature, determined from surface air meteorological data using adiabatic ascent of air from the ground surface. The level p' is defined by:

$$p' = p_s - (p_s - p_t)/4$$ (4)

where p_s is the cloud bottom pressure computed from heat–adiabatic ascent and p_t is the cloud top pressure.

Using the cloud updraft velocity, v, the condensation inflow rate $f[u(t)]$ was obtained by:

$$f[u(t)] = \Delta W \, \rho_m \, v \, dA$$ (5)

where ΔW is the mass of liquid water resulting from condensation during the pseudo–adiabatic ascent of a unit mass of moist air; ρ_m is a vertically averaged (from cloud top to cloud bottom) density of moist air; and dA represents the unit area measure.

Cloud microphysics transforms the discrete mass flux of the falling precipitation drops to the continuous precipitation rate through the use of an exponential particle size distribution:

$$n(D) = N_o \exp(-D/\varepsilon_2)$$ (6)

where N_o is a parameter expressed as a function of the model state variable x_p; D is the drop diameter; and ε_2 is a free model parameter representing the average diameter of hydrometeors in the cloud column.

Since the emphasis on flood, especially flash flood, prediction is in areas of $10^2 - 10^3$ km^2, the use of the station precipitation model requires surface meteorological data sampled at points separated by distances of similar spatial scale. On the average, the surface meteorological data stations in the U.S. have separation distances greater than 100 km. An interpolation procedure is thus necessary to provide input to the precipitation model for all the possible locations of interest. Georgakakos (1986) developed and tested such an interpolation procedure. The variables interpolated were surface air temperature, pressure, and dewpoint temperature. The interpolation accounts for altitude–varying terrain. For flat terrain the procedure reduces to a linear spatial interpolation (weights inversely proportional to distance). Average interpolation errors of about 1°K were obtained from tests in Oklahoma and about 2°K from tests in Montana for the surface air temperature variables. The pressure-variable errors were about 1 mbar. Station distances were greater than 150 km in the test cases. Complemented by the input interpolation procedure, the precipitation model can be used to predict mean areal precipitation for any drainage basin, given only surface air meteorological data at a few stations.

Several flash floods are the result of orographic enhancement of precipitation followed by the resultant fast–moving flood waves on steep slopes. The previously described spatial interpolation methodology can produce estimates of surface air meteorological variables in mountainous terrain. Through the terrain variation of the aforementioned meteorological variables that are used as an input to the precipitation prediction model, the orographic effects are partially taken into account. However, if accurate prediction of precipitation in mountainous areas is to be obtained, the most important effect of orography, namely, the enhancement of the updraft velocity and of the mass influx into the clouds, should also be simulated by the model.

Several studies have examined the issues concerning the modeling of orographic enhancement, given operationally available meteorological data (e.g., Elliott and Shaffer, 1962; Bell, 1978; Rhea, 1978). The basic idea is that the updraft velocity enhancement which is due to orography is equal to the inner product of the horizontal wind vector $\vec{W}$ and the local topographic gradient vector, $\Delta \vec{H}$,

$$V_o = \vec{W}\, \Delta \vec{H} \tag{7}$$

where V_o is the magnitude of the orographic component of the updraft velocity. The operational models presented in the literature to date vary in the way they treat the vertical distribution of the updraft velocity enhancement and the conditions under which enhancement can occur.

In an ongoing research effort, the stochastic–dynamic model discussed above has been modified to incorporate orographic enhancement of the updraft velocity. When only surface meteorological data are available (e.g., in cases when the forecast lead time is less than 6–hours), Equation (7) is used to estimate the surface updraft enhancement and a linear reduction of the orographic updraft velocity component with height is assumed with the vanishing level at the cloud top level. No orographic enhancement is computed in cases when the horizontal surface wind is less than 2.5 m/sec or when the surface relative humidity is less than 65 percent. Thus, given that the aforementioned criteria for the existence of orographic enhancement are fullfilled, the vertically averaged updraft velocity (see Equation (3)) is incremented by the amount of $V_o/2$, while the condensation input rate at cloud bottom (see Equation (5)) is incremented by the amount $\Delta W\, \rho_m\, V_o\, dA$.

The problem of spatial interpolation of the wind vector arises in cases where orographic enhancement is significant. In the present developments, the wind vector observed at an upwind station is assumed to be representative of the wind at the drainage basin of interest. In cases where the drainage basin is divided into more than one orographic zone, a precipitation prediction model (of the type previously described) can be formulated for each orographic zone. Then the drainage basin mean area precipitation at time $t_k[\bar{y}_p(t_k)]$ can be computed by:

$$\bar{y}_p(t_k) = \sum_{i=1}^{L} \frac{A_i}{A} \; y_{pi}(t_k) \tag{8}$$

where A_i is the area of the i-th orographic zone; A is the total area of the drainage basin; and $y_{pi}(t_k)$ is the precipitation prediction at time t_k computed for the i-th orographic zone. The areas of the L orographic zones satisfy:

$$\sum_{i=1}^{L} A_i = A \tag{9}$$

Previous results obtained by the stochastic dynamic precipitation model (see Georgakakos and Bras, 1984a,b; and Georgakakos, 1984) appear encouraging in that the prediction residuals have smaller variance than the precipitation observations while at the same time physically realistic values were obtained for model parameters such as updraft velocity, cloud–averaged drop diameter, and cloud top pressure. Experience in using the model with data from various storm types points to some model weaknesses which should be addressed if improved model predictions are to be obtained. One of the characteristics of the model predictions (especially the deterministic predictions) is the under–estimation of excessive rainfall rates. Several model assumptions contribute to such an underestimation. The assumptions we consider to be the most important in that respect are:

(1) The updraft and rainfall occur uniformly in the space of the column of the precipitation model. In actual storms, regions of updraft are separate from regions of downdraft where drops fall with respect to the ground surface.

(2) The terminal group velocity of falling drops during intense rainfall might be significantly different from the terminal velocity of isolated drops which is what the model uses. The model would tend to underestimate not only high rainfall intensities but also rainfall mass since evaporation in the subcloud layer would be acting for shorter time intervals when group velocity is considered.

(3) Mass advection from the sides of the column is not taken into consideration. A two–dimensional representation is needed.

In an ongoing research effort (e.g., Georgakakos and Lee, 1987), the model formulation has been modified to include a two–dimensional representation, a group velocity component, and a separation of updraft and downdraft regions in each computational grid square. Sensitivity analysis of various, mainly convective, storms is underway. Preliminary results indicate improvement in the performance of the hourly predictions of the deterministic model.

3. Conclusions

The last decade has witnessed significant advancements in the areas of space–time precipitation modeling and forecasting. In terms of stochastic precipitation models, it seems that the present modeling developments outperform our ability to use them in an operational framework. The missing link is reliable parameter estimation. For example, the Waymire et al. (1984) model has fourteen parameters to be estimated. As expected, identification problems may arise when all of these parameters are attempted to be estimated from the time-aggregated data at a sparse raingage network. It is desirable that some of these parameters (such as the size of the cells and the parameters of the rainfall distribution within the cells, etc.) and also some of the model components (such as the functional form of the spread function within the cells) are estimated from data at very fine spatial scales or are physically inferred. This approach would also guarantee the physical meaning of the parameters, whereas parameters estimated from aggregated data and based on hypothesized (and not data–identified) model components should be viewed with caution. For example, the Eagleson et al. (1987) study seems to favor the quadratic exponential spread function, while studies of cell properties based on radar data (e.g., Konrad, 1978; Drufuca, 1977) seem to indicate that the exponential decay function may be more appropriate. Although both models may give similar results in terms of describing the overall spatial rainfall structure, the parameters of the model may change significantly (e.g., compare the values of Table 3 of Rodriguez–Iturbe et al., 1987b); therefore, the physical meaning of the model parameters may be questionable.

It may be that better estimation of stochastic model parameters requires employment of radar or satellite data in conjunction with ground observations. However, the simultaneous use of multiple sensors is not yet developed to the point that they can be used for estimation purposes. One of the main problems is the different properties of the errors of each one of those sensors (see Krajewski, 1987; Krajewski and Georgakakos, 1985).

Concerning the developed stochastic–dynamic quantitative precipitation prediction models, several improvements are feasible in the near future. Some of these improvements are:

(1) Incorporation of data from remote sensors within the state space formulation of the models. A major challenge in this area is the characterization of the errors in data from remote sensors.

(2) Refinement of the physical basis of the models. This aspect would benefit from the training of hydrologists in the area of hydrometeorology.

(3) Coupling of the precipitation models with the large–scale Numerical Weather Prediction Models (NWPM). As a first step, use of NWPM forecasts as the precipitation model input in real time, and assessment of performance in both deterministic and probabilistic predictions can be accomplished.

Although one can argue that operational hydrology needs simpler models, it should be acknowledged that further understanding of the underlying space–time rainfall structure cannot be gained unless physically realistic rainfall models are extensively studied and tested with good quality, high resolution space–time data. Until these models are fully tested, however, the hydrologist needs to exercise judgement to best use the scientific developments and to select a model both appropriate for the modeling objectives and consistent with the available data.

4. Acknowledgements

The work of Efi Foufoula-Georgiou was supported by the National Science Foundation, under Grant CES-8708825. The National Science Foundation also supported the work of K. P. Georgakakos, under Grant ECE-8657526. This support is gratefully acknowledged.

5. References

American Geophysical Union Hydrology Section, Committee on Precipitation: 1984, 'A New Interdisciplinary Focus on Precipitation Research,' *EOS Transactions* **65**(23), 377, 380.

Amorocho, J., and J. Slack: 1970, *'Simulation of Cyclonic Storm Fields for Hydrologic Modelling,'* Amer. Geophys. Union Meeting, Wash., DC.

Amorocho, J., and D. Morgan: 1974, 'Convective Storm Field Simulation for Distributed Catchment Models,' *Proc. Warsaw Symposium on Mathematical Models in Hydrology*, 1971, IAHS, 598–607.

Amorocho, J.: 1981, 'Stochastic Modeling of Precipitation in Space and Time,' in V. P. Singh (ed.), *Statistical Analysis of Rainfall and Runoff*, Water Resour. Publ., Fort Collins, CO.

Amorocho, J., and B. Wu: 1977, 'Mathematical Models for the Simulation of Cyclonic Storm Sequences and Precipitation Fields,' *J. of Hydrology* **32**, 329–345.

62

Austin, P. M., and R. A. Houze, Jr.: 1972, 'Analysis of the Structure of Precipitation Patterns in New England,' *J. Appl. Meteorology* **11**, 926–935.

Bell, R. S.: 1978, 'The Forecasting of Orographically Enhanced Rainfall Accumulations Using 10-Level Model Data,' *Meteorological Magazine* **107**, 113–124.

Bras, R. L., and I. Rodriguez-Iturbe: 1976, 'Rainfall Generation: A Nonstationary Time-Varying Multidimensional Model, *Water Resour. Res.* **12**(1), 450–456.

Buishand, T. A.: 1978, 'Some Remarks on the Use of Daily Rainfall Models,' *J. Hydrology* **36**, 295–308.

Carey, D. I., and C. T. Haan: 1978, 'Markov Process for Simulating Daily Point Rainfall,' *J. Irrig. & Drainage Div., ASCE* **104**(IRI), 111–125.

Caskey, J. E.: 1963, 'A Markov Chain Model for the Probability of Precipitation Occurrence in Intervals of Various Lengths,' *Mon. Weather Rev.* **91**(6), 298–301.

Chang, T. J., M. L. Kavvas, and J. W. Delleur: 1984, 'Daily Precipitation Modeling by Discrete Autoregressive Moving Average Processes,' *Water Resour. Res.* **20**(5), 565–580.

Chin, E. H.: 1977, 'Modeling Daily Precipitation Occurrence Process with Markov Chain,' *Water Resour. Res.* **13**(6), 949–956.

Cho, H-R.: 1985, 'Stochastic Dynamics of Precipitation: An Example,' *Water Resour. Res.* **21**(8), 1225–1232.

Cho, H-R., and D. S. T. Chan: 1987, 'Mesoscale Atmospheric Dynamics and Modeling of Rainfall Fields,' *J. of Geophys. Res.* **98**(D8), 9687–9692.

Court, A.: 1979, 'Precipitation Research, 1975–1978,' *Rev. of Geophys. and Space Phys.* **17**(6), 1165–1175.

Cox, D. R., and V. Isham: 1980, *Point Processes*, Chapman and Hall, London.

Cox, D. R., and P. A. W. Lewis: 1978, *The Statistical Analysis of Series of Events*, Methuen, London.

Drufuca, G.: 1977, 'Radar-derived Statistics on the Structure of Precipitation Patterns,' *J. Appl. Meteor.* **16**(10), 1920–1035.

Eagleson, P. S.: 1978, 'Climate, Soil and Vegetation, 2, The Distribution of Annual Precipitation Derived from Observed Storm Sequences,' *Water Resour. Res.* **14**(5), 713–721.

Eagleson, P. S.: 1984, 'The Distribution of Catchment Coverage by Stationary Rainstorms,' *Water Resour. Res.* **20**(5), 581–590.

Eagleson, P. S., and O. Wang: 1985, 'Moments of Catchment Storm Area,' *Water Resour. Res.* **21**(8), 1185–1194.

Eagleson, P. S., N. M. Fennessey, W. Qinliang, and I. Rodriguez-Iturbe: 1987, 'Application of Spatial Poisson Models to Air Mass Thunderstorm Rainfall,' *J. of Geophys. Res.* **92**(D8), 9661–9678.

Elliott, R. D., and R. W. Shaffer: 1962, 'The Development of Quantitative Relationships Between Orographic Precipitation and Air-Mass Parameters for Use in Forecasting and Cloud Seeding Evaluation,' *J. of Appl. Meteorology* **1**, 218–228.

Feyerherm, A. M., and L. D. Bark: 1967, 'Goodness of Fit of a Markov Chain Model for Sequences of Wet and Dry Days,' *J. Appl. Meteorol.* **6**, 770–773.

Foufoula-Georgiou, E., and D. P. Lettenmaier: 1986, 'Continuous-time Versus Discrete-time Point Process Models for Rainfall Occurrence Series,' *Water Resour. Res.* **22**(4), 531–542.

Foufoula-Georgiou, E., and P. Guttorp: 1986, 'Compatibility of Continuous Rainfall Occurrence Models with Discrete Rainfall Observations,' *Water Resour. Res.* **22**(8), 1316–1322.

Foufoula-Georgiou, E., and D. P. Lettenmaier: 1987, 'A Markov Renewal Model for Rainfall Occurrences,' *Water Resour. Res.* **23**(5), 875–884.

Foufoula-Georgiou, E., and P. Guttorp: 1987, 'Assessment of a Class of Neyman-Scott Models for Temporal Rainfall,' *J. of Geophys. Res.* **92**(D8), 9679–9682.

Gabriel, K. R., and J. Neumann: 1957, 'On a Distribution of Weather Cycles by Lengths,' *Quart. J. R. Meteor. Soc.* **83**, 375–380.

Gabriel, K. R., and J. Neumann: 1962, 'A Markov Chain Model for Daily Rainfall Occurrences at Tel Aviv,' *Quart. J. R. Meteor. Soc.* **88**, 90–95.

Galloy, E., A. Le Breton, and S. Martin: 1981, 'A Model for Weather Cycles Based on Daily Rainfall Occurrences,' in M. Cosnard et al. (eds.), *Rhythms in Biology and Other Fields of Application, Springer Lecture Notes in Biomathematics*, Vol. 49, pp. 303–318.

Georgakakos, K. P.: 1984, 'State Estimation of a Scalar Dynamic Precipitation Model from Time-Aggregate Observations,' *Proceedings of the 16th IEEE Southeastern Symposium on System Theory*, March 25–27, Mississippi, pp. 111–115.

Georgakakos, K. P., and R. L. Bras: 1984a, 'A Hydrologically Useful Station Precipitation Model, 1, Formulation,' *Water Resour. Res.* **20**(11), 1585–1596.

Georgakakos, K. P., and R. L. Bras: 1984b, 'A Hydrologically Useful Station Precipitation Model, 2, Case Studies,' *Water Resour. Res.* **20**(11), 1597–1610.

Georgakakos, K. P., and L. Brazil: 1987, 'Real–Time Flood and Flash–Flood Forecasting Using Hydrometeorological Models and Modern Estimation Theory Techniques,' in L. Bergal, and J. Dolz (eds.), *Flood Warning Systems*, Imprime Cooperativa de Publicaciones del Colegio de Ingenieros de Caminos, Canales y Puertos, Almagro, 42–28010 Madrid, pp. 184–227.

Georgakakos, K. P., and M. D. Hudlow: 1984, 'Quantitative Precipitation Forecast Techniques for Use in Hydrologic Forecasting,' *Bull. Amer. Meteor. Society* **65**(11), 1186–1200.

Georgakakos, K. P., and M. L. Kavvas: 1987, 'Precipitation Analysis, Modeling, and Prediction in Hydrology,' *Reviews of Geophysics* **25**(2), 163–178.

Georgakakos, K. P., and T. H. Lee: 1987, *'Estimation of Mean Areal Precipitation Fields Using Operationally Available Hydrometeorological Data and a Two–Dimensional Precipitation Model,'* Iowa Inst. of Hydraulic Research, The University of Iowa, IIHR Report No. 315, 80 pp.

Georgakakos, K. P.: 1986, 'State Estimation of a Scalar Dynamic Precipitation Model from Time–Aggregate Observations,' *Water Resour. Res.* **22**(5), 744–748.

Georgakakos, K. P.: 1987, 'Real Time Flash Flood Prediction,' *J. Geophys. Res.* **92**(D8), 9615–9629.

Grace, R. A., and P. S. Eagleson: 1966, *'The Synthesis of Short–Time–Increment Rainfall Sequences,'* Hydrodynam. Lab. Rep. 91, Dept. of Civil Engr., Mass. Inst. of Techn., Cambridge, MA.

Green, J. R.: 1964, 'A Model for Rainfall Occurrence,' *J. R. Statist. Soc.*, B, **26**, 345–353.

Green, J. R.: 1965, 'Two Probability Models for Sequences of Wet or Dry Days,' *Monthly Weather Review* **93**, 155–156.

Gupta, V. K., and L. Duckstein: 1975, 'A Stochastic Analysis of Extreme Droughts,' *Water Resour. Res.* **12**(2), 221–228.

Gupta, V. K., and E. Waymire: 1979, 'A Stochastic Kinematic Study of Subsynoptic Space–Time Rainfall,' *Water Resour. Res.* **15**(3), 637–644.

Gupta, V. K., and E. Waymire: 1987, 'On Taylor's Hypothesis and Dissipation in Rainfall,' *J. of Geophys. Res.* **92**(D8), 9657–9660.

Guttorp, P.: 1986, 'On Binary Time Series Obtained from Continuous Time Point Processes Describing Rainfall,' *Water Resour. Res.* **22**(6), 897–904.

Haan, T. N., D. M. Allen, and J. O. Street: 1976, 'A Markov Chain Model of Daily Rainfall,' *Water Resour. Res.* **12**(3), 443–449.

Hamlin, M. J.: 1983, 'The Significance of Rainfall in the Study of Hydrological Processes at Basin Scale,' *J. of Hydrology* **65**, 73–94.

Hobbs, P. V.: 1978, 'Organization and Structure of Clouds and Precipitation on the Mesoscale in Cyclonic Storms,' *Rev. Geophys. and Space Phys.* **16**(4), 741–755.

Hobbs, P. V., and J. D. Locatelli: 1978, 'Rainbands, Precipitation Cores and Generating Cells in a Cyclonic Storm,' *J. Atmos. Sci.* **35**, 230–241.

Hoel, P. G.: 1954, 'A Test for Markov Chains,' *Biometrika* **41**, 430–433.

Hopkins, J. W., and P. Robillard: 1964, 'Some Statistics from the Canadian Prairie Provinces,' *J. Appl. Meteorology* **3**, 600–602.

Houze, R. A., Jr.: 1981, 'Structure of Atmospheric Precipitation Systems: A Global Survey,' *Radio Science* **16**(5), 671–689.

Ison, N. T., A. M. Feyerherm, and L. D. Bark: 1971, 'Wet Period Precipitation and the Gamma Distribution,' *J. Appl. Meteorology* **10**(4), 658–665.

Jacobs, P. A., and P. A. W. Lewis: 1978a, 'A Discrete Time Series Generated by Mixtures I: Correlational and Run Properties,' *J. R. Statist. Soc.*, B, **40**(1), 94–105.

Jacobs, P. A., and P. A. W. Lewis: 1978b, 'A Discrete Time Series Generated by Mixtures II: Asymptotic Properties,' *J. R. Statist. Soc.*, B, **40**(2), 222–228.

Katz, R. W.: 1977, 'An Application of Chain–Dependent Processes to Meteorology,' *J. Appl. Prob.* **14**, 598–603.

Kavvas, M. L., and J. W. Delleur: 1981, 'A Stochastic Cluster Model for Daily Rainfall Sequences,' *Water Resour. Res.* **17**(4), 1151–1160.

Kavvas, M. L., and K. R. Herd: 1985, 'A Radar–based Stochastic Model for Short–Time–Increment Rainfall,' *Water Resour. Res.* **21**(9), 1437–1455.

Kavvas, M. L., P. S. Puri, and M. N. Saquib: 1987a, 'On a Stochastic Description of the Time–Space Behavior of Extratropical Cyclonic Precipitation Fields,' *J. of Stochastic Hydrology and Hydraulics.*

Kavvas, M. L., P. S. Puri, and M. N. Saquib: 1987b, 'Feasibility Studies on a Stochastic Geometric Model for Precipitation Fields,' *J. of Stochastic Hydrology and Hydraulics*.

Keenan, D. M.: 1980, *'Time Series Analysis of Binary Data,'* Ph.D. Thesis, The Univ. of Chicago, IL.

Khanal, N. N., and R. L. Hamrick: 1974, 'A Stochastic Model for Daily Rainfall Data Synthesis,' *Proceedings of the Symposium of Statistical Hydrology*, Tucson, AZ, U.S. Dept. of Agric., Publ. No. 1275, 197–210.

Konrad, T. G.: 1978, 'Statistical Models of Summer Rainstorms Derived from Fine Scale Radar Observations,' *J. of Appl. Meteorology* **17**, 171–188.

Krajewski, W. F.: 1987, 'Radar Rainfall Data Quality Control by the Influence Function Method,' *Water Resour. Res.* **5**, 837–844.

Krajewski, W. F., and K. P. Georgakakos: 1985, 'Synthesis of Radar Rainfall Data,' *Water Resour. Res.* **21**, 764–768.

Lovejoy, S., and B. B. Mandelbrot: 1985, 'Fractal Properties of Rain, and a Fractal Model,' *Tellus* **37A**, 209–232.

Lovejoy, S., and D. Schertzer: 1985, 'Generalized Scale Invariance in the Atmosphere and Fractal Models of Rain,' *Water Resour. Res.* **21**(8), 1233–1250.

Mielke, P. W., Jr.: 1973, 'Another Family of Distributions for Describing and Analyzing Precipitation Data,' *J. Appl. Meteorol.* **10**(2), 275–280.

Mielke, P. W., Jr., and E. S. Johnson: 1974, 'Some Generalized Beta Distributions of the Second Kind Having Desirable Application Features in Hydrology and Meteorology,' *Water Resour. Res.* **10**(2), 223–236.

Milly, P. C. D., and P. S. Eagleson: 1982, *'Infiltration and Evaporation at Inhomogeneous Land Surfaces,'* Tech. Rep. 278, Ralph M. Parsons Lab., Dept. of Civil Engr., Mass. Inst. of Tech., Cambridge.

Milly, P. C. D., and P. S. Eagleson: 1988, 'Effect of Storm Scale on Surface Runoff Volume,' *Water Resour. Res.* **24**(4), 620–624.

Nicks, A. D.: 1982, 'Space–Time Quantification of Rainfall Inputs for Hydrological Transport Models,' *J. of Hydrology* **59**, 249–260.

Obeysekera, J. T. B., and R. D. Salas: 1983, 'GAR Modeling of Streamflow Series,' Presented at the AGU Fall Meeting, San Francisco, *AGU Abstract: EOS* **63**(45), 935.

Obeysekera, J. T. B., G. O. Tabios III, and J. D. Salas: 1987, 'On Parameter Estimation of Temporal Rainfall Models,' *Water Resour. Res.* **23**(10), 1837–1850.

Petterssen, S.: 1969, *Introduction to Meteorology*, 3rd Ed., McGraw-Hill, New York.

Raudkivi, A. J., and N. Lawgun: 1972, 'Generation of Serially Correlated Non–normally Distributed Rainfall Durations,' *Water Resour. Res.* **8**(2), 398–409.

Rhea, J. O.: 1978, *'Orographic Precipitation Model for Hydrometeorological Use,'* Atmospheric Science Paper No. 287, Dept. of Atmospheric Sci., Colorado State Univ., Fort Collins, CO, 198 pp.

Restrepo–Posada, J. P., and P. S. Eagleson: 1982, 'Identification of Independent Rainstorms,' *J. of Hydrology* **55**, 303–319.

Richardson, C. W.: 1981, 'Stochastic Models of Daily Precipitation, Temperature, and Solar Radiation,' *Water Resour. Res.* **17**(1), 182–190.

Rodriguez–Iturbe, I., V. K. Gupta, and E. Waymire: 1984, 'Scale Considerations in the Modeling of Temporal Rainfall,' *Water Resour. Res.* **20**(11), 1611–1619.

Rodriguez–Iturbe, I., D. R. Cox, and P. S. Eagleson: 1986, 'Spatial Modeling of Total Storm Rainfall,' *Proc. R. Soc. Lond.*, A, **403**, 27–50.

Rodriguez–Iturbe, I., D. R. Cox, and V. Isham: 1987a, 'Some Models for Rainfall Based on Stochastic Point Processes,' *Proc. R. Soc. Lond.*, A, **410**, 269–288.

Rodriguez–Iturbe, I., B. Febres de Power, and J. B. Valdes: 1987b, 'Rectangular Pulses Point Process Models for Rainfall: Analysis of Empirical Data,' *Jour. of Geophys. Res.* **92**(D8), 9645–9656.

Rodriguez–Iturbe, I., D. R. Cox, and V. Isham: 1987c, 'A Point Process Model for Rainfall: Further Developments,' *Proc. Royal Soc. London*, A.

Rodriguez–Iturbe, I., B. F. de Power, M. B. Sharifi, and K. P. Georgakakos: 1989, 'Chaos in Rainfall,' *Water Resour. Res.* **25**(7), 1667–1675.

Roldan, J., and D. A. Woolhiser: 1982, 'Stochastic Daily Precipitation Models, 2, A Comparison of Distributions of Amounts,' *Water Resour. Res.* **18**(5), 1461–1468.

Schertzer, D., and S. Lovejoy: 1987, 'Physical Modeling and Analysis of Rain and Clouds by Anisotropic Scaling Multiplicative Processes,' *J. of Geophys. Res.* **92**(D8), 9693–9714.

Sharifi, M. B., K. P. Georgakakos, and I. Rodriguez-Iturbe: In Press, 'Evidence of Deterministic Chaos in the Pulse of Storm Rainfall,' *Journal of Atmospheric Sciences*.

Sivapalan, M., and E. F. Wood: 1987, 'A Multidimensional Model of Non-stationary Space-Time Rainfall at the Catchment Scale,' *Water Resour. Res.* 23(7), 1289–1299.

Skees, P. M., and L. R. Shenton: 1974, 'Comments on the Statistical Distribution of Rainfall Per Period Under Various Transformations,' *Proceedings of the Symposium on Statistical Hydrology*, Tucson, AZ, U.S. Dept. of Agr., Misc. Publ. No. 1275, 172–196.

Small, M. J., and D. J. Morgan: 1986, 'The Relationship Between a Continuous-Time Renewal Model and a Discrete Markov Chain Model of Precipitation Occurrence,' *Water Resour. Res.* 22(10), 1422–1430.

Smith, E. R., and H. A. Schreiber: 1973, 'Point Processes of Seasonal Thunderstorm Rainfall, Part 1, Distribution of Rainfall Events,' *Water Resour. Res.* 9(4), 871–884.

Smith, J. A.: 1987, 'Statistical Modeling of Daily Rainfall Occurrences,' *Water Resour. Res.* 23(5), 855–893.

Smith, J. A., and A. F. Karr: 1983, 'A Point Process Model of Summer Season Rainfall Occurrence,' *Water Resour. Res.* 19(1), 95–103.

Smith, J. A., and A. F. Karr: 1985a, 'Statistical Inference for Point Process Models of Rainfall,' *Water Resour. Res.* 21(1), 73–80.

Smith, J. A., and A. F. Karr: 1985b, 'Parameter Estimation for a Model of Space-Time Rainfall,' *Water Resour. Res.* 21(8), 1251–1257.

Smith, J. A., and W. F. Krajewski: 1987, 'Statistical Modeling of Space-Time Rainfall Using Radar and Rain Gage Observations,' *Water Resour. Res.* 10, 1893–1900.

Stern, R. D., and R. Coe: 1984, 'A Model Fitting Analysis of Rainfall Data (With Discussion),' *J. Roy. Stat. Soc. Ser. A.* 147, 1–34.

Tabios, G. O., J. T. B. Obeysekera, and H. W. Shen: 1987, *'The Influence of Storm Movement on Streamflow Hydrograph Through Space-Time Rainfall Generation and Hydraulic Routing,'* preprint.

Taylor, G. I.: 1938, 'The Spectrum of Turbulence,' *Proc. R. Soc. London*, A, **164**, 476–490.

Thom, H. G.: 1958, 'A Note on the Gamma Distribution,' *Monthly Weather Review* 86(4), 117–122.

Todorovic, P., and D. A. Woolhiser: 1971, 'Stochastic Model of Daily Rainfall,' *Proc. of the USDA-IASPS Symposium on Statist. Hydrology*, Tucson, AZ.

Todorovic, P., and V. Yevjevich: 1969, *'Stochastic Processes of Precipitation,'* Colorado State Univ., Hydro. Paper 35, 1–61.

Tong, H.: 1975, 'Determination of the Order of a Markov Chain by Akaike's Information Criterion,' *J. Appl. Prob.* 12, 488–497.

Valdes, J. B., I. Rodriguez-Iturbe, and V. K. Gupta: 1985, 'Approximations of Temporal Rainfall from a Multidimensional Model,' *Water Resour. Res.* 21(8), 1259–1270.

Waymire, E.: 1985, 'Scaling Limits and Self-Similarity in Precipitation Fields,' *Water Resour. Res.* 21(8), 1271–1281.

Waymire, E., and V. K. Gupta: 1981a, 'The Mathematical Structure of Rainfall Representations, 1, A Review of Stochastic Rainfall Models,' *Water Resour. Res.* 17(5), 1261–1272.

Waymire, E., and V. K. Gupta: 1981b, 'The Mathematical Structure of Rainfall Representations, 2, A Review of the Theory of Point Processes,' *Water Resour. Res.* 17(5), 1273–1285.

Waymire E., and V. K. Gupta: 1981c, 'The Mathematical Structure of Rainfall Representations, 3, Some Applications of the Point Process Theory to Rainfall Processes,' *Water Resour. Res.* 17(5) 1287–1294.

Waymire, E., V. K. Gupta, and I. Rodriguez-Iturbe: 1984, 'A Spectral Theory of Rainfall Intensity at the Meso-β Scale,' *Water Resour. Res.* 20(10), 1453–1465.

Weiss, L. L.: 1964, 'Sequences of Wet and Dry Days Described by a Markov Chain Model,' *Monthly Weather Review* 92, 169–176.

Wilson, C. B., J. B. Valdes, and I. Rodriguez-Iturbe: 1979, 'On the Influence of the Spatial Distribution of Rainfall on Storm Runoff,' *Water Resour. Res.* 15(2), 321–328.

Wiser, E. H.: 1965, 'Modified Markov Probability Models of Sequences of Precipitation Events,' *Monthly Weather Review* 93, 511–516.

Woolhiser, D. A., and G. G. S. Pegram: 1979, 'Maximum Likelihood Estimation of Fourier Coefficients to Describe Seasonal Variations of Parameters in Stochastic Daily Precipitation Models,' *J. Appl. Meteorol.* 8(1), 34–42.

Zawadzki, I.: 1973, 'Statistical Properties of Precipitation Patterns,' *J. Appl. Meteor.* 12, 459–472.

Chapter 4

The Formulation of Evaporation from Land Surfaces

Wilfried Brutsaert
Cornell University
School of Civil and Environmental Engineering
Hollister Hall
Ithaca, New York 14853–3501 U.S.A.

Abstract. Most procedures currently in use to describe evaporation in hydrologic models involve several parameters, characterizing the turbulent transport in the lower atmosphere and/or the moisture availability in the soil and the vegetation; it is still very difficult to determine these parameters a priori. Research in a number of areas will lead to a better understanding of the relevant physical mechanisms.

1. Introduction

Hydrologic modeling is concerned with the efficient and robust description of the different aspects of the hydrologic cycle on land.

Two major pathways in this cycle are precipitation and evaporation. That precipitation is a crucial link in the system is generally agreed upon; for this reason it has been and continues to be the focal point of much of the research in hydrology. Evaporation, on the other hand, has not received as much attention. Yet, while on a global base over land precipitation is of the order of 0.80 m/y, evaporation amounts to roughly 0.50 m/y. In this light it would seem that the quality of a hydrologic model depends critically on its ability to simulate evaporation. This is especially the case for longer term budget calculations at the catchment and basin scale, water resources planning and management for water supply, and irrigation and drainage of agricultural lands. The need for information on evaporation is exacerbated in situations of drought since it is the main depletion mechanism of the available water resources. Therefore, it is an awkward twist of fate that the difficulty of determining evaporation increases as the regional evaporation rate decreases, and as the aridity of the environment increases. In detailing rainfall–runoff phenomena, it is important to capture the "runoff-readiness" of the catchment; even in extreme situations related to floods, this depends primarily on the initial soil–moisture content of the basin, and thus on the antecedent evaporative conditions. For example, Hewlett et al. (1977) have found that in forested regions, storm flows are due almost entirely to gross storm rainfall volume and antecedent moisture storage within the basin, but storm flows are practically unaffected by hourly (or shorter) rainfall intensity. In the same vein, it has been shown (e.g. Liu and Brutsaert, 1978; Loague and Freeze, 1985) that the performance of many rainfall–runoff models at the catchment scale, is sensitive to the methods used to specify the effective precipitation (i.e. the "losses") and the antecedent moisture conditions. This initial state of the catchment is the direct result of the evaporation and soil drainage between storm events.

67

D. S. Bowles and P. E. O'Connell (eds.), Recent Advances in the Modeling of Hydrologic Systems, 67–84.
© 1991 *Kluwer Academic Publishers.*

2. Discussion

2.1. Some Procedures in Current Practice

2.1.1. Concept of Proportional Vapor Fluxes with Soil Moisture "Bucket"

A common method which follows work by Budyko (1955, 1974) and Thornthwaite and Mather (1955) is based on the following proportionality (see Figure 1):

$$E = \beta E_p \tag{1}$$

where E is the rate of actual evaporation, E_p a potential evaporation rate, and β a reduction or moisture availability factor. The potential evaporation is usually taken to represent (e.g. Thornthwaite, 1948) the evaporation from a large area covered completely and uniformly by an actively growing vegetation with adequate moisture supply. This definition is somewhat ambiguous and in practice in the context of Equation (1), there have been two general classes of definitions of potential evaporation. In the first class E_p (denoted by E_{p1}) is assumed to be the evaporation which would take place if the surface were wet, but under the actual (i.e. prevailing) conditions of radiation, air temperature, and wind speed. Thus one way to determine it makes use of the expression:

$$E_{p1} = CeV_r \rho \ (q_s^* - q_a) \tag{2}$$

where Ce is a bulk transfer coefficient for water vapor, V_r the mean wind speed at some reference level z_r above the surface, ρ the density of the air, q_a the mean specific humidity at a reference level z_a (which may but need not be the same as z_r), and q_s^* the saturation specific humidity at the mean temperature of the surface $T = T_s$. Because this surface temperature T_s is often unknown (or the surface difficult to define as in the case of vegetation), an alternative procedure to calculate E_{p1} is that suggested by Penman (1948); it requires information only on wind speed, temperature, humidity at one level, and on the available energy flux at the surface $Q_{ne} = R_{ne} - G_e$, where R_{ne} is the net radiation and G_e the heat flux into the ground (expressed in units of evaporation). In a general form it can be written as:

$$E_{p1} = \frac{\Delta}{\Delta + \gamma} \ Q_{ne} + \frac{\gamma}{\Delta + \gamma} CeV_r \rho \ (q_a^* - q_a) \tag{3}$$

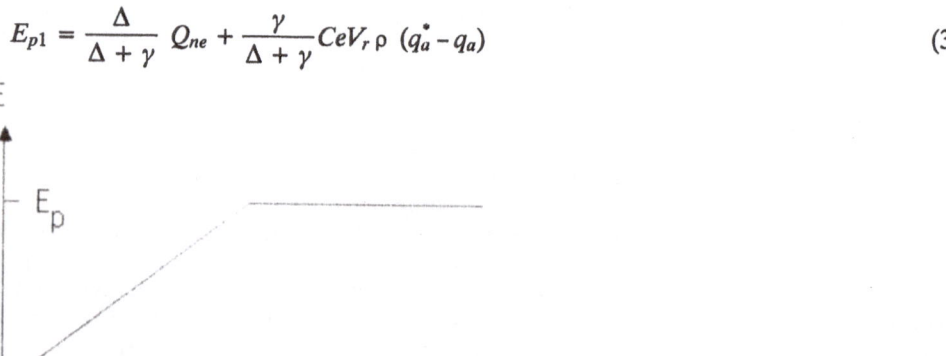

Figure 1. Sketch of the proportional relationship expressed in (1) and (5).

where $\Delta = dq^*/dT$ at the temperature of the air T_a, q_a^* the saturation specific humidity at the same temperature, $\gamma = c_p/L_e$ the psychrometric constant, c_p the specific heat of air at constant pressure, and L_e the latent heat of vaporization.

In the second class of definitions, E_p (denoted here by E_{p2}) is the evaporation that can result from the available energy at the surface under certain conditions of advection at the regional scale. The "certain" refers to either average or minimal regional advection. Many empirical equations that were developed to determine irrigation requirements fall in this category; also in this category is the well–known equation of Priestley and Taylor (1972) as an extension of the equilibrium evaporation of Slatyer and McIlroy (1961, p. 73). It can be written as:

$$E_{p2} = a \frac{\Delta}{\Delta + \gamma} Q_{ne} \tag{4}$$

where α is ideally a constant of the order of 1.26 to 1.28 for extensive water surfaces and for short vegetation such as grass, which is not wet but well supplied with water; in reality α is not a constant and tends to vary in space and in time (e.g. Brutsaert, 1982, p. 220–221).

The application of Equation (1) with Equations (2), (3), or (4) requires knowledge of a number of parameters. The transfer coefficient Ce in Equations (2) and (3) can in principle be formulated on the basis of known similarity theories for the atmospheric boundary layer (e.g. Brutsaert, 1982); however, such theories have mostly been validated either in micrometeorological settings (with characteristic length scales between about 1 and 100 m) or over relatively smooth and uniform terrain; therefore, at the watershed or basin scale (i.e. 1 to 100 km) this coefficient Ce is often simply specified by calibration [for example as a function of wind speed $(a + b V_r)/V_r$], or even taken as a single fixed value.

The reduction factor β is often assumed to be a function of soil water content. In the application of Equation (1) with Equations (2) and (3) a common assumption has been (Figure 2):

$$\begin{aligned} \beta &= 1 && \text{for } w > w_o \\ \beta &= w/w_o && \text{for } w \le w_o \end{aligned} \tag{5}$$

where w_o is a critical soil water content above which E equals E_p. The value of w can be determined on the basis of a soil water budget (e.g. Thornthwaite and Mather, 1955; Budyko, 1974, p. 335; Manabe, 1969; Carson, 1982). The value of w_o must be determined by calibration; for a

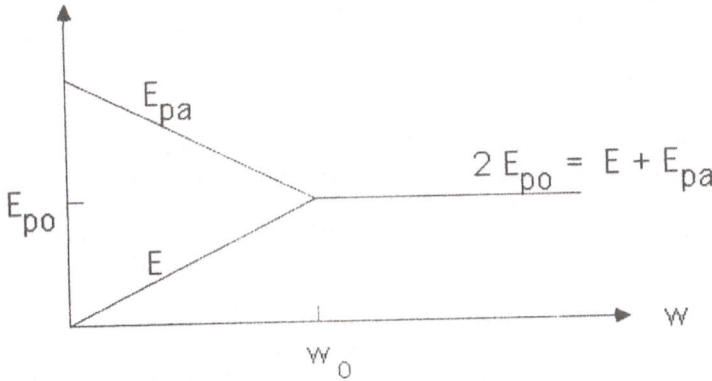

Figure 2. Sketch of the reciprocal relationship expressed in Equation (14).

surface soil layer with typical thickness of about 1 m, w_o is generally taken to be of the order of 10 – 20 cm. In their application of Equation (1) with Equation (4), Davies and Allen (1973) proposed a β which, while nonlinear, was similar to Equation (5). The reduction factor β can also be related to some other surface moisture indices such as the accumulated actual evaporation minus precipitation (Priestley and Taylor, 1972), the soil moisture deficit (e.g. Grindley, 1970), and the antecedent precipitation index (e.g. Mawdsley and Ali, 1985), again through calibration of the model with available data.

2.1.2. Resistance Concept

A second procedure of reducing E_p to E is formally not based on Equation (1). Instead it is based on the concept of a resistance factor r, which accounts for the moisture stress in the vegetation and/or soil and which relates saturation specific humidity q_s^* at the temperature T_s of the evaporating surface to the actual (not saturated) specific humidity, q_s, at the surface. Several such resistance parameters have been used for this purpose (e.g. Monteith, 1973). The one given by Thom (1972) is appropriate in the present context; it can be defined as follows:

$$r_s = \rho \ (q_s^* - q_s)/E \tag{6}$$

in which q_s is the actual (not saturated) mean specific humidity at the evaporating surface. For practical use there have generally been two types of evaporation equations based on the resistance concept. In the derivation of the first type, q_s, which is unknown, is eliminated between Equation (5) and the standard mass transfer equation:

$$E = CeV_r\rho \ (q_s - q_a) \tag{7}$$

to yield the expression

$$E = CeV_r(1 + r_s \ CeV_r)^{-1} \rho \ (q_s^* - q_a) \tag{8}$$

In the derivation of the second type, Equation (8) is used [instead of Equation (2)] to obtain an equation analogous to Equation (3), namely:

$$E = \left[\Delta \ Q_{ne} + \gamma \ CeV_r\rho \ (q_a^* - q_a)\right] \left[\Delta + \gamma \ (1 + r_sCeV_r)\right]^{-1} \tag{9}$$

Equation (9) is in the form of the Penman–Monteith equation [see however Monteith, 1973, 1981; Thom, 1975].

Numerous experiments have been conducted to determine resistance values for different types of vegetation. This has mostly been done in the context of equations related to Equation (9). A few examples are beans (Black et al., 1970), sugar beets (Brown and Rosenberg, 1977), eucalyptus forest (Dunin and Greenwood, 1986), pine forest (Gash and Stewart, 1975; Lindroth, 1985) barley (Monteith et al., 1965), sorghum (Szeicz et al., 1973), and fir forests (Tan and Black, 1976). In addition, many attempts have been made to relate resistance parameters with such factors as Bowen ratio, soil moisture suction in the root zone, soil moisture deficit, humidity deficit in the air, solar radiation, temperature, leaf area index, and others (e.g. van Bavel, 1967; Szeicz and Long, 1969; Federer, 1977; Garratt, 1978; Lindroth, 1985; Stewart, 1988; Gash et al., 1989). The relationships developed so far are mainly statistical, so that they are vegetation- and site-dependent. Therefore, the resistance formulation is probably not yet sufficiently general to be practical for predictive purposes, but it has been useful as a diagnostic index and in certain simulation studies (for example, to calculate missing data).

As a note of caution, in various studies the resistance formulation has not always been used with consistent definitions for Ce and r_s (e.g. Thom, 1972; Brutsaert, 1982, p. 111), so intercomparison of different results may not be easy. For instance, in many applications the drag coefficient Cd (or the related so–called aerodynamic conductance) is used instead of Ce, as required in the rigorous derivation of Equation (3). This drag coefficient is defined by:

$$Cd = u_*^2/V_r^2 \tag{10}$$

where $u_*^2 = \tau_0/\rho$ is the friction velocity and τ_0 the surface shear stress; it is well known that Cd is rarely equal to Ce (e.g. Brutsaert, 1975; 1982, p. 88). As a result of this inappropriate use of Cd (instead of Ce), it is not clear how the turbulence aspects of the transport, normally embodied in Ce, can be partitioned or separated from the strictly vegetational and/or soil moisture aspects of the transport expected to be manifested in r_s. This has undoubtedly contributed to the difficulty in deriving general relationships for both Ce and r_s on the basis of Equation (9).

At a first glance, the resistance formulation with r_s or any other canopy or surface resistance parameter may appear conceptually superior and quite different from the reduction factor β. Actually, however, both procedures are practically the same. Indeed Equation (8) is equivalent with Equations (1) and (2) and a reduction factor:

$$\beta = (1 + r_s \; CeV_r)^{-1} \tag{11}$$

Similarly, Equation (9) is equivalent with Equations (1) and (3) and a reduction factor:

$$\beta = [1 + r_s \; CeV_r \; \gamma/(\Delta + \gamma)]^{-1} \tag{12}$$

Equation (9) is also equivalent with Equations (1) and (4) and a reduction factor:

$$\beta = \alpha^{-1}\left[1 + \gamma \; CeV_r \rho \; (q_a^* - q_a) \; /\Delta \; Q_{ne}\right] [1 + r_s \; CeV_r \; \gamma/(\Delta + \gamma)]^{-1} \tag{13}$$

Again, in practical application of Equations (8) or (9), a knowledge of the parameters Ce and r_s is essential.

2.1.3. Concept of Complementary Vapor Fluxes with Advection–Aridity

This concept was first proposed by Bouchet (1963) who postulated, almost in diametrical opposition to Equation (1), a certain complementary relation between a potential evaporation E_{pa} and the actual regional evaporation E, as follows:

$$E + E_{pa} = 2E_{po} \tag{14}$$

More specifically, the term E_{pa} represents an apparent potential evaporation, the evaporation calculated for the prevailing non–potential atmospheric conditions with the surface assumed to be wet. In contrast E_{po} is the actual evaporation which would occur under potential conditions with adequate water supply. Equation (14) may be derived from the following logic. If for one or other reason, independent of the available energy, E decreases below E_{po} a certain amount of energy becomes available, namely:

$$E_{po} - E = h \tag{15}$$

At the regional scale this decrease of E, relative to E_{po}, probably has a relatively small effect on net radiation; it affects primarily the temperature, humidity and turbulence of the air near the ground. This available energy flux, h, causes an increase in the apparent potential evaporation inferred from these drier and warmer conditions. Bouchet (1963) assumed that E_{pa} is increased by exactly this amount, or:

$$E_{pa} = E_{po} + h \tag{16}$$

A combination of Equations (15) and (16) yields immediately Equation (14). Equation (14) has been used mainly in two types of applications. The application by Morton (1975; 1976; 1986) was concerned primarily with the estimation of long–term evapotranspiration for climatological purposes.

In the application by Brutsaert and Stricker (1979) Equation (14) was combined with the ideas on regional advection, advanced by Slatyer and McIlroy (1961, p. 73). Because the excess in apparent potential evaporation E_{pa} in Equation (16) and the deficit in actual evaporation E in Equation (15) provide an indication of the aridity of the region, and thus are related to regional advection, they coined their approach the "advection–aridity" approach. In brief, in Equation (14) E_{pa} was assumed to be equivalent to E_{p1} as given by Equation (3), and E_{po} was assumed to represent the potential evaporation under conditions of minimal regional advection, that is E_{p2}, as given by Equation (4) or some other suitable expression. Substitution of Equations (3) and (4) into Equation (14) yields (Brutsaert and Stricker, 1979):

$$E = (2a - 1)\frac{\Delta}{\Delta + \gamma} Q_{ne} - \frac{\gamma}{\Delta + \gamma} CeV_r \rho (q_a^* - q_a) \tag{17}$$

The main advantage of Equation (17) is that it does not require any information related to soil moisture, canopy resistance, or other measures of aridity because it relies on meteorological parameters only. Its main limitation is that while the underlying Bouchet concept expressed in Equation (14) is simple and plausible, it was not arrived at in a rigorous, theoretical or experimental way. Equation (17) has been applied in a number of studies; for example, in the Netherlands (Brutsaert and Stricker, 1979), West Africa (Jobe, 1982), Long Island, New York (Steenhuis et al., 1985), Japan (Kotoda, 1986), and British Columbia (Grimmond et al., 1986; Ali and Mawdsley, 1987). More recently, Byrne et al. (1988) have attempted to compensate for some of the limitations of Equations (3) and (4) when used for non–wet and tall vegetation. To apply Equation (14) they assumed that E_{pa} and E_{po} can be obtained from the analogs of Equations (3) and (4) but with the inclusion of a resistance of turgid vegetation that is well supplied with water but not wet. Thus for E_{pa} they adopted Equation (9) (unfortunately, with Cd instead of Ce), and for E_{po} the first term on the right of Equation (9) multiplied by α.

2.1.4. A Unified Parametric Formulation

It is readily seen that, as reviewed here, several of the procedures in current practice can be cast in a single formulation as follows:

$$E = \beta \left[a\frac{\Delta}{\Delta + \gamma'} Q_{ne} + b\frac{\gamma}{\Delta + \gamma'} CeV_r \rho (q_a^* - q_a) \right] \tag{18}$$

in which a and b are weighting parameters for the first and second term, respectively, which together with the remaining parameters β and γ' depend on the chosen model. As before, the parameter β is used if actual evaporation is obtained by reduction of potential evaporation E_{p1} or E_{p2}; in the potential evaporation given by Penman's Equation (3), the remaining parameters are $a = b = 1$ and $\gamma' = \gamma$, whereas in that given by Priestley and Taylor's Equation (4) they are $a = \alpha$, $b = 0$, and $\gamma' = \gamma$. In the Penman–Monteith Equation (9), for actual evaporation $\beta = a = b = 1$ and $\gamma' = \gamma(1 + r_s Ce V_r)$; in Equation (17) of Brutsaert and Stricker they are $\beta = 1$, $a = 2\alpha-1$, $b = -1$ and $\gamma' = \gamma$. Equation (18) illustrates that the different models are related, but the corresponding parameters can vary considerably; this parameter space is presently being investigated.

2.2. Possible Alternatives

The procedures described in Section 2 suffer common shortcomings. First, they depend on several parameters, namely a mass transfer coefficient Ce, reduction factors β or r_s, a regional advection parameter α, the weighting parameters a and b, and possibly others depending on the specific implementation. At present it is still very difficult, if not impossible, to determine these parameters a priori. Second, there is a wide consensus among hydrologists, that the methods themselves are unsatisfactory as they involve crude and sometimes questionable physical conceptualizations.

To improve on this unsatisfactory situation, more work will be needed. Certainly, there must be alternative approaches, but what are the available strategies to break the deadlock? In the case of the methods outlined in Section 2, they all contain a mass transfer parameter Ce. This means that any improvement on this aspect of the problem will have to come from a better understanding of transport in the atmospheric boundary layer (ABL) at the appropriate scales. This boundary layer comprises the air in which the turbulent transport is directly affected by the surface.

But the characteristics of the boundary layer are not the result of local surface conditions only; rather, observed profiles in the boundary layer reflect in some integrated way the surface conditions over upwind fetches of the order of tens of kilometers. Indeed, it is not hard to visualize that it takes about that distance before the effect of some surface feature, say a discontinuity, a smoke puff, or the water vapor released by a single tree, reaches the higher levels of the boundary layer, that is about 10^3 m above the ground. Thus, because of its integrating power, the boundary layer contains information about the surface on a regional scale.

Clearly, the boundary layer is one of the key elements in the understanding and parameterization of land surface processes. While progress on the stably stratified boundary layer has been relatively difficult, in recent years considerable advances have been made regarding the convective boundary layer. The latter is of practical interest in hydrology because evaporation tends to be largest under unstable conditions. In what follows a brief overview is presented of several approaches which are currently being investigated for the unstable ABL.

2.3. Boundary Layer Parameterizations

Transport of momentum, sensible heat, water vapor, and other admixtures of the air in the vicinity of the earth's surface almost invariably involves turbulence. Turbulent transport phenomena are commonly studied by making use of similarity, on the basis of dimensional analysis.

2.3.1. Flux–Profile Relationships

These provide expressions for the mean (in the turbulence sense) wind speed, V, the mean potential temperature, θ, and the mean specific humidity, q, in dimensionless form, as functions of dimensionless height above the ground and of other dimensionless variables. In the development of similarity formulations for turbulent boundary layers, both in wind tunnels and in natural flows, it is convenient to consider several sublayers or regions, depending on the relative importance of the physical variables. The roughness sublayer refers to the flow region closest to the surface where the turbulence is strongly affected by – among other factors – the detailed geometry of the surface roughness. The turbulence in the outer region of the ABL is affected directly by – among other factors – the large–scale pressure gradient, the Coriolis effect of the rotating Earth, and the entrainment due to boundary layer growth. Between the roughness sublayer and the outer region lies the inner region.

This sublayer which is also referred to as the surface sublayer, or the wall layer, is commonly assumed to occupy the lower 10 percent or so of the ABL over a uniform and level surface. The flux–profile relationships are generally accepted to obey Monin–Obukhov (M-O) similarity; in integral form they can be written as follows:

$$V = \frac{u_*}{k}\left[\ln\left(\frac{z-d}{z_o}\right) - \Psi_m\left(\frac{z-d}{L}\right)\right] \qquad (19)$$

$$\theta_1 - \theta = \frac{H}{k\,u_*\,\rho\,c_p}\left[\ln\left(\frac{z-d}{z_1-d}\right) - \Psi_s\left(\frac{z-d}{L}\right) + \Psi_s\left(\frac{z_1-d}{L}\right)\right] \qquad (20)$$

$$q_1 - q = \frac{E}{k\,u_*\,\rho}\left[\ln\left(\frac{z-d}{z_1-d}\right) - \Psi_v\left(\frac{z-d}{L}\right) + \Psi_v\left(\frac{z_1-d}{L}\right)\right] \qquad (21)$$

where $k(=0.4)$ is von Karman's constant, c_p the specific heat of air at constant pressure, d the zero–plane displacement height, z_o the roughness height, the subscript 1 refers to a reference level $z = z_1$ within the surface sublayer and L is the Obukhov stability length. It is defined by:

$$L = \frac{-u_*^3}{k(g/T_1)(H_v/\rho c_p)} \qquad (22)$$

where $H_v = (H + 0.61\,T_1 c_p E)$ is the virtual sensible heat flux at the surface and g the acceleration of gravity. The symbols $\Psi_m(\ )$, $\Psi_s(\ )$, and $\Psi_v(\ )$ denote the (integral) stability functions for momentum, sensible heat, and water vapor, respectively; these functions have been determined experimentally (e.g. Dyer, 1974; Yaglom, 1977; Businger, 1988; Högström, 1988; Brutsaert, 1982). Equations (19) through (22) form a closed system and with measurements of V, θ and q the surface fluxes u_*, H and E can be determined; this is the basis of the profile method.

Over uniform and even surfaces the stability functions for sensible heat and water vapor can usually be assumed to be equal, that is:

$$\Psi_s\left(\frac{z-d}{L}\right) = \Psi_v\left(\frac{z-d}{L}\right) \qquad (23)$$

Note that the assumed equality shown in Equation (23) is the basis of the well known Bowen ratio method to determine E and H from a surface energy budget.

More often than not, natural land surfaces are not level and uniform, so that the applicability of surface layer similarity is not always obvious. Recent experimental studies over irregular and complex surfaces, such as pine forest with clearings and hilly prairie, have begun to shed some light on this issue. It was observed that under unstable conditions profiles of V, θ and q measured with radiosondes can usually be scaled well in terms of M–O similarity (e.g. Brutsaert and Parlange, 1991; Brutsaert and Sugita, 1990; Sugita and Brutsaert, 1990a,b). The Businger–Dyer version of the M–O functions was found to be satisfactory for this purpose. The observed height range of the surface sublayer extended on average between a lower limit of around $(z-d)/z_0 = 40$ to 50, and an upper limit of around $(z-d)/z_0 = 100$ to 130. In these experiments, the horizontal scales of the heterogeneities of the surface were on the order of around 10^2 to 10^3m. Thus these findings strongly suggest that for heterogeneities or nonuniformities of this order of magnitude the ABL can be considered as quasi–uniform from the regional point of view. In other words, within the inner region defined by this height range the law of the wall of M–O similarity can be assumed to be valid, provided the surface roughness parameters, z_0 and d, the surface shear stress, u_*, and the surface fluxes of sensible and latent heat, H and E, are taken as representative regional values.

Although these results are encouraging, several aspects of the ABL above natural terrain will require further investigation. For instance, it is still not clear how rugged the surface can be before the applicability of Equations (19) through (22) breaks down. An experiment in the Swiss Fore–Alps has suggested a marked dissimilarity between θ and q profiles in the surface sublayer, as just defined above (Brutsaert and Kustas, 1987). The terrain of this experiment is characterized by hills of the order of 100 m above the mean valley elevations and by distances between ridges of the order of 1 km; the area, which lies 20 to 30 km north from the Alps, is largely covered by pasture with only 23 percent occupied by patches of forest. While the terrain is quite variable and complex in the usual micrometeorological sense, at the regional scale in the prevailing westerly wind direction it may still be termed "statistically uniform". As mentioned earlier, the inner region extends normally over the approximate range $45 < (z-d)/z_0 < 120$ (where $z_0 = 3.8$ m and $d = 46$ m, approximately, in the present case, see Kustas and Brutsaert, 1986). Under unstable conditions over this range, it was impossible to obtain good agreement between the observed θ–profiles and Equation (20), and between the observed q–profiles and Equation (21). Examples of such profiles are given in Figures 3 and 4. The lack of vertical resolution of the measurements may be one cause of the failure to detect M–O similarity; another possible explanation may be that as the regional roughness of rugged terrain is larger than that of more regular surfaces, the inner region would also have to extend to higher elevations. At these higher elevations in the unstable ABL, the potential temperature profile is usually fairly uniform, whereas the specific humidity still displays a gradient. This probably follows from the fact that the air which is entrained into the ABL from aloft is generally not only warmer, but often drier than that below. Therefore, it is conceivable that the inner region is "squeezed" and eliminated between the expanded roughness sublayer from below and the entrainment from above. This is but one example of an unresolved issue; further research will be needed for a complete understanding of the surface sublayer above rugged terrain.

The outer region of the convective ABL is also commonly referred to as the mixed layer. It is now recognized, however, that well mixed conditions apply mainly to the profile of potential temperature, θ, and not necessarily to other admixtures of the air, such as humidity q (see Figures 3 and 4; see also Mahrt, 1976). Thus θ is usually fairly uniform, whereas q often displays a gradient. This follows from the fact that the convective activity entrains air from outside the ABL which is generally not only warmer but also drier.

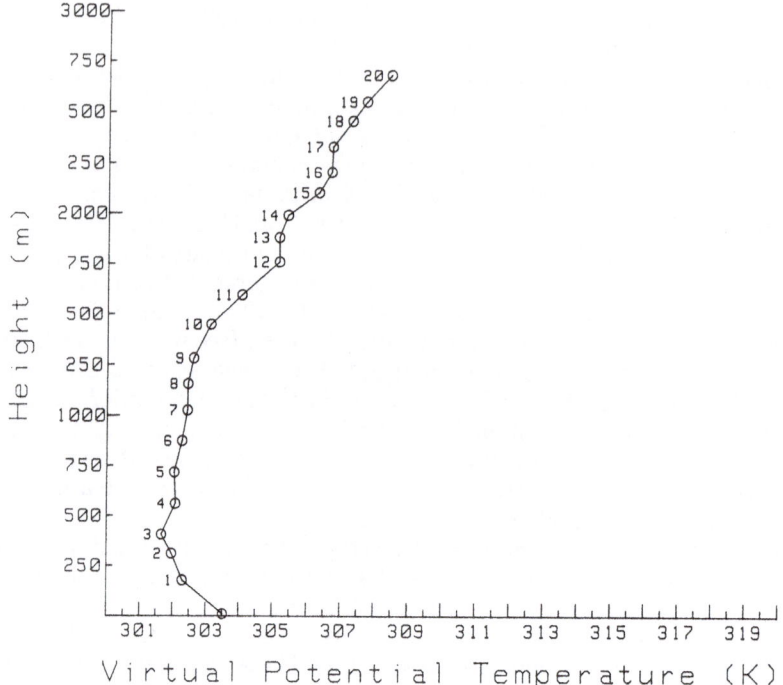

Figure 3. Example of virtual potential temperature profile under stable conditions above rugged terrain. The top of the mixed layer is at approximately 1400 m (from Brutsaert and Kustas, 1987).

The profile of q in the mixed layer has received much attention recently, but several studies have led to similar results for the mean gradient. For example, following ideas of Andre et al. (1979), Driedonks and Tennekes (1984) and Wyngaard and Brost (1984), Kustas and Brutsaert (1987a) adopted the expression:

$$< \frac{\partial q}{\partial z} > = a_1 \frac{\overline{q'w'}_h}{\sigma_h \ell_h} + a_2 \frac{E/\rho}{\sigma_o \ell_o} \tag{24}$$

where $< >$ denotes the mean over the entire mixed layer, $\overline{q'w'}_h$ the water vapor flux at the bottom of the inversion where $z = h_i$, a_1, and a_2 are constants, and σ_h, σ_o, ℓ_h, and ℓ_o are turbulent velocity and length scales for the inversion and surface flux, respectively. Various length and velocity scales were tried in the analysis, but it was found that Equation (24) with $\ell_h = \ell_o = h_i$ and $\sigma_o = u^*$ and $\sigma_h = (w_*^3 + 8u_*^3)^{1/3}$ gave the best representation of their data from the Fore–Alps, with a multiple correlation of $R = 0.51$. The values of the regression constants were found to be $a_1 = 11$ and $a_2 = 2.5$. The variable w_* represents the convective velocity scale:

$$w_* = \left(\frac{g H_v h_i}{T_1 \rho c_p} \right)^{1/3} \tag{25}$$

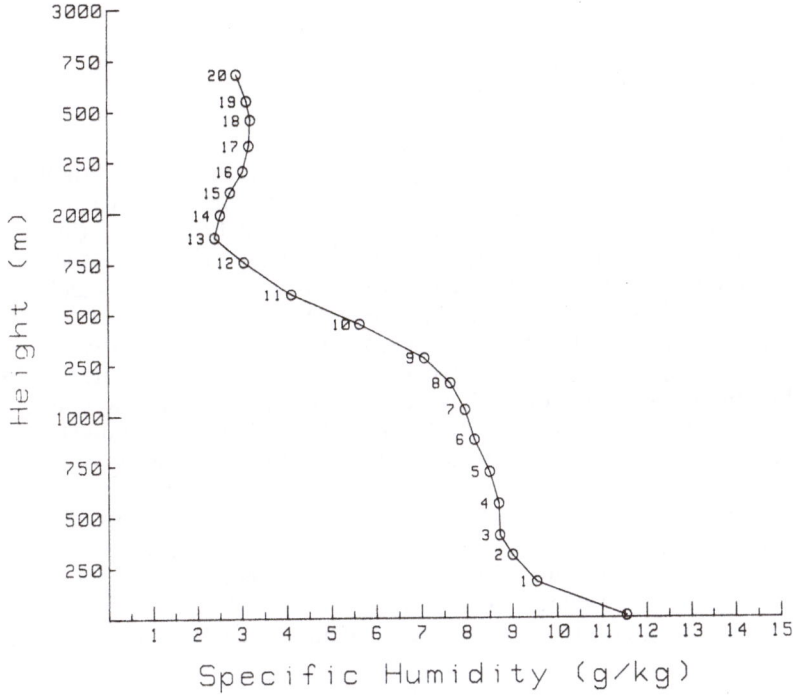

Figure 4. Same as Figure 3 for specific humidity.

It was found that Equation (24) can be useful as a predictor ($R \approx 0.7$) for the inversion flux $\overline{q'w'}_h$. Although the correlations obtained so far with expressions like Equation (24) (e.g. Driedonks and Tennekes, 1984, p. 99) are not impressive, it is clear that in the outer region the entrainment flux is an important variable. One outstanding issue is whether the similarity expression, Equation (24), is sufficiently general so that it is applicable to other scalar admixtures of the air, including potential temperature.

2.3.2. Bulk Mass and Heat Transfer Relations

Progress in the formulation of flux–profile relationships for the outer region has been slow and difficult. The main reason for this is undoubtedly (in contrast to the inner region) that the structure of the outer region in the ABL is rarely the result of equilibrium conditions and is affected by large–scale atmospheric dynamics and ever–changing weather patterns. Therefore, most of the available information gathered in the past has not been collated in the form of profile equations but rather in the form of bulk transfer expressions. In one approach, following work by, among others, Kazanski and Monin (1961), Zilitinkevich and Deardorff (1974), Arya and Wyngaard (1975), Yamada (1976), and Brutsaert and Mawdsley (1976), the surface fluxes for water vapor and sensible heat can be written in a general form as follows:

$$E = k\, u_* \,\rho \,(q_1 - q_b) \left[\ln\left(\frac{h_b - d}{z_1 - d} \right) + \Psi_v\left(\frac{z_1 - d}{L} \right) - D \right]^{-1} \tag{26}$$

and

$$H = k \; u_* \; \rho \; c_p(\theta_1 - \theta_b)\left[\ln\left(\frac{h_b - d}{z_1 - d}\right) + \Psi_s\left(\frac{z_1 - d}{L}\right) - C\right]^{-1} \tag{27}$$

respectively. The subscript 1 refers to a reference z_1 within the inner region; the variable h_b is a characteristic height scale of the ABL which in different applications of the bulk approach has been taken variously as proportional to (u_*/f), the thickness of an Ekman layer (in which f is the Coriolis parameter), h_i the height of the top of the mixed layer, (i.e. the bottom of the overlying inversion layer), or h_c the depth of convective mixing; the variables q_b and θ_b are characteristic specific humidity and characteristic potential temperature for the ABL which have been taken variously as the values measured at around $z = 0.25 \, u_*/f$, at $z = h_i$, at $z = h_c$ or as the mean taken over the whole or over part of the ABL. D and C are similarity functions of several variables but until now the dependence on (h_b/L) is the only one that appears to have been practical (e.g. Brutsaert, 1982, pp. 78–85); recently, it has been suggested [see also Equation (24)] that C and D may depend upon the entrainment fluxes at the top of the ABL (Brutsaert, 1988).

Equations (26) and (27) can be used to calculate E and H from measurements of q and θ. The friction velocity u_*, which is required in the solution, can be calculated by means of a boundary layer drag equation in terms of wind velocity measurements. The formulation appears to be most effective (e.g. Yamada, 1976; Garratt et al., 1982) if the mixed layer average wind velocity $<V>$ (the integral of V over the thickness of the mixed layer h_i, divided by h_i) is used, namely:

$$u_* = k < V > \left\{\left[\left(\frac{h_i - d}{z_o}\right) - B\right]^2 + A^2\right\}^{-1/2} \tag{28}$$

where A and B are similarity functions analogous to C and D.

Several determinations of the functions $A(h_b/L)$, $B(h_b/L)$, and $C(h_b/L)$ on the basis of experimental data for different definitions of h_b, θ_b, and V_b have appeared in the literature (e.g. Melgarejo and Deardorff, 1974; Wyngaard et al., 1974; Arya, 1977; Yamada, 1976; Garratt and Francey, 1978). Unfortunately, however, although the different determinations are generally in fair agreement, the calculated data exhibit a large degree of scatter. The water vapor function, $D = D(h_b/L)$, has not received as much attention. The only determinations to date appear to be those of Ishijima (1977) and Brutsaert and Chan (1978) from analyses of data over the ocean. A few observations were obtained by Myrup et al. (1985), but in view of the small number and the scatter of the data, the implications of the latter results remain unclear. No other independent determinations are available for land surfaces.

The use of bulk Equations (26), (27), and (28), to determine the fluxes E, H, and u_*, is the analog of the profile method for the surface layer by means of Equations (19), (20), and (21). As indicated elsewhere (Brutsaert, 1982, p. 214) the ABL bulk transfer equations can also be used in an energy budget approach, analogous to the method applied by Stricker and Brutsaert (1978) for the surface sublayer. Thus a surface energy budget equation with Equations (27) and (29) constitute the necessary three equations to solve for the three unknowns u_*, H, and E.

Until now Equations (26), (27), and (28) have remained essentially untested in a predictive mode over land, mainly because of a scarcity of suitable data sets. Their applicability and the determination of the most effective characteristic variables h_b, q_b, and θ_b will require further study.

2.3.3. Slab Approach for the ABL

In certain applications, useful results can be obtained by assuming that, except for a relatively thin layer near the ground, the wind speed, V, the potential temperature, θ, and the specific humidity, q, are practically uniform over the entire depth of the ABL. This ABL is assumed to be capped by a very thin and sharp inversion. Thus in the simplest i.e. zero–order form of this approach, which is a good approximation under unstable conditions, the ABL is considered as a perfectly mixed "slab" of thickness, h_i, above the negligibly thin surface layer (see Figure 5). The budget equations for the mean specific humidity, q_m, and potential temperature, θ_m, represent the evolution of storage of latent and sensible heat in such a slab; they can be written as:

$$h\frac{Dq_m}{Dt} = E/\rho_a - \overline{q'w'}_h \tag{29}$$

$$h\frac{D\theta_m}{Dt} = H/\rho_a c_p - \overline{\theta'w'}_h - Q_R/\rho_a c_p \tag{30}$$

where the subscript m denotes mean values for the slab; D/Dt $(= \partial/\partial t + V_m \cdot \nabla_{xy})$ is the horizontal substantial derivative with mean wind V_m and mean horizontal gradient ∇_{xy}; ρ_a is the density of the air near the ground; and Q_R is the radiative heating loss rate from the slab.

For special conditions, namely in the absence of advection and radiative cooling (i.e. $V_m \cdot \nabla_{xy} q_m = 0, V_m \cdot \nabla_{xy} \theta_m = 0$ and $Q_R = 0$), the slab approach has been useful to describe the development of the thickness, h, and of the mean concentration, q_m and θ_m, (e.g. Driedonks, 1982). This approach has also served in attempts to simulate evaporation and to gain more insight into existing parameterizations for E such as Equations (4) and (14) (e.g. McNaughton, 1976; Perrier, 1982; DeBruin, 1983; McNaughton and Spriggs, 1986).

The slab approach has also been used to study the turbulent fluxes at the top of the mixed layer. For sensible heat flux it was confirmed (e.g. Kustas and Brutsaert, 1987b; Brutsaert, 1987) that it is proportional to the surface flux as proposed earlier by, among others, Carson (1973) and Tennekes (1973):

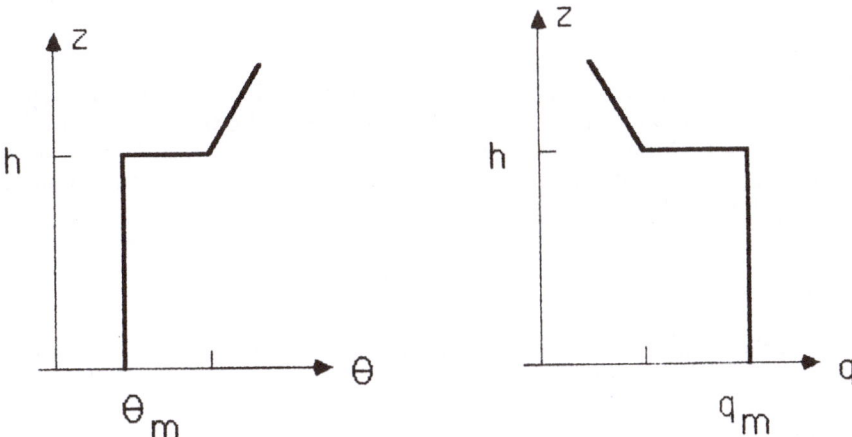

Figure 5. Sketch of zero–order slab model for atmospheric boundary layer.

$$\overline{\theta'w'}_h = -A\,H/\rho_a c_p \qquad (31)$$

where A is a constant around 0.2 for strongly convective conditions and somewhat larger, around 0.5, when mechanical turbulence is also present. For such conditions a better parameterization, suggested by Tennekes (1973), was found to be:

$$\overline{\theta'w'}_h = -A\,H/\rho_a c_p - B\,T_a\,u_*^3/g\,h_i \qquad (32)$$

where A and B are on the order of 0.2 and 5, respectively.

Water vapor is a more passive scalar than sensible heat, so that $\overline{\theta'w'}_h$ cannot be related to E by simple analogs of Equations (31) and (32). For a slab with a mean humidity gradient, $s = <\partial q/\partial z>$, it can be shown by means of an order–one slab model (Brutsaert, 1987) that the humidity flux at $z = h_i$ can be given by:

$$\overline{q'w'}_h = \beta\,E/\rho_a - \frac{\Delta q - s\,\Delta h}{\Delta\theta}\left[\beta\,H/(\rho_a c_p) - \overline{\theta'w'}_h\right] \qquad (33)$$

where $\beta = \Delta h/(\Delta h + h_i)$ and Δq and $\Delta\theta$ the changes in q and θ, respectively, across the inversion of thickness Δh. Equation (33) was found to produce useful results in calculating q_m and θ_m budgets over the ocean surface by means of Equations (29) and (30).

Clearly, however, even with the assumption of the ABL as a well–mixed slab, the closure of the budget Equations (29) and (30) is not without difficulties. For example, a systematic error of 1 K in the mean temperature θ_m of a slab of $h_i = 1,000$ m derived from soundings taken three hours apart, would yield an error in the budget of about 100 W m^{-2}; this is of the order of the sensible heat flux H. In addition, it is usually not straightforward to account properly for the advective terms $V_m \cdot \nabla_{xy}\theta_m$ and $V_m \cdot \nabla_{xy}q_m$ and for the radiative source term Q_R; for this reason these terms are often neglected. The effect of advection over land can be minimized, however, by averaging budgets over several days (e.g. Betts, 1976); it may also be possible to estimate the advection by means of mesoscale atmospheric modeling (Andre et al., 1988). Radiative heating effects depend on the type and amount of aerosols in the ABL (e.g. Glazier et al., 1976; Moores et al., 1979). The solution of all these complications will require more research.

3. Conclusions

One of the major uncertainties in the performance of hydrologic models results from the unknown reliability of the procedures used to describe regional evaporation and related land–surface processes. A better understanding of turbulent transfer in the atmospheric boundary layer, at appropriate temporal and spatial scales, will lead to improved parameterization techniques for the solution of practical problems.

4. Acknowledgements

Part of the work leading to this chapter has been supported by the National Science Foundation through grants ATM-8601115 and ATM-8619193.

5. References

Ali, M. F., and J. A. Mawdsley: 1987, 'Comparison of Two Recent Models for Estimating Actual Evapotranspiration Using Only Regularly Recorded Data,' *J. Hydrology* **93**, 257–276.

André, J.-C., P. Lacarrére, and L. J. Mahrt: 1979, 'Sur la Distribution Verticale de l'humidité dans une Couche Limite Convective,' *J. Rech. Atmos.* **13**, 135–146.

André, J.-C., J.-P. Goutorbe, A. Perrier, et al.: 1988, 'Evaporation Over Land Surfaces: First Results from HAPEX–MOBILHY Special Observing Period,' *Annales Geophysicae (Eur. Geophys. Soc.)* **6**(5), 477–492.

Arya, S. P. S.: 1977, 'Suggested Revisions to Certain Boundary Layer Parameterization Schemes Used in Atmospheric Circulation Models,' *Monthly Weather Rev.* **105**, 215–227.

Arya, S. P. S., and J. C. Wyngaard: 1975, 'Effect of Baroclinicity on Wind Profiles and the Geostrophic Drag Law for the Convective Planetary Boundary Layer,' *J. Atmos. Sci.* **32**, 767–778.

Betts, A. K.: 1976, 'Modeling Subcloud Layer Structure and Interaction with a Shallow Cumulus Layer,' *J. Atmos. Sci.* **33**, 2363–2382.

Black, T. A., C. B. Tanner, and W. R. Gardner: 1970, 'Evapotranspiration from a Snap Bean Crop,' *Agronomy J.* **62**, 66–69.

Bouchet, R. J.: 1963, 'Evapotranspiration Réele et Potentielle,' *Signification Climatique*, Publ. 62, pp. 134–142, Int. Assoc. Sci. Hydrol., Gentbrugge, Belgium.

Brown, K. W., and N. J. Rosenberg: 1977, 'Resistance Model to Predict Evapotranspiration and its Application to a Sugar Beet Field,' *Agronomy J.* **65**, 341–347.

Brutsaert, W.: 1975, 'A Theory for Local Evaporation (or Heat Transfer) from Rough and Smooth Surfaces at Ground Level,' *Water Resour. Res.* **11**, 543–550.

Brutsaert, W.: 1982, *Evaporation into the Atmosphere: Theory, History and Applications*, 299 pp., D. Reidel, Hingham, MA.

Brutsaert, W.: 1987, 'Nearly Steady Convection and the Boundary Layer Budgets of Water Vapor and Sensible Heat,' *Boundary–Layer Meteor.* **39**, 283–300.

Brutsaert, W.: 1988, 'The Parameterization of Regional Evaporation – Some Directions and Strategies,' *J. Hydrology* **102**, 409–426.

Brutsaert, W., and F. K. F. Chan: 1978, 'Similarity Functions D for Water Vapor in the Unstable Atmospheric Boundary Layer,' *Boundary–Layer Meteor.* **14**, 441–456.

Brutsaert, W., and W. P. Kustas: 1985, 'Evaporation and Humidity Profiles for Neutral Conditions over Rugged Hilly Terrain,' *J. Climate Appl. Meteor.* **24**, 915–923.

Brutsaert, W., and W. P. Kustas: 1987, 'Surface Water Vapor and Momentum Fluxes under Unstable Conditions from a Rugged–Complex Area,' *J. Atmos. Sci.* **44**, 421–431.

Brutsaert, W., and J. A. Mawdsley: 1976, 'The Applicability of Planetary Boundary Layer Theory to Calculate Regional Evapotranspiration,' *Water Resour. Res.* **12**, 852–858.

Brutsaert, W., and M. B. Parlange: 1991, 'The Unstable Surface Layer Above Forest: Regional Evaporation and Heat Flux,' *Water Resour. Res.*

Brutsaert, W., and H. Stricker: 1979, 'An Advection–Aridity Approach to Estimate Actual Regional Evapotranspiration,' *Water Resour. Res.* **15**, 443–450.

Brutsaert, W., and M. Sugita: 1990, 'The Extent of the Unstable Monin–Obukhov Layer for Temperature and Humidity above Complex Hilly Grassland,' *Boundary–Layer Meteor.* **51**, 383–400.

Budyko, M. I.: 1955, 'On the Determination of Evaporation from the Land Surface,' *Meteorol. Gidrol.* **1**, 52–58 (in Russian).

Budyko, M. I.: 1974, *Climate and Life*, Academic Press, NY, 508 pp.

Businger, J. A.: 1988, 'A Note on the Businger-Dyer Profiles,' *Boundary–Layer Meteor.* **42**, 145–151.

Byrne, G. F., F. X. Dunin, and P. J. Diggle: 1988, 'Forest Evaporation and Meteorological Data: A Test of a Complementary Theory Advection-Aridity Approach,' *Water Resour. Res.* **24**, 30–34.

Carson, D. J.: 1973, 'The Development of a Dry Inversion-Capped Convectively Unstable Boundary Layer,' *Quart. J. Roy. Met. Soc.* **99**, 450–467.

Carson, D. J.: 1982, 'Current Parameterizations of Land-Surface Processes in Atmospheric General Circulation Models,' in P. S. Eagleson (ed.), *Land Surface Processes in Atmospheric General Circulation Models*, Cambridge Univ. Press, New York, pp. 67–108.

Davies, J. A., and C. D. Allen: 1973, 'Equilibrium, Potential and Actual Evaporation from Cropped Surfaces in Southern Ontario,' *J. App. Meteor.* **12**, 649–657.

DeBruin, H. A. R.: 1983, 'A Model for the Priestley–Taylor Parameter α,' *J. Climate Appl. Meteor.* **22**, 572–578.

Driedonks, A. G. M.: 1982, 'Models and Observations of the Growth of the Atmospheric Boundary Layer,' *Boundary–Layer Meteor.* **23**, 283–306.

Driedonks, A. G. M., and H. Tennekes: 1984, 'Entrainment Effects in the Well–Mixed Atmospheric Boundary Layer,' *Boundary–Layer Meteor.* **30**, 75–105.

Dunin, F. X., and E. A. N. Greenwood: 1986, 'Evaluation of the Ventilated Chamber for Measuring Evaporation from a Forest,' *Hydrol. Process.* **1**, 47–61.

Dyer, A. J.: 1974, 'A Review of Flux–Profile Relationships,' *Boundary–Layer Meteor.* **7**, 363–372.

Federer, C. A.: 1977, 'Leaf Resistance and Xylem Potential Differ Among Broadleaved Species,' *Forest Sci.* **23**, 411–419.

Garratt, J. R.: 1978, 'Transfer Characteristics for a Heterogeneous Surface of Large Aerodynamic Roughness,' *Quart. J. Roy. Meteor. Soc.* **104**, 491–502.

Garratt, J. R., and R. J. Francey: 1978, 'Bulk Characteristics of Heat Transfer in the Unstable, Baroclinic Atmospheric Boundary Layer,' *Boundary Layer Meteor.* **15**, 399–421.

Garratt, J. R., J. C. Wyngaard, and R. J. Francey: 1982, 'Winds in the Atmospheric Boundary Layer, Prediction and Observation,' *J. Atmos. Sci.* **39**, 1307–1316.

Gash, J. H. C., W. J. Shuttleworth, C. R. Lloyd, J.–C. André, J.–P. Goutorbe, and J. Gelpe: 1989, 'Micrometeorological Measurements in Les Landes Forest During HAPEX–MOBILHY,' *Agric. For. Meteor.* **46**, 131–147.

Gash, J. H. C., and J. B. Stewart: 1975, 'The Average Resistance of a Pine Forest Derived from Bowen Ratio Measurements,' *Boundary–Layer Meteor.* **8**, 453–464.

Glazier, J., J. L. Monteith, and M. H. Unsworth: 1976, 'Effects of Aerosol on the Local Heat Budget of the Lower Atmosphere,' *Quart. J. Roy. Meteor. Soc.* **102**, 95–102.

Grimmond, C. S. B., T. R. Oke, and D. G. Steyn: 1986, 'Urban Water Balance 1. A Model for Daily Totals,' *Water Resour. Res.* **22**, 1397–1403.

Grindley, J.: 1970, 'Estimation and Mapping of Evaporation,' in *Symposium on the World Balance*, Vol. I, Publ. 92, Intern. Assoc. Hydrol. Sci., Gentbrugge, Belgium, pp. 200–213.

Hewlett, J. D., J. C. Fortson, and G. B. Cunningham: 1977, 'The Effect of Rainfall Intensity on Storm Flow and Peak Discharge from Forest Land,' *Water Resour. Res.* **13**, 259–266.

Högström, U.: 1988, 'Non–dimensional Wind and Temperature Profiles in the Atmospheric Surface Layer: A Re–evaluation,' *Boundary–Layer Meteor.* **42**, 55–78.

Ishijima, S.: 1977, 'Observational Studies on the Similarity Functions C and D of the Convective Planetary Boundary Layer over the Ocean,' *J. Meteor. Soc. Japan* **55**, 449–456.

Jobe, L. O.: 1982, *An Investigation of Evapotranspiration in the Sudano–Sahelian Region of West Africa,* M.S. Thesis, Cornell University, Ithaca, NY, 259 pp.

Kazanski, A. B., and A. S. Monin: 1961, 'On the Dynamic Interaction Between the Atmosphere and the Earth's Surface,' *Bull. (Izv.) Acad. Sci.*, USSR, Geophys. Ser., 5 (Engl. Edn. AGU), 514–515.

Kotoda, K.: 1986, *Estimation of River Basin Evapotranspiration,* Environmental Research Center Paper no. 8, Univ. of Tsukuba, Ibaraki, Japan, 66 pp.

Kustas, W. P., and W. Brutsaert: 1986, 'Wind Profile Contants in a Neutral Atmospheric Boundary Layer Over Complex Terrain,' *Boundary–Layer Meteor.* **34**, 35–54.

Kustas, W. P., and W. Brutsaert: 1987a, 'Budgets of Water Vapor in the Unstable Boundary Layer Over Rugged Terrain,' *J. Climate Appl. Meteor.* **26**, 607–620.

Kustas, W. P., and W. Brutsaert: 1987b, 'Virtual Heat Entrainment into the Mixed Layer Over Very Rough Terrain,' *Boundary–Layer Meteor.* **38**, 141–157.

Lindroth, A.: 1985, 'Canopy Conductance of Coniferous Forests Related to Climate,' *Water Resour. Res.* **21**, 297–304.

Liu, C. C. K., and W. Brutsaert: 1978, 'A Nonlinear Analysis of the Relationship Between Rainfall and Runoff for Extreme Floods,' *Water Resour. Res.* **14**, 75–83.

Loague, K. M., and R. A. Freeze: 1985, 'A Comparison of Rainfall–Runoff Modeling Techniques on Small Upland Watersheds,' *Water Resour. Res.* **21**, 229–248.

Mahrt, L.: 1976, 'Mixed Layer Moisture Structure,' *Mon. Weath. Rev.* **104**, 1403–1418.

Manabe, S.: 1969, 'Climate and the Ocean Circulation, 1. The Atmospheric Circulation and the Hydrology of the Earth's Surface,' *Monthly Weather Review* **97**, 739–774.

Mawdsley, J. A., and M. F. Ali: 1985, 'Estimating Nonpotential Evapotranspiration by Means of the Equilibrium Evaporation Concept,' *Water Resour. Res.* **21**, 383–391.

McNaughton, K. G.: 1976, 'Evaporation and Advection I: Evaporation from Extensive Homogeneous Surfaces,' *Quart. Jour. Roy. Meteor. Soc.* **102**, 181–191.

McNaughton, K. G., and T. W. Spriggs: 1986, 'A Mixed-Layer Model for Regional Evaporation,' *Boundary-Layer Meteor.* **34**, 243–262.

Melgarejo, F. W., and J. W. Deardorff: 1974, 'Stability Functions for the Boundary-Layer Resistance Laws Based Upon Observed Boundary-Layer Heights,' *Jour. Atmos. Sci.* **31**, 1324–1333.

Monteith, J. L.: 1973, *Principles of Environmental Physics*, American Elsevier, New York, 241 pp.

Monteith, J. L.: 1981, 'Evaporation and Surface Temperature,' *Quart. J. Roy. Soc.* **107**, 1–27.

Monteith, J. L., G. Szeicz, and P. E. Waggoner: 1965, 'The Measurement and Control of Stomatal Resistance in the Field,' *J. Appl. Ecol.* **2**, 345–355.

Moores, W. H., S. J. Caughey, C. J. Readings, J. R. Milford, D. A. Mansfield, S. Abdulla, T. H. Guymer, and W. B. Johnson: 1979, 'Measurements of Boundary Layer Structure and Development over SE England using Aircraft and Tethered Balloon Instrumentation,' *Quart. J. Roy. Meteor. Soc.* **105**, 397–421.

Morton, F. I.: 1975, 'Estimating Evaporation and Transpiration from Climatological Observations,' *J. Appl. Meteor.* **14**, 488–497.

Morton, F. I.: 1976, 'Climatological Estimates of Evapotranspiration,' *J. Hydraul. Div., Proc. ASCE* **102**(HY3), 275–291.

Morton, F. I.: 1986, 'Comment on "Estimating Nonpotential Evapotranspiration by Means of the Equilibrium Evaporation Concept" by J.A. Mawdsley and M.F. Ali,' *Water Resour. Res.* **22**, 2115–2118.

Myrup, L. O., C. D. Johnson, and W. O. Pruitt: 1985, 'Measurement of the Atmospheric Boundary-Layer Resistance Law for Water Vapor,' *Boundary-Layer Meteor.* **33**, 105–111.

Penman, H. L.: 1948, 'Natural Evaporation from Open Water, Bare Soil, and Grass,' *Proc. R. Soc. London*, Ser. A, **193**, 120–146.

Perrier, A.: 1982, 'Land Surface Processes: Vegetation,' in P. S. Eagleson (ed.), *Land Surface Processes in Atmospheric General Circulation Models*, Cambridge Univ. Press, pp. 395–448.

Priestley, C. H. B., and R. J. Taylor: 1972, 'On the Assessment of Surface Heat Flux and Evaporation Using Large-Scale Parameters,' *Mon. Weather Rev.* **100**, 81–92.

Slatyer, R. O., and I. C. McIlroy: 1961, *Practical Microclimatology*, 310 pp., CSIRO, Melbourne, Australia.

Steenhuis, T. S., C. D. Jackson, S. K. J. Kung, and W. Brutsaert: 1985, 'Measurement of Groundwater Recharge on Eastern Long Island, New York, U.S.A.,' *J. Hydrology* **79**, 145–169.

Stewart, J. B.: 1988, 'Modeling Surface Conductance of Pine Forest,' *Agric. For. Meteor.* **43**, 19–37.

Stricker, H., and W. Brutsaert: 1978, Actual Evapotranspiration over a Summer Period in the "Hupsel Catchment", *J. Hydrol.* **39**, 139–157.

Sugita, M., and W. Brutsaert: 1990a, 'How Similar are Temperature and Humidity Profiles in the Unstable Boundary Layer?,' *J. Appl. Meteor.* **29**, 489–497.

Sugita, M., and W. Brutsaert: 1990b, 'Regional Surface Fluxes from Remotely Sensed Skin Temperature and Lower Boundary Layer Measurements,' *Water Resour. Res.* **26**, 2937–2944.

Szeicz, G., and I. F. Long: 1969, 'Surface Resistance of Crop Canopies,' *Water Resour. Res.* **5**, 622–633.

Szeicz, G., C. H. B. Van Bavel, and S. Takami: 1973, 'Stomatal Factor in the Water Use and Dry Matter Production of Sorghum,' *Agric. Meteor.* **12**, 361–389.

Tan, C. S., and T. A. Black: 1976, 'Factors Affecting the Canopy Resistance of a Douglas-Fir Forest,' *Boundary-Layer Meteor.* **10**, 475–489.

Tennekes, H.: 1973, 'A Model for the Dynamics of the Inversion Above a Convective Boundary Layer,' *J. Atmos. Sci.* **30**, 558–567.

Thom, A. S.: 1972, 'Momentum, Mass and Heat Exchange of Vegetation,' *Quart. J. Roy. Meteor. Soc.* **98**, 124–134.

Thom, A. S.: 1975, 'Momentum, Mass and Heat Exchange of Plant Communities,' in J. L. Monteith (ed.), *Vegetation and the Atmosphere, Vol. I, Principles*, Academic Press, Orlando, FL, pp. 57–109.

Thornthwaite, C. W.: 1948, 'An Approach Toward a Rational Classification of Climate,' *Geograph. Rev.* **38**, 55–94.

Thornthwaite, C. W., and J. R. Mather: 1955, '*The Water Balance, Publications in Climatology,*' **8**, No. 1, Lab. of Climatology, Centerton, NJ, 86 pp.

van Bavel, C. H. M.: 1967, 'Changes in Canopy Resistance to Water Loss from Alfalfa Induced by the Soil Water Depletion,' *Agric. Meteor.* **4**, 165–176.

Wyngaard, J. C., S. P. S. Arya, and O. R. Coté: 1974, 'Some Aspects of the Structure of Convective Planetary Boundary Layers,' *J. Atmos. Sci.* **41**, 747–754.

Wyngaard, J. C., and R. A. Brost: 1984, 'Top–down and Bottom–up Diffusion of a Scalar in the Convective Boundary Layer,' *J. Atmos. Sci.* **41**, 102–112.

Yaglom, A. M.: 1977, 'Comments on Wind and Temperature Flux–Profile Relationships,' *Boundary–Layer Meteor.* **11**, 89–102.

Yamada, T.: 1976, 'On the Similarity Functions A, B and C of the Planetary Boundary Layer,' *J. Atmos. Sci.* **33**, 781–793.

Zilitinkevich, S. S., and J. W. Deardorff: 1974, 'Similarity Theory for the Planetary Boundary Layer of Time–Dependent Height,' *J. Atmos. Sci.* **31**, 1449–1452.

Chapter 5

Physics-Based Models of Snow

E. M. Morris
British Antarctic Survey
Ice and Climate Division
High Cross, Madingley Road
Cambridge, CB3 0ET, United Kingdom

Abstract. This chapter is a review of current physics–based techniques for modeling snowmelt. Mixture theory is used to develop the equations of conservation of mass, momentum and energy for snow treated as a two–component, three–phase mixture of ice, water, water vapor, and air. The constitutive laws and boundary conditions required to complete a general snowmelt model are then described, with particular attention to the energy balance at the upper boundary of a snowpack.

1. Introduction

Modern physics–based snowmelt modeling began with Anderson (1976) who replaced the simple index methods then in use by an energy budget approach. Instead of calculating snowmelt from meteorological data (usually only air temperature) by simple linear empirical equations he calculated all the components of the energy input to the snow surface, the energy losses from the surface, and hence the residual energy available to melt the snow. Although Anderson used empirical equations to describe processes within the snow pack, i.e. heat and water flow, it soon became clear that the energy budget approach could be much improved by physics–based modeling of internal as well as boundary processes. The appropriate theory already existed in the hydrological literature. Colbeck realized that the equations of flow for soils could also be applied to snow since both were porous media. In a series of papers (Colbeck, 1972, 1974, 1977) he established the techniques for modeling water flow in wet (or 'ripe') snow. Meanwhile, other authors (Sulakvelidze, 1959; Yen, 1962; Navarre, 1977; Obled and Rosse, 1977; Palm and Tveitereid, 1979) were applying established equations for heat flow in porous media to the problem of modeling temperature variations in cold snow. Morris and Godfrey (1979) made the first attempt to construct a physics–based model which could be used for both cold and ripe snow and which took into account changes in the structure of the snow pack. Since then, similar models have been developed not only for hydrological purposes but also for avalanche prediction and glacier mass balance research (e.g. Morris, 1983; Akan, 1984; Greuell and Oerlemans, 1986).

Recently the mathematical basis of these snowmelt models has been critically examined by Kelly and his coworkers (Kelly et al., 1986; Kelly, 1987). Several interesting mathematical problems have arisen and have provoked a reexamination of the physics of the problem. This chapter will describe the 'state-of-the-art' snow melt model equations in detail, drawing attention to the assumptions which have been made about the physics and their consequences.

So far, physics–based models have only been written for 'unpolluted' snow, that is for a mixture of ice, water, water vapor, and air. However, it is now clear that a major application of

D. S. Bowles and P. E. O'Connell (eds.), Recent Advances in the Modeling of Hydrologic Systems, 85–112.
© 1991 *Kluwer Academic Publishers.*

snowmelt modeling will be simulation of the preferential elution of impurities at the onset of snowmelt. For this reason the chapter contains rather more discussion of the effect of impurities than is strictly required. Further work is needed on the complex physical chemistry of concentrated ionic solutions in thin films before a working program for polluted snow can be written, but it is useful now to define the required mathematical structure of the model even if all the equations are not known.

2. Discussion

2.1. Mixture Theory

The basis for physics–based snowmelt models is the concept that snow can be regarded as a mixture containing two components: air; and water in up to three of its phases, ice, liquid water, and water vapor. Each volume element of snow is assumed to contain all the constituents dispersed within it. This approach has been used by many authors (e.g. Yen, 1962; Colbeck, 1972; Obled and Rosse, 1977; Navarre, 1977; Morris and Godfrey, 1979; Morris, 1983; Akan, 1984) although different degrees of simplification have been proposed. To date, a working 2–component, 3–phase snowmelt model has not been achieved but considerable progress towards defining the necessary equations has been made by Kelly (1987). In this chapter the model equations will be developed in one–dimension for ease of notation but it is not difficult to extend them to three–dimensions.

Let the independent variables defining the state of the bulk phases be the pressures, p^α, the temperatures, T^α, and the number of moles, n_k^α, per unit volume of snow and velocities, w_k^α, of each of the components. Let $\alpha = 1$ for the solid phase, $\alpha = 2$ for the liquid phase, and $\alpha = 3$ for the gas. The two components are water ($k = 1$) and air ($k = 2$). Similarly, let the independent variables defining the state of the interfaces be the surface tensions, σ^i, the temperatures, T^i, and the number of adsorbed moles per unit volume of snow, n_k^i, and velocities, w_k^i, of each component. Three types of interface are possible: $i = 1$ for an interface between solid and liquid phases, $i = 2$ for a liquid–gas interface, and $i = 3$ for a solid–gas interface.

2.1.1. The Conservation Equations

Conservation of Mass. Changes in the system with time can be described by the conservation equations for mass, energy, and momentum. Let N_k^α be the number of moles of component k gained by bulk phase α per unit time per unit volume of snow. Then conservation of mass gives:

$$\frac{d(n_k^\alpha)}{dt} = \frac{\partial(n_k^\alpha)}{\partial t} + w_k^\alpha \frac{\partial(n_k^\alpha)}{\partial z} = N_k^\alpha \qquad (1)$$

where t is time and z the vertical direction.

A similar equation may be written for N_k^i, the number of moles of component k gained by interface i per unit time per unit volume of snow. Assuming that there are no chemical reactions which transform component k into a different component, the number of moles of k per unit volume of snow is conserved:

$$\sum_\alpha N_k^\alpha + \sum_i N_k^i = 0 \qquad (2)$$

If ρ_k^a is the density of component k in phase α Equation (1) may be written:

$$\frac{\partial}{\partial t}(\rho_k^a \; \theta_k^a) + w_k^a \frac{\partial}{\partial z}(\rho_k^a \; \theta_k^a) = N_k^a \; M_k \tag{3}$$

where θ_k^a is the volume fraction of component k in phase α and M_k the molecular weight of component k. The volume fractions are related by the equation:

$$\sum_a \sum_k \theta_k^a = \sum_a \theta^a = 1 \tag{4}$$

and the density of the snow is defined as:

$$\rho_s = \sum_a \rho^a \; \theta^a = \sum_a \sum_k \rho_k^a \; \theta_k^a \tag{5}$$

A weighted velocity for phase α can be defined as:

$$w^a = \sum_k n_k^a \; w_k^a \; / \; \sum_k n_k^a \tag{6}$$

Conservation of Momentum. The equations of conservation of momentum for each bulk phase are:

$$\sum_k \left(\rho_k^a \; \theta_k^a \; (\frac{\partial w_k^a}{\partial t} + w_k^a \frac{\partial w_k^a}{\partial z}) \right) = \sum_k \left(\frac{\partial}{\partial z}(\theta_k^a \; \tau^a) - \rho_k^a \; \theta_k^a \; g + \rho_s \; \beta^a \right) \tag{7}$$

where β^α are the interaction body forces acting on phase α per unit mass. By definition:

$$\sum_k \beta^a = 0 \tag{8}$$

τ^α are the vertical components of intrinsic stress for each phase and gravity g is assumed to be the only external body force. The correct definition of the β^α terms poses considerable problems. In most snow melt models these are skated over by substitution of the empirical Darcy Law for the momentum equations. It is worthwhile stating explicitly what assumptions are being made. Suppose that:

$$\rho_s \; \beta^a = p^a \frac{\partial \theta^a}{\partial z} - \sum_k \frac{\rho_k^a \; \theta_k^a}{K^a} g \left(\theta_k^a \; w_k^a - (1 - \theta_k^1) \; w_k^1 \right) \tag{9}$$

where superscript 1 denotes the solid ice phase and K^α is the hydraulic conductivity of phase α. Then assume that:
(i) the stresses τ^α are purely hydrostatic so that $\tau^\alpha = -p^\alpha$
(ii) the inertia terms in Equation (7) are negligible
(iii) the ice matrix is static so that $w_k^{(1)} = 0$.

Equations (7) and (9) then reduce to:

$$\sum_k \rho_k^a \; \theta_k^a \; w_k^a \; \theta_k^a = -\sum_k \frac{K^a \; \theta_k^a}{g} \left(\frac{\partial p^a}{\partial z} + \rho_k^a \; g \right) \tag{10}$$

which is the Darcy Law.

The relative velocity of component k with respect to the weighted velocity w^α is given by an empirical diffusion equation:

$$w_k^a - w^a = - \frac{D_k^a}{\rho_k^a \; \theta_k^a} \frac{\partial}{\partial z} (\rho_k^a \; \theta_k^a) \tag{11}$$

where D_k^a is the diffusion coefficient of k in α. At this stage it seems reasonable to define the velocities of the components in the interfaces by the equation:

$$w_k^i = w_j^a \tag{12}$$

where α is the more viscous of the bulk phases in contact with interface i, and j is the main component. So, for example, the material in the meniscus between water and moist air inclusions in snow is assumed to move at the same velocity as the water. Note this does not mean that the i interface itself moves at velocity w_j^a; the velocity of the interface between two phases depends also on the rate of phase change.

Conservation of Energy. The derivation of the equation for conservation of energy is not as straightforward as it might seem. If H^α and H^i are the enthalpies of phase α and interface i and Q is the heat per unit volume of snow, the energy conservation equation may be written:

$$\frac{dQ}{dt} = \sum_a \frac{dH^a}{dt} + \sum_i \frac{dH^i}{dt} - \sum_a \theta^a \frac{dp^a}{dt} - \sum_i \Omega^i \frac{d\sigma^i}{dt} \tag{13}$$

Ω^i is the area of interface i per unit volume of snow and σ^i is the surface energy per unit area. To preserve the analogy with pressure in the bulk phases, σ will be defined to be positive for compression and negative for tension. The surface tension of interface i is therefore $-\sigma^i$.

The heat input dQ/dt arises from conduction in the solid phase $\alpha = 1$, absorption of radiation, $\partial R_n / \partial z$, and stress working, $\mathbb{Q}$. Thus:

$$\frac{dQ}{dt} = \frac{\partial}{\partial z} \left(\kappa \frac{\partial T^1}{\partial z} \right) + \frac{\partial R_n}{\partial z} + \mathbb{Q} \tag{14}$$

κ is the thermal conductivity of the solid phase. The enthapy differentials can be expanded in terms of the $k+2$ independent thermodynamic variables so that:

$$\frac{dH^a}{dt} = \frac{\partial H^a}{\partial T^a} \frac{dT^a}{dt} + \frac{\partial H^a}{\partial p^a} \frac{dp^a}{dt} + \sum_k \frac{\partial H^a}{\partial n_k^a} \frac{dn_k^a}{dt}$$

$$\frac{dH^i}{dt} = \frac{\partial H^i}{\partial T^i} \frac{dT^i}{dt} + \frac{\partial H^i}{\partial \sigma^i} \frac{d\sigma^i}{dt} + \sum_k \frac{\partial H^i}{\partial n_k^i} \frac{dn_k^i}{dt} \tag{15}$$

The change in specific enthalpy of component k when it changes from phase β to phase α is defined as:

$$\ell_k^{\alpha\beta} = \frac{\partial H^\alpha}{\partial n_k^\alpha} - \frac{\partial H^\beta}{\partial n_k^\beta} \tag{16}$$

where $\ell_k^{\alpha\beta}$ is either a latent heat or a heat of solution. Similarly the heat of adsorption of component k onto interface i from phase β is:

$$\ell_k^{i\beta} = \frac{\partial H^i}{\partial n_k^i} - \frac{\partial H^\beta}{\partial n_k^\beta} \tag{17}$$

Using Equations (1) and (2) the latent heat and adsorption terms in (15) may be written:

$$\sum_a \left(\frac{\partial H^\alpha}{\partial n_k^\alpha} \frac{dn_k^\alpha}{dt} \right) + \sum_i \left(\frac{\partial H^i}{\partial n_k^i} \frac{dn_k^i}{dt} \right) = \sum_a \ell_k^{\alpha\beta} N_k^\alpha + \sum_i \ell_k^{i\beta} N_k^i \tag{18}$$

If $(C_p)^\alpha$, $(C_p)^i$ are the specific heats of phase α and interface i then by definition:

$$\frac{\partial H^\alpha}{\partial T^\alpha} = (C_p)^\alpha \quad ; \quad \frac{\partial H^i}{\partial T^i} = (C_p)^i \tag{19}$$

Using the weighted velocity w^α,

$$\frac{dT^\alpha}{dt} = \frac{\partial T^\alpha}{\partial t} + w^\alpha \frac{\partial T^\alpha}{\partial z} \tag{20}$$

A similar equation holds for T^i. The energy equation (13) may then be written:

$$\sum_a (C_p)^\alpha \left(\frac{\partial T^\alpha}{\partial t} + w^\alpha \frac{\partial T^\alpha}{\partial z} \right) + \sum_i (C_p)^i \left(\frac{\partial T^i}{\partial t} + w^i \frac{\partial T^i}{\partial z} \right) =$$

$$-\sum_a \ell_k^{\alpha\beta} N_k^\alpha - \sum_i \ell_k^i N_k^i + \frac{\partial}{\partial z} \left(\kappa \frac{\partial T^1}{\partial z} \right) + \frac{\partial R_n}{\partial z} \tag{21}$$

$$+ Q - \sum_a \frac{\partial H^\alpha}{\partial p^\alpha} \frac{dp^\alpha}{dt} - \sum_i \frac{\partial H^i}{\partial \sigma^i} \frac{d\sigma^i}{dt} + \sum_a \theta^\alpha \frac{dp^\alpha}{dt} + \sum_i \Omega^i \frac{d\sigma^i}{dt}$$

The terms involving the interfaces and all the terms on the third line have been considered to be negligible by previous authors. To define the various parameters in this equation it is necessary to consider the particular physical properties of the snow mixture. That is, the equations of state of each phase and interface must be defined.

2.1.2. The Equations of State

For each phase and interface the pressures p^α and surface energies σ^i are related to temperatures T^α, T^i and concentrations n_k^α, n_k^i by equations of state. In general if the Gibbs function, $G^a(p^a, T^a, n_k^a)$ for the phase α or $G^i(\sigma^i, T^i, n_k^i)$ for the interface i can be defined, the other thermodynamic variables, specific volume, entropy and chemical potential may be calculated from the definitions:

$$\theta^a = \frac{\partial G^a}{\partial p^a} \quad ; \quad \Omega^i = \frac{\partial G^i}{\partial \sigma^i}$$

$$S^a = -\frac{\partial G^a}{\partial T^a} \quad ; \quad S^i = -\frac{\partial G^i}{\partial T^i} \tag{22}$$

$$\mu_k^a = \frac{\partial G^a}{\partial n_k^a} \quad ; \quad \mu_k^i = \frac{\partial G^i}{\partial n_k^i}$$

The enthapy $H^\alpha = G^\alpha + T^\alpha S^\alpha$, with a similar equation for interface i, and the internal energy $U^\alpha = H^\alpha - p^\alpha \theta^\alpha$ or $U^i = H^i - \sigma^i \Omega^i$.

The Gibbs function of a multi–component bulk phase α may be written in the form:

$$G^a = \sum_k n_k^a \left((\mu_k^a)_o + \Re\, T^a \ln a_k^a \right) \tag{23}$$

where $(\mu_k^a)_o$ is a reference potential for component k, $\Re$ is the gas constant, and a_k^a is the activity of k in the phase α. For ideal solutions $(\mu_k^a)_o$ is simply the chemical potential of pure component k.

Equation of State for the Gas Phase. The gas phase (moist air) can be regarded as such an ideal solution. The perfect gas law applies to both the dry air and water vapor components and the mixture is inert. The activity of each component k is then simply the mole fraction $x_k^a = n_k^a / \sum n_k^a$ and the reference potential is the chemical potential of the pure gas:

$$(\mu_k^a)_o = \Re\, T^a \left(\phi_k + \ln \frac{p^a}{p_o} \right) \qquad a = 3, \qquad k = 1, 2 \tag{24}$$

where:

$$\phi_k = \frac{(h_k)_o}{\Re T} - \frac{1}{\Re} \int \left(\frac{\int (c_p)_k\, dT}{(T)^2} \right) dT - \frac{(s_k)_o}{\Re} \qquad k = 1, 2 \tag{25}$$

and p_o is a reference pressure, usually 1 bar. The constants $(h_k)_o$ and $(s_k)_o$ and the variation of the molar specific heats with temperature are known for air and water vapor. Differentiation of (24) with respect to p^α gives the familiar equation of state for a mixture of perfect gases:

$$p^a \theta^a = \sum_k n_k^a \Re\, T^a \qquad a = 3, \qquad k = 1, 2 \tag{26}$$

Equation of State for the Liquid Phase. If it is assumed that the liquid phase in unpolluted snow is a very dilute solution of air in water, the activities can again be replaced by the mole fractions. The reference potential for the solvent (water) can be taken to be the chemical potential of pure water:

$$(\mu_k)_o = \Re \, T^\alpha \, \phi_k + \frac{p^\alpha M_k}{\rho_k^\alpha} \qquad a = 2, \; k = 1 \tag{27}$$

where ρ_k^α is the density. The isothermal compressibility of water is very small ($\sim 10^{-5}$ bar^{-1}) so it may be assumed that ρ_k^α is not a function of p^α. The variation with temperature (measured in °C) is given by empirical equations (Pruppacher and Klett, 1978, p. 86):

$$\rho_k^\alpha = \sum A_n \, (T^\alpha)^n/(1 + B_n \, T^\alpha) \quad 0°C \le T^\alpha \le 100°C \;\; a = 2, \quad k = 1$$
$$\rho_k^\alpha = \sum C_n \, (T^\alpha)^n \qquad\qquad\qquad -50°C \le T^\alpha \le 0°C \tag{28}$$

with coefficients as shown in Table 1.

Clearly for wet snow with a temperature close to 0°C the liquid water density can be regarded as a constant.

The reference potential for the solute (dry air) is given by the expression:

$$(\mu_k)_o = \Re \, T^\alpha \, (\phi_k + \ln \frac{k}{p_o}) \qquad a = 2, \quad k = 2 \tag{29}$$

where k is the empirical Henry's Law constant and is of the order of 10^4 bar. If the partial pressure of the air above the liquid is of the order of 1 bar, i.e. the air in the snow is near atmospheric pressure, the mole fraction of air dissolved in the water is about 10^{-4}. For most modeling purposes it is reasonable to neglect this.

Equation of State for the Solid Phase. The solubility of air in solid ice is similarly extremely low at near atmospheric pressures; thus for many applications it may be assumed that both ice and liquid water may be treated as pure substances. The Gibbs functions then are simply:

$$G^\alpha = \Re \, T^\alpha \, \phi_k + \frac{p^\alpha \, M_k}{\rho_k^\alpha} \qquad a = 1,2 \quad k = 1 \tag{30}$$

The isothermal compressibility of ice is small ($\sim 10^{-6}$ bar^{-1}) so the density can be assumed to be independent of pressure. The variation with temperature at 1 bar pressure, in the range $-150°C \le T \le 0°C$, is given by an empirical equation of the form of Equation (28) with coefficients as shown in Table 2 (Pruppacher and Klett, 1978, p. 86).

Again, for modeling purposes it is reasonable to treat the density of ice as a constant.

Table 1. Coefficients in the equation for liquid water density.

n	A_n/kg m^{-3} K^{-n}	B_n/10^{-3} K^{-1}	C_n/kg m^{-3} K^{-n}
0	999.8396		999.84
1	18.224944	18.159725	860 10^{-4}
2	−7.922210 10^{-3}		−108 10^{-4}
3	−55.44846 10^{-6}		
4	149.7562 10^{-9}		
5	−393.2952 10^{-12}		

Table 2. Coefficients in the equation for ice density.

n	$C_n/kg \ m^{-3} \ K^{-n}$
0	916.7
1	–0.175
2	–5.0 10^{-4}

Equation of State for the Interfaces. In previous snowmelt models it has not been necessary to define the Gibbs functions of the interfaces because: (i) the terms in the energy conservation equation involving the interfaces have been assumed to be negligible, and (ii) the adsorption of impurities has not been considered. Under these circumstances only the surface tensions need be defined. They have always been taken to be constant although there is in fact a dependence on temperature and the curvature of the interface. Pruppacher and Klett (1978, p. 104) give an empirical expression for the variation of the surface tension of a water/moist air interface with temperature in the range $0°C \leq T^i \leq 40°C$:

$$-\sigma^i = (76.1 \ 10^{-3} \ J \ m^{-2}) - (0.155 \ J \ m^{-2} \ K^{-1}) \ T^i \quad i = 2 \tag{31}$$

Their calculation (p.108) of the effect of curvature is based on an adsorption $n_k^i/\Omega^i = 0.87 \ 10^{-5}$ mol m^{-2} ($i = 2, k = 2$) for air on a water surface. The reduction in surface tension becomes significant (i.e. greater than 1%) for surface radii r^i less than 10^{-8} m.

It is clearly reasonable to regard σ^i as a constant when modeling wet snow near 0°C. However, it is important to remember the curvature effect when the transition to dry snow is investigated.

The definition of surface tensions for ice/moist air and ice/water interfaces is complicated by the crystalline structure of ice, which suggests that different faces of an ice crystal should have different surface tensions, and the complex surface structure which may develop. There is also evidence that a quasi–liquid layer may exist at the ice–air interface. Pruppacher and Klett (1978, p. 121) suggest values of:

$$\begin{aligned} -\sigma^i &= 100 \ mJ \ m^{-2} \quad \text{(basal face)} \quad i = 3 \\ &= 106 \ mJ \ m^{-2} \quad \text{(prism face)} \end{aligned} \tag{32}$$

Measurements by Walford et al. (1987) suggest a value of:

$$-\sigma^i = 34 \ mJ \ m^{-2} \quad i = 1 \tag{33}$$

for the ice/water interface, and Pruppacher and Klett (1978, p. 130) suggest the empirical equation:

$$-\sigma^i = 28.5 \ 10^{-3} \ J \ m^{-2} + 0.25 \ J \ m^{-2} \ K^{-1} \ T^i \quad i = 1 \tag{34}$$

in the range $-50°C \leq T^i \leq 0°C$.

Although explicit expressions for the surface areas of the interfaces have not been provided in previous models, the Ω^i have in fact been defined implicitly by specifying the radii of the interfaces and the arrangement of ice grains in the system. It is possible to develop equations for the Gibbs functions of the surfaces by analogy with Equation (23) (see for example Defay et al., 1966, pp. 165–170). This will be important for models of polluted snow, in which it will be necessary to represent chemical processes on the surfaces in detail, but for unpolluted snow it is reasonable to use the simplified approach of previous authors and define surface tension alone.

Latent Heats. Having defined the Gibbs functions and hence the enthalpies for the various phases and interfaces it is possible to look more closely at the terms in the energy conservation equation (Equation (21)). The latent heats of fusion, vaporization, and sublimation for water have been assumed to be constants in all previous models. However, it is clear from the discussion above that $\mathcal{L}$ depends on temperature, on the pressure difference between the two phases where there is a curved interface, and on the concentration of other constituents in each phase. Pruppacher and Klett (1978, pp. 137–142) show that the error produced by substituting the latent heats, L, for pure water changing phase across a plane interface for the true latent heats, $\mathcal{L}$, is less than 3 percent for solutions of concentration less than about five moles per liter and interfaces with radius of curvature greater than about 10^{-8}m. Thus, for normal modeling purposes it is reasonable to use L. It is only when the transition to cold snow, a two phase mixture of ice and moist air, is being investigated that it is important to define the latent heats correctly.

Table 3 shows the latent heats of evaporation, melting and sublimation for pure water, a plane interface, and standard atmospheric pressure at various temperatures.

They have been calculated using the Goff–Gratch formulation of the low pressure properties of water (Smithsonian Meteorological Tables (SMT) p. 343).

Specific Heats. Using Equations (19) and (23) it may be shown that the specific heats at constant pressure for moist air may be obtained by summing the molar values for each pure constituent, i.e.:

$$(C_p)^a = \sum_k n_k^a \, (c_p)_k \qquad a = 3 \tag{35}$$

The specific heats $(c_p)_k$ depend on temperature but, as a first approximation, can be given constant values of 29.14 J mol^{-1} K^{-1} for dry air and 33.26 J mol^{-1} K^{-1} for water vapor.

Empirical equations for the specific heats of pure ice and pure water at standard pressure are given by Pruppacher and Klett (1978, p. 89):

$$(c_p)_k = 2.106 \ 10^3 \text{ J kg}^{-1} \text{ K}^{-1} + 7.33 \text{ J kg}^{-1} \text{ K}^{-2} \ T \tag{36}$$
$$-40°C \leq T \leq 0°C \quad a = 1, \quad k = 1$$

$$(c_p)_k = 4.187 \ 10^3 \text{ J kg}^{-1} \text{ K}^{-1} + 1.30 \ 10^{-2} \text{ J kg}^{-1} \text{ K}^{-3} \ (T - 35°C)^2 \tag{37}$$
$$+ 1.59 \ 10^{-5} \text{ J kg}^{-1} \text{ K}^{-5} \ (T - 35°C)^4$$
$$0°C \leq T \leq 35°C \quad a = 2, \quad k = 1$$

Table 3. The latent heats of water.

T/°C	L_e/kJ kg^{-1}	L_m /kJ kg^{-1}	L_s/kJ kg^{-1}
0	2500.7	333.7	2834.4
−10	2524.6	312.3	2836.5
−20	2549.3	288.9	2838.2

The specific heats of the interfaces can be calculated from the Gibbs functions if these are defined. However, if the simplified approach as previously explained is followed, values of $(C_p)_i$ must be estimated. Normally they are considered to be negligible.

2.1.3. Mass Transfer Between Phases and Interfaces

To complete the model it is necessary to find equations for the rates of mass transfer N_k^α, N_k^i in terms of the thermodynamic variables. Strictly the mixture theory approach requires that the equations should arise from consideration of the properties of snow on a macroscopic scale. However, there are so few suitable experimental data available that in practice a microscopic approach has to be used. This means that the shape, size and distribution of the water inclusions and ice grains have to be specified rather than just the specific volumes as in the classic mixture theory approach.

Thermodynamic Equilibrium Equations. Several authors have suggested an approach based on the assumption that a snowpack will remain in thermodynamic equilibrium whatever the changes in temperature and water content produced by changes in the mass and energy inputs. This is essentially saying that changes of phase occur very much more rapidly than heat and mass flow in the snowpack. In this case equations for thermodynamic equilibrium may be used to complete the model. If there is thermal equilibrium between phases α and β and interface i:

$$T^\alpha = T^\beta = T^i \tag{38}$$

The equations for chemical equilibrium are:

$$\mu_k^\alpha = \mu_k^\beta = \mu_k^i \tag{39}$$

The chemical potentials for the ice, water, and moist air are known; the requirement for an independent definition of μ_k^i can be avoided by specifying that $N_k^i = 0$. This will not be a satisfactory approach when modeling polluted snow but will suffice when air is the only adsorbed component. Finally, the equations for mechanical equilibrium are:

$$p^\alpha - p^\beta = \sigma^i \left(\frac{1}{r_{i1}} + \frac{1}{r_{i2}} \right) \tag{40}$$

where r_{i1} and r_{i2} are the two radii of curvature of interface i separating phases α and β. These are the Laplace equations.

The Pendular and Funicular Regimes. Figure 1 schematically shows the two basic configurations of a three phase, unpolluted snow mixture.

If there are two types of interface, between ice and water and between water and gas, the snow is said to be in the funicular regime (Colbeck, 1975). If there is also an ice–gas interface, the snow is in the pendular regime. The snow grains have been drawn as spheres but this is of course a gross approximation for natural snow. Cold snow and saturated snow, which are both two phase mixtures with one interface, have one configuration: ice spheres surrounded by moist air or water.

In the pendular regime suppose that the ice–gas interface ($i = 3$) is spherical. Then the two principal radii are equal:

$$r_{31} = r_{32} \tag{41}$$

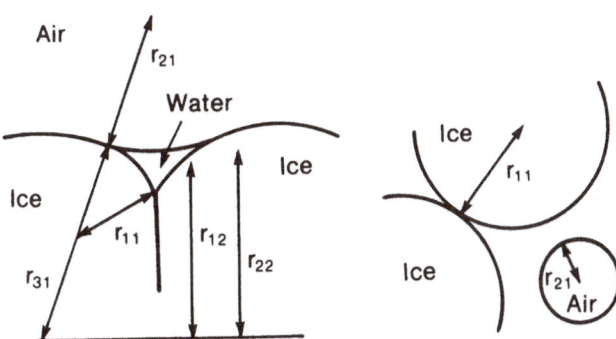

Figure 1. The two basic configurations of an unpolluted snow mixture showing the radii of the various interfaces.

The ice grains are slightly compressed at their points of contact to maintain mechanical equilibrium (Colbeck, 1979) so that the radius of the ice–water interface $r_{11} < r_{31}$. If Ξ is the dihedral angle:

$$(r_{31} - r_{21})^2 = (r_{22} - r_{21})^2 + \frac{(r_{11} \cos(\Xi/2))^2 (r_{31} - r_{21})^2}{(r_{11} - r_{21})^2} \tag{42}$$

and

$$r_{12} = \frac{1}{2}\left(r_{21} + r_{31}\frac{(r_{22} - r_{21})}{(r_{31} - r_{21})} \right) \tag{43}$$

Equations (41) to (43) and the Laplace equations serve to define the six principal radii for the system. The best experimental value of the dihedral angle is that of Walford et al. (1987) who give $\Xi = 33.6° \pm 0.7°$.

In the funicular regime both ice grains and gas bubbles may be considered spherical. Thus $r_{11} = r_{12}$ and $r_{21} = r_{22}$. With the two Laplace equations for this configuration these equations define the four principal radii.

Volume Fraction Laws. It has already been remarked that the areas of the interfaces Ω^i are defined implicitly by the radii of curvature provided that the arrangement of ice grains is known. The same is true of the specific volumes of the bulk phases. Thus, if the equilibrium approach is accepted, any volume fraction laws used in the snowmelt model must be compatible with Equations (38) to (40). Morris and Godfrey (1979) introduced the volume fraction law for water:

$$\theta^{(2)} = (1 - \theta^{(1)})\left(\frac{p^{(2)} - p^{(3)}}{p_o} \right)^s \tag{44}$$

by analogy with the soil moisture characteristic equation and suggested values of $s = -0.93 \pm 0.12$, $p_o = -206 \, (+692 \text{ or } -161)$ Pa for the index s and air entry pressure p_o on the basis of the scant experimental data available (Figure 2).

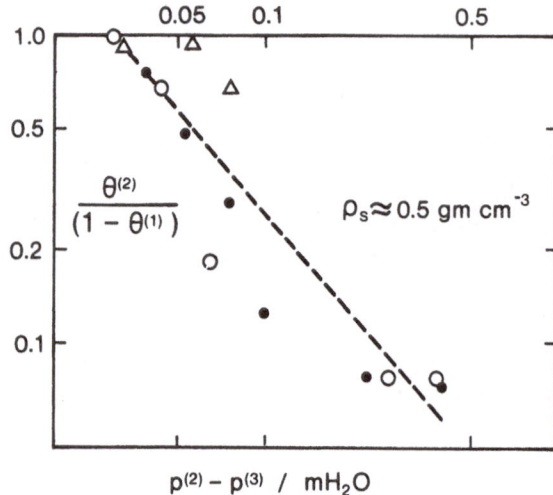

Figure 2. The water content of snow as a function of capillary pressure.

Morris and Kelley (1990) have derived the exact volume fraction equations for simple arrays of ice grains and show that Equation (44) is a reasonable approximation with $s = -1.64 \pm 0.02$ and:

$$\psi_a = \frac{(-10^{-3}\text{Pa m})}{R_{13}} \left(3.34 \ \exp(1.43 \ n_3) \frac{\theta^{(1)}}{1-\theta^{(1)}} \right)^{0.61} \tag{45}$$

where n_3 is the coordination number i.e. the number of contacts with other grains per grain. This is encouraging support for the microphysical approach. Although in general two independent volume fraction laws are required for a three phase mixture, previous snowmelt models have used only one because it has been arbitrarily assumed that the velocity of the ice grains is zero. This assumption is acceptable in hydrological modeling but in other fields, especially avalanche research, compaction of the snowpack must be taken into account.

2.1.4. Further Constitutive Equations

In section 2.1.1 the hydraulic conductivities K^α and diffusivities D_k^α were introduced. Constitutive equations are required to define these for the particular material being considered. A macroscopic approach is usual for snow but must be compatible with the microphysical equations already derived.

Hydraulic Conductivity. By analogy with the equation commonly used in soil physics, Colbeck (1972) defined the hydraulic conductivity of water in snow as a power function of the water content. Colbeck and Anderson (1982) give the equation:

$$K^{(2)} = K_{sat}^{(2)} \left(\frac{\theta^{(2)} - 0.07}{1 - \theta^{(1)} - 0.07} \right)^3 \tag{46}$$

based on field experiments on a natural snowpack. The saturated hydraulic conductivity $K_{sat}^{(2)}$ is related to the dry density and grain size of the snow. Figure 3 shows some experimental data and the variation predicted by the Shimizu equation:

$$K_{sat}^{(2)} = 0.42 \ 10^6 \ \text{m}^{-1} \ \text{s}^{-1} \ (2 \ r_{31})^2 \ \exp(-7.8 \ \rho^{(1)} \ \theta^{(1)}/\rho^{(2)}) \tag{47}$$

and the Bender equation:

$$K_{sat}^{(2)} = 2.24 \ 10^3 \ \text{m}^{-0.63} \ \text{s}^{-1} \ (2 \ r_{31})^{1.63} \ (1 - \theta^{(1)}) \tag{48}$$

Neither of these empirical equations are particularly satisfactory; the Shimizu equation has been more widely used.

Diffusion of Water Vapor in Air. The diffusivity D_k^α which appears in Equation (11) will not necessarily be equal to the molecular diffusivity of k in a large volume of α, given that the bulk phases in snow are mixed and sometimes discontinuous. However, the value of $0.226 \ 10^{-4} \ \text{m}^2$ s^{-1} for water vapor in a large volume of air at $0°C$ and 1 bar pressure (SMT, 1971, p. 395) can be used as a first estimate. D_k^α should not be confused with the effective diffusivity for snow referred to by some authors e.g. Morozov (1967). This quantity is derived from measurements on bulk snow samples and is defined for snow, not for the gas phase as is D_k^α. In general the effective diffusivity is larger than the molecular diffusivity of water vapor in air. Figure 4 shows the variation of the effective diffusivity for water vapor in snow with temperature and density measured by Mozorov (1967).

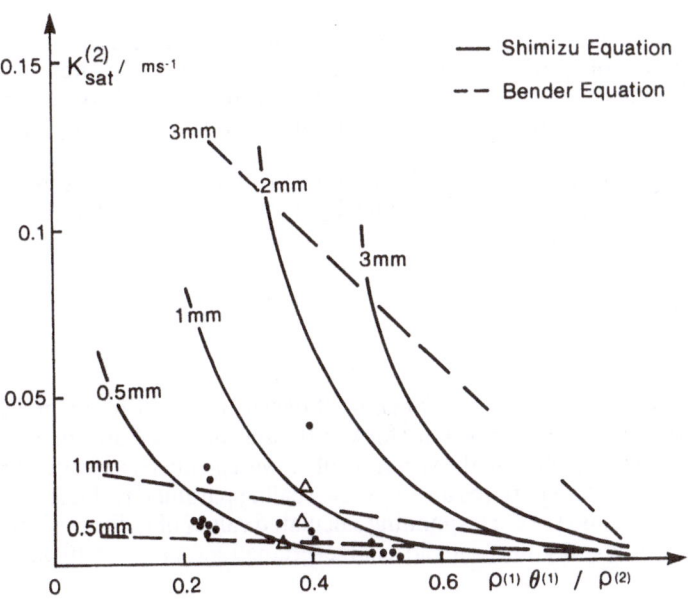

Figure 3. The saturated hydraulic conductivity of snow as a function of its relative density.

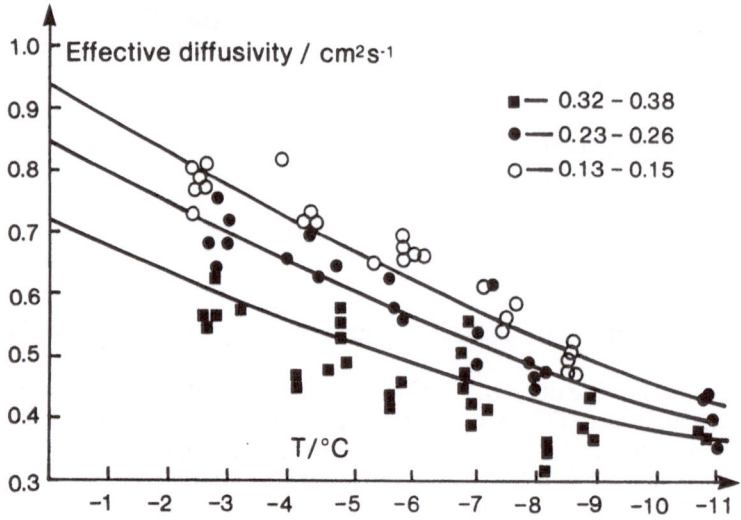

Figure 4. The effective diffusivity of snow as a function of temperature (after Morozov, 1967).

Radiation Absorption. The radiation absorption rate may be represented by the Bouger–Lambert equation:

$$\frac{\partial R_\lambda}{\partial x_1} = -v_\lambda R_\lambda \tag{49}$$

where v_λ is the extinction coefficient for wave length λ. Mellor (1977) has summarized the data available on the variation of v_λ with wave length, grain size, and water content of the snow. For short wave radiation $v_S \sim 10 \, \text{m}^{-1}$, whereas long–wave radiation is absorbed much more rapidly with $v_L \sim 250 \, \text{m}^{-1}$. Figure 5 shows the variation of v_λ with λ for samples of wet and dry snow measured by Manz (1974).

A theoretical analysis is given by Bohren and Barkstrom (1974), who treat the snow as a collection of transparent ice spheres of diameter d which both transmit and scatter incoming radiation. They calculate that for visible wavelengths:

$$v_s = 0.84 \left(\frac{\kappa_i}{d}\right)^{1/2} \left(\frac{\rho_s}{\rho_i}\right) \tag{50}$$

where $\kappa_i(\lambda)$ is the absorption coefficient of ice. For snow of density $\rho_s = 0.5 \, 10^3 \, \text{kg m}^{-3}$ and grain size $d = 1 \, \text{mm}$ equation (49) gives $v_\lambda = 4.1 \, \text{m}^{-1}$ for $\lambda = 0.5 \, \mu\text{m}$ and $\kappa_i = 8 \, 10^{-2} \, \text{m}^{-1}$. This is a little low, but the equation does simulate the variation of v_λ with ρ_s and d correctly so it is worth using in a snowmelt model with perhaps an empirical scaling parameter. Table 4 gives values of κ_i (Hobbs, 1974, p. 207) and the corresponding calculated values of v_λ for fresh, fine-grained snow ($\rho_s = 10^2 \, \text{kg m}^{-3}$, $d = 0.1 \, \text{mm}$) and old, coarse-grained snow ($\rho_s = 5 \, 10^2 \, \text{kg m}^{-3}$, $d = 0.1 \, \text{mm}$).

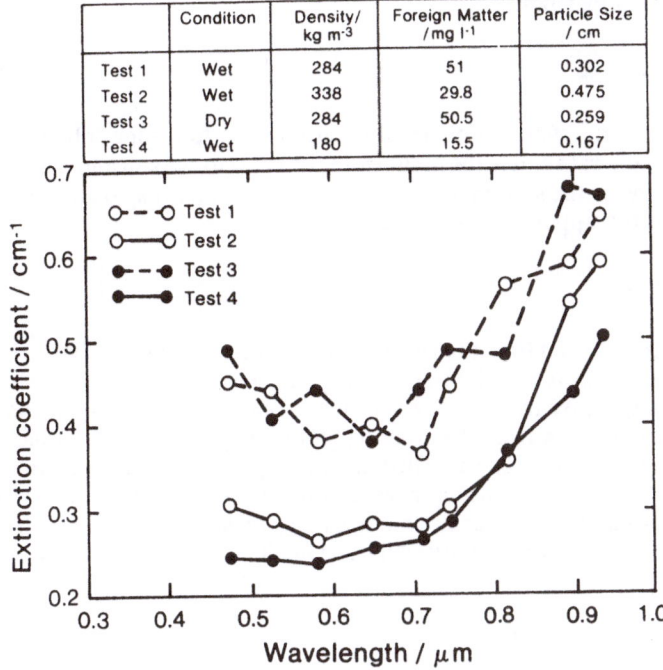

	Condition	Density/ kg m^{-3}	Foreign Matter / mg l^{-1}	Particle Size / cm
Test 1	Wet	284	51	0.302
Test 2	Wet	338	29.8	0.475
Test 3	Dry	284	50.5	0.259
Test 4	Wet	180	15.5	0.167

Figure 5. Extinction coefficient of snow as a function of wave–length (after Manz, 1974).

Table 4. Calculated values of the extinction coefficient for snow.

$\lambda/\mu m$	κ_i/m^{-1}	ν_λ/m^{-1} fresh snow	ν_λ/m^{-1} old snow
0.315	0.11	3.05	4.82
0.400	0.04	1.83	2.91
0.500	0.08	2.60	4.11
0.600	0.20	4.11	6.50
0.700	0.60	7.12	11.26
0.800	1.80	12.33	19.50

2.2. Modeling the Boundary Conditions

The boundary conditions at the upper and lower surfaces of the snow are specified by defining the mass and energy fluxes $\mathcal{M}_{bk}$ and $\mathcal{B}$. Most models for unpolluted snow assume that only liquid water and water vapor are transmitted across these boundaries although some take account of removal of ice at the upper surface as blowing snow and the penetration of air into the snow because of atmospheric pressure variations. When ice losses are negligible the flux of water substance at the upper boundary is:

$$\mathcal{M}_{b_k} = E - I \qquad k = 1, \; z = \mathcal{Z} \tag{51}$$

where I is the rainfall rate and E the evaporation or sublimation rate. At the lower boundary $z = 0$, $\mathcal{M}_{b_k}$ is the water production rate and is the input from the snow to the soil and overland flow systems. It is not the same as the snowmelt rate since the melt water may be delayed in the pack. The energy flux at the upper boundary may be written:

$$\mathcal{E} = H - R_n + P \qquad z = \mathcal{Z} \tag{52}$$

where H is the sensible heat transferred from the snow by turbulence in the atmosphere, R_n is the net radiation (positive downwards by micrometeorological convention), and P the heat convected by rain or snowfall (positive upwards). At the lower boundary the components of $\mathcal{E}$ are the heat conducted into the snow from the ground (geothermal heating), net radiation absorbed by the ground, and heat convected by infiltration or seepage. The various terms in Equations (50) and (51) can be calculated using measured meteorological data. It is convenient to distinguish these calculations, which are made using a *meteorological model*, from the calculation of the fluxes at the lower boundary which are made using the *snowmelt model* described in section 2.1.

2.2.1. Radiation

The net radiation may be divided into short–wave, solar radiation and long–wave, thermal radiation, S and L respectively. Dividing each of these into incoming and outgoing components leads to the equation:

$$R_n = S\downarrow + S\uparrow + L\downarrow + L\uparrow \tag{53}$$

The incoming solar radiation flux may be further divided into direct, Q, and diffuse, D, components. The direct solar radiation is attenuated as it passes through the atmosphere and may be estimated if necessary from the equation (Garnier and Ohmura, 1968):

$$Q = Q^* f \left(\cos \Pi \, \cos \Omega \, \cos Y + \sin \Pi \, \sin Y \right) \tag{54}$$

where Π is the latitude, Ω the hour angle, Y the sun's declination, and Q^* the solar constant. The transmission factor f varies with the zenith angle of the sun and decreases with altitude. Dozier (1978) gives a more complete treatment of the methods of estimating direct solar radiation.

The incoming diffuse radiation on a horizontal slope depends not only on the nature of the atmosphere, for example the amount of cloud, but also on the nature of the slope surface. Gardiner (1987) has shown that multiple reflection of radiation between clouds and a snow surface with high albedo can increase $S\downarrow$ by a factor of 2 or more in overcast conditions. He has successfully modeled $S\downarrow$ for a two year period, using observed cloud heights and amounts from an Antarctic research station. The albedo of the surrounding surface varied with the extent of sea ice cover. This is a promising technique but does require sufficient data on the clouds for the transmission factors to be accurately estimated. For many hydrological applications such data will not be available, and D must be estimated as a simple fraction of Q.

Outgoing shortwave radiation of wavelength λ is related to incoming radiation of the same wavelength by the equation:

$$(S \uparrow)_\lambda = a_\lambda \, (S \downarrow)_\lambda \tag{55}$$

where α_λ is the reflectance and depends on the direction of the incoming radiation and the physical properties of the snow. O'Brien and Munis (1975) have measured α_λ, for snow in the range $0.6\ \mu m \le \lambda \le 2.5\ \mu m$ (Figure 6). Values are high at the visible end of the spectrum, decreasing with increasing λ in the range $0.3\ \mu m$ to $0.7\ \mu m$. There have been many theoretical studies of snow reflectance (e.g. Bohren and Barkstrom, 1974; Warren, 1982) which clarify the different roles played by the properties of the snow (grain size, density and water content) and the external factors (cloud type, inclination of the sun, etc.).

In many snowmelt models α_λ is taken to be constant over the short-wave range so that

$$S \uparrow = a \, S \downarrow \tag{56}$$

where α is the short-wave albedo. Simple empirical equations relating α to cloud cover (if known) and the age of the snow have proven useful; although it would be clearly better to use the physics-based equations for α_λ if at all possible.

The incoming long-wave radiation may be estimated using Stefan's Law for emitted radiation:

$$L \downarrow = \varepsilon_a \, \sigma \, (T_a^*)^4 \tag{57}$$

where ε_a is the emissivity of the air, T_a^* its absolute temperature, and σ is the Stefan-Bolzmann constant. A common expression for the emissivity is given by Brutsaert (1975a):

$$\varepsilon_a = 0.642\ \mathrm{Pa}^{-1/7}\ \mathrm{K}^{1/7}\ (p_v \, / T_a^*)^{1/7} \tag{58}$$

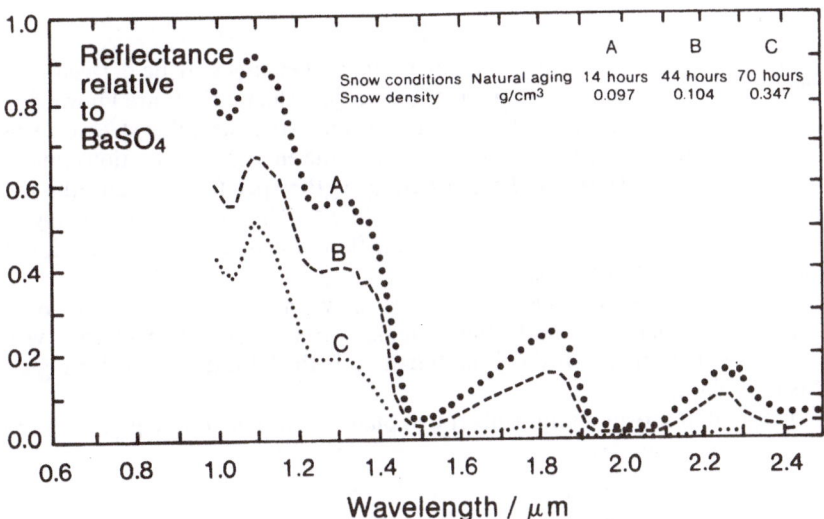

Figure 6. Reflectance of snow as a function of wave-length (after O'Brien and Munis, 1975).

where p_v is the partial pressure of water vapor in the atmosphere. The outgoing long–wave radiation depends on the temperature of the snow surface T_z^* :

$$L \uparrow = \varepsilon_s \, \sigma \, (T_z^*)^4 \tag{59}$$

The emissivity of the snow ranges from 0.97 – 0.99.

2.2.2. Turbulent Transfer

A survey of values of the components of the energy input to a melting snow cover measured over various periods of time in various different parts of the world, shows that the sensible heat input can be up to twice the net radiation input and is very often around 40 percent of it. The latent heat required to produce the measured evaporation can be of the same order as the net radiation although it is usually around 10 percent. Therefore, it is necessary for snowmelt modelers to measure or calculate the sensible heat flux and evaporation (which are both produced by turbulent transfer processes) with at least the same level of precision as is achieved for net radiation. Unfortunately, estimation of sensible heat and evaporation are two of the most difficult aspects of snowmelt modeling.

Flow in the Atmospheric Boundary Layer. Suppose that the instantaneous velocity components, u_i ($i = 1,3$), pressure, p, temperature, T, density, ρ, and humidity, q, in the atmosphere are each composed of a temporal mean plus a fluctuation. That is, for variable ξ:

$$\xi = \bar{\xi} + \xi' \tag{60}$$

The temporal means are defined by averaging over a period τ which includes a large number of fluctuations. Thus:

$$\bar{\xi} = \frac{1}{\tau} \int_t^{t+\tau} \xi \, dt \tag{61}$$

A set of equations of conservation of mass, momentum, and energy may be derived for the fluctuations. The momentum conservation equations for the fluctuations are known as the Reynolds equations and contain terms of the form $-\rho \, (\overline{u_i' u_j'})$ (the Reynolds stresses) which arise from the turbulent transfer of momentum. Similarly, the energy conservation equations contain turbulent heat fluxes $H = c_p \rho (\overline{T' u_j'})$, where c_p is the specific heat, and the water vapor equations contain turbulent vapor fluxes $E = \rho (\overline{q' u_j'})$. This set of mass, energy, and momentum equations is not sufficient to determine the flow. Further equations, to close the set, are provided by turbulence models.

The momentum equations for the fluctuations may be written in terms of the velocity vector $v = (u_1, u_2, u_3)$ and the angular frequency of rotation vector $\Omega = (\omega \cos \alpha \cos \phi, \omega \sin \alpha \sin \phi, \omega \sin \phi)$ of the Earth. ϕ is the latitude and α the angle between the x axis and North.

For steady horizontal flow over a uniformly rough plane the momentum equations reduce to:

$$\frac{\partial}{\partial x} \, (\overline{u_1' u_3'}) = f \, (\bar{u}_2 - v_g) + \nu \frac{\partial^2 \bar{u}_1}{\partial x_3^2} \tag{62}$$

$$\frac{\partial}{\partial x_3} \left(\overline{u_2' u_3'}\right) = -f \left(\overline{u}_1 - u_g\right) + \nu \frac{\partial^2 \overline{u}_2}{\partial x_3^2} \tag{63}$$

where the horizontal variation of turbulent fluxes is assumed to be negligible; u_g and v_g are the components of the geostrophic wind and $f = 2\,\omega \sin \phi$ is the Coriolis parameter. A further approximation may be made in the lowest layer of the atmosphere where the direction of flow does not change significantly with height. In this case the coordinate system may be defined in such a way that $\overline{u}_2 = 0$. Then Equation (62) can be integrated to give an expression for the Reynolds shear stress:

$$-\rho\left(\overline{u_1' u_3'}\right) = f \rho \, v_g \, x_3 - \mu \frac{\partial \overline{u}_1}{\partial x_3} + \tau \tag{64}$$

where $\mu = \nu\rho$ is the dynamic viscosity of the air. If the surface is solid the fluctuations u_3' must be zero at $x_3 = 0$ so the integration constant τ (the surface shear stress) is:

$$\tau = \mu \left(\frac{\partial \overline{u}_1}{\partial x_3}\right)_{x_3 = 0} \tag{65}$$

$u_* = (\tau/\rho)^{1/2}$ is the friction velocity.

In the turbulent boundary layer the surface shear stress is much greater than the geostrophic wind term and the momentum flux is constant, that is:

$$-\rho\left(\overline{u_1' u_3'}\right) = \tau \tag{66}$$

A surface is defined as 'rough' if it has elements which protrude into the turbulent flow layer. A surface is 'smooth' if a laminar sublayer can develop.

For variable ξ it is supposed that:

$$-\rho\left(\overline{\xi' u_j'}\right) = \Gamma_{\xi, x_j} \frac{\partial \overline{\xi}}{\partial x_j} \tag{67}$$

where Γ_{ξ, x_j} is called the turbulent exchange coefficient for ξ in direction x_j.

The Neutral Boundary Layer. Prandtl suggested the expression:

$$\Gamma_{\overline{u_1}, x_3} = \rho k^2 \, (x_3)^2 \left|\frac{\partial \overline{u}_1}{\partial x_3}\right| \tag{68}$$

for the momentum exchange coefficient for flow in the lower part of a neutral boundary layer lying over a flat plane $x_3 = 0$. Here k is von Kármán's constant. Using Equations (66), (67), and (68) a logarithmic expression for the velocity in the turbulent boundary layer:

$$\overline{u}_1 = \frac{u}{k} * \ln\left(\frac{x_3}{z_o}\right) \tag{69}$$

is obtained. The aerodynamic roughness length z_0 is an integration constant.

Values of z_0 may be determined by measuring the wind profile $\bar{u}_1(x_3)$ in neutral conditions. Table 5 gives a selection of values obtained at sites where the condition of steady horizontal flow over a uniformly rough plane is most likely to have been achieved.

The Monin–Obukhov Functions. The term neutral indicates that the turbulence is not affected by changes in density of the air produced by temperature gradients. When turbulent transfer is suppressed by the temperature gradient (i.e. cold air lies underneath warm air), the boundary layer is referred to as stable. Conversely, turbulent transfer is increased in unstable boundary layers with cold air lying over warm. In these non–neutral conditions Equation (68) is not suitable.

One method for specifying the transfer coefficients in non–neutral conditions that is very widely used in hydrology is derived from Monin–Obukhov similarity theory. The coefficient for vertical transfer for variable ξ is defined as:

$$\Gamma_{\xi,x_3} = \frac{\rho k u_* x_3}{\varphi_\xi (x_3/\Lambda)} \tag{70}$$

where the Monin–Obukhov functions φ_ξ depend only on height above the surface and the Monin–Obukhov scaling length Λ. If scaling parameters T_* and q_* for temperature and humidity are defined by the equations:

$$T_* u_* = -(\overline{T'u_3'}) = -\frac{H}{\rho c_p} \tag{71}$$

$$q_* u_* = -(\overline{q'u_3'}) = -\frac{E}{\rho} \tag{72}$$

then:

$$\Lambda = \frac{t u_*^2}{kg(T_* + 0.61\ T\ q_*)} \tag{73}$$

Table 5. Aerodynamic roughness length of flat snow and ice–covered areas determined from profile measurements in neutral conditions.

Site	Reference	z_0/mm
Britannia Sø (melting ice)	Grainger and Lister (1966)	0.4
Maudheim ice shelf (snow)	Liljequist (1953)	0.1
(level snow)	Konstantinov (1966)	0.2 – 0.8
(average snow)	"	1 – 5
(friable snow)	"	5 – 20
Finse (snow)	Harding (1986)	0.2 – 4
Brunt ice shelf (snow)	King (In press)	0.11 ± 0.01

The Monin–Obukhov length acts as a measure of the stability of the atmosphere; $\Lambda < 0$ for unstable flow and $\Lambda > 0$ for stable flow. The general behavior of φ_ξ is an increase with increasing stability. The form of the Monin–Obukhov functions can be determined from wind, temperature, and humidity profile measurements in non–neutral (sometimes referred to as diabatic) conditions.

For variable ξ Equation (66) can be integrated to give:

$$\xi = \frac{\xi_*}{k}\left(\ln(x_3/z_\xi) + \Phi_\xi(x_3/\Lambda)\right) \tag{74}$$

where:

$$\Phi_\xi(x_3/\Lambda) = -\int_{z_\xi}^{x_3} (1 - \varphi_\xi(x_3/\Lambda)) \frac{dx_3}{x_3} \tag{75}$$

and z_ξ is the scaling length for parameter ξ (except that by convention the aerodynamic roughness length is written z_o not z_u). When x_3/Λ is very small, i.e. in near–neutral conditions, the function Ψ_ξ can be replaced by the linear term ($\alpha\, x_3/\Lambda$). The profiles of ξ then have a log–linear form. Although many expressions for the Monin–Obukhov functions under various stability conditions have been derived, little of the work has been based on measurements made over snow. Granger (1977) reports that for stable, but near neutral, conditions with $0 \leq x_3/\Lambda \leq 0.1$ the ratio $\varphi_T/\varphi_u = 1$ and $\varphi_q/\varphi_u \simeq 2$. Thus from Equation (70) the heat transfer coefficient is the same as and the water vapor transfer coefficient is about half the momentum transfer coefficient. For unstable conditions $\varphi_T/\varphi_u = (1 - 58x_3/\Lambda)^{-1/4}$ and φ_q/φ_u decreases with increasing stability. Later Granger and Male (1978) suggested the expression $\varphi_T/\varphi_u = (1 + 7x_3/\Lambda)^{0.1}$ for the stable range $x_3/\Lambda > 0.1$. Figure 7 shows how these expressions compare with those derived from studies over other surfaces (Swinbank, 1968; Businger et al., 1971; Dyer and Hicks, 1970).

There are few data available on the roughness lengths z_T and z_q over snow, and it is often assumed that these are the same as z_o. However, Brutsaert (1975) suggests on theoretical grounds that z_T and z_q should be considerably smaller than z_o. Andreas (1986) has developed expressions for z_ξ/z_o in terms of the roughness Reynolds Number $R_* = u_* z_o/\eta$. Figure 8 shows his predicted values for flow over aerodynamically smooth and rough snow and ice surfaces. The values for transition surfaces are interpolated.

The success of the profile equations in predicting H and E from limited meteorological data at a point site depends on: (i) how far the site and weather conform to the conditions for which the theory is valid, and (ii) how accurately the parameters such as z_ξ can be specified. Table 5 shows that even in good micrometeorological conditions, the range of measured values of z_o is two orders of magnitude. Luckily, the corresponding uncertainty in H and E is less; if all z_ξ were increased from 1 mm to 10 mm, predicted values of H and E would increase by a factor of 2.25, and similarly a decrease in z_ξ to 0.1 mm would decrease H and E by a factor of 0.56.

Eddy Correlation Methods. Turbulent fluxes may be measured directly by recording time series for the vertical wind velocity component, air temperature, and humidity and computing the covariances $\overline{T'u_j'}$ and $\overline{q'u_j'}$ of values of the functions measured at the same time and place. The problems which arise with this eddy correlation method arise from the finite response time of the instruments, which means that high frequency fluctuations cannot be recorded, and the fact that all the sensors cannot physically be in the same place. Slow response of the sensors can be remedied by increasing the measurement height, z, and most eddy correlation measurements over snow have been made at $z = 3$ to 5 m. However, this may mean that the

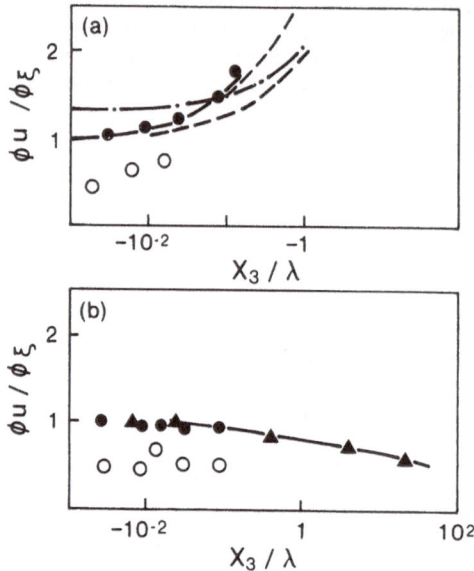

(a) unstable range: – o – Granger (1977) φ_u/φ_T , – – Swinbank (1968),
– • – Businger et al. (1971), – – – Dyer and Hicks (1970),
o Granger (1977) φ_u/φ_q

(b) stable range: o Granger (1977) φ_u/φ_T, – $\triangle$ – Granger and Male (1978),
o Granger (1977) φ_u/φ_q

Figure 7. Ratios of the Monin–Obukhov functions as a function of stability.

instruments are above the constant flux boundary layer and will not produce good values of H and E at the snow surface, which is, of course, what is required by the hydrologist. Nevertheless, several authors report successful use of eddy correlation measurements in energy budget studies.

Blowing Snow. At wind speeds greater than about 7 m s^{-1} grains of snow at a cold snow surface will start to move. Near the surface a saltation layer of about 10 cm thickness develops. Above this layer the snow is transported by turbulent transfer.

The presence of blowing snow in the atmosphere affects the turbulent transfer of momentum, heat, and water vapor in several ways. The roughness of the snow surface is no longer fixed; deposition and erosion by the wind produces features such as sastrugi and snow–waves, so z_0 becomes a function of $\bar{u}_1$. Chamberlain (1983) has shown that the roughness lengths measured in blowing snow conditions fit Charnock's equation:

$$z_o = 0.016 \ u_*^2/g \qquad (76)$$

for the roughness length of the sea. Wind profiles measured in blowing snow very often have the logarithmic form expected for neutral conditions over a fixed surface (Male, 1980), but deviations indicating stable conditions have also been reported for moderate wind speeds (Schmidt, 1982). In practical hydrological modeling applications, it is probably reasonable to use standard profile equations, such as Equation (74), for the wind speed at all times, but use

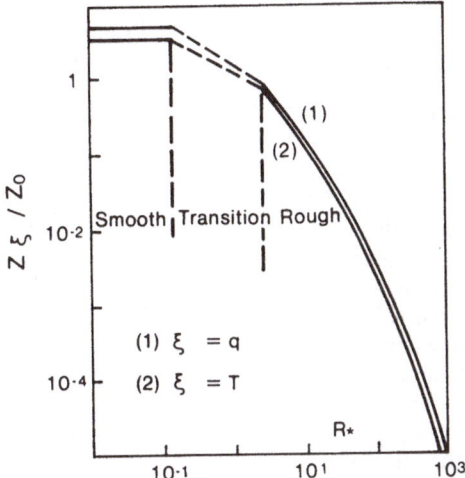

Figure 8. Model predictions of z_ξ/z_0 over snow and sea ice (after Andreas, 1986).

a variable value of aerodynamic roughness length (calculated using Charnock's equation) when the measured wind speed at a reference height exceeds a certain level.

Profiles of temperature and humidity in blowing snow are affected by sublimation from the ice particles which act as sources of humidity and sinks of latent heat in the atmosphere. H and E are therefore not constant within the blowing snow layer. There are few field measurements, although Schmidt (1982) has measured humidity profiles over snow covered prairie and found a linear dependence of q on x_3 from 0.2 to 1 m above the surface. However, there has been considerable theoretical work which allows horizontal mass transport, sublimation, and heat transfer rates to be calculated. Schmidt describes a method of computing the vertical profile of sublimation rate as a function of wind speed, humidity, and snow particle concentration; this in turn can be used to derive temperature and humidity profiles.

Sublimation from blowing snow is an important hydrological process which cannot be neglected; Schmidt (1982) has estimated that in his experiments "blowing snow sublimation was at least two orders of magnitude greater than surface evaporation would have been had the surface remained stable." However, the major importance of blowing snow processes for the hydrologist is as a mechanism for the redistribution of snowfall. Unfortunately, theoretical work on this problem is still in an early stage. It is possible to predict the build–up of snow drifts around obstacles of simple form using the equations of flow, but prediction of the pattern of redistribution of snow within a catchment for a given synoptic wind is not yet possible.

Katabatic Flow. On sloping snow and ice surfaces local pressure gradient (katabatic) forces act in addition to the synoptic–scale pressure gradient. Prandtl (1952) has suggested that for katabatic flow the downslope velocity profile can be represented by the empirical equation:

$$\bar{u}_1 = C \sin\left(\frac{\pi x_3}{4H}\right) \exp\left(-\frac{\pi x_3}{4H}\right) \tag{77}$$

where H is the height at which $\bar{u}_1$ is a maximum. H can be very small; de la Casiniere (1974) reports values of $H \simeq 25$ cm over a snow slope of about $5°$ for example. In the Prandtl theory the thickness of the katabatic flow layer is related to the slope angle, the surface roughness, and the temperature of the free atmosphere. Thus, if these values are known the parameter H can be estimated and used to calculate turbulent transfers (Kuhn, 1978).

In principle, it is possible to imagine that a "classical" boundary layer could develop at the base of a katabatic flow layer. In this case the boundary layer equations might still be used to calculate fluxes at the snow surface. Munro and Davies (1977) found log–linear profiles up to a height of 1 m over a glacier in conditions of katabatic flow. They used the maximum velocity height H instead of the Monin–Obukhov length Λ as a scaling factor. This is a promising result from the point of view of the hydrological modeler since it implies that, provided the input meteorological data are collected at a sufficiently low level, there does not have to be a different set of model equations for katabatic flow conditions.

The danger of using boundary layer equations to analyze profile data at a sloping site is that if the lowest sensor happens to be above the boundary layer, the roughness lengths deduced will be too high. This explains why some of the values given in the literature for mountainous areas are greater than would be expected from the nature of the surface.

2.3. Effective Values of Snowpack Model Parameters

If measured values of an output variable are available then, working "backwards," it is possible to derive "effective" values of the parameters which appear in the meteorological and snowmelt model equations. These "effective" values are those which produce the best simulations of the output variable for the particular site and model equations chosen. Ideally, the effective values should be close to the actual values that would be determined from measurements at the site and thus independent of the structure of the models. However, in practice the effective values may be influenced by errors in the input data or over–simplifications in the model equations. So far the only model parameters which have been thoroughly investigated are the roughness lengths for turbulent transfer.

2.3.1. Effective Values of the Roughness Lengths

Harding et al. (1987) used measurements of snowmelt and evaporation in the firn area of the Hintereisferner to optimize parameters z_H and z_E of an energy–budget based snowmelt model. These parameters are defined such that $\Phi_E{}^2 = \Phi_u \Phi_q$ and $\Phi_H{}^2 = \Phi_u \Phi_T$ where the function $\Phi(z_\xi)$ is given by Equation (75). They are the scaling lengths for sensible heat transfer and evaporation. The effective values were found to be $z_H = 40 \pm 20$ mm and $z_E \simeq 5$ mm. This illustrates the difficulty of using a priori estimates of turbulent transfer parameters in snowmelt models. The a priori estimate of z_0 for a smooth flat snow cover would be $\simeq 1$ mm (from Table 5). The best estimates for z_T and z_q, and hence z_E and z_H, would then also be $\simeq 1$ mm. The effective values of the model parameters would, therefore, be underestimated in this case.

2.4. Modeling Snowpack Processes on the Catchment Scale

The equations derived in the previous sections have been used by many authors in hydrological models to calculate the amount of snowmelt in a catchment (WMO, 1986). The basic assumption has been that the catchment area can be divided up into a mosaic of elements each of which can be treated independently as a point site. Snowmelt from each element is then routed through the catchment using a distributed hydrological model. The problem with this method is that input meteorological data must be provided for each catchment element.

2.4.1. Distribution of Air Temperature and Humidity

Air temperature is perhaps the easiest of meteorological variables to distribute. It is usually assumed that it varies only with the altitude h so that:

$$T(h) = T(0) = \gamma h \tag{78}$$

The lapse rate γ can be set to a constant value, e.g. 6.5 K km^{-1} the dry adiabatic lapse rate. Humidity is more difficult; it is often assumed that the relative humidity remains constant over the whole catchment.

2.4.2. Distribution of Radiation Data

Net radiation data cannot be provided for a wide area by a simple transformation of net radiation data recorded at one point. Each component of the radiation budget must be corrected in a different way. The simplest correction to make is that for the direct solar radiation. Suppose that a catchment element has a slope of ζ degrees and faces A degrees from north. That is the vector perpendicular to the slope has an azimuth, A, and zenith angle ζ. The direct solar radiation perpendicular to the surface is:

$$Q = Q * f \, (cos Y (sin \Pi \, \cos \Omega \, \cos A \, \sin \zeta - \sin \Omega \, \sin A \, \sin \zeta +$$
$$\cos \Pi \, \cos \Omega \, \cos \zeta) + \sin Y \, (\sin \Pi \, \cos \zeta - \cos \Pi \, \cos A \, \sin \zeta \,)) \tag{79}$$

The transmission factor f will vary with altitude since it depends on the optical air mass of the atmosphere between the sun and the catchment element. One empirical equation which has been used is:

$$f(h) = (f_o) \, m_o (1 - h/10 \text{ km}) \tag{80}$$

where f_o is the mean transmissivity along the zenith path, and the index $m_o(1 - h/10$ km$)$ is the optical air mass at altitude h.

The direct radiation must also be corrected for shadowing. This can be done in a catchment model by defining a horizon equation relating the angle of elevation of the horizon to the azimuth. The corrections for diffuse radiation are more complex. Factors which must be taken into account include reflected radiation from surrounding surfaces and the anisotropic distribution of the diffuse radiation itself. Dozier (1978) has produced a successful model of solar radiation for snow surfaces in mountainous areas which includes these factors.

Incoming radiation is corrected for the effect of altitude by correcting the emissivity ε_a which depends (Equation [58]) on the air temperature and humidity. It must also be corrected for the effect of a restricted sky area in mountainous regions. The horizon equation can be used for this.

2.4.3. Distribution of Wind Speed and Direction

The current approach in hydrological models is to assume that wind speed and direction are the same for each element. This may be adequate for flat areas with only one type of vegetation, but in mountainous areas the synoptic scale wind is channeled through the valleys and the distribution of wind will no longer be spatially homogeneous. Temporal variability in the spatial distribution will be produced by heating and cooling effects. These give rise to valley winds which reverse direction over the course of a day. The solution must be to improve the meteorological element of distributed hydrological models so that wind, temperature, and humidity patterns over the whole of a catchment may be deduced from such measured data as are available.

There have been numerous theoretical studies of boundary layer flow over surfaces with varying elevation and/or surface roughness. Some of these have considered highly simplified cases. However, a few models have been developed for sufficiently realistic conditions to be of potential use in catchment hydrology. For example, Walmsley et al. (1986) have developed a three–dimensional model for neutral boundary layer flow over moderately steep terrain with slopes up to about 0.3 and z_0 varying by a factor of up to 100. The technical problems of interfacing such a distributed meteorological model with a distributed hydrological model are considerable but must be faced if any substantial progress is to be made in snowmelt modeling.

3. Conclusions

Many of the problems that have been described in this chapter are not confined to snowmelt modeling. All aspects of the meteorological model are relevant to the calculation of mass and energy inputs to soils and vegetation although particular effects arise because of the high albedo and smooth surface of snow. The snowmelt model equations are relevant to soils; indeed, the interface processes touched on in this chapter are particularly important once chemical reactions on the surfaces of soil particles are considered, for example in studies of acid rain. The problem of distribution of input data is an obstacle to progress in all aspects of catchment hydrology. What is special about snowmelt modeling is the strong feed–back in the system being modeled. The meteorological inputs both control and are controlled by the snow processes. It is this complex sensitivity which demands a physics–based approach and which has forced snowmelt modelers to use the most advanced techniques available for hydrological modeling.

4. References

Akan, A. O.: 1984, 'Simulation of Runoff from Snow Covered Hillslopes,' *Water Resources Research* **20**(6), 707–713.

Anderson, E. A.: 1976, *'A Point Energy and Mass Balance Model of a Snow Cover,'* NOAA Technical Report, National Weather Service 19, US Department of Commerce.

Andreas, E. L.: 1986, *'Theory for the Scalar Roughness and the Scalar Transfer Coefficients Over Snow and Sea Ice,'* US Army Corps of Engineers, Cold Regions Research and Engineering Laboratory, Report 86–9, 19pp.

Bohren, C. F., and B. R. Barkstrom: 1974, 'Theory of the Optical Properties of Snow,' *Journal of Geophysical Research* **79**(30), 4527–4535.

Brutsaert, W.: 1975, 'The Roughness Length for Water Vapour, Sensible Heat and Other Scalars,' *J. Atmos. Sci.* **32**, 2028–2031.

Brutsaert, W.: 1975a, 'On a Derivable Formula for Long–Wave Radiation from Clear Skies,' *Water Resources Research* **11**(5), 742–744.

Businger, J. A., J. C. Wyngaard, Y. Izumi, and E. F. Bradley: 1971, 'Flux Profile Relationships in the Atmospheric Surface Layer,' *J. Atmos. Sci.* **28**, 181–189.

de la Casiniere: 1974, 'Heat Exchange over a Melting Snow Surface,' *J. Glaciol.* **13**(67), 55–72.

Chamberlain, A. C.: 1983, 'Roughness Length of Sea, Sand and Snow,' *Boundary Layer Met.* **25**(4), 405–410.

Colbeck, S. C.: 1972, 'A Theory of Water Percolation in Snow,' *Journal of Glaciology* **11**(63), 369–385.

Colbeck, S. C.: 1974, 'On Predicting Water Runoff From a Snow Cover,' in *Advanced Concepts in the Study of Snow and Ice Resources*, US National Academy of Sciences, Monterey, 1973, pp. 55–66.

Colbeck, S. C.: 1975, 'Grain and Bond Growth in Wet Snow,' in *Snow Mechanics – Symposium – Mécanique de la Niege,* Proceedings of the Grindelwald Symposium, April 1974, IAHS–AISH Pub. No. 114, pp. 51–61.

Colbeck, S. C.: 1977, 'Short Term Forecasting of Water Runoff From Snow and Ice,' *Journal of Glaciology* 19(81), 571–588.

Colbeck, S. C.: 1979, 'Grain Clusters in Wet Snow,' *Journal of Colloid and Interface Science* 72(3), 371–384.

Colbeck, S. C., and E. A. Anderson: 1982, 'The Permeability of a Melting Snow Cover,' *Water Resources Research* 18(4), 904–908.

Defay, R., I. Prigogine, A. Bellemans, and D. H. Everett: 1966, *Surface Tension and Adsorption,* Longmans, Green and Co. Ltd.

Dozier, J.: 1978, 'A Solar Radiation Model for a Snow Surface in Mountainous Terrain,' in S. C. Colbeck, and M. Ray (eds.), *Modeling of Snow Cover Runoff,* US Army Cold Regions Research and Engineering Laboratory, Hanover, NH, September 1978, pp. 144–153.

Dyer, A. J., and B. B. Hicks: 1970, 'Flux Gradient Relationships in the Constant Flux Layer,' *Quart. J. Royal Met. Soc.* 96, 715–721.

Gardiner, B. G.: 1987, 'Solar Radiation Transmitted to the Ground Through Cloud in Relation to Surface Albedo,' *Journal of Geophysical Research* 94(D4), 4010–4018.

Garnier, B. J., and A. Ohmura: 1968, 'A Method of Calculating the Direct Shortwave Radiation Income of a Slope,' *Journal of Applied Meteorology* 7, 796–800.

Grainger, M. E., and H. Lister: 1966, 'Wind Speed, Stability and Eddy Viscosity Over Melting Ice Surfaces,' *J. Glaciol.* 6, 101–129.

Granger, R. J.: 1977, *'Energy Exchange During Melt on a Prairie Snowcover,'* MSc Thesis, Department of Mechanical Engineering, University of Saskatchewan.

Granger, R. J., and D. H. Male: 1978, 'Melting of a Prairie Snowpack,' *J. Applied Meteorology* 17(12), 1833–1842.

Greuell, W., and J. Oerlemans: 1986, 'Sensitivity Studies With a Mass Balance Model Including the Temperature Profile Calculations Inside the Glacier,' *Zeitschrift für Gletscherkunde und Glazialgeologie* **Band 22** Heft 2, s.101–124.

Harding, R. J.: 1986, 'Exchanges of Energy and Mass Associated With a Melting Snowpack,' in *Modelling Snowmelt–Induced Processes,* Proceedings of the Budapest Symposium, July 1986, IAHS Publ. No. 155, pp. 3–15.

Harding, R. J., H. Escher-Vetter, A. Jenkins, G. Kaser, M. Kuhn, E. M. Morris, and G. Tanzer: 1987, 'Energy and Mass Balance Studies in the Firn Area of the Hintereisferner,' *Proc. Symposium on Glacier Fluctuations and Climatic Change,* Reidel.

Hobbs, P. V.: 1974, *Ice Physics,* Clarendon Press, Oxford.

Kelly, R. J.: 1987, *'Mathematical Models of Multi–phase Snowmelt,'* Ph.D. Thesis, University of East Anglia.

Kelly, R. J., L. W. Morland, and E. M. Morris: 1986, 'A Three–Phase Mixture Model for Melting Snow,' in E. M. Morris (ed.), *Modelling of Snowmelt Induced Processes,* Proceedings of the Budapest Symposium July 1986, IAHS Publication No. 155, pp. 17–26.

King, J. C.: In Press, 'Measurements of Turbulence in the Stably–Stratified Surface Layer Over an Antarctic Ice Shelf,' *Boundary Layer Meteorology.*

Konstantinov, A. R.: 1966, *'Evapouration in Nature,'* Israel Program for Scientific Translation, Jerusalem.

Kuhn, M.: 1978, 'The Height of Maximum Speed in Drainage Winds as Parameters of the Energy Budget of a Glacier,' in *Arbeiten aus der Zentralanstalt für Meteorologie und Geodynamik* 31 Wien 1978.

Liljequist, G. H.: 1953, 'Radiation and Wind and Temperature Profiles Over an Antarctic Snowfield – A Preliminary Note,' *Proc. Meteorological Conference Toronto,* 1953, Am. Met . Soc., Royal Met. Soc., pp. 78–87.

Male, D. H.: 1980, 'The Seasonal Snowcover,' in S . C . Colbeck (ed.), *Dynamics of Snow and Ice Masses,* Academic Press, pp. 305–395.

Male, D. H., and R. J. Granger: 1981, 'Snow Surface Energy Exchange,' *Water Resources Research* 17(3), 609–627.

Manz, D. H.: 1974, *'Interaction of Solar Radiation With Snow,'* M.Sc. Thesis, Department of Agricultural Engineering, University of Saskatchewan, Saskatoon.

Mellor, M.: 1977, 'Engineering Properties of Snow,' *Journal of Glaciology* 19(81), 15–65.

112

Morozov, G. A.: 1967, 'Computation of the Change in the Density of Snow Cover Under the Influence of Water Vapour Diffusion, Convection, Sublimation and Condensation in it,' *Soviet Hydrology Selected Papers* **3**, 314–319.

Morris, E. M.: 1983, 'Modelling the Flow of Mass and Energy Within a Snowpack for Hydrological Forecasting,' *Annals of Glaciology* **4**, 198–203.

Morris, E. M., and J. Godfrey: 1979, 'The European Hydrological System Snow Routine,' in S. C. Colbeck, and M. Ray (eds.), *Modeling of Snow Cover Runoff*, US Army Cold Regions Research and Engineering Laboratory, Hanover, NH, September 1978, pp. 269–278.

Morris, E. M., and R. J. Kelley: 1990, 'A Theoretical Determination of the Characteristic Equation for Snow in the Pendular Regime,' *Journal of Glaciology* **36**(123), 179–187.

Munro, D. S., and J. A. Davies: 1977, 'An Experimental Study of the Glacier Boundary Layer Over Melting Ice,' *J. Glaciol.* **18**(80), 425–436.

Navarre, J. P.: 1977, 'Modele Unidimensionel d'evolution de la Neige Deposee,' *Meteorologie Applie*, 109–120.

Obled, C., and B. Rosse: 1977, 'Mathematical Models of a Melting Snowpack at an Index Plot,' *Journal of Hydrology* **32**(1/2), 139163.

O'Brien, H. W., and R. H. Munis: 1975, 'Red and Near–Infrared Spectral Reflectance of Snow,' in A. Rango (ed.), *Workshop on Operational Applications of Satellite Snowcover Observations*, NASA SP–391, pp. 319–334.

Palm, E., and M. Tveitereid: 1979, 'On Heat and Mass Flux Through Dry Snow,' *Journal of Geophysical Research* **84**(C2), 745–749.

Prandtl, L.: 1952, *Essentials of Fluid Dynamics*, Hafner Publ. Co., New York, 452 pp.

Pruppacher, H. R., and J. D. Klett: 1978, *Microphysics of Clouds and Precipitation*, D. Reidel Publishing Company, Dordrecht, Holland.

Schmidt, R. A.: 1982, 'Vertical Profiles of Wind Speed, Snow Concentration and Humidity in Blowing Snow,' *Boundary Layer Met.* **23**, 233–246.

Smithsonian Meteorological Tables: 1971, 6th Revised Edition, Smithsonian Institution Press, City of Washington.

Sulakvelidze, G. K.: 1959, 'Thermoconductivity Equation for Porous Media Containing Saturated Vapour, Water and Ice,' *Bulletin of the Academy of Sciences of the USSR, Geophysics Series*, 186–188.

Swinbank, W. C.: 1968, 'A Comparison Between Predictions of Dimensional Analysis for the Constant Flux Layer and Observations in Unstable Conditions,' *Quart. J. Royal Met. Soc.* **94**, 460–467.

Walford, M. E. R., D. W. Roberts, and I. Hill: 1987, 'Optical Measurements of Water Lenses in Ice,' *Journal of Glaciology* **33**(114), 159–161.

Walmsley, J. L., P. A. Taylor, and T. Keith: 1986, 'A Simple Model of Neutrally Stratified Boundary Layer Flow Over Complex Terrain with Surface Roughness Modulations (MS3DJH/3R),' *Boundary Layer Met.* **3**(1/2), 157–186.

Warren, S. G.: 1982, 'Optical Properties of Snow,' *Reviews in Geophysics and Space Physics* **20**(1), 67–89.

WMO: 1986, 'Intercomparison of Models of Snowmelt Runoff,' *Operational Hydrology Report No. 3*, WMO–No.646 Secretariat of the World Meteorological Organisation, Geneva, Switzerland.

Yen, Y. C.: 1962, 'Effective Thermal Conductivity of Ventilated Snow,' *Journal of Geophysical Research* **67**, 1091–1098.

Yen, Y. C.: 1963, 'Heat Transfer by Vapour Transfer in Ventilated Snow,' *Journal of Geophysical Research* **68**(4), 1093–1101.

Chapter 6

A Physically Based Snowcover Model

Bertel Vehviläinen
National Board of Waters and the Environment
Water and Environment Research Institute
P.O. Box 436
00101 Helsinki, Finland

Abstract. The aim of this chapter is to present the development of a physically based snowcover model, for simulating snow accumulation and snowmelt. The model can be used for operational purposes in large river basins with minimum daily input data. It may be expected that a physically based model would provide better predictions than the common degree–day snow models normally used in operational catchment hydrology. However, such an advantage has not yet been achieved with the model presented or with any other physical snowcover model; although they do have the advantage of providing information on the different processes included in snow accumulation and snowmelt phenomenon, such as snow density and depth, providing information for studies on soil frost simulation or for calculation of snow albedo. This information can be used in developing simpler snow models.

1. Introduction

The physically based snowcover model presented in this chapter is comprised of two independent model components: a snowpack energy balance model, and a snowpack mass balance model. The energy balance model simulates snowmelt and refreezing. The mass balance model uses the output (potential melt or refreezing) from the energy balance model to simulate snow accumulation/ablation, compaction of snowpack, liquid water storage in snowpack, and water yield from snowpack. The main characteristics of snowpack (i.e. snow water equivalent, density, depth, and albedo) are simulated in the snowpack mass balance model). Albedo is especially critical in the energy balance model for simulation of net shortwave radiation. Shortwave radiation can be simulated with or without direct shortwave measurements. Additionally, the effect of forestation is taken into account in shortwave and longwave simulation. The energy balance model is designed to be run with normal meteorological data (daily values of cloudiness, air temperature, precipitation, wind velocity, and air moisture) so that it can be linked to operational watershed models which are used to forecast snowmelt induced floods in real time.

The energy and mass balance model components are described in the two following subsections of this chapter. As part of this presentation results are presented for application of the model in the Loimijoki and Tujuoja basins.

The Loimijoki basin is located in Southern Finland (61°N 23°E). The basin covers 1980 km^2 within a relatively flat altitude range of 50–100m. About half of the basin (55 percent) consists of open areas (mainly cultivated land and marshlands) and 45 percent is forest. Only 0.5 percent of the basin is lakes, and thus it has a fairly short response area for Finnish conditions. The response time for rainfall or snowmelt is less than 24 hours.

The Tujuoja research basin is a small experimental basin near the western coast of central Finland 64°N, 25°E. The area covers 20.6 km^2, of which 82 percent is forest (canopy density

113

D. S. Bowles and P. E. O'Connell (eds.), Recent Advances in the Modeling of Hydrologic Systems, 113–135.
© 1991 *Kluwer Academic Publishers.*

114

30 percent), 12 percent field, 4 percent bog, and 2 percent urban area. No lakes are in the area and the mean slope is 2.3 percent.

Figures 1–8 present the main input data and simulation results for the winter 1971–1972 (in verification period 1970–1976) from the Loimijoki basin.

2. Discussion

2.1. Energy Balance Snowmelt Models

Temperature index models are not very stable. The best degree–day factor values vary from spring to spring and within each spring. By using an energy balance snowpack model, one expects to get a model, which is stable from spring to spring and without errors due to variable model parameters.

The energy balance of snowpack (RTOT) which determines the amount of snowmelt is presented in Equation (1):

$$RTOT = RSN + RLN + RLAT + RSEN + RP + RG - CO \tag{1}$$

where RSN is shortwave radiation balance; RLN is longwave radiation balance; $RSEN$ is sensible heat; $RLAT$ is latent heat from evaporation and sublimation; RP is heat from liquid precipitation; RG is heat exchange at ground surface; and CO is heat deficit of snowpack.

2.1.1. Shortwave Radiation Balance

Shortwave radiation (wavelength less than 3 μm) originally comes from the sun, but a part comes as a diffused component reflected from clouds and terrain. Heavy cloud cover can reduce shortwave radiation to 25 percent of the clear–sky value.

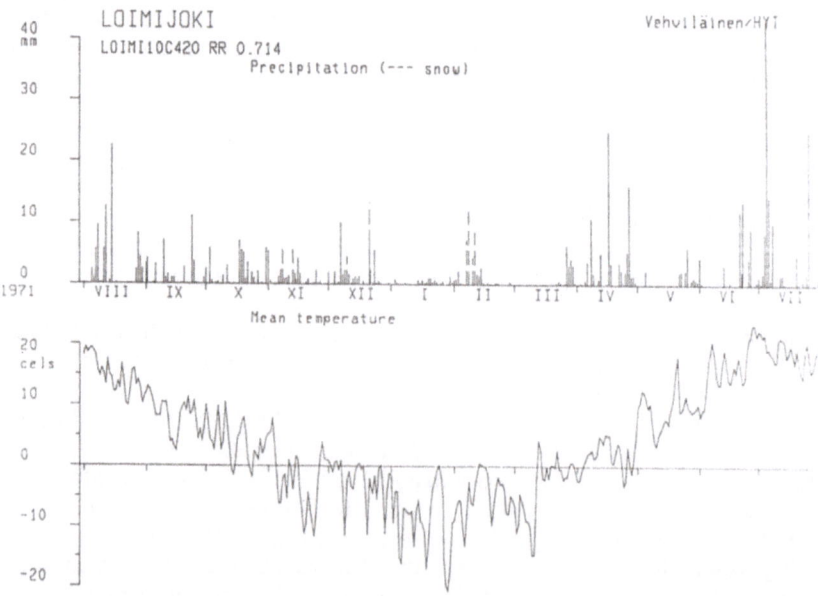

Figure 1. Daily precipitation and mean temperature at the Loimijoki Basin in the winter 1971–1972.

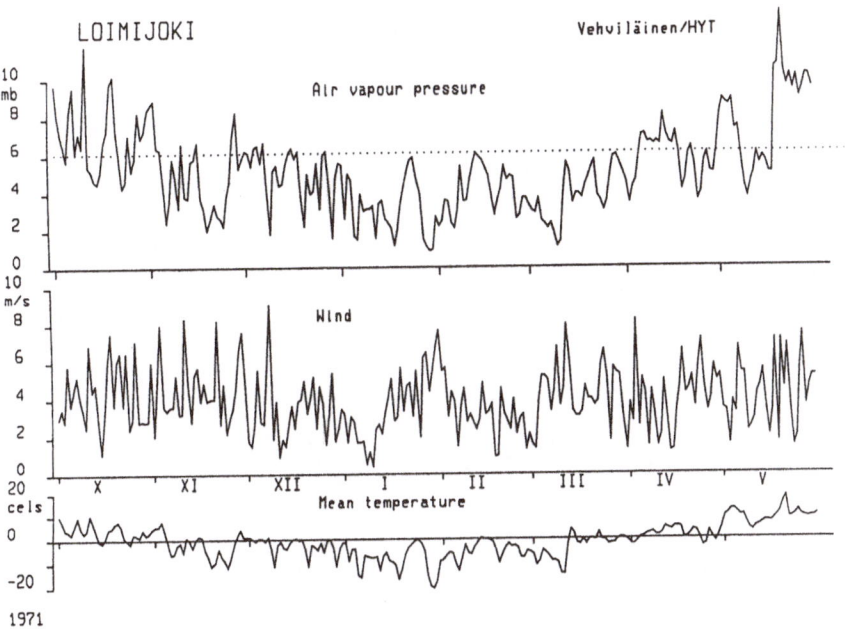

Figure 2. Air vapor pressure, wind, and mean temperature at the Loimijoki Basin, in the winter 1971–1972.

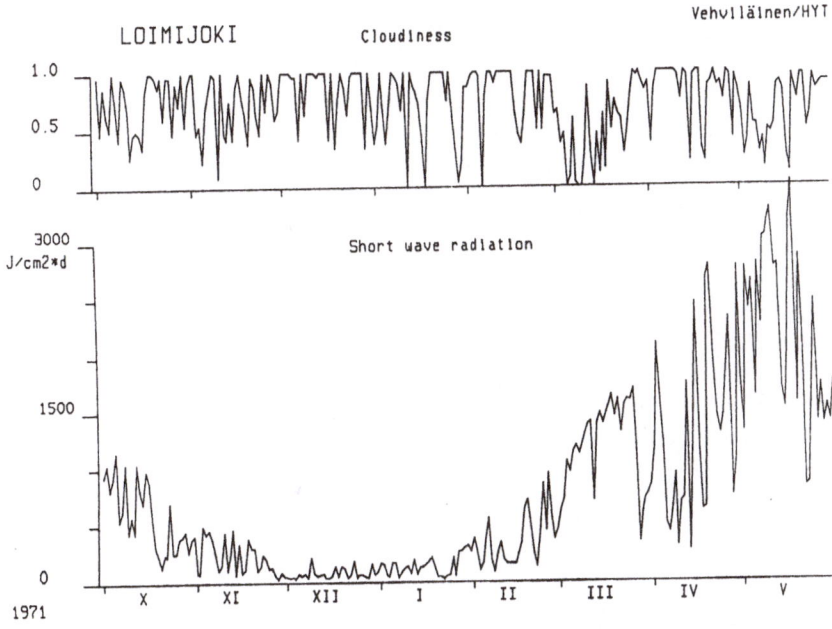

Figure 3. Cloudiness and measured shortwave radiation at the Loimijoki Basin in the winter 1971–1972.

116

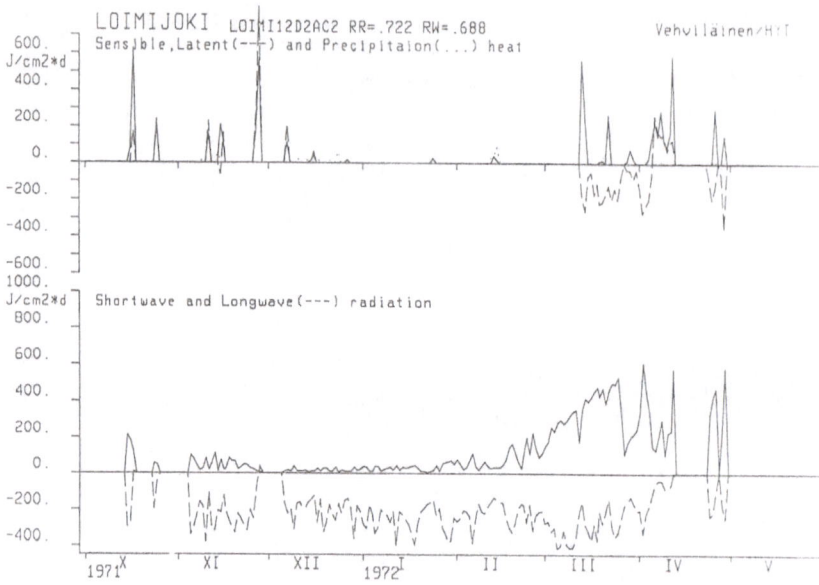

Figure 4. Simulation of different energy balance terms on snow surface: shortwave and longwave radiation, sensible and latent heat.   Simulation of precipitation heat of whole snowpack.

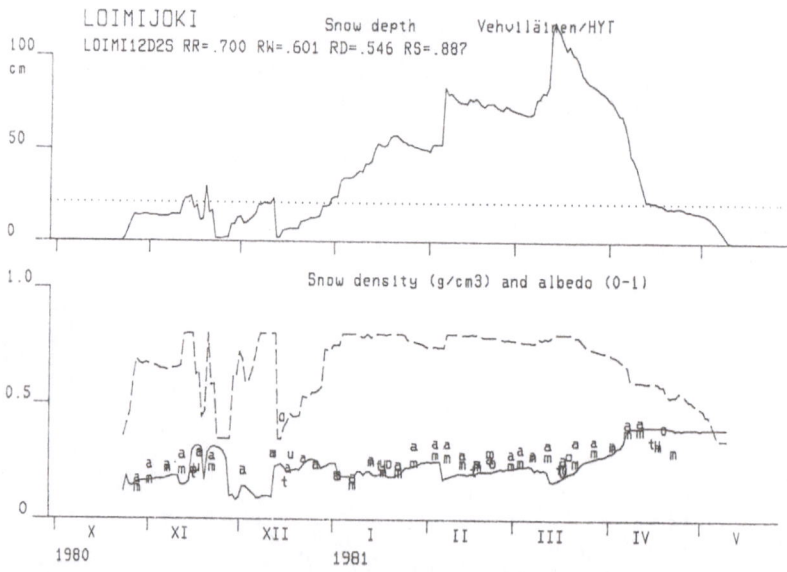

Figure 5. Simulated snow depth and density and measured density (letters) in the Loimijoki Basin in the winter 1971-1972.

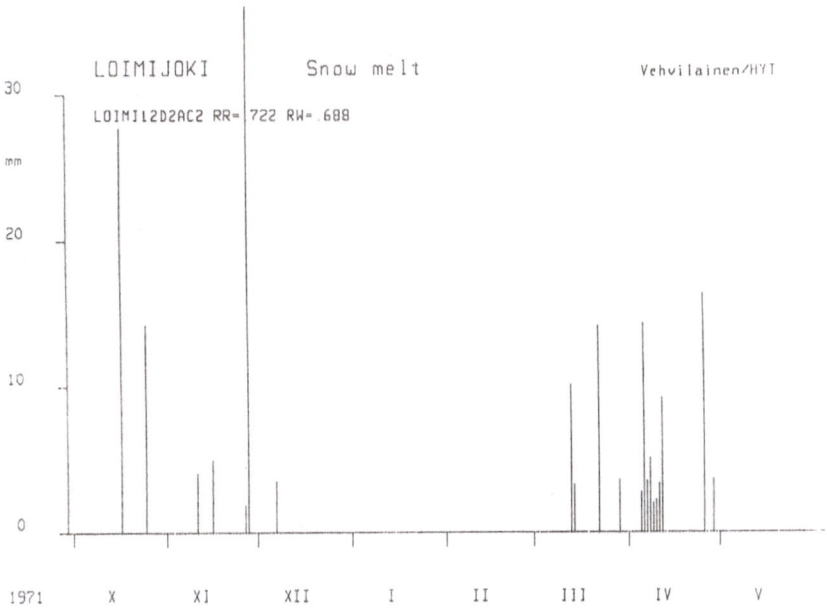

Figure 6. Simulated daily snowmelt values at the Loimijoki Basin in the winter 1971–1972.

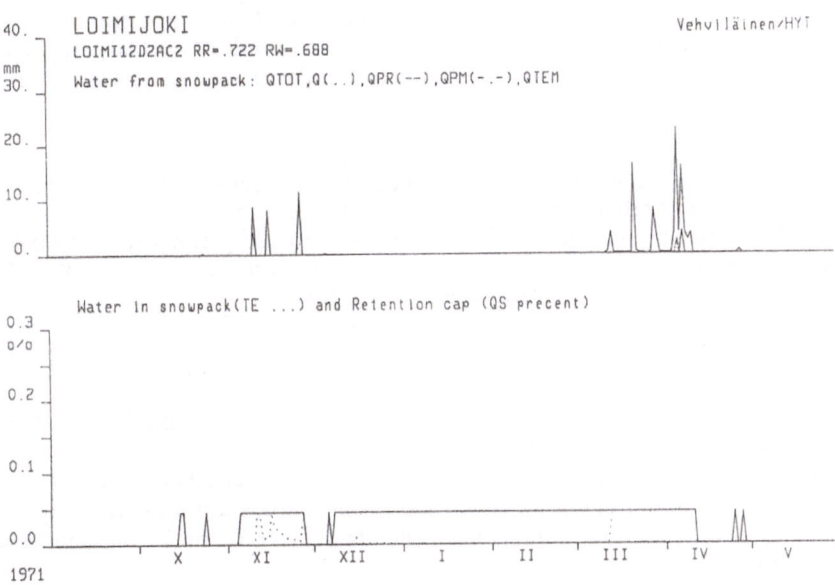

Figure 7. Simulated water yield from snowpack, retention capacity and water in snowpack at the Loimijoki Basin in the winter 1971–1972.

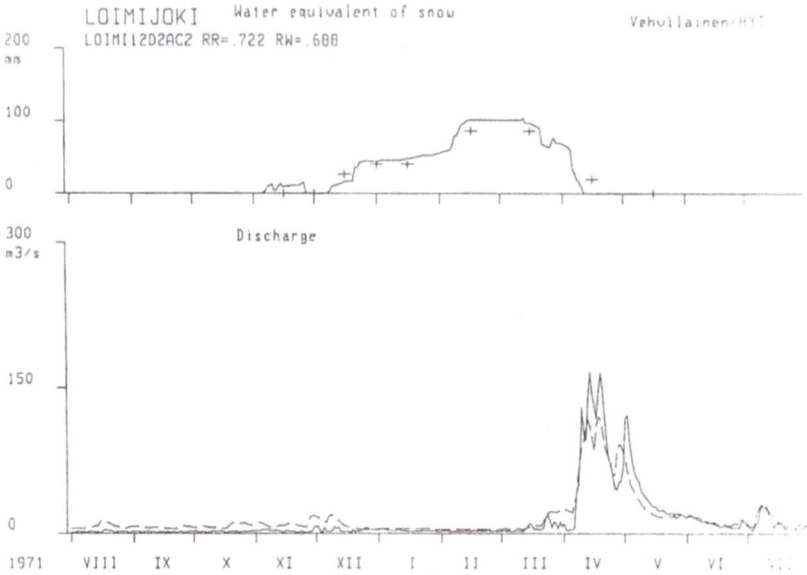

Figure 8. Simulated and observed areal water equivalent and discharge at the Loimijoki Basin in the winter 1971–1972.

A large amount of incoming shortwave radiation is reflected by snowcover. This reflected portion, a function of snow albedo, depends on the wave length, the slope and aspect of the snow surface, and especially on the snow structure. New snow reflects over 80 percent whereas coarse–grained old snow reflects only 30–40 percent of the incoming shortwave radiation.

2.1.2. Incoming Shortwave Radiation

The incoming shortwave radiation can be either measured or calculated from information on local latitude, date, time, and cloudiness. Measurements of shortwave radiation are made by a Bellani type pyranometer at Tujuoja and by a Moll–Gorzynski pyranometer at Loimijoki.

For calculating direct solar radiation above clouds incident upon a horizontal surface (RS_i), the following function is used (Kuchment et al., 1983):

$$RS_i = SLR \; \sin a - SLR2 \; \sin a^{0.5} \tag{2}$$

where SLR is solar constant, about 8.4 J/cm^2 min; a is the angle of the radiation with the horizontal; and $SLR2$ is parameter.

The solar constant is the average intensity of solar radiation received on a plane of unit area normal to the incident radiation at the outer limit of the earth's atmosphere.

The following function from spherical trigonometry is used for calculation of sin a (Eagleson, 1970):

$$\sin a = \sin d \; \sin f + \cos d \; \cos f \; \cos h \tag{3}$$

where d is declination, f is local latitude, and h is sun's hour angle.

The effect of cloud is taken into account with the following equation:

$$RS = RS_i (1 - CS \ CL) \qquad (4)$$

where CL is cloudiness (cloudless sky $= 0$, overcast sky $= 1$), and CS is parameter.

The complex processes of atmospheric absorption, scattering, and reflection (cloudless sky situation) are not considered; however, they are included in the parameter values of this simple model.

In Equations (2), (3), and (4), the incoming solar radiation model has been calibrated against the measured incoming solar radiation at Tujuoja and Loimijoki basins. Model performance is evaluated according to R^2-criteria, as follows:

$$R^2 = (F_o^2 - F^2)/F_o^2 \qquad (5)$$

in which:
$$F_o^2 = (X_{obs} - X_{mean,obs})^2 = \text{variance}$$
$$F^2 = (X_{obs} - X_{sim})^2 = \text{mean square error}$$

where X_{obs} is observed value of incoming solar radiation, $X_{mean,obs}$ is mean observed value, and X_{sim} is simulated value.

Table 1. R^2 for calibration and verification periods.

Basin	Calibration 1976–1981	Verification 1970–1976
Tujuoja	0.822	0.750
Loimijoki	0.901	0.888

Table 2. Parameter values obtained from calibration.

Basin	SLR	SLR2	CS
Tujuoja	6.7 J/cm^2 min	0.00	0.67
Loimijoki	9.6 J/cm^2 min	0.96	0.71

The calibrated values of the solar constant, SLR, were less at Tujuoja and higher at Loimijoki than the 8.4 J/cm^2 min value proposed by Eagleson (1970). In this incoming solar radiation model, the energy balance snowmelt model is used without measuring shortwave radiation. Figure 9 provides an example from the verification period of simulation of incoming shortwave radiation by this model.

Incoming Solar Radiation in Forest. Snowcover in a forest receives reduced solar radiation. The reduction factor, forest transmission coefficient (*CF*), is a function of forest canopy density.

Canopy density is the percentage of forested area that is covered by a horizontal projection of the vegetation canopy. This coefficient, *CF*, is calculated by the equation (WMO, 1974):

$$CF = 1 - CF1 \ [1 - (1 - CD)^2]^{0.5} \qquad (6)$$

120

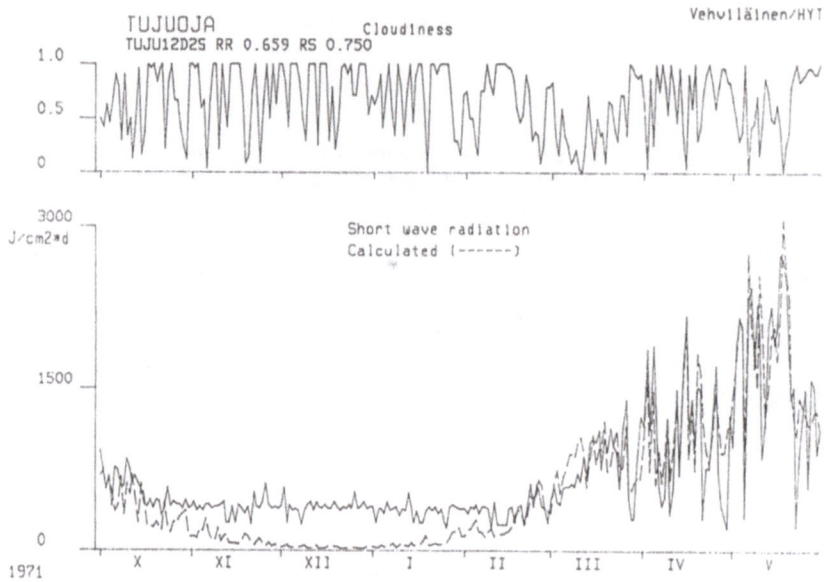

Figure 9. The simulation of incoming shortwave radiation compared to measured values at Tujuoja basin during the winter 1974–1975.

where $CF1$ is transmission coefficient with 100 percent canopy density, dependent on forest type (about 0.9); and CD is canopy density.

The incoming solar radiation through forest (RS_f) is now calculated by equation:

$$RS_f = CF\ RS \tag{7}$$

The calculated values of canopy density for the whole area are about 30 percent for Tujuoja and about 10 percent for Loimijoki. The values of the transmission coefficient with 100 percent canopy density $(CF1)$ are slightly lower than 0.9: about 0.85 for Tujuoja, and about 0.75 for Loimijoki. (The exact values from calibrations are presented later.) With these average values, the transmission coefficient is 0.40 for Tujuoja and 0.67 for Loimijoki according to Equation (6). These values are near the values read from the Figure 10 according to canopy density, CD, only.

Snow Albedo. The shortwave radiation balance on snowcover can be presented as (Mantis, 1951):

$$RS = RSN + RSR + RSB \tag{8}$$

where RS is incident radiation on snow; RSN is absorbed radiation; RSR is reflected radiation; and RSB is transmitted radiation.

It is usually assumed that $RSB = 0$, and the net incoming radiation (RSN) is calculated according to the snow albedo (AL) which is the ratio of reflected radiation (RSR) to the incident radiation on snowcover (RS):

$$AL = RSR/RS \tag{9}$$

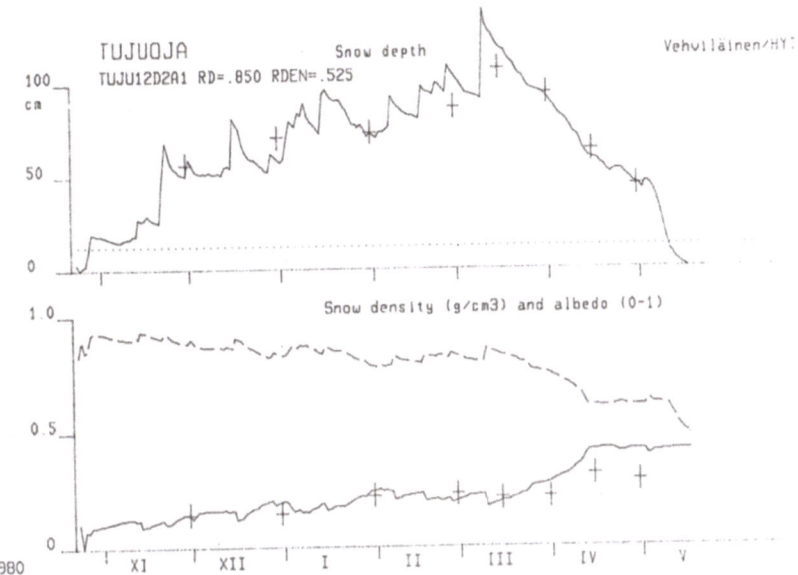

Figure 10. An example of simulation of snow albedo, snow depth, and snow density for the Tujuoja basin (+ = observed value).

For a forested area the net shortwave radiation RSN is thus:

$$RSN = (1 - AL) \, RS_f \tag{10}$$

Snow albedo is easily measured at a point, but daily measurement of albedo for a large watershed is difficult, often impossible. Therefore, some empirical functions can be used for calculating watershed albedo during snowcover. Snow albedo is a function of snow structure and, therefore, a function of snow density. According to Kuchment et al. (1983), the following equation can be used for snow albedo simulation:

$$AL = C_{a1} - D_s \tag{11}$$

where C_{a1} is parameter (about 1.03); and D_s is snow density (g/cm³).

The depth of snowcover also affects snow albedo. When snow depth thins sufficiently, a portion of incoming shortwave radiation is transmitted through the snowcover and absorbed by the underlying ground. The effect of the transmitted radiation (RSB) can be taken into account with the following functions:

$$AL = A1 \, (1 - A2) + AL \, A2 \tag{12}$$

$$AL = AL_{min} \qquad , \text{if } AL < AL_{min}$$

where $A1$ is $C_{a2} - D_s$; $A2$ is H/H_{lim}; C_{a2} is parameter (about 0.9); H is depth of snowcover; H_{lim} is threshold value for snow depth; and AL_{min} is albedo of ground surface (0.20 – 0.35).

In Figures 10 and 11 examples from Tujuoja and Loimijoki are presented, where snow albedo is simulated with Equations (11) and (12). Snow density and snow depth are simulated by the physically based snowpack model presented later.

122

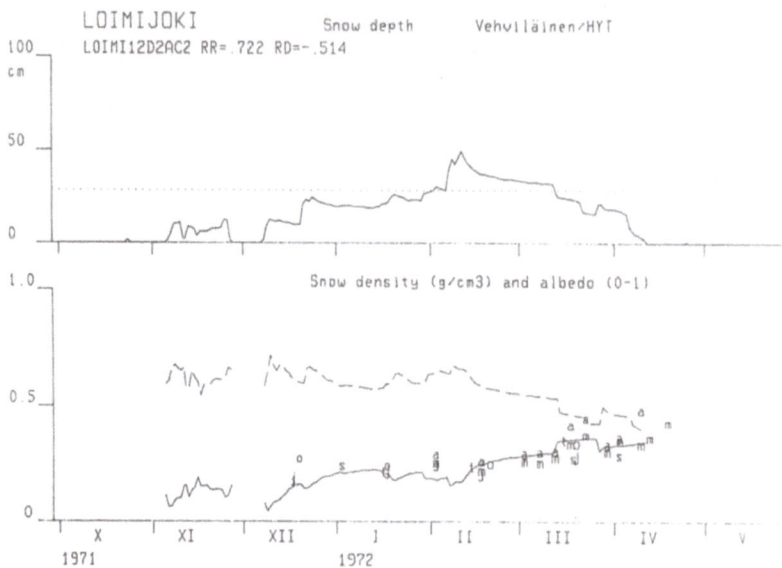

LOIMIJOKI   Snow depth   Vehviläinen/HYT
LOIMI12D2AC2 RR=.722 RD=-.514

Figure 11. An example of simulation of snow albedo, depth, and density for the Loimijoki basin. a, m, t, o, u = observed densities.

Albedo is often calculated as a function of the age of snow (USCE, 1956). Snow albedo can also be calculated as a function of water equivalent (Kuusisto, 1984). This method has been used in Equation (13):

$$AL = AL_{max} (W/WMAX)^{eal}$$

$$AL = AL_{min} \quad , if AL < AL_{min}$$

(13)

where AL_{max} is snow albedo before melt; AL_{min} is albedo at the end of snowmelt period; eal is parameter (1.0 for Loimijoki); W is water equivalent of snow; and $WMAX$ is maximum water equivalent of snowpack.

2.1.3. Longwave Radiation Balance

The net longwave energy RLN for snowpack is the difference between downward longwave radiation (RLD) and upward longwave radiation (RLU):

$$RLN = RLD - RLU$$

(14)

With snow, the upward longwave energy is the sum of emitted $RLDE$ and reflected $RLDR$ energy by snow.

$$RLU = RLUE + RLUR$$

(15)

Because snow albedo for longwave radiation is only about 0.03 (Dunkle et al., 1949), the reflection of longwave radiation by snowcover is omitted in longwave radiation calculations by assuming $RLUR = 0$.

Assuming snow is a near–perfect black body in the longwave portion of spectrum, the upward longwave emission by snow is calculated with an emissivity, E_s, between 0.97 – 1.00 (Kondratyev, 1969):

$$RLUE = E_s \, SF \, T_s^4 \tag{16}$$

where E_s is emissivity of snow surface (0.97 –1.00); SF is Stefan–Boltzmann constant; and T_s is snow surface temperature (°K).

T_s is approximated by air temperature at 2 m level when no snowmelt, because the time step is 24 hours. $T_s = 273$ °C when snow is melting.

In a clear–sky situation, downward longwave radiation comes mostly from the lower 100 m of the atmosphere and, of that, 2 percent is emitted by ozone, 17 percent by carbon dioxide, and 81 percent by water vapor after Geiger (1961). Thus variation of downward longwave radiation is mainly dependent on the variation of moisture and temperature in the 100 m layer of atmosphere above ground. In clear–sky conditions, this downward radiation is calculated by:

$$RLD = E_a \, SF \, T_K^4 \tag{17}$$

where E_a is atmospheric emissivity; and T_K is air temperature (°K).

Many empirical equations exist for atmospheric emissivity, which are functions of water vapor pressure in air and air temperature. One common function, first presented by Brunt (1952), is:

$$E_a = a + b \, e_a^{0.5} \tag{18}$$

This equation gave quite consistent predictions of atmospheric emissivity according to Hatfield et al. (1983). Kuzmin (1961) has suggested parameter values of $a = 0.62$ and $b = 0.05$ based on thorough studies in the steppe and forest regions of Russia.

For a cloudy sky the atmospheric emissivity is also a function of cloud thickness, density, type, and altitude. For a partly cloudy sky, estimating downward longwave radiation is difficult and many empirical functions have been proposed for it. Kuzmin (1961) used the following function:

$$RLD = E_c \, SF \, T_K^4 \tag{19}$$

$$E_c = [1 + C_L \, (CL + CL_1)]E_a \tag{20}$$

where C_L is parameter, about 0.12; CL is total cloudiness (0 – 1); and CL_1 is cloudiness of the lower layer (ranges from 0 – 1).

A solid forest canopy is an efficient absorber and emitter of longwave radiation. The longwave absorptivity and emissivity are at least 0.90 (Brooks, 1959), and perhaps as high as 0.97 (Baumgartner, 1967). Thus a forest canopy can be treated as a blackbody at ambient air temperature during windy periods. In the absence of wind, the temperature of the forest canopy rises above the air temperature during clear–sky, daytime conditions. At night, the situation is reversed.

The downward longwave radiation in a forest is calculated by equation (Eagleson, 1970):

$$RLD = E_f \, SF \, T_K^4 \tag{21}$$

$$E_f = CD_f + (1 - CD_f) E_a \tag{22}$$

where CD_f is forest canopy density for longwave radiation.

2.1.4. Sensible Heat Exchange

Sensible heat exchange is the product of the turbulent exchange processes of heat in the layer above snow surface. This turbulent heat flux can be directly measured by eddy correlation techniques, which requires sophisticated instrumentation.

A more common method is to evaluate sensible heat exchange based on wind and temperature profiles above the snow surface (Prandtl, 1932):

$$RSEN = -C_a D_a K_h \, dT_{pot}/dz \tag{23}$$

where C_a is specific heat of air (kJ/kg °C); D_a is air density (kg/m^3); K_h is eddy diffusivity for convective energy transfer (m^2/s); and T_{pot} is potential temperature (i.e. temperature at the surface pressure (°C)).

In practice the eddy diffusivity is calculated from wind profile assuming that the eddy diffusivity for convective heat transfer K_h equals eddy diffusivity for momentum K_m which is usually valid during snowmelt (Anderson, 1976). The sensible heat exchange is calculated by equation:

$$RSEN = D_a C_a k^2 (K_h/K_m)(U_2 - U_1)(T_{pot2} - T_{pot1})/(ln z_2/z_1)^2 \tag{24}$$

where k is von Karman's constant (0.4); K_m is eddy diffusivity for momentum (m^2/s); U_1 is wind velocity (m/s) at the snow surface (usually extrapolated); U_2 is wind velocity (m/s) at the height of measurement device; z_1 is the height (m) of extrapolated wind velocity near snow surface; and z_2 is the height (m) of measurement device.

This equation is used with standard meteorological observations, assuming $U_1 = 0$, and $T_{pot1} = 0$, and $K_h/K_m = 1$ during snowmelt.

This method of profiles is only valid for open areas, where well-defined wind and temperature profiles exist. In forested areas the vegetation prevents the formation of well-defined wind and temperature profiles 2 – 3 meters above snow. Therefore it is reasonable to use simplified equations for sensible heat exchange in the forested basins of Tujuoja and Loimijoki. These equations are usually of the form:

$$RSEN = CSEN \, U \, (T - T_s) \tag{25}$$

where $CSEN$ is bulk transfer coefficient for sensible heat exchange [kJ/(m^3 °C)], the value of which varies from 0.001 to 0.015 (Male and Gray, 1981).

The values of bulk transfer coefficients are usually averages over relative long periods and Equation 84 should not be applied over intervals of less than 24 hours, at least with the presented coefficient values.

2.1.5. Latent Heat Exchange

Turbulent mixing of the air layers is the main process causing heat and moisture transfer above the snow surface. Because the origin of these two exchange processes is the same, they occur simultaneously. Sensible and latent heat exchange need the gradient of temperature and moisture, on one hand, and wind, on the other hand, to exist. Moisture and temperature gradients are developed due to the energy supply of shortwave and longwave radiation. Wind eliminates gradients simultaneously increasing turbulent heat absorption/intake and moisture evaporation/condensation. Without wind, heat and moisture transfer ceases although gradients exist; there is no turbulent transport process. Conversely, wind can easily eliminate weak gradients and again stop the transfer of heat and moisture.

The same method of profiles (or gradients) used for sensible heat exchange is also valid for latent heat exchange (Prandtl, 1932):

$$RLAT = -HV\, D_a\, K_e\, de_a/dz \tag{26}$$

where HV is heat of vaporization (335 kJ/kg); K_e is eddy diffusivity of latent energy transfer (m^2/s); and e_a is air moisture (mb).

The eddy diffusivity of momentum K_m is first calculated from wind profile. Assuming K_e/K_m is unity during snowmelt, latent heat exchange can be evaluated based on measurements on two levels z_1 and z_2:

$$RLAT = D_a\, HV(K_e/K_m)(U_2 - U_1)(e_{a2} - e_{a1})/(\ln z_1/z_2)^2 \tag{27}$$

The values of the ratio K_e/K_m reported in the literature have a wide range, and the assumption of a value of unity from this ratio during snowmelt is questionable.

The same reasons, absence of well-defined wind and moisture profiles near surface, which led to the use of simplified equations of sensible heat in forested areas are also valid for latent heat exchange. For the simulation of latent heat exchange in forested basins of Tujuoja and Loimijoki, a similar equation is used for sensible heat:

$$RLAT = CLAT\, U(e_a - e_s) \tag{28}$$

where $CLAT$ is bulk transfer coefficient for latent heat exchange [kJ/(m^3/mb)]; e_s is vapor pressure on the snow surface; and e_a is vapor pressure in air over snow.

The values of $CLAT$ vary from 0.002 to 0.025 kJ/m^3 mb according to Male and Gray (1981).

2.1.6. Precipitation Heat

Precipitation heat is the heat or energy delivered to a snowpack, considering the difference between the temperature of precipitation and the temperature of snowpack. If the temperature of snowpack is under 0°C, the liquid precipitation freezes and delivers the latent heat of fusion to snowpack. With no freezing, the energy supply delivered to the snowpack (RP) is:

$$RP = D_w\, C_w(T_p - T)P/1000 \tag{29}$$

where D_w is density of water (kg/m^3); C_w is specific heat of water [kJ/(kg °C)]; T_p is temperature of rain (°C); and P is depth of rain (mm/d).

The energy supplied by this process is small compared to other components of the energy balance equation.

If all liquid precipitation is frozen in the snowpack, the energy supplied from liquid precipitation becomes important relative to other energy balance terms. The latent heat of fusion of water (335kJ/kg) is only possible at the beginning of snowmelt, when the temperature of snowpack is below 0°C.

When the temperature of snowpack is −10 °C, the energy supplied from 1 mm of rain at 5 °C air temperature is about 6 J/cm^2 [Equation (28)]. When the water is frozen the additional energy supply to the snowpack is 34 J/cm^2.

With abundant rainfall on the snow, the heat from liquid precipitation is rapidly distributed into the entire snowpack. Rainfall may very quickly change the structure of snow compared to normal melting process.

2.1.7. Heat Exchange at Ground Surface

The main form of heat transfer in the ground is molecular heat conductivity. The heat flux from the ground to the snowpack (RG) is given by the equation:

$$RG = -k_g \, dT_g/dz \tag{30}$$

where T_g is temperature of ground surface layer (°C); and k_g is thermal conductivity of ground surface layer.

Thermal conductivity of the ground is strongly dependent on soil moisture. Thermal conductivity increases with increasing soil moisture content. The thermal conductivity of frozen soil is higher than the thermal conductivity of an unfrozen soil because the thermal conductivity of ice is approximately four times that of water.

During the snowmelt period, the heat exchange depends on the condition of the soil. If the underlying soil is not frozen, the melting snow receives heat from the soil. When ground frost exists the heat flux is positive to the ground and the snow loses energy. In both cases the heat flux is usually in the limits of −40 to + 40 J/cm^2 (Kuzmin, 1961), indicating snowmelt or freezing of 0.1 mm of water. This value is negligible compared to other terms in the energy balance equation.

In Finnish studies Sucksdorff (1982) obtained 31 J/cm^2 for ground heat flux below the maximum depth of soil frost. Data from the Finnish Meteorological Institute yielded values of 8–35 J/cm^2 d during the winter 1979 (Kuusisto, 1984); the maximum value occurred at the beginning of winter and the minimum at the end of winter.

2.1.8. Internal Energy

Before snow begins melting, the temperature of snow must rise to 0 °C. This negative energy storage, called cold content (CO) can be calculated by:

$$CO = H(D_i C_i + D_w C_w + D_v C_v)T_s \tag{31}$$

where H is snow depth; D_i, D_w, D_v is density of ice, water and vapor; C_i, C_w, C_v is specific heat of ice, water and vapor; and T_s is mean snowpack temperature.

The energy content of water vapor is negligible and, for practical calculations, the equation is used in the form:

$$CO = C_s W T_s \tag{32}$$

where C_s is specific heat of snow (2 kJ/kg °C); W is $D_s H$ = water equivalent of snow; and D_s is density of snow.

The correct simulation of internal energy requires the simulation of heat transfer by conduction and mass transfer into snowpack. A simpler method is to calculate the cold content directly from snow surface energy balance by assuming a minimum temperature for snowpack as a limit for cold content. Braun (1985) used a similar constant temperature approach for estimating of cold content in his snowmelt studies.

2.1.9. The Contribution of Different Energy Balance Terms

Tables 3 through 6 present the monthly contribution of different energy balance terms on the energy balance of snow surface over the whole simulation period 1970–1981. The results are expressed in J/cm^2 month and also as a percentage from the sum of positive energy balance terms by which we can better compare the importance of different positive energy balance terms possible snowmelt.

The net shortwave radiation RSN is the most dominant positive energy source for snow-cover in both basins over the whole year and especially during spring, although in the last days or week, during the melting period the change of negative to positive net longwave radiation

Table 3. Monthly means of energy balance terms on snowcover in J/cm^2 month over the period of 1970–1981 in the basin of Tujuoja (Frk = number of months).

Month	Frk	RSN	RLN	RSEN	RLAT	RP	RTOT
		J/cm^2	J/cm^2	J/cm^2	J/cm^2	J/cm^2	J/cm^2
January	11	313	-3772	120	17	108	-3214
February	11	902	-3698	41	-13	95	-2672
March	11	2917	-3403	191	-174	153	-317
April	11	4857	-1570	529	-679	28	3166
May	11	1121	565	282	-49	8	1928
June	1	89	130	60	12	2	293
July	–	–	–	–	–	–	–
August	–	–	–	–	–	–	–
September	7	88	14	62	12	5	182
October	10	419	-834	143	56	87	-128
November	11	371	-2361	142	43	199	-1607
December	11	336	-3467	225	30	226	-2650

Table 4. The relative contribution of energy balance terms on total energy balance on snowcover expressed as percentage from the sum of positive energy balance terms per month in the basin of Tujuoja during period of 1970–1981 (Frk = number of months included).

Month	Frk	RSN	RLN	RSEN	RLAT	RP
		%	%	%	%	%
January	11	54	-828	20	2	24
February	11	88	-365	4	-1	9
March	11	89	-105	6	-5	5
April	11	90	-29	10	-13	1
May	11	56	29	15	-2	0
June	1	31	44	20	4	1
July	–	–	–	–	–	–
August	–	–	–	–	–	–
September	7	49	8	34	7	3
October	10	59	-118	20	8	12
November	11	48	-324	19	6	27
December	11	43	-639	23	3	20

Table 5. Monthly means of energy balance terms on snowcover in J/cm^2 month over the period of 1970–1981 in the basin of Loimijoki (Frk = number of months included).

Month	Frk	RSN	RLN	RSEN	RLAT	RP	RTOT
		J/cm^2	J/cm^2	J/cm^2	J/cm^2	J/cm^2	J/cm^2
January	11	294	−2783	101	62	201	−2125
February	11	1144	−2932	67	4	150	−1567
March	11	2463	−2316	233	−250	148	278
April	9	1722	−106	290	−350	6	1562
May	1	850	350	156	−75	5	1286
June	−	−	−	−	−	−	−
July	−	−	−	−	−	−	−
August	−	−	−	−	−	−	−
September	−	−	−	−	−	−	−
October	6	171	−343	13	18	37	−104
November	10	238	−1058	163	162	108	−387
December	11	154	−2046	152	53	125	−1562

Table 6. The relative contribution of energy balance terms of total energy balance on snowcover expressed as percentage from the sum of positive energy balance terms in month in the basin of Loimijoki during period of 1970–1981 (Frk = number of months included).

Month	Frk	RSN	RLN	RSEN	RLAT	RP
		%	%	%	%	%
January	11	45	−231	15	9	30
February	11	84	−222	4	0	11
March	11	87	−81	8	−9	5
April	9	85	−5	15	−17	0
May	1	63	26	11	−5	0
June	−	−	−	−	−	−
July	−	−	−	−	−	−
August	−	−	−	−	−	−
September	−	−	−	−	−	−
October	6	73	−160	5	5	17
November	11	35	−157	24	24	16
December	11	32	−431	31	11	26

diminishes the share of shortwave radiation. In one case at Tujuoja in June, net longwave radiation exceeds net shortwave radiation. Also in midwinter, when the temperature in snowpack is below zero and cold content exists, liquid precipitation contributes markedly to the positive energy balance through freezing of rain in snowpack, which diminish the share of shortwave radiation.

The absolute values of net shortwave radiation peak in spring, because incoming shortwave radiation strongly increases from March to May. The absolute values in spring are about ten times higher than in the autumn or winter.

The net longwave radiation is dominately negative throughout the entire snowcover period except during the first days of snow in autumn and the last days of snow in spring. Air temperature and hence forest canopy temperature can be high during these periods causing abundant downward longwave radiation which changes the longwave radiation balance to positive.

Sensible heat is the second most important positive energy balance term. In winter its contribution is under 10 percent of the sum of positive terms because of cold air and typically weak winds. But with higher air temperatures in spring, the share of sensible heat rises to between 10 percent and 20 percent. During early winter or late autumn, when air temperature can rise considerably, as is associated with high wind speed during the passing of low pressure centers typical of frontal weather systems, the contribution of sensible heat can rise to 30 percent of the total positive energy balance terms.

Latent heat exchange on a snow surface is predominately negative during the melting period of March to May, which means evaporation and sublimation from snowcover. The lowest values during this period were –350 and –679 J/cm² month for the Loimijoki and Tujuoja, respectively, the evaporation of water consumes an additional 250 J/mm, and the sublimation of ice 33 J/mm. So the average energy loss of –350 to –679 J/cm² month indicates only 1–2 mm of evaporation/sublimation per month at most in spring. In autumn and into early winter, condensation prevails in the latent heat exchange because of high air vapor pressure and frequently high wind speeds. The amounts of condensation are also very low; 56 and 162 J/cm² month indicating maximum condensation of 0.2 – 0.6 mm/month.

Precipitation heat strongly contributes to the sum of positive energy terms in winter because liquid precipitation freezes in the cold snowpack. During the melting period, when the temperature of snowpack is 0°C, the share of precipitation heat strongly diminishes. The energy contributed by precipitation is due only to the heating effect, in which the temperature of rain lowers to the temperature of snow according to Equation (28) without latent heat of freezing. During winter precipitation heat contributes 20 – 30 percent of the sum of positive energy balance terms.

Total energy balance is positive in spring and autumn. The amount of snowmelt in millimeters can be calculated by dividing the total energy balance *RTOT* by the latent heat of freezing (33 J/mm).

The absolute energy amounts per month are higher at Tujuoja because of a longer snowcover period and higher values of water and snow equivalents.

2.2. Mass Balance Snowpack Model

The simulation of the snow albedo by Equations (11) and (12) requires information on the density and depth of snowcover. In model simulation work the only reasonable way to obtain the density and depth of snowcover is to simulate it using a computer model. A physically based snowcover model (Motovilov and Vehviläinen, 1987), which simulates water equivalent, ice concentration, density, depth, and liquid water in the snowpack (Figure 12) is presented here. The inputs for the snowcover model are precipitation, evaporation from snowcover, and potential melt/freezing simulated by the energy balance model presented above.

In this snowcover model snow is divided into three phases: air, water, and ice. The processes which change the concentrations of different snow phases are precipitation, snow evaporation, melt/freezing, and compaction of snow.

2.2.1. The Effect of Precipitation and Snow Evaporation on Snowcover

The density of snow in precipitation D_p (g/cm³) is a function of air temperature (°C) after Kuchment et al. (1983):

$$D_p = 0.13 + 0.0135\,T + 0.00045\,T^2 \tag{33}$$

130

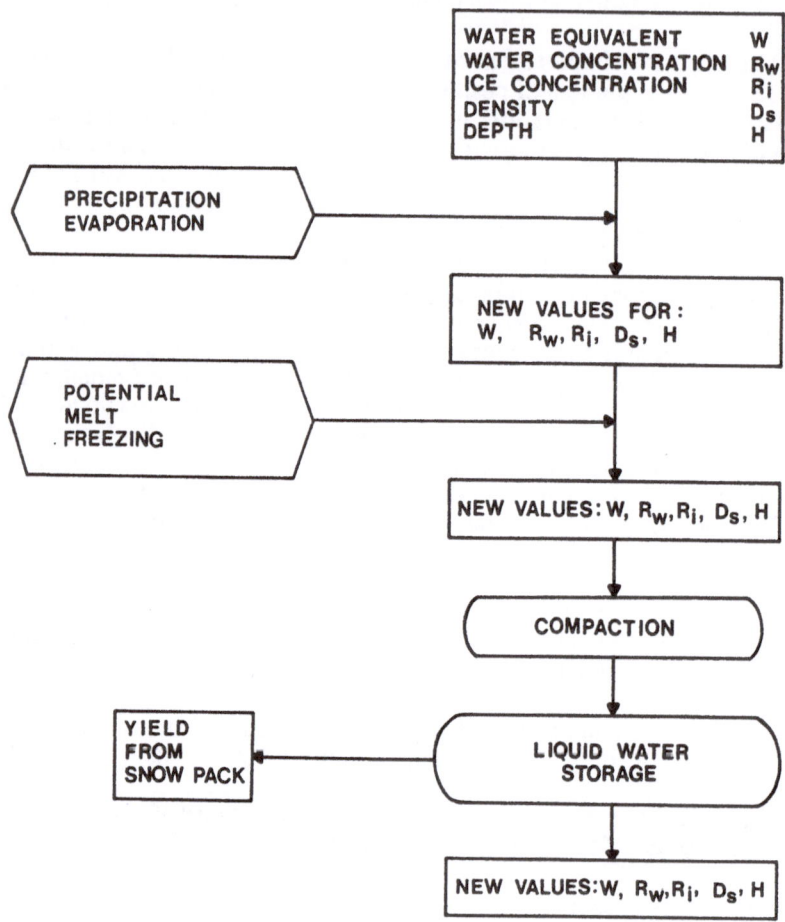

Figure 12. The structure of snowcover model.

Snow precipitation PS (cm) and evaporation from snowcover ESC (cm) cause changes in depth, water equivalent, ice concentration, water concentration, and density of snow. Possible liquid precipitation will change water equivalent, water concentration, and density of snow.

The change of the depth of snow, H (cm), due to solid precipitation, PS, and the evaporation from snowcover, ESC, is calculated by equation:

$$dH_p = PS\ D_w/D_p - ESC\ D_w/D_s \tag{34}$$

where D_w is the density of water (g/cm^3); and D_s is the density of snowpack (g/cm^3).

The change of the water equivalent of snow W (cm) is then:

$$dW_p = PS + PL - ESC \qquad (35)$$

The change of ice concentration R_i of snow is:

$$dR_i = (PS - ESC) \, D_w/D_i \, 1/(H + dH_p) \qquad (36)$$

where R_i is the ice concentration of snow in volumetric units (cm^3/cm^3); D_i is the density of ice (g/cm^3); and H is the depth of snowpack (cm).

The change of water concentration R_w needs to be calculated, if liquid precipitation PL (cm) occurs by following equation:

$$dR_w = PL/(H + dH_p) \qquad (37)$$

where R_w is the water concentration of snow in volumetric units (cm^3/cm^3).

The average snow density D_s (g/cm^3) for snowpack is now:

$$D_s = (W + dW_p)/(H + dH_p) \qquad (38)$$

2.2.2. The Effect of Snowmelt on Snowcover

The potential snowmelt M (cm) is calculated by snow surface energy balance model presented before. The change of snow depth due to melt is calculated by equation:

$$dH_m = MD_w/D_i \, 1/R_i \qquad (39)$$

The change of water concentration is now:

$$dR_w = M/(H - dH_m) \qquad (40)$$

and the average density for snowcover D_s (g/cm^3) is calculated by:

$$D_s = W/(H - dH_m) \qquad (41)$$

2.2.3. The Effect of Freezing of Liquid Water in a Snowpack on Snow Characteristics

The potential freezing of water F (cm) in a snowpack is calculated by the energy balance model previously presented. The upper limit of back-freezing of water is the storage of liquid water in snowpack. The change of ice concentration of snow due to freezing is:

$$dR_i = F \, D_w/D_i \, 1/H \qquad (42)$$

The change of water concentration is:

$$dR_w = F/H \qquad (43)$$

132

2.2.4. Compaction of Snowcover

Density changes of snowcover are due to the wind effect on snowcover, compaction of over-lying snow, and metamorphic changes in the structure of snowcover.

Kojima (1967) and Yosida (1963) have presented similar quantitative expressions for com-paction of snowcover. According to them the compaction of snowcover can be expressed as:

$$1/D_s \, dD_s/dt = W/n \tag{44}$$

where n is viscosity coefficient, which is dependent on temperature and density of snow (cm h).

According to Kojima (1967) the relationship between n and the density of snow can be expressed as:

$$n = n_c \exp(C_2 \, D_s) \tag{45}$$

where n_c is the hypothetical viscosity coefficient with zero density (cm h); and C_2 is parameter (cm^3/g) to be determined from observations, according to Kojima (1967): 21 cm^3/g.

Now we can express Equation (44) in the following form using Equation (45):

$$1/D_s \, dD_s/dt = C_1 \, W \exp(-C_2 \, D_s) \tag{46}$$

where C_1 is $1/n_c$ = the fractional increase in density per centimeter water equivalent of load per hour, according to Kojima (1967): 0.026 – 0.069 1/(cm h).

Kojima (1967) reported that Equation (46) performed generally well against the observed data except for cases of low–density new snow, wind–packed snow, and depth–hoar layers. Low–density snow and wind–packed snow layers increased in density at a faster rate than Equation (45) expressed.

Mellor (1964) have presented another equation for the viscosity coefficient n, which takes into account the temperature and type of snow by:

$$n/n_o = \exp[[AC/R][(T_c - T)/T \, T_c]] \tag{47}$$

where T_c is 0°C temperature in Kelvin degrees (273 °K); n_o is the viscosity coefficient at temperature of 0°C; AC is the activation energy of the snow (4 10^4 J/mol); and R is the gas constant, about 8 J/mol °K.

For a normal temperature range during the snowcover period, Equation (47) can be presented as:

$$n/n_o = \exp[C_3 * (T_c - T)] \tag{48}$$

where C_3 is $AC/(R \cdot T \, T_c)$ = about 0.08 1/°K

Equation (46) will now be after temperature dependency of Equation (48) has been taken into account:

$$1/D_s \, dD_s/dt = C1 \exp[-0.08(T_c - T)] \, 0.1W \exp(-C_2 D_s) \tag{49}$$

According to Mellor (1964), the value of C_1 varies because of differences in temperature and snow type as can be seen from Equation (49): instead of constant C_1, which Kojima (1967) used in his Equation 105, we have no C_1 exp $[-.08 (T_c - T)]$. Equation (49) is used in a snow-cover simulation model for simulation of compaction. Instead of snow density, D_s, the concentration of ice in snow, R_i, is used in Equation (49).

The change of depth due to compaction is calculated now:

$$dH_c = R_i/(R_i - dR_i)H \tag{50}$$

The change of water concentration in snowpack due to compaction is:

$$dR_w = R_w \, dH_c/H \tag{51}$$

2.2.5. Water Yield from Snowcover

Before any liquid water is released from snowcover, the retention capacity of water must be fulfilled. The maximum retention capacity of snow WHL (in volume percentage) can be calculated by a function presented by Kuzmin (1957):

$$WHL = 0.11/D_s - 0.11 \tag{52}$$

When retention capacity is expressed as maximum water concentration in snow R_{wm}, Equation (52) is (Kuchment et al., 1983):

$$R_{wm} = \frac{0.11(1 - D_i/D_w \, R_i)}{1.11 - 0.11(D_w/D_i \, R_i)} \tag{53}$$

Water yield from snowcover WQ (cm) is now:

$$
\begin{aligned}
WQ &= 0 && , \text{if } R_w < R_{wm} \\
WQ &= (R_w - R_{wm})H && , \text{if } R_w > R_{wm}
\end{aligned} \tag{54}
$$

The water yield, WQ, from snowpack diminishes the water equivalent, W, by the same value, and the density of snow, D_s (g/cm^3), after water yield and compaction is:

$$D_s = (W - WQ)/(H + dHc) \tag{55}$$

3. Conclusions

Two versions of the physically based snowcover model were compared against the best temperature index snow model used in the operational watershed models in Finland (Vehviläinen, 1986). In the first version only the energy balance model portion of the physically based snowcover model is used. In the second version the entire snowcover model, which simulates the main characteristics of snowcover, is used. The model performance criteria, R^2, is used to compare simulated and observed areal snow water equivalent (WE) in the verification period (ver) 1970–1976 in the Loimijoki Basin. The observed areal snow water equivalent is based on snow line measurements.

Additionally, the results obtained with the complete operational watershed model, which is a modified version of HBV-model (Bergström, 1976) are presented. The watershed model is tested using different versions of snow models against the observed discharges (Q) both in the calibration period (cal) 1976–1981 and in the verification period (ver) 1970–1976. Only the results against runoff (q) from the Tujuoja basin are available:

	Tujuoja		Loimijoki		
	R^2 (cal)	R^2 (ver)	R^2 (cal)	R^2 (ver)	
	q	q	Q	Q	WE
Energy balance model	.821	.708	.821	.670	.539
–with snowcover model	.850	.710	.842	.758	.833
Temperature index	.847	.737	.856	.780	.837

In the Loimijoki basin, the results from the energy balance model were clearly worse than the operationally used temperature index model. The entire physically based snowcover model presented in this text and the operationally used temperature index snow model performed equally well when compared using areal snow water equivalent models.

Similar results were obtained when the whole watershed model was run with different snow models. The watershed model with the complete mass balance snowpack model performed better than just the energy balance approach, especially in the Loimijoki basin. Still the temperature index model is superior in both approaches, when comparing the results in the verification period. The improvement in the verification results in Loimijoki was moderate.

If we take into account only the task of operational flood forecasting due to snowmelt, there is no reason to use the complex, and data demanding physical snow model in place of the simpler temperature index snow model. However, the physically based snow model is a valuable tool when there is a need to simulate continuous snow depth or density values, such as in a soil frost simulation model.

4. References

Anderson, E. A.: 1976, 'A Point Energy and Mass Balance Model of a Snowcover,' NOAA Tech. Rep. NWS-19, U. S. Dept. Commer., Washington, DC.

Baumgartner, A.: 1967, 'Energetic Bases for Differential Vaporization From Forest and Agricultural Land,' Proc. Internat. Sympos. Forest Hydrol., pp. 381–390, Pergamon Press.

Bergström, S.: 1976, 'Development and Application of a Conceptual Runoff Model for Scandinavian Catchments,' SMHI. Nr RHO 7, Norrköping.

Braun, N. L.: 1985, 'The Simulation of Snowmelt-Runoff in Lowland and Lower Alpine Regions of Switzerland,' Züricher Geographische Schriften no 21, Geographisches Institut. Zürich.

Brooks, F. A.: 1959, An Introduction to Physical Microclimatology, 264 pp. illus., Davis, CA, Assoc. Students Store.

Brunt, D.: 1952, Physical and Dynamical Meteorology, Cambridge Univ. Press, Cambridge, MA.

Dunkle, R. V., et al.: 1949, 'Non Selective Radiometer for Hemispherical Irradiation and Net Radiation Interchange Measurements,' Univ. Calif. Dept. Engin., Thermal Radiation Proj. 9, CA.

Eagleson, P. S.: 1970, Dynamic Hydrology, McGraw-Hill, NY.

Geiger, R.: 1961, Das Klima der bodennahen Luftschicht (The Climate Near the Ground), (Scripta Technica Inc., Trans.), Harvard University Press, Cambridge, MA.

Hatfield, J. L., R. J. Reginato, and S. B. Idso: 1983, 'Comparison of Long-Wave Radiation Calculation Methods Over the United States,' Water Resources Research 19(1), 285–288.

Kojima, K.: 1967, 'Densification of Seasonal Snow Cover, Physics of Snow and Ice,' Proceedings of International Conf. on Low Temperature Science, Sapporo, Vol. I, Part 2, The Institute of Low Temperature Science, Hokkaido University, Sapporo, pp. 929–952.

135

Kondratyev, K. Y.: 1969, *Radiation Energy of the Sun,* Gidromet., Leningrad.

Kuchment, L. S., V. N. Demidov, and Y. G. Motovilov: 1983, *Formirovanie Rechnogo Stoka (River Runoff Formation Physically Based Models),* Nauka, Moskva.

Kuusisto, E.: 1984, *'Snow Accumulation and Snowmelt in Finland,'* Publications of Water Research Institute no. 55, Helsinki.

Kuzmin, P. P.: 1957, *'Fizitsheskije Svojstva Sneznogo Pokrova (The Physical Properties of Snow Cover),'* Gidrometeoizdat, 179 pp.

Kuzmin, P. P.: 1961, *Melting of Snow Cover,* Leningrad, (Israel Program for Scientific Translations, Jerusalem 1972).

Male, D. H., and D. M. Gray: 1981, 'Snowcover Ablation and Runoff,' *Handbook of Snow,* Pergamon Press.

Mantis, H. T.: 1951, *'Review of the Properties of Snow and Ice,'* SIPRE Report 4.

Mellor, M.: 1964, *'Properties of Snow,'* Cold Regions Science and Engineering Monograph III–A1, Cold Regions Research and Engineering Laboratory, Hanover, NH, 105 pp.

Motovilov, J., and B. Vehviläinen: 1987, 'Snow Cover and Snowmelt Runoff Model in the Forest Zone,' *Proceedings of the International Soviet–Finnish Symposium in Water Research in Moscow, 1986,* VYH Monistesarja Nro 27., Helsinki.

Prandtl, L.: 1932, 'Meteorologische Anwendung der Strömungslehre,' *Beitr. Phys. d. freien Atmos.* **19**, 188–202.

Sucksdorff, Y.: 1982, *'Lämpötilavaihtelut Maaperän Pintakerroksessa (The Variation of Temperature in the Upper Soil Layers),'* MSc-Thesis, Univ. of Helsinki, Dept. of Geophysics, 55 pp.

USCE: 1956, *'Snow Hydrology; Summary Report of the Snow Investigations,'* North Pacific Division, Corps of Engineers U.S. Army, Portland, OR, 437 pp.

Vehviläinen, B.: 1986, 'Operational Spring Time Forecasting Difficulties and Improvements,' *Nordic Hydrology* **17**, 363–370.

WMO: 1974, *Guide to Hydrological Practices,* Third Edition, WMO-No.168, Geneva.

Yosida, Z.: 1963, 'Physical Properties of Snow,' in W. D. Kingery (ed.), *Ice and Snow,* M.I.T. Press, Cambridge, MA, pp. 485–527.

Chapter 7

Infiltration, Soil Moisture, and Unsaturated Flow

Keith Beven
University of Lancaster
Centre for Research on Environmental Systems
Institute of Environmental and Biological Sciences
Lancaster, LA1 4YQ, United Kingdom

Abstract. Strategies and progress in modeling the unsaturated zone are described and, in particular, limitations on the use of the Richards equation. Problems of identifying soil parameters in the face of heterogeneity and preferential flow pathways are discussed. The advantages of parametric simplicity for certain modeling purposes are stressed.

1. Introduction

In 1931 L. A. Richards proposed a description of single phase water flow in soils that has since come to dominate physical representations of water flow in the unsaturated zone. Richards combined an equation of continuity with Darcy's law, under the assumption that the linear relationship between mass flux density and hydraulic gradient remains valid for unsaturated flows. For isothermal, incompressible flow the Richards' equation may be written in the form:

$$\frac{\partial \Theta}{\partial t} = \nabla(K \nabla h) + \frac{\partial h}{\partial z} \pm S \tag{1}$$

where ∇ is the differential operator; h is the soil capillary potential; Θ is the volumetric soil water content; S is a source/sink function to account, for example, for uptake of water by roots; and K is the hydraulic conductivity of the soil which, for unsaturated conditions, is expected to be a nonlinear and hysteretic function of soil moisture status at any point in the soil but is independent of the local hydraulic potential gradient.

This partial differential equation requires that the relationship between the variables K, Θ, and h (known as the soil characteristic curves) must be specified a priori before a solution can be obtained. These relationships are generally highly nonlinear for unsaturated flows, and analytical solutions of Equation (1) are known only for special cases with simple boundary conditions. For hydrological modeling purposes, it will generally be necessary to use a numerical solution of the Richards equation, in which values of Θ or h are calculated at a number of node points in space (Figure 1). A number of numerical solution techniques are available, as will be outlined below, but until recently it has been quite uncommon for this physical description of unsaturated flows to be incorporated into hydrological models as a result of the computational expense of the nonlinear distributed calculations.

Thus, while recognizing that the Richards equation has represented the best physical description of soil physics available, most hydrological modelers have chosen to use more

D. S. Bowles and P. E. O'Connell (eds.), Recent Advances in the Modeling of Hydrologic Systems, 137–151.

138

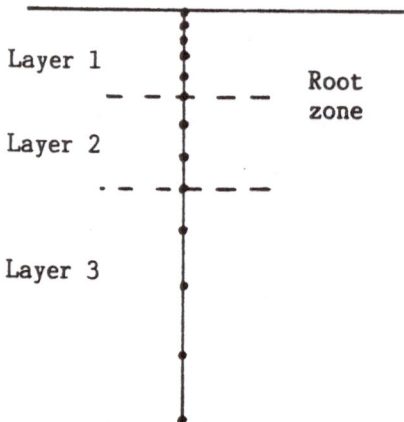

Figure 1. Models of the unsaturated zone.

conceptual representations of the unsaturated zone. Some of these have a good physical justification, at least in part, such as the various infiltration equations based on simplifications of the one–dimensional Richards equation of Philip (1957), Morel–Seytoux and Khanji (1974), and Smith and Parlange (1978), all of which have been incorporated into hydrological models at the catchment scale. Other components, notably representations of the root zone, have been based on simple storage elements with no direct physical justification (Figure 2). There are numerous models in this category, ranging from the early lumped models of Crawford and Linsley (1966) and Dawdy and O'Donnell (1965) to the semi–distributed models of Beven and Kirkby (1979) and Ledoux et al. (1984). These simple water balance models are used operationally for predicting soil water status such as in the U.K. Meteorological Office MORECS system (Gardner and Bell, 1980). Such models are justified only empirically, by being shown to give reasonable results on a variety of catchments. The variety of formulations of such storage models will not be considered here, but the success of such models will be discussed further later.

2. Discussion

In what follows we shall examine the progress that has been made in modeling the unsaturated zone using solutions of the Richards equation. Problems with using Richards

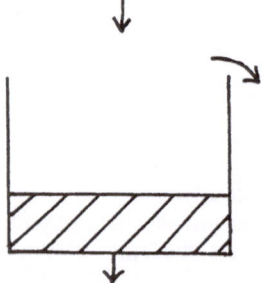

Figure 2. Simple conceptual storage model.

equation–based models will then be considered, leading on to a discussion of model calibration and the estimation of predictive uncertainty. Finally a strategy for structuring future applications of unsaturated zone models is presented.

2.1. Progress in Modeling the Unsaturated Zone

Flow of water in the unsaturated zone is a complex phenomenon involving transfers of water, air, and vapor through dynamic flow pathways under the influence of hydraulic, temperature, density, and osmotic gradients in a compressible porous medium. Equation (1) represents only the simplest form of Darcian description, formulated as a single–phase isothermal and incompressible flow of water under the influence of a gradient of hydraulic potential. It would be possible to formulate the problem as a multi–phase flow process in a deformable medium with a coupled thermal transport model (see for example Green et al., 1970; Narasimhan and Witherspoon, 1977; and Milly, 1982).

The resulting, more complex, equations could also be solved by approximate numerical methods without undue difficulty but at the expense of increased computer time relative to the single phase Equation (1). More significantly, the additional complexity would also be manifest in greater requirements for parameter and input data. These would include characteristic curves for the transport of air and vapor, thermal capacity and diffusivity, densities of water and air, and upper boundary conditions for air pressures, humidity, and temperature in addition to water fluxes. Both laboratory and simulation experiments have shown that under certain conditions the effects of air, vapor, and thermal fluxes may be important in the unsaturated zone. Theoretical reasoning would therefore suggest that the use of Equation (1) as a simulator of the unsaturated zone should be expected to be inaccurate in some circumstances. Limiting conditions for the use of Equation (1) for simulating unsaturated zone processes in the field have not, however, been established and it is usually assumed, with some justification, that the error resulting from ignoring the multi–phase nature of the flow process may be negligible, relative to the problems of estimating the single–phase flow parameters for field situations.

There are a number of approximate numerical techniques available for the solution of Equation (1), including finite difference, finite element, and integrated finite difference techniques. These techniques are now well established and, with the growing availability of powerful dedicated desktop computers, are becoming viable as routine techniques for modeling the unsaturated zone in catchment scale models. Examples are the one–dimensional unsaturated components of the SHE model and the USDA Opus model (Smith and Ferreira, 1989); the two–dimensional components of Freeze (1972), Neuman (1973), Beven (1977), Feddes et al. (1975), Rulon et al. (1985) and the Institute of Hydrology Distributed Model (Beven et al., 1987); and the three–dimensional variably saturated models of Freeze (1971), Huyakorn et al. (1986), Sharma et al. (1987), Binley et al. (1989), and my discussion in Chapter 17. Success in reproducing field measurements of soil water potentials and water contents using a Richards equation–based model has been reported, for example by Jensen and Jønch–Clausen (1982, Figure 3).

2.2. Problems in Modeling the Unsaturated Zone

A major problem in applying these models to field situations is in the estimation of the parameters of the soil characteristic curves. A number of different functional relationships have been used to describe the relationships between K, Θ, and h (see Table 1). The majority of these functional forms are single–valued relationships, but models for hysteretic relationships have also been proposed with varying degrees of success in reproducing measured curves (eg. Mualem, 1977, 1984).

140

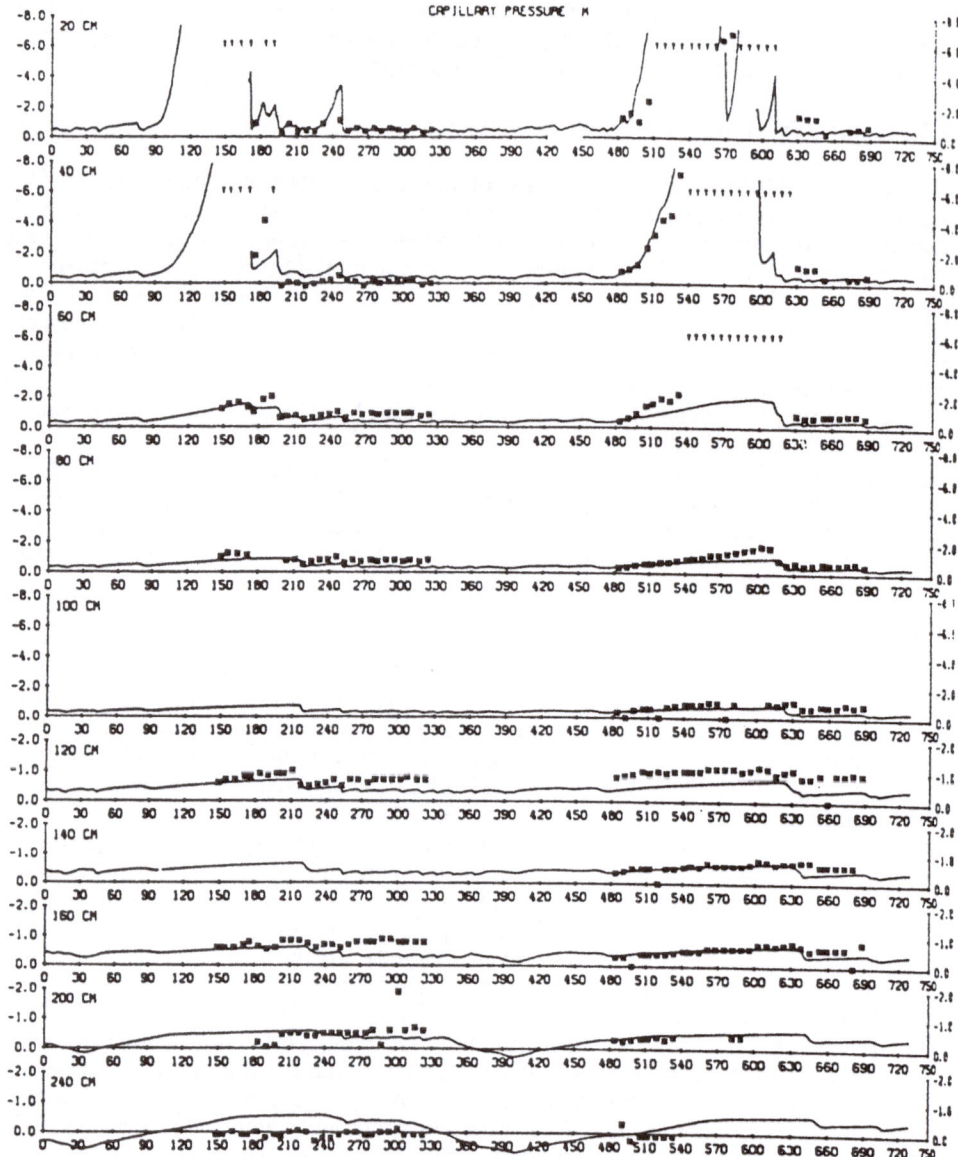

Figure 3. Comparison of observed and predicted soil moisture contents and potentials using a 1-dimensional finite difference model (after Jensen and Jønch–Clausen, 1982).

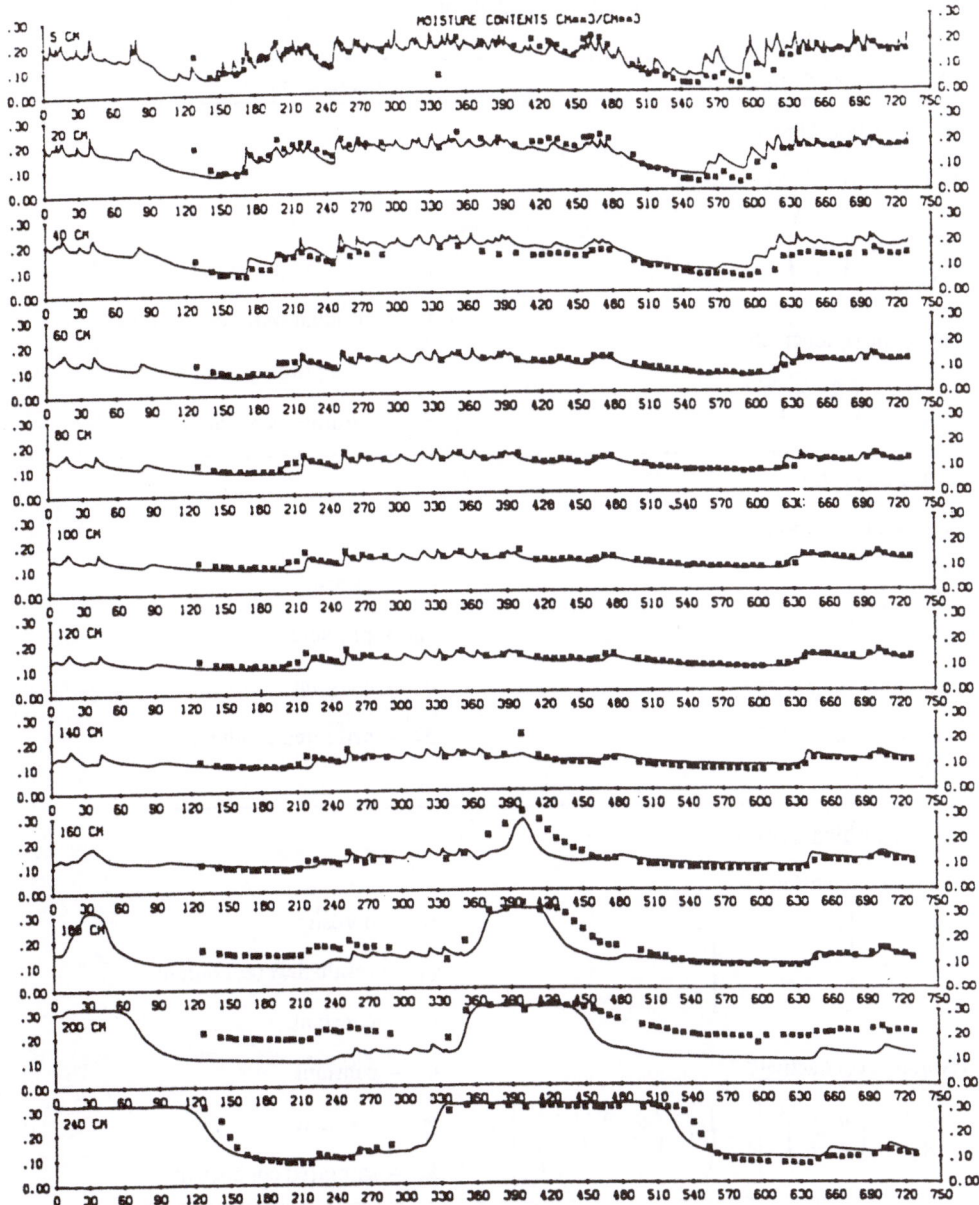

Figure 3. (Continued)

Table 1. Soil water retention and hydraulic conductivity relationships (after Rawls and Brakensiek, 1988).

Hydraulic soil characteristics	Parameters

Brooks–Corey (1964)

Soil water retention

$$\frac{\theta - \theta_r}{\phi - \theta_r} = \left(\frac{\varphi_b}{\varphi}\right)^{\lambda}$$

λ = pore size index

φ_b = bubbling pressure

θ_r = residual water content

Hydraulic conductivity

ϕ = total porosity

$$\frac{K(\theta)}{K_s} = \left(\frac{\theta - \theta_r}{\phi - \theta_r}\right)^{3-2/\lambda}$$

K_s = saturated conductivity

Campbell (1974)

Soil water retention

$$\frac{\theta}{\phi} = \left(\frac{H_b}{\varphi}\right)^{1/b}$$

ϕ = porosity

H_b = pressure

Hydraulic conductivity

b = constant

$$\frac{K(\theta)}{K_s} = \left(\frac{\theta}{\phi}\right)^{2b+3}$$

K_s = saturated conductivity

Van Genuchten (1980)

Soil water retention

$$\frac{\theta - \theta_r}{\phi - \theta_r} = \left[\frac{1}{1 + (a\ \varphi)^n}\right]^m$$

ϕ = porosity

θ_r = residual water content

α = constant

Hydraulic conductivity

n = constant

$$\frac{K(\theta)}{K_s} = \left(\frac{\theta - \theta_r}{\phi - \theta_r}\right)^{1/2}\left[1 - \left(1 - \left(\frac{\theta - \theta_r}{\phi - \theta_r}\right)^{1/m}\right)^m\right]^2$$

m = constant

K_s = saturated conductivity

θ = water content; φ = capillary suction (cm)
$K(\theta)$ = hydraulic conductivity for given water content (cm/hr)

Field application of Richards equation models requires that the parameters of the soil moisture characteristic relationships included in the model be specified a priori, or calibrated by comparing simulated and observed variables until some "optimum" set of parameters is decided upon. Some progress has been made in the estimation of the parameters of Richards equation–based models. Work carried out by Rawls and Brakensiek (1982; 1988; and Rawls et al., 1982), Clapp and Hornberger (1978), and Cosby et al. (1984) has attempted to relate soil hydraulic characteristics measured on a large number of soil samples to soil texture classes (eg. Figure 4) and other soil variables that are more usually measured during soil surveys. This approach would allow soil hydraulic parameters to be estimated a priori. However, it is worth noting that these estimates will be dependent on the measurement and sample collection techniques used in providing the data that was used in the regression analyses. The saturated hydraulic conductivity measurements carried out by the USDA, for example, have been routinely made on "fist sized" fragments and must consequently reflect the effects of macropores on saturated conductivity only poorly. In addition, the quoted estimates associated with a given texture class, or estimated by a regression equation using other variables, must be associated with a standard error of estimation, representing the uncertainty in the estimate. This uncertainty will often be large.

It is also worth noting that if Equation (1) is used as a basis for determining parameters from field or laboratory experiments, the neglect of additional factors, such as air and thermal transport, will mean that the parameter values derived in this way will be subject to the experimental conditions unless it can be shown that a single–phase flow model is adequate. Thus, these physical models are consequently not free of interaction between model parameters and model structure. This problem is compounded by the use of different numerical solution schemes for Equation (1), with different space and time steps (see for example, Calver and Wood, 1989).

The fact remains that nearly all the commonly used catchment–scale hydrological models use a conceptual storage–based approach to modeling the unsaturated zone. There is some justification for using such an approach. Most hydrological models require that their parameters be calibrated against an observed data set, usually discharge records. The discharge variable is usually the most readily available indicator of catchment response processes. It has some advantages in this respect in that it is a truly integrative variable, representing runoff production processes over the whole catchment. It is also often, but not always, the variable that is required to be predicted by a hydrological model. However, it should always be remembered that in many climates the discharge variable is not the major output variable from a catchment area, evapotranspiration being the dominant output component of the water balance. There is, of course, a significant link between evapotranspiration and unsaturated zone flow processes, and it would clearly be better to calibrate a model of the unsaturated zone on measurements of actual evapotranspiration, if they were available. However, they are not, except in some recent research studies, and the discharge variable must usually be used alone.

A discharge record may be considered as a series of hydrographs. Modeling a single hydrograph is not very difficult. All that is required is a loss function and a time transformation. This can be done with two or three parameters. Modeling a number of hydrographs will require some representation of the effect of initial conditions and rainstorm profiles on discharge production, which may justify the use of another two or three parameters. However, there will usually only be sufficient information in a discharge record to justify the calibration of four or five parameters. Adding more parameters will add increasing degrees of freedom for fitting the discharge curve but will be sure to result in identifiability problems in calibration, manifest in intercorrelated parameter values, and valleys, plateaux, and local minima in the response surface.

This implies that for hydrological models calibrated in this way, the description of the unsaturated zone must be parsimonious, i.e. it must represent the processes using only a small number of parameters. Models based on physical theory, which may involve four or five parameters in defining the soil moisture characteristic curves for the unsaturated zone component alone (even assuming a homogeneous porous medium), are not immune from this

144

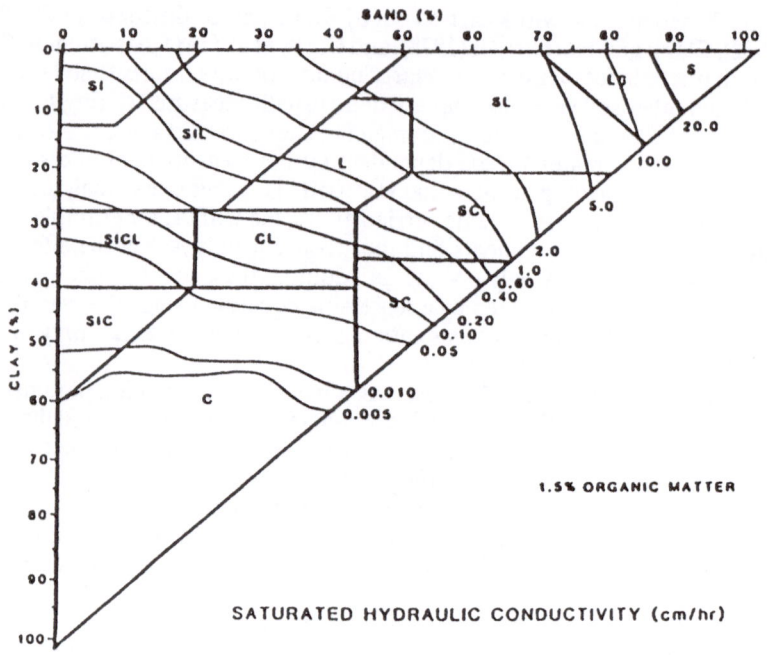

SATURATED HYDRAULIC CONDUCTIVITY (cm/hr)

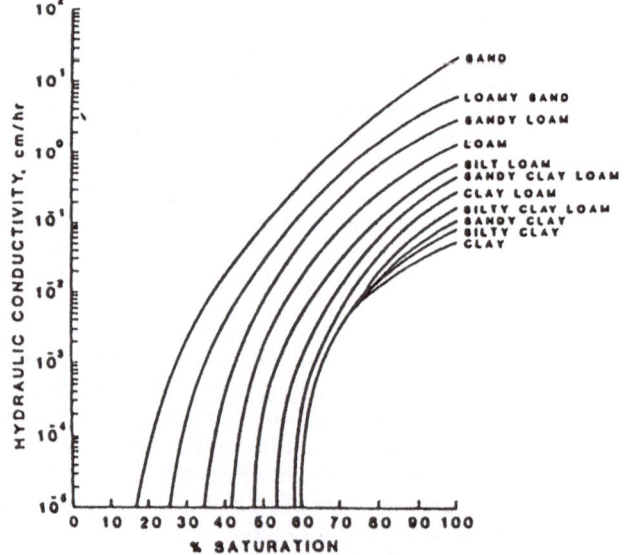

Unsaturated hydraulic conductivity curves
classified by soil texture.

Figure 4. Estimation of soil hydraulic characteristics from soil texture (after Rawls and Brakensiek, 1988).

conclusion if they are calibrated in this way (see the discussion in Beven, 1989). They may only avoid this problem if at least some of the parameters are defined a priori from physical information about the soil. It is not yet clear how successful such a priori estimates will be in making predictions at the catchment scale, but the case for using a parsimonious conceptual approach is clear. Such models have only few parameters, so identification of parameter values by comparison of observed and predicted discharges, given that the input data may not be error free, may be reasonably robust. Clearly the storage–based models also have the advantage of computational efficiency which may be an important criterion in making multiple runs for calibration or in simulating long periods of record.

Most conceptual storage models can, however, be shown to be incompatible with certain physical processes in the unsaturated zone that can be predicted by the Richards equation, such as upward capillary movement from a shallow water table or the effects of surface crusting on infiltration rates.

It would appear then that what is needed is a computationally efficient, parametrically parsimonious description of the unsaturated zone that incorporates much of the physical basis of the Richards equation, at least in an approximate way. Indeed, a number of attempts to produce such a model have been published (see for example Clapp, 1982; Milly, 1986; Charbeneau, 1988; Morel–Seytoux, 1988). Most of these studies have resulted in models that are simplified in terms of the computations required rather than the number of parameters involved.

2.3. Reality in the Unsaturated Zone

There is also a further problem that is not often addressed in attempts to model the unsaturated zone: the nature of reality in the unsaturated zone.

In Chapter 17, I outline a perceptual model of catchment response. A perceptual model is a subjective and personal set of impressions of the relevant hydrological processes. In general, any perceptual model will be too complex to describe using formal mathematics and consequently cannot be used for other than qualitative predictions. Let us consider the parts of this author's perceptual model that may be relevant to models of the unsaturated zone in an undisturbed soil profile (Figure 5).

The major inputs to the unsaturated zone generally occur as fluxes at the soil surface. The effect of a vegetation cover will often give rise to marked spatial heterogeneity in the input intensities over the soil surface due to the effects of stemflow and local variations in throughfall. Under some circumstances, at the soil surface the penetration of water into the soil may be governed by a very thin layer at the boundary which is associated with the surface litter, allowing only local penetration or surface crusting and resulting in restricted penetration. These sources of variability in the infiltration of water into the soil will interact with the local soil structure to determine flow pathways for the infiltrating water. For some soils, a significant part of the infiltrating water may follow flow pathways associated with soil macropores (cracks, root channels, earthworm channels and other continuous large voids) leading to rapid penetration to depth and bypassing of the bulk soil matrix. It is worth noting that where this occurs it may be particularly important for the transport of solutes and pollutants in the soil, as well as for the flow of water itself. Even within the soil matrix there may be distinct pathways of preferential flow at a number of different scales associated with continuous larger capillary size pores and local zones of higher hydraulic conductivity.

Following infiltration, patterns of redistribution will reflect both the heterogeneous patterns of infiltration and the influence of plant roots in extracting water to satisfy evapotranspiration needs. The roots introduce a further source of heterogeneity: the patterns of active roots varying in depth, space, and time. It must be remembered that the root system is not static, even for perennial vegetation covers, and that it may be possible for the roots to

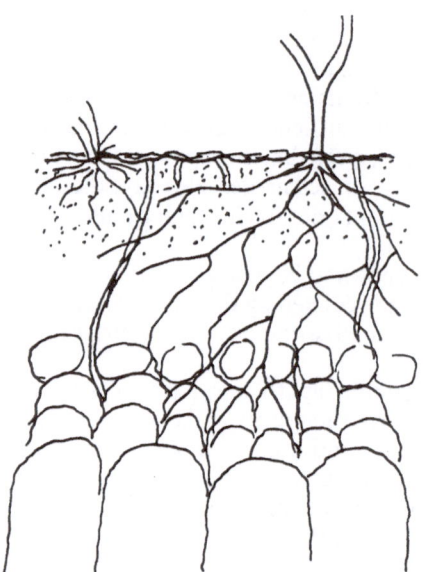

Figure 5. The soil profile.

grow towards wetter soil at a rate faster than Darcian matrix flow will allow the water to flow towards the roots.

2.4. Prospects for Modeling the Unsaturated Zone

This picture of reality in the unsaturated zone is one of complexity and heterogeneity. It is also one that does not conform to the assumptions underlying the formulation of the Richards' equation. There is no doubt that the Richards' equation is an adequate descriptor of unsaturated flow in homogeneous soil in the laboratory where the hydraulic properties of the soil can be adequately characterized (eg. Luthin et al., 1975; Nieber and Walter, 1981). It is not so clear, however, under what range of conditions it can be properly applied to describe the flow of water in heterogeneous field soils. The studies that have been done have almost invariably involved parameter calibration against observed state data which, as noted above, cannot result in an unequivocal justification of the model formulation.

What of the complex processes alluded to in the previous section? It is certain that there are some circumstances in which a Darcian description of the unsaturated zone is inadequate (see for example Luxmoore et al., 1981). Physically–based descriptions of these more complex processes, however, will inevitably involve further parameters, many of which would be very difficult to define or measure on any particular area of interest (such as: the distribution of input intensities at the ground surface for different vegetation types, rainstorm profiles, and wind speeds; the distributions, connectivities, and geometries of macropores and the way in which they change with time, especially in swelling clay soils with cracks; and the changing patterns of active roots, particularly in a growing crop). Consequently the problem of parameterization would be inevitably greatly increased.

To summarize the state of play so far, on the one hand we have physically–based models of the unsaturated zone that may not be adequate descriptions of the complex heterogeneous processes operating in the unsaturated zone but that may be overparameterized when the parameters are calibrated against measurements of bulk output variables such as stream discharge. The calibrated parameter values will then reflect both the model structure and errors in the observations so that any physical interpretation and extrapolation to other conditions must be made with great care. On the other hand, there are conceptual models with few parameters that may be parametrically parsimonious but which represent the hydrological processes of the unsaturated zone in only a crude functional way. Parameter values will then be very closely interlinked to model structure and will be site–specific so that extrapolation to other sites may not be possible.

What then of the prospects for modeling the unsaturated zone? Exploratory exercises into more complex physically–based descriptions have been made, including multiphase flow models (Green et al., 1970), models that incorporate the effects of macropores and preferential pathways (Hoogmoed and Bouma, 1980; Beven and Germann, 1981; Beven and Clarke, 1986; and Davidson, 1985) and models that attempt to reproduce the effects of spatial variability of soil properties (Dagan and Bresler, 1983; Yeh et al., 1985, Sharma et al., 1987; and Binley et al., 1989). Theoretically, there are few restrictions on incorporating further complexity into models of unsaturated flow. Numerical solutions may be necessary to obtain solutions to the descriptive equations, but as computer power increases this becomes less of a limitation. However, for practical purposes, the difficulties of obtaining sufficient data to characterize the parameters of the more complex descriptions are paramount and will not be overcome in the foreseeable future.

There have also been some interesting exercises in applying simple conceptual models of the unsaturated zone. Calder et al. (1983), for example, have applied a number of different model structures to predict soil moisture deficits for six different grassland sites in the UK. The model structures differed in the equations for estimating potential evapotranspiration (including the Penman, Priestley–Taylor, and Thom–Oliver formulations) and the functional relationships between actual and potential evapotranspiration rates. They found that the use of a "climatological mean" potential evapotranspiration rate (a simple annual sinusoidal variation about a mean rate constant for all sites and all years), together with a single parameter root constant model was successful for all the sites over all the years—even the extreme drought year of 1976 (Figure 6). Additional site specific meteorological data and the use of the more complex evapotranspiration equations did not prove to yield improved estimates of the measured soil moisture deficits. They conclude:

"To some extent the optimized root constant values will be incorporating the effects of interception, soil moisture drainage, deep abstraction and evaporation from the soil surface. It is the belief of the authors that until these effects are further investigated, and incorporated explicitly in SMD models, little advantage can be gained from further developments and increased sophistication of the meteorological aspects of these models." (p.351)

In view of the discussion above, the same conclusions might be drawn about increased sophistication in the modeling of the soil physics —*if only discharge or bulk soil moisture deficit data are available for use in calibration*. More complex parameterizations may only be justified where there are sufficient measured data available to allow calibration with acceptable confidence. Thus, the choice of a model structure for a particular application requiring prediction of actual data (rather than a purely hypothetical study) should take into account the type of data available for calibration. Use of a complex model, however physically–based, may not be justified without appropriate and adequate calibration data. It should also not be expected that physically–based models will necessarily give more accurate results than a simpler model structure if only bulk output variables (discharge or recharge rates) are of interest, especially if there is no independent means of assessing the magnitude of the variable (as in the case of recharge).

148

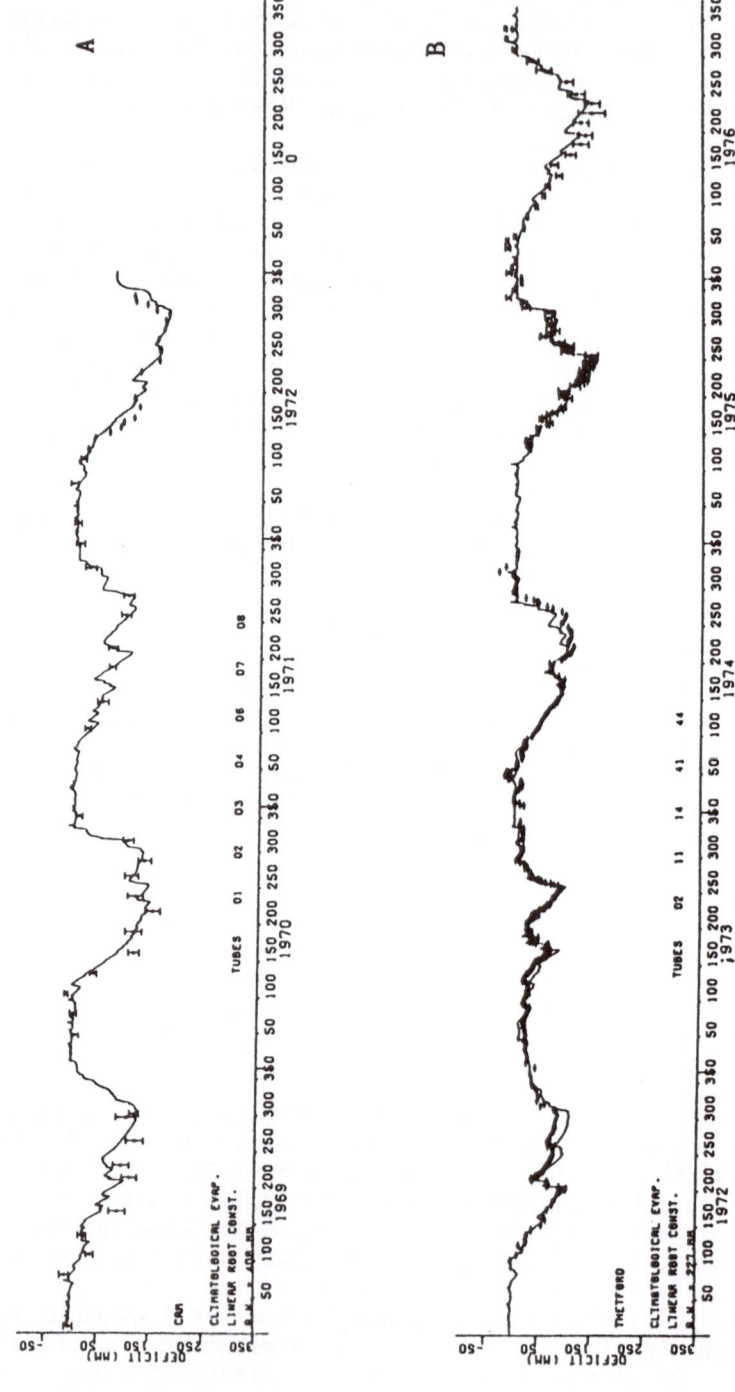

Figure 6. Comparison of predicted and observed soil moisture deficits for A. Cam, B. Thetford sites in U.K. (after Calder et al., 1983).

3. Conclusions

It is possible to subsume all the considerations of appropriate model structures discussed here into a unifying framework by allowing that all model predictions must be associated with some degree of uncertainty. This uncertainty will arise from a number of causes, including the inadequacies of a given model structure, errors in the input data, and errors in the data used for calibration. Modeling studies of hydrological processes, particularly for the unsaturated zone with all its complexities, must start to make realistic assessments of predictive uncertainty. Overparameterization of complex models calibrated against relatively few measured data will then be seen as a dramatic increase in predictive uncertainty. The provision of further data, whether measurements of the hydraulic properties of soils or time series of state variables, could be evaluated in terms of the costs and benefits of constraining the uncertainty associated with the predictions. In Chapter 17, I outline a general procedure to allow an assessment of uncertainties for nonlinear models in terms of likelihood functions in the context of predicting runoff production. It would be equally applicable to the problem of modeling the unsaturated zone, whether at the scale of a single profile, the field or hillslope, or complete catchments.

4. References

Beven, K. J.: 1977, 'Hillslope Hydrographs by the Finite Element Method,' *Earth Surface Processes* **2**, 13–28.

Beven, K. J.: 1989, 'Changing Ideas in Hydrology: The Case of Physically–Based Models,' *J. Hydrology* **105**, 157–172.

Beven, K. J., and M. J. Kirkby: 1979, 'A Physically–Based Variable Contributing Area Model of Basin Hydrology,' *Hydrological Sciences Bulletin* **24**, 303–325.

Beven, K. J., and P. F. Germann: 1981, 'Water Flow in Soil Macropores. II. A Combined Flow Model,' *J. Soil Science* **32**, 15–29.

Beven, K. J., and R. T. Clarke: 1986, 'On the Variation of Infiltration into a Homogeneous Soil Matrix Containing a Population of Macropores,' *Water Resources Research* **22**, 383–388

Beven, K. J., A. Calver, and E. M. Morris: 1987, *'The Institute of Hydrology Distributed Model,'* Institute of Hydrology Report No. 97, Wallingford, Oxon.

Binley, A. M., J. Elgy, and K. J. Beven: 1989, 'A Physically–Based Model of Heterogeneous Hillslopes. I. Runoff Production,' *Water Resources Research* **25**(6), 1219–1226.

Brooks, R. H., and A. T. Corey: 1964, *'Hydraulic Properties of Porous Media,'* Colorado State University Hydrology Papers, 3, 27 pp.

Calder, I. R., R. J. Harding, and P. T. W. Rosier: 1983, 'An Objective Assessment of Soil Moisture Deficit Models,' *J. Hydrology* **60**, 329–355.

Calver, A., and W. L. Wood: 1989, 'On the Discretisation and Cost Effectiveness of the Finite Element Solution for Hillslope Subsurface Flow,' *J. Hydrology* **110**, 165–179.

Campbell, G. S.: 1974, 'A Simple Method for Determining Unsaturated Conductivity from Moisture Retention Data,' *Soil Science* **117**, 311–314.

Charbeneau, R. J.: 1988, 'Liquid Moisture Redistribution: Hydrologic Simulation and Spatial Variability,' in H. J. Morel-Seytoux (ed.), *Unsaturated Flow in Hydrologic Modelling*, NATO ASI Series C, v 275, Reidel, 127–160.

Clapp, R. B.: 1982, *'A Wetting–Front Model of Soil Water Dynamics,'* Unpublished PhD Thesis, University of Virginia, Charlottesville.

Clapp, R. B., and G. M. Hornberger: 1978, 'Empirical Equations for Some Soil Hydraulic Properties,' *Water Resources Research* **14**, 601–604.

Cosby, B. J., G. M. Hornberger, R. B. Clapp, and T. R. Ginn: 1984, 'A Statistical Exploration of the Relationships of Soil Moisture Characteristics to the Physical Properties of Soils,' *Water Resources Research* **20**, 682–690.

Crawford, N. H., and R. K. Linsley: 1966, *'Digital Simulation in Hydrology, Stanford Watershed Model IV,'* Department of Civil Engineering, Stanford University, Technical Report No. 39.

Dagan, G., and E. Bresler: 1983, 'Unsaturated Flow in Spatially-Variable Fields. 1. Derivation of Models of Infiltration and Redistribution,' *Water Resources Research* **19**, 413–420.

Davidson, M. R.: 1985, 'Asymptotic Behaviour of Infiltration in Soils Containing Cracks or Holes,' *Water Resources Research* **21**, 1345–1353.

Dawdy, D. R., and T. O'Donnell: 1965, 'Mathematical Models of Catchment Behaviour,' *J. Hydraul. Div., ASCE* **91**, 123–137.

Feddes, R. A., S. P. Neuman, and E. Bresler: 1975, 'Finite Element Analysis of Two–Dimensional Flow in Soils Considering Water Uptake by Roots. II. Field Applications,' *Soil Sci. Soc. Amer. Proc.* **39**, 231.

Freeze, R. A.: 1971, 'Three Dimensional Transient Saturated-Unsaturated Flow in a Groundwater Basin,' *Water Resources Research* **7**, 347–366.

Freeze, R. A.: 1972, 'Role of Subsurface Flow in Generating Surface Runoff. 2. Upstream Source Areas,' *Water Resources Research* **8**, 1272–1283

Gardner, C. M. K., and J. P. Bell: 1980, *'Comparison of Measured Soil Moisture Deficits With Estimates by MORECS,'* IAHS Publication No. 130, 337–341.

Green, D. W., H. Dabiri, C. F. Weinaug, and R. Prill: 1970, 'Numerical Modelling of Unsaturated Groundwater Flow and Comparison of the Model to a Field Experiment,' *Water Resources Research* **6**, 862–874.

Hoogmoed, W. D., and J. Bouma: 1980, 'A Simulation Model for Predicting Infiltration into Cracked Soil,' *Soil Sci. Soc. Amer. J.* **44**, 458–461.

Huyakorn, P. S., E. P. Springer, V. Guvanasen, and T. D. Wadsworth: 1986, 'A Three-Dimensional Finite-Element Model for Simulating Water Flow in Variably Saturated Porous Media,' *Water Resources Research* **22**, 1790– 1808.

Jensen, K. H., and T. Jønch-Clausen: 1982, 'Unsaturated Flow and Evapotranspiration Modelling as a Component of the European Hydrological System (SHE),' in V. P. Singh, (ed.), *Modelling Components of the Hydrologic Cycle*, Water Resource Publications (P.O. Box 2841, Littleton, CO 80161–2841, U.S.A.), 235–252.

Ledoux, E., G. Girard, and J. P. Villeneauve: 1984, 'Proposition d'un Modèle Couplé Pour la Simulation Conjointe des Écoulements de Surface et des Écoulements Souterrains Sur Un Bassin Hydrologique,' *La Houille Blanche* 1/2, 1984.

Luthin, J. N., A. Orhun, and G. S. Taylor: 1975, 'Coupled Saturated-Unsaturated Transient Flow in Porous Media: Experimental and Numeric Model,' *Water Resources Research* **11**, 973–978.

Luxmoore, R. J., T. Grizzard, and M. R. Patterson: 1981, 'Hydraulic Properties of Fullerton Cherty Silt Loam,' *Soil Sci. Soc. Amer. J.* **45**, 692–698.

Milly, P. C. D.: 1982, 'Moisture and Heat Transport in Hysteretic Inhomogeneous Porous Media: A Matric Head-based Formulation and a Numerical Model,' *Water Resources Research* **18**, 489–498.

Milly, P. C. D.: 1986, 'An Event-based Simulation Model of Moisture and Energy Fluxes at a Bare Soil Surface,' *Water Resources Research* **22**, 1680–1692.

Morel-Seytoux, H. J.: 1988, 'Recipe for Simple But Physically-based Modeling of the Infiltration and Local Runoff Processes,' in H. J. Mortel-Seytoux, and D. G. DeCoursey (eds.), *Proc. 8th AGU Front Range Branch Hydrology Days*, 226–247, Hydrology Days Publications, Fort Collins, CO.

Morel-Seytoux, H. J., and J. Khanji: 1974, 'Derivation of an Equation of Infiltration,' *Water Resources Research* **10**, 794–800.

Mualem, Y.: 1977, 'Extension of the Similarity Hypothesis Used for Modelling Soils Characteristics,' *Water Resources Research* **13**, 773–780.

Mualem, Y.: 1984, 'Prediction of the Soil Boundary Wetting Curve,' *Soil Science* **137**, 379–390

Narasimhan, T. N., and P. A. Witherspoon: 1977, 'Numerical Model for Saturated-Unsaturated Flow in Deformable Porous Media. I. Theory,' *Water Resources Research* **13**, 657–664.

Neuman, S. P.: 1973, 'Saturated-Unsaturated Seepage by Finite Elements,' *J. Hydraul. Div., ASCE* **99**, 2233–2250.

Nieber, J. L., and M. F. Walter: 1981, 'Two-dimensional Soil Moisture Flow in a Sloping Rectangular Region: Experimental and Numerical Studies,' *Water Resources Research* **17**, 1722–1730.

Philip, J. R.: 1957, 'The Theory of Infiltration. 4. Sorptivity and Algebraic Infiltration Equations,' *Soil Science* **84**, 257–264.

Rawls, W. J., and D. L. Brakensiek: 1982, 'Estimating Soil Water Retention From Soil Properties,' *J. Irrig. Drain., ASCE* **108**, 166–171.

Rawls, W. J., and D. L. Brakensiek: 1988, 'Estimation of Soil Hydraulic Properties,' in H. J. Morel-Seytoux (ed.),*Unsaturated Flow in Hydrologic Modelling*, Reidel, 275–300.

Rawls, W. J., D. L. Brakensiek, and K. E. Saxton: 1982, 'Estimation of Soil Water Properties,' *Trans. Amer. Sco. Agric. Eng.* **25**, 1316–1320, 1328.

Richards, L. A.: 1931, 'Capillary Conduction of Liquids Through Porous Mediums,' *Physics* **1**, 318–333.

Rulon, J. J., R. Rodway, and R. A. Freeze: 1985, 'The Development of Multiple Seepage Faces on Layered Slopes,' *Water Resources Research* **21**, 1625– 1636.

Sharma, M. L., R. J. Luxmoore, R. De Angelis, R. C. Ward, and G. T. Yeh: 1987, 'Subsurface Water Flow Simulated for Hillslopes With Spatially Dependent Soil Hydraulic Characteristics,' *Water Resources Research* **23**, 1523–1530.

Smith, R. E., and V. A. Ferreira: 1989, 'A Comparison of Unsaturated Zone Model Components,' in H. J. Morel–Seytoux (ed.), *Unsaturated Flow in Hydrologic Modelling*, NATO ASI Series C, v 275, Reidel, 391–412.

Smith, R. E., and J-Y. Parlange: 1978, 'A Parameter–Efficient Hydrologic Infiltration Model,' *Water Resources Research* **14**, 533–538.

Van Genuchten, M. Th.: 1980, 'Predicting the Hydraulic Conductivity of Unsaturated Soils,' *Proc. Soil Sci. Soc. Amer.* **44**, 892–898.

Yeh, T. C. J., L. W. Gelhar, and A. L. Gutjhar: 1985, 'Stochastic Analysis of Unsaturated Flow in Heterogeneous Soils (2 Papers),' *Water Resources Research* **21**, 447–464.

Chapter 8

Overland Flow: A Two-Dimensional Modeling Approach

Ezio Todini and **M. Venutelli**
E. Todini & Partners, S. r. l.
Via Zanardi, 16
40131 Bologna, Italy

Abstract. The paper deals with the problem of overland flow by introducing the equations that describe the two–dimensional motion of water on a plane and the hypotheses that allow it to reach simplified models. The differential equations obtained in continuous form are then discretized by means of spatial discretization methods such as the finite difference, the integrated finite difference, or the finite element methods. Finally a comparative analysis of the performances of different numerical algorithms used for solving the overall system of equations, obtained after discretization in time and space, is presented in order to evaluate their speed and their computer memory requirements.

1. Introduction

The aim of this chapter is to describe the equations governing two–dimensional overland flow and, in particular, the numerical schemes used to resolve them.

The establishment of a plan and the definition of possible actions aimed to the improvement of the hydraulic conditions of a given flood prone area, in order to mitigate the risk and the effects of flooding, requires the knowledge of the hydraulic levels that could be reached during extreme flood events. In view of its importance, many studies – which will be referred to later – have been devoted to this subject.

This chapter focuses primarily on the assumptions that have led to a class of simplified models, the parabolic models, which seem quite adequate for the solution of most problems encountered in practice.

2. Discussion

Sections 2.1 and 2.2 of this chapter include basic equations in the continuum, according to the Eulerian theory, for the description of phenomena occurring in fluids. The aim of the Eulerian approach is to determine the characteristics of flow (dependent variables) by investigating fixed positions at each point in the space occupied by a fluid; thus velocity and pressure, and – where relevant – density, are a function of the coordinates of the investigation point and of time (independent variables). This concept is set against the other theory, propounded by Lagrange, according to which the phenomena are studied by analyzing the behavior of a given fluid particle in its motion through space; i.e. the coordinates of the particle (dependent variables) are defined as functions of given initial values of time (independent variables).

D. S. Bowles and P. E. O'Connell (eds.), Recent Advances in the Modeling of Hydrologic Systems, 153–166.
© 1991 *Kluwer Academic Publishers.*

Sections 2.3 through 2.6 of the chapter is concerned with an examination of numerical methodologies, in particular the methods of finite differences, integrated finite differences, and finite elements for solving the approximate model referred to.

In Section 3, a discussion is made of algorithms used to resolve the resulting system of equations, considering the possible tradeoffs between the use of larger computer memory and computer time requirements.

2.1. Basic Equations of Mass and Momentum Balance

The indefinite equations of mass and momentum balance, expressing respectively the principle of the conservation of mass and the conditions of dynamic equilibrium at each point in time for an incompressible fluid are, according to the classic literature (Citrini and Noseda, 1975; Marchi and Rubatta, 1981; and others), as follows:

$$Div \ \mathbf{v} = 0 \tag{1}$$

$$\frac{d\mathbf{v}}{dt} = \mathbf{F} - \frac{grad\mathbf{p}}{\rho} + \nu \ \nabla^2 \ \mathbf{v} \tag{2}$$

Equation (2) is commonly known as the Navier–Stokes equation. Applying the rule of Eulerian derivation, Equations (1) and (2), in terms of components, can be written as:

$$\frac{\partial u}{\partial x} + \frac{\partial v}{\partial y} + \frac{\partial w}{\partial z} = 0 \tag{3}$$

$$\frac{\partial u}{\partial t} + u\frac{\partial u}{\partial x} + v\frac{\partial u}{\partial y} + w\frac{\partial u}{\partial z} = F_1 - \frac{1}{\rho}\frac{\partial p}{\partial x} + \nu \ \nabla^2 \ u \tag{4}$$

$$\frac{\partial v}{\partial t} + u\frac{\partial v}{\partial x} + v\frac{\partial v}{\partial y} + w\frac{\partial v}{\partial z} = F_2 - \frac{1}{\rho}\frac{\partial p}{\partial y} + \nu \ \nabla^2 \ v \tag{5}$$

$$\frac{\partial w}{\partial t} + u\frac{\partial w}{\partial x} + v\frac{\partial w}{\partial y} + w\frac{\partial w}{\partial z} = F_3 - \frac{1}{\rho}\frac{\partial p}{\partial z} + \nu \ \nabla^2 \ w \tag{6}$$

where u,v,w and F_1, F_2, F_3 represent, respectively, the components of velocity and mass force referred to the mass unit in directions x, y and z; p, ν, ρ and t are respectively pressure, kinematic viscosity, fluid density and time; the symbol ∇^2 is the Laplace operator. The indefinite equation of momentum balance (2), completed by the equation of continuity of mass (1), by the equation of state (which for an incompressible fluid is ρ = cost), by the relations between stresses and strains which—still under the assumption of an incompressible fluid—make it possible to arrive at the equation of motion in the form (2), and by the boundary and initial conditions typical of the specific process of motion considered enables the determination of the characteristic elements of motion at each point in time, providing a complete and detailed description.

This result, which is certainly the most complete solution of the dynamic problem, is extremely difficult to achieve because of the many obstacles encountered in the integration of the system of differential equations with partial derivatives on real world domains. On the other hand it should also be noted that such a detailed description of the phenomena is generally not required vis-à-vis the solution of many practical problems.

2.2. Partial Differential Equations of Two-Dimensional Free Surface Flow

The basic partial differential equations on which this chapter concentrates may be derived from the Equations (3) to (6) describing mass and momentum balance.

Integration of the Equation (3) over the liquid height h of a generic cross section and applying the Leibnitz rule, as in Pinder and Gray (1977), Chen and Chow (1971), Chow and Ben–Zvi (1973), Lai (1986), Di Silvio (1978), leads to:

$$\frac{\partial h}{\partial t} + \frac{\partial(Uh)}{\partial x} + \frac{\partial(Vh)}{\partial y} = q \tag{7}$$

where possible inflows and/or outflows (rainfall, infiltration, evaporation, etc...) were considered and where U and V represent the mean velocities along the vertical depth in the x and y direction, respectively.

In the case of almost horizontal flows, the vertical component of velocity and acceleration can be ignored, and under this assumption (Lai, 1986) Equation (6) reduces to the form:

$$\rho \, F_3 - \frac{\partial p}{\partial z} = 0 \quad ; \quad F_3 = -g \tag{8}$$

Integrating with respect to z and selecting the atmospheric pressure p_0 over the water surface as the integration constant, leads to:

$$p = \rho g \, (H - z) + p_0 \tag{9}$$

where H is the water surface elevation. Then, using Equation (9) and assuming p_0 to be everywhere constant, the following expressions can be found:

$$\frac{\partial p}{\partial x} = \rho \, g \, \frac{\partial H}{\partial x}$$

$$\tag{10}$$

$$\frac{\partial p}{\partial y} = \rho \, g \, \frac{\partial H}{\partial y}$$

Equations (4) and (5) can be integrated over the liquid height h, as with the equation of continuity of mass, and applying once more the Leibnitz rule and taking into account Equation (10) with the further assumption $F_1 = F_2 = 0$, the equations of momentum balance become:

$$\frac{\partial(Uh)}{\partial t} + \frac{\partial(U^2h)}{\partial x} + \frac{\partial(UVh)}{\partial y} + g \, h \, \frac{\partial h}{\partial x} = g \, h \, (S_{0x} - S_{fx}) + D_{1x} \tag{11}$$

$$\frac{\partial(Vh)}{\partial t} + \frac{\partial(V^2h)}{\partial y} + \frac{\partial(UVh)}{\partial x} + g\,h\,\frac{\partial h}{\partial y} = g\,h\,(S_{0y} - S_{fy}) + D_{1y} \tag{12}$$

where S_{0x} and S_{0y} are the components of the bottom slope in the two directions x and y, respectively; S_{fx} and S_{fy} are the head loss components in the same directions; D_{lx} and D_{ly} account for possible lateral inflows and/or outflows (Akanbi and Katopodes, 1988).

A detailed examination of the equations of mass (7) and of momentum balance (11) and (12) are given, for example, in Abbott (1979), Cunge (1975), Abbott and Cunge (1975), Daubert and Graffe (1967), Akan and Yen (1981), and Hromadka II et al. (1986).

Moreover, the problem of numerical solution of these equations has been addressed by many authors. An extensive study of the various methods is provided by Lai (1986) and Katopodes and Strelkoff (1978, 1979). The method of finite differences can be found in Grubert's study (1976) and more recently in Fennema and Chaudhry (1989). The method of integrated finite differences can be found in Narasimhan and Witherspoon (1976), Narasimhan et al. (1978), Pacciani (1987), Calo' (1987). The method of finite elements can be found not only in classic texts (Pinder and Gray, 1977; Gallagher et al., 1975), or in a general overview paper such as the one by Cheng (1978), but also in Fugazza and Gallati (1977), Katopodes (1980), Katopodes (1984), Hosseinipour and Amein (1984), Vansnick and Zech (1984), and Akanbi and Katopodes (1988).

In the present chapter, to provide a general overview of the problem while reducing the complexity of the derivation, following the works undertaken by Xanthopoulos and Koutitas (1976), and by Hromadka II et al. (1985) and J.D. De Vries et al. (1986), Equations (7), (11) and (12) have been simplified; in particular, in the equations of continuity of momentum per unit of weight, Equations (11) and (12) the dynamical terms (i.e. the derivatives of momentum with respect to time) as well as the kinetic terms (i.e. the derivatives of velocity with respect to x and y) can been neglected without a great loss of accuracy in the description of the two dimensional flow pattern. By ignoring, without loss of generality, the terms D_{lx} and D_{ly} resulting from possible inflows and/or outflows, a simplified form of the equation of momentum can be found in which the surface water slope coincides with the friction slope.

$$\frac{\partial h}{\partial x} = S_{0x} - \frac{n_x^2 U^2}{h^{4/3}} \tag{13}$$

$$\frac{\partial h}{\partial y} = S_{0y} - \frac{n_y^2 V^2}{h^{4/3}} \tag{14}$$

Obtaining from these latter two equations the velocity components U and V and replacing them in the equation of mass balance (7), the following final expression is obtained:

$$\frac{\partial}{\partial x}\left(K_x \frac{\partial H}{\partial x}\right) + \frac{\partial}{\partial y}\left(K_y \frac{\partial H}{\partial y}\right) + q = \frac{\partial H}{\partial t} \tag{15}$$

where:

$$K_x = \frac{\frac{h^{5/3}}{n_x}}{\left(\frac{\partial H}{\partial x}\right)^{1/2}} \qquad ; \qquad K_y = \frac{\frac{h^{5/3}}{n_y}}{\left(\frac{\partial H}{\partial y}\right)^{1/2}} \tag{16}$$

with q representing external net flow exchanges (inflows–outflows)

In this chapter the solution of eq. (15) or of the equivalent system (7), (13), (14) are based upon the method of finite differences, the method of finite integrated differences and the method of finite elements.

2.3. The Method of Finite Differences

Under this method, the domain is discretized in a finite number of points which form a regular mesh grid in which the approximate value of the solution is determined, as shown in the classic literature (Abbott, 1979; Lapidus and Pinder, 1982). In the following case, since the unknown h varies with the spatial coordinates x and y and with time t, the scheme is the one shown in Figure 1 in which the spatial coordinates are indicated with the symbol i for the x direction and symbol j for the y direction; time is counted with n multiples at interval Δt.

Referring to the system (7), (13), (14), the equation of continuity (7), using a general form of the finite temporal differences, integrated between the instant $n.\Delta t$ and the instant $(n+1).\Delta t$, takes the form:

$$\frac{h_{i,j}^{n+1} - h_{i,j}^{n}}{\Delta t} + \varepsilon \left(\frac{\partial(Uh)}{\partial x} + \frac{\partial(Vh)}{\partial y} - q \right)^{n+1} = (1-\varepsilon) \left(\frac{\partial(Uh)}{\partial x} + \frac{\partial(Vh)}{\partial y} - q \right)^{n} = 0 \quad (17)$$

with q accounting for external inflows and/or outflows.

The method becomes explicit for $\varepsilon = 0$; in this case the unknown height h at interval $(n+1)$ Δt, can be expressed explicitly as a function of known quantities at time n Δt; nevertheless the procedure requires that certain stability conditions be satisfied. On the other hand, if $\varepsilon > 0$, the unknown height is defined implicitly; this method involves more complex computations but it does offer the advantage of furnishing stable solutions (with $0.5 \leq \varepsilon \leq 1$). Lastly, in the case of $\varepsilon = 0.5$ the centered implicit scheme (Crank–Nicolson) is obtained, in which the disadvantage of relatively complicated calculations is compensated by an enhanced degree of accuracy of the solution.

For the replacement of spatial partial derivatives with finite incremental relations, again with reference to Figure 1, one of the following forms can be used:

In these latter expressions there is a truncation error connected with the substitution of the tangent at point (i,j) with the chord; in Equation (18) this is centered around the point itself; in

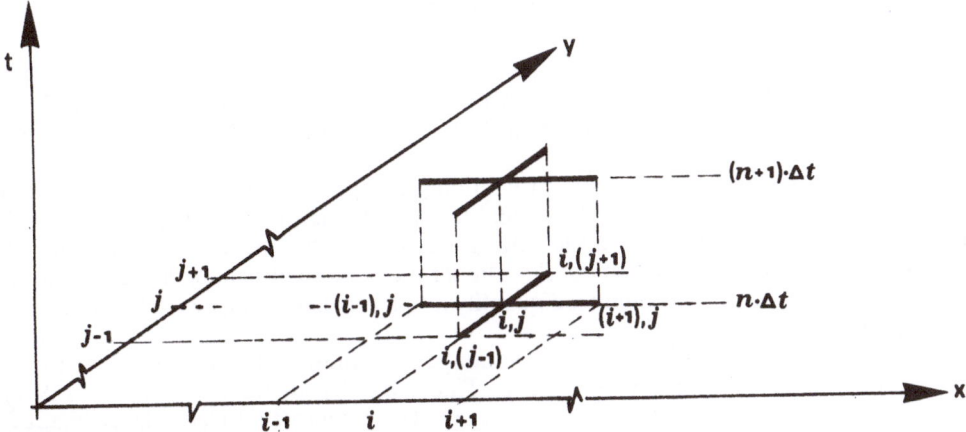

Figure 1. The finite differences mesh.

Equation (19) it refers to the previous Δx interval, and, lastly, in Equation (20) it refers to the next interval. Similar expressions may be written for the other spatial coordinate y:

$$\frac{\partial(Uh)}{\partial x} \cong \frac{1}{2\Delta x}[(Uh)_{i+1,j} - (Uh)_{i-1,j}] \tag{18}$$

$$\frac{\partial(Uh)}{\partial x} \cong \frac{1}{\Delta x}[(Uh)_{i,j} - (Uh)_{i-1,j}] \tag{19}$$

$$\frac{\partial(Uh)}{\partial x} \cong \frac{1}{\Delta x}[(Uh)_{i+1,j} - (Uh)_{i,j}] \tag{20}$$

The values of the velocity components U_{ij} and $V_{i,j}$ can be obtained, once more by applying one of schemes, (18), (19), and (20), from Equations (13) and (14). As far as the water free surface depth values h are concerned, the average values within the interval can be assumed.

2.4. The Method of Integrated Finite Differences

Equation (15) is integrated over a surface element with an area S_i and boundary H.

$$\int_{S_i}\left[\frac{\partial}{\partial x}\left(K_x\frac{\partial H}{\partial x}\right) + \frac{\partial}{\partial y}\left(K_y\frac{\partial H}{\partial y}\right)\right] dS_i + \int_{S_i} q \ dS_i = \frac{\partial}{\partial t}\int_{S_i} H \ dS_i \tag{21}$$

Applying the theorem of divergence, the first term on the left part of Equation (21) is transformed into a line integral. For the remaining two elements in the expression it is assumed that q_i and H_i are the mean values assumed by the respective quantities over S_i, thus obtaining:

$$\int_L\left(K_x\frac{\partial H}{\partial x}\mathbf{i} + K_y\frac{\partial H}{\partial y}\mathbf{j}\right) dL + q_i \ S_i = S_i\frac{\partial H_i}{\partial t} \tag{22}$$

The basic aim of integrated finite differences is to discretize the total domain of the flow in subdomains or conveniently small "elements" and to assess the mass balance in each element as shown in Equation (22).

In physical terms, the line integral in Equation (22) represents the total flow through the boundary of the element, whereas the second term, on the left hand side in Equation (22), represents the external flow exchanges. Therefore, the first element represents, globally, the accumulation of mass within element S_i. The term on the right hand side transforms, on element S_i, represents the accumulated quantity into a corresponding mean variation of the water potential.

In order to illustrate the method, the whole domain is divided into polygonal cells [Dirichlet tassellation, better known to hydrologists as Thiessen polygons (Sloan, 1987; Bear, 1979)], whose mean properties are associated with the nodal point i inside the element (see Figure 2). To ensure the highest degree of precision, the interfaces between the elements must be

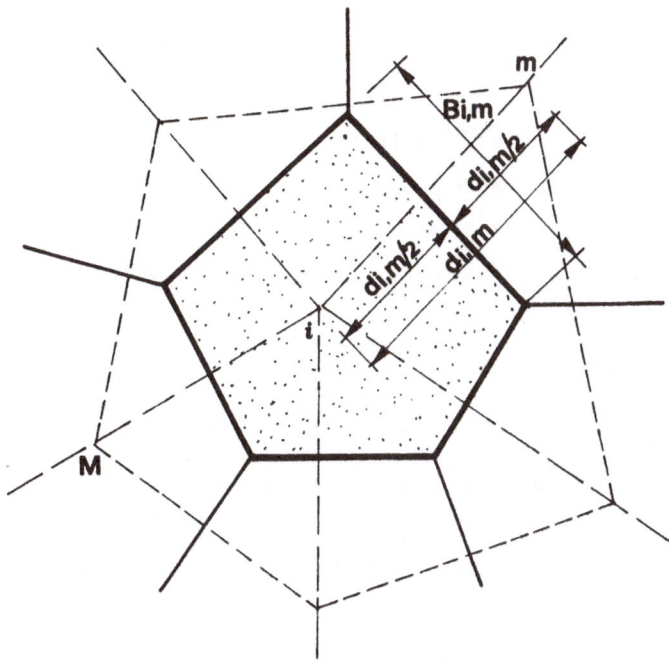

Figure 2. The integrated finite differences mesh.

perpendicular to the lines joining the nodal points and they must intersect such lines at their median point (Bear, 1979; Narasimhan and Witherspoon, 1976).

It is further assumed that the function H changes linearly among the nodal points and that its point value coincides with the mean value. Therefore Equation (22) becomes:

$$\sum_{m=1}^{M} K_{i,m}\frac{H_m - H_i}{d_{i,m}}B_{i,m} + q_i \ S_i = S_i\frac{\partial H_i}{\partial t} \tag{23}$$

where $d_{i,m}$ is the distance between the nodes i and m, $K_{i,m}$ is the inverse of the resistance along the line $d_{i,m}$, and $B_{i,m}$ is the breadth of the interface between the cell i and the cell m.

Equation (23) can be rewritten as:

$$\sum_{m=1}^{M} U_{i,m}(H_m - H_i) + q_i \ S_i = S_i\frac{\partial H_i}{\partial t} \tag{24}$$

where $U_{i,m} = (K_{i,m} \ B_{i,m})/d_{i,m}$ is the conductivity of the interface of separation between the elements i and m and represents the quantity of fluid transferred for a unit potential difference between the nodal points i and m (Narasimhan and Witherspoon, 1976).

With regard to discretization over time, Equation (24) takes the following general form:

$$\sum_{m=1}^{M} [\varepsilon \ U_{i,m}^{t+\Delta t} + (1-\varepsilon) \ U_{i,m}^{t}]$$
$$\left\{ [\varepsilon \ H_m^{t+\Delta t} + (1-\varepsilon) \ H_m^{t}] - [\varepsilon \ H_i^{t+\Delta t} + (1-\varepsilon) \ H_i^{t}] \right\} + \tag{25}$$
$$+ [\varepsilon \ q_i^{t+\Delta t} + (1-\varepsilon) \ q_i^{t}] \ S_i = \frac{S_i}{\Delta t} (H_i^{t+\Delta t} - H_i^{t})$$

With $\varepsilon = 0$, $\varepsilon = 1$ and $\varepsilon = 0.5$, eq. (25) leads to previously cited schemes: explicit, fully implicit, and centered (Crank–Nicolson).

2.5. The Method of Finite Elements

According to this method, as abundantly described in the classic literature (Zienkiewicz, 1977; Dhatt and Touzot, 1985; Becker et al., 1986), the domain of definition of the problem is divided into a certain number of subdomains or elements; the intersection points of the division lines (mesh) are called nodes. At these points the approximate value of the solution is determined; with this value, by means of linear combinations, it is possible to obtain the value of the unknown function, again in approximate form, at every point in the domain. Equation (15) can be written in the form:

$$f(H) = \frac{\partial}{\partial x}\left[K_x \ (H)\frac{\partial H}{\partial x} \right] + \frac{\partial}{\partial y}\left[K_y \ (H)\frac{\partial H}{\partial y} \right] + q - \frac{\partial H}{\partial t} \tag{26}$$

and, defining with:

$$H^*(x,y,t) = \sum_{j=1}^{N} H_j \ (t) \ \Phi_j \ (x,y,) \tag{27}$$

an unknown approximation of the solution sought $H(x,y,z)$, given N known base functions $\Phi_j(x,y)$, and the N unknown nodal values H_j, then the application of the Galerkin method leads to the imposition of orthogonality condition between the residue $f(H^*)$ and the N base functions:

$$\int_{\Omega} f(H^*) \ \Phi_i \ d\Omega = 0 \qquad\qquad i = 1, 2,..., N \tag{28}$$

which can be written in full as:

$$\int_{\Omega}\left\{ \frac{\partial}{\partial x}\left[K_x \ (H^*)\frac{\partial H^*}{\partial x} \right] + \frac{\partial}{\partial y}\left[K_y \ (H^*)\frac{\partial H^*}{\partial y} \right] + q - \frac{\partial H^*}{\partial t} \right\} \ \Phi_i \ d\Omega = 0 \tag{29}$$
$$i = 1, 2,..., N$$

Figure 3 shows the general base function $\Phi_i(x,y)$ for the i–th node and all the functions taking the form $\Psi_{i,\ell}(x,y)$ which constitute it (Becker et al., 1986). The example in the figure refers to triangular linear elements which offer the advantage, over quadrangular elements for

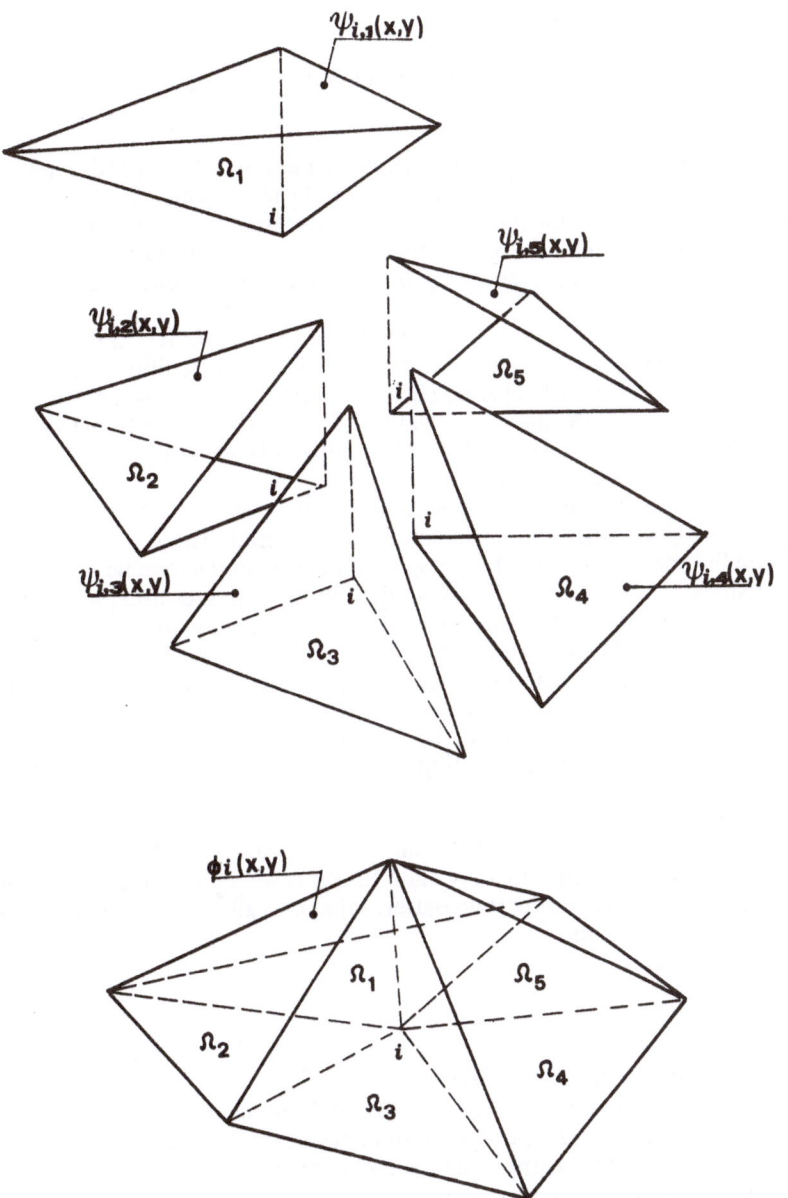

Figure 3. The finite elements base function representation.

example, of complete interpolating polynomials to the maximum degree which increases the approximation of the method as described in Becker et al. (1986) and Reddy (1986).

Resuming the procedure, first the formula of integration by parts (1st identity of GREEN) is applied to Equation (29) to eliminate the second derivatives and, then in view of the non-linearity, the equation is resolved by using one of the interative methods, such as Newton-Raphson or others (Huyakorn and Pinder, 1983).

2.6. Solution of the Resulting System of Nonlinear Equations

From a mathematical point of view, the integration in time of the system of partial differential equations describing mass and momentum balance, is generally transformed, at each step in time, into the solution of a system of linear ordinary equations (explicit solution for $\varepsilon = 0$), or nonlinear ordinary equations (implicit solution for $0 < \varepsilon \le 1$), via one of the above mentioned approaches, namely the finite differences, the integrated finite differences, or the finite elements.

In general, for two dimensional problems, the implicit scheme is preferred to the explicit since it allows for larger integration time steps, without giving rise to stability problems.

Therefore, when dealing with implicit schemes since a direct solution for systems of non-linear ordinary equations is generally not available, the usual approach is to transform the original problem into a sequence of linear problems, the solution of which will iteratively converge to the solution of the nonlinear system.

As a result, for solving a system of nonlinear equations it is necessary to solve a number of linear systems; consequently, very efficient linear solvers are required. These algorithms will be mainly based upon the knowledge both of topological and structural characteristics (thus taking advantage of sparsity, symmetry and positive definition) of the resulting matrix.

For the solution of a finite differences scheme, which is based on square or rectangular elements, many algorithms have been developed in the past. The interested reader may find in Abbott (1979) a deeper insight in the numerical methods used for the solution of the systems of equations resulting from finite differences schemes.

However, more recent approaches tend to concentrate on the description of two-dimensional problems by means of integrated finite differences or finite element schemes which are usually based upon a mesh of nodes connected as to form irregular triangular elements (Hromadka II et al., 1985; Katopodes, 1984).

The topology of the resulting matrix, extremely similar in the case of both the integrated finite differences and the finite elements, does not allow for the relatively easy methods found in the case of finite differences, but shows certain patterns that may allow for the definition of the most suitable solution algorithms.

The resulting matrix, in fact, is:
· Sparse
· Symmetric
· Positive definite
· Stieltjes type for the integrated finite differences

The rank of the matrix is equal to the number of nodes and the number of independent non zero elements equals the number of nodes plus the number of sides of the triangular elements.

The sum of the absolute value of the extra–diagonal terms is smaller–equal to the absolute value of the corresponding principal diagonal term.

Moreover, in the case of the matrix resulting from the integrated finite differences scheme, all the extra–diagonal terms are smaller–equal to zero, while the diagonal terms are positive (which defines a Stieltjes matrix) (Ortega and Rheinboldt, 1970).

From this very brief analysis the following conclusions can be derived:

(1) The method used for the solution of the resulting linear equations system must account for sparsity, symmetry, and positive definition.

(2) The method can be an iterative method (Jacoby, Gauss–Seidel). (Although these methods proved to be not particularly efficient even when optimal over–relaxation coefficients are used to accelerate convergence [Stoer and Burlish, 1980].)

(3) The conjugate gradient method (and in particular the Incomplete Cholesky Modified Conjugate Gradient) which also resorts into an iterative scheme is found to be the most efficient with respect to computer memory requirements (Ajiz and Jennings, 1984; Kershaw, 1978).

(4) The direct methods (Gauss, Cholesky Factorization), provided that an internal re-numbering of the nodes takes place, in order to minimize the fill-in of the original matrix (the number of non–zero elements may greatly increase, thus reducing the benefits of sparcity) are the most efficient from the point of view of computational time requirements (Duff and Reid, 1982; Duff and Reid, 1982; Salgado, 1988).

Incomplete Cholesky Modified Conjugate Gradient (ICMCG) and the MA27 – Harwell Library subroutine based on a Complete Cholesky factorization preceded by a renumbering of nodes--are shown in Figures 4 and 5, respectively.

3. Conclusions

Overland flow problems can be subdivided into two main categories:
- Runoff concentration on catchment slopes during rain events
- Flooding of flood prone areas

The two problems can be studied by means of simplified parabolic models as opposed to the fully dynamic hyperbolic descriptions. In particular, two dimensional routing of surface runoff due to excess rainfall, which considers a very thin layer of water, must be studied by means of parabolic models to avoid enhanced stability problems.

The spatial description can be obtained via one of the classical schemes, namely the finite differences, the integrated finite differences, and the finite elements; although, in recent years a preference has been placed by many authors on the last two, which can better represent

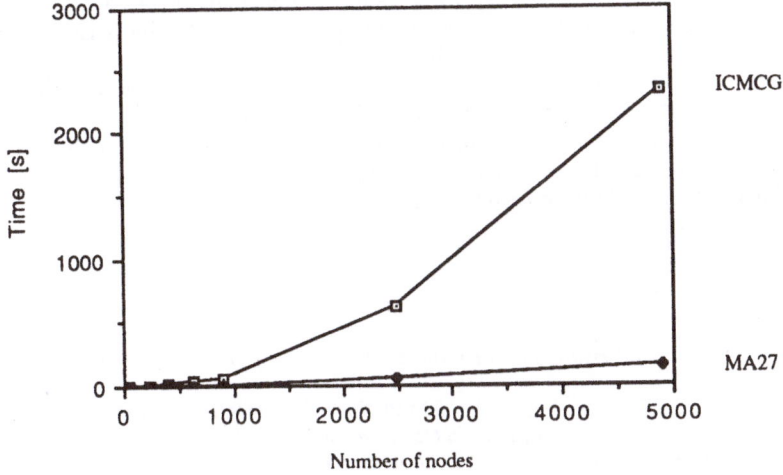

Figure 4. Comparison of time requirements.

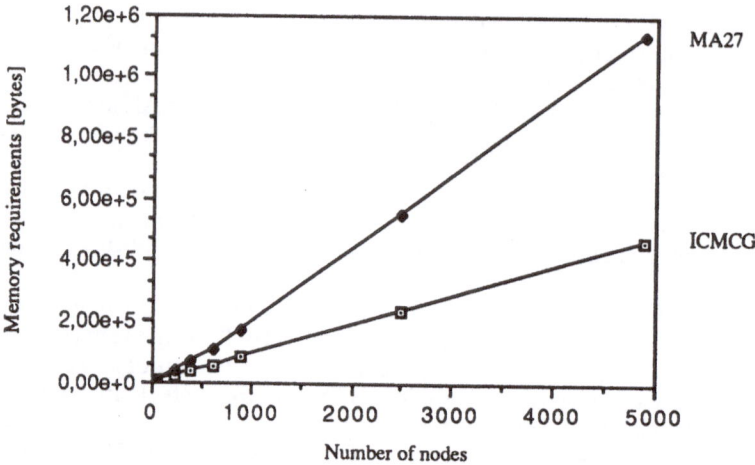

Figure 5. Comparison of size requirements.

non–homogeneous and non–hysotropical conditions deriving from the topographical and the geomorphological nature of the integration domain.

The application of integrated finite differences or of finite elements requires the use of linear systems solvers particularly suited to the topological and structural characteristics of the matrices originated by the selected scheme. In particular, direct methods combined to an internal node renumbering show to be the most suitable in terms of time consumption, while the incomplete Cholesky factorization modified Conjugate Gradient show to be the most suitable in terms of computer memory requirements.

With the advent of parallel processing at low cost (100 MFlops with 8 parallel processors for less than 25,000 $US), however, these latter statements could be overruled by new system solvers or even new integration techniques for differential equations, such as the "molecular automata" techniques that proved to be extremely efficient on parallel computers (Higuera et al., 1989).

This is probably the domain in which larger efforts should be placed in the nearby future to increase the size of problems that can be solved, also taking into account the increased availability of satellite imagery combined to Geographical Information Systems.

The use of these techniques on the one hand will improve and simplify the description of the integration domain and its characteristics, and on the other hand will enhance the understanding of the results that can be immediately displayed on maps or graphically overimposed to digital terrain model outputs, thus obtaining three–dimensional representations of possible flood prone areas. This information can be extremely useful to planners or decision makers.

4. References

Abbott, M. B.: 1979, *Computational Hydraulics. Elements of the Theory of Free Surface Flow*, Pitman Publishing, London.

Abbott, M. B., and J. A. Cunge: 1975, 'Two–Dimensional Modeling of Tidal Deltas and Estuaries,' in K. Mahmood, and V. Yevjevich (eds.), *Unsteady Flow in Open Channels*, Vol. II, 763–812, Fort Collins.

Ajiz, M., and A. Jennings: 1984, 'A Robust Incomplete Cholesky Conjugate Gradient Algorithm,' *Int. Jour. for Numerical Methods in Engineering* **20**, 949–966.

Akan, A. O., and B. C. Yen: 1981, 'Diffusion–Wave Flood Routing in Channel Networks,' *J. Hydr. Div. ASCE* **107**(HY6), 719–732.

Akanbi, A. A., and N. D. Katopodes: 1988, 'Model for Flood Propagation on Initially Dry Land,' *J. Hydr. Div. ASCE* **114**(HY7), 689–706.

Bear, J.: 1979, *Hydraulics of Groundwater*, McGraw–Hill.

Becker, E. B., G. F. Carey, and J. T. Oden: 1986, *'Finite Elements,' An Introduction*, vol. I, Prentice–Hall.

Calo', P.: 1987, *'Analisi delle Conseguenze dello Scarico Rapido della Prevista Diga di Elvas (BZ) Mediante un Modello di Propagazione Bidimensionale,'* Tesi di Laurea, Universita di Bologna.

Chen, C., and V. T. Chow: 1971, 'Formulation of Mathematical Watershed–Flow Model,' *J. Mech. Div. ASCE* **97**(EM3), 809–828.

Cheng, R. T.: 1978, 'Modeling of Hydraulic Systems by Finite–Element Methods,' *Advances in Hydroscience* **11**, 207–284.

Chow, V. T., and A. Ben–Zvi: 1973, 'Hydrodynamic Modeling of Two–Dimensional Watershed Flow,' *J. Hydr. Div. ASCE* **99**(HY11), 2023–2040.

Citrini, D., and G. Noseda: 1975, *Idraulica*, Casa Editrice Ambrosiana, Milano.

Cunge, J. A.: 1975, 'Two–Dimensional Modeling of Flood Plains,' in K. Mahmood, and V. Yevjevich (eds.), *Unsteady Flow in Open Channels*, vol. II, 705–762, Fort Collins.

Daubert, A., and O. Graffe: 1967, 'Quelques Aspects des Écoulements Presque Horizontaux a deux Dimensions en Plan et Non Permanents Application Aux Estuaires,' *La Houille Blanche* **8**, 847–860.

De Vries, J. D., T. V. Hromadka, and A. J. Nestlinger: 1986, 'Applications of a Two–Dimensional Diffusion Hydrodinamic Model,' Hydrosoft 86–Hydraulic Engineering Software, *Proc. 2nd. Int. Conf.*, Southampton, 393–412.

Dhatt, G., and G. Touzot: 1985, *The Finite Element Method Displayed*, John Wiley.

Di Silvio, G.: 1978, 'Modelli Matematici per lo Studio della Propagazione di Onde Lunghe e del Trasporto di Materia Nei Corsi d'Acqua e Nelle Zone Costiere,' in Pàtron (ed.), *Metodologie Numeriche per la Soluzione di Equazioni Differenziali dell'Idrologia e dell'Idraulica*, Bologna.

Duff, I. S., and J. K. Reid: 1982, *'The Multifrontal Solution of Indefinite Sparse Symmetric Linear Systems,'* Rep. n. CSS 122, Computer Science and System Division, AERE Harwell Laboratory, Oxfordshire.

Duff, I. S., and J. K. Reid: 1982, *'MA–27 – A Set of FORTRAN Subroutines for Solving Sparse Symmetric Sets of Linear Equations,'* Rep. n. R–10533, AERE Harwell Laboratory, Oxfordshire.

Fennema, R. J., and M. H. Chaudhry: 1989, 'Implicit Methods for Two–Dimensional Unsteady Free–Surface Flows,' *Journal of Hydraulic Research* **27**(3), 321–332.

Fugaza, M., and M. Gallati: 1977, *A Finite Element Numerical Model of the Two Dimensional Shallow Water Motion*, Baden–Baden, IAHR vol. 2, 191–196.

Gallagher, R. H., J. T. Oden, C. Taylor, and O. C. Zienkiewicz: 1975, *Finite Elements in Fluids*, vol. 1, John Wiley.

Grubert, J. P.: 1976, 'Numerical Computation of Two–Dimensional Flows,' *J. Wat. Harbors and Coastal Div. ASCE* **102**(WW1), 1–12.

Higuera, F. G., S. Succi, and R. Benzi: 1989, *'Lattice Gas Dynamics with Enhanced Collisions.'*

Hosseinipour, Z., and M. Amein: 1984, 'Finite Element Computation of Two–Dimensional Unsteady Flow for River Problems,' *Proc. 5th Int. Conf. on Finite Elements on Water Resources*, Burlington, VT, 457–466.

Hromadka II, T. V., C. E. Berenbrock, J. R. Freckleton, and G. L. Guymon: 1985, 'A Two–Dimensional Dam–Break Flood Plain Model,' *Adv. Water Resour.* **8**, 7–14.

Hromadka II, T. V., A. J. Nestlinger, and J. D. De Vries: 1986, 'Comparison of Hydraulic Routing Methods for One–Dimensional Channel Routing Problems,' *Hydrosoft 86–Hydraulic Engineering Software, Proc. 2nd Int. Conf.*, Southampton, 85–98.

Huyakorn, P. S., and G. F. Pinder: 1983, *Computational Methods in Subsurface Flow*, Academic Press, London.

Katopodes, N. D.: 1980, 'Finite Element Model for Open Channel Flow Near Critical Conditions,' *Proc. 3rd Int. Conf. on Finite Elements on Water Resources*, Oxford, MS, 5.37–5.46.

Katopodes, N. D.: 1984, 'Two–Dimensional Surges and Shocks in Open Channels,' *J. Hydr. Div. ASCE* **110**(HY6), 794–812.

Katopodes, N. D., and T. Strelkoff: 1978, 'Computing Two–Dimensional DamBreak Flood Waves,' *J. Hydr. Div. ASCE* **104**(HY9), 1269–1288.

166

Katopodes, N. D., and T. Strelkoff: 1979, 'Two–Dimensional Shallow WaterWave Models,' *J. Mech. Div. ASCE* **105**(EM2), 317–334.

Kershaw, D.: 1978, 'The Incomplete Cholesky–Conjugate Gradient Method for the Iterative Solution of Systems of Linear Equations,' *Journal of Comp. Physics* **26**, 43–65.

Lai, C.: 1986, 'Numerical Modeling of Unsteady Open–Channel Flow,' *Advances in Hydroscience* **14**, 161–333.

Lapidus, L., and G. F. Pinder: 1982, *Numerical Solution of Partial Differential Equations in Science and Engineering*, John Wiley.

Marchi, E., and A. Rubatta: 1981, *'Meccanica dei Fluidi. Principi e Applicazioni Idrauliche,'* UTET, Torino.

Narasimhan, T. N., and P. A. Witherspoon: 1976, 'An Integrated Finite Difference Method for Analyzing Fluid Flow in Porous Media,' *Water Resour. Res.* **12**(1), 57–64.

Narasimhan, T. N., P. A. Witherspoon, and A. L. Edwards: 1978, 'Numerical Model for Satured–Unsatured Flow in Deformable Porous Media. 2. The Algorithm,' *Water Resources Research* **14**(2), 255–261.

Ortega, J. M., and W. C. Rheinboldt: 1970, *Iterative Solution of Non–Linear Equations in Several Variables*, Academic Press.

Pacciani, M.: 1987, *'Analisi Comparativa di Metodi Impliciti ed Espliciti per il Calcolo della Propagazione Bi–Dimensionale di Onde di Piena,'* Tesi di Laurea, Università di Bologna.

Pinder, G. F., and W. G. Gray: 1977, *Finite Element Simulation in Surface and Subsurface Hydrology*, Academic Press, London.

Reddy, J. N.: 1986, *Applied Functional Analysis and Variational Methods in Engineering*, McGraw-Hill.

Salgado, R. O.: 1988, *'Computer Modelling of Water Supply Distribution Networks Using the Gradient Method,'* Ph.D. Thesis, University of Newcastle upon Tyne.

Sloan, S. W.: 1987, 'A Fast Algorithm for Constructing Delaunay Triangulations in the Plane,' *Advanced Engineering Software* **9**(1), 34–55.

Stoer, J., and R. Burlish: 1980, *Introduction to Numerical Analysis*, Springer Verlag, New York.

Vansnick, M., and Y. Zech: 1984, 'Fourier Analysis for Testing a Finite Elements Method in Shallow Water Problems,' *Proc. 5th Int. Conf. on Finite Elements on Water Resources*, Burlington, VT, 467–476.

Xanthopoulos, T. H., and C. H. Koutitas: 1976, 'Numerical Simulation of Two Dimensional Flood Wave Propagation Due to Dam Failure,' *Journal of Hydraulic Research* **14**(4), 321–331.

Zienkiewicz, O. C.: 1977, *The Finite Element Method*, 3rd Ed., McGraw-Hill, New York.

Part III – State–of–the–Art in Modeling Linked Components of the Hydrologic Cycle

Chapter 9

Real Time Coupling of Hydrologic and Meteorological Models for Flood Forecasting

Konstantine P. Georgakakos
The University of Iowa
Department of Civil and Environmental Engineering
Iowa Institute of Hydraulic Research
Iowa City, Iowa 52242 U.S.A.

and

Efi Foufoula–Georgiou
University of Minnesota
Department of Civil and Mineral Engineering
St. Anthony Falls Hydraulic Laboratory
Minneapolis, Minnesota 55414 U.S.A.

Abstract. Recent advances in real time hydrometeorology are critically reviewed. Presented are the principles of flood forecasting with models that couple meteorological with hydrologic components and that are complemented with state estimators for real time state updating from precipitation and streamflow observations. The nature of the real–time coupling of these stochastic–dynamical hydrometeorological models is investigated. The cases for which such coupling of meteorological and hydrologic components appears to be most beneficial to flood forecasting are identified. Some promising research areas are also indicated.

1. Introduction

Stochastic–dynamical hydrologic models have been used in the past for the real time forecasting of flood flows (e.g. Kitanidis and Bras, 1980a,b). Mathematical models of runoff production and flood routing were complemented by state estimators that updated the model states in real time from available observations. In those models precipitation was treated as an input with given mean and variance. With the advent of a stochastic–dynamical precipitation model (Georgakakos and Bras, 1984a,b) and of a procedure for the spatial interpolation of the meteorological input, precipitation could be treated as a dynamical process with a state variable that could be added on to the state variables of the hydrologic models. A stochastic–dynamical hydrometeorological model was the result (Georgakakos and Hudlow, 1984; Georgakakos, 1986a,b; Georgakakos, 1987).

Some of the characteristics of the stochastic–dynamical hydrometeorological models that make them particularly useful in real–time flood and flash flood forecasting are:

(1) Extended forecast lead time, especially in cases of flash floods, because of the quantitative precipitation forecast capability.

(2) Better distribution of the precipitation volume in the various soil zones because of the existence of a dynamical equation for the precipitation state.

D. S. Bowles and P. E. O'Connell (eds.), Recent Advances in the Modeling of Hydrologic Systems, 169–184.
© 1991 *Kluwer Academic Publishers.*

(3) Improved modeling of uncertainty through the incorporation of a precipitation state in the state vector.

(4) Capability for updating the upper soil states from only precipitation observations.

Real time operation of the stochastic–dynamical hydrometeorological models for the prediction of flood flows involves the following steps:

(1) Collection of meteorological data from stations located nearby the watershed of interest.

(2) Spatial interpolation of the meteorological data and determination of spatially-averaged values for the watershed at each time.

(3) Using a set of initial mean and covariance values for the states of the hydrometeorological model and the interpolated meteorological input values, the model dynamical equations and the state estimator variance equations are solved for the forecast lead time interval. Predictions of the watershed–average precipitation rate and of the flow at the watershed outlet are obtained from the predicted states of the system.

(4) Collection of raingauge data and stage data at observation time. Computation of watershed mean areal (watershed–average) precipitation rate and discharge from collected data.

(5) Use of the state estimator update equations for the determination of improved estimates for the mean and covariance of the states of the hydrometeorological model at the observation time.

(6) Use of the updated state mean and covariance as initial conditions for the next time step with steps 1 through 5 repeated.

Georgakakos (1986a,b) used the modified Sacramento hydrologic model and the Georgakakos and Bras (1982) flood routing model as the soil and channel components of the hydrometeorological model. Georgakakos (1987) successfully used a simple Antecendent Precipitation Index (API) model for runoff generation together with the previously mentioned flood routing model as the basis of an Integrated Hydrometeorological Forecast System (IHFS) for flood and flash–flood prediction.

2. Discussion

In following sections, we present the basis of hydrometeorological modeling for real–time forecast application. The nature of the real–time coupling of hydrologic with meteorological models is examined in detail. A preliminary assessment of the usefulness of hydrometeorological models—based on past and on–going research—is attempted, and potentially fruitful research avenues are identified.

2.1. Forecast Problem Solution Using Real–Time Hydrometeorological Databases

Operational streamflow forecasting has been the subject of extensive research in recent years. The use of mathematical models that simulate the dynamics of the precipitation and terrestrial runoff processes and the use of improved monitoring systems have increased our capability to reliably predict floods and flash floods. Along these lines of research, conceptual mathematical models for soil moisture accounting have been devised and used (e.g., Peck, 1976) in real–time flood forecasting. State estimators that process the observations of discharge and automatically update the model states (model variables that summarize past inputs) have been designed based on concepts borrowed from Modern Estimation Theory (e.g., Kitanidis and Bras, 1980a,b; Georgakakos, 1986a,b). The estimators use information on the magnitude of the uncertainty that enters the mathematical model (due to inaccurate specification of input, model parameters, and model structure) and produce forecasts of the expected flood discharges and associated estimation–error variances. They also use observations of discharge and associated observation error statistics, and they produce improved initial conditions for the state variables in a sequential manner (i.e., real–time updating) for improved forecasts. Thus, the conceptual models complemented by state estimators use available data either as input or as observations for state updating in a way that takes into account the expected errors in each data type. For this reason it is expected that coupled hydrologic and meteorological models complemented by state estimators can form the basis of flood and flash–flood warning systems that utilize all available information pertinent to flood occurrence. An outline of the mathematical formulation that leads to the solution algorithm of the operational flood forecasting problem is presented next.

Denote by $\mathbf{f}(\mathbf{x}; \mathbf{u} \, \mathbf{a})$ the nonlinear function that signifies the derivative of the state vector $\mathbf{x}$ with respect to time t for a hydrometeorological model. It is assumed that dependence of $\mathbf{f}$ on t is through the state vector $\mathbf{x}$, the vector of input variables $\mathbf{u}$, and the vector of parameters $\mathbf{a}$. The water content of clouds and the water content of channel reaches can be two of the elements of the state vector. Air temperature and evapotranspiration can be two of the elements of the input vector, while cloud–drop diameter and ground water depletion rates can be some of the elements of the parameter vector. The mathematical expression for the derivative vector function $\mathbf{f}$ will be referred to as the conceptual model. If the model errors (in input, model parameters and structure) can be represented as an additive sequence of random vector disturbances $\mathbf{w}(t)$, then:

$$\frac{d\mathbf{x}(t)}{dt} = \mathbf{f}\big(\mathbf{x}(t); \mathbf{u}(t), \mathbf{a}\big) + \mathbf{w}(t) \tag{1}$$

Consider an observation vector $\mathbf{z}(t_k)$ composed of the mean areal precipitation over a catchment and of catchment streamflow at time instant t_k, and let a vector function $\mathbf{h}(\mathbf{x}; \mathbf{u} \, \mathbf{a})$ model the relationship between the model state vector $\mathbf{x}(t_k)$ and the input vector $u(t_k)$ at time t_k on the one hand and the observation at the same time on the other. If $\mathbf{v}(t_k)$ signifies the observation error vector at time t_k, then:

$$\mathbf{z}(t_k) = \mathbf{h}\big(\mathbf{x}(t_k); \mathbf{u}(t_k), \mathbf{a}\big) + \mathbf{v}(t_k) \tag{2}$$

where it has been assumed that the drainage basin of interest is a headwater one with the only available observations of streamflow located at the outlet of the basin (a realistic assumption in the operational environment).

The set of Equations (1) and (2) represents the state–space mathematical formulation of the precipitation–streamflow process. Conceptualization and parameterization of the component processes, such as precipitation, overland flow, groundwater flow, and flood–wave propagation, gives expressions for the functions $\mathbf{f}$ and $\mathbf{h}$.

The Kalman Filter (e.g., Gelb, 1974) applied to the nonlinear system (1)–(2) (the algorithm is called Extended Kalman Filter in this case) results in the following set of recursive differential and algebraic equations for step $[t_{k-1}, t_k]$:

$$\frac{d\hat{\mathbf{x}}(t)}{dt} = \mathbf{f}\big(\hat{\mathbf{x}}(t)^-; \mathbf{u}(t), \mathbf{a}\big); \quad t_{k-1} \le t \le t_k \tag{3}$$

$$\frac{dP(t)^-}{dt} = F(t)P(t)^- + P(t)^- F(t)^T + Q(t); \quad t_{k-1} \le t \le t_k \tag{4}$$

$$K(t_k) = P(t_k)^- H(t_k)^T \left[H(t_k)P(t_k)^- H(t_k)^T + R(t_k) \right]^{-1} \tag{5}$$

$$\hat{\mathbf{x}}(t_k)^+ = \hat{\mathbf{x}}(t_k)^- + K(t_k) \left[\mathbf{z}(t_k) - \mathbf{h}(\hat{\mathbf{x}}(t_k)^-; \mathbf{u}(t_k), \mathbf{a}) \right] \tag{6}$$

$$P(t_k)^+ = \left[1 - K(t_k)H(t_k) \right] P(t_k)^- \tag{7}$$

where:

$$[F(t)]_{i,j} = \frac{\partial f_i(\hat{\mathbf{x}}(t)^-; \mathbf{u}(t), \mathbf{a})}{\partial x_j(t)} \tag{8}$$

$$[H(t)]_{i,j} = \frac{\partial h_i(\hat{\mathbf{x}}(t)^-; \mathbf{u}(t), \mathbf{a})}{\partial x_j(t)} \tag{9}$$

and with the following symbol definitions:

- f_i the ith element of the vector function **f**,
- h_i the ith element of the vector function **h**,
- x_j the jth element of the state vector **x**,
- $\hat{\mathbf{x}}(t)^-$ the unbiased minimum variance estimate of $\mathbf{x}(t)$ given all observations before time t,
- $\hat{\mathbf{x}}(t)^+$ the unbiased minimum variance estimate of $x(t)$ given all observations up to and including time t,
- $P(t)^-$ the covariance matrix of the state estimation error vector given all observations before time t,
- $P(t)^+$ the covariance matrix of the state estimation error vector given all observations up to and including time t,
- $Q(t)$ the covariance parameter matrix of the model error process $w(t)$,
- $R(t)$ the variance function of the observation error sequence $v(t)$.

It is noted that the derivatives in Equations (8) and (9) are functions of the state vectors evaluated at the best estimate of the state vector before the observation $\mathbf{z}(t_k)$ at time t_k, becomes available.

The set of Equations (3)–(7) can be used recursively to: a) process the observations $\mathbf{z}(t_k)$ ($k = 1,2...,M$) as they become available in real time, and b) produce forecasts of the state vector $\mathbf{x}(t_k)$ $(\hat{\mathbf{x}}(t_k)^-)$ and associated estimation–error covariance matrix $P(t_k)$ $(P(t_k)^-)$. (Initial conditions $\mathbf{x}_0$ and estimation–error covariance matrix P_0 are required for the initiation of the algorithm.) Given these forecasts, forecasts of mean areal precipitation and streamflow $\hat{z}(t_k)$ and associated estimation–error covariance $P_z(t_k)^-$ can be produced by:

$$\hat{z}(t_k) = \mathbf{h}(\hat{\mathbf{x}}(t_k)^-; \mathbf{u}, \mathbf{a}) \tag{10}$$

$$P_z(t_k)^- = H(t_k)P(t_k)^- H(t_k)^T \tag{11}$$

If desired, use of a Gaussian assumption can be made for the sequence of streamflow forecast errors resulting in capability for probabilistic flood occurrence predictions. Thus, if ϕ

denotes the discharge corresponding to a prespecified flood stage, and the Gaussian probability density function of the streamflow $\zeta(t_k)$ at time t_k is denoted by $p[\zeta(t_k); \hat{z}_s(t_k), P_{z_s}(t_k)^-]$, then the probability $Pr(\phi, t_k)$ that the threshold ϕ will be exceeded at time t_k is given by:

$$Pr(\phi, t_k) = \left[1 - \int_{-\infty}^{\phi} p\left(\zeta(t_k); \hat{z}_s(t_k), P_{z_s}(t_k)^- \right) d\zeta(t_k) \right] \tag{12}$$

where $\hat{z}_s$ is the mean–of–streamflow element of $\hat{z}$ and P_{z_s} the variance–of–streamflow element in P_z.

The key assumptions that lead to the numerically convenient form of Equations (3)–(7) are that the continuous–time error process $w(t)$ is a Gaussian white–noise stochastic process with mean vector equal to the zero vector and covariance parameter matrix equal to $Q(t)$, and that the discrete–time error sequence $v(t_k)$ is a Gaussian independent sequence with mean equal to zero and covariance matrix equal to $R(t_k)$. The initial vector $x(t_0)$ is also assumed Gaussian with mean vector x_0 and covariance matrix P_0.

Prior to implementation of the estimation algorithm for a particular conceptual model, it is required that values for the model parameters, a, and for the estimator parameters, $Q(t)$ and $R(t_k)$ (for all t and k), be identified from available morphometric and hydrologic historical data from the site of interest. Thus, before the application of algorithm in real–time streamflow forecasting, a parameter identification study is required.

A detailed survey of parameter estimation techniques and applications in the area of real–time flow forecasting has recently been presented by Rajaram and Georgakakos (1987) and will not be attempted here. Some conclusions of that review relevant to the present study will be discussed in the following.

Even though estimation techniques have been extensively studied in the model parameter estimation context, very few methodologies have been proposed and tested for the estimation of the estimator parameters $Q(t)$ and $R(t_k)$. Of particular importance is the identification of the elements of $Q(t)$ since very little a priori information is available for it. The variance $R(t_k)$ of the streamflow observation error can be reliably estimated in most cases (especially in headwater basin with no looped rating curves) since it depends on the characteristics of the measuring device. The stochastic approximations technique has been used (e.g., Kitanidis and Bras, 1978; Georgakakos, 1984) as the basis of a methodology for the determination of selected elements of the matrix $Q(t)$ assumed independent of time t. The idea of the methodology is to enforce at all times consistency between the actual statistics of the one–step–ahead predicted residuals (residual = observed – predicted streamflow) and those statistics as they are predicted by the estimator in real time. This methodology is well suited to low dimensional Q matrices. It fails to perform well in cases of high dimensional systems and in cases when the model error is highly nonstationary (the matrix Q depends on time t). Recognizing the deficiency of existing methods to provide a general solution to the problem of determining values for the estimator parameters, Rajaram and Georgakakos (1987) proposed a conceptual framework leading to an identification methodology that does not suffer from the aforementioned deficiencies.

Rajaram and Georgakakos (1987) proposed the decomposition of the model error process into two component processes: the model–parameter error process, and the input error process. The conceptual model equations and a priori estimates of the model–parameter and input estimation errors were used to parameterize the statistics of the component processes. For model parameter calibration by automated techniques, the final parameter error covariance matrix can be used as an a priori estimate of the model–parameter estimation error. For manual calibration of the model parameters, the Hydrologists' degree–of–belief estimates of the parameter error variances can be used instead. The proposed methodology required the identification of only two free parameters from historical data. Because of the small number of free parameters, almost any statistically–sound parameter estimation technique can be used for their identification. Previous application of these techniques to large–scale hydrologistical (Georgakakos et al., 1988), and hydrologic–chemical models that simulate watershed response to acidic deposition (Rajaram and Georgakakos, 1987) has given encouraging results.

Using the concepts previously presented and design requirements of national flash–flood warning systems (see Georgakakos, 1986c), Georgakakos (1987) designed a prototype IHFS for the real time forecasting of floods and flash floods. Figure 1 presents the IHFS configuration. IHFS is presented here as an example of a hydrometeorological flood warning system.

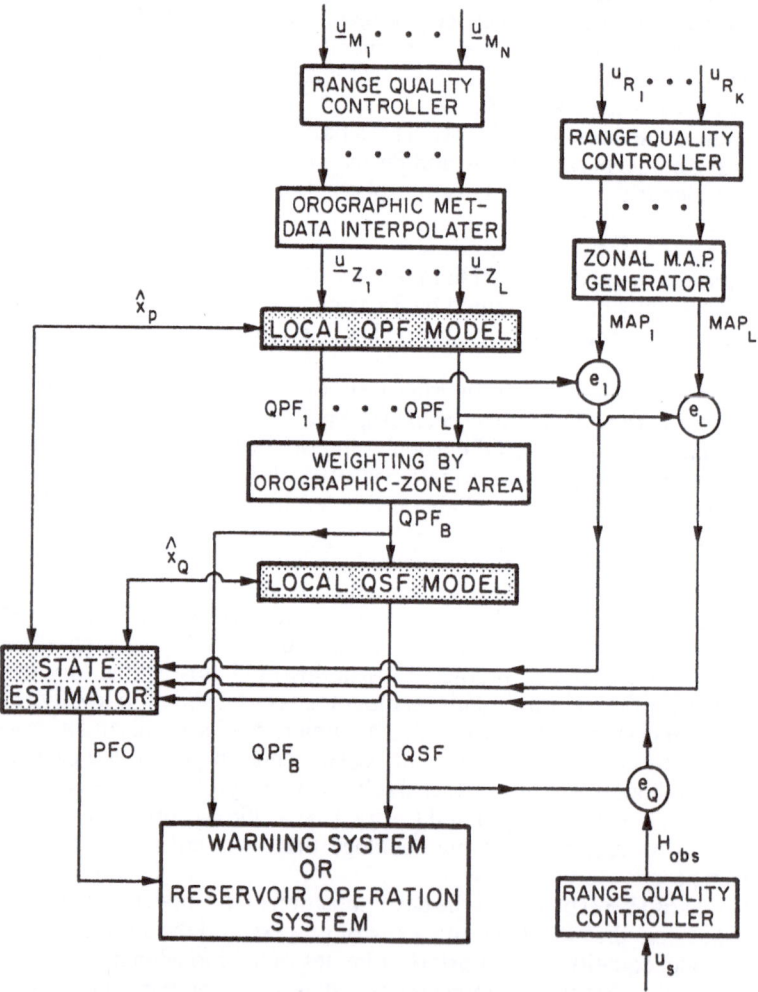

Figure 1. Systems design of an integrated hydrometeorological forecast system (after Georgakakos, 1987).

In Figure 1 $\underline{u}_{M_1}$ through $\underline{u}_{M_N}$ represent the input vectors of the meteorological variables from several meteorological sensors located near the drainage basin of interest. Once the meteorological input enters the system, consistency checks are initiated to discard erroneous data. At the present stage of development, IHFS includes only checks on the magnitude of the input variables (range quality control). The meteorological data that pass the quality control are interpolated in space to arrive at the vector of meteorological variables that characterize the basin of interest. In cases where mountainous terrain is part of the drainage basin of interest, the basin is divided in several (L) orographic zones of increasing altitude, and quality controlled meteorological input is spatially interpolated at the centroid of each orographic zone. Georgakakos' (1986a) spatial interpolation procedure that accounts for mountainous terrain is used. The result of the interpolation is a set of meteorological variable vectors $\underline{u}_{Z_1}$ through $\underline{u}_{Z_L}$ that characterize each one of the L orographic zones. These vectors constitute the input to the stochastic–dynamical quantitative precipitation forecast (QPF) model which generates mean areal precipitation forecasts QPF_1 through QPF_L for each orographic zone. A mean areal precipitation forecast QFP_B for the whole drainage basin is produced as a weighted average of the zonal precipitation forecasts. Each zonal forecast is weighted by the area of the corresponding orographic zone. The drainage basin mean areal precipitation forecast feeds a spatially lumped, hydrologic, quantitative stage forecast (QSF) model (e.g., an API procedure or a conceptual hydrologic model) that: 1) computes and subtracts from the input mean areal precipitation forecast the infiltration losses to the ground, 2) routes the remaining input as overland flow and channel runoff to the point where the flood forecast is needed, and 3) converts the flow discharge to river stage at the point of interest. The output of the hydrologic model is the QSF. Both the basin mean areal precipitation forecast QPF_B and the stage forecast QSF are IHFS system outputs and they feed the flood warning system (see Figure 1).

IHFS processes observations of point precipitation from several (K) raingages (if any exist) located within or near the basin and/or observations of flow stage at the point of interest. At the time that the raingage observations u_{R_1} through u_{R_K} enter the IHFS system and after they pass a range check for validity (range quality controller in Figure 1), they are divided in categories based on raingage elevation, one category for each orographic zone. Subsequently, mean areal precipitation estimates for each category are computed: MAP_1 through MAP_L. The observed MAPs are compared to the model produced QPFs that correspond to the same time and orographic zone; a set of errors e_1 through e_L are computed, one for each orographic zone. Through a similar process, in cases where a river stage observation is available, an error e_Q is produced by comparison of the QSF model predicted QSF and the quality controlled observation H_{obs} (Figure 1). The computed errors from all the observations and the predicted state variables of the QPF model x_p and of the QSF model x_Q are input to a state estimator. The estimator: 1) updates the predicted state variables based on the input errors and returns the updated state variables to their respective models so that the next forecasts will be based on updated initial conditions, 2) generates variances for the forecasts QPF_B and QSF through a set of variance equations and during the forecast stage (no observations are yet available), and 3) generates forecasts of the probability that a certain given critical flood stage will be exceeded at various times in the future based on Equation (12) (thus, the probability of flood occurrence PFO in Figure 1). The probability PFO is a IHFS system output, and it feeds the warning system together with the forecasts QPF_B and QSF.

The structure of the IHFS system is capable of accommodating various real–time conditions ranging from sparse data cases to cases with several data sensors and historical datasets. Upper air data from radiosondes can be used when available to feed the QPF component. However, IHFS can produce forecasts with only surface meteorological observations. Orographic enhancement of precipitation is performed on option. The model component that generates surface runoff from rainfall can be a simple Antecedent

Precipitation Index (API) model or the rather sophisticated Sacramento soil moisture accounting model (see Georgakakos, 1986a), depending on data availability for calibration. The state estimator is designed in such a way that it can accommodate observation vectors with time varying composition. IHFS can be used both in headwater and in first–order–tributary basins. Georgakakos et al. (1988) present a User's Manual for the IHFS computer program.

2.2. Real Time Coupling

It is desired to identify the means by which meteorological and hydrologic components are coupled in stochastic–dynamical hydrometeorological models. Real–time coupling of the hydrologic with the meteorological components is illustrated in the following by way of analysis of the state estimator equations for a headwater drainage basin hydrometeorological model.

Define by x_p the n_1–dimensional vector of the precipitation model states, and by x_h the n_2–dimensional vector of the hydrologic model states. Dependence on time t is not explicitly shown for notational convenience; it is noted, however, that such dependence exists and in most cases it is implicit through dependence on input data.

Form the composite n–dimensional vector x as:

$$\mathbf{x} = \begin{bmatrix} \mathbf{x}_p \\ \cdots \\ \mathbf{x}_h \end{bmatrix} \text{ with } n_1 + n_2 = n \tag{13}$$

Assume that the precipitation model dynamical equations are:

$$\frac{d\mathbf{x}_p}{dt} = \mathbf{f}_p(\mathbf{x}_p; \mathbf{a}_p, \mathbf{u}_p) \tag{14}$$

and the hydrologic model dynamical equations are:

$$\frac{d\mathbf{x}_h}{dt} = \mathbf{f}_h(\mathbf{x}_h, \mathbf{x}_p; \mathbf{a}_h, \mathbf{a}_p, \mathbf{u}_p, \mathbf{u}_h) \tag{15}$$

with attendant observation equations:

$$\mathbf{z}_p = \mathbf{h}_p(\mathbf{x}_p; \mathbf{a}_p, \mathbf{u}_p) \tag{16}$$

$$\mathbf{z}_h = \mathbf{h}_h(\mathbf{x}_h; \mathbf{a}_h) \tag{17}$$

where t is time, $\mathbf{f}_p(\bullet)$ is a functional form of the precipitation model temporal state derivative vector, $\mathbf{f}_h(\bullet)$ is a functional form of the hydrologic model temporal state derivative vector, $\mathbf{h}_p(\bullet)$ is a functional form of the algebraic equation that describes the relationship between precipitation state and precipitation observation, $\mathbf{h}_h(\bullet)$ is a functional form of the algebraic equation that describes the relationship between hydrologic–model state and outflow observation, $\mathbf{a}_p$ is a k_1–dimensional vector of precipitation model parameters, $\mathbf{a}_h$ is a k_2–dimensional vector of hydrologic model parameters, $\mathbf{u}_p$ is a ℓ_1–dimensional vector of precipitation model input, $\mathbf{u}_h$ is a ℓ_2–dimensional vector of hydrologic model input, $\mathbf{z}_p$ is a m_1–dimensional vector of precipitation observations, and $\mathbf{z}_h$ is a m_2–dimensional vector of drainage basin outflow observations.

Dependence on input and parameter vector is explicitly shown in the previous formulation based on the specific models used in Georgakakos (1986a,b; 1987). The analysis to follow, however, is independent of the specific models used.

Making use of the composite vector **x**, the state–space form of Equations (14) through (17) can be written alternatively as in Eqs. (1) and (2), with:

$$\mathbf{f} = \begin{bmatrix} \mathbf{f}_p \\ \cdots \\ \mathbf{f}_h \end{bmatrix} \tag{18}$$

$$\mathbf{h} = \begin{bmatrix} \mathbf{h}_p \\ \cdots \\ \mathbf{h}_h \end{bmatrix} \tag{19}$$

$$\mathbf{a} = \begin{bmatrix} \mathbf{a}_p \\ \cdots \\ \mathbf{a}_u \end{bmatrix} \tag{20}$$

$$\mathbf{u} = \begin{bmatrix} \mathbf{u}_u \\ \cdots \\ \mathbf{u}_h \end{bmatrix} \tag{21}$$

$$\mathbf{z} = \begin{bmatrix} \mathbf{z}_p \\ \cdots \\ \mathbf{z}_h \end{bmatrix} \tag{22}$$

The filter Equations (3) through (9) can be used for propagation and updating of the state vector between and at the times of observation respectively. Coupling of the meteorological with the hydrologic components is first realized in Equations (14) and (15) and secondly in Equations (5) through (7). In the former set of equations, coupling is due to the conservation of mass law: the output mass flux of the precipitation model is equal to the mass flux input into the hydrologic component. It is a one-way coupling—from the precipitation to the hydrologic model. The later set of equations [(5), (6) and (7)] couples the precipitation components with the hydrologic components through the feedback terms in Equations (6) and (7). Namely, the terms:

$$K(\mathbf{z} - \mathbf{h}(\hat{\mathbf{x}}^- ; \ \mathbf{a}, \mathbf{u}))$$

and

$$KHP^-$$

Next, we partition the composite vectors and matrices into their components corresponding to the precipitation and hydrologic processes to gain insight into the factors of the feedback coupling in the state mean.

At first, partition matrix K as follows:

$$K = \begin{bmatrix} K_{pp} & \cdot & K_{ph} \\ \cdots & \cdots & \cdots \\ K_{hp} & \cdot & K_{hh} \end{bmatrix} \tag{23}$$

where K_{pp} is a $(n_1 x m_1)$–dimensional matrix, K_{ph} is a $(n_1 x m_2)$–dimensional matrix, K_{hp} is a $(n_2 x m_1)$–dimensional matrix, and K_{hh} is a $(n_2 x m_2)$–dimensional matrix. Then, using Equations (6), (13), (19), (22) and (23) we obtain:

$$\hat{x}_p^+ = \hat{x}_p^- + K_{pp}\big(z_p - h_p(\hat{x}^-; \, a, u)\big) + K_{ph}\big(z_h - h_h(\hat{x}^-; \, a, u)\big) \tag{24}$$

and

$$\hat{x}_h^+ = \hat{x}_h^- + K_{hp}\big(z_p - h_p(\hat{x}^-; \, a, u)\big) + K_{hh}\big(z_h - h_h(\hat{x}^-; \, a, u)\big) \tag{25}$$

It is clear then that feedback coupling is due to non–zero gain matrices: K_{ph} and K_{hp}. Also, it is noted that forecast errors in precipitation are used to correct hydrologic model states, and forecast errors in outflow are used to correct precipitation model states—a two–way coupling.

To further examine the nature of the feedback coupling, it is necessary to express the matrices K_{ph} and K_{hp} as functions of state and noise uncertainty via Equation (5). For that purpose partitioning of the covariance matrix P is necessary:

$$P = \begin{bmatrix} P_{pp} & \cdot & P_{ph} \\ \cdots & \cdots & \cdots \\ P_{hp} & \cdot & P_{hh} \end{bmatrix} \tag{26}$$

where P_{pp} is a $(n_1 x n_1)$–dimensional symmetric covariance matrix, P_{hh} is a $(n_2 x n_2)$–dimensional symmetric covariance matrix, P_{ph} is a $(n_1 x n_2)$–dimensional matrix, and P_{hp} is a $(n_2 x n_1)$–dimensional matrix with:

$$P_{ph} = P_{hp}^T \tag{27}$$

At this point and for simplification of the derivations involved, we set:

$$m_1 = 1 \tag{28}$$

$$m_2 = 1 \tag{29}$$

Such an assumption is not a binding one in cases of headwater areas without pronounced relief (differences in elevation less than 500 meters) where the two observation variables are mean areal precipitation over the whole basin and outflow discharge (or stage) (see

Georgakakos, 1986a,b, and Georgakakos, 1987, for examples). Under assumption–equations (28) and (29), the inverse matrix in Equation (5) is a two–dimensional matrix.

Partition the observation gradient matrix H as follows:

$$H = \begin{bmatrix} H_{pp} & \vdots & H_{ph} \\ \cdots & \cdots & \cdots \\ H_{hp} & \vdots & H_{hh} \end{bmatrix} \tag{30}$$

where H_{pp} is a $(1xn_1)$–dimensional matrix, H_{ph} is a $(1xn_2)$–dimensional matrix, H_{hp} is a $(1xn_1)$–dimensional matrix, and H_{hh} is a $(1xn_2)$–dimensional matrix. In addition H_{ph} and H_{hp} are matrices whose elements are all equal to zero. Also, partition the $(2x2)$–dimensional matrix R as:

$$R = \begin{bmatrix} R_1 & 0 \\ 0 & R_2 \end{bmatrix} \tag{31}$$

assuming no linear dependence between the errors in observing precipitation and the errors in observing discharge.

The inverse $(2x2)$–dimensional matrix in Equation (5) can then be written as:

$$(HP^-H^T + R)^{-1} = \frac{1}{D} \begin{bmatrix} R_2 + H_{hh}P_{hh}H_{hh}^T & \vdots & -H_{pp}P_{ph}H_{hh}^T \\ \cdots & \cdots & \cdots \\ -H_{pp}P_{ph}H_{hh}^T & \vdots & R_1 + H_{pp}P_{pp}H_{pp}^T \end{bmatrix} \tag{32}$$

with the scalar determinant D defined by :

$$D = R_1R_2 + R_1H_{hh}P_{hh}H_{hh}^T + R_2H_{pp}P_{pp}H_{pp}^T + H_{pp}P_{pp}H_{pp}^T H_{hh}P_{hh}H_{hh}^T \tag{33}$$

Then, from Equations (5) and (23) it follows that:

$$K_{ph} = \left[P_{ph}H_{hh}^T R_1 + P_{ph}H_{hh}^T H_{pp}P_{pp}H_{pp}^T - P_{pp}H_{pp}^T H_{pp}P_{ph}H_{hh}^T \right] \frac{1}{D} \tag{34}$$

and

$$K_{hp} = \left[P_{hp}H_{pp}^T R_2 + P_{hp}H_{pp}^T H_{hh}P_{hh}H_{hh}^T - P_{hh}H_{hh}^T H_{pp}P_{ph}H_{hh}^T \right] \frac{1}{D} \tag{35}$$

The factors that contribute to the initiation of feedback coupling between the meteorological (precipitation) and the hydrologic components can be obtained from Equations (34) and (35) by setting:

$$P_{ph} = 0 \tag{36}$$

where 0 represents an $(n_1 \times n_2)$–dimensional matrix whose elements are all equal to zero. Using Equations (36) and (27) in Equations (34) and (35), one obtains:

$$K_{ph} = 0 \quad \text{and} \quad K_{hp} = 0 \tag{37}$$

Thus, coupling is due to the presence of non–zero P_{ph}. Such non–zero P_{ph} is obtained during the propagation step of the Extended Kalman Filter (Equation (4)) because of the mass coupling of the meteorological hydrologic components. That is:

$$\frac{dP_{ph}}{dt} = F_{pp}P_{ph} + P_{ph}F_{hh}^T + F_{ph}P_{hh} + P_{pp}F_{hp}^T \tag{38}$$

assuming that there is no dependence between model errors in the precipitation model and model errors in the hydrologic model, and using:

$$F_{pp} = \frac{\partial \mathbf{f}_p}{\partial \mathbf{x}_p} \tag{39}$$

$$F_{ph} = \frac{\partial \mathbf{f}_p}{\partial \mathbf{x}_h} \tag{40}$$

$$F_{hh} = \frac{\partial \mathbf{f}_h}{\partial \mathbf{x}_h} \tag{41}$$

$$F_{hp} = \frac{\partial \mathbf{f}_h}{\partial \mathbf{x}_p} \tag{42}$$

Were $F_{ph} = 0$ and $F_{hp} = 0$, Equation (38) would reduce to:

$$\frac{dP_{ph}}{dt} = F_{pp}P_{ph} + P_{ph}F_{hh}^T \tag{43}$$

For zero initial matrix P_{ph}, Equation (43) yields zero solution for all times. It is then clear that feedback coupling is due to the existence of non–zero F_{ph} or F_{hp}. State derivative dependence is shown in Equations (14) and (15). It follows that F_{hp} is non–zero due to the terms in $\mathbf{f}_h$ that set the input to the hydrologic model equal to the output of the precipitation model.

In summary, mass conservation is responsible for the feedback coupling by the state estimator. An appealing result!

2.3. Previous Case Studies and Summary of Results

Georgakakos (1986a,b) established the improved performance of flood forecasting systems when precipitation prediction components are coupled to hydrologic runoff components via mass conservation equations and state–updating feedback. His results indicate improved performance in predicting the timing and magnitude of flood–peaks compared to the performance of traditionally–used hydrologic prediction models (models without precipitation prediction components). Thus, for the 2344 km^2 hydrologic drainage basin of Bird Creek with outlet near Sperry, Oklahoma, USA, a performance comparison study was undertaken involving stochastic–dynamical hydrologic models that used precipitation as an input, and stochastic–dynamical hydrometeorological models. The forecast lead time was equal to the data observation interval of 6 hours. Histograms of the time difference between predicted and observed peak and the error in predicting the magnitude were constructed for both types of model and for 30 significant flood events over a five year period. The results showed the models that coupled meteorological with hydrologic components had a better performance in forecasting both the peak timing and the peak magnitude.

Sensitivity with respect to the location of the meteorological station supplying input data to the precipitation component was examined by Georgakakos (1987) for a recent event that caused excessive flooding in West Virginia, USA. He found that for the case of updating from both stage and precipitation, the IHFS predictions were not significantly sensitive to the location of the meteorological station. Noticeable differences were observed in the predictions using input from various meteorological stations for cases of no updating and updating from only stage. The location of the station with respect to the storm moisture inflow was a significant factor in those cases.

A significant finding for cases of short–term (compared to catchment response time) flash flood prediction was that the predictions were insensitive to the soil–moisture accounting component of the system for all cases of updating examined (see Georgakakos, 1987). In addition, it was found that best results were obtained when real–time updating from both precipitation and stage data was performed. Perhaps more importantly, Georgakakos (1987) showed that in such a case, the predictions appear to be insensitive to the spatial configuration of the meteorological stations and to the runoff–generating model parameters. This last finding is particularly significant for the real time implementation of IHFS and for the design of sensor networks for localized flash flood predictions. It indicates that for robust behavior, real–time data from both precipitation and stage (or discharge) sensors are necessary.

In a recently reported research effort (Georgakakos et al., 1988) data from three drainage basins of different areas and hydrologic characteristics were used in an extensive sensitivity analysis with a goal to ascertain the effects on real time stage predictions of errors in observed data that are processed by IHFS. Simple persistence of the input meteorological data was used to produce forecast input values so that IHFS prediction could be causal, simulating real time operation. The results of this study can be useful in assessing the worth of each type of data to real time flood forecasting and, therefore, in identifying the cases for which coupling of meteorological with hydrologic models is most beneficial. Preliminary results of this study are itemized in the following for each flood event examined.

For the July '76 North Fork Big Thompson River in Colorado, USA, flash flood (an area of about 218 km^2 with response time less than 1 hour, and data interval and forecast lead time equal to 1 hour) the following were important conclusions:

(1) Dew–point temperature errors had pronounced effects on the IHFS stage and probability of flooding predictions.

(2) The IHFS component that generates orographic enhancement of rainfall was very important for this flash flood. Significant underestimation of stage results without orographic enhancement of rainfall.

For the May '84 Tug Fork in West Virginia, USA, flood (an area of about 1215 km^2 with response time of 13 hours, and data interval and forecast lead time of 3 hours) the main conclusions were:

(1) In case of state updating from both stage and rainfall, errors in the stage data are most detrimental to accurate IHFS stage predictions.

(2) In case of state updating from precipitation only, errors in precipitation were most influential to IHFS stage and probability of flooding predictions.

(3) Errors in the dew–point data of the meteorological station near the storm moisture inflow area have relatively greater effects on the IHFS stage predictions than similar errors in the data of stations removed from that area.

In the case of the April '88 West Otter Creek in Iowa, USA, flood (an area of about 47 km^2 with response time greater than 24 hours, with a data interval and forecast lead time of 1 hour) the main conclusions were:

(1) The rating curve and the estimate of the basin response time affect the stage predictions of IHFS significantly.

(2) Errors in stage data have the most pronounced effects on the accuracy of the IHFS stage predictions.

The results summarized previously provide a quantitative basis for the assessment of the utility of data of various types in real–time flood prediction. That is, as the ratio of forecast lead time to basin response time decreases to values near 0.1, the utility of reliable stage data increases while the utility of the meteorological data and rainfall data decreases. On the other hand, as the ratio of forecast lead time to basin response time approaches the value of 1, the utility of the meteorological data increases and that of the stage data and of precipitation data decreases. Figure 2 quantifies data–type importance based on the IHFS sensitivity study. In addition to the above mentioned comments, it was clear from the analysis that use of all data types gave the best IHFS performance for all the floods examined and that the stage predictions of IHFS were very good when no errors were introduced in the data.

3. Conclusions

Stochastic–dynamical hydrometeorological models that couple meteorological and hydrologic components have proven useful to real–time flood forecasting for headwater drainage basins. Conservation of mass couples the aforementioned components through both the dynamical equations of the models (a one–way coupling) and the state–estimator feedback (a two–way coupling). When simple persistence schemes are used to supply meteorological-input (air temperature, pressure, and dew–point temperature) forecasts for real–time operation, coupling is most important for floods for which the ratio of forecast lead time to drainage basin response time is near 1. The importance of the coupling should increase for smaller ratios when more accurate meteorological input forecasts are used by the hydrometeorological models (e.g., use of forecasts of meteorological–input produced by operational large–scale numerical weather prediction models).

Extension of the concepts of coupling to large basins with several headwater areas, using possibly two–dimensional precipitation prediction models and two–dimensional catchment models, appears feasible. Such an extension would facilitate the use of radar and satellite data by the forecast system in a physically–based manner. Furthermore, such an extension could result in coupling forecast systems and reservoir control systems in real time for optimal performance using all available data. The very recent work of Lee and Georgakakos (1990) in the development of two dimensional stochastic–dynamical rainfall prediction models is a first step in that direction.

It appears essential that studies should continue that could lead to the generation of values which would complete Figure 2 for several ratios between 0.1 and 1. Such results could further our understanding of the worth of each type of hydrometeorological data to real–time flood forecasting and, consequently, could lead to improved methods of observation network design.

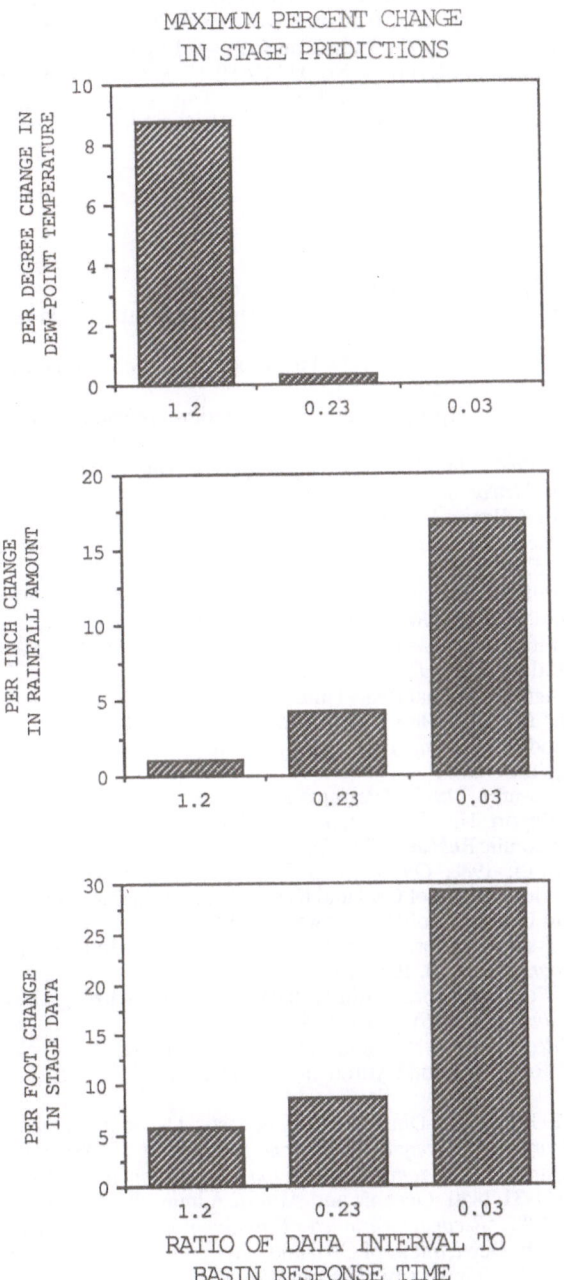

Figure 2. Sensitivity of IHFS predictions to input data type.

4. Acknowledgements

The research and development work of the first author reported herein has been supported, at various stages, by the Hydrologic Research Laboratory, U.S. National Weather Service, NOAA, through Cooperative Agreement No. NA86AA-H-HY12, by the U.S. Army Corps of Engineers, Cold Regions Research and Engineering Laboratory, through Contract No. DACA89-87-K-004, and by the U.S. National Science Foundation through Grant No. CES-8657526. This support is gratefully acknowledged. The work of the second author was supported by the National Science Foundation through Grant No. CES-8708825.

5. References

Gelb, A. (ed.): 1974, *Applied Optimal Estimation*, MIT Press, Cambridge, MA.

Georgakakos, K. P., and R. L. Bras: 1982, 'Real Time, Statistically Linearized Adaptive Flood Routing,' *Water Resour. Res.* **18**(3), 513–524.

Georgakakos, K. P., and R. L. Bras: 1984a, 'A Hydrologically Useful Station Precipitation Model, 1. Formulation,' *Water Resour. Res.* **20**(11), 1585–1595.

Georgakakos, K. P., and R. L. Bras: 1984b, 'A Hydrologically Useful Station Precipitation Model, 2. Case Studies,' *Water Resour. Res.* **20**(11), 1597–1610.

Georgakakos, K. P., and M. D. Hudlow: 1984, 'Quantitative Precipitation Forecast Techniques for Use in Hydrologic Forecasting,' *Bull. Am. Meteor. Soc.* **65**(11), 1186–1200.

Georgakakos, K, P.: 1984, 'Model–Error Adaptive Parameter Determination of a Conceptual Rainfall Prediction Model,' *Proc. IEEE Southeast, Symp. on System Theory, 16th*, IEEE Computer Society Press, Silver Spring, MD, pp. 111–115.

Georgakakos, K. P.: 1986a, 'A Generalized Stochastic Hydrometeorological Model for Flood and Flash-flood Forecasting, 1. Formulation,' *Water Resour. Res.* **22**(13), 2083–2095.

Georgakakos, K. P.: 1986b, 'A Generalized Stochastic Hydrometeorological Model for Flood and Flash-flood Forecasting, 2. Case Studies,' *Water Resour. Res.* **22**(13), 2096–2106.

Georgakakos, K. P.: 1986c, 'On the Design of National Real-Time Warning Systems with Capability for Site-Specific Flash Flood Forecasts,' *Bull. Am. Meteorol. Soc.* **67**(10), 1233–1239.

Georgakakos, K. P.: 1987, 'Flash Flood Prediction,' *J. of Geophysical Research* **92**(D8), 9615–9629, Copyright by the American Geophysical Union.

Georgakakos, K. P., S. Li, and C. Viswesvaran: 1988, *'Integrated Hydrometeorological Forecast System (IHFS), A User's Manual'* (IIHR Report No. 322), Department of Civil and Environmental Engineering and Iowa Institute of Hydraulic Research, The University of Iowa, Iowa City, IA, 70 pp.

Georgakakos, K. P., H. Rajaram, and S. Li: 1988, *'On Improved Operational Hydrologic Forecasting of Streamflows'* (IIHR Report No. 325), Department of Civil and Environmental Engineering and Iowa Institute of Hydraulic Research, The University of Iowa, Iowa City, IA, 162 pp.

Kitanidis, P. K., and R. L. Bras: 1980a, 'Real Time Forecasting with a Conceptual Hydrologic Model, 1. Analysis of Uncertainty,' *Water Resour. Res.* **16**(6), 1025–1033.

Kitanidis, P. K., and R. L. Bras: 1980b, 'Real Time Forecasting with a Conceptual Hydrologic Model, 2. Applications and Results,' *Water Resour. Res.* **16**(6), 1034–1044.

Kitanidis, P. K., and R. L. Bras: 1978, *'Real Time Forecasting of River Flows'* (Technical Report No. 235), R. M. Parsons Laboratory for Water Resources and Hydrodynamics, Dept. Civil Engineering, MIT, 324 pp.

Lee, T. H., and K. P. Georgakakos: 1990, 'A Two–Dimensional Stochastic–Dynamical Quantitative Precipitation Forecasting Model,' *Journal of Geophysical Research–Atmospheres* **95**(D3), 2113–2126.

Peck, E. L.: 1976, *'Catchment Modeling and Initial Parameter Estimation for the National Weather Service River Forecast System,'* NWS HYDRO-31, Natl. Oceanic and Atmos. Admin., Washington, DC.

Rajaram, H., and K. P. Georgakakos: 1987, *'Recursive Parameter Estimation of Conceptual Watershed Respond Models'* (IIHR Report No. 319), Department of Civil and Environmental Engineering and Iowa Institute of Hydraulic Research, The University of Iowa, 169 pp.

Chapter 10

Modeling of Saturated Flow and the Coupling of Surface and Subsurface Flow

Børge Storm
Danish Hydraulic Institute
Agern Allé 5, Dk–2970
Hørsholm, Denmark

Abstract. Saturated flow may play an important role in the description of the flow regime in catchments. Most prominent examples are runoff generation from humid catchment with shallow soils and catchments which are subject to major groundwater development. Groundwater flow has been subject to extensive research in particular with respect to numerical solution techniques, and today models are available which may cover a variety of field problems. The interaction between surface and subsurface water is a problem of significant importance in water resource engineering, and approaches have been developed to accommodate these problems under the constraints of computer technology and field data availability. A brief review of methods developed in the recent years will be presented, along with examples from two case studies.

1. Introduction

Saturated subsurface flow plays a significant role in the hydrological cycle. During drought periods it provides and sustains streamflow through baseflow, while during storm events it may contribute significantly to the stormflow as well as influence the magnitude of the overland flow from a rising water table. Groundwater abstractions for water supply and irrigation may influence the natural recharge and discharge properties and, as such, change the flow regime in the catchment. Comprehensive and ill–managed irrigation schemes may produce areas suffering from waterlogging, a phenomenon which is becoming an increasing problem in many command areas.

Saturated and partly–saturated flow in a porous medium is a broad subject which may include a variety of specialized problems, such as flow in fractured rocks, land subsidence, sea water intrusion, multiaquifer systems, multiphase flow, runoff generation, groundwater development, etc. In connection with groundwater contamination, development of field techniques and statistical methods for assessment of the basic aquifer parameters have been the subject of enhanced research.

It is not intended to address all the above mentioned subjects relevant to the modeling of saturated flow in this presentation. The main emphasis will be put on a review of some of the modeling efforts which have been made to analyze and understand the role of saturated flow in the hydrological cycle in recent years. The chapter will also present some modeling approaches for a description of the interaction between surface and subsurface flow in different hydrological regimes.

D. S. Bowles and P. E. O'Connell (eds.), Recent Advances in the Modeling of Hydrologic Systems, 185–203.
© 1991 *Kluwer Academic Publishers.*

2. Discussion

Development of numerical solution techniques to describe flow of water in saturated and partly saturated porous media has been the subject of considerable study in the past decades. A plethora of numerical models of various complexity have been produced (see e.g. Van der Heijde et al., 1985), and the mathematical foundation for solving the most important boundary value problems in groundwater hydrology is available (Bear and Verruijt, 1987, or other standard textbooks).

Three–dimensional modeling of unsaturated–saturated transient flow in nonhomogeneous, anisotropic aquifers was introduced by Freeze (1971). Since then a series of similar studies have been published—for example see Huyakorn et al. (1986b), among others. However, the computer technology, in terms of memory requirements and CPU times, may still be one of the serious impediments for application of these rigorous three–dimensional approaches in water resource engineering.

Therefore, studies have been mainly concerned with the description of two–dimensional time–variant subsurface flows either as a cross–sectional analysis on hillslopes (profile models), including the unsaturated and saturated zones, or as an areal analysis of horizontal saturated flow in confined or unconfined aquifers.

The use of groundwater aquifers as a resource and as a place for disposal of hazardous waste has implied a need for groundwater models for prediction of groundwater flow in regional and local systems. These "traditional" models, e.g. the Tyson–Weber model (Thomas, 1973), the PLASM model (Prickett and Lonnquist, 1971), or the USGS–MOC model (Konikow and Bredehoeft, 1978), consider the surface water and unsaturated water as boundaries, i.e. discharge to the stream or recharge from the unsaturated zone.

The modeling of hillslope flows has been motivated by, in particular, a desire to describe and predict the runoff from humid catchments with shallow sloping soils. The runoff during storm events from these catchments mainly results from the water–table response in zones contiguous and noncontiguous to the stream and from the formation of seepage areas. Near-field analysis of the migration of contaminant plumes is another example where a profile model may be appropriate.

In many cases, the purely two–dimensional approach is inadequate when the interaction between the surface water and the subsurface water is important. The two–dimensional model can conveniently be extended to a quasi–three–dimensional model by including a vertical component. Examples of such models for unconfined aquifers are the SHE model (Abbott et al., 1986a,b), the IHDM model (Morris, 1980), or PREDIS (Gilding, 1983). For example, in the SHE the subsurface flow regime is a coupled one–dimensional unsaturated and a two–dimensional saturated system, whereas the IHDM model represents the catchment as a series of hillslope planes linked to a one–dimensional surface water system. Numerous models have been developed to account for multilayer systems where the horizontal flows in the aquifers are linked by vertical flow through an aquitard, e.g. Chorley and Frind (1978), Refsgaard and Hansen (1982a).

2.1. Governing Equations

Three–dimensional flow of water in a variable–saturated, porous medium may be described by the following equation:

$$\frac{\partial}{\partial x_i} K_{ij}(\frac{\partial \psi}{\partial x_j} + z_j) = S_s \frac{\partial \psi}{\partial t} + q \quad i,j = 1, 2, 3 \tag{1}$$

where x_i are Cartesian coordinates; t is time; ψ is the pressure head; z_j is the unit vector in vertical direction; $K_{ij} = K_{ij}(\psi)$ is the hydraulic conductivity tensor; q is the volumetric flow rate via source or sinks per unit volume of the medium, and S_s is the specific storage given by:

$$S_s = S\rho g(\alpha + n\beta) + n\frac{\partial S}{\partial \psi} \tag{2}$$

where α and β are the compressibility coefficients of the water and the porous medium respectively; ρ is the density; n is the total porosity; and S is the degree of saturation. Equation (1) is valid under the general set of assumptions of: Darcian flow, isothermic system, incompressible grains, linearly and reversibly elastic soil, only vertical deformation, and negligible osmotic and chemical concentration gradients as well as negligible internal forces and velocity heads in the saturated zone (Freeze, 1971).

Equation (1) may be solved for the following initial and boundary conditions in the domain D:

$$\psi = \psi_0 \; in \; D \quad t = 0 \tag{3}$$

$$q = q_B \; on \; B_1 \quad t > 0 \tag{4}$$

$$\psi = \psi_B \; on \; B_2 \quad t > 0 \tag{5}$$

where ψ_0 is the initial pressure head distribution; q_B is the specified flux normal to a segment of the external boundary B_1; and ψ_B is the specified pressure head on a segment of the external boundary B_2. In addition to Equations (3)–(5), the condition $\psi = 0$ is considered at the seepage face.

For incompressible water and soil, Equation (2) may be reduced to:

$$S_s = C(\psi) \tag{6}$$

where $C(\psi)$ is the specific moisture capacity. The two–dimensional version of Equation (1) in combination with Equations (2) or (6) for a cross–section has been derived and applied in many studies, among others e.g. Rubin (1968), Verma and Brutsaert (1970), Neuman (1973), Cooley (1983), Nieber and Walter (1981), and Hillman and Verschuren (1988).

In traditional areal analysis with groundwater models, a two–dimensional form of Equation (1) is used for essentially horizontal–saturated flow. This is obtained by integration over the depth, yielding:

$$\frac{\partial}{\partial x_i}\left(K_{ij}b\frac{\partial h}{\partial x_j}\right) = S_s b\frac{\partial h}{\partial t} + Q \quad i,j = 1,2 \tag{7}$$

where b is the local saturated thickness of the aquifer, and K_{ij} and S_s represent effective values. Q is the local volume flux per unit area. Equation (7) uses the facts that most aquifers are thin relative to their horizontal dimensions and the flow is essentially horizontal, which is equivalent to assuming that the vertical gradient in the piezometric head $\partial \phi/\partial x_3 = 0$. ϕ is given by $\phi = z + p/\rho g$, where p is the pressure potential. This is referred to as the Dupuit assumption which is the alibi for replacing ϕ with the hydraulic head h. Rushton and Rathod (1985) show that vertical flow components can indeed be inferred from application of the hydraulic head h. Equation (7) may also be used in profile applications by setting $b = 1$ m and changing Q to an area flux per unit area.

2.1.1. Unconfined Aquifers

For unconfined groundwater flow, the state variable h is the position of the water table which defines the upper boundary of the flow domain. Equation (7) is known as Boussinesq's equation. In groundwater applications the specific yield S_y (or drainable porosity) is traditionally employed for the storage coefficient S, assuming that the effects of water and aquifer compressibility are negligible. Brutsaert and El–Kadi (1984; 1986) analyze the effects of partial saturation above the water table and the compressibility effects of the water and aquifer. They propose that the ratio $-\psi_t/D$ can be used as a criterion to assess the relative importance of the flow in the partially saturated zone. The variable $-\psi_t$ is "a representative or characteristic capillary pressure, expressed as height of water column, which is needed to reduce the saturation of a soil by a certain substantial fraction," and D is "a characteristic vertical dimension of the flow region or of the aquifer." They show that for small ratios, the compressibility of the aquifer may become important under conditions of large sudden pressure changes in the water. In situations where the effect of partly saturated flow is not negligible, the compressibility terms can safely be neglected.

The specific yield S_y is in general defined as the volume of water drained from a soil column of a unit horizontal area extending from the water table to the ground surface, per unit decline of the water table. It corresponds to the difference in water content between effective saturation θ_{es} and the residual water content θ_r retained by the soil against gravity.

The specific yield is not a static term, and employing it purely by its definition given above may be erroneous. In situations of relatively fast lowering of the water table, the drainage from the unsaturated zone may lag behind, depending on the soil properties and the actual moisture content; therefore, specific yield is time–dependent. In cases of infiltration the rise in the water table depends on its actual position. Gillham (1984) showed the effect of capillary fringe on the water table rise and found highly disproportionate responses of shallow water tables to rainfall input. This effect is often referred to as the Wieringermeer effect (Hooghoudt, 1947). A second phenomenon which may introduce extreme water table responses is the so–called Lisse effect (Hooghoudt, 1947), caused by air entrapment and an increase in the gas–phase pressure during an infiltrating wetting front. Heliotis and De Witt (1987) have shown field evidence for both phenomena in a wetland area.

In groundwater studies of regional deep aquifers, it may be legitimate to apply $S_y = \theta_{es} - \theta_r$ in the flow equation Equation (7). Temporal variations in the water content distribution above the water table may be small; the time steps applied in the models are usually sufficiently large to assume successive steady state conditions in the partly saturated zone. The vertical recharge to the water table may, in these cases, be calculated empirically.

A more important role is played by the moisture content distribution in the unsaturated zone in areas of shallow aquifers or in situations where irrigation may cause water logging problems. Infiltration may cause significant changes in the soil moisture distribution; the water table response is then highly dependent on the flow of water above the water table. Therefore, a proper solution requires an accurate description of both the specific yield term S_y and the recharge term Q in Equation (7). The recharge Q may be estimated from a mass balance calculation of the unsaturated zone by:

$$Q = \frac{\partial}{\partial t} \int_{z_0}^{z_g} \theta(z)dz + \Sigma q \tag{8}$$

where $\theta(z)$ is the volumetric moisture content; z_0 is a datum; z_g is ground surface elevation; and Σq is the net inflows to the unsaturated zone. The moisture content distribution may be estimated from the one–dimensional version of Richards' equation:

$$C(\psi)\frac{\partial \psi}{\partial t} = \frac{\partial}{\partial z}\left(K_z(\psi)\frac{\partial \psi}{\partial z} + 1\right) - A(z) \tag{9}$$

where C and K are the specific moisture capacity and the capillary conductivity, respectively; A is a sink term; and z is the vertical Cartesian coordinate. The moisture content can be obtained from the moisture retention curve. The approximate solution to Equation (1), found by coupling Equations (7) and (9), should be considered as a practical approach which is required for simulation on catchment sizes of several square kilometers and for simulation periods which exceed just the length of the individual storm events.

2.1.2. Multilayer Systems

In multilayer aquifer systems, the flow through the connecting aquitard may be described by:

$$\frac{\partial}{\partial z}\left(K\frac{\partial h}{\partial z}\right) = S_s\frac{\partial h}{\partial t} \tag{10}$$

where K, S_s and h are aquitard properties. Equations (7) and (10) are coupled via the leakage flow term Q at the aquifer–aquitard interfaces. The solution to Equation (10) can be found analytically, see e.g. Frind (1979), involving a convolution procedure which eliminates spatial discretization of the aquitard. By employing Equation (10) the storage and the transient development of the pressure distribution in the aquitard is taken into account.

In some cases steady–state flow through the aquitard may be appropriate. A simple check of the validity of this assumption is given by de Marsily (1986):

$$\exp\left(-\frac{\pi^2 K \Delta t}{S_s e^2}\right) << 0.5 \tag{11}$$

where e is the aquitard thickness and Δt is the length of time step of the transient calculation in the aquitard. For steady–state flow, the leakage flux may be described directly by Darcy's law. However, storage in the aquitard may still be accounted for by adding half of the storage coefficient to the storage coefficient in each of the adjacent aquifers.

2.2. Coupled Surface–Subsurface Flow in Catchment Models

One of the goals in catchment modeling is to simulate the complex interaction of surface– and subsurface water flow. Runoff generation from humid catchments with shallow soils and unconfined aquifers is a classical problem in hydrology and has been the subject of numerous studies and modeling attempts.

190

The concept of subsurface and surface flow from variable contributing zones is generally recognized. These zones may expand and contract both seasonally and during rainstorms. The actual extent of the zones is determined by physical characteristics of the catchment, as well as the hydrological conditions (see Figure 1). The dispute concerning runoff generation is mainly about the question of the extent to which the Darcian approach to subsurface flow is realistic or not. There have been many studies providing field evidence for either case (Pearce et al., 1986).

Several studies have been concerned with finding practical alternatives to the three-dimensional flow equation, Equation (1), which can be applied to catchment-scale problems with a tolerable loss of accuracy. A promising approach is to separate the unsaturated and saturated domains. The flow in the unsaturated zone is considered to be vertical, modeled by Richard's equation, Equation (9), an assumption which is indeed acceptable in most situations. The horizontal groundwater flow is modeled by the one- or two-dimensional version of Boussinesq equation, Equation (7); the water table acts as an internal boundary, and the recharge to the water table is calculated by Equation (8).

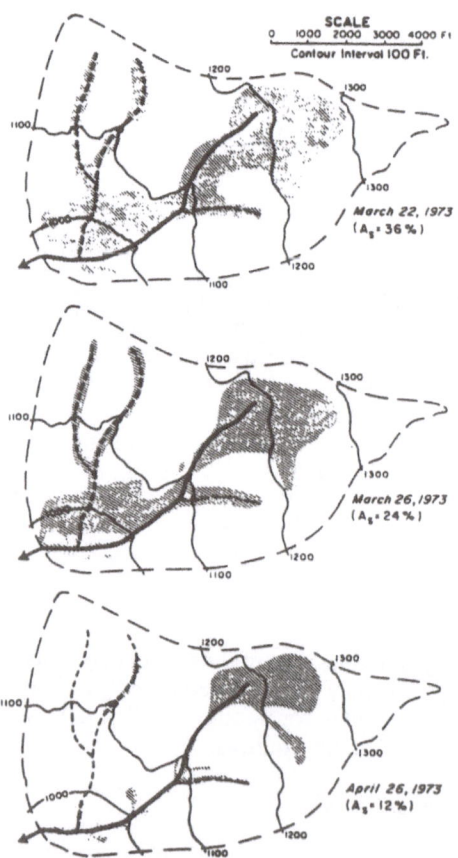

Figure 1. Seasonal variation of the saturated zone in a catchment at Randboro, Quebec (from Dunne et al., 1975).

This approach was proposed by Pikul et al. (1974). Similar principles have been adopted in the PREDIS model (Gilding, 1983) and the SHE model (Abbott et al., 1986a,b). Vachaud and Vauclin (1975) questioned this approach mainly for two reasons: 1) a horizontal flow component exists in the capillary fringe, and 2) it is complex to determine the storage coefficient S (the specific yield S_y in Equation (7)) properly in the coupled system. The first point cannot be refuted but may be of minor importance on a catchment scale with aquifers that have low watertable gradients (Pikul et al., 1975). From a computational point of view, the physical meaning of S_y as it is introduced in Equation (7) becomes irrelevant in a coupled system. A central point in the coupled system approach is that the water table response can only be calculated by employing an iterative procedure, and the estimation of the storage coefficient may influence the number of iterations to be carried out. This point is elaborated below in a brief introduction to the coupling procedure applied in the SHE model.

2.2.1. Method of Coupling Unsaturated and Saturated Zones in the SHE Model

The inherent difficulty in the description of the transient state of a two–domain system arises from the fact that the two domains are solved in parallel in two separate calculation steps. The following method adopted in the SHE model was originally proposed by Belmans et al. (1983) for a transient one–dimensional, finite–difference soil water–root uptake model with a groundwater flux as bottom boundary condition.

As water is recharging the water table, a change in water table position and a redistribution of water simultaneously take place. In a parallel running of the two components, the unsaturated zone (UZ) and the saturated zone (SZ), the water table which is lower boundary in UZ is fixed while the UZ calculations are advanced one time–step. Using the calculated recharge rate Q (from Equation (8)) in the SZ calculations, a mass balance error will be introduced even with a nearly perfect estimate of the storage coefficient S and steady–state groundwater flow.

The coupling procedure in the SHE model uses the fact that if the horizontal flow in the saturated zone are perfectly known (e.g. zero flow), the dynamics of the water table can be calculated by iteratively changing the water table position until an acceptable error in the mass balance is obtained. The algorithm for this procedure is briefly outlined below:

Let us consider grid square (i,j) in the model grid after the UZ calculations have advanced from time n to time $n+1$:

(1) The accumulated mass balance error for column (i,j) at time $n+1$ is:

$$E_{cum}^{n+1} = \int_{z_0}^{z_g} \left[\theta^{n+1}(z) - \theta^n(z) \right] dz + \Sigma q + E_{cum}^n$$

where Σq includes infiltration rate, evapotranspiration loss rates, and the horizontal saturated net outflow rate from column calculated in SZ in the previous time step; z_0 is a datum; and z_g is ground surface elevation.

(2) If $E_{cum}^{n+1} < 0$, too little water is stored in the soil column, and too much is stored if E_{cum}^{n+1} is positive. If abs $(E_{cum}^{n+1}) < E_{max}$, where E_{max} is a tolerance criteria, the mass balance error is acceptable and corrections are not required. The solution proceeds to the next SZ–time step $n->n+1$ using the recharge rate Q obtained from Equation (8).

(3) If abs $(E_{cum}^{n+1}) > E_{max}$ the following corrective steps are carried out:

(a) The UZ calculations are repeated for the four lowest node points using a zero–flux boundary at the top node of this mini UZ system.

(b) For a negative value of E_{cum}^{n+1} the water table position n^n is raised in prescribed intervals and the pressure head profile is calculated. A new value of E_{cum}^{n+1} is obtained. If the change in water table has reduced the error to an acceptable level, the iterations are stopped.

192

(4) A new recharge rate Q is calculated taking the adjustments in the water table into account:

$$Q = -(h^{n^*} - h^n)S_y/\Delta t$$

where h^n and h^{n^*} are the water table elevation at time n and after the iterations in UZ, respectively.

For a positive value of E_{cum}^{n+1} step 3b is performed by lowering the water table position h^n.

The procedure is further illustrated in Figure 2. Assuming zero horizontal SZ flow, it appears that during the time intervals $n -> n+m$, V_1 mm of water has been "lost" while V_2 mm has been "gained." The difference is caused due to an imperfect storage coefficient S (in this case a too large value) and the rise in the water table is, therefore, only balanced to a minor extent. However, by adjusting the water table position in the time interval $n+m-1$ to $n+m$, the mass balance error is corrected until the two volumes V_1^* and V_2^* in Figure 2, column b, are approximately equal.

To illustrate the coupling procedure further, Figure 3 shows one year of simulation on a single soil column. It is assumed that the net horizontal flow to the column is zero. The figure illustrates the moisture conditions in the unsaturated zone and the movement of the water table. The temporal variation of E_{cum} is also shown for two estimates of the specific yield S_y. The general tendency in the variation in E_{cum} is that correction of the water table is required when the recharge to the water table is significant. Note that during a rising water table E_{cum} is negative while during the summer period when capillary rise takes place E_{cum} is positive. It is also noteworthy that recharge in the wet autumn produces a larger rise in the water table than obtained from just employing S_y and without considering the actual water content in the unsaturated zone. Using a specific yield of $(\theta_{es}-\theta_r)/10$ reduces the number of times iteration in the coupling is required, indicating that in this specific case this estimate would be more appropriate than $S_y = \theta_{es}-\theta_r$. One important point, however, is that this one-year simulation requires approximately ten thousand time steps to avoid mass balance errors in the UZ

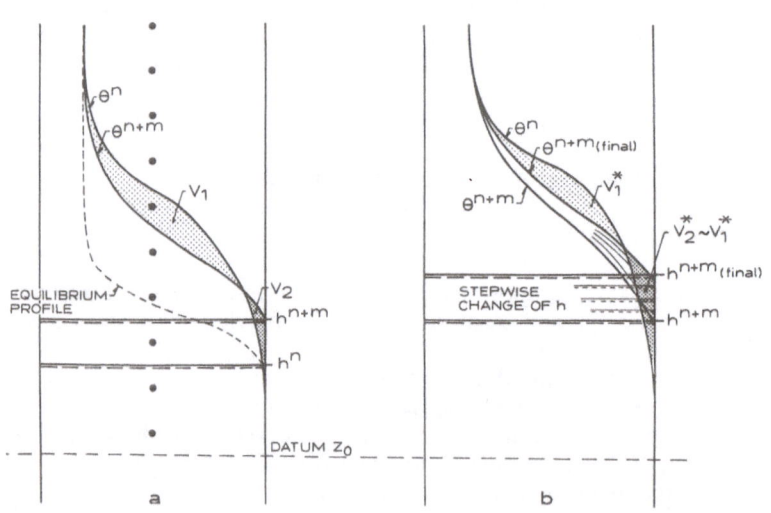

Figure 2. Illustration of the iterative adjustments of the water table in the coupling procedure.

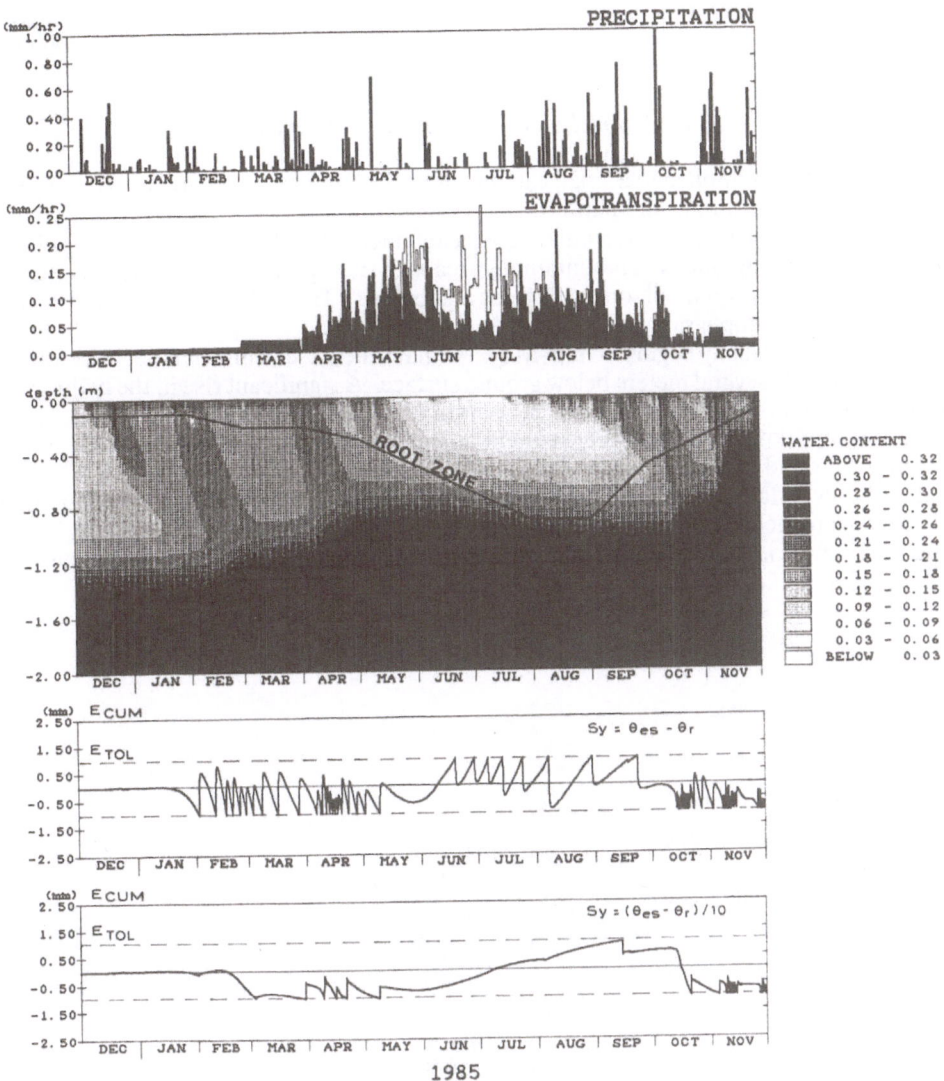

Figure 3. Single column simulation of the actual evapotranspiration--the soil moisture distribution in the unsaturated zone and the water table fluctuation over one year. The iteration frequency in the coupling procedure for two estimates of the specific yield S_y is illustrated.

calculations. It is seen that adjustment of the water table is required approximately a hundred times in the worst case, and as such does not damage the CPU requirements significantly.

2.2.2. Application of the SHE Model on a Hilly Upland Catchment

The SHE model was applied to a small catchment in Thailand. This section will briefly illustrate a few results from the simulations.

The Khlong Yang catchment is a hilly, sloped catchment of 45 sq.kms with shallow soils. It can be divided into two areas: a mountainous area with steep gradients and covered by dense forest, and lower scattered hill tops with grass and bushes. The soil is predominantly loamy. Figure 4 shows the topography.

The wet season usually begins in mid–May. At that time the catchment is very dry and the phreatic surface is several meters below ground surface. A significant rise in the hydrograph due to overland flow is only observed in heavy rain during the early period of the wet season. Later, overland flow becomes more frequent as the soils saturate and the phreatic surface rises to the ground surface.

At mid–July a state of saturation is observed in the low–lying areas which even in periods with no rain are fuelled by subsurface flow from the high elevation areas in the catchment. In these high elevation areas the phreatic surface level is generally below the ground surface.

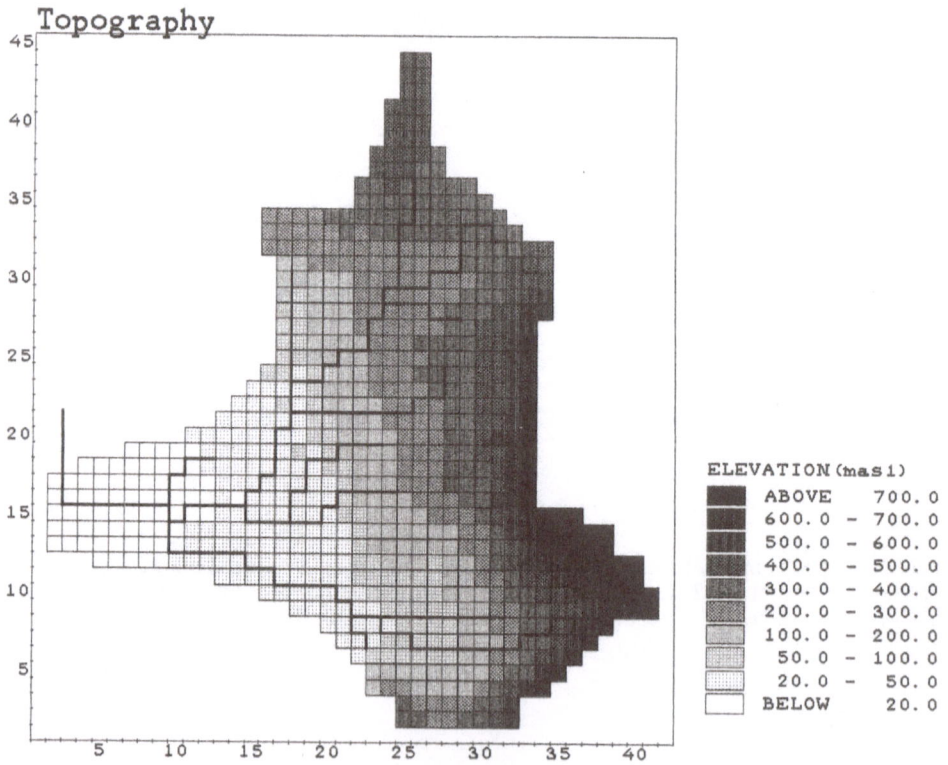

Figure 4. Gray scale representation of the topography in the Khlong Yang catchment.

The SHE model was calibrated on a three-months period, June–August 1986, using observed runoff from the catchment.

The calibration results shall not be commented upon here, but Figures 5 and 6 below serve to illustrate the information which can be extracted from the simulation results with the SHE

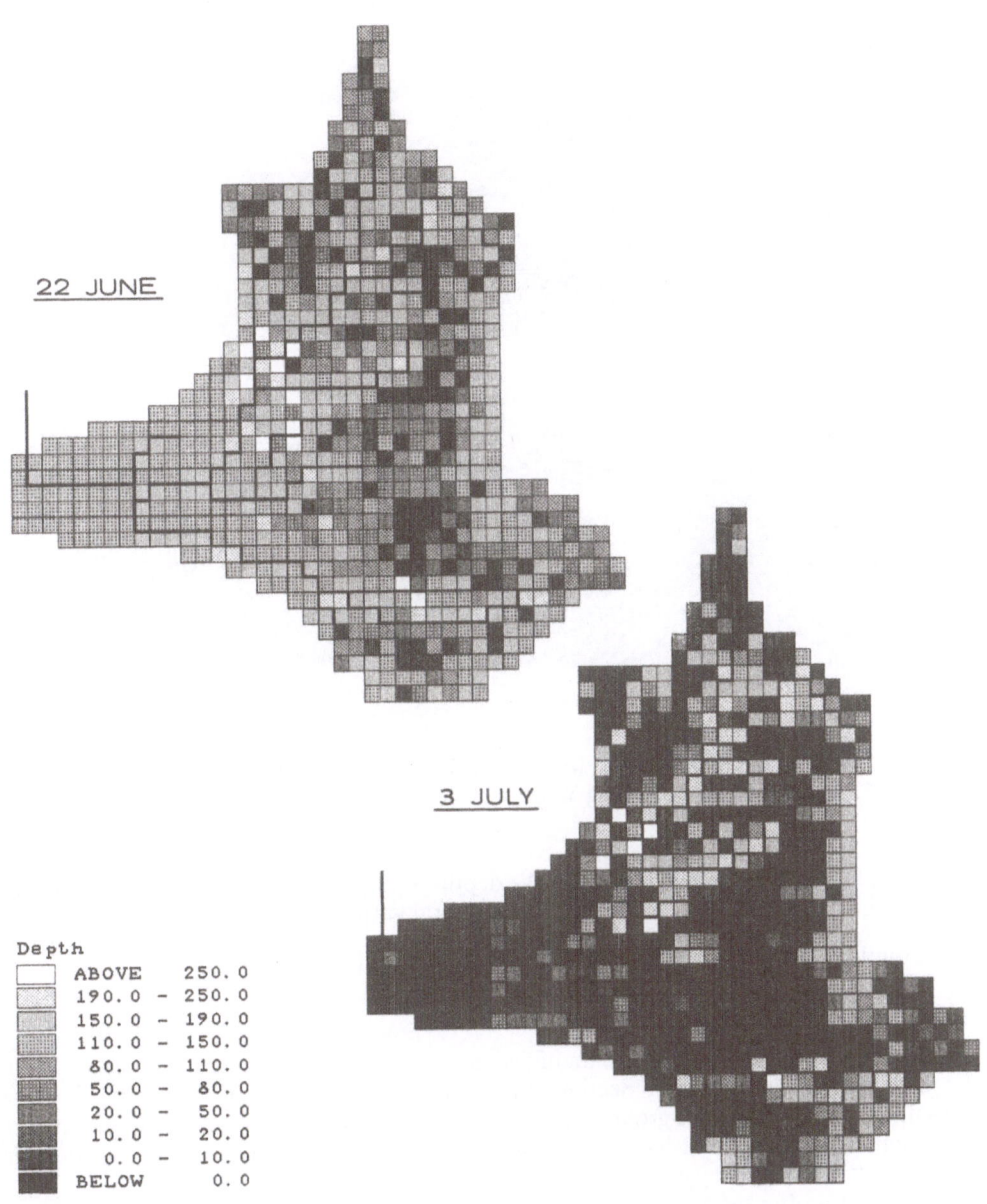

Figure 5. Spatial distribution of the depth to phreatic surface at the two dates; 22 June and 3 July 1986 in the Khlong Yang Catchment.

196

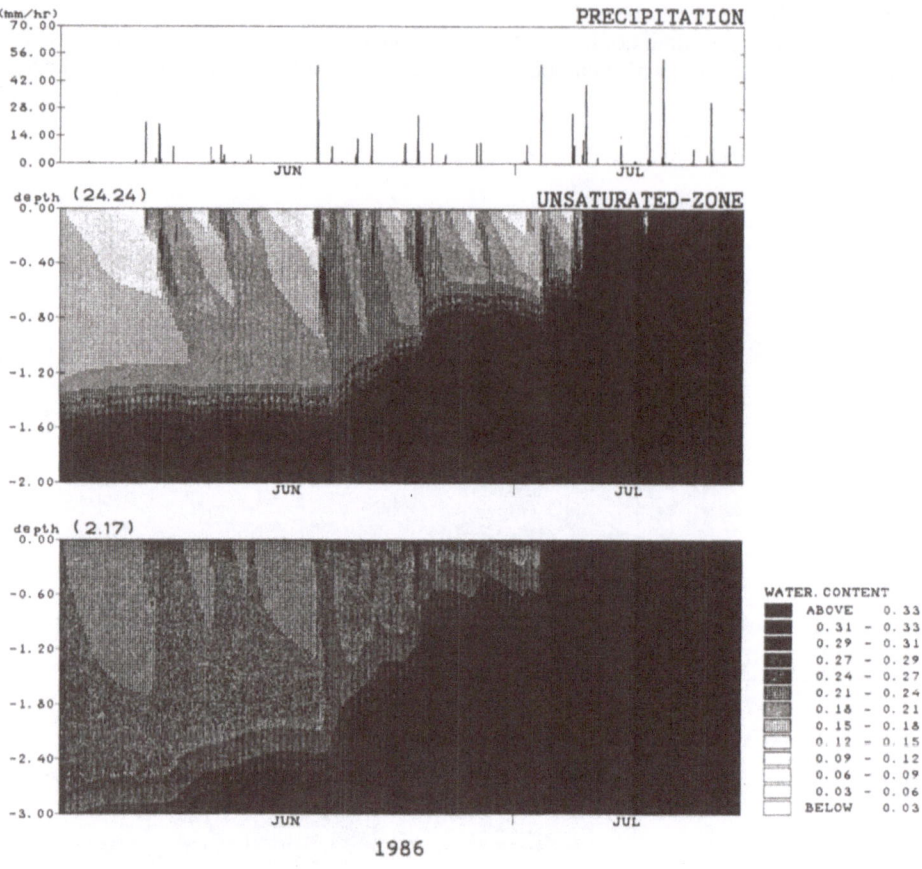

Figure 6. Temporal variation in soil moisture distribution and water table position at two sites; grid square (2,17) and (24,24) in the Khlong Yang Catchment.

model. In Figure 5 the spatial pattern of the depth to the groundwater table is illustrated for two specific dates, and Figure 6 shows the temporal variations for two specific sites.

2.3. Coupled Surface and Subsurface Flow in Multiaquifer Catchments

2.3.1. *Conceptual Model*

The flow regime in catchments with multilayer systems may be considered as an extension to the flow regime described in Section 2.1. Such systems may consist of locally-lying unconfined aquifers near the ground surface, underlain by several layers of confined aquifers and aquitards. There is a hydraulic connection between the layers, but a considerable permeability contrast may exist.

Such systems offer the advantage that the flow in the aquitards may be considered vertical and one-dimensional while the flow in the aquifers is approximately horizontal, an assumption which is generally recognized (Herrera and Yates, 1977; Chorley and Frind, 1978;

Huyakorn et al., 1986a; Refsgaard and Stang, 1981) in the development of a quasi three-dimensional model. The flow regime in the upper unconfined aquifer may be modeled under the assumptions already mentioned.

Multilayer systems may be a result of depositional and erosional processes. A common geological feature in many Danish catchments are quaternary sediments underlain by Danian and Silandian calcareaous sediments which were deposited in connection with glacier and meltwater activities.

Figure 7 shows a schematic cross–section through a river valley and the upstream creek in such a catchment.

The conceptual model for such an environment is a regionally confined aquifer with high permeabilities, overlain by a low permeable aquitard of morain clay. The thickness may range from ten to a hundred meters. The phreatic water table in the glacial deposits is observed to be highly dependent on the topography and is generally lying a few meters below the ground surface. Zones of vertical recharge are found in high elevation areas while discharge areas are situated in river valleys where the piezometric head in the confined aquifer is higher than the watertable and the streams. Lateral flow occurs in both aquifers.

In a catchment which is subject to groundwater development from a regionally confined aquifer, the groundwater abstraction may introduce changes in the flow regime and influence the seasonal variation in the streamflow.

The physical mechanism can briefly be outlined as follows (Refsgaard and Hansen, 1982b):

(1) A groundwater abstraction induces drawdowns of the hydraulic heads, H_l, of the confined aquifer.

(2) The drawdowns in the confined aquifer increase the hydraulic gradient $(H_u-H_l)/L$ between the phreatic and confined aquifers having hydraulic heads H_u and H_l, respectively, and separated by an aquitard of thickness L. This change causes an increase of the average recharge from the phreatic aquifers.

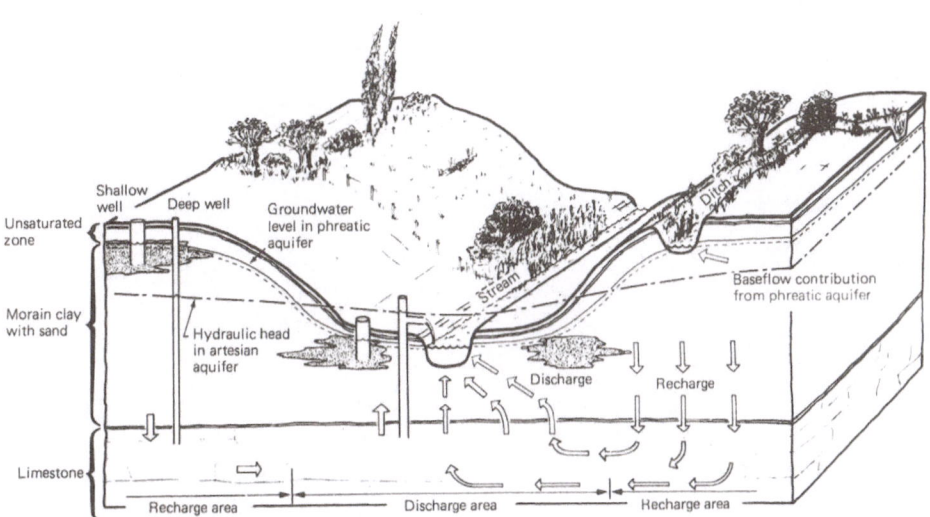

Figure 7. Schematic description of groundwater movements, recharge and discharge areas (from Hansen and Dyhr–Nielsen, 1983).

(3) A larger average recharge rate through the aquitard causes the phreatic water table to decline faster.

(4) This means that the average elevation of the phreatic surface is lowered, relative to the natural and artificial drains responsible for a horizontal routing of water towards the streams in the phreatic aquifers. Thus, compared with the undisturbed situation, a smaller part of the water percolating out of the root zone into the phreatic aquifers will be diverted to the streams as drainflow and baseflow.

2.3.2. *Application of a Catchment Model on a Catchment with a Multilayer Aquifer System*

To verify the above mentioned point a catchment model was developed for the Suså catchment as part of a major Danish International Hydrological Programme (IHP) research project. The model area, the topographic divides, and the groundwater polygonal mesh are shown in Figure 8.

The overall structure of the model is outlined in Figure 9. As can be seen, the model consists of four separate components for the confined regional aquifer, the aquitard, the phreatic aquifers and the root zone, respectively. The time steps in the calculations are one day in all parts of the model.

The nearly horizontal groundwater flow in the confined regional aquifer is accounted for by a traditional, spatially distributed, two-dimensional model of the integrated finite difference type (Thomas, 1973).

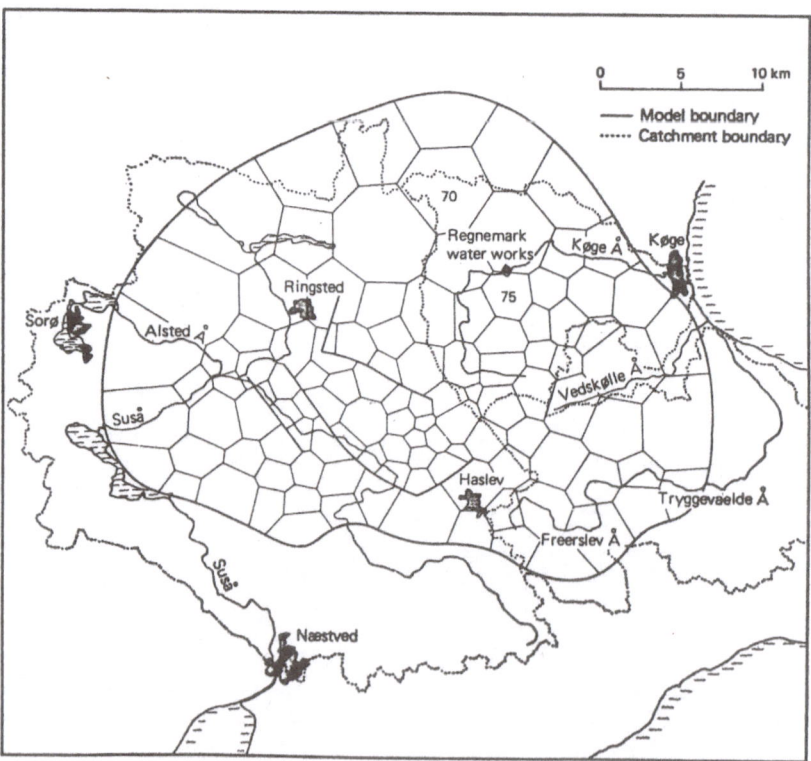

Figure 8. Model area, catchment boundaries, and groundwater polygonal mesh.

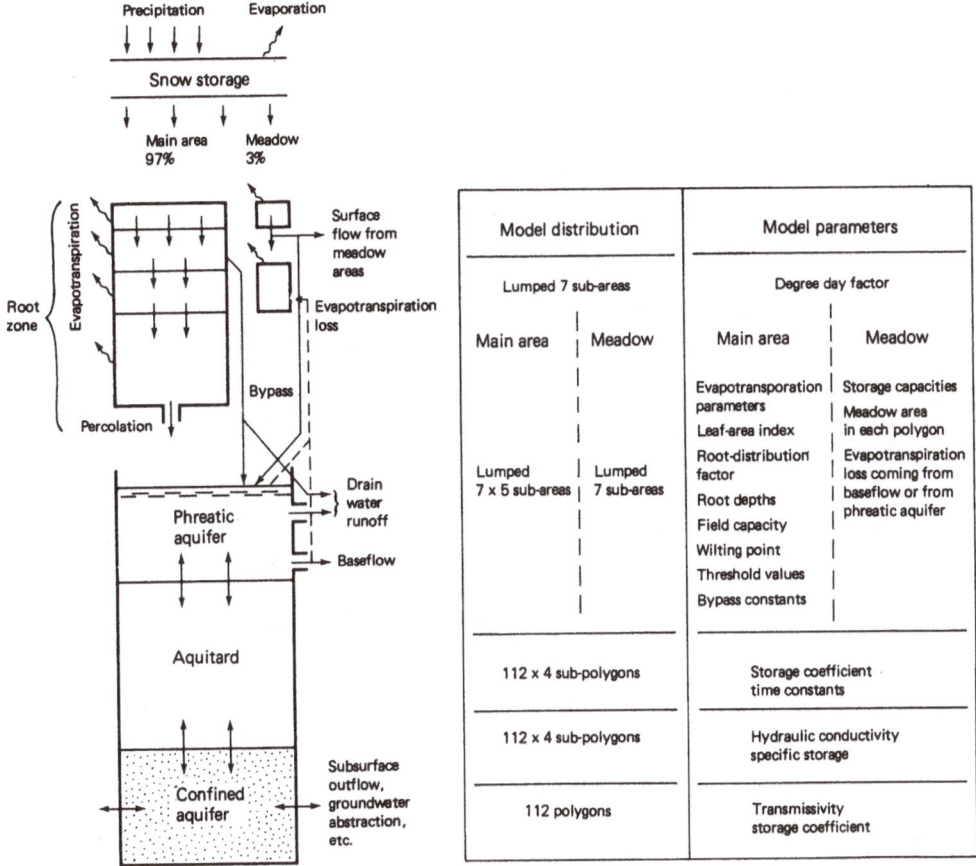

Figure 9. The vertical structure, spatial distribution, and parameters of the hydrological model.

The local vertical exchange of water through the aquitard between the phreatic and the confined aquifer depends on the height difference between the water table and the piezometric head. The nonstationary aquitard flows are calculated by numerical integration of the analytical aquitard response functions (Refsgaard and Stang, 1981).

As previously mentioned, it is a characteristic feature of the phreatic aquifer that the water table elevation, although not constant in time, is highly dependent on the topography. The local vertical exchange of water between the phreatic and the confined regional aquifer depends on the height difference between the water table and the piezometric head of the regional aquifer. From hypographic curves, for example see Figure 10, it is evident that the exchange rate may vary considerably within a polygon. Recharge areas, having downward flow from the phreatic to the confined aquifer, as well as discharge areas, characterized by upward flows, are often experienced within the same polygon. As the discharge to the phreatic aquifers gives rise to the main part of the baseflow contributions to the streams in dry summer periods, it is of the utmost importance for a correct simulation (especially for low streamflows) that the water exchange mechanisms are appropriately modeled.

200

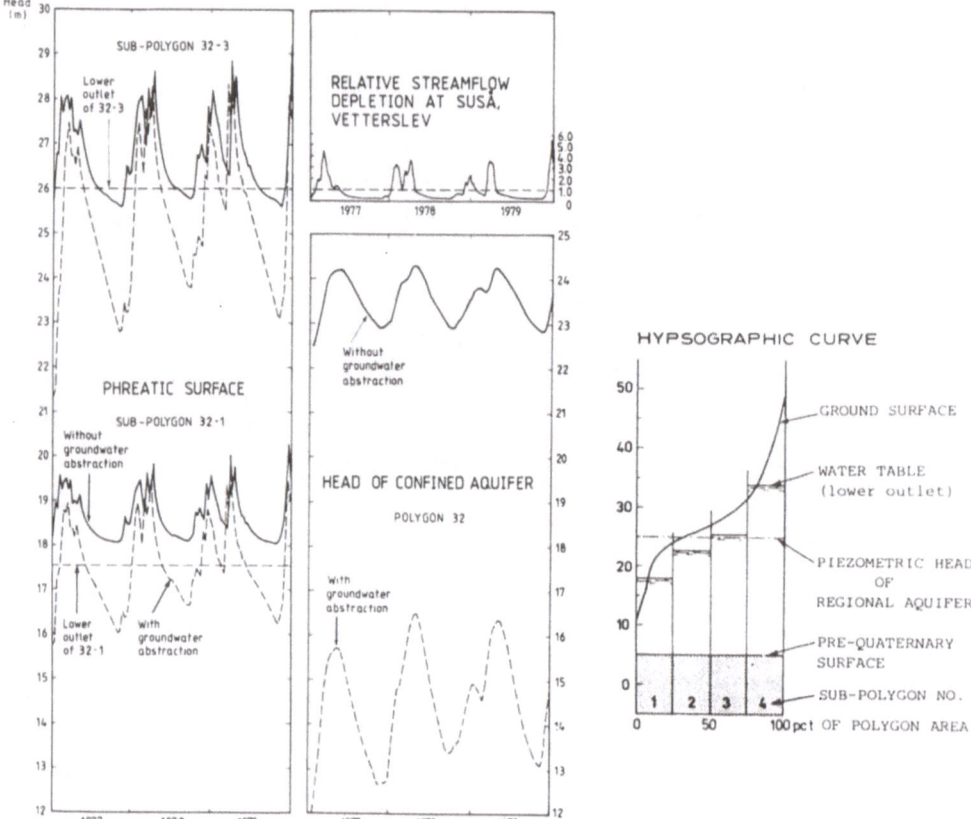

Figure 10. A relative streamflow depletion curve together with the variations of hydraulic heads of the confined and two of the phreatic aquifers in polygon 32, both with and without a groundwater abstraction (from Refsgaard and Hansen, 1982b).

Consequently, on the basis of surface topography, the phreatic aquifers in each polygon have been divided into four equally large sub–areas, assuming homogeneous recharge and discharge conditions within each sub–area.

The horizontal water movement in the phreatic aquifer is modeled as a direct routing of runoff to the rivers, linearly dependent on the height of the water table elevation above a base–flow and a drain water outlet (elevation of tile drain pipes).

For the root zone calculations the catchment is divided into seven sub–areas denoted as precipitation areas with separate precipitation input and soil parameters. Spatial variation in vegetation is accounted for by subdividing each of the seven precipitation areas further into six different vegetation areas in accordance with the actual vegetation distribution within each precipitation area.

Calibration and validation of the model against historical time series of hydraulic heads in the confined aquifer, streamflows, and soil moisture contents in the root zone are described in Refsgaard and Hansen (1982a,b). These results confirm that the proposed model set–up may describe the hydrological regime in the Suså catchment.

The consequences of a major groundwater abstraction on the hydraulic heads are illustrated in Figure 10 together with a relative streamflow depletion curve from the same area. The groundwater abstraction in question is a 21 million m³/year abstraction from the central part of the Suså catchment.

From Figure 10 it is noticed that the recessions in hydraulic heads during the summer season are increased considerably as a consequence of the groundwater abstraction. For sub–polygon 32–3 (the second highest elevation – interval of polygon 32) this means that a larger volume of water must be replenished in the phreatic aquifer before the horizontal water flow as baseflow and drainflow begins in the winter season. For instance, the lowest point of the water table in 1977 is 0.4 m below the lower outlet in the nonabstraction situation, but 3.2 m below in the abstraction situation.

Furthermore, it is seen that the condition of sub–polygon 32–1 has changed from a groundwater discharge situation (upward flow in the aquitard) to a recharge situation (downward flow in the aquitard). The consequence is that there is no longer streamflow contributions from sub–polygon 32–1 in the low flow periods.

Generally, the streamflow depletions from the groundwater discharge areas are more evenly distributed throughout the year than the streamflow depletions from the recharge areas.

3. Conclusions

This presentation has attempted to provide a brief overview of some of the modeling efforts which have taken place in the recent years concerning saturated flow. Particular emphasis has been made on modeling studies in which the saturated zone has been introduced as an integrated part of catchment modeling.

A review of research publications reveals that an array of three–dimensional, finite element models are available for handling groundwater flow and solute transport in unconfined and multilayer aquifer systems. Numerical solution techniques have been developed which have enhanced the computational performance in terms of CPU times and memory requirements. However, for application in practical studies in water resource engineering, these approaches are still not widely used because of limited computer facilities generally available in such studies as well as the difficulties in obtaining adequate hydrogeological data bases.

Less comprehensive approaches are two–dimensional models which may represent the horizontal and vertical time–variant flow on vertical cross–sections, or the horizontal time–variant flow in regional aquifers. In the former type of study the models have been extensively applied in studying the generation of seepage areas in catchments with shallow soils, and the migration of contaminant plumes near waste disposal sites. Regional groundwater models have almost become standard tools in groundwater management and planning. They are widely used to forecast the effects of groundwater exploitation and in the analysis of the migration of contaminants. In fact the major obstacle to the application of two–dimensional models is the definition of the spatially varying aquifer parameters.

Many applications on catchment scale in which the interaction between surface and subsurface flow or the interaction between different layers in the vertical flow may call for a three–dimensional approach. This has long been recognized in connection with groundwater development of multilayer aquifer systems in which a representation of the horizontal flow only may be inappropriate. In predicting runoff from humid catchments the description of the water table position is important and requires a description of the flow conditions in the unsaturated zone as well. Approximate techniques which couple the unsaturated and saturated zones have been developed which may prove useful in many types of water resource studies. This chapter demonstrated the application of two quasi three–dimensional models in two case studies.

202

4. Acknowledgements

Thanks are given to my colleague, J.C. Refsgaard, for his helpful comments on this manuscript.

5. References

Abbott, M. B., J. C. Bathurst, J. A. Cunge, P. E. O'Connell, and J. Rasmussen: 1986a, 'An Introduction to the European Hydrological System – Système Hydrologique Européen, "SHE", 1: History and Philosophy of a Physically–Based, Distributed Modelling System,' *J. Hydrol.* **87**, 45–69.

Abbott, M. B., J. C. Bathurst, J. A. Cunge, P. E. O'Connell, and J. Rasmussen: 1986b, 'An Introduction to the European Hydrological System – Système Hydrologique Européen, "SHE", 2: Structure of a Physically–Based, Distributed Modelling System,' *J. Hydrol.* **87**, 61–77.

Bear, J., and A. Verruijt: 1987, *Modelling Groundwater Flow and Pollution*, D. Reidel, Dordrecht.

Belmans, C., J. G. Wesseling, and R. A. Feddes: 1983, 'Simulation Model of the Water Balance of a Cropped Soil: SWATRE,' *J. Hydrol.* **63**, 271–286.

Brutsaert, W., and A. I. El-Kadi: 1984, 'The Relative Importance of Compressibility and Partial Saturation in Unconfined Groundwater Flow,' *Water Resour. Res.* **20**, 400–408.

Brutsaert, W., and A. I. El-Kadi: 1986, 'Interpretation of an Unconfined Groundwater Flow Experiment,' Technical Note, *Water Resour. Res.* **22**(3), 419–422.

Chorley, D. W., and E. O. Frind: 1978, 'An Iterative Quasi-Three-dimensional Finite Element Model for Heterogeneous Multiaquifer Systems,' *Water Resour. Res.* **14**(5), 943–952.

Cooley, R. L.: 1983, 'Some New Procedures for Numerical Solution of Variably Saturated Flow Problems,' *Water Resour. Res.* **19**(5), 1271–1285.

Dunne, T., T. R. More, and C. H. Taylor: 1975, 'Recognition and Prediction of Runoff Producing Zones in Humid Regions,' *Hydrol. Sci. Bull.* **20**, 305–327.

Freeze, R. A.: 1971, 'Three-Dimensional, Transient, Saturated–Unsaturated Flow in a Groundwater Basin,' *Water Resour. Res.* **7**(2), 347–366.

Frind, E. O.: 1979, 'Exact Aquitard Response Functions for Multiple Aquifer Mechanics,' *Adv. in Water Resources* **2**, 77–82.

Gilding, B. H.: 1983, 'The Soil–Moisture Zone in a Physically–Based Hydrologic Model,' *Adv. in Water Resources* **6**, 36–43.

Gillham, R. W.: 1984, 'The Capillary Fringe and its Effects on Water Table Response,' *J. Hydrol.* **67**, 307–324.

Hansen, E., and M. Dyhr-Nielsen: 1983, 'The Suså Project: Modelling for Water Resources Arrangement', *Nature and Resource* **19**(3), 10–18.

Van der Heijde, P., Y. Bachmat, J. D. Bredehoeft, B. Andrews, D. Holtz, and S. Sebastian: 1985, *Groundwater Management: The Use of Numerical Models. 2 Edn.*, American Geophysical Union, Washington, DC.

Heliotis, F. D., and C. B. DeWitt: 1987, 'Rapid Water Table Responses to Rainfall in a Northern Peatland Ecosystem,' *Water Resources Bulletin* **23**(6).

Herrera, I., and R. Yates: 1977, 'Integrodifferential Equations for Systems of Leaky Aquifers and Applications, 3. A Numerical Method of Unlimited Applicability,' *Water Resour. Res.* **13**(4), 725–732.

Hillman, G. R., and J. P. Verschuren: 1988, 'Simulation of the Effects of Forest Cover, and its Removal on Subsurface Water,' *Water Resour. Res.* **24**(2), 305–314.

Hooghoudt, S. B.: 1947, *'Waarnemingen van Grondwaterstanden voor de Landbouw,'* Verslagen Technische Bijeenkomsten 1–6, Commissie voor Hydrologisch Onderzoek TNO. The Hague, 94–110 (published in 1952).

Huyakorn, P. S., B. G. Jones, and P. F. Andersen: 1986a, 'Finite Element Algorithms for Simulating Three-dimensional Groundwater Flow and Solute Transport in Multilayer Systems,' *Water Resour. Res.* **22**(3), 361–374.

Huyakorn, P. S., E. V. Springer, V. Guvanasen, and T. D. Wadsworth: 1986b, 'A Three-dimensional Finite Element Model for Simulating Water Flow in Variably Saturated Porous Media,' *Water Resour. Res.* **22**(13), 1790–1808.

Konikow, L. F., and J. D. Bredehoeft: 1978, 'Computer Model of Two-dimensional Solute Transport and Dispersion in Groundwater,' *U.S. Geol. Survey Techniques of Water-Resources Inv.*, Book 7, Chap. C2.

de Marsily, D.: 1986, *Quantitative Hydrogeology*, Academic Press, Inc.

Morris, E. M.: 1980, 'Forecasting Flood Flows in Grassy and Forested Basins Using a Deterministic Mathematical Model,' *Hydrological Forecasting (Proc. Oxford Symp., April 1980)*, IAHS Publ. No. 129, pp. 247–255.

Nieber, J. L., and M. F. Walter: 1981, 'Two–dimensional Soil Moisture Flow in a Sloping Rectangular Region: Experimental and Numerical Studies,' *Water Resour. Res.* **17**(6), 1722–1730.

Neuman, S. P.: 1973, 'Saturated–Unsaturated Seepage by Finite Elements,' *J. Hydraul. Div. Amer. Soc. Civil Eng.* **99**, 2233–2251.

Pearce, A. J., M. K. Stewart, and M. G. Sklash: 1986, 'Storm Runoff Generation in Humid Catchments 1. Where Does the Water Come From,' *Water Resour. Res.* **22**(8), 1263–1272.

Pikul, M. F., R. L. Street, and I. Remson: 1974, 'A Numerical Model Based on Coupled One–dimensional Richards and Boussinesq Equations,' *Water Resour. Res.* **10**(2), 295–302.

Pikul, M. F., R. L. Street, and I. Remson: 1975, 'Reply,' *Water Resour. Res.* **11**(3), 510.

Prickett, T. A., and C. G. Lonnquist: 1971, 'Selected Digital Computer Techniques for Groundwater Evaluation,' *Illinois State Water Survey Bulletin 55*.

Refsgaard, J. C., and O. Stang: 1981, *'An Integrated Groundwater/Surface Water Hydrological Model,'* Danish Committee for Hydrology, Report Suså H 13, 122 pp.

Refsgaard, J. C., and E. Hansen: 1982a, 'A Distributed Groundwater/Surface Water Model for the Suså Catchment, Part 1: Model Description,' *Nordic Hydrology* **13**(5), 299–310.

Refsgaard, J. C., and E. Hansen: 1982b, 'A Distributed Groundwater/Surface Water Model for the Suså Catchment, Part 2: Simulation of Streamflow Depletion Due to Groundwater Abstraction,' *Nordic Hydrology* **13**(5), 311–322.

Rubin, J.: 1968, 'Theoretical Analysis of Two–dimensional, Transient Flow of Water in Unsaturated and Partly Unsaturated Soils,' *Soil Sci. Soc. Amer. Proc.* **32**(5), 607–615.

Rushton, K. R., and K. S. Rathod: 1985, 'Horizontal and Vertical Components of Flow Deduced from Groundwater Heads,' *J. Hydrol.* **79**, 261–278.

Thomas, R. G.: 1973, *'Groundwater Models. Irrigation and Drainage,'* Spec. Pap. Food Agricultural Organis., No. 21, U.N., Rome.

Vachaud, G., and M. Vauclin: 1975, 'Comments on "A Numerical Model Based On Coupled One–dimensional Richards and Boussinesq Equations" by Mary F. Pikul, Robert L. Street, and Irwin Remson,' *Water Resour. Res.* **11**(3), 506–509.

Verma, R. D., and W. Brutsaert: 1970, 'Unconfined Aquifer Seepage by Capillary Flow Theory,' *J. Hydraul. Div. Am. Soc. Civ. Eng.* **96**(HY6), 1331–1344.

Chapter 11

Environmental Features Important in Nonpoint Source Models - Microclimatology

Donn G. DeCoursey
USDA–ARS
Hydro–Ecosystem Research Unit
243 Federal Building
301 South Howes, P.O. Box E
Fort Collins, Colorado 80522 U.S.A.

Abstract. Models designed to aid research in fields such as nonpoint source pollution must be designed with the ability to realistically simulate the effect of management changes on the soil physical, chemical, and biological environment. This and the next chapter describe features of some of the more complex models that have been developed to aid in analyzing nonpoint source pollution. These features are microclimatology, plant or crop growth, and soil properties as influenced by management and tillage. The discussion of microclimatology in this chapter covers the general heat budget of crop canopies and techniques for calculating energy mass fluxes, surface radiation balance and its components, soil heat flux, and the aerial and crop resistances affecting energy transport. The discussion of crop growth in Chapter 12 describes methods of calculating photosynthesis and light interception, respiration, translocation, root activity and nutrient uptake, plant development and morphogenesis, and some examples of crop growth models. Also in Chapter 12, I discuss how soil properties are influenced by management. It covers estimates of interception and depression storage, soil bulk density and hydraulic properties.

1. Introduction

Recent advances in use of hydrologic models extends more and more frequently into other disciplines where they provide a support function rather than the main focus of attention. One example is analysis of groundwater quality problems, in particular identifying the fate of nonpoint source contaminants such as pesticides, nutrients, and acid rain. Point source problems of groundwater contamination are generally analyzed using the tools of groundwater hydraulics with solute transport processes piggybacked on them. Frequently the solution to point source problems involves insertion of a curtain wall into the aquifer and pumping of a well field. In some cases increased levels of biological degradation are induced. However, nonpoint sources of groundwater contamination are usually so extensive that treatment of the aquifer is prohibitively expensive, and treatment, if it is to be effected, must be in the root zone where we have some degree of control on the fate of the contaminants before they move out.

Up until the last few years research to evaluate the effect of land use change or management on the disposition and movement of water was through the use of experimental plots. They are still used, but because of their expense are supplemented by model studies. There are several reasons besides the issue of cost. Problems of temporal and spatial variability make it virtually impossible, using sampling techniques, to detect subtle differences in response of

205

D. S. Bowles and P. E. O'Connell (eds.), Recent Advances in the Modeling of Hydrologic Systems, 205–240.

the soil environment that come about as a result of management changes. It is also possible to analyze many more alternatives using models than it is using experimental data.

Models designed to aid research in fields such as nonpoint source pollution must be designed with the ability to realistically simulate the effect of management changes on the soil physical, chemical, and biological environment. If the model does not have the ability to do this, then parameters of the model must be "calibrated" to field data that realistically respond to the change. We have already stated that this may not be possible. In this chapter emphasis is placed on three features of some of the more complex models that have been developed to aid in analysis of problems such as nonpoint source pollution. These are microclimatology, plant or crop growth models, and soil properties as influenced by management and tillage.

Water movement in the soil is another component of these models that is extremely important. If water movement is not described in physically realistic terms, i.e. driven by tensiometric gradients in the unsaturated state and consideration given to macropore structure when water is ponded on the surface, then erroneous estimates of water and solute movement are possible. Other chapters deal extensively with this subject, thus it is not discussed here.

To illustrate the significance of microclimatology, plant growth, and soil property features on water and solute movement, assume that our objective is to compare conventional and limited or minimal tillage of corn on the potential for nitrate contamination of groundwater. In this case we must know how subsurface water movement would be affected by the two different management scenarios. In this example, the mulch provided by the dead cover material in the limited tillage corn will lead to a change in the soil temperature regime and distribution of soil water content from that of conventional tillage. This is accomplished by reducing surface evaporation and increasing shading, thus increasing water available for transpiration by the corn and a change in evaporation and movement of soil water in the profile. The cover material and reduced tillage can change the infiltration rates because of increased macropore structure and different soil moisture distribution in the surface layers. Tillage also affects infiltration rates in conflicting ways. It can break up macropore structure, such as worm holes and old root cavities, but it can also create large macropores such as those observed in a freshly plowed field. Reconsolidation of soil after tillage can lead to dramatic changes in infiltration rate over a few days or weeks' time. In one case (DeCoursey, 1980) two orders of magnitude change in infiltration rates were observed over one month as tilled soil was reconsolidated by rainfall.

The conflicting nature of the effect of no-tillage farming practices is also exemplified by 1988 weather conditions. In Nebraska, no-till corn produced better than conventional tillage because there was less water loss under no-till conditions. However, in Iowa and Illinois no-till crops had reduced yields because cover and shading by residue produced cool wet soils, inducing shallow rooted crops that ran out of water later in the season. The effects just described may be either reduced or amplified by soil type, climate, type of crop, topographic features, and management practice; thus analysis by any other means than detailed mathematical models is almost impossible.

In the following presentation, aspects of microclimatology are discussed. Crop growth and the effects of management on soil properties are discussed in Chapter 12. The degree to which these processes are incorporated into a model depends upon the objective of the model. In some models these three processes may be treated very superficially, and in other cases they may be simulated in great detail. I cannot begin to cover the wealth of information that exists on these topics in the space permitted; thus, I hope to identify a few sources of information that can lead users into as much detail as they would like. Because of numerous references and use of existing figures, nomenclature is not consistent; variables are defined following their use.

Microclimatology is the study of climate within a few meters of the ground surface. In this region of the Earth's atmosphere energy, temperature, and moisture gradients are greater than in any other region of the atmosphere. These gradients, combined with rapid changes in time and elevation, make microclimate much different from the more moderate and stable climate of the upper atmosphere. It is in this interesting layer of the atmosphere that we live, grow our crops, and raise our livestock. Since plants play such an important role in the disposition of water and solutes in the soil surface, we need to understand the microclimate in which they grow. Much of my presentation on the subject is taken from an ASAE Monograph, *Modification of the Aerial Environment of Plants* by Barfield and Gerber (1979); Section 3 – Energy Flow in the Plant Environment System. I supplement their material with frequent reference to Rosenberg et al. (1983) and other authors.

2. Discussion

Solar radiation provides almost all of the energy received at the Earth's surface. Some of the radiant energy is reflected back to space, and the Earth reradiates some energy back in the thermal waveband. The radiant energy remaining at the Earth's surface is the net radiation R_n. Net radiation drives the physical processes responsible for plant growth, heating the soil, heating the air, and driving evapotranspiration. The energy balance of these processes is expressed as:

$$R_n + G + A + \lambda ET + P + M = 0 \tag{1}$$

in which G is the flux of heat into or out of the soil, A is the flux of sensible heat between the air and the soil or plant surfaces, λET is the latent heat required to vaporize water through surface evaporation (E) and plant transpiration (T), P is the energy fixed in plants by photosynthesis, ET is evapotranspiration in units of LT^{-1}, λ is the latent heat of vaporization, and M is the energy involved in miscellaneous processes such as heat storage in the plant canopy and respiration.

For purposes of this presentation terms for photosynthesis and respiration are negligible and will not be discussed. Radiation balance and estimates of its components, sensible and latent heat fluxes and their calculations, and soil heat flux and its simulation will be discussed. Resistances affecting energy transport will also be described.

3. Radiation Energy Balance

The radiation balance that produces net radiation, R_n, required to drive plant growth and related processes is given by:

$$R_n = R_{s_i} - R_{s_o} + R_{T_i} - R_{T_o} \tag{2}$$

in which the R_S terms and the R_T terms are solar or shortwave radiation and thermal or longwave radiation, respectively. Specifically, R_{S_i} is total incoming solar radiation, R_{S_o} is outgoing or reflected solar radiation, R_{T_i} is incoming thermal radiation and R_{T_o} is outgoing thermal radiation (Shaw and Decker, 1979) made up of reflected thermal radiation and Earth's radiated longwave energy.

Effective incoming solar radiation is dominated by wavelengths between 0.15 and 5.0 microns (Wallace and Hobbs, 1977) and is predominantly the ultraviolet (4 percent), visible (44 percent), and short infrared radiation (52 percent) (Berry et al., 1945). It consists of direct beam and diffuse components. Diffuse radiation is received from all parts of the sky with slightly more coming from the vicinity of the sun. When the sky is overcast and at night, diffuse radiation accounts for 100 percent of incoming solar radiation. When the sun is high and the sky clear, diffuse radiation accounts for a little over 15 percent of incoming solar radiation. The flux of direct beam radiation is a function of latitude, clearness of the sky, time of day, and season (Shaw and Decker, 1979).

Solar radiation reaching the surface is absorbed, reflected, or transmitted. Albedo, which is the proportion of the solar radiation that is reflected, is wavelength dependent. Albedo values for crops range from 10 to 25 percent, for dry grass 20 to 30 percent, and for snow 50 to 95 percent (see Table 1.3 on page 44 of Rosenberg et al., 1983).

Radiation is directly related to the temperature of the radiating body, frequently referred to as a blackbody. A blackbody is a hypothetical mass that absorbs all incident radiation and emits the maximum possible radiation in all wavelength bands and all directions. The gain or loss of energy, ΔE, of a blackbody at an absolute temperature, T_1, (degrees Kelvin), placed in an enclosure at absolute temperature, T_2 is given by:

$$\Delta E = \sigma \, (T_1^4 - T_2^4) \tag{3}$$

in which σ is the Stefan–Boltzman constant, $5.67 \times 10^{-8} \, Jm^{-2} s^{-1} \, {}^\circ K^{-4}$. The energy flux density, E, of radiation from a body is:

$$E = \epsilon \, \sigma \, T^4 \tag{4}$$

in which ϵ is the emissivity, the fraction of radiant energy emitted by the body (for a blackbody, $\epsilon = 1.0$). The Wien displacement law describes the wavelength of maximum emission for a blackbody in terms of its Kelvin temperature:

$$\lambda_{max} = \frac{2897.}{T} \tag{5}$$

Thus the sun, at approximately $6100\,{}^\circ K$, emits its maximum radiation at about .5 μm, which is in the visible range. The soil, plants, and their atmospheres, are much cooler than the sun; thus, they emit their radiation in the longer–waved infrared band of the spectrum. At these wavelengths CO_2 and H_2O vapor absorb the energy, thus preventing its escape from the Earth and trapping the longwave radiation. This trapping is responsible for the greenhouse effect that is claimed to be increasing global temperatures.

From the microclimate perspective clouds reflect, absorb, and transmit solar radiation thus considerably affecting the incoming solar energy. Reflection varies from 45–70 percent for thin stratus clouds to 80–90 percent for heavy cumulonimbus clouds. Absorption can vary from 1–8 percent for thin stratus to 10–20 percent for the heavy cumulonimbus clouds.

Figure 1, from Shaw and Decker (1979), shows the radiation balance over a 24–hour period for a fescue meadow near Columbia, MO. The radiation flux densities are averages for the period July 1–15 during the years 1965–74 and were measured on a horizontal surface. If values shown on Figure 1 are added algebraically, the result would be net solar radiation. The value would be negative for nighttime hours, but if the sky were overcast, net radiation would approach zero. Similar radiation balance components from other areas of the globe vary considerably from Figure 1 as a result of snow cover and seasonal variation in the albedo and moisture content of the crop and soil surface. Figure 2 shows how the radiation balance changes over a year. It was taken from the same field as Figure 1 and represents average five day periods for the years 1965–74. For more discussion of solar energy, see Chapter 1 of Rosenberg et al. (1983).

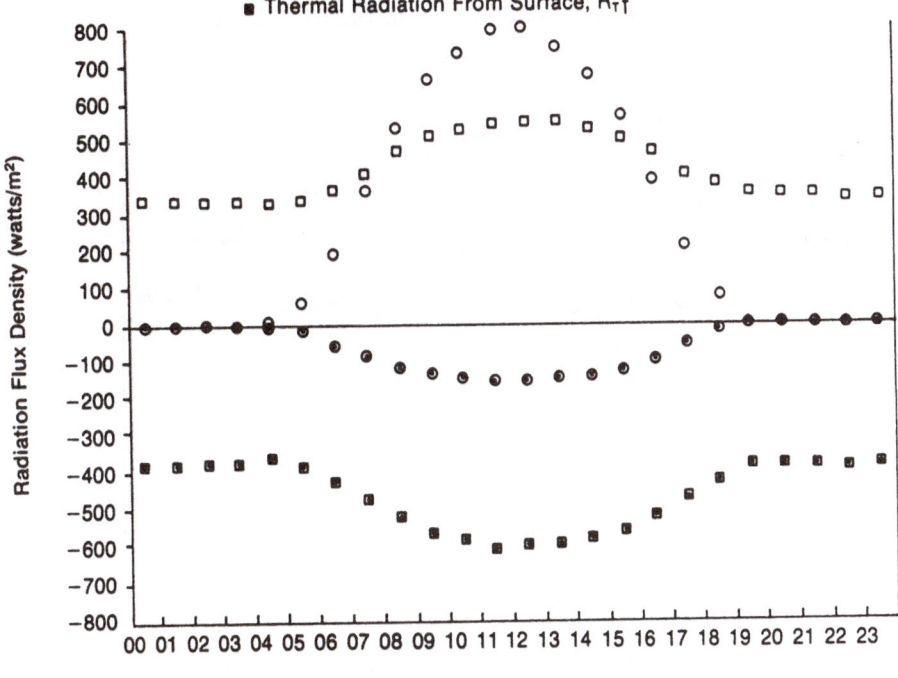

Figure 1. Hourly radiation balance for a fescue meadow near Columbia, MO. The radiation flux densities are averages for July 1–15 during the years 1965–1974 from measurements on a horizontal surface (Shaw and Decker, 1979).

3.1. Estimating Radiation Balance and its Components

Methods of estimating radiation balance and its components were abstracted from Davies and Idso (1979).

Solar energy reaching the Earth's surface is controlled by the atmosphere's ability to emit, absorb, and scatter radiation. Equations for describing radiative transfer were derived by Kondrat'yev (1969). The expression for incoming shortwave radiation, R_{s_i}, is:

$$\frac{\cos\theta}{\rho}\frac{dI_\lambda}{dz} = \frac{s_\lambda}{4\pi}\int_{\omega=4\pi} I_\lambda(z,r')\,\gamma_\lambda\,(z,r',r)\,d\omega - (k_\lambda + s_\lambda)\,I_\lambda\,(z,r) \tag{6a}$$

and for thermal or longwave radiation, R_{T_i}, in cloudless skies with no fog or pollutants is:

$$\frac{\cos\theta}{\rho}\frac{dI_\lambda}{dz} = k_\lambda(E_\lambda - I_\lambda) \tag{6b}$$

In these equations θ is the angle of incidence of the radiant beam, I_λ the monochromatic radiation intensity at wavelength λ, ρ density of air, z height, k_λ the mass absorption

210

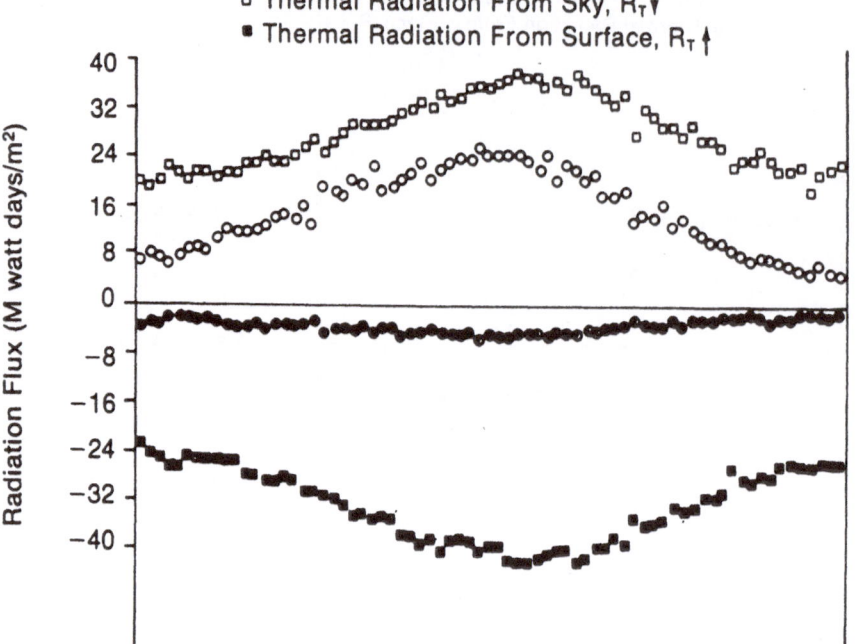

○ Solar Radiation From Above (Direct and Diffuse), $R_s \downarrow$
● Solar Radiation Reflected From Surface, $R_s \uparrow$
□ Thermal Radiation From Sky, $R_T \downarrow$
■ Thermal Radiation From Surface, $R_T \uparrow$

Figure 2. Radiation balance for a fescue meadow near Columbia, MO. The radiation flux densities are averaged for five–day periods for 1965–1974 from measurements on a horizontal surface (Shaw and Decker, 1979).

coefficient, s_λ the mass scattering coefficient, E_λ the monochromatic thermal radiation intensity, $(1/4\pi)\gamma\gamma_\lambda(z, r', r)$ a scattering function which accounts for scattering of an impinging ray from direction r' into the direction r (the direction of I_λ), and ω is solid angle.

Solution of 6a is more difficult than 6b because of complexity of the scattering function and the wavelength dependence of the absorption and scattering coefficients. In a cloudless sky the main absorbers are water vapor, ozone, and aerosols.

3.1.1. Incoming Shortwave Radiation

Solution of the shortwave equation, 6a, for cloudless sky conditions is obtained by simplifications that produce a general numerical procedure that produces consistently satisfactory estimates. Davies and Idso (1979) show derivation of the following solution for incoming shortwave radiation R_{s_i} :

$$R_{s_i} = I(z_T) \cos \theta (\psi_{0_3} - a_w)\psi_{aa}[\psi_R \psi_{as} + (1 - \psi_R \psi_{as})B] \qquad (7)$$

in which $I(z_T)$ is the radiation intensity at the top of the atmosphere, a_w is the absorptance of water vapor, B is the ratio of forward to total scatter, and the ψ terms are transmittances through the absorbing water vapor, ozone, and aerosols. See Davies and Idso (1979) for a discussion of expressions to evaluate the ψ and B terms. Even with all the simplifications described, atmospheric transmittance can be calculated to within 1.5 percent; see Figure 3, taken from Idso (1969). Many researchers use this technique for estimating incoming solar radiation where actual measurements are not available.

The effects of clouds on incoming solar energy must be incorporated into solar radiation models. The problem here is that cloud patterns can change drastically from hour to hour and observations do not include enough information on extent, number of layers, cloud depth, or type to permit short time estimates. Many linear and nonlinear empirical relationships have been developed. Perhaps the best method requires cloud type and amount at low, middle, and high levels, with total transmittance being the product of transmittance values at all levels. Results using these methods show considerable variability for daily estimates, but for 5- and 10-day mean values estimates are within 10 percent of measured estimates; see Figure 4.

3.1.2. Incoming Longwave Radiation

Longwave radiation coming from the radiating atmosphere is obtained by integrating Equation (6b) between the top of the atmosphere and the ground surface. This incoming flux can be represented by the sum of contributions from the cloudless sky, high clouds, middle clouds, and low clouds. Fluxes from each of these components are estimated graphically from

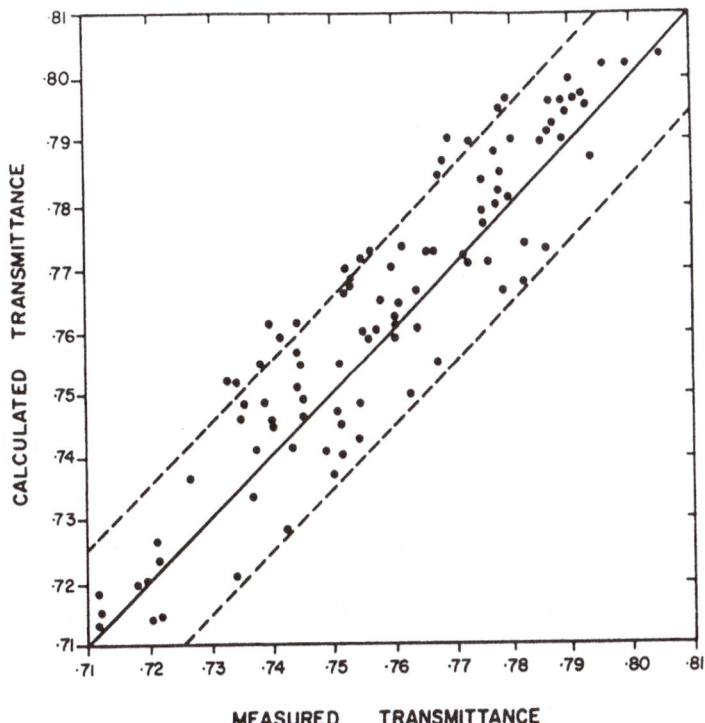

Figure 3. Correlation between calculated and measured atmospheric transmittance for solar radiation on individual cloudless days at Phoenix, AZ. Dashed lines depict a departure of ± 1.5 percent from 1:1 correspondence (Davies and Idso, 1979).

212

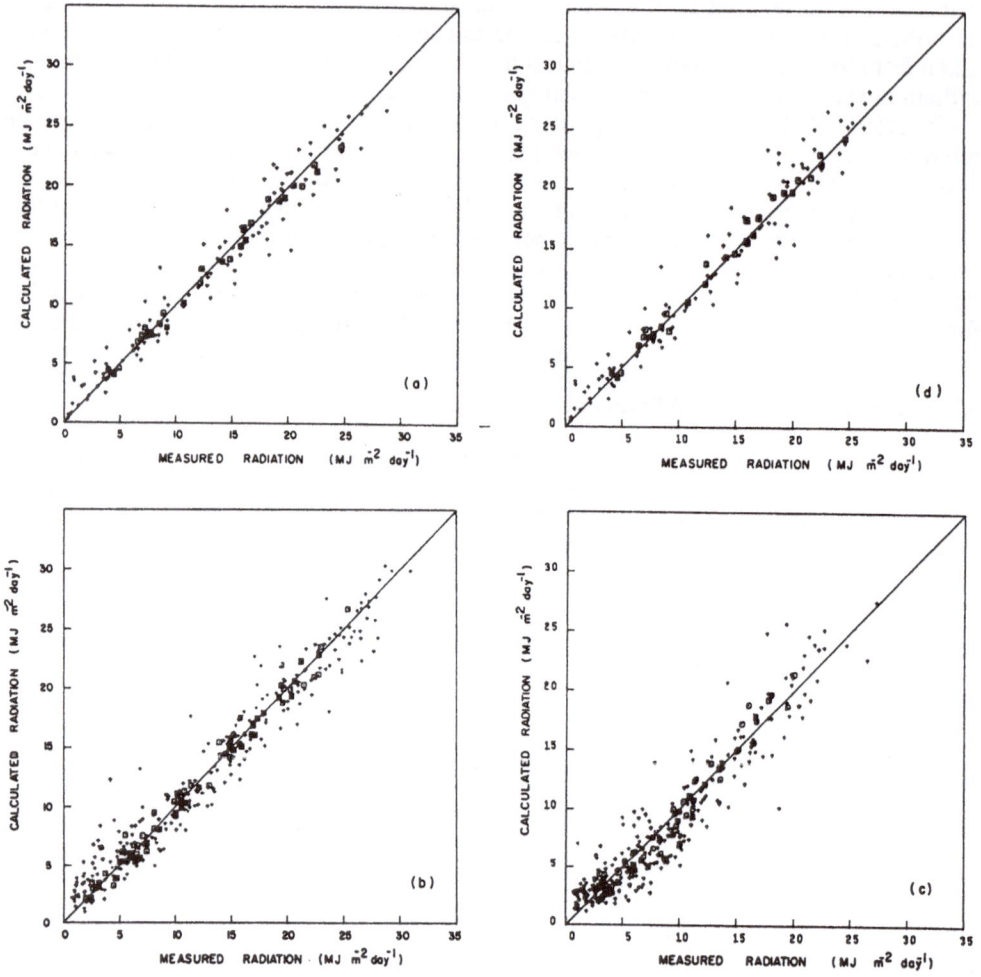

Figure 4. Correlation between calculated and measured values of global radiation for daily [+], five-day mean [□] and ten-day mean [■] periods in southern Ontario:

[a] Grimsby [July to Nov. 1969]

[b] Trenton [July 1973 to July 1974] } *Davies, Schertzer and Nunez [1975]*

[c] Trenton [Aug 1973 to March 1976]   *Atwater and Ball [1974]*

[d] Grimsby [July to Nov. 1969]

Calculation assumes no aerosol (Davies and Idso, 1979).

radiosonde data. However, because of the lack of such data and instability in estimating parameter values, Idso and Jackson (1969) found that an empirical relationship between cloudless sky longwave radiation, $R_{T_{io}}$, nocturnal screen temperature, T_a, and the Stefan-Bolzman constant, σ, could be used successfully to predict cloudless sky longwave radiation:

$$R_{T_{io}} = \sigma T_a^4 \left[1 - 0.261 \exp[- 7.77 x 10^{-4}(273 - T_a)^2] \right] \tag{8}$$

Paltridge (1970), in a modification of the Swinbank (1963) expression, showed that the effect of clouds can be obtained by adding a component to $R_{T_{io}}$ to give an expression for incoming longwave radiation R_{T_i} that includes the effect of clouds:

$$R_{T_i} = R_{T_{io}} + (1 - 0.7)\epsilon_C \, \sigma \, T_C^4 \, C \tag{9}$$

in which σ is the Stefan-Bolzman constant, ϵ_C is the emissivity contributed by cloud amount, C, at temperature, T_C. Daily values of measured and calculated values using the Swinbank and Paltridge models show most calculated values are within two percent of measured values and nearly all are within four percent (Davies and Idso, 1979).

3.1.3. Outgoing Energy Fluxes

The outgoing energy is made up of two components: reflected shortwave radiation and emission of longwave radiation from the surface. Reflected radiation is a function of the intensity of incoming solar radiation and the surface albedo. Albedo varies widely from 0.05 for a dark, wet clay surface to 0.95 for fresh snow cover. Albedo values for soil are a function of color and moisture conditions and can vary by a factor of five as a function of soil type only. Table 1, taken from Davies and Idso (1979), lists typical values for a range of soil type. Albedo also

Table 1. Albedo (non-normalized) of wet and dry soils and their ratios (from Idso and Reginato, 1974) (Davies and Idso, 1979).

Reference Source	Type of Soil	Wet	Dry	Wet/Dry
Van Wijk (1963)	Dark clay	0.05	0.16	0.31
Portman (1954)	Unspecified	0.06	0.16	0.38
Chia (1967)	Black soil	0.07	0.14	0.50
Kondrat'yev (1954)	Black soil	0.09	0.14	0.57
Ekern (1965)	Bare latosol	0.08	0.14	0.57
Ångström (1925)	Black mould	0.08	0.14	0.57
Ångström (1925)	Grey sand	0.09	0.18	0.50
Piggin & Schwerdtfeger (1973)	Red-brown clay loam	0.10	0.20	0.50
Kondrat'yev (1954)	Grey soil	0.11	0.27	0.41
Monteith (1959)	Clay loam with flints	0.11	0.18	0.61
Rijks (1967)	Dark brown clay	0.11	0.17	0.65
Graham & King (1961)	Grey brown loam	0.12	0.21	0.57
Idso et al. (1975)	Avondale clay loam	0.14	0.30	0.47
Kalma & Bredham (1972)	Tippera clay loam (red)	0.14	0.23	0.61
Budyko (1956)	Grey soil	0.15	0.27	0.56
Kondrat'yev (1954)	Blue loam	0.16	0.23	0.70
Büttner & Sutter (1935)	Dune sand	0.24	0.37	0.65
Average	All types	0.111	0.205	0.65

varies with type of crop. Numerous experiments have been reported that identify the albedo of a variety of crop surfaces. Some of these are reported in Table 2, also taken from Davies and Idso (1979). Use of these values of albedo is satisfactory for daily estimates of outgoing solar energy, but they cannot be used for short–term estimates because of a dependence on the solar zenith angle. Albedo tends to increase with an increase in the zenith angle, see Davies and Idso (1979) for a discussion of this dependence.

The intensity of outgoing longwave radiation is a function of surface temperature and emissivity, as described previously. The emissivity value of a soil is a function of its surface water

Table 2. Albedo of crops and vegetation (Davies and Idso, 1979).

Reference Source	Surface Type	Albedo
	Wheat	
Monteith (1965)		0.26
Davies and Buttimor (1969)		0.22
Denmead (1976)		0.2(4)*
Fritschen (1967)		0.21
Piggin & Schwerdtfeger (1973)		0.23
	Barley	
Monteith (1965)		0.23
Fritschen (1967)		0.23
Piggin & Schwerdtfeger (1973)		0.26
	Oats	
Impens and Lemeur (1969)		0.24
Fritschen (1967)		0.23
	Grassland	
Nkemdirim (1972)		0.23
Ripley and Redmann (1976)		0.17
Monteith (1965)		0.24
Davies and Buttimor (1969)		0.24
Arnfield (1975)		0.28
Stanhill (1970)		0.25
Oguntoyinbo (1970)		0.21
	Corn	
Arnfield (1975)		0.29(11)*
	Tropical Vegetation	
Oguntoyinbo (1970)	Sahel savanna	0.21
	Sudan savanna	0.20
	Guinea savanna	0.19
	Derived savanna	0.16
	Rain forest	0.13
	Swamp forest	0.12
	Temperate vegetation	
Jarvis, James & Landsberg (1976)	Coniferous	0.08–0.13
Stanhill (1970)	Hardwood	0.10–0.12
	Sub–Artic surfaces	
Davies (1963)	Tundra	0.17
	Burn	0.12
	Open woodland	0.11
	Bog and Muskeg	0.10
	Close crown forest	0.09
	(mixed deciduous coniferous)	

* *Listed albedo value is the mean from the number of studies in brackets.*

content. In general soils have values ranging from 0.89 to 0.97. Water has a value of about 0.97. For vegetation emissivity depends on the surface configuration. Most crops have values that lie between 0.96 and 1.0. The surface temperature required to estimate the outgoing long-wave radiation is generally measured using infrared thermometry. Measures of outgoing longwave radiation require careful calibration and experimentation. See Davies and Idso (1979) for references on sources of error.

The instrumentation, measurements, and estimates required to calculate net radiation on a term–by–term basis has discouraged many from using this technique, and several simple regression equations have been developed. The most common methods relate net flux of all radiant energy to incoming shortwave radiation (correlation coefficients of 0.99 for hourly data and 0.95 or better for daily data). Since net radiation is a function of surface texture, albedo, zenith angle, and other features, care must be taken in use of some of the simpler expressions. See Davies and Idso (1979) for a discussion of some of these issues.

4. Sensible and Latent Heat Fluxes

Sensible and latent heat fluxes are components A and λET in Equation (1). Sensible heat is carried into and out of the canopy by parcels of air moving through the canopy. The flux density of sensible heat in the layer of air above the canopy is considered positive (toward the surface) when the temperature increases with height. The magnitude of the flux density is a function of the number and magnitude of air parcels moving into the canopy. This is given by the eddy diffusivity or Austausch coefficient which increases with wind speed, surface roughness, and temperature gradient. Air turbulence is also responsible for the transmission of water vapor through the canopy. Both the soil and crop canopy serve as sources and sinks for energy depending upon whether they are warmer or cooler than the air parcels. As such they are considered in the energy balance.

The rate of sensible heat transfer between the atmosphere and the soil or plant canopy is dominated by moisture exchange almost to the point that sensible heat transfer is the residual after satisfying latent heat demands. If there is adequate soil moisture, a full–plant canopy may utilize 5 or 6 mm of water per day or about 75 percent of net radiation. On cloudy days there can be almost no sensible heat transfer. In some cases, generally in irrigated regions, warm air from surrounding nonirrigated areas can be advected into the field in the form of sensible heat, and latent heat exchange can exceed net radiation. In one case in central California latent heat exchange exceeded net radiation by 50 percent.

Latent heat exchange from a surface is the sum total of energy of evaporation (E) from the soil surface and transpiration (T) from the plant canopy. In most cases they are combined as evapotranspiration (ET) and the flux of water is expressed as mass of water vapor per cm^2 of surface or in energy terms of W/m^2. Atmospheric properties of net radiation, turbulent exchange, and water vapor gradients determine the capacity for latent heat exchange. The maximum amount of evapotranspiration possible under conditions of unlimited water availability is called potential evapotranspiration. Under limited water conditions, extremely high temperatures, and biological control, actual evapotranspiration is less than potential. As mentioned previously latent heat exchange tends to use most of the net radiation available. This is illustrated in Figure 5 where total solar radiation, net solar radiation, and the latent heat energy of evapotranspiration are shown for a rye grass field on May 29 and June 1. Note that during the night and early morning hours ET energy exceeds net radiation as the plant canopy uses sensible heat advected from the soil. Also the ET rates for June 1 show that the plants used more energy during the entire day than was available from net solar radiation, thus indicating that wind very likely advected sensible heat from the surrounding area into the field.

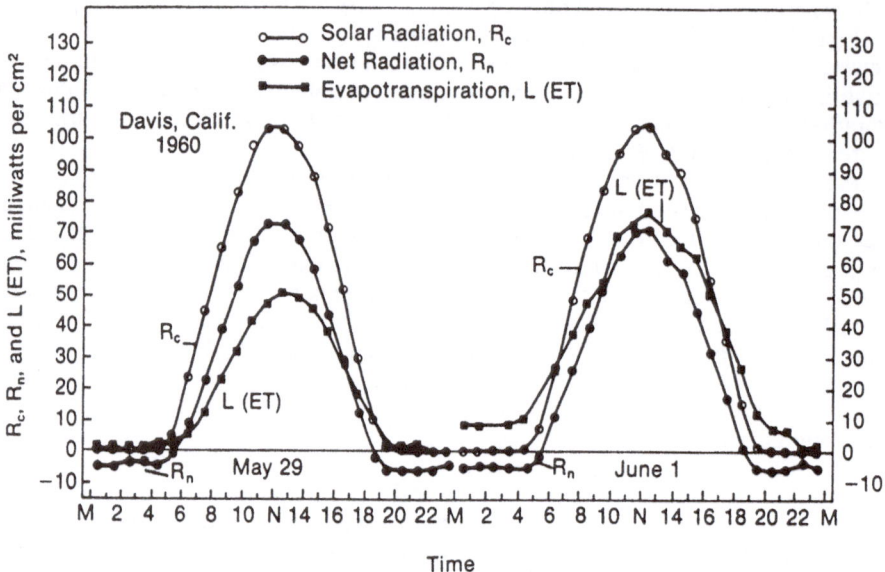

Figure 5. Hourly latent heat energy of evapotranspiration L(ET) for rye grass compared with solar (R_c) and net (R_n) radiation (see Pruitt and Angus, 1961; Shaw and Decker, 1979).

Soil heat transfer is relatively small compared to latent and sensible heat exchange. The soil profile gains heat in the early spring reaching a maximum soil heat flux density in midsummer, then declining until by late summer there is a net daily transfer of energy to the atmosphere. On a daily basis there is a diurnal cycle of heat gain during the daylight hours and loss to the atmosphere at night, with very little net change in profile temperature from day to day. Figure 6 shows the average daily heat flux density under a fescue meadow near Columbia, MO.

The principles described in early sections of this chapter on energy balance and the radiative and nonradiative components also apply to plant surfaces. The mechanisms are more complex, however, because leaves within the canopy receive less energy than those on the surface and may have different orientations. Shaw and Decker (1979) discuss the energy balance of plant surfaces with reference to several specific studies.

4.1. Calculating Energy and Mass Fluxes

A large number of methods have been derived for calculating energy and mass flux. In almost all of these, the application has been to soil surface evaporation (E) and plant transpiration (T) combined as evapotranspiration (ET). Tanner (1967) classifies these as (a) hydrological, (b) climatological, and (c) micrometeorological. The classification is generally based on data required to drive the models. Since the emphasis of this chapter is on microclimatology, discussion will be confined generally to the micrometeorological methods and a few combination equations that are generally classified as climatological methods. However, a few of the other methods will be described to place the computational methods in perspective. Much of the material presented in this discussion of computational methods follows that of Rosenberg et al. (1983). For a more complete discussion of these and other methods see Tanner (1967) and Jensen (1974).

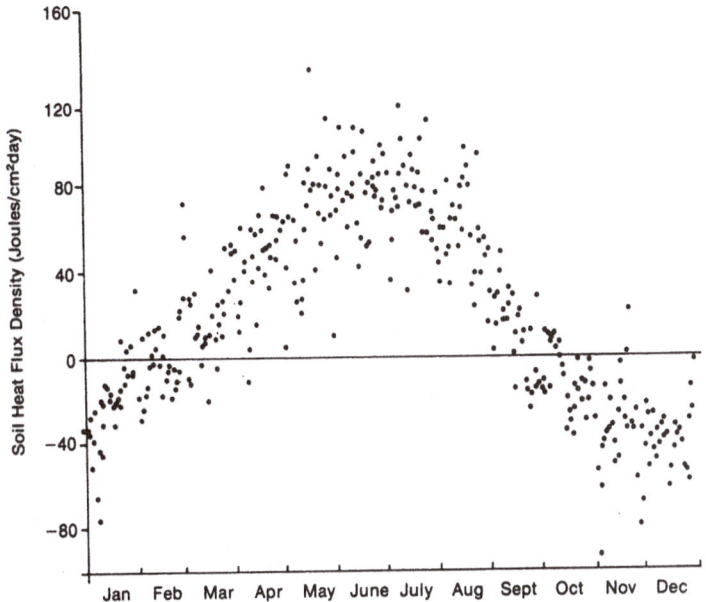

Figure 6. Average daily heat flux density under a fescue meadow near Columbia, MO [1965-74] (Shaw and Decker, 1979).

4.1.1. Hydrologic Methods

Hydrologic methods for estimating ET are based on water balance. The following expression defines the general equation as:

$$P + \Delta SW \pm RO - D - ET = 0 \tag{10}$$

in which P is precipitation, ΔSW is change in soil water, RO is runoff (positive if runon), D is deep percolation (beyond the root zone), and ET is evapotranspiration. In using Equation (10) ET is the residual with all other components measured or estimated. The method is generally easy to use but is not accurate enough for daily estimates. Rosenberg et al. (1983) list several applications of the method.

4.1.2. Climatologic Methods

Climatologic methods are generally divided into temperature-based methods, solar energy-based methods, and combination methods. Temperature-based methods rely almost solely on air temperature, thus estimates are reasonable for periods of several weeks or longer but not for short periods of a few days. The most commonly used models are Thornthwaite (1948), Blaney and Criddle (1950), Hargreaves (1974), and Linacre (1977). Since solar radiation contributes most of the energy for evapotranspiration, one would expect a good correlation between solar energy and evapotranspiration and thus good predictive capability (see Figure 7). Several authors have attempted to develop such relationships. The most notable are Makkink (1957), Jensen and Haise (1963), and Idso et al. (1975; 1977). Linacre (1967) discusses some of the strengths and weaknesses of the methods.

218

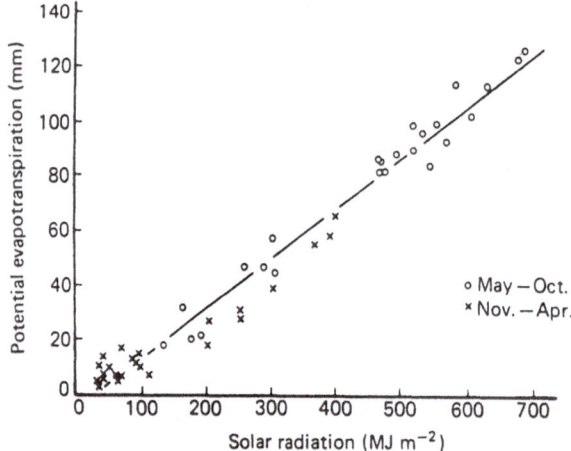

Figure 7. Relation of monthly total potential evapotranspiration to monthly totals of solar radiation (after Aslyng, 1974, Evapotranspiration and Plant Production Directly Related to Global Radiation Nordic Hydrology Vol. 4–6) (Microclimatology, The Biological Environment, Rosenberg, N.J., B.L. Blad, S.B. Verma, Copyright 1983 by John Wiley & Sons, Inc., Reprinted by permission of John Wiley & Sons, Inc.).

4.1.3. Micrometeorologic Methods

Micrometeorologic methods (sometimes known as gradient methods) for estimating ET are based on vapor pressure, wind, and temperature gradients in and above a crop canopy. The first application of the method was by Dalton (about 1800) in predicting evaporation, E, as a function of vapor pressure gradient:

$$E = C\,(e_o - e_d) \tag{11}$$

in which C is an empirical coefficient and e_o and e_d are water vapor pressure at the surface and in the air at some point above the surface, respectively. It can only be applied with some degree of confidence over water or other saturated surface where the vapor pressure at a given temperature is known. Variations of the expression show C to be a function of wind speed. The method is most widely used to estimate lake evaporation because data requirements are relatively simple (relative humidity, temperature, and wind speed).

A variation of the gradient method is the aerodynamic method. Thornthwaite and Holzman (1942) proposed a relationship for calculating evapotranspiration, λET, involving the gradients of water vapor pressure, e, and the logarithmic wind profile, u:

$$\lambda ET = \rho_a k^2 \frac{(e_2 - e_1)(u_2 - u_1)}{\ln(z_2/z_1)^2} \tag{12}$$

in which ρ_a is the density of moist air, k is the von Karman constant and z is the height of measurement. Other versions of Equation (12) have been proposed. In general they require accurate estimates of wind speed and water vapor pressure at several heights above the surface and estimates of surface roughness elements. See Kanemasu et al. (1979) for more information on these methods. In general the methods are popular because they are based on classical fluid dynamics theory.

Another popular version of the micrometeorological approach is the Bowen ratio–energy balance (BREB) method. The Bowen ratio, β, is the ratio of sensible heat, A, to latent heat, λET:

$$\beta = \frac{A}{\lambda ET} = \gamma \frac{K_h\ \partial T/\partial z}{K_w\ \partial e/\partial z} \tag{13}$$

in which γ is the psychometric constant, K_h is the turbulent heat exchange coefficient, K_w is the exchange coefficient for water, T is air temperature, e is the water vapor pressure, and z is vertical distance. By assuming $K_h = K_w$ and $(\partial T/\partial z\ /\ \partial e/\partial z) = \Delta T/\Delta e$:

$$\beta = \gamma \frac{\Delta T}{\Delta e} \tag{14}$$

Substituting $A = \beta \lambda ET$ into Equation (1), assuming P and M to be negligible, yields the (BREB) method for estimating the latent heat λET:

$$\lambda ET = -\left[\frac{R_n + G}{(1 + \beta)(\Delta T/\Delta e)}\right] \tag{15}$$

The BREB method has been found to be in good agreement with lysimeter measurements in nonadvective conditions (Tanner, 1960; Pruitt and Lourence, 1968; Denmead and McIlroy, 1970). Figure 8 compares results from advective and nonadvective experiments. *LE* in this figure is the same as λET in Equation (15). In advective conditions K_h has been shown to be

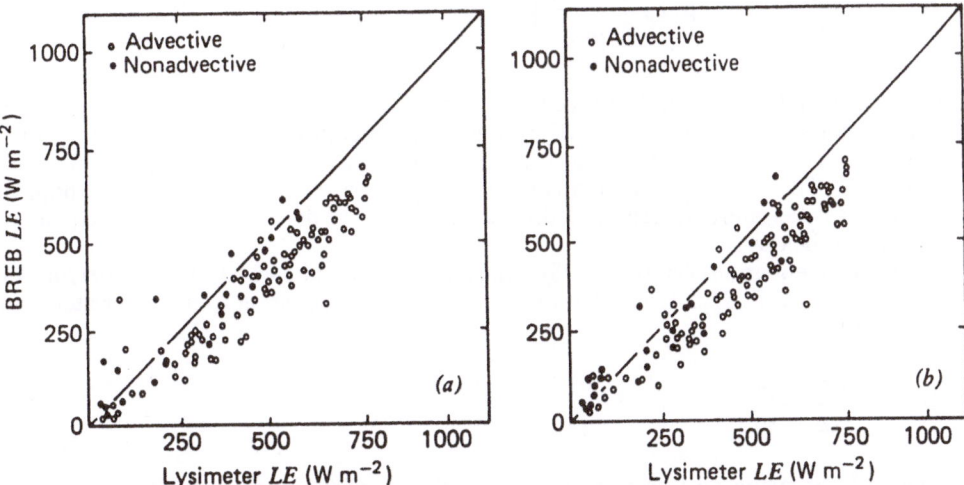

Figure 8. Relation between latent heat flux LE calculated with the Bowen ratio energy balance (BREB) method and lysimetrically measured LE. Temperature and vapor pressure data for BREB calculations were from (a) 1.00 to 1.25 m level and (b) 1.25 to 1.50 m level (adapted from Blad and Rosenberg, 1974, Journal of Applied Meteorology, American Meteorological Society) (Microclimatology, The Biological Environment, Rosenberg, N.J., B.L. Blad, S.B. Verma, Copyright 1983 by John Wiley & Sons, Inc., Reprinted by permission of John Wiley & Sons, Inc.).

greater than K_w, and prediction can be improved using actual ratios and modifying Equations (14) and (15) to include the ratio K_h/K_w (Blad and Rosenberg, 1974). The Bowen ratio method is advantageous for field application because instrumentation requirements are relatively simple.

4.1.4. Eddy Correlation Methods

Eddy correlation methods of calculating λET are based on continuous monitoring of the movement of air parcels as they move up and down in the crop canopy. The equation for calculating λET is:

$$\lambda ET = \frac{\epsilon}{P} \rho_a \overline{w'e_a'} \tag{16}$$

in which ϵ is the ratio of the molecular weight of water to that of dry air M_w/M_a, P is atmospheric pressure, ρ_a is the density of moist air, and w' and e_a' are instantaneous departures from the mean vertical wind speed and mean water vapor pressure in air, respectively; the overbar on their product represents a time average. The method provides accurate estimates of λET but requires very sensitive instrumentation.

4.1.5. Combination Method

Combination formulas are those that combine features of both the climatologic and micrometeorologic methods. In one such method λET is calculated as the residual in the energy balance equation [Equation (1)], assuming P and M to be negligible, with sensible heat flux, A, estimated using an aerodynamic equation:

$$\lambda ET = -\left[R_n + G + \rho_a c_p \frac{(T_a - T_s)}{r_a} \right] \tag{17}$$

in which T_a and T_s are air and surface temperatures, respectively, r_a is the boundary layer resistance, and c_p is the specific heat of air at constant pressure. Since surface temperature cannot be measured directly, Equation (17) is modified to eliminate $(T_a - T_s)$ (see Kanemasu et al., 1979). The modified equation is a fundamental expression from which the popular Penman, van Bavel, and Slatyer and McIlroy equations can be derived. These three equations are discussed below.

The Penman equation (Penman, 1948) is the most popular and widely used model for estimating potential ET, ET_p. It was developed to estimate evaporation from open water, bare soil, and grass:

$$\lambda ET = \frac{\Delta R_n + \gamma E_a}{(\Delta + \gamma)} \tag{18}$$

in which λET is evapotranspiration or evaporation if over open water, Δ is the slope of the saturation vapor pressure curve at the mean wet bulb temperature of the air, γ is the psychometric constant $= c_p P/\epsilon\lambda$, and R_n is the net radiation, in this case, over open water. E_a is given by:

$$E_a = f(U)(e_a - e_d) \tag{19}$$

In which e_a is the saturation water vapor pressure, e_d is the actual water vapor pressure of the air, and $f(U)$ is a linear function of daily wind run. Equation (18) has been modified to reflect reduced evaporation over cropland. Generally the modification is an empirical multiplier that lies between 0 and 1.0; values of 0.8 for summer and 0.6 for winter have been recommended but these are only very crude estimates; some researchers have argued that Equation (18) predicts ET_p directly and that no modification for crop lands is necessary. In this case R_n + G from a cropland surface is used in place of R_n. The Penman equation is not accurate for conditions of strong sensible heat advection. See Rosenberg et al. (1983) for a discussion of the effects of advective energy and references.

Slatyer and McIlroy (1961) developed an expression for potential ET similar to Penman by replacing the vapor pressure deficit with the wet–bulb depression:

$$\lambda ET_p = \frac{\Delta}{\Delta + \gamma}(R_n + G) - \rho_a c_p \frac{D_o - D_z}{r_a} \tag{20}$$

in which D_o and D_z are the wet-bulb depression at the surface and some distance z above the surface, respectively, and r_a is the aerial resistance. Other terms are as previously defined.

van Bavel (1966) introduced an expression similar to Penman that he considered an improvement in that it involved no empirical constants or functions but included a reference elevation of wind and temperature measurements, z, a reference roughness, z_o, the von Karman constant, k, and wind speed, u:

$$\lambda ET = \frac{\Delta(R_n + G) + \gamma \lambda B_v(e_a - e_d)}{\Delta + \gamma} \tag{21}$$

in which λ is the latent heat of vaporization and B_v is given by:

$$B_v = \frac{\rho_a \epsilon k^2}{P} \frac{u}{(\ln(z/z_o))^2} \tag{22}$$

in which all terms are previously defined. See Rosenberg et al. (1983) for limitations on use of these expressions; they are very sensitive to wind.

About the same time that van Bavel introduced his expression for the logarithmic eddy diffusion function into the wind term of the Penman equation, Montieth (1965; 1973) derived a very similar combination equation (Penman/Montieth) of the form:

$$\lambda ET = \frac{\Delta(R_n - G) + \rho c_p(e_a - e_d)/r_a}{\Delta + \gamma (1 + r_c/r_a)} \tag{23}$$

in which λET is the vapor flux density, R_n is net radiation flux density to the plant canopy, G is soil heat flux density, Δ is the slope of the saturation vapor pressure curve, ρ is the density of air, c_p is the specific heat of the air, e_a is the saturation vapor pressure at current air temperature, e_d is the saturation vapor pressure at the dewpoint temperature, r_a is the aerodynamic resistance to vapor and heat diffusion, γ is the psychometric constant, and r_c is the bulk stomatal (canopy) resistance. This equation is an improvement over the others discussed because the stomatal resistance term produces an equation that simulates actual ET rather than just potential, ET_p.

The expression for the aerodynamic resistance, r_a in Equation (23) is given by:

$$r_a = \frac{\left[\ln\left(\frac{z_m - d}{z_{om}}\right)\right]\left[\ln\left(\frac{z_h - d}{z_{oh}}\right)\right]}{k^2 \, u_z} \tag{24}$$

in which z_m is the height of wind measurements, z_h is the height of air and humidity measurements above the ground surface, d is the zero plane displacement height (a reference elevation where the logarithmic wind velocity approaches zero), z_{om} is a roughness length for momentum transfer, z_{oh} is a roughness length of vegetation for vapor and heat transfer, k is the von Karmon constant for turbulent diffusion, and u_z is the wind speed at height z.

Section 4.2 of this chapter presents more information on the resistance features of Equation (23) and similar expressions. Montieth (1973) is also a particularly good reference for concepts imbedded in Equation (23).

Allen et al. (1989) compares five Penman–type equations using lysimeter data from eleven locations around the world on reference crops of grass and alfalfa. In this comparison the Penman/Montieth equation was superior to the others. The Penman/Montieth equation has also been found to be stable for daily estimates of ET, but one must be careful in units used. See Allen et al. (1989) for a good discussion of units and expressions used to obtain values for the various parameters. Of particular note is his discussion of the relative size of z_{om} and z_{oh}, which are frequently assumed equal. It is now assumed that the momentum roughness length, z_{om}, is at least ten times as large as that of the vegetation roughness length for vapor and heat transfer, z_{oh}. Recent personal communication with M.E. Jensen would indicate that z_{om} could be as large as 100 times that of z_{oh}.

Kanemasu et al. (1979) in discussing these and other methods of calculating the latent heat loss compared the advantages and disadvantages of several general classes of the micrometeorological methods in Table 3. In general the Bowen ratio energy balance is the most popular for routine measurements of latent and sensible heat fluxes. The eddy–correlation method is probably superior in terms of accuracy and flexibility, but instrumentation requirements are severe.

Three–dimensional forms of the energy balance equation have been developed. The Penman/Montieth equation has also been partitioned to canopy layers. Space does not permit discussion of these topics; see Kanemasu for a discussion of the 3-D methods and Shuttleworth (1979) for canopy partitioning.

Kanemasu et al. (1979) compared many of the methods described above using data from a grass canopy field experiment at the University of California at Davis in 1967. The results are shown in Figures 9–11. Results shown as "Difference" are based on aerodynamic methods in which an iterative scheme is used to calculate the friction velocity of the air and a temperature scaling parameter. The results shown as "Bulk" are solutions of aerodynamic methods where measures of surface resistance, such as canopy stomatal resistance are required but unavailable. Solutions here are also iterative. Figure 9 shows three methods of estimating the friction velocity of the air, $u*$, in meters per second. The eddy correlation technique is considered superior. Figure 10 compares five measures of sensible heat flux, and Figure 11 compares four measures of latent heat flux or ET. On these plots, instrumentation was such that the Bowen–ratio energy balance approach was considered superior. Vegetation growing in the lysimeter was not as vigorous as the surroundings, thus lysimeter estimates were not considered accurate.

Table 3. Comparison of various methods for estimating latent or sensible heat fluxes (after Kanemasu et al., 1979).

Technique	Primary Advantages	Primary Disadvantages
Eddy–correlation	Direct measurement of flux, independent of surface conditions.	Instrumentation can be complex (fast response sensors are required), particularly for deployment at heights less than 4 m. The demands for fast response can be excessive during stable conditions.
Aerodynamic	Instrumentation is relatively simple.	Adequate fetch is critical. Surface parameters usually are required. High degree of accuracy in measurements is needed. Methods sometime fail during light winds.
Bowen ratio	Instrumentation is relatively simple.	Measurements during environmental conditions in which gradients are small can be in error. Surface energy balance is usually required.
Combination	Instrumentation is extremely simple.	Knowledge of heat transfer coefficient or surface resistance is required over a surface not saturated with water.
Three–dimensional	Necessary form of energy balance during conditions of strong horizontal advection.	Large number of measurements.

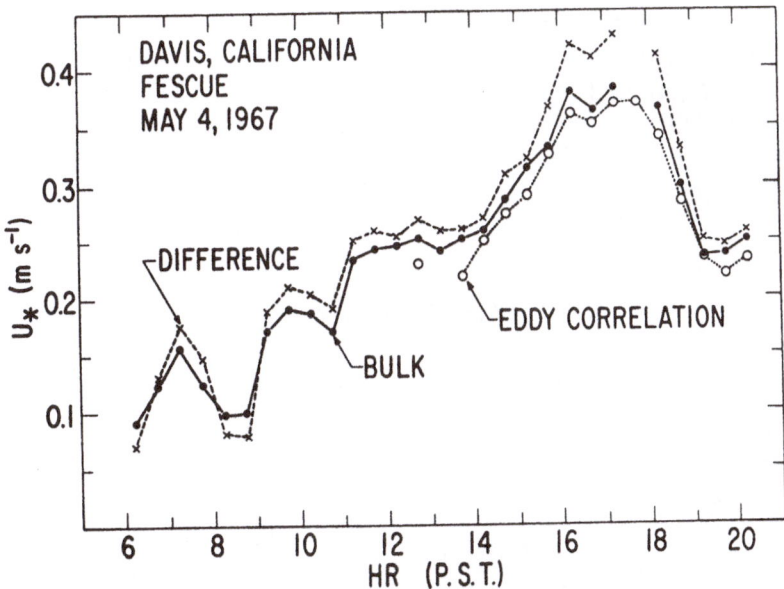

Figure 9. A comparison of results from three methods of measuring friction velocity (Kanemasu et al., 1979).

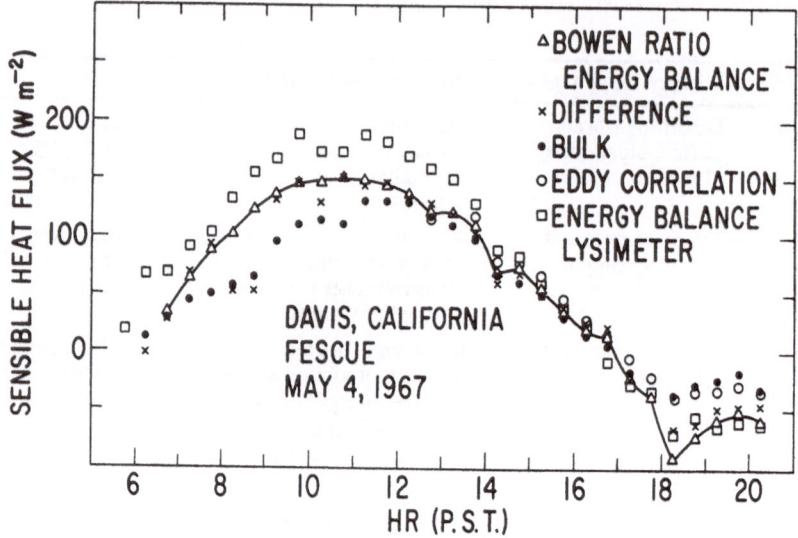

Figure 10. A comparison of results from five methods of measuring sensible heat flux (Kanemasu et al., 1979).

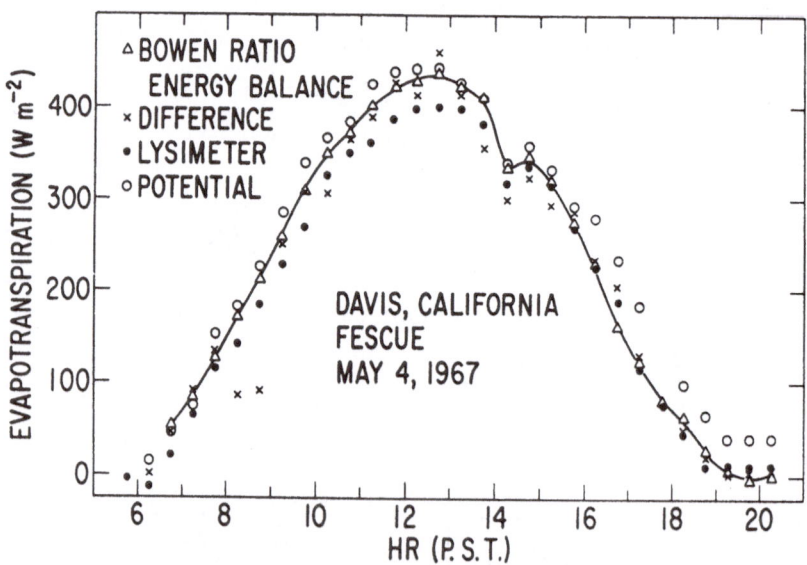

Figure 11. A comparison of results from four methods of measuring evapotranspiration (Kanemasu et al., 1979).

4.2. Aerial and Crop Resistances Affecting Energy Transport

Since the effect of wind and resistance factors are so significant in estimating actual ET, a discussion of aerial and crop resistances needed for the aerodynamic and combination methods is in order. Resistance, as used here, consists of resistance to flow of water vapor through stomates of plant leaves and resistance to flow of air and water vapor through the crop canopy and the turbulent boundary layer above the crop surface. Much of the information presented here is abstracted from Verma and Barfield (1979) and Montieth (1973).

The transport of sensible and latent heat in the atmosphere above a crop canopy depends upon their respective gradients and resistances to flow. For sensible heat the gradient consists of the difference between the temperature of the crop or soil surface and the air above the surface. Latent heat transfer is driven by differences in water vapor concentration in the crop leaves and in the air above the leaf surface. Equations for flux of sensible and latent heat are similar: for sensible heat the expression for the heat flux, A_c, between the crop or soil surface and the atmosphere is given by:

$$A_c = \frac{\rho_a \; c_p (T_s - T_a)}{r_{as}} \tag{25}$$

in which ρ_a is the density of air, c_p is the specific heat of air, T_s is the mean surface temperature of the crop, T_a is the air temperature at a given distance above the crop, and r_{as} is the aerial diffusion resistance to heat transport. The comparable expression for latent heat flux, E_c, between the crop or soil surface and the atmosphere is given by:

$$E_c = \frac{\rho_a \; \lambda (\chi_{vs} - \chi_{vz})}{r_{av} + r_c} \tag{26}$$

in which λ is the latent heat of vaporization, χ_{vs} is the saturation water vapor pressure at the surface temperature and χ_{vz} is the water vapor pressure of the free air, r_{av} is the air resistance to water vapor movement, and r_c is the crop resistance to water vapor movement (through the stomata). Thus, resistance to water vapor movement is made up of two components: resistance to movement in the air, and resistance to movement through the stomata to the air surface.

The movement of air across a surface imposes a drag on the surface that transports momentum from the air to the surface. An expression for the momentum flux, τ is:

$$\tau = \frac{\rho_a u}{r_{am}} \tag{27}$$

in which u is the mean wind speed at a given elevation and r_{am} is the aerial resistance to momentum transport.

The shear stress or velocity, $u*$, is a term frequently used in the description of resistance. It is defined as:

$$u* = \sqrt{\tau / \rho_a} \tag{28}$$

thus, the resistance to momentum transfer can be defined:

$$r_{am} = u / u*^2 \tag{29}$$

The shear velocity can be measured or estimated from the log velocity profile expression:

$$u* = (ku)/\left[\ln\left(\frac{z-d}{z_o}\right) + \psi_m\right]$$
(30)

in which k is the von Karman constant, d is the zero plane or crop displacement height (in practice usually considered to be 2/3 crop height), z_o is crop roughness height, and ψ_m is a diabatic stability correction parameter (if neutral conditions are assumed as is often the case, ψ_m is zero).

Equations (25)–(30) are presented because they are fundamental in the estimation of resistance to energy and mass transport. If r_{as} and r_{av} can be related to r_{am}, then it is possible to calculate sensible and latent heat transfer. It can be seen from Equation (29) that this will require measured or calculated values of $u*$. Since measured values of $u*$ are not generally available, one will need values of z_o and d. The balance of Section 4.2 is devoted to the relation of r_{am} to r_{as} and r_{av}, and methods of estimating r_c, z_o and d.

The transfer of sensible and latent heat from the crop surface to the atmosphere is not the same as the transfer of momentum because the transfer of heat depends upon the ability of the plant to provide the amount called for by the gradient that exists. For example, if the plant cannot provide the water vapor called for because physical conditions have forced the stomates to be nearly closed, then the transport is limited by what is available through the stomata. This is not the case for the transfer of momentum which is entirely dependent upon the shear stress. Thus a dimensionless parameter B^{-1} has been introduced to account for the increased resistance:

$$B^{-1} = r_{as} - r_{am}$$
(31)

Thus it is possible to express the resistance to transfer of sensible and latent heat in terms of the resistance to momentum transfer. Verma and Barfield (1979) show that B^{-1} can be estimated from the momentum roughness height z_o. The relation between aerodynamic roughness, z_{oh}, and momentum roughness, z_{om}, expressed in Equation (24) is an alternate derivation by Montieth (1973).

4.2.1. Crop Roughness and Displacement Height

The momentum roughness height, z_o, is defined as the height at which the neutral wind profile extrapolates to zero wind speed. It is generally accepted to be a function of wind velocity, and the height, shape, and distribution of roughness elements, but there does not appear to be a unique relationship (see Montieth, 1973). Much more data for a variety of crops and different wind speeds are needed to define the relationship. Verma and Barfield (1979) discuss some of the empirical relations that have been proposed. One of them, by Szeicz et al. (1969), relates z_o at a wind speed of 2.5 m/s as:

$$z_{o,2.5} = 0.105\ H \qquad \text{for maize and corn}$$
(32a)

$$z_{o,2.5} = 0.03\ H^{1.3} \qquad \text{for all other crops}$$
(32b)

in which H is the height of the crop. Figure 12 shows a relationship proposed by Szeicz et al. (1969) for estimating z_o for wind speeds other than 2.5 m/sec. As mentioned above there is no generally accepted approach. Allen et al. (1989), in their discussion of Equation (24), suggest the best predictors of z_o and d, consistent with experimental results is:

$$z_o = 0.123 \, H \tag{33}$$

and

$$d = 0.67 \, H \tag{34}$$

4.2.2. Stomatal and Crop Resistance

In the introduction to this section of the chapter the resistance to latent heat transport is defined as being made up of two components: one is the air resistance to water vapor movement, r_{av}, and the other is the crop resistance to water vapor movement, r_c. Frequently the crop resistance component is expressed fundamentally in terms of a leaf resistance, r_s. Values

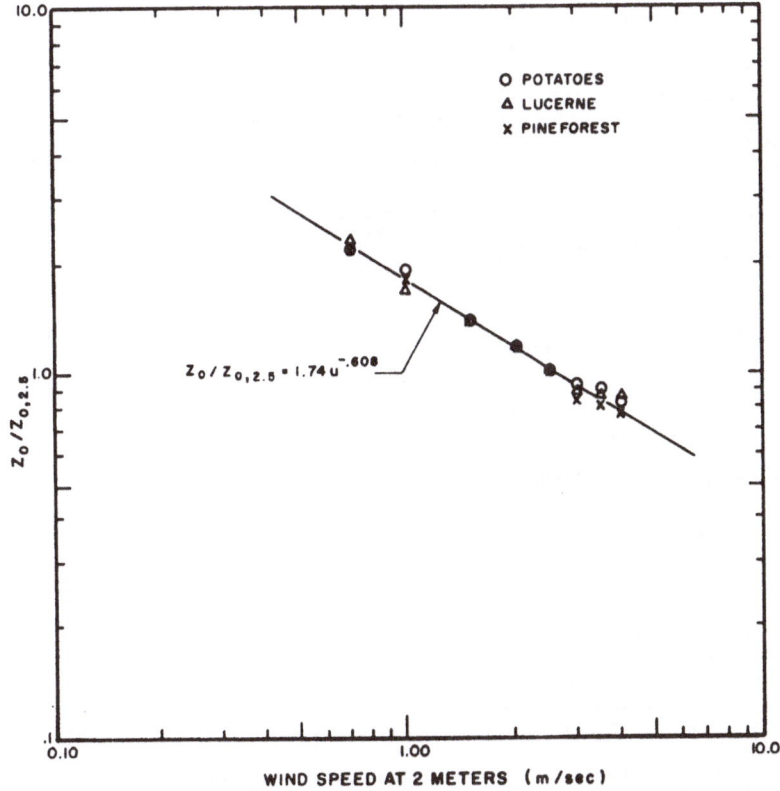

Figure 12. The effect of windspeed on roughness height as proposed by Szeicz et al., 1969 (Verma and Barfield, 1979).

228

of r_{av} and r_s are illustrated in Figure 13 along with representative temperature and water vapor profiles T and χ, respectively. Verma and Barfield (1979) identify several references on the subject of crop resistance. A variety of relationships have been proposed to estimate r_s; it is a function of solar energy, air temperature, and leaf water potential. Figure 14 shows the effect of leaf water potential on the value of r_s for upper surfaces (adaxial) and lower surfaces (abaxial) of snap beans. Expressed as a crop resistance, rather than an individual stomatal resistance, numerous expressions have been proposed. Perhaps the most direct, although an extreme simplification of a very complex process is:

$$r_c = r_s/LAI_e \tag{35}$$

in which LAI_e is the effective leaf area index.

Verma and Barfield (1979) describe application of the above and similar expressions in calculating r_{as} and latent energy, LE, fluxes for a variety of assumptions concerning the availability of data. Figure 15 shows one example (Heilman and Kanemasu, 1976) applied to a soybean field and a sorghum field. In addition to the calculated latent energy, net radiation,

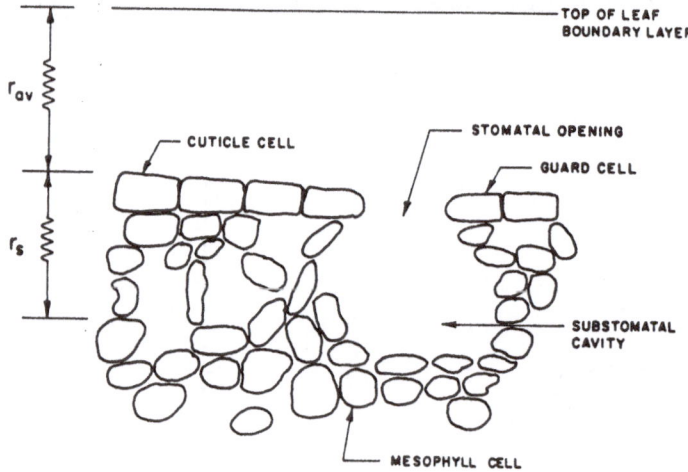

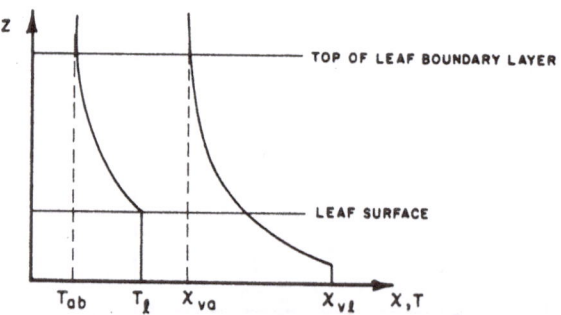

Figure 13. Schematic of leaf showing stomatal cavity, stomatal resistance, and assumed temperature (T) and water vapor concentration (χ) profiles. Stomatal resistance as used here is actually the sum of stomatal pore, mesophyll cell wall, and internal air pore resistance (Verma and Barfield, 1979).

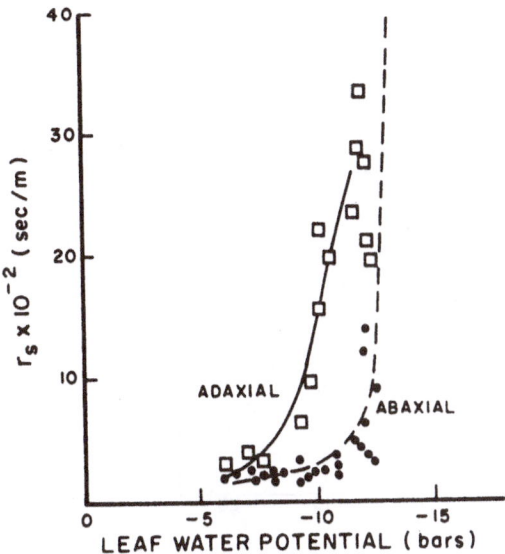

Figure 14. Effect of water stress on stomatal resistance of the upper (adaxial) and lower (abaxial) surfaces of snap beans. Data from Kanemasu and Tanner (1969) and curves from simulations by DeMichelle and Sharpe (1973) (Verma and Barfield, 1979).

R_n, and latent energy, based on lysimeter data, are also shown. Figure 16 is another example. In this case the surface temperature of the crop and evaporation rates are calculated on two different dates, using two different assumptions of LAI_e in relating r_s to r_c in Equation (35). It is presented to show accuracy of these type predictive equations; in this case the best form of LAI_e is $LAI/2$. The difference in evaporation rates for these two assumptions of LAI_e show the significance of r_c in the expressions.

4.3. Partial Canopy Cover

The material presented to this point presumes the surface to be uniform (soil, open water, or complete canopy cover). Frequently a model of ET from partial canopy cover is desired. This could be in the form of forest ET in which the latent heat loss is partitioned between the trees and understory or it could be in estimating the ET from fields with incomplete cover. Several papers, over the years, have addressed this problem. Two early references, Black et al. (1970) and Ritchie (1972), discuss partial cover of row crops. Shuttleworth (1979), in a theoretical analysis of the Penman/Montieth equation, derives a general expression for partitioning ET into multiple canopy layers. Shuttleworth and Wallace (1985) apply the general model of Shuttleworth (1979) to an agricultural crop over bare ground (or water) where as Kelliher and Black (1986) apply Shuttleworth's model to estimating ET from a Douglas Fir stand in which the understory was removed. Another subject of interest to those studying ET is that of sensitivity analysis. Beven (1979), in analyzing the Penman/Montieth equation, presents a good summary of the subject. It is interesting to note that r_c was the most sensitive parameter in the equation.

230

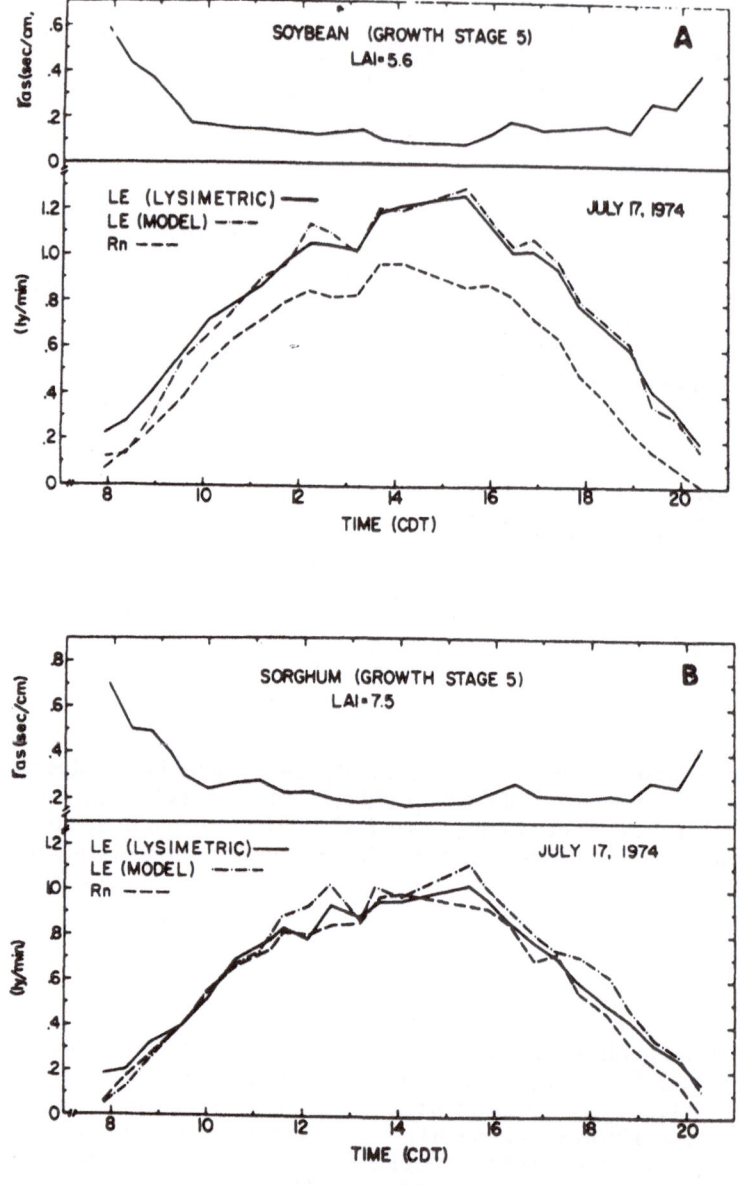

Figure 15. Estimates of r_{as} and comparisons of model LE, lysimetric LE, and R_n for soybean [A] and sorghum [B] on July 17, 1974 (from Heilman and Kanemasu, 1976) (Verma and Barfield, 1979).

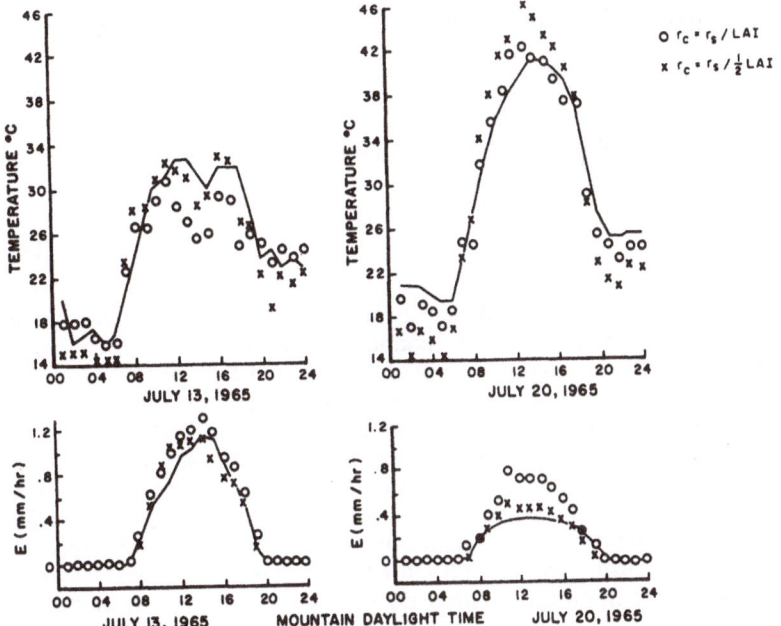

Figure 16. Comparison of predicted and observed (solid line) crop temperatures and evaporation for Arizona sorghum data using Case II model (data from Ehrler and van Bavel, 1967) (Verma and Barfield, 1979).

5. Soil Temperature and Heat Flux

The soil is the only major heat storage reservoir in the soil–plant–atmosphere system, thus it is important as a source/sink term in thermal energy balance models. Soil temperature also plays a dominant role in plant growth and chemical and biological processes. Thus it is important to understand how soil temperatures fluctuate, how to calculate soil heat flux, and what effect various soil features have on soil heat flux. Much of the material presented in this chapter is taken from Rosenberg et al. (1983) and Kimball and Jackson (1979).

5.1. Definitions and Fundamental Equations

The heat capacity of soil, C, is the ratio of heat absorbed or released to a corresponding rise or fall in soil temperature. The mass specific heat, C_s, is the amount of heat required to raise 1 kg of soil 1°C. The volume specific heat C_v is the amount of heat required to raise a unit volume of soil 1°C. Mass specific heat and volume specified heat are related by:

$$\rho \, C_s = C_v \tag{36}$$

in which ρ is the density of the soil. The soil temperature gradient and thermal conductivity, K_g, determine the soil heat flux, G:

$$G = -K_g \frac{dT}{dz} \tag{37}$$

232

in which T and z are soil temperature and depth, respectively. In a nonsteady–state condition where energy is being stored or taken from a soil:

$$\frac{dT}{dt} = \frac{1}{C_v}\frac{dG}{dz} \qquad (38)$$

substituting Equation (37) into Equation (38) and defining the thermal diffusivity as $D_T = K_g/C_v$ yields:

$$\frac{dT}{dt} = D_T\frac{d^2T}{dz^2} \qquad (39)$$

Values of C_s and C_v are functions of soil bulk density and moisture content, thus they can vary greatly from soil to soil and over time. Thermal conductivity, K_g, is a function of porosity, texture, moisture content, and organic matter content of the soil; thus K_g varies with soil conditions. Figure 17 shows how thermal conductivity varies with moisture content, texture, and soil matrix potential; see Rosenberg et al. (1983) for references. Table 4, also taken from Rosenberg et al. (1983), shows a few representative values for soil density, specific heat, thermal conductivity, and thermal diffusivity. Soil thermal diffusivity is primarily a function of the pore space and fraction of this space that is filled with water. It is a parabolic function of moisture content: a small amount of water isolates the pore, decreasing diffusivity; more water, replacing air, increases diffusivity; but still more water increases the heat capacity, again reducing diffusivity. For more information on the thermal properties of soil see van Wijk (1964 and 1965).

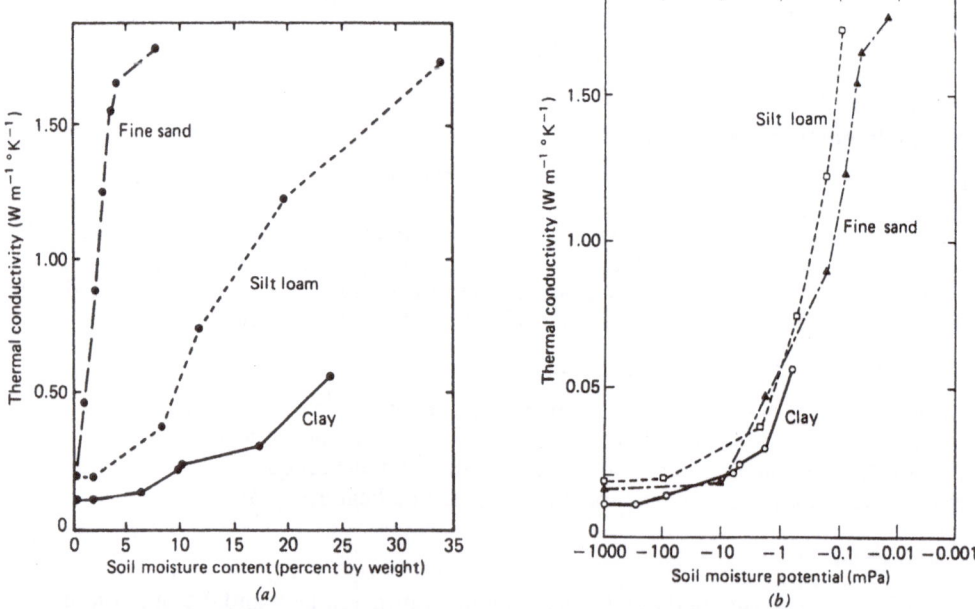

Figure 17. Thermal conductivity of three soils related to (a) to moisture content and (b) to moisture potential (after Al–Nakshabandi and Kohnke, 1965, Journal of Applied Meteorology, American Meteorological Society, Elsevier Science & Publishers, Physical Sciences & Engineering Division) (Microclimatology, The Biological Environment, Rosenberg, N.J., B.L. Blad, S.B. Verma, Copyright 1983 by John Wiley & Sons, Inc., Reprinted by permission of John Wiley & Sons, Inc.).

Table 4. Density and thermal properties of some soil and reference materials[a] (Microclimatology, The Biological Environment, Rosenberg, N.J., B.L. Blad, S.B. Verma, Copyright 1983 by John Wiley & Sons, Inc., Reprinted by permission of John Wiley & Sons, Inc.).

	Density, ρ (kg 1^{-1})	Specific Heat		Thermal Conductivity, K_g (W m^{-1} $°K^{-1}$)	Thermal Diffusivity, D_T (m^2 s^{-1})
		Mass C_s ($J°K^{-1}$ kg^{-1})	Volume, C_v ($J°K^{-1}$ 1^{-1})		
Water @ 20°C	1.00	4.188×10^3	4.188×10^3	5.862×10^{-1}	1.4×10^{-7}
Dead air	0.013	1.005×10^3	1.307×10^1	2.089×10^{-2}	1.6×10^{-5}
Silver metal	10.50	0.251×10^3	2.636×10^3	4.523×10^2	1.72×10^{-4}
Clay	1.80	3.350×10^3	6.030×10^3	1.206	2.0×10^{-7}
Light soil with roots	0.30	1.256×10^3	3.768×10^2	1.131×10^{-1}	3.0×10^{-4}
Wet sandy soil	1.60	1.675×10^3	2.680×10^3	6.398×10^{-3}	1.0×10^{-6}

[a]Based on data from Sutton (1953) and others.

5.2. Surface and Subsurface Temperature Features

Soil surface temperature changes drastically with time of day in response to solar energy. But its change is damped by the insulating effect of vegetative cover and high soil moisture conditions. The change is also much less in the winter than in the summer due to much less solar energy input. Even though daily surface temperature variations can have rather extreme values, up to as much as 35–40 °C in desert conditions, subsurface temperatures are damped considerably because of the relatively low thermal diffusivity of soil. Soil thermal diffusivity ranges from about 20 to 100 m^2s^{-1} whereas dry, dead air is about 1600 m^2s^{-1}. Figure 18, taken from Rosenberg et al. (1983), illustrates the change in diurnal response with season, the insulating effect of surface cover, and the damping with soil depth. The amplitude of daily sinusoidal fluctuations is a function of soil texture as illustrated in Figure 19. These figures show that very little energy is stored in the soil profile from day to day. Most energy absorbed during daylight hours is radiated out during nighttime hours. Even though very little energy is stored from day to day, compared to the amount entering and leaving, there is continual redistribution with a net downward flux during the summer. Figure 20 (Rosenberg et al., 1983) shows typical temperature profiles for different periods of the day, illustrating the rapid change that takes place near the surface. Figure 18 also shows the lag in timing of the occurrence of the peak of the diurnal maximum temperature. At great depths, the highest temperature may be reached in midwinter and lowest values in the summer. A "rule–of–thumb" for estimating the depth at which the annual cycle is extinguished is to calculate the product of the depth of the diurnal wave and $(365)^{1/2}$ or 19. In Figure 18 the diurnal wave depth is about 0.4 m, thus the depth of the annual wave is about 7.6 m. Since the amplitude of the fluctuation decreases exponentially with depth, a sine wave is fitted to the surface temperature fluctuation, then it is possible to use the thermal diffusivity and parameters of the fitted expression to estimate the rate of damping with depth, z_D, in the soil profile (Kimball and Jackson, 1979). A second rule of thumb is that the constant temperature beneath the depth of the annual wave is approximately the average annual air temperature.

5.3. Soil Heat Flux

Equation (37), previously defined, describes the rate of heat movement into the soil profile. Kimball and Jackson (1979) show that if a sinusoidal function is fitted to the soil surface temperature as a boundary condition, then it is possible to define a function of the soil properties called the thermal inertia, I_T:

$$I_T = \sqrt{\omega \, C_v \, K_g} \tag{40}$$

234

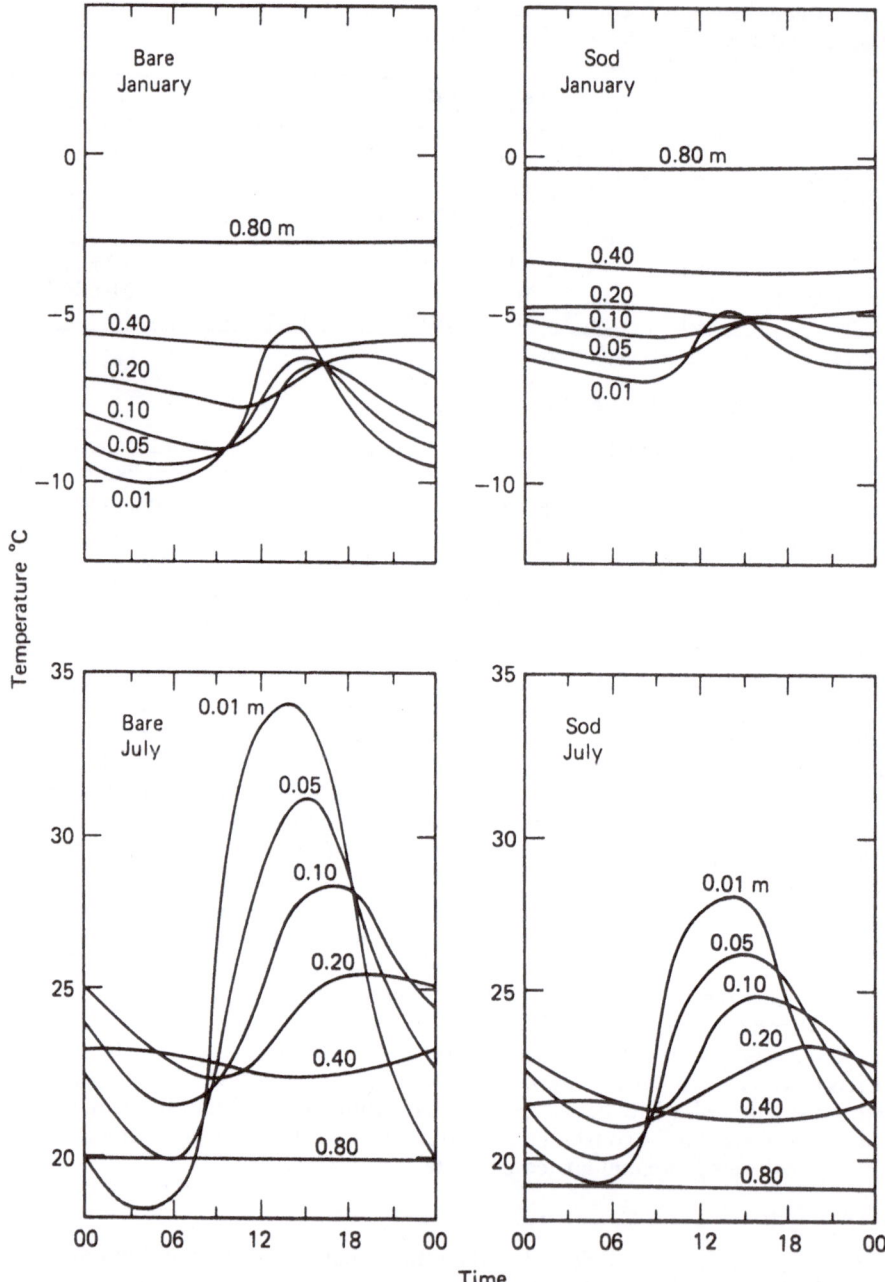

Figure 18. Average hourly soil temperature under bare and sod–covered soil at St. Paul, Minnesota in January (top) and July (bottom). Soil depth is shown in m (after Baker, 1965, Minnesota Farm and Home Science, and the University of Minnesota Agricultural Experiment Station) (Microclimatology, The Biological Environment, Rosenberg, N.J., B.L. Blad, S.B. Verma, Copyright 1983 by John Wiley & Sons, Inc., Reprinted by permission of John Wiley & Sons, Inc.).

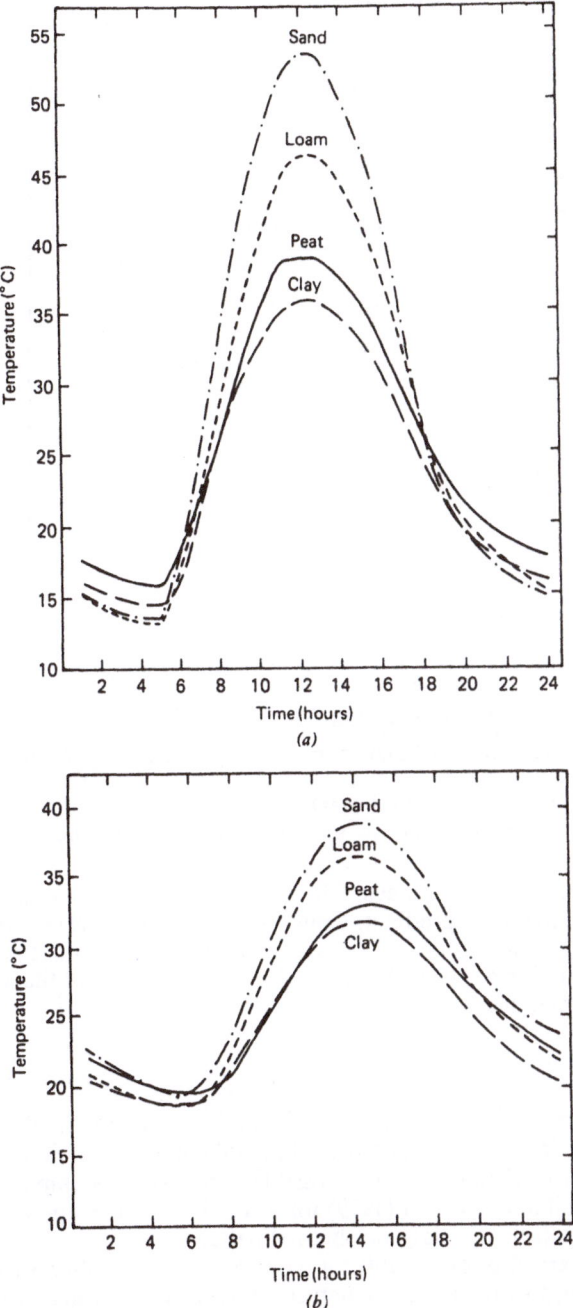

Figure 19. Daily course of temperature (a) at the surface and (b) at a depth of 50 mm on clear summer days at Sapporo, Japan (after Yakuwa, 1946) (<u>Microclimatology, The Biological Environment</u>, Rosenberg, N.J., B.L. Blad, S.B. Verma, Copyright 1983 by John Wiley & Sons, Inc., Reprinted by permission of John Wiley & Sons, Inc.).

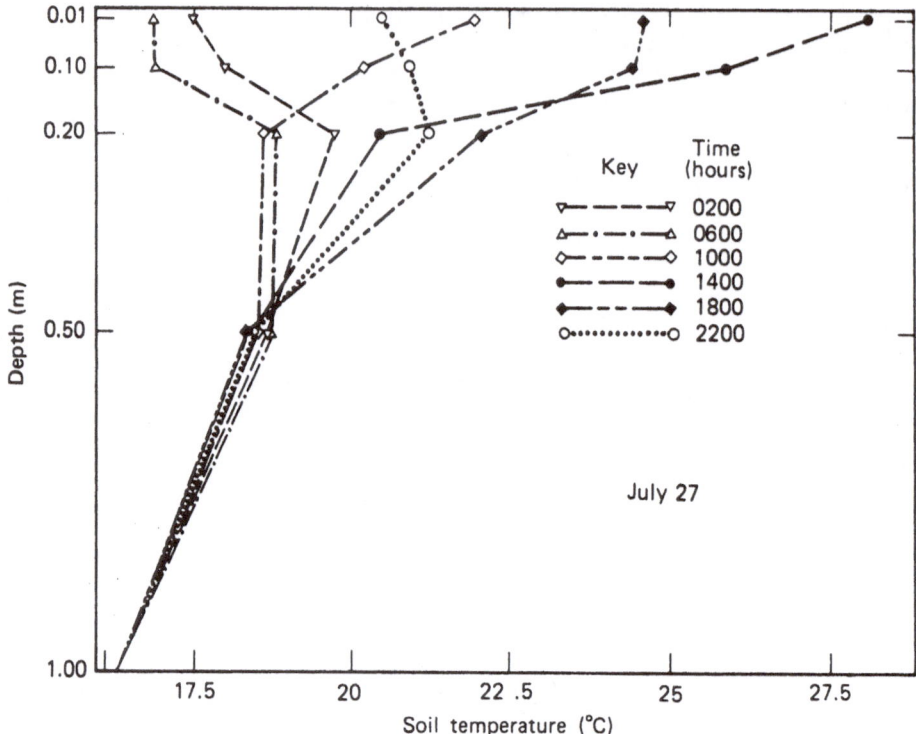

Figure 20. Vertical temperature profiles in soil during the course of a typical summer day at Argonne, Illinois (after Carson and Moses, 1963, Journal of Applied Meteorology, American Meteorological Society) (Microclimatology, The Biological Environment, Rosenberg, N.J., B.L. Blad, S.B. Verma, Copyright 1983 by John Wiley & Sons, Inc., Reprinted by permission of John Wiley & Sons, Inc.).

in which ω is the angular velocity in fitting the sine wave. The larger the value of I_T the larger will be the soil heat flux for a given temperature gradient. Conversely, for a given soil heat flux the amplitude of the temperature wave will be smaller for a larger thermal inertia. The ratio of the amplitudes of temperature fluctuation of two soils having the same surface heat flux will equal the inverse ratio of their thermal inertias:

$$A_1/A_2 = I_{T_2}/I_{T_1} \qquad (41)$$

as an example, the difference in thermal inertias between dry clay and wet sand (21.5/4.9) will cause the temperature fluctuation of dry sand to be about 4.4 times that of wet clay. Values of I_T, K_g, C_v, and Z_D change as soil is tilled; therefore, it is possible to describe the impact of tillage on soil heat flux. See Kimball and Jackson (1979) for a discussion of temperature fluctuation in layered soils and the effects of tillage on thermal patterns.

Temperature gradients in the soil are also responsible for moving soil water. Water vapor moves from areas of high kinetic energy to areas of low kinetic energy, i.e. from warm areas to cool areas. Therefore, temperature gradients force water vapor to flow within the soil profile. Observed values vary from about 0.4 to 2 mm per day. This compares with evaporation values that vary from zero to 12 mm per day, thus temperature-induced vapor flow is not negligible. Temperature gradients shown in Figure 20 indicate that thermally induced water flow will be reversed a number of times each day, especially in the surface layers.

5.4. Properties Affecting Soil Heat Flux

The volumetric heat capacity, C_v, as previously described, is made up of the heat capacities of individual soil constituents. De Vries (1963), in a review of other workers' data, concluded that the heat capacities of minerals, organic matter, and water are 1.9, 2.5, and 4.2 MJ m^{-3} °C^{-1}, respectively. Thus for unfrozen soil:

$$C_v = 1.9 \, X_m + 2.5 \, X_o + 4.2 \, X_w \tag{42}$$

in which X_m, X_o or X_w are the volume fractions of minerals, organic matter, and water, respectively.

De Vries (1963) also described a method for calculating the thermal conductivity, K_g, based on electrical field theory for calculating the dielectric constant of a granular material.

Bulk density affects heat transmission because it is a measure of the relative volume of minerals, organic matter, water and air. Thus Equation (42) can be used to show the effect of tillage on bulk density.

Water content affects heat flow in many ways. First it affects the heat capacity as described in Equation (42). Second, immediately following the inception of rain, infiltrating water advects heat into the soil medium. Third, following a rainfall event the surface will dry, creating a steep water content gradient that in turn creates a vapor transfer that advects both water and heat from one region to another.

6. Conclusions

The subject of microclimatology is frequently glossed over in traditional approaches used in modeling hydrologic systems because the scale of simulation generally precludes definition of the systems at a level that is meaningful from the microclimatological perspective. Microclimatology has primarily been the tool of agriculturalists. With increasing interest in water quality, it has become apparent that more emphasis needs to be placed on the physical processes influencing the movement and disposition of water in the root zone.

In this chapter the subject of microclimatology has been described very briefly. The emphasis has been to introduce the subject and lead the reader to numerous references that can be used in further study. The first topic discussed is radiation energy balance. Methods of calculating incoming shortwave and longwave radiation are presented along with albedos required to estimate outgoing energy fluxes. Net solar radiation is partitioned into five major processes: the flux of heat into or out of the soil, the flux of sensible heat between the air and the soil or plant surfaces, the latent heat used in vaporizing water through surface evaporation and plant transpiration, energy used by plants in photosynthesis, and energy used in miscellaneous processes such as heat storage in the plant canopy and plant respiration. Energy used in photosynthesis and in miscellaneous processes is minor compared to that used in sensible and latent heat and at times in soil heating.

Various methods of calculating the sensible and latent heat fluxes are presented. These include hydrologic, climatologic, micrometeorologic, eddy correlation and combination methods. More emphasis is placed on the combination methods because they are generally superior and describe the fundamental processes more completely. Since resistance to the flux of energy, water vapor and momentum are so important in estimating the flux rates, these topics are described. These include the resistance to wind movement created by the plant geometry, resistance to water vapor transport through the stomata on plant leaves and the influence of partial canopy cover on all processes.

Finally, the subject of soil temperature and heat flux is presented. The effects of soil type, moisture content, and thermal properties on the cyclic patterns of soil temperature with depth in the profile are described.

238

7. References

Allen, R. G., M. E. Jensen, J. L. Wright, and R. D. Burnam: 1989, 'Operational Estimates of Reference Evapotranspiration,' *Agron. J.* **81**, 650–662.

Al–Nakshabandi, G., and H. Kohnke: 1965, 'Thermal Conductivity and Diffusivity of Soils as Related to Moisture Tension and Other Physical Properties,' *Agric. Meteorol.* **2**, 271–279.

Ångström, A.: 1925, 'The Albedo of Various Surfaces of Ground,' *Geografiska Annaler.* **7**, 321–342.

Arnfield, A. J.: 1975, 'A Note on the Diurnal, Latitudinal and Seasonal Variation of the Surface Reflection Coefficient,' *J. Appl. Meteorol.* **14**(8), 1603–1608.

Aslyng, H. C.: 1974, 'Evapotranspiration and Plant Production Directly Related to Global Radiation,' *Nordic Hydrol.* **5**, 247–256.

Baker, D. G.: 1965, 'Factors Affecting Soil Temperature,' *Minn. Farm Home Sci.* **22**, 11–13.

Barfield, B. J., and J. F. Gerber (eds.): 1979, *'Modification of the Aerial Environmental of Plants,'* ASAE Monograph #2 in series, Am. Soc. of Agricultural Engineers, St. Joseph, MI 49805, 538 pp.

Berry, F. A., E. Bollay, and N. R. Beers: 1945, *Handbook of Meteorology*, McGraw Hill Book Co., NY, pp. 292, 297.

Beven, K.: 1979, 'A Sensitivity Analysis of Penman–Montieth Actual Evapotranspiration Estimates,' *Journal of Hydrology* **44**, 169–190.

Black, T. A., C. B. Tanner, and W. R. Gardner: 1970, 'Evapotranspiration from a Snap Bean Crop,' *Agron J.* **62**, 66–69.

Blad, B. L., and N. J. Rosenberg: 1974, 'Lysimetric Calibration of the Bowen Ratio–Energy Balance Method for Evapotranspiration Estimation in the Central Great Plains,' *J. Appl. Meteorol.* **13**, 227–236.

Blaney, H. F., and W. D. Criddle: 1950, *'Determining Water Requirements in Irrigated Areas from Climatological and Irrigated Data,'* USDA–SCS Tech. Paper No. 96, 48 pp.

Budyko, M. I.: 1956, *'The Heat Balance of the Earth's Surface,'* English Trans. Stepanova, N.A. 1958. Office of Technical Services, U.S. Dept. of Commerce, Washington, DC.

Büttner, K., and E. D. Sutter: 1935, 'Abkuhl, Gross Ind. Dunen. Ruckstrahl Verschied,' Bodenbedeck, f. uv. u. gesante Sonnenstrahl, *Strahlentherapie* **54**, 156–173.

Carson, J. E., and H. Moses: 1963, 'The Annual and Diurnal Heat–Exchange Cycles in Upper Layers of Soil,' *J. Appl. Meteorol.* **2**, 397–406.

Chia, L. S.: 1967, 'Albedos of Natural Surfaces in Barbados,' *Quart. J. Roy. Meteorol. Soc.* **93**(395), 116–120.

Davies, J. A.: 1963, 'Albedo Investigations in Labrador–Ungava,' *Arch. Meteorol. Geophys. Bioklim Ser. B.* **13**(1), 137–150.

Davies, J. A., and P. H. Buttimor: 1969, 'Reflection Coefficients, Heating Coefficient and Net Radiation at Simcoe, Southern Ontario,' *Ag. Meteorol.* **6**(3), 373–386.

Davies, J. A., and S. B. Idso: 1979, 'Estimating the Surface Radiation Balance and Its Components,' in B. J. Barfield, and J. E. Gerber (eds.), *Modification of the Aerial Environment of Plants, ASAE Monograph #2 in Series*, Am. Soc. of Agric. Engrs., pp. 183–210.

DeCoursey, D. G.: 1980, 'Runoff, Erosion and Crop Yield Simulation for Landuse Management,' *Trans. ASAE* **23**(2), 374–386.

DeMichele, D. W., and P. J. H. Sharpe: 1973, 'An Analysis of the Mechanics of Guard Cell Motion,' *J. Theoretical Biol.* **41**, 77–96.

Denmead, O. T.: 1976, 'Temperate Cereals,' in J. L. Monteith (ed.), *Vegetation and the Atmosphere, Vol. 2*, Academic Press, New York.

Denmead, O. T., and C. C. McIlroy: 1970, 'Measurements of Non–Potential Evaporation from Wheat,' *Agric. Meteorol.* **7**, 285–302.

De Vries, D. A.: 1963, 'Thermal Properties of Soils,' in W. R. van Wijk (ed.), *Physics of Plant Environment*, John Wiley and Sons Inc., New York, NY, pp. 210–235.

Ehrler, W. L., and C. van Bavel: 1967, 'Sorghum Foliar Response to Soil Water Content,' *Agron. J.* **59**, 243–246.

Ekern, P. C.: 1965, 'The Fraction of Sunlight Retained as Net Radiation in Hawaii,' *J. Geophys. Res.* **70**(4), 785–793.

Fritschen, L. J.: 1967, 'Net and Solar Radiation Relations Over Irrigated Field Crops,' *Agric. Meteorol.* **4**(17), 55–62.

Graham, W. G., and K. M. King: 1961, 'Short–Wave Reflection Coefficient for a Field of Maize,' *Quart. J. Roy. Meteorol. Soc.* **87**(373), 425–428.

Hargreaves, G. H.: 1974, 'Estimation of Potential and Crop Evapotranspiration,' *Trans. ASAE* **17**, 701–704.

Heilman, J. L., and E. T. Kanemasu: 1976, 'An Evaluation of a Resistance Form of the Energy Balance to Estimate Evapotranspiration,' *Agron. J.* **68**, 607–612.

Idso, S. B.: 1969, 'Atmospheric Attenuation of Solar Radiation,' *J. Atmos. Sci.* **26**(4), 1088–1095.

Idso, S. B., and R. D. Jackson: 1969, 'Thermal Radiation from the Atmosphere,' *J. Geophys. Res.* **74**(23), 5397–5403.

Idso, S. B., R. D. Jackson, and R. J. Reginato: 1975, 'Estimating Evaporation: A Technique Adaptable to Remote Sensing,' *Science* **189**, 991–992.

Idso, S. B., R. D. Jackson, R. J. Reginato, B. A. Kimball, and F. S. Nakayama: 1975, 'The Dependence of Bare Soil Albedo on Soil Water Content,' *J. Appl. Meteorol.* **14**(1), 109–113.

Idso, S. B., and R. J. Reginato: 1974, 'Assessing Soil Water Status Via Albedo Measurement,' *Hydrology and Water Resources in Arizona and the Southwest* **4**, 41–54.

Idso, S. B., R. J. Reginato, and R. D. Jackson: 1977, 'An Equation for Potential Evaporation from Soil, Water, and Crop Surfaces Adaptable to Use by Remote Sensing,' *Geophys. Res. Letters* **4**, 187–188.

Impens, I., and R. Lemeur: 1969, 'The Radiation Balance of Several Field Crops,' *Arch. Meteor. Geophys. Bioklim.* **B17**, 261–268.

Jarvis, P. G., G. B. James, and J. J. Landsberg: 1976, 'Coniferous Forest,' in J. L. Monteith (ed.), *Vegetation and the Atmosphere*, Vol. 2, Academic Press, New York, pp. 171–240.

Jensen, M. E.: 1974, *Consumptive Use of Water and Irrigation Water Requirements,* Rep. by the Tech. Comm. Irrig. Water Requ. Irrig. and Drain. Div., Am. Soc. of Civil Engrs.

Jensen, M. E., and H. R. Haise: 1963, 'Estimating Evapotranspiration from Solar Radiation,' *J. Irrig. Drainage Div. ASCE* **89**, 15–41.

Kalma, J. D., and R. Bredham: 1972, 'The Radiation Balance of a Tropical Pasture, I. The Reflection of Short–Wave Radiation,' *Agric. Meteorol.* **10**(4), 251–259.

Kanemasu, E. T., and C. B. Tanner: 1969, 'Stomatal Diffusion Resistance of Snap Beans I. Influence of Leaf Water Potential,' *Plant Physiol.* **44**, 1547–1552.

Kanemasu, E. T., M. L. Wesely, B. B. Hicks, and J. L. Hielman: 1979, 'Techniques for Calculating Energy and Mass Fluxes,' in B. J. Barfield, and J. F. Gerber (eds.), *Modification of the Aerial Environment of Crops, ASAE Monograph #2 in Series*, Am. Soc. of Agric. Engr., pp. 156–182.

Kelliher, F. M., and T. A. Black: 1986, 'Estimating the Effects of Understory Removal from Douglas Fir Forest Using a Two–Layer Canopy Evapotranspiration Model,' *Water Res. Res.* **22**(13), 1891–1899.

Kimball, B. A., and R. D. Jackson: 1979, 'Soil Heat Flux,' in B. J. Barfield, and J. E. Gerber (eds.), *Modification of the Aerial Environment of Plants, ASAE Monograph #2 in Series*, Am. Soc. Agric. Engrs., pp. 211– 229.

Kondrat'yev, K. Ya: 1969, *Radiation in the Atmosphere*, Academic Press, New York, 912 pp.

Kondrat'yev, K. Ya: 1954, 'Albedo of the Underlying Surface and Clouds,' in *The Radiant Energy of the Sun*, Leningrad, Chapter 9, English Summary by A. Kurlent and P. Larson, McGill Univ.

Linacre, E. T.: 1967, 'Climate and the Evaporation from Crops,' *J. Irrig. Drainage Div. ASCE* **93**, 61–79.

Linacre, E. T.: 1977, 'A Simple Formula for Estimating Evapotranspiration Rates in Various Climates, using Temperature Data Alone,' *Agric. Meteorol.* **18**, 409–424.

Makkink, G. F.: 1957, 'Ekzameno de la Formula de Penman,' *Neth. J. Agric. Sci.* **5**, 290–305.

Montieth, J. L.: 1959, 'The Reflection of Short–Wave Radiation by Vegetation,' *Quart. J. Roy. Meteorol. Soc.* **85**(366), 386–392.

Montieth, J. L.: 1965, 'Evaporation and Environment,' *Symp. Soc. Exp. Bio.* **19**, 205–234.

Montieth, J. L.: 1973, *Principles of Environmental Physics*, American Elsevier Publ. Co., New York, 241 pp.

Nkemdirim, L. C.: 1972, 'A Note on the Albedo of Surfaces,' *J. Appl. Meteorol.* **II**(5).

Oguntoyinbo, J. S.: 1970, 'Reflection Coefficient of Natural Vegetation Crops and Urban Surfaces in Nigeria,' *Quart. J. Roy. Meteorol. Soc.* **96**(409), 430–441.

Paltridge, G. W.: 1970, 'Daytime Longwave Radiation from the Sky,' *Quart. J. Roy. Meteorol. Soc.* **96**(410), 645–653.

Penman, H. L.: 1948, 'Natural Evapotranspiration from Open Water, Bare Soil and Grass,' *Proc. R. Soc. London Ser. A.* **193**, 120–145.

240

Piggin, I., and P. Schwerdtfeger: 1973, 'Variations in the Albedo of Wheat and Barley Crops,' *Arch. Meteorol. Geophys. Bioklim. Ser. B* **21**(4), 365–391.

Portman, D. J.: 1954, 'The Measurement of Radiation,' *Publs. Clim. Johns Hopkins Univ.* **7**, 289–292.

Pruitt, W. O., and D. E. Angus: 1961, 'Comparison of Evapotranspiration with Solar and Net Radiation and Evaporation with Water Surfaces,' *Chapter VI, First Annual Report*, U.S. Army Electronic Proving Grounds Contract No. DA-36-039-SC-80334, Univ. of Calif.-Davis.

Pruitt, W. O., and F. J. Lourence: 1968, *'Correlation of Climatological Data with Water Requirement for Crops,'* Dep. of Water Sci. and Engr. Paper No. 9001, Univ. of Cal.-Davis, 59 pp.

Rijks, D. A.: 1967, 'Water Use by Irrigated Cotton I. Reflectance of Short-Wave Radiation,' *J. Appl. Ecol.* **4**, 561–568.

Ripley, E. A., and R. E. Redmann: 1976, 'Grassland,' in J. L. Monteith (ed.), *Vegetation and the Atmosphere, Volume 2, Case Studies*, Academic Press, New York, pp. 349–398.

Ritchie, J. T.: 1972, 'Model for Predicting Evaporation From a Row Crop with Incomplete Cover,' *Water Resources Research* **8**, 1204–1213.

Rosenberg, W. J., B. L. Blad, and S. B. Verma: 1983, *Microclimate, The Biological Environment, 2nd Edition*, John Wiley and Sons, New York, NY, 495 pp.

Shaw, R. H., and W. L. Decker: 1979, 'The General Heat Budget of Canopies,' in B. J. Barfield, and J. E. Gerber (eds.), *Modification of Aerial Environment of Plants, ASAE Monograph #2 in Series*, Amer. Soc. of Agric. Eng., pp. 141–155.

Shuttleworth, J. W.: 1979, 'Below-Canopy Fluxes in a Simplified One-Dimensional Theoretical Description of the Vegetation-Atmosphere Interaction,' *Boundary Layer Meteorology* **17**, 315–331.

Shuttleworth, J. W., and J. V. Wallace: 1985, 'Evaporation from Sparce Crops-An Energy Combination Theory,' *Royal Met. Soc.* **111**, 839–855.

Slatyer, R. O., and I. C. McIlroy: 1961, *Practical Microclimatology*, CSIRO, Australia and UNESCO.

Stanhill, G.: 1969, 'A Simple Instrument for the Field Measurement of Turbulent Diffusion Flux,' *J. Appl. Meteorol.* **8**(4), 509–513.

Stanhill, G.: 1970, 'Some Results of Helicopter Measurements of Albedo,' *Solar Energy* **13**, 59–66.

Sutton, O. G.: 1953, *Micrometeorology*, McGraw-Hill, New York, 333 pp.

Swinbank, W. C.: 1963, 'Longwave Radiation from Clear Skys,' *Quart. J. Roy. Meteorol. Soc.* **89**(381), 339–348.

Szeicz, G., G. Endrodi, and S. Tackman: 1969, 'Aerodynamic and Surface Factors in Evaporation,' *Water Res. Res.* **5**(2), 380–394.

Tanner, C. B.: 1960, 'Energy Balance Approach to Evapotranspiration from Crops,' *Soil Sci. Soc. Am. Proc.* **24**, 1–9.

Tanner, C. B.: 1967, 'Measurement of Evapotranspiration,' in R. M. Hagan, H. R. Haise, and T. W. Edminster (eds.), *Irrigation of Agricultural Lands*, Am. Soc. of Agron., Madison, WI, pp. 534–574.

Thornthwaite, C. W.: 1948, 'An Approach Toward a Rational Classification of Climate,' *Geogr. Rev.* **38**, 55–94.

Thornthwaite, C. W., and B. Holzman: 1942, 'Measurement of Evaporation from Land and Water Surface,' *USDA Tech. Bull. 817*, 75 pp.

van Bavel, C. H. M.: 1966, 'Potential Evaporation: The Combination Concept and Its Experimental Verification,' *Water Resources Res.* **2**, 455–467.

van Wijk, W. R. (ed.): 1963, *Physics of Plant Environment*, North-Holland Pub. Co., Amsterdam, 382 pp.

van Wijk, W. R.: 1964, 'Two New Methods for the Determination of the Thermal Properties of Soil Near the Surface,' *Physics* **30**, 387–388.

van Wijk, W. R. P: 1965, 'Soil Microclimate, its Creation, Observation, and Modification,' Chapter 3, in P. E. Waggoner (ed.), *Am. Meteorol. Soc. Meteorological Monograph* **6**, 59–73.

Verma, S. B., and B. J. Barfield: 1979, 'Aerial and Crop Resistances Affecting Energy Transport,' in B. J. Barfield, and J. E. Gerber (eds.), *Modification of the Aerial Environment of Plants, ASAE Monograph #2 in Series*, Am. Soc. Agric. Engrs., pp. 230–248.

Wallace, J. M., and P. V. Hobbs: 1977, *Atmospheric Science an Introductory Survey*, Academic Press, New York, 467 pp.

Yakuwa, R.: 1946, 'Über die Bodentemperaturen in den verschiedenen Bodenarten in Hokkaido (On Soil Temperature in the Different Soil Types of Hokkaido),' *Geophys. Mag. (Tokyo)* **14**, 1–12.

Chapter 12

Environmental Features Important in Nonpoint Source Models - Crop Growth and Influence of Management on Soil Properties

Donn G. DeCoursey
USDA-ARS
Hydro-Ecosystems Research Unit
243 Federal Building
301 South Howes, P. O. Box E
Fort Collins, Colorado 80522 U.S.A.

Abstract. The movement of nutrients and pesticides in the root zone is profoundly affected by both vegetation and state of the soil surface. Thus it is important that effects of vegetation and soil state on water and solute movement be accurately represented in dynamic mathematical models of the soil environment. Two global problems that have long-term impacts on our environment, groundwater quality, and climate change are uniquely related to the soil-plant-atmosphere interface. This and the preceding chapter discuss elements of the system that must be adequately represented in simulation models if they are to be effective tools of management. In this chapter concepts of plant growth processes and the effects of land use management and tillage on soil properties are discussed. Plant or crop growth processes described include photosynthesis and light interception, respiration, translocation, root growth, nutrient uptake, plant development, and morphogenesis. Some examples of crop growth models are presented. The discussion of soil properties covers estimates of interception, depression storage, soil bulk density, and hydraulic properties as influenced by management.

1. Introduction

The movement of nutrients and pesticides is environmentally important for several reasons. Perhaps most important, at the present time, is the concern over the potential for groundwater contamination. However, a farm manager may also be concerned about residual pesticides that may be in the root zone of future crops and could harm their growth. In most agricultural areas a large portion of precipitation is transpired back to the atmosphere by plants. The location in the soil where the plant gets its moisture affects the movement of water and associated solutes (pesticides, salts and nutrients). Thus good plant growth models are needed to estimate nutrient and pesticide fate. From the perspective of nutrients, a reliable plant growth model could aid in estimating future application rates. The removal of water, nutrients and pesticides influences the concentration and disposition of chemicals in the root zone. In many cases the plant may have more impact than any other component. Thus the need to consider their presence in simulating water and solute movement.

The movement of water and solutes is also affected by tillage. Order of magnitude change in hydraulic conductivity can be caused by the operation of a tillage tool and the gradual reconsolidation resulting from rainfall and microorganism activity. Distribution of

D. S. Bowles and P. E. O'Connell (eds.), Recent Advances in the Modeling of Hydrologic Systems, 241–268.

macropore soil structure and the resulting decrease in infiltration can both enhance and retard the movement of nutrients and pesticides depending upon their physical characteristics, the method of application, soil properties, and rainfall intensity. Thus, it is very important that the effects of tillage on soil properties be incorporated into hydrologic models being used in simulating soil water and solute movement.

2. Discussion

2.1. Crop Growth Modeling

Space does not permit an in–depth description of all the processes of crop growth, but there are numerous references on the subject. Two recent references have provided much of the material described in this chapter; they are *Mathematical Models in Agriculture* by France and Thornley (1984) and *Modeling Plant Growth and Development* by Charles–Edwards, Doley, and Rimmington (1986). Following is a brief description of the processes that are simulated to varying degrees in most crop growth models. Both books give numerous references to specific processes.

2.1.1. Plant Growth Processes

Plants acquire their energy from the sun, and their food source from the air and from nutrients in soil water. In a process called photosynthesis, plants use energy in the visible waveband (.4–.7μm) to convert water and carbon dioxide molecules in the air to oxygen and a carbohydrate moiety, $(CH_2O)_n$. As light energy is absorbed by chlorophyll in leaf tissues, electrons emitted by the chlorophyll effect a reduction of gaseous carbon dioxide to carbohydrate through a sequence of reactions catalyzed by proteins in the leaf tissues:

$$CO_2 + H_2O + light \rightarrow (CH_2O)_n + O_2 \tag{1}$$

The dependence of the photosynthetic process upon these proteins means that mineral elements from the soil such as nitrogen, phosphorus, sulfur, and other minor elements, which are essential constituents of the proteins, are essential to the proper functioning of the leaf. Thus the root system and its environment is as important as sunlight to a healthy plant.

Some of the carbohydrates produced by photosynthesis are then consumed within the plant for new growth and by respiration for maintenance of existing plant tissues. Since respiration occurs in all plant parts, the carbohydrate moiety is split and transported to all plant parts. The carbohydrate partitioned to roots is used to maintain the existing roots and grow new roots in regions of the soil profile where food, water, and the atmosphere are best. Growth of the above ground parts causes the plant to become larger until it reaches anthesis, when it starts growing reproductive parts. At this time larger and larger fractions of the carbohydrate are partitioned to the reproductive parts. The plant continues to grow, developing its fruit, until at maturity senescence begins. During senescence plant tissues deteriorate and die. In some plants this happens to all the leaves at the same time, on others it is a progressive thing starting with the older, more shaded leaves and moving progressively up the plant. During this process, some plants may mobilize nutrients from the dying parts and translocate them to new growing parts. Some plants, also periodically drop or abscise their leaves as they become less productive. In the following parts of this chapter, these individual processes are described.

Photosynthesis and Light Interception. Many of the basic principles of photosynthesis are not yet understood, but observations of the process at the leaf, plant, and crop levels provide enough information to simulate the process. Many different expressions have been proposed for leaf photosynthetic rate, P_ℓ, but the most common expression is:

$$P_\ell = \frac{a\,I_\ell\,P_{max}}{a\,I_\ell + P_{max}} \tag{2}$$

in which α is the photosynthetic efficiency; I_ℓ is the light flux density on the leaf; and P_{max} is the photosynthetic rate at saturating light intensity. P_{max} is frequently expressed as:

$$P_{max} = \tau\,C_a \tag{3}$$

in which τ is the CO_2 conductance and C_a is the CO_2 concentration.

The canopy photosynthetic rate is obtained by integrating Equation (2) over the leaf area of the plant. However, since the light flux density on leaves within the canopy is less than on the surface, solution to the integral equations will depend upon assumptions about the structure of the canopy, leaf orientation, and leaf area index. Obviously, every crop has its own pattern, but Monsi and Saeki (1953) assumed that the light flux density, I, on a horizontal plane at any point in the canopy can be expressed as:

$$I = I_o\,e^{-kL} \tag{4}$$

in which I_o is the light flux density at the surface of the canopy, k is an extinction coefficient, and L is the leaf area index for the total canopy above the point represented by I. Figure 1 shows values of I for two different values of k over a range of leaf area indices. It can be shown that I_ℓ and I are related by:

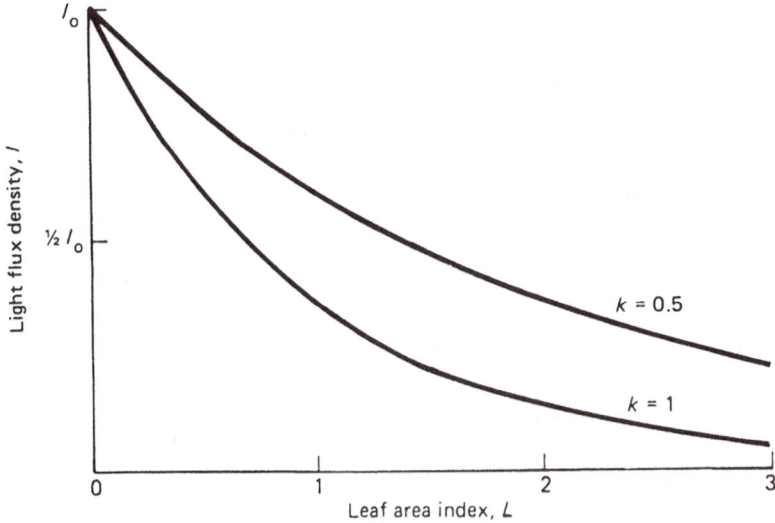

Figure 1. The Monsi–Saeki equation (4). Two values of the extinction coefficient k are shown (Mathematical Models in Agriculture, Butterworths, 1984, courtesy J. France and J.H.M. Thornley).

$$I_\ell = \left(\frac{k\,I_o}{1-m}\right)e^{-kL} \tag{5}$$

in which m is the leaf transmission coefficient. Substituting Equation (5) in Equation (2) and integrating across the canopy yields the canopy photosynthetic rate, P_c:

$$P_c = \frac{P_{max}}{k}\left\{\ell n\left[\frac{k\,a\,I_o + P_{max}(1-m)}{k\,a\,I_o\,e^{-kL} + P_{max}(1-m)}\right]\right\} \tag{6}$$

Solution of Equation (6) is shown in Figure 2 for representative values of P_{max}, k, m, and α. Light flux density is expressed as radiant energy in the wavelength that plants use i.e. photosynthetically active radiation (PAR) which is about 0.38 – 0.71 μm.

Respiration. Respiration in plants is the conversion of sugars, stored carbohydrates, and nitrogen into plant parts or plant proteins. It is also the use of some of the carbohydrates in maintaining plant status quo. There is no consensus on how to best divide respiration into growth and maintenance components. However, France and Thornley (1984) discuss four different approaches:

(1) A phenomenological approach in which the total respiration, R, is a linear combination of photosynthetic rate, P_c and plant weight W:

$$R = kP_c + cW \tag{7}$$

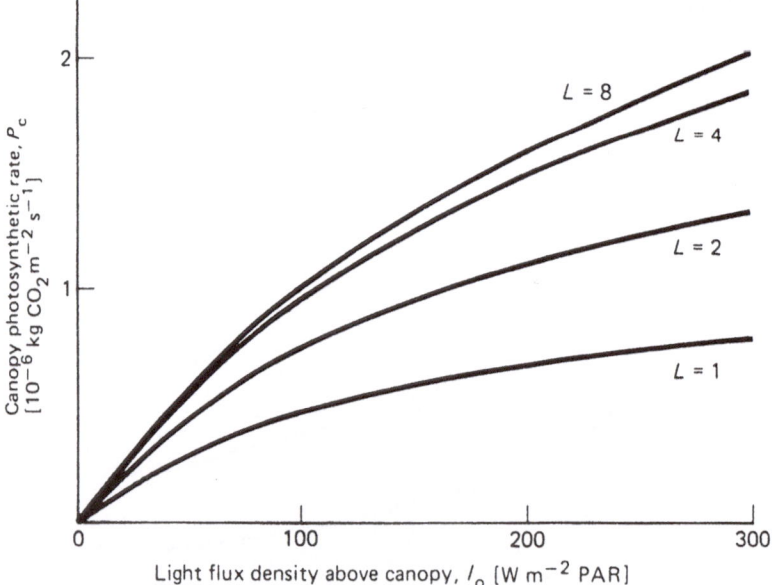

Figure 2. Crop photosynthetic response curve to light according to Equation (6), for the leaf area indices L shown. Parameter values are $P_{max} = 1.2 \times 10^{-6}$ kg CO_2 m^{-2} s^{-1}, $\alpha = 13 \times 10^{-9}$ kg CO_2 (JPAR)$^{-1}$, $k = 0.8$ and $m = 0.1$ (Mathematical Models in Agriculture, Butterworths, 1984, courtesy J. France and J.H.M. Thornley).

in which k and c are coefficients. In general, models of this form in which the maintenance requirement is a constant do not function well.

(2) The second approach is a biochemical approach in which knowledge about pathways of reactions, stoichiometry, and energetics are used to predict efficiencies of the growth and maintenance processes. Very wide ranges in values obtained by this method indicate that much remains to be done before the method is useful.

(3) The third method is a substrate–balance analysis of respiration that is based on an experimental evaluation of the disposition of a quantity of substrate supplied during a time interval. This approach leads to an expression that requires estimates of dry matter accumulation and specific respiration rate. See France and Thornley (1984) for a discussion of expressions to represent the processes.

(4) The fourth method is a recycling model of growth and maintenance respiration. In this method both growth and maintenance are assumed produced by the same synthetic process, but a component of the structural dry matter is degraded and can be resynthesized, producing a maintenance–like respiration component. The equation that results from this approach is considerably different from the one that is derived from the assumptions in the third approach.

It is obvious that much remains to be done if crop growth models are to be able to specifically simulate respiration with any degree of reliability. One approach, very similar to Equation (7), being used to estimate maintenance respiration (Thornley, 1977; Barnes and Hole, 1978) is based on the amount of proteinaceous material contained in plant tissues. The proteinaceous material is directly proportioned to elemental nitrogen content, N. The expression is:

$$R = a_o P_c + b_o N \tag{8}$$

in which R is the daily integral of total respiration rate, P_c is the daily integral of the rate of total photosynthesis, a_o and b_o are coefficients. The $b_o N$ term is the maintenance respiration which may be approximated by a constant.

Other assumptions lead to different expressions. One assumption leads to an expression similar in form to Equation (8); the daily respiration is related directly to the amount of light energy, $b_1 J$, received. Thus $b_o N$ in Equation (8) is replaced by $b_1 J$. Yet another assumption is that respiration can be based on the protein content of plants. Researchers making these assumptions are able to support their arguments from data they have collected. An example of the research data collected in such experiments is shown in Figure 3 where the total daily respiration is compared to total daily photosynthesis. Note in Figure 3(b) the difference between values for a clear day following dull days, and a cloudy day following clear days. This could be justification for the assumption that respiration is a function of light energy. Data scatter and the very likely spurious correlation that exists in the data support the need for additional research across a wide variety of crops and assumptions.

Translocation, the Movement of Plant Sugars. Plant growth is a very complex process with growth taking place in roots, stems, leaves, etc. all at the same time. Also, a wide variety of protein and carbohydrate products are required by these growing tissues and these products are developed in the leaves during the photosynthetic process. Translocation is the movement of this photosynthate from leaves to roots, to new shoots, or to places of temporary or permanent storage until they are needed elsewhere.

Many different hypotheses have been formulated to describe the translocation process, but no single description of the process is accepted as complete. Ordinary diffusion is too slow to account for the observed rates of movement of sugar between the different plant tissues. It is difficult to envision how diffusive flow could be facilitated to the extent required to explain observed rates at which sugars are translocated between plant parts. Another hypothesis

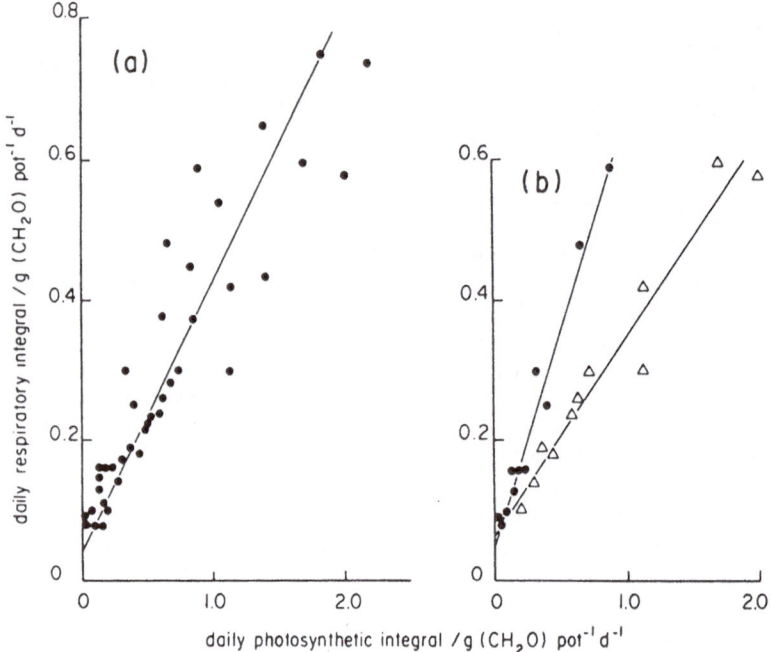

Figure 3. (a) Observed relationship between the daily dark respiratory integral and the daily photosynthetic integral for seedlings of rye grass (Lolium multiflorum). (b) Observed relationships between the daily respiratory integral and the daily photosynthetic integral for seedlings of Lolium perenne on a dull day following three successive bright days (0) and on a bright day following three successive dull days (Δ) (Charles–Edwards et al., 1986).

assumes that mass flow is driven by osmotic pressures. This process is simple and could sustain the high rates of translocation of sugars that have been observed; however, it depends upon a controversial subject of open sieve plate pores required to move the volume of phloem sap. Other mechanisms, which do not depend upon open sieve plate pores, require complex cellular machinery and the existence of a concentration gradient between the source and sink organs. Such gradients have been observed. France and Thornley (1984) describe the mathematics of some of these processes.

More simplistic approaches to the translocation process have been proposed for partitioning carbohydrates between plant organs. One such process assumes that a balance is maintained between mass of plant parts (roots and shoots, for example). Assume that two organs, A and B, have masses of M_A and M_B, then a growth ratio G_R is maintained such that:

$$G_R = \frac{M_A}{M_B} \tag{9}$$

Another method is to assume that a substrate pool of a given concentration, S_c, is assigned to plant parts A and B, using expressions for relative growth as follows:

$$G_A = \frac{k_A S_c}{K_A + S_c} \quad \text{and} \quad G_B = \frac{k_B S_c}{K_B + S_c} \tag{10}$$

in which G_A and G_B are the relative growth rates of organs A and B and k's and K's are constants. Functions such as those in Equation (10) can model both a fixed system of assignment priorities, i.e. the k's and K's remain in the same ratio, or changing priorities if the k's and K's are functions of the age or degree of maturity of the plant. Figure 4 shows how the equations may be used to give priority to organ A at low substrate concentration but to organ B at higher concentrations.

Root Activity and Nutrient Uptake. The roots of a plant are made up of existing root tissues and new growth. Obviously, the rate of root growth is dependent upon the amount of new assimilate or labile carbohydrates translocated to the roots. Generally, maintenance of existing root tissues and their function has the highest priority and the production of new shoots above ground has lowest priority.

Charles–Edwards et al. (1986) describes a simple root growth model in balance with the above ground shoot growth. The concentrations of labile carbohydrate in the shoot and root tissues are A_S and A_R, respectively, and the mass of shoot and root materials are W_S and W_R, respectively. The rate of growth of shoots, A_S, and roots, A_R, is defined as:

$$dA_S/dt = [\nabla F - \phi(A_S - A_R) - kA_S]/W_S \tag{11}$$

and

$$dA_R/dt = [\phi(A_S - A_R) - \beta - kA_R]/W_R \tag{12}$$

in which ∇F is the daily net assimilation rate of the plant, k is a growth constant, ϕ is translocation conductance, and β is the rate of assimilate use in the process of nutrient uptake. The growth rates of shoots and roots are kA_S and kA_R, respectively. Charles–Edwards et al. (1986) states that the number of growing points, n_R, in the roots that can be sustained is given by:

$$n_R = [\phi(A_S - A_R) - \beta]a_R \tag{13}$$

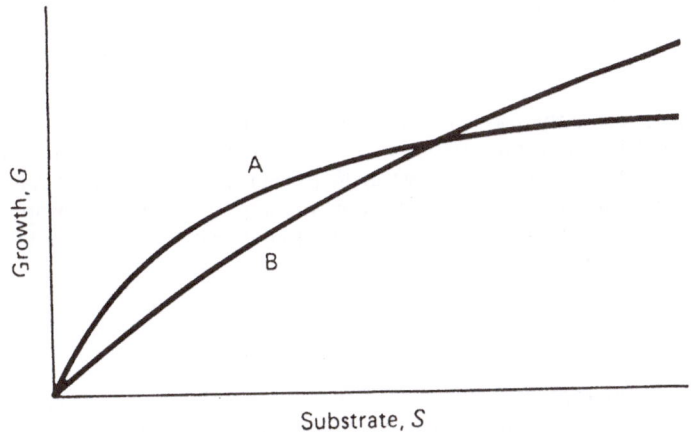

Figure 4. Growth responses, G_A and G_B, to a substrate pool of concentration S_c for two organs, A and B, according to equation (10) (Mathematical Models in Agriculture, Butterworths, 1984, courtesy J. France and J.H.M. Thornley).

in which a_R is the flux of assimilate required to sustain one root growing point. If we assume a steady-state growth pattern, i.e. $dA_S/dt = dA_R/dt$, we get:

$$n_R = [k \phi \nabla F - k \beta(\phi + k)]/[\phi^2 + (\phi + k)^2] \tag{14}$$

This expression shows that the number of new roots will increase as the plant grows, i.e., as ∇F increases, and that the number will decrease as the rate of assimilate required to take up nutrients, β, increases, i.e., as the nutrients in the surrounding soil decrease. If the model is applied to each growth point, it will be seen that growth can take place in regions of high mineral and nutrient concentrations. This pattern of root growth is very frequently observed.

Root length to diameter and weight can be used to estimate surface area and thus nutrient uptake. Root dry weight, W; surface area, A; volume, V; length, L; and radius, r are simply related as:

$$V = W/\rho \tag{15}$$

in which ρ is the root dry matter density.

$$V = \pi r_2 L \tag{16}$$

and

$$A = 2\pi r L \tag{17}$$

From these expressions, it is possible to estimate root surface area needed, along with concentrations, to estimate uptake of nutrients.

Nutrient ion movement toward a plant root takes place by: (i) mass flow of soil water, and (ii) by diffusion in which the concentration gradient between the soil and root surface is the driving mechanism. The latter is expressed simply as:

$$F_i = \lambda C_i \tag{18}$$

in which F_i is the flux of ion, i, across the root surface; C_i is the ion concentration in the soil solution and λ is an uptake coefficient. If the coefficient is large, the ion concentration at the root surface will be low and uptake may be limited by the diffusion rate. However, if λ is low the ion concentration at the root surface will be nearly constant, and uptake will be proportional to λ. If nutrient uptake is controlled more by water flux into the root, then the flux of the ion, i, to the root surface is given by:

$$F_i = VC_i \tag{19}$$

in which V is the water flux into the root. If the rate of uptake of the ion by the root is greater than the supply rate, the ion concentration in the solution at the root surface will decrease and the concentration gradient will cause diffusive movement toward the root. The opposite will happen if the uptake rate is less than the supply rate. These facts show why it is important to know the growth pattern and surface area of active roots. For a more detailed discussion of root growth and soil water movement, see Huck and Hillel (1983).

Plant Development and Morphogenesis. The rate at which a plant progresses from seed into a seedling, from seedling into a mature vegetative plant, or from a vegetative plant into a reproductive one is dependent upon physiological and environmental factors. Plant development in this context is different from plant growth in terms of dry matter production.

The rate of change in the development state of a plant often appears to be sensitive to both temperature and photoperiod or day length. The age and size of the plant are also important. If water, nutrient, or temperature stress has prevented a plant from becoming a mature vegetative plant, then it may not have a leaf canopy large enough to assimilate enough material to produce fruit (grain) or else the yield will be considerably reduced. Frequently plant development is a function of degree days; if the mean ambient temperature has been low, for example, it may take considerably longer for the plant to mature to the point of setting seed. Leaf development is one way that plant maturity is evaluated. The effect of temperature on rate of maturity is directly linked to the energy provided by warmer temperatures. However, there is a range of physiological temperatures within which growth can take place. Charles–Edwards et al. (1986) describes a chemical rate constant, k, that is frequently used to express growth rate in plants as:

$$k = k_R (1 + b \, \Delta T) \tag{20}$$

in which ΔT is the difference between the temperature T and a reference temperature R which is crop specific; these temperatures are usually expressed in °K; b is defined as equal to $a_1 TR$. k_R, a reference value of k at temperature, R, and a_1, a coefficient, are based on the Arrhenius equation:

$$ln(k) = a_o - a_1/T \tag{21}$$

in which a_o and a_1 are constants.

Charles–Edwards et al. (1986) shows how Equation (20) is used to derive the general degree/day expression:

$$t_{AB} = a_n + b \, \Sigma \, (\Delta T) \tag{22}$$

in which t_{AB} is sum of the degree–days needed to move from plant state A to state B, b was previously defined and a_n is a function of plant state, b, and k_R. This expression is frequently used to describe the rate of plant development.

Plant flowering is perhaps the most critical time in the simulation of plant growth; thus, it has received a lot of attention. Some plants flower most rapidly if the duration of the light period is short, others flower if it is long. Some plants will not flower at all if the day length is either too long or too short. Within these extremes lie plants that seem to be independent of day length.

A very comprehensive soybean model SOYMOD (Meyer et al., 1979) illustrates very well the growth of various plant parts. Figure 5 shows results using SOYMOD; part (a) shows the development of leaves, stems, roots and the fruit, part (b) shows the leaf area over the growing season, and part (c) shows the total dry matter produced over the growing season. Space does not permit a detailed discussion of the model; see Meyer et al. (1979) for more information. For references to other examples see France and Thornley (1984).

2.1.2. Examples of Some Models

Some of the first models to be developed were empirical attempts to relate crop growth (or more likely yield) directly to various aspects of climate, weather, and environment. The

250

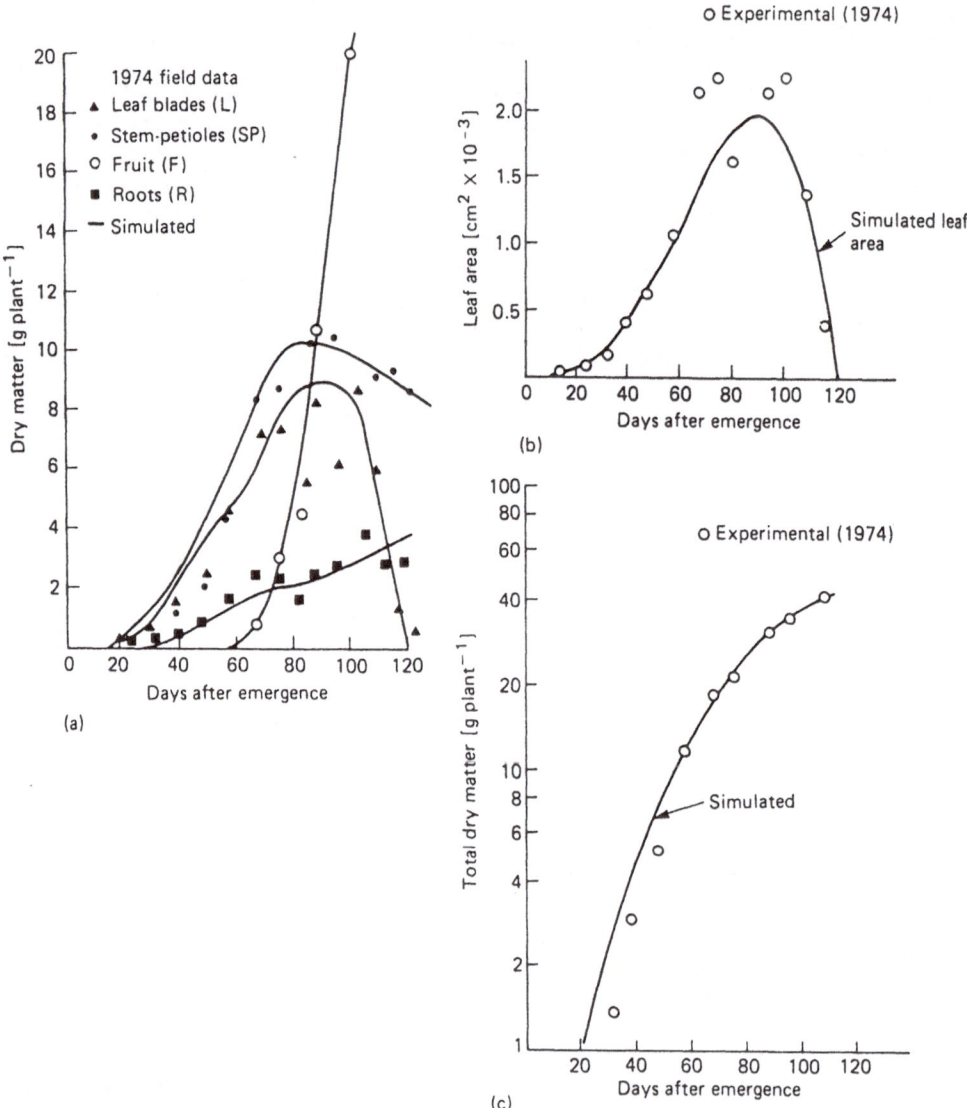

Figure 5. Simulated and experimental data for SOYMOD (after Meyer et al., 1979). (a) Dry matter for leaf blades, stem, petiole, fruit and roots. (b) Total leaf area. (c) Total dry matter (Mathematical Models in Agriculture, Butterworths, 1984, courtesy J. France and J.H.M. Thornley).

objectives were to discover which factors affect yield most greatly and then to consider whether management can increase yields or decrease costs by manipulating these factors. The features most often considered were temperature, solar radiation, day length, fertilizer and nutrient status, and CO_2 concentration. Table 1 taken from France and Thornley (1984) lists a few empirical models and some of their features.

Table 1. Some published empirical crop models. Environmental or management factors included: I = irradiance; W = water status (rainfall, humidity, wind); T = temperature; D = day length (Mathematical Models in Agriculture, Butterworths, 1984, courtesy J. France and J.H.M. Thornley).

Crop	Factors	Method	Time Interval	Reference
Grass	I, W, T, D	Growth and development rates are modified	7 days	Angus et al. (1980)
Rice	I, T	Regression of climatic factors on yield	Phases of development	Murata (1975)
Wheat	W, T	Stepwise multiple regression on yield	Day	Bridge (1976)
Wheat	W	Linear regression of yield on seasonal rainfall	Season	Seif and Pederson (1978)
Wheat	T, W	Regression of yield on a derived index	Month	Sakamoto (1978)
Wheat	T, W	Multiple curvilinear regression	Month	Pitter (1977)
Wheat	T, W	Regression equations for daily growth rate	Day	Haun (1974)

Crop/growth models used as components of hydrologic or other models are of necessity simpler in structure than physiological models. In the following paragraphs two such models are described.

Opus Crop Growth Model. The crop growth model in Opus, an advanced simulation model for nonpoint source pollution transport at the field–scale (Ferreira and Smith, 1990) illustrates a plant growth model needed as a component in a model for environmental assessment of pollutant transport. The model simulates incremental (daily) plant material growth as:

$$\Delta P = C_e \, F_e \, F_s \, F_m \, S \, L \, \Delta t \tag{23}$$

in which C_e is a plant sensitive photosynthetic conversion coefficient; F_e is an energy conversion efficiency factor, a function of the leaf area index; F_s is a plant age factor; F_m is a plant size factor; S is a stress factor considering water, nutrients, and temperature stresses; L is daily radiation; and Δt is a time interval. F_s is scaled 0 to 1.0 and slows the plant growth as degree/day age approaches maturity. F_m is similar in that it slows the plant growth as a maximum genetic size is reached (see Figure 6). The water component of the stress factor, τ_w, (zero for max stress to 1.0 for no stress) assumes a value of 0 for soil water contents with tension greater than 15–bar, θ_{15}, and increases linearly from that point to a value of 1.0 at $1.2 \, \theta_{15}$. The nitrogen stress factor, τ_N is:

$$\tau_N = 2.0 \frac{actual \, N_{use}}{potential \, N_{use}} - 1.0 \tag{24}$$

in which potential use is that required to maintain a normal plant nitrogen concentration that varies with plant size. The temperature stress factor, τ_T is a function of the ratio of the ambient temperature and the range for optimal growth.

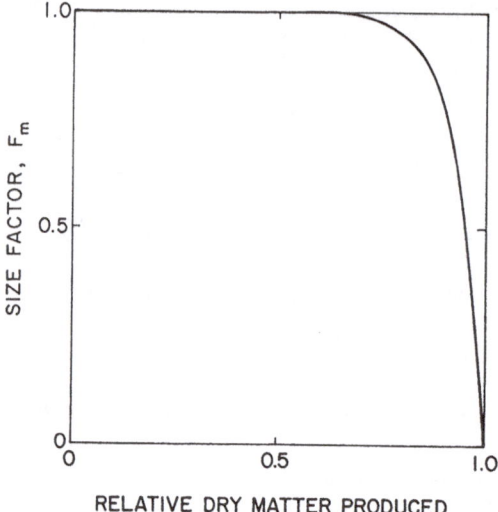

Figure 6. The relative growth factor in the mechanistic plant model as a function of the plant's genetic size (Smith and Ferreira, in preparation).

Plant material produced by Equation (23) is stoichiometrically divided between roots, leaves, stalk, and fruit (seed) according to age and degree/days. The leaf area index of the plant is a nonlinear function of the mass of plant material above ground. Root penetration into the soil is also a nonlinear function of root mass up to a maximum potential depth. The mass of roots, in relation to a normal balance between roots and above ground biomass, and the leaf area index determine the energy conversion factor, F_e which is shown in Figure 7.

Performance of the model is shown in Figures 8–10. Figure 8 shows a simulated corn crop without water stress. Figure 9 shows the same crop under a severe water stress. Figure 10 shows a perennial grass with no stress but two harvest operations.

NTRM Crop Growth Model. NTRM, A Soil–Crop Simulation model for Nitrogen, Tillage, and Crop–Residue Management (Shaffer and Larson, 1987), was designed to provide management assistance at the farm, extension, and engineering levels. It simulates physical, chemical, and biological processes in the soil–water–plant continuum using integrated submodels for soil temperature, carbon, and nitrogen transformations, unsaturated flow of water, crop and root growth, evaporation and transpiration, tillage, interception and infiltration, chemical equilibria processes, solute transport, and crop residue decomposition. It serves as an example of an application of hydrologic principles to an important area of farm management that includes the potential for solute movement through the root zone.

Specific chapters in the report describe soil temperature, interception, transpiration, unsaturated flow, solute transport and other processes. These specific chapters describe the development of models simulating some of the processes discussed earlier in the chapter. The crop growth component of NTRM is more comprehensive than the one in Opus. The version described herein simulates corn growth.

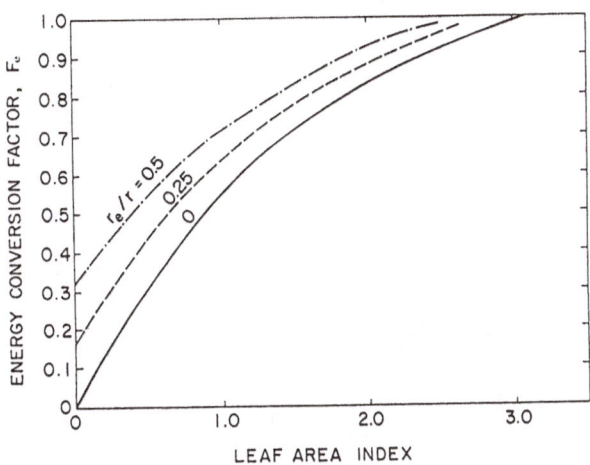

Figure 7. The plant efficiency in capturing solar energy is a function, as shown, of its leaf area index. The factor is modified in cases where roots exist in excess of those normally present for the plants age or size. An example of such an effect is the rapid regrowth of hay or alfalfa after harvesting (Smith and Ferreira, in preparation).

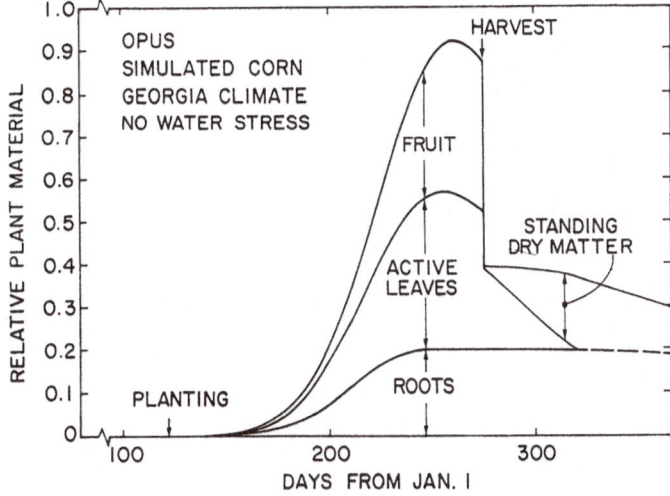

Figure 8. Growth of a corn crop in Georgia, simulated by Opus, in the absence of water or nutrient stresses (Smith and Ferreira, in preparation).

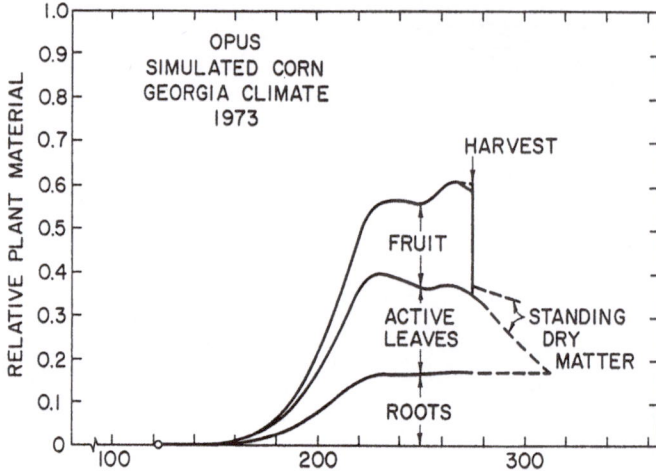

Figure 9. Growth of a corn crop as in Figure 8 except that water stress was included, reducing the growth and yield (Smith and Ferreira, in preparation).

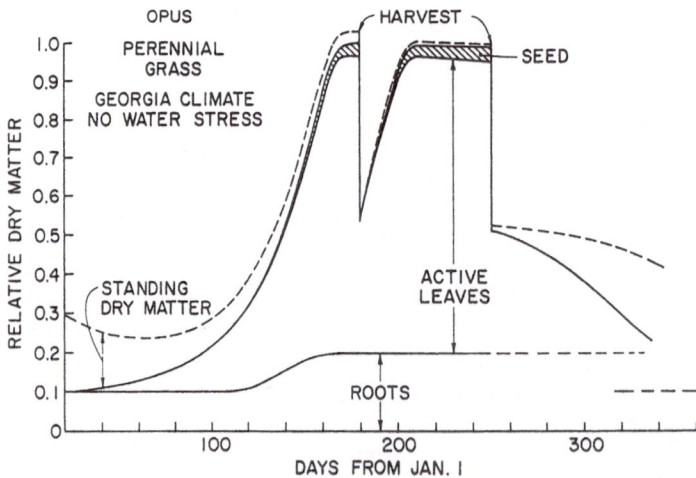

Figure 10. An Opus plant model simulation of the growth of a multiply-harvested grass crop, illustrating the rapid regrowth feature, where root resources are available (Smith and Ferreira, in preparation).

Space does not permit a complete description of the crop growth submodel; however, the functional aspects of the above–ground components are shown in Figure 11 (Shaffer, 1987). Photosynthesis, which produces the carbohydrates responsible for growth, is a function of the solar radiation, percent of leaf area that is sunlit, available carbohydrate, and transpiration. Carbohydrates generated by photosynthesis are consumed by respiration processes for growth and plant maintenance. Maintenance respiration is a function of air temperature and accumulated dry matter produced. Total respiration is calculated as a function of nitrogen uptake/demand and root mass produced. Growth respiration is the difference between total and maintenance respiration. The difference between the total photosynthate produced and that used in maintenance is stored as available carbohydrate. The production of dry matter is a scalar function of growth respiration. With respect to corn, dry matter produced prior to tasseling goes into leaves and stems. After that date, which is degree day dependent, all dry matter produced is converted into seed. In both cases (production of leaves and stems before tasseling and grain or seed after tasseling) the amount of dry matter produced is a function of total leaf area. Expressions for calculating leaf area are different for the two conditions; prior to tasseling leaf area expansion is function of air temperature, respiration, a crop coefficient, and an air temperature sensitivity coefficient. After tasseling the leaf area continues to increase to a maximum and then decline as a function of tasseling date, maturity date, and

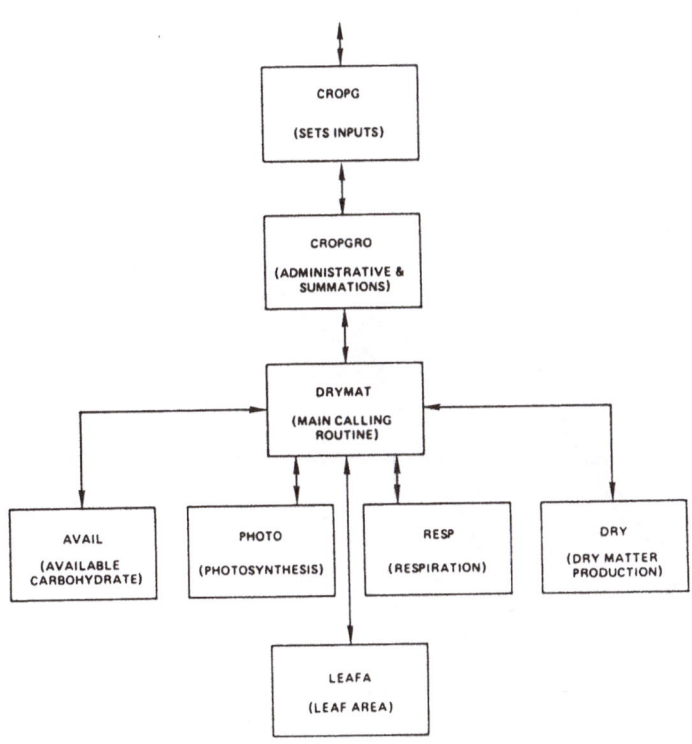

Figure 11. The above ground component of crop growth in NTRM (Shaffer, 1987).

accumulated degree days. Nitrogen uptake is that required to maintain pre–established concentrations in the dry matter that decrease with time. Figures 12 and 13 show how the model output compares with observed data on leaf area, dry matter production, and yield from the Sandhill area of Nebraska. Figure 14 shows how the model output compares to data from four sites in Midwestern United States.

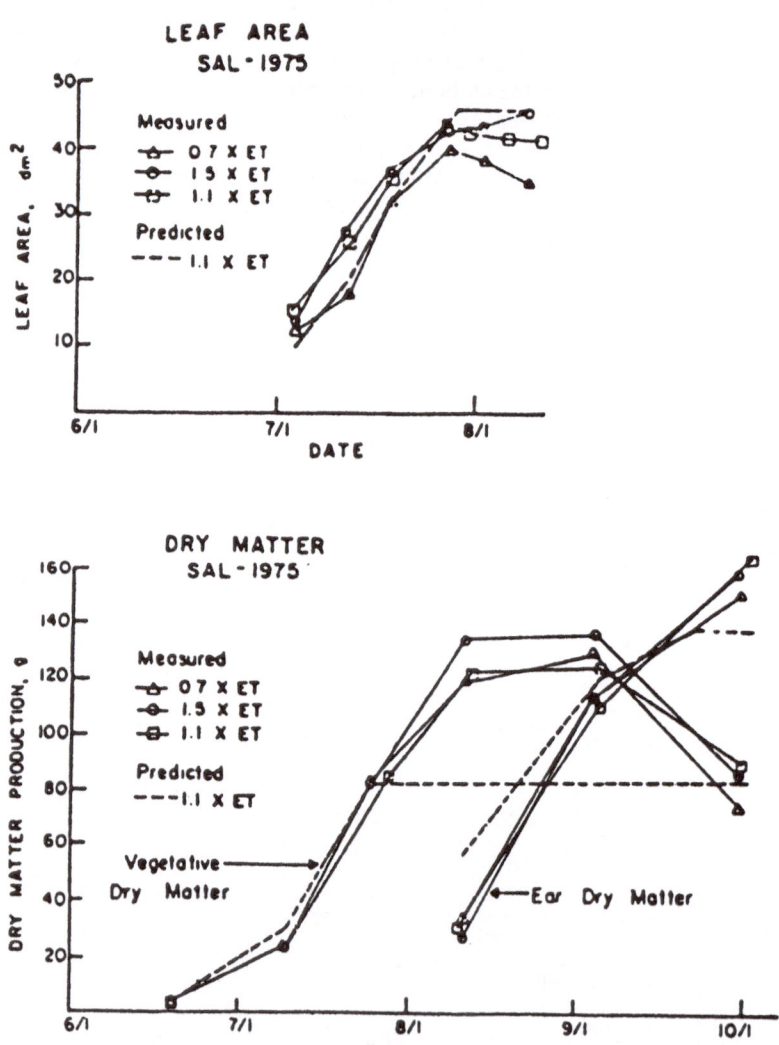

Figure 12. Actual and predicted corn leaf area and dry matter production from the Sandhills Agricultural Laboratory, Nebraska, 1975 (Shaffer, 1987).

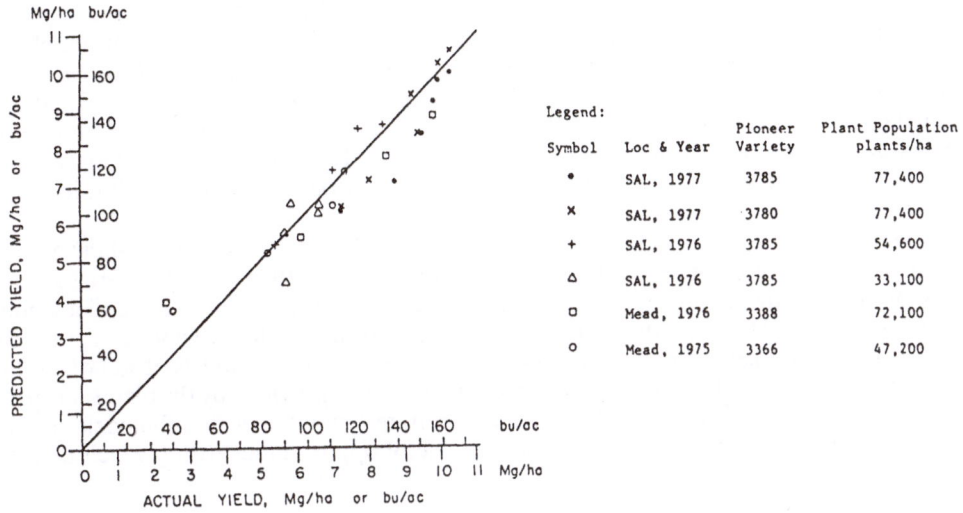

Figure 13. Yield comparisons of the corn growth submodel for various irrigation treatments and varieties at Mead and Sandhills Agricultural Laboratories, Nebraska (Shaffer, 1987).

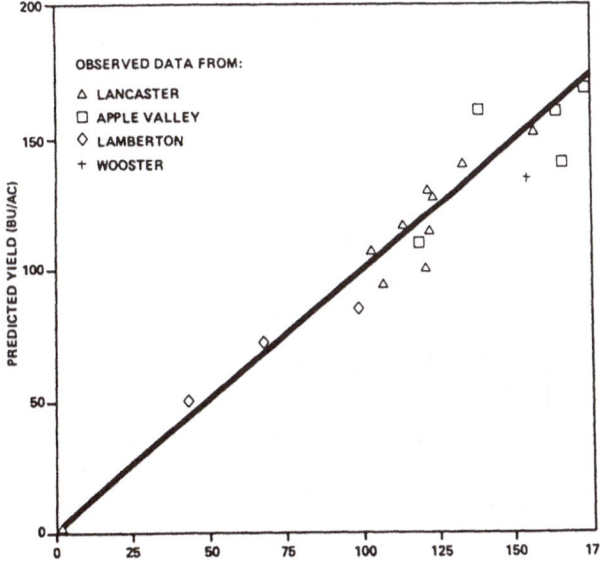

Figure 14. Comparison of the NTRM -- corn submodel with yields from midwestern states (Shaffer, 1987).

258

A general flow chart for the root growth component of Shaffer's corn plant is shown in Figure 15. This figure taken from Shaffer and Clapp (1987), shows the degree of complexity needed to simulate root growth. Since space does not permit a discussion of all the features of root growth, readers should refer to the author's paper. Figure 16 shows observed and predicted corn root densities at three different dates on a site in Rosemount, MN. Figure 17 shows changes in root distribution over a growing season. Each line represents the fraction of roots in each succeeding 15–cm depth increment starting from the surface (top line in Figure 17).

Numerous crop growth models have been developed for specific purposes. Many of them are detailed physiological descriptions of the growth process. In general these are too complex and require too much data to be of value in studies of solute movement and nonpoint source pollution. However, they serve as examples of problems to which we can apply the concepts of hydrology and theory of soil water movement. Just as models of soil water flow will not be accurate without the sinks of water and nutrients provided by the plants, the plant physiologist's model of crop growth will not be accurate without realistic submodels of water and solute movement. Thus there is a wide range of opportunities within which to apply knowledge of hydrology and hydraulics.

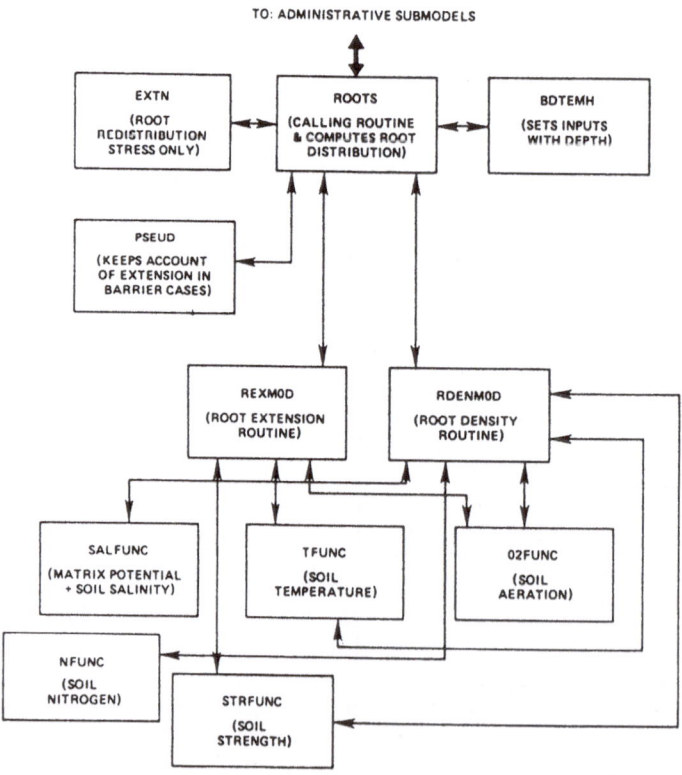

Figure 15. Root growth submodel diagram (Shaffer and Clapp, 1987).

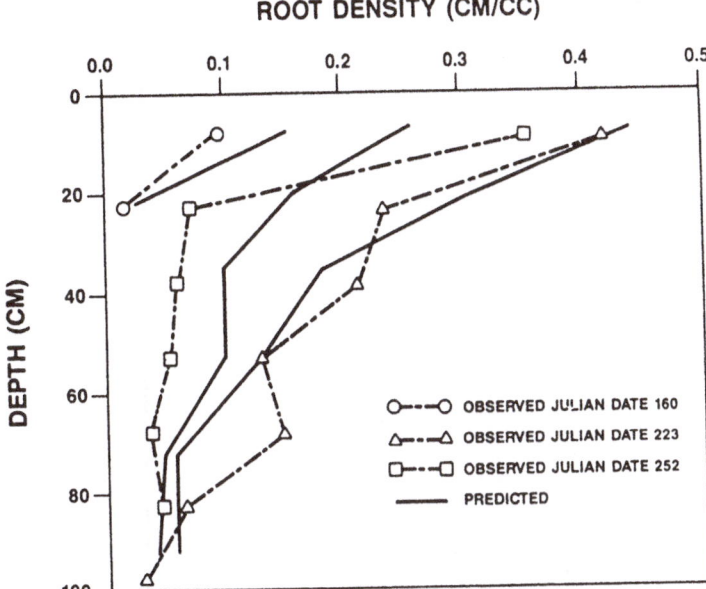

ROOT DENSITY (CM/CC)

Figure 16. Observed and calculated corn root densities, Rosemont, MN (Shaffer, 1987).

For those interested in pursuing this in more depth, refer to Norman (1979), Norman and Campbell (1983), Charles–Edwards et al. (1986), France and Thornley (1984), Baker (1980), and Legg (1981).

2.2. Soil Properties as Affected by Management

Soil properties change continuously with time, the forces of nature, and with management (primarily tillage). The major factors responsible for this change are the forces of raindrop impact, freeze/thaw cycles, the loosening and compaction of tillage tools, crop management, and soil biological processes. All of these processes in one way or another change the texture, water–holding capacity and hydraulic features of the soil. In this part of the chapter some of the more recent work in this area is reviewed because changes in soil attributes can change, by orders of magnitude, the rates of water movement throughout the growing season. For a recent discussion (from which much of this material comes) and numerous references, see Onstad and Voorhees (1987).

The energy of raindrops on exposed soil surfaces causes the soil particles to be rearranged, forming a more dense matrix. If the soil has little or no aggregation, rearrangement is rapid and the hydraulic conductivity of the "skin" that develops is very low. If the soil is well aggregated, the raindrop energy packs the primary and aggregate soil particles into a more dense matrix and erodes individual particles from the aggregates. The net result is reduced hydraulic conductivity. Figure 18 shows decrease in hydraulic conductivity of four tilled soils as a function of time until ponding. Part of the decrease in hydraulic conductivity may have been the intensity and amount of water applied. Movement of water through the soil matrix reduces the hydraulic conductivity of internal soil layers that may have been tilled. Ignoring

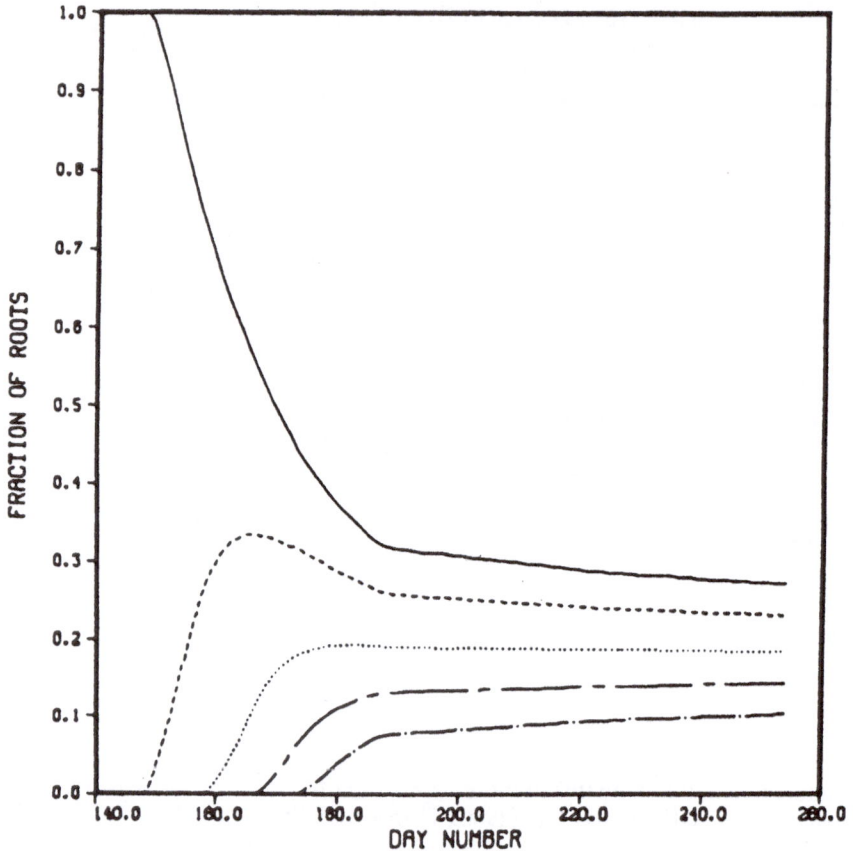

Figure 17. Corn root growth as a function of depth and time (Shaffer and Clapp, 1987).

the problems of swelling and shrinking soils, the moisture tension that increases as soil dries will tend to consolidate the soil–reducing hydraulic conductivity. At the time of this writing expressions for predicting the effect that soil moisture content has on bulk density were being developed by a team of scientists working on a new soil erosion model. A draft of the documentation describes preliminary expressions (Lane and Nearing, 1988).

Rainfall energy expended on the soil surface reduces the amount of water infiltrating a soil by eroding tillage marks and random roughness elements responsible for interception storage. The WEPP team expresses this as an exponential function of accumulated rainfall (Lane and Nearing, 1988). See Onstad (1984), Onstad et al. (1984), and Onstad and Voorhees (1987), for further discussion.

Freezing and thawing usually modify soil structure and aggregation. The effect of freezing and thawing cycles on a soil is dependent on the size of soil aggregates, the antecedent soil water content, and the level of ambient freezing temperature. In some cases the soil–water content of aggregates increases, the size of aggregates changes, and their bulk density decreases. The effect of freezing and thawing on hydraulic conductivity is even harder to

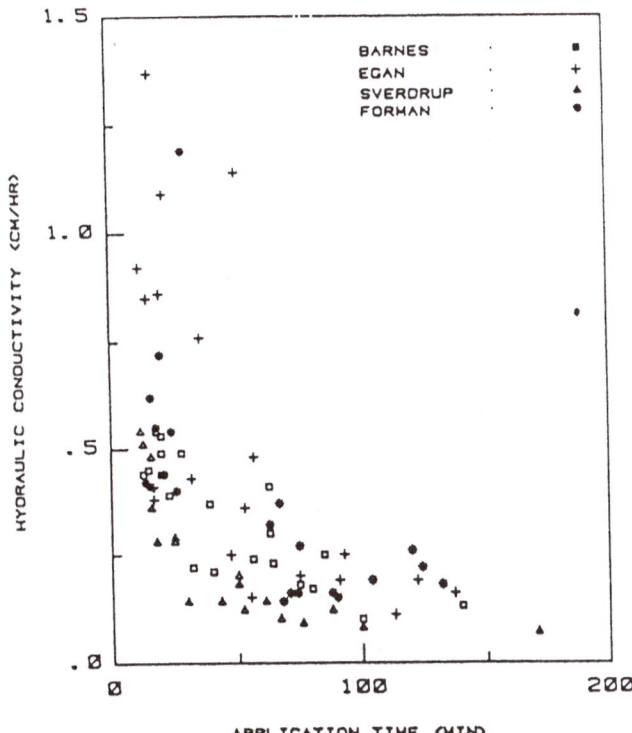

Figure 18. Hydraulic conductivity at the time of ponding versus application time for four soils (Bosch, 1986) (Onstad and Voorhees, 1987).

predict. See Onstad and Voorhees (1987). Figure 19 shows the change in relative hydraulic conductivity as a function of freezing and thawing soil cores at different soil water contents. Even though the data show a lot of scatter, the magnitude of change in hydraulic conductivity with moisture content at freezing is significant. If the soils are dry, freeze/thaw activity can increase hydraulic conductivity, but the opposite is true for wet soils. These results of change in soil features due to freezing and thawing do not reflect rapid reconsolidation that is likely to occur in the first few rain storm events following the last thaw, especially if the soil surface is bare.

The soil tillage required, in different farming operations, has a major impact on soil structure. Its effect is a function of the degree of tillage, soil type, degree and weight of wheel and animal traffic, and moisture content at the time of tillage. In general primary tillage (deep plowing, etc.) increases roughness and porosity of the soil, but secondary tillage decreases roughness and porosity. See Table 2 from Voorhees et al. (1981). Also, it is generally accepted that most tillage practices affect the macropore structure more than they do the mesopore or micropore structure. See Onstad and Voorhees (1987) for additional references on this subject.

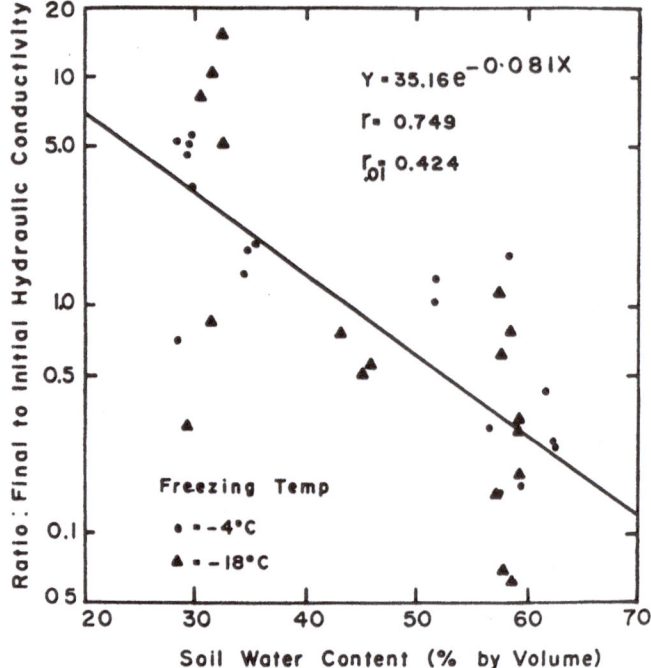

Figure 19. Semilog plot of the ratio of final to initial soil hydraulic conductivity after freezing and thawing soil cores at different soil water contents (Benoit, 1973). (Reproduced from Soil Science Society of America Proceedings, Volume 37, No. 1. Jan–Feb, 1973, pp. 3–5, by permission of the Soil Science Society of America, Inc.)

Crop management, as it is used here, refers to crop rotations and management practices as to their impact on the soil structure and its ability to absorb water. Farming practices that are identified as conventional use a variety of tillage tools to prepare the soil, plant, cultivate, and harvest the crop. In general the effects of these tillage practices are described very briefly in the previous paragraph. Conservation tillage and related limited or minimum tillage systems have a different affect on soil structure. These tillage systems contrast with conventional systems in that they leave much more crop residue on the surface and disturb the surface of the soil much less. This means that there is more infiltration in conservation–tilled systems because the surface residue breaks up the energy of rain drop impact, thus preventing the development of a surface skin or seal. Also, soil that is disturbed very little is likely to be more aggregated and have a larger and a more stable macropore structure. This is created by the presence of more earth worms, root channels, cracks around the base of stems, etc. Also, the presence of increased surface residue reduces surface evaporation, reduces soil temperatures slowing biological and chemical reactions, and thus increases the levels of organic matter and produces a better water holding capacity and better soil structure and soil environment.

In the following parts of this chapter some of the above features are discussed in more detail.

Table 2. Random roughness and total fractional porosity (vol. of pores/ vol. soil) of tilled layer as affected by tillage systems (Voorhees et al., 1981) (after Onstad and Voorhees, 1987).

Tillage Method		Random Roughness Measured after Spring Tillage (cm)		Total Fractional Porosity Measured after Spring Tillage	
Fall	Spring	Range	Mean	Range	Mean
Plow	—	1.96–7.26	3.24	0.54–0.96	0.72
Tandem disk	—	1.75–1.83	1.79	0.54–0.69	0.61
Chisel	—	1.35–1.68	1.52	0.68–0.69	0.68
Plow	—	1.30–3.63	2.22	0.48–0.68	0.60
Plow	Disk and harrow	0.89–1.52	1.27	0.58–0.66	0.61
None	Plow	1.17–3.61	2.28	0.55–0.84	0.66
None	Plow, disk and harrow	0.69–2.16	1.15	0.53–0.82	0.67
None	Plow and wheel tracked	0.93–1.15	1.06	0.52–0.56	0.54
None	Rotary till	1.55–1.75	1.65	0.64–0.67	0.66
None	None	0.46–0.86	0.59	0.52–0.53	0.52

2.2.1. Interception and Depression Storage

The interception of rainfall by plants and crop residue is not a soil feature, but it relates more to management than it does to a crop growth model, thus it is described here. The canopy interception of rainfall by plants and crop residues reduces the amount available for infiltration and surface runoff. This water is retained on the leaves and stems until it evaporates. It does not cover the surface uniformly nor is it stored on the leaves at uniform depths. These features are functions of the type of leaf tissue and structure of the plant. In some plants, corn for example, rainfall is directed both toward the main stalk and away from the plant. In soybeans the plant structure creates a drip line around the outer perimeter of leaves.

The maximum amount of water that can be retained on a plant, P_{ICM}, is given by Ritchie (1972) as:

$$P_{ICM} = 0.08L - W_c \tag{25}$$

in which L is the leaf area index, W_c is the water on the canopy at the beginning of the event in cm, and P_{ICM} is in cm.

Interception by crop residues or mulch, P_{IRM}, is given by a similar empirical expression:

$$P_{IRM} = 0.08M - W_R \tag{26}$$

in which M is the dry weight of residues in metric tons/ha and W_R is the amount of water on the surface at the beginning of the event.

Depression storage on the soil surface is a function of the microrelief created by tillage tools and the degree of smoothing caused by rainfall impact. It is also a function of the soil slope; the greater the slope, the smaller the depression storage. Linden (1979) shows, in the Figure 20, the depression storage, D, as a function of the random roughness, RR, index

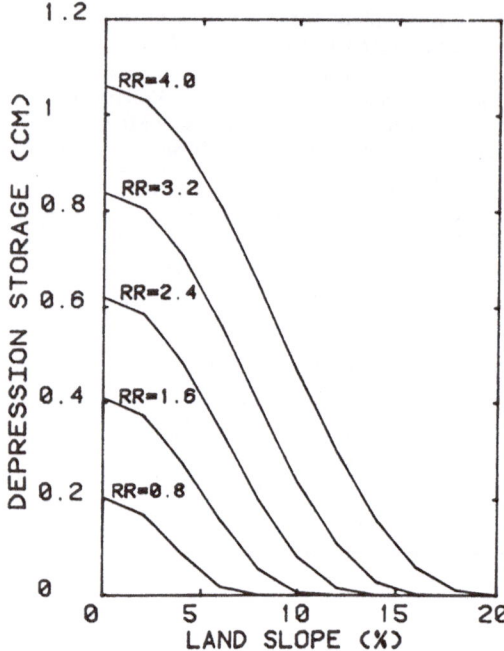

Figure 20. The upper limit to depression storage versus general land slope for five values of random roughness (RR) in cm (Linden, 1979) (after Linden and Van Doren, 1987).

(standard deviation of height measurements in cm), and land slope, L. The smoothing effect of rainfall on the surface random roughness is given by:

$$RR = RR_I - (RR_I - 0.8)\,(1 - exp(-1.5B(cos\,\phi/A)E - 0.015P)) \tag{27}$$

in which RR_I is the initial roughness, B is the exposed fraction of the soil surface, ϕ is the average angle of inclination of the surface, A is the surface contact area per unit of horizontal area, E is the rainfall energy, and P is the amount of precipitation. Table 2 shows values of relative roughness for selected tillage operations. See Linden (1979) for more specific information.

2.2.2. Soil Bulk Density, Porosity and Hydraulic Properties

The tilled layers of the soil experience large changes in bulk density, primarily due to the effect of tillage operations on the macropore structure of the soil. These changes were discussed in previous paragraphs; the magnitude of the change is presented in Table 2. The reconsolidation that occurs as a result of the settling action of rainfall is described by Linden and Van Doren (1987) as:

$$\alpha = \alpha_I - (\alpha_I - \alpha_c)(1 - exp(-1.5E - 0.015P)) \tag{28}$$

in which α is the soil porosity, α_I is the initial porosity, α_c the stable porosity, E the rainfall energy, and P the amount of rainfall. This is an exponential decrease in porosity that is somewhat slower than that of the erosion of surface features but is a function of rainfall energy and amount.

Rawls and Brakensiek analyzed the hydraulic and texture properties of a large number of soils to determine the effect of texture and tillage on hydraulic properties. In their papers (Rawls et al., 1983; Rawls and Brakensiek, 1985a) they present a method of estimating the Brooks and Corey soil water retention parameters and Green and Ampt infiltration parameters as functions of the percent sand, silt, clay, and organic matter. The parameter values are shown plotted on the traditional soil texture triangle. In companion papers (Brakensiek and Rawls, 1983; Rawls and Brankensiek, 1985b) they discuss procedures for estimating the Green and Ampt infiltration equation parameters for tilled soils. The analyses include estimates of hydraulic conductivity as influenced by soil crusting. The parameters of the soil crust conductivity function are related to the soil water properties and soil surface cover and roughness. Since the effects of tillage on infiltration are primarily on soil bulk density (or porosity), the impact on infiltration is imputed from change in porosity. Figure 21 shows the effect of moldboard plowing on porosity as a function of texture. Numbers by the lines in the figure show the initial increase in porosity as a result of plowing and the subsequent decrease (reconsolidation) during the growing season, in parentheses. Rawls and

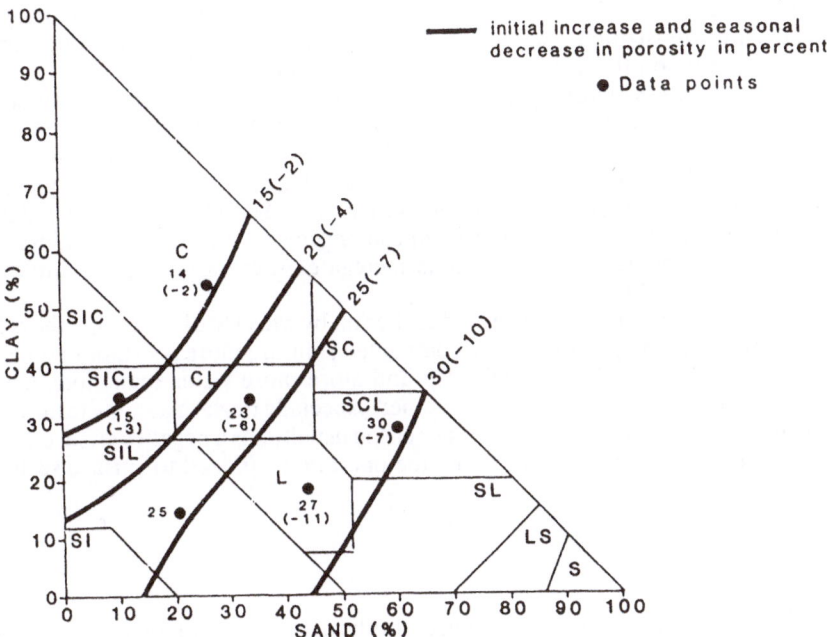

Figure 21. Percent porosity increases due to a moldboard plow (percent decreases during the growing season) as a function of soil texture (Rawls et al., 1983).

Brakensiek (1985b) also developed an expression for calculating saturated hydraulic conductivity, K_s

$$K_s = 19.5\phi - 0.0282 \ C_f + 0.000181 \ S_f^2 - 0.00941 \ C_f^2$$
$$- 8.40 \ \phi^2 + 0.0777 \ \phi \ S_f - 0.00298 \ S_f^2 \ \phi^2 - 0.0195 \ C_f^2\phi^2$$
$$+ 0.0000173 \ S_f^2 \ C_f + 0.0273 \ C_f^2 \ \phi + 0.00143 \ S_f^2 \ \phi \tag{29}$$
$$- 0.0000035 \ C_f^2 \ S_f - 8.97$$

in which ϕ is the soil porosity, C_f is the clay fraction, and S_f is the sand fraction. Similar figures and expressions are used for other soil properties. Presentation of all the expressions is beyond the intent of this chapter; see their papers for other figures and expressions.

Ahuja et al. (1989) also analyzed nine different soils and developed the empirical expression:

$$K_s = 765 \ \phi_e^{3.29} \tag{30}$$

in which ϕ_e is the effective porosity. Efforts are also underway to use the macropore features of the soil to calculate the movement of ponded water from the surface in assumed cylindrical macropores or planar cracks. Using these features the maximum flow-rate capacity of the macropores is calculated using conventional unit-gradient gravity flow with absorption of macropore solution by the soil matrix. Other chapters describe this effort.

It is obvious from this discussion that the effects of tillage on hydraulic conductivity can be calculated since the parameter that changes most in response to tillage is the porosity; porosity is the soil feature most often equated to a change in saturated hydraulic conductivity. For additional information on this subject see Onstad and Voorhees (1987).

Many conservation practices incorporate various amounts of surface residues into the surface layers. Those practices that leave large amounts of material at or above the soil surface provide protection against the erosive forces of flowing water as well as against raindrop impact. They also provide numerous entryways for large volumes of water and frequently prevent all surface runoff for several months or even years. The incorporated residues increase the organic matter content as they decompose improving soil structure and water holding capacity; the residues also form more stable aggregates; thus they change the infiltration characteristics considerably.

Soil structure is changed by energy transmitted to the soil by wheel traffic. During the off-season (from fall through early spring) clods of compacted soils are more resistant to breakdown (Onstad and Voorhees, 1987) and will trap and store more water than clods from uncompacted soils. Not all soils respond in this manner, especially those of a sandy texture or lower in organic matter. Traffic during the planting season usually has a negative effect on soil surface conditions because instability created by the operation can lead to surface sealing and lower porosity.

3. Conclusions

The effects of nonpoint sources of contamination on both surface and ground waters is now one of the most important water quality issues. In Chapter 11, microclimatology was described because of its importance. In this paper emphasis was placed on crop growth and the effects of management on soil hydraulic properties. These two topics are presented because they play such important roles in the disposition of large volumes of water and nutrients.

Simulation of plant growth processes described include photosynthesis and light interception, plant respiration, translocation and the movement of plant sugars, root activity and nutrient uptake, and plant development and morphogenesis. Examples of two crop growth models that have been developed for use with hydrologic models are also described along with some examples of their response. Soil properties which are so important in estimating hydraulic conductivity can cause order of magnitude changes in this parameter. The soil properties discussed in regard to management operations include hydraulic conductivity, interception, depression storage, bulk density, and porosity.

4. References

Ahuja, L. R., D. K. Cassel, R. R. Bruce, and B. B. Barnes: 1989, 'Evaluation of Spatial Distribution of Hydraulic Conductivity Using Effective Porosity Data,' *Soil Sci.* **148**(6), 404–411.

Angus, J. F., A. Kornher, and B. W. R. Torssell: 1980, *'A Systems Approach to Estimation of Swedish Ley Production'*, Progress Report 1979/80, Report 85, Uppsala: Swedish University of Agricultural Sciences.

Baker, D. N.: 1980, 'Simulation for Research and Crop Management', in F. T. Corkin (ed.), *Proceedings of World Soybean Research Conference II*, Boulder, CO, Westview Press, pp. 533–546.

Barnes, A., and C. C. Hole: 1978, 'A Theoretical Basis of Growth and Maintenance Respiration,' *Ann. Bot.* **42**, 1217–1221.

Benoit, G. R.: 1973, 'Effect of Freeze-Thaw Cycles in Aggregate Stability and Hydraulic Conductivity of Three Soil Aggregate Sizes,' *Soil Sci. Soc. Am. Proc* **37**(1), 3–5.

Bosch, D. D.: 1986, *'The Effects of Rainfall on the Hydraulic Conductivity of Soil Surfaces,'* Unpublished M.S. Thesis, University of Minnesota, 161 pp.

Brakensiek, D. L., and W. J. Rawls: 1983, 'Agricultural Management Effects on Soil Water Processes Part II: Green and Ampt Parameters for Crusting Soil,' *Trans. ASAE* **26**, 1753–1757.

Bridge, D. W.: 1976, 'A Simulation Model Approach for Relating Effective Climate to Winter Wheat Yields on the Great Plains,' *Agricultural Meteorology* **17**, 185–194.

Charles-Edwards, D. A., D. Doley, and G. M. Rimmington: 1986, *Modeling Plant Growth and Development*, Academic Press, 235 pp.

Ferriera, V. A., and R. E. Smith: 1990, 'Opus: An Advanced Simulation Model for Nonpoint-Source Pollution Transport at the Field Scale – An Overview,' in D. G. DeCoursey (ed.), *Proceedings of the International Symposium on Water Quality Modeling of Agricultural Non-Point Sources*, Logan, UT, U.S. Dep. of Ag., Agricultural Research Service, ARS-81, pp. 823–834.

France, J., and J. H. M. Thornley: 1984, *Mathematical Models in Agriculture*, Butterworths, 335 pp.

Haun, J. R.: 1974, 'Prediction of Spring Wheat Yields from Temperature and Precipitation Data,' *Agronomy Journal* **66**, 405–409.

Huck, M. G., and D. Hillel: 1983, 'A Model of Root Growth and Water Uptake Accounting for Photosynthesis, Respiration, Transpiration and Soil Hydraulics,' in D. Hillel (ed.), *Advances in Irrigation, Vol. 2*, Academic Press, pp. 273–333.

Lane, L. V., and M. A. Nearing: 1988, *'WEPP Profile Model Documentation,'* USDA-ARS, Tucson, AZ, Draft Only.

Legg, B. J.: 1981, 'Aerial Environment and Crop Growth,' in D. A. Rose, and D. A. Charles-Edwards (eds.), *Mathematics and Plant Physiology*, Academic Press, London, pp. 129–149.

Linden, D. R.: 1979, *'A Model to Predict Soil Water Storage as Affected by Tillage Practices,'* Ph.D. Thesis, Dept. of Soil Sci., Univ. Minn., St. Paul, MN.

Linden, D. R., and D. M. Van Doren Jr.: 1987, 'Chapter 14 Simulation of Interception, Surface Roughness, Depression Storage, and Soil Settling,' in M. J. Shaffer, and W. E. Larsen (eds.), *NTRM, a Soil-Crop Simulation Model for Nitrogen, Tillage, and Crop-Residue Management*, USDA, ARS, Cons. Res. Report 34-1, pp. 90–93.

Meyer, G. E., R. B. Curry, J. G. Streeter, and H. J. Mederski: 1979, *'A Dynamic Simulator of Soybean Growth, Development, and Seed Yield, I. Theory, Structure and Validation,'* Research Bulletin of the Ohio Agricultural Research and Development Center, No. 1113, 36 pp.

Monsi, M., and T. Saeki: 1953, 'Uber den Lichtfaktor in den Pflanzengesellschaften und seine Bedeutung für die Stoffproduktion,' *Japanese Journal of Botany* **44**, 603–613.

Murata, Y.: 1975, 'Estimation and Simulation of Rice Yield from Climatic Factors,' *Agricultural Meteorology* **15**, 117–131.

Norman, J. M.: 1979, 'Modeling the Complete Crop Canopy,' in B. J. Barfield, and J. F. Gerber (eds.), *Modification of the Aerial Environment of Plants, ASAE Monograph, No. 2 in a Series*, pp. 249–277.

Norman, J. M., and G. Campbell: 1983, 'Application of a Plant-Environment Model to Problems of Irrigation,' in D. Hillel, *Advances in Irrigation*, Academic Press, **2**, pp. 155–188.

Onstad, C. A.: 1984, 'Depressional Storage on Tilled Soil Surfaces', *Trans. ASAE* **27**:3, 729–732.

Onstad, C. A., and W. B. Voorhees: 1987, 'Chapter 5 Hydraulic Soil Parameters Affected by Tillage,' in T. J. Logan, J. M. Davidson, J. L. Baker, and M. R. Overcash (eds.), *Effects of Conservation Tillage on Groundwater Quality, Nitrates and Pesticides*, Lewis Publishers, pp. 95–112.

Onstad, C. A., M. L. Wolfe, C. L. Larson, and D. C. Slack: 1984, 'Tilled Soil Subsidence During Repeated Wetting,' *Trans. ASAE* **27**(3), 733–736.

Pitter, R. L.: 1977, 'The Effect of Weather and Technology on Wheat Yields in Oregon,' *Agricultural Meteorology* **18**, 115–131.

Rawls, W. J., D. L. Brakensiek, and B. Soni: 1983, 'Agricultural Management Effects on Soil Water Processes. Part I: Soil Water Retention and Green and Ampt Infiltration Parameters,' *Trans. ASAE* **26**, 1747–1752.

Rawls, W. J., and D. L. Brakensiek: 1985a, 'Agricultural Management Effects on Soil Water Retention,' in D. G. DeCoursey (ed.), *Proceedings Natural Resources Modeling Symposium*, Pingree Park, CO, Oct. 16–21, 1983, USDA-ARS-30, pp. 115–117.

Rawls, W. J., and D. L. Brakensiek: 1985b, 'Prediction of Soil Water Properties for Hydrologic Modeling,' in *Proceedings ASCE Watershed Management Symposium*, Denver, CO, 297 pp.

Ritchie, J. T.: 1972, 'Model for Predicting Evaporation From a Row Crop with Incomplete Cover,' *Water Resources Research* **8**, 1204–1213.

Sakamoto, C. M.: 1978, 'The Z-index as a Variable for Crop Yield Estimation,' *Agricultural Meteorology* **19**, 305–313.

Seif, E., and D. G. Pederson: 1978, 'Effect of Rainfall on the Grain Yield of Spring Wheat, with an Application,' *Australian Journal of Agricultural Research* **29**, 1107–1115.

Shaffer, M. J.: 1987, Chapter 7 'Crop Growth Submodel (Corn),' in M. J. Shaffer, and W. E. Larson (eds.), *NTRM, A Soil-Crop Simulation Model for Nitrogen, Tillage, and Crop–Residue Management*, USDA, ARS, Cons. Res. Report 34–1, pp. 57–62.

Shaffer, M. J., and C. E. Clapp: 1987, Chapter 10, 'Root Growth Submodel,' in M. J. Shaffer, and W. E. Larson (eds.), *NTRM, A Soil–Crop Simulator Model for Nitrogen, Tillage, and Crop–Residue Management*, USDA, ARS, Cons. Res. Report 34–1, pp. 63–72.

Shaffer, M. J., and W. E. Larson, (eds.): 1987, *NTRM, A Soil–Crop Simulation Model for Nitrogen, Tillage, and Crop–Residue Management,* USDA, ARS, Cons. Res. Report 34–1, 103 pp.

Smith, R. E., and V. A. Ferreira: 1990, *'Opus, An Advanced Simulation Model for Non–Point Source Pollution Transport at the Field Scale,'* Model Documentation, Draft.

Thornley, J. H. M.: 1977, 'Growth, Maintenance, and Respiration: A Reinterpretation,' *Ann. Bot.* **41**, 1191–1203.

Voorhees, W. B., R. R. Allmaras, and C. E. Johnson: 1981, 'Alleviating Temperature Stress,' in G. F. Arkin, and H. M. Taylor (eds.), *Modifying the Root Environment to Reduce Crop Stress*, ASAE Monograph, pp. 217–266.

Chapter 13

Framework for Erosion and Sediment Yield Modeling

James C. Bathurst
University of Newcastle upon Tyne
Department of Civil Engineering
Natural Environment Research Council
Water Resource Systems Research Unit
Newcastle upon Tyne, NE1 7RU, United Kingdom

and

Jon M. Wicks
Sir William Halcrow and Partners Ltd.
Swindon, SN4 0QD, United Kingdom

Abstract. The physically–based distributed modeling approach to linking on–site rates of soil erosion in a basin to sediment yield at the basin outlet is examined. Processes considered are: hillslope erosion by raindrop impact, overland flow, gullying and mass movement; overland flow sediment routing; channel erosion, bed sediment infiltration and storage; and channel sediment routing. The sophistication of model numerical solution techniques is not matched by that of available data provision methodologies or the current understanding of the relevant processes.

1. Introduction

The relevance and aims of erosion and sediment yield modeling are established by consider-ing the sediment–related problems which are of current concern and which, in many cases, are exacerbated or caused completely by human activity. Frequently these arise in one part of a river basin as a result of activity in other parts of the basin; therefore, a full understanding of sediment–related problems can be attained only by taking an integrated view of the sedimen-tation processes (e.g. Simons et al., 1979; Sundborg, 1983). This is illustrated by the following examples, using the framework of Schumm's (1977) fluvial system to link the source area for erosion, the channel system transporting the load, and the sink area for deposition:

(1) *Source area*. Poor farming practices expose the soil to erosion by wind, rain and over-land flow, resulting in loss of topsoil and nutrients, reduced crop productivity and desertifica-tion (e.g. Sundborg, 1983). Deforestation of hillslopes can precipitate increased erosion by landslides, resulting in loss of life, property and livelihood (e.g. Simons, 1988).

(2) *Channel system*. The sediment released into the river system can promote channel instability (with implications for loss of land and maintenance of structures), cause bed aggradation (hindering navigation and increasing flooding) and clog fish spawning grounds (both ecologically and economically damaging) (e.g. Pain, 1987).

(3) *Sink area*. Deposition of the sediment creates problems of siltation and sedimentation. Reservoirs commonly silt up two or more times faster than expected at the time of design, often because little consideration is given to the impacts of upstream land–use change on

269

D. S. Bowles and P. E. O'Connell (eds.), Recent Advances in the Modeling of Hydrologic Systems, 269–288.
© 1991 *Kluwer Academic Publishers.*

sediment supply (e.g. Tejwani, 1984). Siltation reduces reservoir capacity, promotes upstream flooding, leads to abrasion of turbine blades in hydroelectric plants and threatens the operation of irrigation and transport canal systems (e.g. Larson and Albertin, 1984).

As an additional area of concern, adsorption to sediment particles is increasingly found to provide a mechanism for the transport of contaminants derived from agricultural fertilizers and industrial wastes such as heavy metals and radionuclides (e.g. Ongley, 1982; Walling, 1983, 1988). Deposition of sediments carrying such loads, in the channel or on the floodplain, can have detrimental consequences, particularly for ecology and agriculture.

Amelioration of these problems requires means of: predicting, on a basin–scale, the erosion, transport, deposition and storage of sediment arising from proposed courses of action; and identifying the control measures and methods of land management which can minimize adverse impacts. Such aims can be achieved only through a sound understanding of the relevant erosion and sediment yield processes and the degree to which they are affected by human activities. This is the province of erosion and sediment yield modeling, the essence of which is the linking of on–site rates of erosion and soil loss in a basin to the sediment yield at the basin outlet. However, the linkages and feedbacks which relate sediment erosion, transport, deposition and storage are complex, vary with spatial and temporal scale and are currently not always well expressed quantitatively or even qualitatively. The need to improve our understanding in this area has often been stressed (e.g. Wolman, 1977; Sundborg, 1983; Walling, 1983).

This chapter examines progress in the understanding of various aspects of the sediment system and their representation by mathematical models. The advantages and disadvantages of different modeling approaches have been covered in a number of recent reviews (e.g. Rice, 1982; Walling, 1983; Pickup, 1988; Singh et al., 1988; Lane et al., 1988) and, therefore, will not be repeated in detail here. The emphasis is on the physically–based approach since this is felt to be of greatest relevance to the solution of current sediment related problems through its ability to incorporate the impacts of man's activities and to indicate the links between the different components of the sediment system.

2. Discussion

2.1. The Sediment System

To be appropriate to a basin, a sediment yield model must match its components with the physical processes determining sediment sources, erosion and linkages within the basin. These may be considered as follows.

2.1.1. Sediment Sources

In general, sediment is identified as coming from hillslope or channel sources (e.g. Dunne and Leopold, 1978, pp. 506–589; Simons et al., 1979). The three major components of the former are sheet and rill erosion caused by raindrop impact and overland flow, gullying, and mass movement. The two components of the latter are channel bed and bank erosion. Other sources which are less general but which can have important influences in individual basins include volcanos, glaciers, wind erosion, uprooting of trees, burrowing by animals, freeze–thaw effects, biological production and chemical weathering. A large number of natural factors determine the impact and balance of these processes in any basin but the greatest single influence is probably human activity (e.g. Dunne and Leopold, 1978, p. 510; Sundborg, 1983; Simons et al., 1979).

The overall sediment yield of a basin may be derived from a single type of source or from a combination. Generally it is held that the sediment yield from upland headwater basins is determined primarily by hillslope processes while channel sources become progressively more important as basin size increases (e.g. Simons et al., 1979). However, this pattern is not always observed.

2.1.2. Sediment Transport

Transport occurs in particulate or solute form. Particulate transport can be further classified according to mechanism as suspended load (maintained in suspension by the flow turbulence) and bed load (moving in contact with the bed). For channels an alternative classification is according to source as wash load (derived from outside the channel) and bed material load (derived from the channel bed). There is a degree of interchange between the two: wash load generally moves in suspension while bed material load moves in both bed and suspended form. In sand–bed channels or in lowland rivers, bed load may account for only 5–20 percent of total particulate load (e.g. Lane and Borland, 1951) but the proportion may rise to 50 percent or more in mountain rivers with coarse bed material (e.g. Moosbrugger, 1957; Lauffer and Sommer, 1982). The ratio of solute to particulate load is also variable (e.g. Slaymaker, 1977). However, solute transport is not considered further in this chapter.

2.1.3. Sediment Storage and Sediment Delivery Ratio

Generally the sediment delivery ratio, or the ratio of sediment delivered at the basin outlet to the gross erosion within the basin, is rather less than unity. Various impediments, in the form of temporary or permanent storage, prevent much of the mobilized sediment from reaching the outlet, at least within timespans relevant to human activity. The magnitude of the delivery ratio for a particular basin can be related to a wide range of geomorphological and environmental factors but there is a lack of information on the relative importance of the various losses within the hillslope–channel network (e.g. Walling, 1983). An investigation in the Oka Basin, central European USSR, reported by Golubev (1982) indicates that only 10 percent of the gross soil erosion reaches the larger rivers: 60 percent is deposited on the lower parts of hillslopes, 20 percent in ephemeral channels and 10 percent in the minor streams of the network. Goswami (1985) estimates that, for the 607-km Assam reach of the Brahmaputra, during 1971–79 about 70 percent of the suspended sediment inflow to the reach was being retained in the channel. Seasonal variations in the sediment delivery ratio are caused by the erosion (e.g. winter processes) and transport (e.g. summer storm runoff) being out of phase (e.g. Duijsings, 1986).

2.1.4. Sediment Yield Dynamics

The yield at the basin outlet is a function of sediment supply and runoff generation. The sediment concentration and flood waves are often out–of–phase, both in single events and seasonally, and their relationship describes hysteresis loops of either direction (e.g. Walling and Webb, 1982; Bathurst, 1987). The amount of sediment available for transport may become progressively depleted during events, sequences of events and between seasons. This is especially noticeable in flow regimes with large seasonal variations, for example snowmelt regimes (Østrem et al., 1971) and monsoon regimes (Gole and McManus, 1988; Kattan et al., 1987). The source of the runoff also affects the yield. A snowmelt flood may carry less sediment than does a rainfall flood of the same magnitude in the same river because the latter is more heavily supplied by the erosion products of rainfall impact and overland flow from outside the channel (e.g. Bathurst, 1987). A given channel discharge can therefore carry a wide range of sediment loads in the same river.

2.1.5. Spatial Distribution

Usually a large proportion of the sediment yield is derived from only a small proportion of the basin (e.g. Gregory and Walling, 1973). This is particularly true where the supply comes from landslides feeding directly into the channel. Even where supply is by overland flow transport, though, the proportion of the basin affected is limited by the development of variable source or partial contributing areas for the overland flow (Walling, 1983; Dunne and Leopold, 1978, pp. 262–274).

2.2. Model Requirements and Approaches

2.2.1. Model Requirements

Walling and Webb (1982) suggest that any meaningful sediment yield model should include the following:

(1) Means of reproducing storm runoff as the driving agent for sediment yield. To this may be added snowmelt runoff.

(2) Means of reproducing the typical hysteretic behavior of sediment concentration.

(3) Allowance for the temporal supply and depletion of sediment within multiple events and sequences of events.

(4) Allowance for variable source areas, i.e. spatial distribution.

Another important requirement for the practical sediment yield model is that it should be able to incorporate the impact of human activities on the natural sediment system, both detrimental and remedial, with particular relevance to changes in land use. The extent to which available modeling approaches can meet these requirements is reviewed briefly in the following sections.

2.2.2. Empirical Models

Such models usually depend upon establishing a relationship between concurrent time series of an input and the output, calibrated from existing hydrometeorological and sediment yield records and containing no consideration of the physical processes involved. The simplest example is the sediment rating curve, relating sediment concentration, C, (usually only suspended load) to water discharge, Q, at some point in a stream:

$$C = a\,Q^b \tag{1}$$

where a and b are empirical constants. For lack of choice this method has been widely applied; however, the scatter of data about the relationship can involve one or more orders of magnitude, the relationship can exhibit both systematic and random variation and the method is thus subject to error (e.g. Walling and Webb, 1982; Pickup, 1988). Its usefulness is limited by the assumption that a single–valued relationship between suspended load concentration and water discharge can exist. The suspended load, at least during rainstorms, consists to a greater or lesser extent of wash load derived from outside the channel. The river can usually carry wash load in quantities well in excess of those supplied. Hence it is not channel discharge which determines the sediment load but the rate of erosion within the basin and the rate of delivery to the channel, neither of which is necessarily related to channel discharge.

2.2.3. Conceptual Models

These models are based on some consideration of the physical nature of the delivery process. However, the physical significance is not so clear that the model parameters can be assessed from direct measurements; instead they must be calibrated from concurrent input and output time series. The available models retain varying degrees of empiricism and, when applied to a basin, are spatially lumped.

Examples of yield models include modifications to the rating curve approach in order to account for sediment supply and depletion effects (Haddock, 1978; VanSickle and Beschta, 1983), a unit sediment graph for the wash load produced by storm runoff (Rendon–Herrero, 1978) and a dynamic basin yield model incorporating sediment availability, removal and translation functions (Moore, 1984).

The Universal Soil Loss Equation (USLE) (Wischmeier and Smith, 1978; and numerous textbooks, e.g. Dunne and Leopold, 1978, pp. 523–531) is a gross soil erosion model which has formed the basis of various approaches to the calculation of net sediment yield (e.g. Walling, 1983; Hrissanthou, 1986). It is able to account for land management activities but retains an empirical basis. Recently the United States Department of Agriculture, in its Water Erosion Prediction Project, has sought to develop a significant conceptual improvement over the USLE, based on up-to-date hydrological and soil erosion understanding and computer capabilities (Foster, 1987).

2.2.4. Stochastic Models

These models have some component of a random character, having a distribution in probability through time. Identical inputs may result in different outputs if run through the model under identical conditions. Their use in sediment yield modeling is a recognition of the complexity of the sediment yield process, the apparently random nature of sediment yield supply and the occurrence of different sediment transport rates for similar flow conditions in a river. The basis of the approach is that sediment yield can be described by a stochastic process and associated probability distribution functions. Detailed reviews are given by Woolhiser and Renard (1980) and Singh et al. (1988) and recent applications are presented by VanSickle (1982) and Griffiths (1980).

Particular advantages of this approach are that it can provide insight into the expected variability of sediment yield over various time periods and can force the consideration of the inherent uncertainty in the governing processes, model input and model output. However, stochastic models require the availability of a lengthy sediment yield record for their development and, when used for prediction, may not be able to account for nonstationarity in sediment yield arising from land–use change (e.g. VanSickle, 1982).

2.2.5. Physically–based Distributed Models

These models are based on our understanding of the physics of the sediment yield processes and describe the sediment system using equations governing the transfer of mass, momentum and energy (e.g. Abbott et al., 1986; Beven, 1985). In principle they can be applied outside the range of conditions used for calibration and, as their parameters have physical meaning, they can be evaluated from direct measurement and without the need for a lengthy hydrometeorological record. Limited only by the relevance of the physical laws on which they are based, they can be applied to any type of basin and can simulate the full sediment yield regime, providing multiple outputs on a spatially distributed basis. However, they also have heavy computer requirements, require the evaluation of a large number of parameters on a spatially distributed basis, incorporate spatial scaling problems of currently unknown impact and depend for their accuracy upon the accuracy of the hydrological model used to generate the surface flows which transport the eroded sediment.

The relative transposability of such models gives them an advantage in predicting the impacts of land–use change and of development in basins with short or nonexistent hydrometeorological and sediment yield records. Simpler models are based on response functions which require lengthy records for calibration or are too simple to represent the complex changes in system behavior associated with land–use change. However, current conceptual understanding of sediment yield processes is not always sufficient or capable of being expressed mathematically. Therefore, as currently deployed, physically–based distributed yield models describe the erosion and transport arising from raindrop impact, overland flow and channel flow only. Recent examples include the Colorado State University model (Simons et al., 1982), the Modified ANSWERS (Beasley et al., 1982; Park et al., 1982b), FESHM (Ross et al., 1980), SWAM (DeCoursey, 1982), the Modified USDAHL–74 (Yoo and Molnau, 1987) and two models based on the SHE hydrological modeling system, SHE–SEM (Nielsen et al., 1986; Storm et al., 1987) and SHESED–UK (Wicks and Bathurst, in press; Wicks et al., 1988).

The physical basis of this type of model and the potential for extending it to wider applicability is examined in the following section.

2.3. Models of Erosion and Sediment Yield Processes

2.3.1. Erosion by Raindrop Impact and Overland Flow

Soil particles may be detached and transported downslope by both raindrop impact and overland flow. In general, though, the greater erosion is caused by raindrop impact and most of the transport is brought about by overland flow. In some instances the erosion and transport processes apply more or less uniformly over the soil surface, the overall effect then being called sheet erosion. Frequently, though, the flow may be concentrated into rills which provide an efficient agency for transport as well as a further mechanism for erosion. The interrill areas are then characterized by sheet erosion which feeds the rill flows. These processes have seen the greatest development in physically–based modeling and there is a plethora of related papers. Therefore, the following covers only the main points.

Erosion by Raindrop Impact. Raindrop impact detaches or breaks down soil aggregates from the soil mass, making them available for transport. The amount eroded depends on raindrop properties (including size distribution and fall velocity), soil properties and system characteristics such as vegetation cover and overland flow depth (Park et al., 1982a). A sophisticated model of the spatial redistribution of soil by rainfall, considering the dispersion of raindrop splash droplets and the entrainment of mineral particles from a disaggregated soil mixture in the droplets, has been explored by Wright (1987). However, most models rely on less detailed representations.

For bare soil the amount of material detached by splash, D_r, is usually given as a function of a soil erodibility coefficient, k_r, and either a physically–based rainfall parameter, such as kinetic energy or momentum, or rainfall intensity raised to an experimentally determined power (e.g. Gilley and Finkner, 1985). Where a physically–based rainfall parameter is used, it may also be expressed as a function of rainfall intensity, so that the general detachment equation is:

$$D_r = k_r I^b \qquad (2)$$

where I = rainfall intensity and b is an exponent, usually in the range of 1 to 2. A theoretical basis for the use of the square of raindrop momentum as the rainfall parameter is provided by Styczen and Høgh–Schmidt (1986); this approach is used in both SHE–SEM and SHESED–UK.

Soil erodibility depends on soil texture and structure (e.g. percent clay, aggregate stability, particle size), surface moisture content, whether the ground is frozen, and other factors.

Dependency on surface moisture and freeze/thaw action implies that erodibility can vary with time through an event (e.g. Park et al., 1982a). Many attempts have been made to relate the raindrop soil erodibility coefficient to the various soil properties (e.g. Pall et al., 1982; Young and Onstad, 1982) but no satisfactory relationship has yet been devised. The coefficient is, therefore, often left as a calibration factor.

Major factors altering the raindrop impact erosion from that applicable to bare soil include ground, canopy and surface water covers. Ground cover includes forest litter, stones, snow-pack and vegetation less than about 0.5m high. This can be allowed for by a proportional area reduction factor. Canopy cover refers to the taller vegetation. This may initially dissipate the kinetic energy of the raindrops but it also allows redistribution and coalescence of the rain into leaf or branch drips (e.g. Mosley, 1982; Moss and Green, 1987; Herwitz, 1987). These drips are generally larger than the incoming raindrops and have a narrower size distribution. Depending on fall height, reinterception within the canopy, proportion of rain entering stem-flow and other factors, drip formation can result in erosion being greater under a vegetation canopy than on bare soil (e.g. Armstrong and Mitchell, 1987; Moss and Green, 1987; Herwitz, 1987; Morgan, 1985). In the SHE the overall process is modeled using the same "momentum squared" approach as for raindrop impact but requires additional information on fall height, drip diameter, percent canopy cover and proportion of intercepted rain reaching the ground as leaf drip.

Surface water acts to reduce the energy imparted to the soil by raindrop impact. Park et al. (1982b) have proposed the following correction factor:

$$F_w = 1 \qquad\qquad \text{if } h \leq d_m$$
$$F_w = e^{(1-h/d_m)} \qquad \text{if } h > d_m \tag{3}$$

where F_w = water depth correction factor; h = water depth; and d_m = median raindrop diameter.

Modifying Equation (2) to allow for the above effects gives a single equation for the soil eroded by raindrop impact. For SHESED–UK this is:

$$D_r = k_r \, F_w \, (1 - C_g) \left[(1 - C_c) \, M_r + M_d \right] \tag{4}$$

where C_g = proportion of ground shielded by ground cover; C_c = proportion of ground shielded by canopy cover; M_r = momentum squared for rain falling directly to the ground; and M_d = momentum squared for leaf drip, spatially averaged. Similar equations are used in other models, with additional terms for slope, wind direction, soil moisture and other factors as needed (e.g. Yoo and Molnau, 1987).

Equations (3) and (4) require information on the size distribution of the raindrops. Where this is not measured, a derived distribution can be used instead: SHESED–UK, for example, incorporates the Marshall–Palmer (1948) distribution. Information on the extent of ground and canopy cover is obtained by various field sampling techniques or by photogrammetric and other remote sensing methodologies.

Erosion by Overland Flow. Soil particles are detached as a result of the shear stress exerted by the flow. The amount of soil eroded depends on both the excess shear stress above the level for initiation of detachment and the extent to which the flow transport capacity exceeds the sediment load being supplied from upstream (e.g. Lane et al., 1988).

The basic erosion equation is:

$$D_f = k_f \, (\tau - \tau_c)^b \tag{5}$$

where D_f = amount eroded; k_f = overland flow soil erodibility coefficient; τ = shear stress; τ_c = critical shear stress; and b = an exponent (= 1 in SHESED–UK). As with the rainfall soil erodibility coefficient, k_f has not been satisfactorily related to the soil properties and is often left as a calibration coefficient.

The critical shear stress depends on the nature of the soil. For noncohesive soils there are a number of approaches such as the Shields relationship (e.g. Simons and Sentürk, 1977) but cohesive materials have proven more difficult to model (e.g. Kelly and Gularte, 1981). Yoo and Molnau (1987) quote the formula of Smerdon and Beasley (1961):

$$\tau_c = 0.0503 \ (10^{0.0183 \ Pc}) \ (1.0 - e^{-s}) \tag{6}$$

where Pc = percent clay content and s = soil moisture index.

Transport capacity of the flow depends on flow and sediment characteristics as formulated in a sediment transport equation. Rather little information is available on sediment transport by shallow overland flow, particularly for aggregates and fine particles, and it is usually necessary to use streamflow equations—for which there is an extensive literature (Walling, 1983). Reviews of suitable equations are presented by Alonso et al. (1981) and Julien and Simons (1985).

More physically–based considerations are examined by Aziz and Prasad (1985) (who give the flow conditions for the determination of the transport capacity as the incipient conditions of instability in the propagation of water and sediment waves) and Novotny (1980) (who indicates the importance of raindrop impact on the overland flow as a source of energy for keeping sediment particles in suspension).

Rill and Interrill System. Coupled models have been devised to account for the combined erosion from rills and the interrill areas (e.g. Foster et al., 1977; Young and Onstad, 1982). However, the grid–scale used to represent spatial variation in catchment models such as SHESED–UK is usually large (e.g. 50–500 m) compared with the scale of individual rill and interrill systems. Given also the current lack of experimental data and the difficulty of distinguishing the separate effects of rill and interrill erosion, such coupled models may be too detailed for realistic calibration at other than a local (subgrid) scale. Means of accounting for rill erosion in catchment models will need to be developed along different lines (perhaps involving effective parameter values) which extrapolate from the subgrid process scale to give the integrated effect at the model grid scale. Currently, though, only an empirical approximation to this approach has been proposed, with erosion estimated as sheet erosion and then modified by a rill susceptibility factor (e.g. Komura, 1976). It remains important to model rilling because: (a) such an approach is more physically based, (b) rilling causes more erosion than does sheet flow, and (c) the source of the erosion (from the surface of an interrill area or from deeper within the tillage layer via a rill) has implications for the transport by particle adsorption of agricultural chemicals (which may be incorporated within the soil or applied to the surface) (Young and Onstad, 1982).

Sediment Routing. Three model components are needed to route the eroded sediment across the basin. The first concerns overland flow as the basic driving agent and computes the spatial and temporal variation of flow velocity and depth, usually with some version of the Saint Venant equations for mass continuity and force–momentum (e.g. Woolhiser, 1975; Li, 1979). Sources of error arise from the questionable relevance of the equations to spatially nonuniform shallow flow and in the calculation of flow resistance as a function of soil and vegetation roughnesses and raindrop impact. Usually the flow resistance coefficient is left as a calibration factor which effectively takes account of such uncertainties. Accuracy also depends on the hydrological model used to determine the net amount of precipitation available for runoff.

The second component computes the spatial and temporal variation in sediment concentration or mass, using the mass conservation equation (e.g. Li, 1979). In the two–dimensional formulation used in SHESED–UK this is:

$$\frac{\partial(ch)}{\partial t} + (1-\lambda)\frac{\partial z}{\partial t} + \frac{\partial g_x}{\partial x} + \frac{\partial g_y}{\partial y} = 0 \tag{7}$$

where h = water depth; c = sediment concentration; λ = soil porosity; z = soil surface elevation or alternatively the depth of loose soil; g_x, g_y = volumetric sediment transport rates per unit width in the x and y directions, respectively; and t = time.

The third component keeps an account of local bed elevation and particle size distribution as they are affected by the erosion, transport and deposition processes. This usually involves some comparison of soil availability, incoming sediment load, transport ability relative to particle size and transport capacity. Where appropriate, the component might also account for the filtering of sediment by vegetation and by soil infiltration (e.g. Novotny et al., 1986). Solution techniques for the routing scheme tend to be numerical (e.g. Li, 1979; Lane et al., 1988). An important aspect of this component is the sediment discharge equation giving the capacity transport rates in terms of the flow conditions. As noted above, though, an appropriate relationship for shallow overland flow has not yet been derived and the use of a stream-flow equation as an alternative may therefore represent a significant source of error. Also, the simulated sediment transfers depend heavily on the flow data supplied by the overland flow component, so any uncertainty in the flow simulation will feed through to the sediment simulation. This can be partially avoided during calibration of the sediment component by using measured rather than calculated flow discharges, for example in connection with rainfall simulator studies (e.g. Wicks et al., 1988).

Routing of sediment by size fraction is especially important when considering the movement of contaminants adhering to soil particles. The erosion and sediment delivery processes act selectively on different soil particle sizes and there is a tendency for suspended sediment loads to be enriched with the finer sizes (clay) and depleted of silt and sand–sized material compared with the source material (Walling, 1983, 1988). A contributory process in the fining of the sediment load is likely to be breakdown of aggregates during wetting by rain (Loch et al., 1988). Contaminant enrichment is closely related to sediment size and concentration and especially accompanies a physical enrichment of the finer particle sizes, particularly clays (e.g. Young and Onstad, 1982). Concentrations of the adsorbed nutrients, metals and organic contaminants can be several orders of magnitude above ambient water concentrations (e.g. Ongley, 1982). It may be noted that the process coincides with the range of sediments (cohesive) for which the erosion and transport processes are least understood.

2.3.2. Gullying

Gullying may be considered as the continuation of rilling on a grander scale. On agricultural land, gullies may be differentiated from rills as being too large to be obliterated by normal cultivation methods but otherwise can vary considerably in size (e.g. Dunne and Leopold, 1978, p. 516). In agricultural catchments where raindrop impact and overland flow erosion processes are active, gullying generally contributes in a relatively minor way to the gross export of soil (Dunne and Leopold, 1978, pp. 516–517; Piest et al., 1975). However, locally (for example in small basins [e.g. Harvey, 1987] or basins undergoing urbanization [e.g. Nicklin et al., 1986]), it can be the dominant source.

In general terms, the export of material depends on both sediment production and sediment removal. The former process may be related, for example, to freeze/thaw, seepage and soil moisture effects and to soil structure and acts typically by causing bank collapse (e.g.

Piest et al., 1975; Nicklin et al., 1986; Harvey, 1987). Moisture content in particular affects bank material strength and shear forces. Removal of the material occurs in runoff events and is a function of flow strength and recurrence interval. The relationship between sediment supply, storage and removal varies with season, climate and flow regime among other factors (e.g. Piest et al., 1975; Faber and Imeson, 1982).

The physically–based approach to modeling gully erosion requires consideration of the geotechnical factors affecting soil stability, the hydraulic factors affecting flow erosion and transport, and the hydrological factors determining soil moisture content and runoff. This kind of approach has been explored by Nicklin et al. (1986) among others but has not yet evolved into a general prediction methodology. Therefore, it is necessary to resort to more empirical approaches, such as modifying the erosion estimated by overland flow with a gully susceptibility factor (e.g. Komura, 1976).

2.3.3. Mass Movement

This is the en masse movement of material, primarily under the influence of gravity. It can range from very fast to imperceptibly slow and from persistently active to episodic, involve a range of solid/water ratios and incorporate any amount of material (e.g. Dunne and Leopold, 1978, pp. 546–558; Megahan and King, 1985). Mass erosion may feed sediment directly into the channel system, with both immediate and longer–term consequences for sediment yield (e.g. Newson, 1980), or may alternatively mobilize sediment within the basin to be worked on by overland flow, raindrop impact or later debris flows (e.g. Benda and Dunne, 1987).

Attempts to predict landslide occurrence and sediment yield have been made with statistical and physically–based approaches (e.g. Megahan, 1986). Statistical approaches include Monte Carlo analysis (Rice, 1982) and multivariate discriminant analysis for predicting the probability of landslides occurring in response to logging (e.g. Rice and Pillsbury, 1982; Furbish and Rice, 1983). Good results have been obtained with the latter method but its potential for general application is uncertain.

The physically–based approach uses accepted geotechnical principles to define a factor of safety, involving the two opposing forces acting on the soil mantle:

$$FS = \frac{Resistance\ of\ soil\ to\ failure}{Downslope\ component\ of\ soil\ weight} \tag{8}$$

Hillslope failure occurs when the downslope weight exceeds the resistance to shear ($FS < 1$). Moisture content is especially important in determining soil strength and weight and a large majority of hillslope failures are triggered by the action of water (e.g. Addison, 1987; Jenkins et al., 1988). Resistance to failure is also reduced by the removal of vegetation and the subsequent decay of roots previously exerting a binding force (e.g. Megahan and King, 1985; Megahan, 1983).

The factor of safety can be calculated for a slope if representative values of relevant soil, vegetation, slope angle and groundwater variables are known. In general, though, they are not well quantified and their natural variability and the lack of knowledge of subsurface conditions introduce considerable uncertainty into the factor. Therefore, attempts have been made to modify the approach with some form of probability scheme (e.g. Ward et al., 1981; Sidle, 1987). The resulting hybrid approach holds considerable promise but is not yet generally applicable. A particular difficulty is posed by the provision of data, especially rainfall and soil moisture. The possibility of using simpler, more empirical techniques (based on Caine's [1980] formula for the rainfall needed to trigger shallow landslides) is reviewed by Megahan and King (1985).

2.3.4. Channel Erosion and Transport

The relative importance of the channel bed and banks as sediment sources depends on channel cross–sectional shape, the erosive power of the flow and the materials of which the bed and banks are composed. Gravel–bed rivers with noncohesive sand or gravel banks have high rates of lateral erosion compared with bed scouring, so a large proportion of the total particulate sediment load comes from the banks. In sand–bed channels with cohesive clay or silt banks, most of the total load comes from the bed (e.g. Osman and Thorne, 1988).

Bank Erosion. The mechanics of bank failure are related to the forces of erosion, the size, geometry and structure of the bank and the engineering properties of the bank material (e.g. Hooke, 1979; Simons and Li, 1982; Thorne, 1982). Thorne (1982) divides the erosion processes into fluvial entrainment (removing material directly from the bank or precipitating gravitational collapse by deepening the adjacent channel) and subaerial/subaqueous weakening and weathering (involving bank strength and stability, particularly as affected by soil moisture). Other factors include freeze/thaw effects, ice gouging, animal and human activities, wind and boat waves and vegetation (e.g. Simons and Li, 1982; Hey and Thorne, 1986; Murgatroyd and Ternan, 1983). Where erosion occurs by gravitational collapse, the failed material acts to protect the bank and the erosion rate is then a function of both the collapse mechanism and the rate of removal of the failed material by fluvial entrainment (Thorne, 1982).

Physically–based determination of bank stability and erosion requires both hydraulic and geotechnical analysis (e.g. Thorne, 1982) and the consideration of seepage and the hydrological factors affecting the soil moisture conditions. Osman and Thorne (1988) present a slope stability analysis for steep banks which takes into account the removal of failed material and which, used in conjunction with a method to calculate lateral erosion distance, enables prediction of bank stability response to lateral erosion or bed degradation. However, this type of approach has still to evolve into a general methodology for predicting sediment yield; for the time being it is necessary to resort to more conceptual approaches such as the supply model of VanSickle and Beschta (1983).

Recent studies of the contribution of bank erosion to channel sediment load, mainly as suspended sediment, are described by Newson and Leeks (1987) and Odgaard (1987).

Bed Erosion and Sediment Routing. Channel sediment routing is the mathematical procedure by which inflowing sediment load is passed along the channel to the catchment outlet during a sequence of flow events. The load is composed of both wash load (from outside the channel, including bank supply) and bed material load (to be determined according to hydraulic scour, deposition and transport criteria).

As for overland flow transport, three model components are required (e.g. Bennett and Nordin, 1977). The first routes the flow of water as the basic driving mechanism for sediment transport. This is accomplished with some form of the Saint Venant equations (e.g. Cunge et al., 1980; Fread, 1985) and again the flow resistance coefficient is often used as a calibration factor.

The second component routes the sediment using the mass conservation equation. Considering total load, this is (e.g. Chen, 1979):

$$\frac{\partial(CA)}{\partial t} + (1-\lambda)\frac{\partial(wz)}{\partial t} + \frac{\partial Q_s}{\partial x} = q_s \tag{9}$$

where C = mean sediment concentration; A = flow cross–sectional area; λ = bed porosity; w = active bed width, in which erosion or deposition occurs; z = local bed elevation; Q_s = volumetric sediment discharge; q_s = lateral sediment inflow per unit length of channel; t = time; and x = distance. It is also possible for the bed and suspended loads to be considered separately, using two versions of the mass conservation equation and accounting for exchanges between the two loads (Bennett and Nordin, 1977).

The third component keeps an account of the local channel–bed elevation and sediment size composition. This accounts for erosion, deposition, armour layer development and the availability of different size fractions on the bed. As understanding increases, this component must also take into account interstorm bed consolidation effects which can increase the resistance to initiation of motion of the bed material and which can thus result in significant differences between the conditions under which initiation and cessation of transport occur during a subsequent flow event (e.g. Reid and Frostick, 1987). As for overland flow transport, an important aspect of this component is the sediment discharge equation linking the capacity transport rate and flow conditions. Generally this applies only to bed material load (not wash load) and is assumed to be a single–valued relationship. A wide range of equations is available (e.g. Vanoni, 1975; White et al., 1975; Alonso et al., 1981; Bathurst et al., 1987). However, all have an empirical component and a given equation should be applied only when it is appropriate to the flow conditions, channel characteristics and sediment types. Thus, for example, an equation developed for lowland sand–bed channels should not be applied to mountain boulder–bed channels.

Routing proceeds according to some comparison of availability of sediment on the bed, incoming sediment load, transport ability relative to particle size and the limiting transport capacity (e.g. Bennett and Nordin, 1977; Chen, 1979; Cunge et al., 1980; Li and Fullerton, 1987; Pickup, 1988, among others). Various schemes exist for routing the sediment according to size fraction but the processes of selective transport, particle interaction and transport of cohesive materials are not yet well understood (e.g. Alonso et al., 1981; Samaga et al., 1986).

2.3.5. *Channel Bed Sediment Infiltration*

Infiltration of sediment (usually fine material) into the channel bed represents an entry into temporary storage. The process is also important because it can clog fish spawning grounds, cause bed cementation and consolidation, create a source for sediment pulses following bed scour, reduce stream/aquifer interflows, reduce dissolved oxygen levels within the bed and provide a sink for adsorbed contaminants (e.g. Woods, 1980; Reid and Frostick, 1987; van't Woudt et al., 1979). Increased clogging has been related to altered sediment input arising from land–use change and to altered flow regime (e.g. Woods, 1980; Petts, 1988).

There is a limited literature on clogging of river sediments. However, the principal determining factors appear to be flow velocity and turbulence at the bed interface and the relationship between infiltrating particle size and the size of bed voids (e.g. Beschta and Jackson, 1979; Carling, 1984; Frostick et al., 1984; Cunningham et al., 1987; Sakthivadivel, 1972). With cohesive sediments, mechanical processes are augmented by electrostatic and electrochemical processes producing aggregation (e.g. Kovács, 1981, p. 351); thus, cohesive sediments are much more effective at clogging than are noncohesive sediments (e.g. Cedergren, 1977, p. 40).

A physically–based numerical model for fine sediment infiltration in gravel–bed streams has been developed by Alonso et al. (1986). Differences in potentiometric head along the longitudinal bed profile drive the intragravel flow which in turn carries part of the suspended sediment load from the channel flow into the gravel bed. Water movement in the gravel is modeled with the Boussinesq equation and the intragravel transport of sediment is described by modifications to the equations governing the dispersion of tracers. Coupling algorithms are included so that, as the fine sediment settles in the interstices of the gravel matrix, the pore space is gradually filled and the gravel permeability and intragravel flow are adjusted accordingly. The approach is promising and further developments along these lines are needed. Improved means of measuring the subsurface particle size distribution and hydraulic conductivity are also needed to provide appropriate test data. (Subsurface particle size distribution is also an important input to sediment transport equations which consider the subsurface, rather than surface, material as the basic source for sediment load.)

2.3.6. Channel Sediment Storage

Storage sites may develop in the channel bed (through infiltration), as channel bars (e.g. Church and Jones, 1982), as slowly moving sediment waves (e.g. Griffiths, 1979) and as over-bank deposition (e.g. Meade, 1982). The average length of time a particle spends in storage (the residence time) varies according to the stability and location of the storage site.

There is an increasing realization that storage has a major impact on the movement of sediment through a channel system (Walling, 1988). Wolman (1977) observes that a sediment yield of several years can be stored in the channel yet remain virtually indiscernible in the system. Similarly, Swanson et al. (1982) note that the volume of temporarily stored material along a channel is commonly more than ten times the average annual export of total particulate sediment. Moderate changes in storage, perhaps arising from an altered flow regime, can thus cause major changes in sediment yield even if there is no change in external supplies. Storage sites may also act initially as sinks for sediments and adsorbed contaminants, later releasing the material either gradually or as a pulse. Therefore, they have important implications for the dispersal, decay or degradation of contaminants and the interaction of the contaminants with ecological and agricultural systems. Finally, the development of a bar storage site on one side of a channel may precipitate, or be directly connected with, bank erosion and the release of sediment on the other. Storage and release can thus proceed in parallel in the same channel reach.

Physically–based simulation of the interaction between storage and sediment yield is not yet feasible at other than short time–scales. There is a lack of the necessary quantitative information on bar formation, consolidation and movement, cycles of bed aggradation and degradation, and long term wave movement. A conceptual model which accounts for variations in sediment discharge arising from imbalances between erosion and deposition has been proposed by Hey (1979, 1987). More recently Kelsey et al. (1986) have presented a stochastic model for the transport of stored sediment along a river. The model is based on the probability of transition of a particle of sediment between four storage reservoirs: active (often mobilized), semiactive, inactive, and stable (hardly ever mobilized). The probabilities are derived from the calculated sediment residence times for each reservoir. A successful application was carried out in a research basin but provision of the data needed to calculate residence times poses a problem for more general application.

2.4. Scale Effects

Spatial distribution in physically–based models is typically represented by a grid network. Each grid element is then a miniature lumped model within which all conditions are spatially uniform. Selection of grid size must balance accurate representation of the basin (requiring a fine grid) against minimization of computer requirements (requiring a coarse grid). The interaction of grid scale with model operation is recognized to be important but relatively little research on the problem has been carried out so far (e.g. Walling, 1983).

Two areas which need consideration are:

(1) Typically parameters are sampled and evaluated at spatial scales significantly different (generally smaller) from those of the model grid scale. The parameter values may not then be representative of conditions as they are simulated in the model. This applies particularly to the soil hydraulic parameters, which may be obtained from cores with no allowance for macropore effects and spatial variability within the grid square. The result can be an erroneous simulation of overland flow and thence sediment transport. Therefore,

measurement techniques should correspond, as far as possible, with the model grid scale. Remote sensing techniques may have an important role to play here.

(2) The model grid scale may typically exceed the spatial scale at which individual erosion processes occur. Thus the assumption of a uniform topography in a grid square may ignore distributions of concave and convex slopes at the subgrid scale and their impact on patterns of erosion and deposition. Similarly, the spatial lumping of rainfall in a grid square may misrepresent significant variations in rainfall intensity and thence soil erosion. Too coarse a grid will also prevent representation of partial contributing area effects. Attempts to compensate for such misrepresentations in, for example, the calibration of the soil erodibility coefficients will reduce the general transposability of the model. One solution may be the development of correction factors linking the simulation output for a given area represented at an appropriately fine grid scale with the outputs obtained at increasingly coarse representations (e.g. Julien and Frenette, 1986). A similar approach may be needed when, within the same basin, erodibility coefficients calibrated at one scale are applied at another scale with different storage and yield characteristics.

3. Conclusions

So far only partial progress has been made in developing physically–based distributed models which provide the integrated view of sediment yield processes needed in tackling the types of problems mentioned in the introduction. Integrated models have been developed for basins dominated by raindrop impact and overland flow erosion but even these involve considerable simplification of the operating processes. Other sediment yield processes have proved less susceptible to modeling (e.g. mass movement, storage) or have only recently come under scrutiny (e.g. river bed sediment infiltration). Where uncertainty in data provision and process description is likely to continue, physically–based models may not be able to provide complete solutions. However, hybrid models involving a probabilistic component offer a promising approach.

Numerical solution techniques have now advanced to the level where they are no longer the principal limiting factor in model application. Their sophistication, though, is not always matched by that of available data provision methodologies, which are not generally capable of supplying the large amounts of data ideally required to develop, test and apply advanced modeling systems. Further developments are therefore required in the application of remote sensing technology, the refinement of representative sampling techniques for channel subsurface sediments and the use of rainfall simulators for calibration simulation amongst other methodologies. These should be aimed particularly at the required spatial scales. Data relevant to all sediment erosion and yield variables should be collected so that calibration and validation can be carried out against more than just time series of sediment output for a basin.

There is also a need for process studies to improve the theoretical basis of models. Particular areas which require study are sediment transport dynamics in overland flow and channel flow, the behavior of cohesive sediments and of sediment mixtures with nonuniform size distributions, storage dynamics, bed sediment infiltration and contaminant adsorption. Detailed studies at small scales should be balanced by sediment budget studies which provide a more integrated view of the sediment–yield process. The basis of the hydrology model which supplies overland flow and soil moisture information should also be carefully reviewed.

On the computing side, future trends seem to revolve around ensuring increased use of sophisticated sediment yield models. On the one hand this involves increasing the ease of use through the development of front–end and trouble–shooting packages, geographic information systems and improved portability based on microcomputer applications. On the other it involves the interpretation of simulation results in a form useful to decision makers through the combination of sediment yield models with ecological, economic and other types of model. Such mega–models will enable, for example, the physical effects of land–use change to be related to economic costs. However, the complexity of advanced models is such that, to avoid their misuse, increasingly lengthy training or apprenticeship periods will be needed to provide a good understanding of the model background together with the ability to interpret model results, including tracing the causes of errors. In a related manner, and to a greater degree than is necessary with simpler models, operators of physically–based distributed models must have a good appreciation of the physical processes of sediment transfer and of the response mechanisms of the basin being simulated.

4. References

Abbott, M. B., J. C. Bathurst, J. A. Cunge, P. E. O'Connell, and J. Rasmussen: 1986, 'An Introduction to the European Hydrological System – Système Hydrologique Européen, "SHE", 1 : History and Philosophy of a Physically–based, Distributed Modelling System; 2 : Structure of a Physically–based Distributed Modelling System,' *J. Hydrol.* **87**(1/2), 45–77.

Addison, K.: 1987, 'Debris Flow During Intense Rainfall in Snowdonia, North Wales : A Preliminary Survey,' *Earth Surf. Proc. Landf.* **12**(5), 561–566.

Alonso, C. V., W. H. Neibling, and G. R. Foster: 1981, 'Estimating Sediment Transport Capacity in Watershed Modeling,' *Trans. Am. Soc. Agric. Engrs.* **24**(5), 1211–1220 & 1226.

Alonso, C. V., C. Mendoza, and G. Q. Tabios: 1986, 'Sediment Intrusion into the Substrate of Gravel Bed Streams,' in *World Water Issues in Evolution*, Proc. Am. Soc. Civ. Engrs. Water Forum '86, Long Beach, California, Vol. 2, pp. 1959–1967.

Armstrong, C. L., and J. K. Mitchell: 1987, 'Transformations of Rainfall by Plant Canopy,' *Trans. Am. Soc. Agric. Engrs.* **30**(3), 688–696.

Aziz, N. M., and S. N. Prasad: 1985, 'Sediment Transport in Shallow Flows,' Proc. Am. Soc. Civ. Engrs., *J. Hydraul. Engrg.* **111**(10), 1327–1343.

Bathurst, J. C.: 1987, 'Measuring and Modelling Bedload Transport in Channels with Coarse Bed Materials,' in K. Richards (ed.), *River Channels: Environment and Process*, Blackwell, Oxford, pp. 272–294.

Bathurst, J. C., W. H. Graf, and H. H. Cao: 1987, 'Bed Load Discharge Equations for Steep Mountain Rivers,' in C. R. Thorne, J. C. Bathurst, and R. D. Hey (eds.), *Sediment Transport in Gravel–bed Rivers*, Wiley, Chichester, pp. 453–477.

Beasley, D. B., L. F. Huggins, and E. J. Monke: 1982, 'Modeling Sediment Yields from Agricultural Watersheds,' *J. Soil Wat. Conservation* **37**(2), 113–117.

Benda, L., and T. Dunne: 1987, 'Sediment Routing by Debris Flow,' in *Erosion and Sedimentation in the Pacific Rim*, Intl. Ass. Hydrol. Sci. Publ. No. 165, pp. 213–223.

Bennett, J. P., and C. F. Nordin: 1977, 'Simulation of Sediment Transport and Armouring,' *Hydrol. Sci. Bull.* **22**(4), 555–569.

Beschta, R. L., and W. L. Jackson: 1979, 'The Intrusion of Fine Sediments into a Stable Gravel Bed,' *J. Fish. Res. Board Canada* **36**, 204–210.

Beven, K.: 1985, 'Distributed Models,' in M. G. Anderson, and T. P. Burt (eds.), *Hydrological Forecasting*, Wiley, Chichester, pp. 405–435.

Caine, N.: 1980, 'The Rainfall Intensity–Duration Control of Shallow Landslides and Debris Flows,' *Geogr. Ann.* **62A**(1–2), 23–27.

Carling, P. A.: 1984, 'Deposition of Fine and Coarse Sand in an Open–Work Gravel Bed,' *Can. J. Fish. Aquat. Sci.* **41**, 263–270.

Cedergren, H. R.: 1977, *Seepage, Drainage and Flow Nets*, 2nd edn., Wiley, New York, 534 pp.

Chen, Y. H. : 1979, 'Water and Sediment Routing in Rivers,' in H. W. Shen (ed.), *Modeling of Rivers*, Wiley, New York, pp. 10.1–10.97.

Church, M., and D. Jones: 1982, 'Channel Bars in Gravel–bed Rivers,' in R. D. Hey, J. C. Bathurst, and C. R. Thorne (eds.), *Gravel–bed Rivers*, Wiley, Chichester, pp. 291–324.

Cunge, J. A., F. M. Holly, and A. Verwey: 1980, *Practical Aspects of Computational River Hydraulics*, Pitman, London, 420 pp.

Cunningham, A. B., C. J. Anderson, and H. Bouwer: 1987, 'Effects of Sediment–Laden Flow on Channel Bed Clogging,' Proc. Am. Soc. Civ. Engrs., *J. Irrig. Drain. Engrg.* **113**(1), 106–118.

DeCoursey, D. G.: 1982, 'ARS' Small Watershed Model, *Proc. Am. Soc. Agric. Engrs. 1982 Summer Meeting*, University of Wisconsin, Madison, WI.

Duijsings, J. J. H. M.: 1986, 'Seasonal Variation in the Sediment Delivery Ratio of a Forested Drainage Basin in Luxembourg,' in *Drainage Basin Sediment Delivery*, Intl. Ass. Hydrol. Sci. Publ. No. 159, pp. 153–164.

Dunne, T., and L. B. Leopold: 1978, *Water in Environmental Planning*, Freeman, San Francisco, 818 pp.

Faber, T., and A. C. Imeson: 1982, 'Gully Hydrology and Related Soil Properties in Lesotho,' in *Recent Developments in the Explanation and Prediction of Erosion and Sediment Yield*, Intl. Ass. Hydrol. Sci. Publ. No. 137, pp. 135–144.

Foster, G. R., L. D. Meyer, and C. A. Onstad: 1977, 'An Erosion Equation Derived from Basic Erosion Principles,' *Trans. Am. Soc. Agric. Engrs.* **20**(4), 678–682.

Foster, G. R. (Compiler): 1987, *'User Requirements: USDA–Water Erosion Prediction Project (WEPP),'* Rep. 1, Natl. Soil Erosion Res. Lab., USDA–ARS, 43 pp.

Fread, D. L.: 1985, 'Channel Routing,' in M. G. Anderson, and T. P. Burt (eds.), *Hydrological Forecasting*, Wiley, Chichester, pp. 437–503.

Frostick, L. E., P. M. Lucas, and I. Reid: 1984, 'The Infiltration of Fine Matrices into Coarse–Grained Alluvial Sediments and its Implications for Stratigraphical Interpretation,' *J. Geol. Soc. London* **141**(6), 955–965.

Furbish, D. J., and R. M. Rice : 1983, 'Predicting Landslides Related to Clearcut Logging, Northwestern California, USA,' *Mountain Res. Dev.* **3**(3), 253–259.

Gilley, J. E., and S. C. Finkner: 1985, 'Estimating Soil Detachment Caused by Raindrop Impact,' *Trans. Am. Soc. Agric. Engrs.* **28**, 140–146.

Gole, S. P., and J. McManus: 1988, 'Sediment Yield in the Upper Krishna Basin, Maharashtra, India,' *Earth Surf. Proc. Landf.* **13**(1), 19–25.

Golubev, G. N.: 1982, 'Soil Erosion and Agriculture in the World: an Assessment and Hydrological Implications,' in *Recent Developments in the Explanation and Prediction of Erosion and Sediment Yield*, Intl. Ass. Hydrol. Sci. Publ. No. 137, pp. 261–268.

Goswami, D. C.: 1985, 'Brahmaputra River, Assam, India: Physiography, Basin Denudation, and Channel Aggradation,' *Wat. Resour. Res.* **21**(7), 959–978.

Gregory, K. J., and D. E. Walling: 1973, *Drainage Basin Form and Process*, Arnold, London, 458 pp.

Griffiths, G. A.: 1979, 'Recent Sedimentation History of the Waimakariri River, New Zealand,' *N.Z.J. Hydrol.* **18**(1), 6–28.

Griffiths, G. A.: 1980, 'Stochastic Estimation of Bed Load Yield in Pool–and–Riffle Mountain Streams,' *Wat. Resour. Res.* **16**(5), 931–937.

Haddock, D. R.: 1978, *'Modeling Bedload Transport in Mountain Streams of the Colorado Front Range,'* MS Thesis, Colorado State University, Fort Collins, CO, 74 pp.

Harvey, A. M.: 1987, 'Sediment Supply to Upland Streams: Influence on Channel Adjustment,' in C. R. Thorne, J. C. Bathurst, and R. D. Hey (eds.), *Sediment Transport in Gravel–bed Rivers*, Wiley, Chichester, pp. 121–146.

Herwitz, S. R.: 1987, 'Raindrop Impact and Water Flow on the Vegetative Surfaces of Trees and the Effects on Stemflow and Throughfall Generation,' *Earth Surf. Proc. Landf.* **12**(4), 425–432.

Hey, R. D.: 1979, 'Dynamic Process–Response Model of River Channel Development,' *Earth Surf. Proc.* **4**, 59–72.

Hey, R. D.: 1987, 'River Dynamics, Flow Regime and Sediment Transport,' in C. R. Thorne, J. C. Bathurst, and R. D. Hey (eds.), *Sediment Transport in Gravel–bed Rivers*, Wiley, Chichester, pp. 17–37.

Hey, R. D., and C. R. Thorne: 1986, 'Stable Channels with Mobile Gravel Beds,' Proc. Am. Soc. Civ. Engrs., *J. Hydraul. Engrg.* **112**(8), 671–689.

Hooke, J. M.: 1979, 'An Analysis of the Processes of River Bank Erosion,' *J. Hydrol.* **42**, 39–62.

Hrissanthou, V.: 1986, 'Mathematical Model for the Computation of Sediment Yield from a Basin,' in *Drainage Basin Sediment Delivery*, Intl. Ass. Hydrol. Sci. Publ. No. 159, pp. 393–401.

Jenkins, A., P. I. Ashworth, R. I. Ferguson, I. C. Grieve, P. Rowling, and T. A. Stott : 1988, 'Slope Failures in the Ochil Hills, Scotland, November 1984,' *Earth Surf. Proc. Landf.* **13**(1), 69–76.

Julien, P. Y., and D. B. Simons: 1985, 'Sediment Transport Capacity of Overland Flow,' *Trans. Am. Soc. Agric. Engrs.* **28**(3), 755–762.

285

Julien, P. Y., and M. Frenette: 1986, 'Scale Effects in Predicting Soil Erosion,' in *Drainage Basin Sediment Delivery*, Intl. Ass. Hydrol. Sci. Publ. No. 159, pp. 253–259.

Kattan, Z., J. Y. Gac, and J. L. Probst: 1987, 'Suspended Sediment Load and Mechanical Erosion in the Senegal Basin – Estimation of the Surface Runoff Concentration and Relative Contributions of Channel and Slope Erosion,' *J. Hydrol.* **92**, 59–76.

Kelly, W. E., and R. C. Gularte: 1981, 'Erosion Resistance of Cohesive Soils,' Proc. Am. Soc. Civ. Engrs., *J. Hydraul. Div.* **107**(HY10), 1211–1224.

Kelsey, H. M., R. Lamberson, and M. A. Madej: 1986, 'Modeling the Transport of Stored Sediment in a Gravel Bed River, Northwestern California,' in *Drainage Basin Sediment Delivery*, Intl. Ass. Hydrol. Sci. Publ. No. 159, pp. 367–391.

Komura, S.: 1976, 'Hydraulics of Slope Erosion by Overland Flow,' Proc. Am. Soc. Civ. Engrs., *J. Hydraul. Div.* **102**(HY10), 1573–1586.

Kovács, G.: 1981, *Seepage Hydraulics*, Elsevier, Amsterdam, 730 pp.

Lane, E. W., and W. M. Borland: 1951, 'Estimating Bed Load,' *Trans. Am. Geophys. Union* **32**(1), 121–123.

Lane, L. J., E. D. Shirley, and V. P. Singh: 1988, 'Modelling Erosion on Hillslopes,' in M. G. Anderson (ed.), *Modelling Geomorphological Systems*, Wiley, Chichester, pp. 287–308.

Larson, C. L., and W. Albertin: 1984, 'Controlling Erosion and Sedimentation in the Panama Canal Watershed,' *Water International* **9**(4), 161–164.

Lauffer, H., and N. Sommer: 1982, 'Studies on Sediment Transport in Mountain Streams of the Eastern Alps,' *Proc. 14th Congress Intl. Commission on Large Dams*, Rio de Janeiro, pp. 431–453.

Li, R-M.: 1979, 'Water and Sediment Routing from Watersheds,' in H. W. Shen (ed.), *Modeling of Rivers*, Wiley, New York, pp. 9.1–9.88.

Li, R-M., and W. T. Fullerton: 1987, 'Investigation of Sediment Routing by Size Fractions in a Gravel–bed River,' in C. R. Thorne, J. C. Bathurst, and R. D. Hey (eds.), *Sediment Transport in Gravel–bed Rivers*, Wiley, Chichester, pp. 421–442.

Loch, R. J., J. L. Cleary, E. C. Thomas, and S. F. Glanville: 1988, 'An Evaluation of the Use of Size Distributions of Sediment in Runoff as a Measure of Aggregate Breakdown in the Surface of a Cracking Clay Soil Under Rain,' *Earth Surf. Proc. Landf.* **13**(1), 37–44.

Marshall, J. S., and W. M. Palmer: 1948, 'The Distribution of Raindrops with Size,' *J. Meteor.* **5**, 165–166.

Meade, R. H.: 1982, 'Sources, Sinks and Storage of River Sediments in the Atlantic Drainage of the United States,' *J. Geol.* **90**(3), 235–252.

Megahan, W. F.: 1983, 'Hydrologic Effects of Clearcutting and Wildfire on Steep Granitic Slopes in Idaho,' *Wat. Resour. Res.* **19**(3), 811–819.

Megahan, W. F.: 1986, 'Recent Studies on Erosion and its Control on Forest Lands in the United States,' in *Forest Environment and Silviculture*, Proc. 18th IUFRO World Congress, Ljubljana, pp. 178–189.

Megahan, W. F., and P. N. King: 1985, 'Identification of Critical Areas on Forest Lands for Control of Nonpoint Sources of Pollution,' *Environmental Management* **9**(1), 7–18.

Moore, R. J.: 1984, 'A Dynamic Model of Basin Sediment Yield,' *Wat. Resour. Res.* **20**(1), 89–103.

Moosbrugger, H.: 1957, 'Le Charriage et le Débit Solide en Suspension des Cours d'Eau de Montagnes,' *Proc. IASH Toronto General Assembly*, Intl. Ass. Sci. Hydrol. Publ. No. 43, pp. 203–231.

Morgan, R. P. C.: 1985, 'Effect of Corn and Soybean Canopy on Soil Detachment by Rainfall,' *Trans. Am. Soc. Agric. Engrs.* **28**(4), 1135–1140.

Mosley, M. P.: 1982, 'The Effect of a New Zealand Beech Forest Canopy on the Kinetic Energy of Water Drops and on Surface Erosion,' *Earth Surf. Proc. Landf.* **7**, 103–107.

Moss, A. J., and T. W. Green: 1987, 'Erosive Effects of the Large Water Drops (Gravity Drops) that Fall from Plants,' *Australian J. Soil Res.* **25**, 9–20.

Murgatroyd, A. L., and J. L. Ternan: 1983, 'The Impact of Afforestation on Stream Bank Erosion and Channel Form,' *Earth Surf. Proc. Landf.* **8**, 357–369.

Newson, M.: 1980, 'The Geomorphological Effectiveness of Floods – a Contribution Stimulated by Two Recent Events in mid–Wales,' *Earth. Surf. Proc.* **5**, 1–16.

Newson, M. D., and G. J. Leeks: 1987, 'Transport Processes at the Catchment Scale,' in C. R. Thorne, J. C. Bathurst, and R. D. Hey (eds.), *Sediment Transport in Gravel–bed Rivers*, Wiley, Chichester, pp. 187–218.

Nicklin, M. E., J. I. Remus, and J. Conroy: 1986, 'Gully Bank Erosion of Loessial Soil in Urbanizing Watersheds,' in *Drainage Basin Sediment Delivery*, Intl. Ass. Hydrol. Sci. Publ. No. 159, pp. 141–151.

Nielsen, S. A., B. Storm, and M. Styczen: 1986, 'Development of Distributed Soil Erosion Component for the SHE Hydrological Modelling System,' Presented at Brit. Hydraul. Res. Ass. Intl. Conf. on Water Quality Modelling in the Inland Natural Environment, Bournemouth, UK.

Novotny, V.: 1980, 'Delivery of Suspended Sediment and Pollutants from Nonpoint Sources During Overland Flow,' Wat. Resour. Bull. 16(6), 1057–1065.

Novotny, V., G. V. Simsiman, and G. Chesters: 1986, 'Delivery of Pollutants from Nonpoint Sources,' in Drainage Basin Sediment Delivery, Intl. Ass. Hydrol. Sci. Publ. No. 159, pp. 133–140.

Odgaard, A. J.: 1987, 'Streambank Erosion Along Two Rivers in Iowa,' Wat. Resour. Res. 23(7), 1225–1236.

Ongley, E. D.: 1982, 'Influence of Season, Source and Distance on Physical and Chemical Properties of Suspended Sediment,' in Recent Developments in the Explanation and Prediction of Erosion and Sediment Yield, Intl. Ass. Hydrol. Sci. Publ. No. 137, pp. 371–383.

Osman, A. M., and C. R. Thorne: 1988, 'Riverbank Stability Analysis. I. Theory; II. Applications,' Proc. Am. Soc. Civ. Engrs., J. Hydraul. Engrg. 114(2), 134–172.

Pain, S.: 1987, 'After the Goldrush,' New Scientist, 20 August, 36–40.

Pall, R., W. T. Dickinson, D. Green, and R. McGirr: 1982, 'Impacts of Soil Characteristics on Soil Erodibility,' in Recent Developments in the Explanation and Prediction of Erosion and Sediment Yield, Intl. Ass. Hydrol. Sci. Publ. No. 137, pp. 39–47.

Park, S. W., J. K. Mitchell, and G. D. Bubenzer: 1982a, 'Splash Erosion Modeling: Physical Analyses,' Trans. Am. Soc. Agric. Engrs. 25, 357–361.

Park, S. W., J. K. Mitchell, and J. N. Scarborough: 1982b, 'Soil Erosion Simulation on Small Watersheds: A Modified ANSWERS Model,' Trans. Am. Soc. Agric. Engrs. 25, 1581–1588.

Petts, G. E.: 1988, 'Accumulation of Fine Sediment within Substrate Gravels along Two Regulated Rivers, UK,' Regulated Rivers: Research and Management 2(2), 141–153.

Pickup, G.: 1988, 'Hydrology and Sediment Models,' in M. G. Anderson (ed.), Modelling Geomorphological Systems, Wiley, Chichester, pp. 153–215.

Piest, R. F., J. M. Bradford, and G. M. Wyatt: 1975, 'Soil Erosion and Sediment Transport from Gullies,' Proc. Am. Soc. Civ. Engrs., J. Hydraul. Div. 101(HY1), 65–80.

Reid, I., and L. E. Frostick: 1987, 'Toward a Better Understanding of Bedload Transport,' in Recent Developments in Fluvial Sedimentology, Proc. 3rd Intl. Fluvial Sedimentology Conference, Fort Collins, CO, pp. 13–19.

Rendon–Herrero, O.: 1978, 'Unit Sediment Graph,' Wat. Resour. Res. 14(5), 889–901.

Rice, R. M.: 1982, 'Sedimentation in the Chaparral: How Do You Handle Unusual Events?,' in Sediment Budgets and Routing in Forested Drainage Basins, Rep. PNW–141, USDA-FS, Pacific Northwest Forest and Range Experiment Station, pp. 39–49.

Rice, R. M., and N. H. Pillsbury: 1982, 'Predicting Landslides in Clearcut Patches,' in Recent Developments in the Explanation and Prediction of Erosion and Sediment Yield, Intl. Ass. Hydrol. Sci. Publ. No. 137, pp. 303–311.

Ross, B. B., V. O. Shanholtz, and D. N. Contractor: 1980, 'A Spatially Responsive Hydrologic Model to Predict Erosion and Sediment Transport,' Wat. Resour. Bull. 16(3), 538–545.

Sakthivadivel, R.: 1972, 'Sediment Transport Through a Porous Column,' in H. W. Shen (ed.), Sedimentation, H. W. Shen, Colorado State University, Fort Collins, CO, pp. 26.1–26.17.

Samaga, B. R., K. G. Ranga Raju, and R. J. Garde: 1986, 'Bed Load Transport of Sediment Mixtures; Suspended Load Transport of Sediment Mixtures,' Proc. Am. Soc. Civ. Engrs., J. Hydraul. Engrg. 112(11), 1003–1035.

Schumm, S. A.: 1977, The Fluvial System, Wiley, New York, 338 pp.

Sidle, R. C.: 1987, 'A Dynamic Model of Slope Stability in Zero–Order Basins,' in Erosion and Sedimentation in the Pacific Rim, Intl. Ass. Hydrol. Sci. Publ. No. 165, pp. 101–110.

Simons, D. B., and R-M. Li: 1982, 'Bank Erosion on Regulated Rivers,' in R. D. Hey, J. C. Bathurst, and C. R. Thorne (eds.), Gravel–bed Rivers, Wiley, Chichester, pp. 717–747.

Simons, D. B., and F. Sentürk: 1977, Sediment Transport Technology, Water Resources Publications, Fort Collins, CO, 807 pp.

Simons, D. B., T. J. Ward, and R-M. Li: 1979, 'Sediment Sources and Impacts in the Fluvial System,' in H. W. Shen (ed.), Modeling of Rivers, Wiley, New York, pp. 7.1–7.27.

Simons, D. B., R-M. Li, T. J. Ward, and L. Y. Shiao: 1982, 'Modeling of Water and Sediment Yields from Forested Drainage Basins,' in Sediment Budgets and Routing in Forested Drainage Basins, Rep. PNW–141, USDA-FS, Pacific Northwest Forest and Range Experiment Station, pp. 24–38.

Simons, P.: 1988, 'Après Ski le Déluge,' *New Scientist*, 14 January, 49–52.

Singh, V. P., P. F. Krstanovic, and L. J. Lane: 1988, 'Stochastic Models of Sediment Yield,' in M. G. Anderson (ed.), *Modelling Geomorphological Systems*, Wiley, Chichester, pp. 259–285.

Slaymaker, O.: 1977, 'Estimation of Sediment Yield in Temperate Alpine Environments,' in *Erosion and Solid Matter Transport in Inland Waters*, Intl. Ass. Hydrol. Sci. Publ. No. 122, pp. 109–117.

Smerdon, E. T., and R. P. Beasley: 1961, 'Critical Tractive Forces in Cohesive Soils,' *Agricultural Engineering* 42(1), 26–29.

Storm, B., G. H. Jørgensen, and M. Styczen: 1987, 'Simulation of Water Flow and Soil Erosion Processes with a Distributed Physically–Based Modelling System,' in *Forest Hydrology and Watershed Management*, Intl. Ass. Hydrol. Sci. Publ. No. 167, pp. 595–608.

Styczen, M., and K. Høgh-Schmidt: 1986, '*A New Description of the Relation between Drop Sizes, Vegetation and Splash Erosion,*' Proc. KOHYNO, Nordic Hydrologic Program, Copenhagen.

Sundborg, Å.: 1983, 'Sedimentation Problems in River Basins,' *Nature and Resources* 19(2), 10–21.

Swanson, F. J., R. J. Janda, and T. Dunne: 1982, 'Summary: Sediment Budget and Routing Studies,' in *Sediment Budgets and Routing in Forested Drainage Basins*, Rep. PNW-141, USDA–FS, Pacific Northwest Forest and Range Experiment Station, pp. 157–165.

Tejwani, K. G.: 1984, 'Reservoir Sedimentation in India – Its Causes, Control and Future Course of Action,' *Water International* 9(4), 150–154.

Thorne, C. R.: 1982, 'Processes and Mechanisms of River Bank Erosion,' in R. D. Hey, J. C. Bathurst, and C. R. Thorne (eds.), *Gravel–bed Rivers*, Wiley, Chichester, pp. 227–259.

Vanoni, V. A. (ed.): 1975, *Sedimentation Engineering*, Am. Soc. Civ. Engrs. Manuals and Reports on Engineering Practice, No. 54, New York, 745 pp.

VanSickle, J.: 1982, 'Stochastic Predictions of Sediment Yields from Small Coastal Watersheds in Oregon, U.S.A.,' *J. Hydrol.* **56**, 309–323.

VanSickle, J., and R. L. Beschta: 1983, 'Supply-based Models of Suspended Sediment Transport in Streams,' *Wat. Resour. Res.* **19**(3), 768–778.

Van't Woudt, B. D., J. Whittaker, and K. Nicolle: 1979, 'Ground Water Replenishment from River Flow,' *Wat. Resour. Bull.* **15**(4), 1016–1027.

Walling, D. E.: 1983, 'The Sediment Delivery Problem,' *J. Hydrol.* **65**, 209–237.

Walling, D. E.: 1988, 'Erosion and Sediment Yield Research – Some Recent Perspectives,' *J. Hydrol.* **100**, 113–141.

Walling, D. E., and B. W. Webb: 1982, 'Sediment Availability and the Prediction of Storm–Period Sediment Yields,' in *Recent Developments in the Explanation and Prediction of Erosion and Sediment Yield*, Intl. Ass. Hydrol. Sci. Publ. No. 137, pp. 327–337.

Ward, T. J., R–M. Li, and D. B. Simons: 1981, 'Use of a Mathematical Model for Estimating Potential Landslide Sites in Steep Forested Drainage Basins,' in *Erosion and Sediment Transport in Pacific Rim Steeplands*, Intl. Ass. Hydrol. Sci. Publ. No. 132, pp. 21–41.

White, W. R., H. Milli, and A. D. Crabbe: 1975, 'Sediment Transport Theories: A Review,' *Proc. Instn. Civ. Engrs.* 59, Part 2, 265–292.

Wicks, J. M., J. C. Bathurst, C. W. Johnson, and T. J. Ward: 1988, 'Application of Two Physically–based Sediment Yield Models at Plot and Field Scales,' in *Sediment Budgets*, Intl. Ass. Hydrol. Sci. Publ. No. 174, pp. 583–591.

Wicks, J. M., and J. C. Bathurst: In press, 'SHESED–UK - a Physically–based, Distributed Sediment Yield Model for the SHE Hydrological Modelling System,' *J. Hydrol.*

Wischmeier, W. H., and D. D. Smith: 1978, '*Predicting Rainfall Erosion Losses – A Guide to Conservation Planning,*' US Dept. Agric., Agricultural Handbook No. 537.

Wolman, M. G.: 1977, 'Changing Needs and Opportunities in the Sediment Field,' *Wat. Resour. Res.* 13(1), 50–54.

Woods, P. F.: 1980, 'Dissolved Oxygen in Intragravel Water of Three Tributaries in Redwood Creek, Humboldt County, California,' *Wat. Resour. Bull.* 16(1), 105–111.

Woolhiser, D. A.: 1975, 'Simulation of Unsteady Overland Flow,' in K. Mahmood, and V. Yevjevich (eds.), *Unsteady Flow in Open Channels*, Water Resources Publications, Fort Collins, CO, Vol. 2, pp. 485–508.

Woolhiser, D. A., and K. G. Renard: 1980, 'Stochastic Aspects of Watershed Sediment Yield,' in H. W. Shen, and H. Kikkawa (eds.), *Application of Stochastic Processes in Sediment Transport*, Water Resources Publications, CO, pp. 3.1–3.28.

Wright, A. C.: 1987, 'A Model of the Redistribution of Disaggregated Soil Particles by Rainsplash,' *Earth Surf. Proc. Landf.* **12**(6), 583–596.

Yoo, K. H., and M. Molnau: 1987, 'Upland Soil Erosion Simulation for Agricultural Watersheds,' *Wat. Resour. Bull.* **23**(5), 819–827.

Young, R. A., and C. A. Onstad: 1982, 'The Effect of Soil Characteristics on Erosion and Nutrient Loss,' in *Recent Developments in the Explanation and Prediction of Erosion and Sediment Yield*, Intl. Ass. Hydrol. Sci. Publ. No. 137, pp. 105–113.

Østrem, G., and others: 1971, '*Slamtransportstudier i Norksa Glaciärälver 1970,*' Forskningsrap. 12, Naturgeografiska Institutionen, Stockholms Universitet, Stockholm, 133 pp.

Chapter 14

Integrated Quantity/Quality Modeling - Root Zone Leaching

Donn G. DeCoursey
USDA–ARS, Hydro–Ecosystem Research
243 Federal Building
301 South Howes, P.O. Box E
Fort Collins, Colorado 80522 U.S.A.

Abstract. In recent years global problems like climate change, acid rain, water quality, food production, and drought dominate discussions of world problems. Hydrology and hydrologic processes play dominant roles in the solution of these problems. In this and the next chapter, the roles of hydrology and hydrologic process models are investigated as they are applied in solving problems of nonpoint source contamination of water. The need for these models, the contaminants considered, and general features are discussed. Because of the importance of rootzone process in solving the global problems, these processes are discussed in this chapter. Nonpoint source water quality modeling of receiving waters (ground water, surface runoff, and surface impoundments) is presented in Chapter 15.

1. Introduction

1.1. Need for Nonpoint Source Water Quality Models

Good water quality is basic to the health of all mankind. Over time man can adjust to local natural contaminants in water supplies. In less well developed countries, inhabitants have adapted to high salt concentrations and the presence of pathogenic organisms in water supplies. However, in general, man cannot adapt to non–natural or even natural pollutants when they are introduced over a very short period of time at concentrations toxic to normal bodily functions.

In the past man polluted his water supplies but was able to find other sources; thus he did not consider it a major problem. However, in the last 50–75 years, or in some cases last two centuries, as populations have grown the availability of these water supplies has diminished and man has been forced to do something about contamination. Initially point sources of river water contamination such as sewer outlets were identified. Consequently most industrialized countries, in attempts to make the water "drinkable" and safe for use in industry or agriculture, imposed strong measures to clean up the rivers and lakes and to keep them clean. As a result of these restrictions on the dumping of contaminants, most municipal and industrial points of contamination were removed, and the quality of streams improved.

However, in the last 20–30 years, and in some cases even less, we have increased the application of pesticides and fertilizers and are now concerned about the concentrations of these materials in both surface and groundwater supplies. Concentrations of a variety of agricultural chemicals have been found in lakes, streams, and groundwater in concentrations high

D. S. Bowles and P. E. O'Connell (eds.), Recent Advances in the Modeling of Hydrologic Systems, 289–322.

enough to be toxic. We have also discovered that old land fills, buried tanks, sewage disposal sites and similar dumping grounds are responsible for extremely dangerous contamination of groundwater supplies (Office of Technology Assessment, 1984).

In the United States a multibillion dollar "Super Fund" has been set up by the U.S. Congress to aid in cleaning up the sites contaminated by "point sources" such as buried land fills. These point sources are extremely expensive to treat, but since the contamination can be traced by the plume that develops, it is possible to clean up such sites. Recently attention shifted to nonpoint sources in a Presidential Initiative on Enhancing Water Quality.

In this and the next chapter, emphasis will be placed on nonpoint sources such as those created by the broad application of pesticides and fertilizers. These are not necessarily agricultural, although agricultural nonpoint source pollution has received the most attention. Other sources are in cities where fertilizer and pesticides are applied to yards, parks, and industrial sites. The biggest difference between problems of point and nonpoint sources is that the clean–up of nonpoint sources is almost impossible, especially contaminated groundwater. Even the clean–up of surface waters can be difficult if the source is polluted groundwater. Direct surface runoff contamination can be controlled by proper management.

Mathematical models, such as those discussed herein, have been helpful in analyzing and obtaining solutions to water quality problems. Model development technology has advanced to the point that it is recognized by the Congress of the United States. In fact the Congress instructed the Office of Technology Assessment (OTA) to assess the nation's ability to use models efficiently and effectively in analyzing water resource problems and to make recommendations for improving the use of available technologies.

The OTA report (Office of Technology Assessment, 1984) contained some interesting comments about mathematical models that are applicable to problems of nonpoint source water pollution.

- Mathematical models are among the most sophisticated tools available for analyzing water resource issues. They can significantly improve the informational background on which decisions are based and substantially reduce the cost of managing water resources.
- Mathematical models have significantly expanded the nation's ability to understand and manage its water resources and have made it feasible to compare quantitatively the likely effects of alternative resource decisions.
- Models are often the least expensive and best available alternative for analyzing complex resource problems.
- Models have the potential to provide even greater benefits for water resource decision-making in the future.

Mathematical models developed to solve most nonpoint source problems are primarily hydrologic models modified to include pollutant transport. This is because the primary transport mechanism is the movement of water. Since most of the chapters in this book describe modeling of the hydrologic system, emphasis in this chapter is on modifications of hydrologic systems to incorporate nonpoint pollutant transport mechanisms. Only enough discussion of the hydraulics of flow will be presented to make the discussion relevant.

In the presentation nonpoint sources of pollutants are those sources that discharge to a watershed in such a way that they depend upon the vagaries of the hydrologic cycle to transport them to the stream system. The water course will be saturated and unsaturated paths through the vadose zone, stream channels, lakes, and groundwater aquifers. Pollutants discussed will consist of plant nutrients, pesticides, salts, bacteria, organic matter, and sediments.

2. Discussion

The root zone is the unsaturated zone of the soil profile extending from the soil–atmosphere interface to the bottom of the region within which most plant roots are found. It is a very chemically and biologically active region of the biosphere. The residence time of chemicals within this region, and the processes operative within it, ultimately determine the degree of groundwater contamination. It is only through a better understanding of the basic physical, chemical, and biological processes active in this unsaturated zone that we can better manage it to minimize ground and surface water contamination and associated risks to the public.

Much of the material in the following presentation is abstracted from four recent state-of-the-art papers (Vachaud et al., 1990; Wagenet et al., 1990; De Willigen et al., 1990; Jones et al., 1990). Because of space constraints it will not be possible to delve very deeply into the processes of water and solute transport in the vadose zone, but the four papers above will provide references that can be used for more in-depth study and numerous references. Dagan et al. (1990) and van Genuchten et al. (1990) also discuss important issues of vadose zone transport. In the material presented in this chapter, I have specifically avoided a very important component of rootzone leaching processes, preferential flow. Keith Beven, in other chapters of this book, discusses this topic rather extensively. Also see Clothier (1990) and DeCoursey and Rojas (1990) for examples and a model that incorporates the process, respectively.

In the following presentation, three general classes of models, some of their strengths and weaknesses, and three model examples are described. Vachaud et al. (1990) and Addiscott and Wagenet (1985) make a key distinction between deterministic and stochastic models. Deterministic models define a system in such a way that the occurrence of a given set of events leads to a uniquely-definable outcome; stochastic models define the system such that it presupposes the outcome to be uncertain and structures the models to account for the uncertainty. Deterministic models are further divided into mechanistic and functional models. Mechanistic models take into account the most fundamental mechanisms of the processes as presently known and understood. Functional models incorporate simplified treatments of complex processes. In the following descriptions the three classes of models will be referred to as mechanistic, functional, and stochastic. Mechanistic models will be discussed first and in more depth because they provide the best approach to a description of the major processes of an integrated model.

2.1. Mechanistic Leaching Models

2.1.1. Fundamental Equations

Mechanistic models describe soil water movement by the Darcian–based Richards' equation (Vachaud et al., 1990):

$$C(h)\frac{\partial h}{\partial t} = \frac{\partial}{\partial z}\left\{K(h)\left(\frac{\partial h}{\partial z}-1\right)\right\} + A(z,t) \tag{1}$$

and the Fickian–based convection–dispersion equation for solute movement:

$$\frac{\partial(\rho S)}{\partial t} + \frac{\partial(\theta C)}{\partial t} = \frac{\partial}{\partial z}\left(\theta D_s \frac{\partial C}{\partial z}\right) - \frac{\partial}{\partial z}(J_w C) + \Sigma\phi_j \tag{2}$$

$\quad$ (a) $\qquad$ (b) $\qquad\qquad$ (c) $\qquad\qquad$ (d) $\qquad$ (e)

Specific equations defining flux and transient and steady state flow that form the basis of mechanistic models are presented in Table 1. These models also implement the movement of interactive and inert solutes subject to transformation if it takes place.

In the above equations, and those in Table 1, h represents the pressure head (positive in the saturated region, negative in the vadose zone); $C(h)$ is the capillary capacity or the slope of the soil water retention curve $\theta(h)$; θ is the volumetric water content; $K(h)$ is the soil hydraulic conductivity; $D(\theta)$ is the capillary diffusivity; J_w is the volumetric water flux density; $A(z,t)$ represents sources and sinks of water, notably those resulting from evaporation and plant water extraction; z is vertical distance, positive downward; t is time; C and S are concentrations of material in solution and on solid phases of the soil, respectively; ρ is the soil bulk density; J_S is the solute flux density; D_S is the apparent dispersion coefficient; $V = J_w/\theta$ is the pore water velocity; and ϕ_j are rates of solute removal or supply. In Equation (2), terms (a), (b), and (c) are the chemical aspects and (c) and (d) are the physical aspects of convection-dispersion. Specifically (a) and (b) describe changes in sorbed and dissolved concentrations with time, respectively, (c) accounts for hydrodynamic dispersion, (d) is the advection of solute and (e) includes transformations (such as precipitation/dissolution, complexation, etc.).

Table 1. Basic equations of mechanistic models (Vachaud et al., 1990).

Flux equations		
Water (J_W)		**Solute (J_S)**
$J_W = -K(h)(\frac{\partial h}{\partial z} - 1)$ /Ia/		$J_S = -\theta D_S \frac{\partial C}{\partial z} + J_w C$ /II/
or		
$J_W = -D(\theta)\frac{\partial \theta}{\partial z} + K(\theta)$ /Ib/		

Transient Transport Equations		
Richards:		
$C(h)\cdot\frac{\partial h}{\partial t} = -\frac{\partial J_W}{\partial z} + A(z,t)$ /IIIa/		$\frac{\partial T}{\partial t} = -\frac{\partial J_S}{\partial z} + \Sigma\phi_j$ /IVa/
		$T = \varrho S + \theta C$ /IVb/
or		S and C are linked through submodels Inert solute; no transformation
Fokker–Planck:		
$\frac{\partial \theta}{\partial t} = -\frac{\partial J_W}{\partial z} + A(z,t)$ /IIIb/		$\frac{\partial(\theta C)}{\partial t} = \frac{\partial}{\partial z}\left(\theta D_S \frac{\partial C}{\partial z}\right) - \frac{\partial}{\partial z}(J_w C)$ /V/

Steady Water Flow Conditions		
$J_W = $ Const. ; $\theta = $ Const.		$\frac{\varrho}{\theta}\cdot\frac{\partial S}{\partial t} + \frac{\partial C}{\partial t} = D_S\cdot\frac{\partial^2 C}{\partial z^2} - V\frac{\partial C}{\partial z} + \frac{1}{\theta}\Sigma\phi_j$ /VI/
		Inert solute, no transformation
		$\frac{\partial C}{\partial t} = D_S\frac{\partial^2 C}{\partial z^2} - V\frac{\partial C}{\partial z}$ /VII/

2.1.2. Transformations and Nutrient Cycling

The transformations included in term (e) are those of chemical precipitation and dissolution, oxidation–reduction, radioactive decay, transformation by microorganisms, plant uptake, etc. These processes apply to nutrient cycling and transformation and to pesticide dissipation. The dissipation rates for pesticides are made up of several independent processes that may or may not take place at the same time. These processes include volatilization, photolysis, hydrolysis, biodegradation, oxidation, and complexation. Volatilization can be an important dissipation process during and immediately after application. Photolysis, the photodecomposition of a chemical, may or may not be important. Hydrolysis (a chemical reaction in which the hydrogen and hydroxyl ions of the reactant are split) of esters and amines to their corresponding acid may occur very rapidly (less than an hour) in moist soil. Oxidation is a dissipation process for some pesticides. Complexation, or bonding, of pesticides is an important phenomenon that for all practical purposes becomes a loss pathway. Biodegradation takes place in the soil matrix as soil microbes use pesticides and other substances as a food source.

The dissipation rates for photolysis, hydrolysis, and oxidation are first order kinetics expressed as:

$$C_{d_{i,j}} = C_{o_{i,j}} \exp(-k_{i,j} d) \tag{3}$$

in which $C_{o_{i,j}}$ and $C_{d_{i,j}}$ are the initial and, d–days later, concentrations of pesticide, j, with dissipation process, i. $k_{i,j}$ is the kinetic dissipation rate for pesticide j, and dissipation process i.

Other dissipation processes, especially those deeper in the soil matrix appear to occur rapidly at first and then slow down. Frequently these rates are expressed as a two compartment model.

$$C_d = C_o \exp(-(k_n + k_r)d) + \\ C_o(k_r/(k_n + k_r - k_d))[\exp(-k_d d) - \exp(-(k_n + k_r)d)] \tag{4}$$

in which C_o is the initial concentration; C_d is the concentration d–days later; k_n, k_r, and k_d are the rapid, transition, and slow dissipation rates, respectively.

Each pesticide and its daughter products have individual features that would indicate what type expression would work best. For most pesticides not enough information is available to identify specific process rates and a general expression is used:

$$C_d = C_o \exp(-K_c d) \tag{5}$$

in which K_c is the sum of all dissipation process rates.

With respect to plant nutrients, space does not permit more than a cursory description of the processes of nutrient cycling. Plant residues and animal and microbial litter on the surface of the soil (or incorporated in surface layers of the soil) decompose as microorganisms use the organic carbon in the residue as a food source. This process mineralizes the nutrients stored in the litter or residue. Nitrogen compounds are mineralized into ammonium; labile (readily decomposed) phosphorus compounds are mineralized into phosphate. The ammonium is converted to nitrite and then into nitrate by microbial processes. Collectively the nitrogen processes are referred to as nitrification; they are usually rather rapid. Nitrate and phosphate are frequently added to the soil by fertilization. Also nitrate is added by

294

symbiotic and nonsymbiotic fixation and by rain water in which lightening has "fixed" the nitrogen. Symbiotic fixation is the production of nitrate by organisms (nodules) attached to the roots of legumes and other similar plant species. Nonsymbiotic fixation is the production of nitrate by organisms in the soil that are not necessarily associated directly with plant roots. Denitrification is a process that removes nitrate from the soil. As in most other processes it is accomplished by microbial action; however, denitrification takes place in anaerobic conditions as microbes, using organic carbon as a food source, take oxygen from nitrate. The process of denitrification releases N in the form of N_2 and NO gases. Another microbial process that removes nitrate from the soil is immobilization. It is the use of nitrate in microbial growth and reproduction, i.e. a conversion of nitrate from the mineral form to organic compounds in the bodies of the microbes. Phosphorus is immobilized in the same way, but phosphorus occurs in different degrees of availability depending upon its forms. The most readily available forms are referred to labile phosphorus. Phosphorus is also readily adsorbed on clay particles in the soil; thus there is an adsorption/desorption process that must be considered in phosphorus availability. Other processes that must be considered in nutrient cycling are plant uptake, leaching, and removal in surface runoff or transport in the adsorbed form on soil or organic particles. The nitrogen cycle is shown in Figure 1.

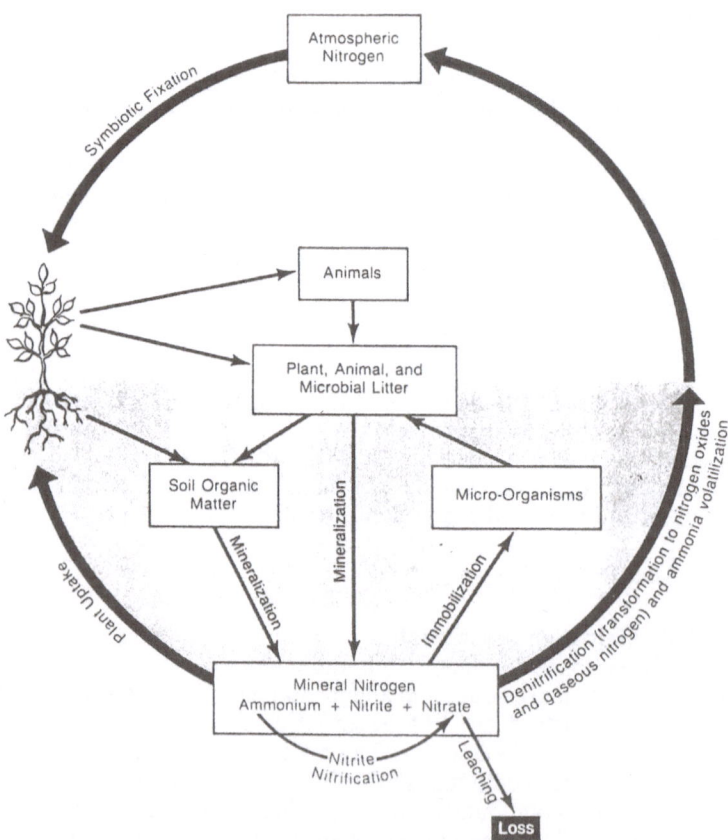

Figure 1. The nitrogen cycle (adapted from Reeder and Sabey, 1987).

2.1.3. Sorption

Sorption is the transfer of solutes from the solution phase to soil particle surfaces. This is accomplished by a variety of chemical mechanisms depending upon the structure of the chemical and the type of soil particle. Desorption is the reverse process of transfer of solutes from the soil to solution phase. Hysteresis may be exhibited in that the desorption rate may not equal the sorption rate. Both the concentrations on the soil and the rates of adsorption and desorption are functions of the surface chemistry. An equilibrium between the sorbed and solution phases is established after a period of time which can be expressed as a ratio of the concentration in solution, C, to the concentration adsorbed on soil particles, S. In many models, the equilibrium is assumed to be instantaneous and various isotherms are used to express the relation between concentrations. The simplest expression is a linear isotherm between S and C:

$$S = K_d C \tag{6}$$

in which K_d is the equilibrium adsorption coefficient. Nonlinear isotherms include the Freundlich equation:

$$S = K_d C^{1/m} \tag{7}$$

and the Langmuir equation:

$$S = \frac{K_d C}{1 + K_d C} \tag{8}$$

in which m is a number less than 1.0 and the other parameters are as previously defined. For additional discussion of adsorption isotherms see Bolt (1979). Since organic matter plays such a dominant role in the sorption process, and since both it and the equilibrium coefficients change with depth in the profile, Equation (6) is frequently expressed as:

$$S = K_{oc} O_c C \tag{9}$$

in which K_{oc} is the organic matter adsorption coefficient and O_c is the organic matter fraction in the soil. Equation (9) implies that all sorption is to the organic matter fraction of the soil, an assumption that is most true for nonionic, nonpolar chemicals. However, to the extent it holds true, it is a very useful predictor of the respective concentrations, S and C, requiring only a knowledge of soil organic matter content. Other phenomena such as hysteresis in the adsorption/desorption process are sometimes present. However, in general there is not enough information on the hysteresis phenomenon to use it in present day models.

In the last few years laboratory and field data have shown breakthrough curves (plots of pesticide concentration vs. time) indicating nonequilibrium adsorption. This feature has been described using chemical and physical nonequilibrium or kinetic models. Chemically controlled kinetic rate reactions are frequently expressed as linear first–order reversible processes, i.e. the adsorption rate is slow enough that equilibrium is achieved only after a very long time:

$$\frac{\partial S}{\partial t} = \alpha(KC - S) \tag{10}$$

in which K is the kinetic adsorption coefficient and α is a first order rate coefficient.

Even though Equation (10) has improved predictive capability, it did so when used in studies carried out at relatively low flow velocities.

Another chemical nonequilibrium model that has improved prediction is the two–site model in which sorption is assumed to consist of two components. The two components are distinguished by the rates at which material is sorbed, type–1 sites, in which concentration of sorbed material (S_1) reaches equilibrium rapidly and type–2 sites, in which concentration of sorbed material (S_2) is very slow in attaining equilibrium. The two sites are thought to be associated with different constituents (minerals, organic matter, etc.). Total sorption is the sum of that sorbed at type–1 and type–2 rates:

$$S_1 = F \, K_d \, C \tag{11}$$

and

$$\frac{\partial S_2}{\partial t} = \alpha\big((1 - F)KC - S_2\big) \tag{12}$$

in which F is the fraction of the sorption that reacts at type 1 rates (van Genuchten et al., 1990).

An alternative to the chemically controlled nonequilibrium model is one in which the reason for the appearance of nonequilibrium is lack of access to sorption sites. In natural soils some of the pores may be isolated or located in regions where pore water velocities may be very slow or zero. This leads to a concept of mobile/immobile water or a bimodal pore-water velocity where only a fraction of the pores (mesopores) may be subject to convective-dispersive transport and thus equilibrium type sorption conditions. In the remaining pores (micropores) the water is considered immobile and the exchange of solutes with the sorbed material is at a rate controlled by diffusion, thus solute exchange occurs much more slowly. This concept of mesopores and micropores should not be confused with that of the relation between macropores and the soil matrix; flow in the mesopores is still considered Darcian; whereas flow in macropores is much more rapid with little interaction between the flowing water and water in the soil matrix. Figure 2 shows pores in a medium–textured soil. It shows what could be considered to be micro, meso, and macropore structure. For additional information on the physical concept of nonequilibrium sorption see van Genuchten and Wierenga (1976) and van Genuchten et al. (1990). It can be shown that the chemical and physical concepts of nonequilibrium adsorption have identical mathematical structures. It is also possible that both concepts may be responsible for observed nonequilibrium sorption conditions.

2.1.4. Solution Techniques

Solutions of Equations (1) and (2) as presented, or in simplified forms, have been achieved analytically, quasi–analytically, and numerically. In order to exact an analytical solution, soil properties and the initial and boundary conditions must be simplified or approximated. Many analytical solutions exist (van Genuchten and Alves, 1982); some are of academic interest but others may be very helpful in verifying numerical solutions, aiding in parameter identification or equation derivation. However, analytical solutions used in the general solution of field problems is very limited because of nonrealistic simplifying assumptions concerning steady–state water flow or boundary conditions.

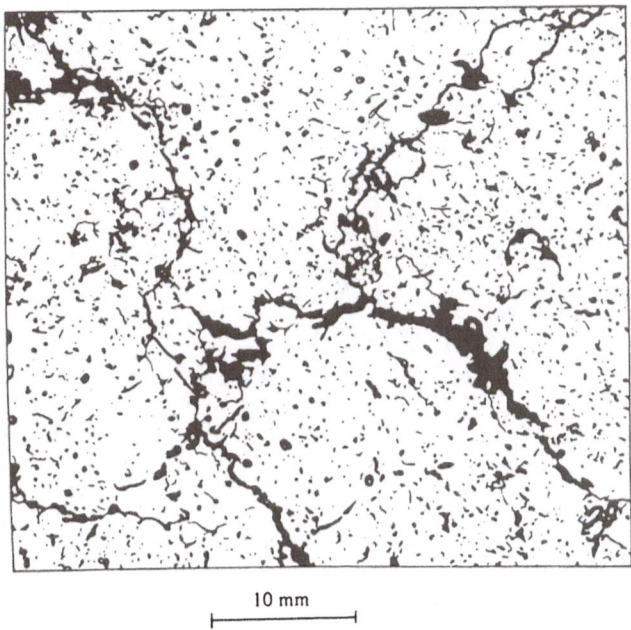

10 mm

Figure 2. Pores (in black) in a cross–section of a soil of medium texture at a depth of 10 cm. Much of the pore space is in pores too small to be represented (Lafeber, 1965, Australian Journal of Soil Research) (Marshall and Holmes, 1979, Soil Physics, Cambridge University Press).

Under normal circumstances, solute leaching occurs under unsteady flow conditions, thus forcing a numerical solution to the simultaneous flow equations. Many models (finite difference, finite element, and integrated finite difference techniques) are available to solve the equations. However, the highly nonlinear terms (see Figure 3 which shows $K(\theta)$ for example) can exact rather stringent time steps that can cause lengthy computer runs. Also, sensitivity analyses show that an accurate estimation of the water flux is necessary to insure good prediction of the solute distribution pattern. Because unsaturated flow theory is being discussed in other chapters, I do not intend to pursue it any further in this chapter, except as needed in discussion of transport processes.

2.1.5. Strengths of Mechanistic Models

Mechanistic models are generally of most interest to the research community because they include the most rigorous solutions to solute transport. As such they enable the user to study the interactions of water flux, moisture content, and soil properties on solute transport. They enable the researcher to uncouple the many interacting processes and study specific processes. They also enable the researcher to obtain solutions to the problem at any point within the domain of the problem area. Other solution techniques are more limited in their solution sites.

298

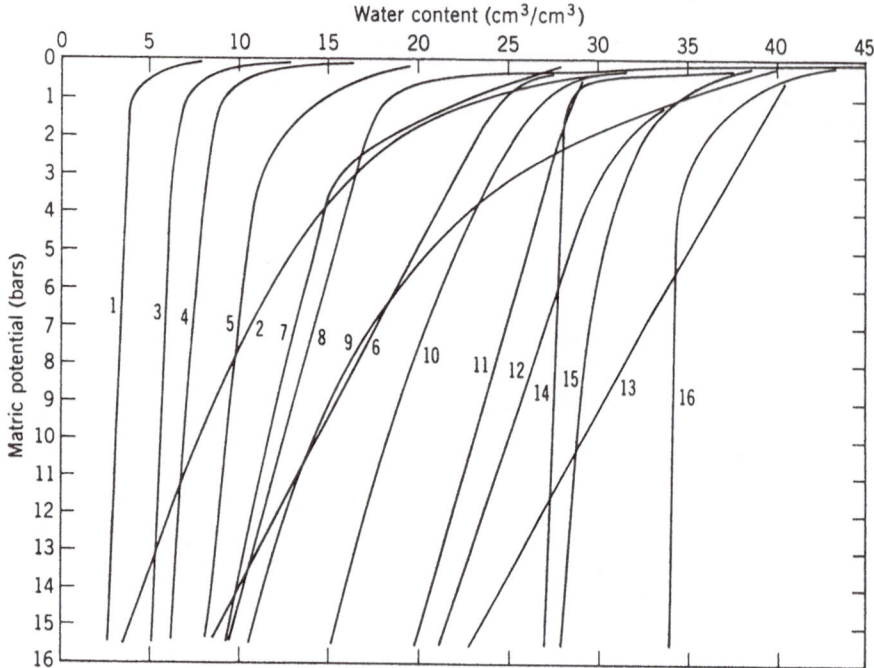

Figure 3. Desorption curves for various soils sketched from data at 0.1, 0.3, 0.6, 3, and 15 bars tension given by Holtan et al., 1968. (1) Continental gravelly sandy loam, Arizona; (2) Sassafras sandy loam, Maryland; (3) Progresso fine sandy loam, New Mexico; (4) Vaucluse sandy loam, Georgia; (5) Albion loam, Oklahoma; (6) Abilene clay loam, Texas; (7) Hartsells loam, Ohio; (8) Palouse silt loam, Washington; (9) Fayette silt loam, Wisconsin; (10) Nellis gravelly loam, New York; (11) Larb–like silty clay loam, South Dakota; (12) Memphis silt loam, Mississippi; (13) Drummer silty clay loam, Illinois; (14) Auston silty clay, Texas; (15) Marshall silty clay loam, Iowa; (16) Bascomlike clay, South Dakota (Baver et al., 1972, Soil Physics, fourth edition, John Wiley and Sons).

2.1.6. Weaknesses

Just as mechanistic models are advantageous to researchers, they are difficult to use in an applied or predictive sense. Values of $K(\theta)$ can vary by orders of magnitude in an area as small as a field. Moisture and tension gradients are also spatially variable to the extent that predicting solute movement at any given point in the field is tenuous at best and is very likely impossible. Mechanistic models are computer intensive, demand a lot of input, and are difficult to transfer between users (Wagenet et al., 1990).

2.2. Functional Models

Functional models simplify the equations of water and solute flow presented in Equations (1) and (2) and in Table 1. See Addiscott and Wagenet (1985) and Wagenet et al. (1990) for a current review of functional models. Functional models from the homogeneous and layered soil perspectives are discussed using the example of Vachaud et al. (1990).

2.2.1. The Soil as One Homogeneous Layer

In this approach, the diffusion and dispersion terms in the water and solute transport equations are ignored and:

$$\theta \frac{\partial C}{\partial t} = -J_w \frac{\partial C}{\partial z} \tag{13}$$

which moves water and solute as if it were piston flow. The location of α, the solute concentration peak, is given by:

$$a = Q/\theta_{fc} \tag{14}$$

in which Q is added water and θ_{fc} is the water content of the soil at field capacity. Other researchers have expanded on the concept, adding diffusion and dispersion effects but keeping the piston flow concept. See Rose et al. (1982). In a flushing type of environment such as in sandy soils or in the leaching of salts, these modifications seem to work satisfactorily. Simple models like this cannot handle heterogeneous soils, or profiles that have solute already in place, without modification that generally leads to layered soil models.

2.2.2. Layered Soil

Using the concept of mass conservation of water and solute, Equation (13) can be applied to a layered soil but it cannot be solved analytically. Thus it is expressed as a progressive solution in which the mass of solute in each layer is equal to the mass of solute added, less the mass drained out. For an entire soil profile this is expressed as (after Wagenet et al., 1990):

$$\sum_i z_i \left(\theta_{fc_i} C_{fc_i} - \theta_{o_i} C_{o_i} \right) = C_{iw} z_{iw} - C_{dw_i} \left[z_{iw} - \sum_i z_i \left(\theta_{fc_i} - \theta_{o_i} \right) \right] \tag{15}$$

in which z_i are the thicknesses of the soil layers; θ_{o_i} is the initial soil water content of layer i; C_{fc_i} is the concentration of solute at field capacity in layer i; and iw and dw are inflowing water and drainage water, respectively. Equation (15) and even simpler versions have been used satisfactorily in the leaching of salts; however, the expression does not consider precipitation/dissolution or ion exchange. Use of Equation (15) results in numerical dispersion which is a function of the number and thicknesses of the assigned layers. Thus the solution suffers from cumulative errors.

One modification of Equation (15) is of interest in the movement of nonconservative, or adsorbed chemicals. This form of Equation (15) assumes that all flow into each layer (water and solute) is uniformly mixed and achieves equilibrium before moving to the next layer. The method, even though it considers adsorption, suffers from the cumulative errors of the previous method. Both of the methods suffer from a lack of dispersion that occurs with movement through soils with a wide distribution of pore sizes and also from the inability to model soil water and solute movement under partially saturated conditions. However, compared to more complete models, models based on Equation (15) were found to be satisfactory for many situations. This is probably due to problems and inaccuracies associated with parameter values for highly variable field conditions, used in the more complete solutions described in the previous section of the chapter.

2.2.3. Strengths and Weaknesses of Functional Models

Further attempts to improve performance of the layered model come through the introduction of diffusion and dispersion coefficients between the sorbed and solution phases of any solute added to the soil. This and previous modifications of Equation (15) are empirical attempts to explain deviations of the model from observed data. Since the model upon which these "adjustments" were made is not physically based, the adjusted models do not provide insight into the basic processes, nor do they allow one to predict the effects of changing any of the soil features such as texture, pore size distribution, etc. The mechanistic models, previously described, that simulate soil water and solute flux must be used if insight into the basic processes is desired. Other weaknesses in use of functional models include use in situations where most of the solute moves in an unsaturated state rather than in response to the flushing type action of large events, and use in layered soils where the layers are physically quite different from one another. These situations are common and even though the model was not designed to be used in these ways, it will be. Thus, there are likely to be significant errors associated with its use under these conditions.

Some of the real strengths in use of functional models come from the fact that soil–water characteristics curves, $K(\theta)$ relationships, and hydraulic conductivity, K_s, are not needed. Representative values of $K(\theta)$ and K_s for given field conditions are very difficult and expensive to obtain as previously discussed. Other advantages of functional models include simple computer codes that are easy to transfer between users, less input data, and the ability to incorporate processes such as snowmelt and precipitation/runoff in much simpler forms than in mechanistic models. In support of the desirability for simple structure, it is interesting to note that De Willigen and Neeteson (1985) in comparing six simulation models of nitrate movement found that the main difficulties were in developing an adequate description of microbiological transformations of nitrogen.

2.3. Stochastic Models

Spatial variability, in some cases over several orders of magnitude, in values of both the physical and chemical properties of soils, even within a field, have created massive sampling problems and estimation errors using mechanistic models. Moreover, the erratic nature of chemical and physical processes in many cases precludes our ability to estimate their values and predict field level response. Thus, many researchers feel that stochastic models offer more potential for field application than either mechanistic or functional models. Vachaud et al. (1990) divides stochastic models into two groups for discussion: (1) approaches in which allowance is made for spatial variability of hydrodynamic soil properties in existing mechanistic models, and (2) approaches focused on the variability of solute and water movement through use of transfer functions, taking no account of the mechanisms.

2.3.1. Mechanistic Models with Stochastic Hydrodynamic Properties

In stochastic hydrodynamic models randomly-distributed input values are used in a mechanistic submodel to produce a distribution of responses representative of a field. The field is assumed to be a collection of uniform homogeneous vertical soil columns with different hydraulic properties. Water and solute movements are described by Equations IIIb and V in Table 1. The unsaturated hydraulic conductivity is viewed as a random stochastic process which is usually considered to have a lognormal probability density function. This infers that capillary diffusivity, the dispersion coefficient, and water flux are also stochastic processes. The best example of a stochastic model of solute movement is one developed over several years by Bresler and Dagan (1979, 1981, 1983a,b) and Dagan and Bresler (1983). Their model will be described in an attempt to present the concepts embodied in stochastic models.

The soil is considered horizontal in the x, y plane with z positive downwards. Water flow takes place as a result of recharge, R, at the soil surface at moisture content, θ, and at a pore water velocity V. R is assumed vertical and flow is steady. Even though the surface is assumed spatially variable, the soil is assumed homogeneous in the vertical direction. Solutes are assumed conservative, i.e. they are not adsorbed, thus Equation VII in Table 1 describes the diffusion/convection equation of solute transport. The concentration of solute, C, is assumed dimensionless and normalized to 1.0. With boundary conditions of: $C = 0$ for $z > 0$ and $t = 0$; and $C = 1$ for $z = 0$ and $t > 0$. Solution of Equation VII in Table 1 with these boundary conditions yields:

$$C(z, t) = 1/2 \left[1 - erf\left(\frac{z - Vt}{z(D_s t)^{\frac{1}{2}}} \right) \right]$$

(16)

in which the apparent dispersion coefficient, D_s, is the sum of molecular diffusion, D_p, and hydrodynamic dispersion, D_h. D_h is defined as:

$$D_h = \lambda V$$

(17)

in which λ is the dispersivity.

In their mechanistic stochastic model (MSM) Bresler and Dagan assume that both V and D vary with x and y in a random way such that each realization (area of a field) is taken from an ensemble of points with the same probability density function (pdf). Their objective is to evaluate C as a function of z and t, $f(z,t; C)$, in the cumulative probability form, $P(z,t; C)$:

$$P(z, t; C) = \int f(z, t; C) dC$$

(18)

P is interpreted as the percent of a field whose concentration is less than C at given z and t. Some of the moments of $P(z,t; C)$ are of special interest: the first moment is the average concentration at depth z and time t; the second moment is the horizontal variation of concentration at z and t; and the third moment is a measure of the symmetry of concentration given z and t. Bresler and Dagan also integrate Equation (16) over a depth span, $z_2 - z_1$, obtaining a closed form solution for the average concentration.

Since the randomness in C is created by the randomness in V and D, solution of Equation (18) depends on the pdf's of V and D. Space does not permit the derivation of these pdf's and the computational scheme other than to say that functional expressions for V and D in terms of λ, K_s, $K(\theta)$ and θ are substituted in Equation (16) which is then substituted in Equation (18) to get the desired cumulative distribution. Spatial variability in λ and K_s are responsible for the variability in V and D. Both λ and $(K_s)^{1/2}$ are assumed to be lognormally distributed. Recharge is also assumed to be a random variable from a uniform (rectangular) pdf in x and y.

Bresler and Dagan illustrate the use of Equation (18) by applying it to two different soils whose properties differ markedly in variability and average hydraulic conductivity. Figure 4 shows the value of $P(z, t; C)$ for the two soils. Figure 4a shows that 50 percent of Panoche soil has a concentration of less than 0.02, 0.3 and 0.6 at depths of 12.5, 6.2, and 2.3 cm respectively, 24 hours after recharge. The Bet Dagan soil (4b) is much more permeable and 50 percent of the field has concentrations of less than 0.01, 0.05, 0.22, 0.6, and 0.75 at depths of 43, 39, 28, 17, and 8 cm respectively after only 5 hours. Figure 5 shows the fraction of the area of Bet Dagan soil whose depth averaged concentration is less than that shown for various thicknesses of soil. For example, 50 percent of the area will have a depth averaged concentration in the top 28 cm less than about 0.65. Figure 6 is another interesting evaluation in that it shows the mean concentration, the variance of concentration, and skewness of concentration as functions of depth in the profile at the end of 24 hrs for the Panoche soil and at the end of 5 hours on the Bet Dagan soil. The comparison shows how different the two soils are.

302

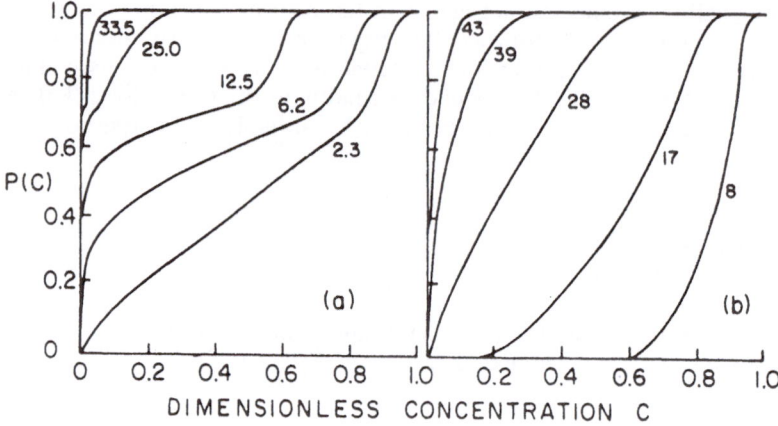

Figure 4. The probability function $P(z, t; C)$ presented as a function of C for various fixed z and t. (a) Panoche soil $\sigma_Y = 1.16, \lambda = 3$ cm, $D_p = 0.02$ cm^2/hr, $r = 1, \sigma_\mu = 0$, $S_R = 0, t = 24$ hours. (b) Bet Dagan soil $\sigma_Y = 0.4, \lambda = 3$ cm, $D_p = 0.02, r = 0.1, \sigma_\mu = 0, S_R = 0, t = 5$ hours. The numbers labeling the curves are values of z in centimeters (Bresler and Dagan, Water Resources Research Vol. 17(6), pg. 1687, 1981 Copyright by the American Geophysical Union).

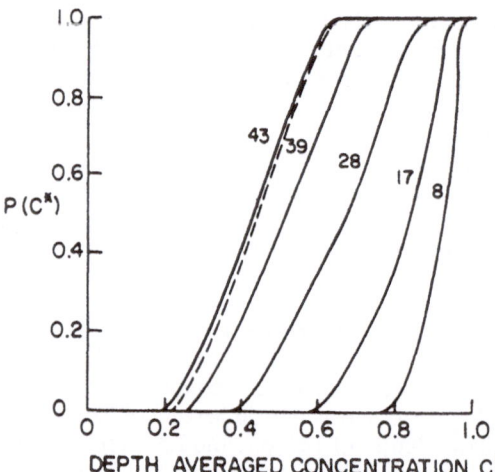

Figure 5. The probability function $P(z_2, t; C^*)$ presented as a function of C^* for fixed z_2 and t in Bet Dagan soil $\sigma_Y = 0.4$, $\lambda = 3$ cm, $D_p = 0.02, r = 0.1, \sigma_\mu = S_R = 0, t = 5$ hours. The numbers labeling the curves are z_2 in centimeters. The dashed line curve is for $\lambda = 0$ (Bresler and Dagan, Water Resources Research Vol. 17(6), pg. 1687, 1981 Copyright by the American Geophysical Union).

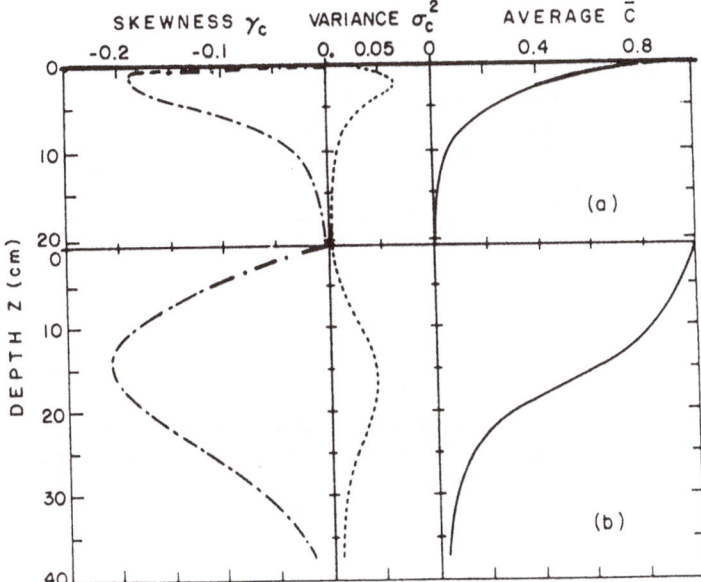

Figure 6. The first ($\overline{C}$), second (σ^2), and third (Y) moments as a function of z for fixed t. (a) Panoche soil σ_y = 1.15, λ = 3 cm, σ_μ = 0, D = 0.02 cm^2/hr, r = 0.2, S_R = 0, t = 24 hours. (b) Bet Dagan soil σ_y = 0.04, λ = 3 cm, σ_μ = 0, D_p = 0.02, r = 0.1, S_R = 0, t = 5 hours (Bresler and Dagan, <u>Water Resources Research</u> Vol. 17(6), pg. 1687, 1981 Copyright by the American Geophysical Union).

Bresler and Dagan go on to evaluate the sensitivity of Equation (18) to internal parameters, to look at average concentration profiles and average values for various layers. They conclude their remarks with the following statement:

The first major conclusion was that even the average concentration profile in the hetero-geneous field, $\overline{C}(z, t)$, cannot be generally modeled as the solution of a convection–diffusion equation with constant coefficients. Thus, the traditional approach to solute transport is not warranted even for this gross characterization of the transport phenomenon, and different tools have to be employed for large fields.

The second conclusion is that if pore scale dispersivity is taken at its usual laboratory-measured value, its influence upon solute distribution is negligible as compared with that of soil variability. If, however, the large field value (which reflects mainly heterogeneity of the profile) is adopted, it has a definite impact upon the solute distribution. Although hetero-geneity combined with convection is the dominant spreading mechanism in Panoche soil, while for Bet Dagan soil it is less pronounced at depths of interest in agricultural applica-tions, accounting for both mechanisms seems to be warranted in most cases.

2.3.2. Stochastic Models that Use Transfer Function Concepts

The use of a transfer function concept applied to field–scale soil water and solute movement was first suggested by Raats (1978). This basic concept has been used in hydrologic studies for many years and is known as the unit response function or unit hydrograph in hydrology. It

304

has also been used to account for lag and attenuation in coefficient type stream channel routing. Jury (1982) implemented the concept by deriving or suggesting appropriate transfer functions for dispersion, spatially variable input rates, and concentration distributions at different depths in the profile of a field with spatially variable hydrodynamic characteristics. The transfer functions integrate convection, dispersion, sorption, radioactive decay, biological transformation, and any other processes that lead to a change in concentrations with depth and time. Jury et al. (1986) visualizes the entrance and exit surfaces of the soil profile as shown in Figure 7. The concept, expressed in equation form for the probability of a tracer, $P_L(I)$, applied at the surface reaching a depth, L, after a net amount of water, I, has been applied, is given by:

$$P_L(I) = \int f_L(I')\, dI' \tag{19}$$

in which $f_L(I')$ is the probability density function, i.e. the probability that the applied tracer will arrive between I and $I + dI$. If we assume for the time being that there is no adsorption or dispersion such that in any one flow path the concentration at the exit of the path is the same as at the entrance, then the average concentration at the exit depth L, $C_L(I)$, is given by:

$$C_L(I) = \int_0^\infty C_{in}(I - I')\, f_L(I')\, dI' \tag{20}$$

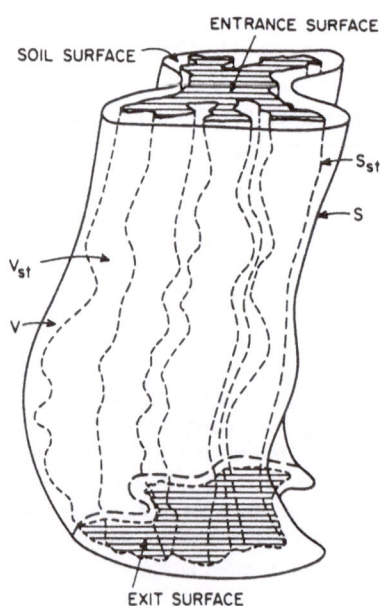

Figure 7. Schematic illustrations of a field soil unit of volume, V, bounded by a surface, S, and enclosing a transport volume, V_{st}, and its bounding surface, S_{st} (Jury et al., Water Resources Research, Vol. 22(2), pg. 244, 1986. Copyright by the American Geophysical Union).

in which $C_{in}(I - I')$ is the concentration of solution discharging at I'. See Jury (1982) for a more complete explanation of the derivation of the transfer function model (TFM). The soil described by Equation (20) can be visualized as an ensemble of twisted capillary paths of different lengths within which fluid flows by piston flow or by a model of soil which assumes piston flow within flow paths of variable water content or capillary diameters. Assuming that the distribution of flow path lengths is lognormally distributed, $f_L(I')$ is replaced by the lognormal distribution

$$f_L(I') = \exp\left[\frac{-(\ln I - \mu)^2/\sigma^2}{(2\pi)^{1/2} \sigma I}\right] \tag{21}$$

in which μ is the mean of the distribution of $\ln I$ and σ^2 is the corresponding variance. Equation (20) is modified to predict the concentration at any depth and Equation (21) substituted prior to solution.

Examples of the output of the model are shown in Figures 8 and 9. Figure 8 simulates the average solute concentration or breakthrough curves at 100 cm depth for input solute pulses of three different lengths and varying recharge volumes. Figure 9 illustrates extensive solute spreading over depth which results from the simulation. For example after a 10–cm pulse has been moved into the soil by 50 cm of water, it is dispersed over a depth of 300 cm and its average maximum concentration is about 15 percent of its initial value at the surface. White et al. (1986) used Jury's transfer function model to analyze field data on bromide and chloride

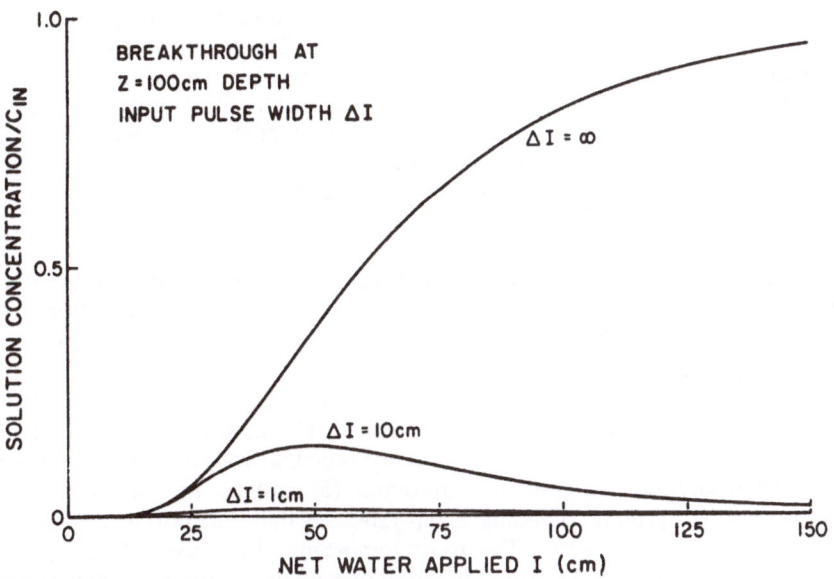

Figure 8. Simulated average solute concentrations at Z = 100 cm as a function of net applied water for a rectangular solute pulse of width ΔI added to soil surface (Jury, <u>Water Resources Research</u>, Vol. 18(2), pg. 366, 1982. Copyright by the American Geophysical Union).

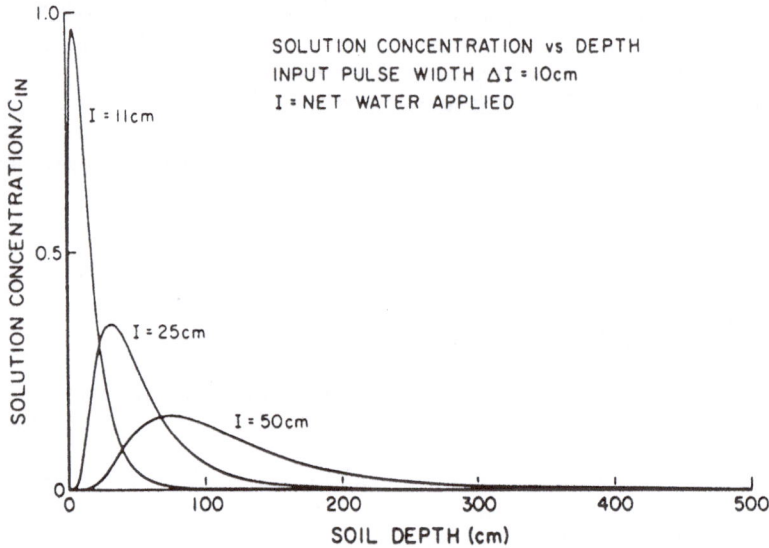

Figure 9. Simulated average solute concentrations versus depth for an inlet pulse width $\Delta I = 10$ cm for several values of net water applied (Jury, Water Resources Research, Vol. 18(2), pg. 366, 1982. Copyright by the American Geophysical Union).

movement. The travel time pdf for unsteady flow in an unsaturated–well–structured clay suggested a two component mechanism comprising fast solute transport through large pores and slow to negligible transport through the remaining pores. The volume of water responsible for the transport varied from 14 to 90 percent of that added. In a companion paper to the Jury et al. (1986) and White et al. (1986) papers, Sposito et al. (1986) developed transfer functions representative of both the rapid and slow features of flow and showed that the effects of convection, dispersion, and linear adsorption processes could also be simulated.

It would appear that the transfer function approach could possibly be developed to the point of being a good management model because it does not suffer from the weaknesses of the mechanistic and functional models. However, the biggest disadvantage of the transfer function approach is its site- and condition–specific nature, manifested in the requirement for calibration and the dependence that this calibration has on boundary conditions. The cost of data sets required for such calibration over the range of soils and boundary conditions needed could preclude its use.

Dagan et al. (1990) in a discussion of spatial variability compare the response of TFM and MSM. Similarities in response of the two systems include: (1) showing that the law of mass balance, interpreted in the context of probability theory, is demonstrated by both; (2) showing that effective dispersivity increases linearly with depth; (3) emphasizing the effect of pore water velocity variations on solute spreading; and (4) showing that transport is dominated by convection in the vertical direction. The most important difference between the two approaches is that travel time in TFM is a function of only net amount of water applied, whereas in MSM, travel time is a function of both the spatial hydraulic properties and the soil surface conditions.

2.4. Examples of Leaching Models

2.4.1. General Status

Many leaching models have been developed over the years. It is difficult to find a suitable way of categorizing these models because they vary so much in structure, type of application, solute included, size of area to which they apply, etc. Earlier in this chapter, papers by Wagenet, Vachaud, Jones and De Willigen and their coauthors are referenced. All of these papers have excellent summaries of these models. Table 2 is a listing of the models they review. Space does not permit an in-depth review of these papers, but their concluding remarks are very revealing. De Willigen et al. (1990) state,

> "Few of the models are at present suitable for solving problems related to management and regulation.... One of the reasons for the limited practical use of such models is the lack of thorough validation of the models.... Moreover, in many cases too many site-specific parameters are used that are not readily available or very difficult to obtain. In order for models to be used as management tools, a prerequisite is that proper documentation is available and that the model code is easy to understand."

The general consensus is that most models described do not satisfy these user needs. With this introduction, two different models that De Willigen et al. consider to be well documented, tested, validated, and suited to use in management, regulation, and planning and one general research model that includes some of the most recent concepts are discussed. Additional research level models are not described because the previous discussion of mechanistic, functional, and stochastic models describes and identifies several other good examples of research level models. Even though the models discussed below are primarily for the planner, they are equally valuable as aids to theoretical and field research. Selection of the following models for discussion does not imply that other models are not equally good. The first model is designed for nitrate leaching and nitrogen transformations, the second model deals with pesticides and the last model is a very general research level model that includes all major processes.

2.4.2. A Nitrate Leaching Model

The nitrate leaching and nitrogen transformation process model presented herein was described by De Willigen et al. (1990). The following information is abstracted from their paper. The model SOILN (Johnson et al., 1987) has both mechanistic and empirical submodels. The model was applied to several sites and data sets: sites where information on processes was extensive, sites where data sets describing relatively long-time series and several crops were available, and sites with long-term trends in nitrate leaching in a whole watershed where information was very sparse. Thus it would appear that SOILN can be used for practical problems related to planning, management, and regulation.

The model consists of two parts which are used in sequence. First, soil moisture and temperature conditions are simulated in a water and heat submodel is used to produce driving variables for a nitrogen submodel. Common to both submodels is a vertical structure that facilitates division of the soil into layers, depending upon the accuracy requested and available information on basic physical and biological characteristics.

Water and Heat Transport. The water and heat submodel (Jansson and Halldin, 1980) is based on two coupled differential equations describing heat and water transport (derived from Fourier's and Darcy's laws, respectively) in a one-dimensional soil profile. Snow dynamics, frost, evapotranspiration, precipitation, groundwater flow, water uptake by the

Table 2. Some currently available leaching models.

Model Name (if any)	Author Reference	Use
SLIM	Addiscott, 1982	Nitrate movement in the field
SWAM	Alonso and DeCoursey, 1985	Small watershed scale–fertilizer, pesticide, sediment & general hydrology
WATCROS	Aslyng and Hansen, 1982	Water balance and crop production
ANIMO	Berghuijs et al., 1985	Nitrogen model
PULSE	Bergstrom et al., 1987	Nitrate leaching
	Boesten, 1986	Pesticide movement
	Burns, 1985	Nitrate leaching
PRZM	Carsel et al., 1984	Solute leaching
TRANS CSM	Chardon, 1984	Cadmium mobility
	Clothier et al., 1988	Ammonium nitrate movement
	Dautrebande and Agneesens, 1985	Pesticide transport
TRANS	DeSmedt et al., 1981	Solute movement (mobile/immobile water)
	De Willigen et al., 1982	Phosphate transport
PESTAN	Enfield et al., 1982	Pesticide transport
	Ferrari and Cuperus, 1973	Non–absorbed ion transport
MORELN	Geng et al., 1987	Nitrate transport
NITCROS	Hansen and Aslyng, 1984	Nitrogen balance and crop products
RZWQM	Hebson and DeCoursey, 1987 DeCoursey and Rojas, 1990	Mgmt. model – fertilizer, pesticide, solute and transport in general
WHNSIM	Huwe and Van der Ploeg, 1987	Plant uptake
SOILN	Johnsson et al., 1987	Nitrogen dynamics
BAM	Jury et al., 1983	Transport of pesticides and other organics
CREAMS	Knisel, 1980	Water quality (general)
RENLEM	Kragt and DeVries, 1987	Solute transport–plant uptake waterflow
GLEAMS	Leonard et al., 1986	Water quality (general)
CALF	Nicholls et al., 1982	Pesticide degradation and movement
CMLS	Nofziger and Hornsby, 1985	A management tool for chemical movement
	Richter et al., 1988	Nitrate processes (general)
SIMUL	Rieu, 1984	Solute transport and precipitation
COFARM	Shaffer et al., 1984	Nitrogen and tillage mgmt at farm level
NTRM	Shaffer and Larson, 1987	Nitrogen tillage, residue mgmt.
OPUS	Ferreira and Smith, 1990	General hydrology – field scale
MOUSE	Steenhuis et al., 1987	Pesticide transport
	Van der Zee and Van Riemsdijk, 1986	Phosphate transport
ONZAT	van Drecht, 1985	Field/Lab treatment of nonlinearly sorbed solutes
	Vinten et al., 1988	Crop response to saline water
LEACHMP	Wagenet and Hutson, 1987	Leaching of nitrogen fertilizers, organic salts, and pesticides
	Zandt and De Willigen, 1981	Nitrate movement

plant, and drainage are included. The model predicts daily values for soil temperature and water content at any level in the soil profile using standard meteorological data as driving variables. The model also simulates transport in the saturated zone above tile drainage systems and the saturated water flow directed towards a stream or ditch. Surface runoff can occur due to limited infiltration capacity or limited permeability of the soil.

Nitrogen Transformations and Plant N–Uptake. Biological and chemical transformations in the model include mineralization, immobilization, nitrification, denitrification, plant uptake, and organic matter decomposition. Biological N–transformations in the model (Figure 10) apply to each soil layer. Mineral–N pools include ammonium and nitrate. Organic–N is distributed among litter, faeces, and humus. The carbon pools for litter and faeces are included to control nitrogen mineralization and immobilization rates. The abiotic response functions regulating decomposition, mineralization, and nitrification are functions of soil temperature and soil water content. Denitrification is calculated as a function of temperature, water content, and nitrate concentration. Plant uptake of nitrogen is controlled by an assumed root distribution and a growth potential logistic curve. Agricultural management characteristics and different crop characteristics can be used to estimate most model parameters controlling inputs and outputs of nitrogen.

Transport of Nitrogen. Only nitrate is considered to be mobile; flows are calculated as the product of water flow and nitrate concentration in the soil layer from which they come. Diffusion/dispersion is not explicitly accounted for, but due to the division of the soil profile into layers of finite thickness numerical dispersion will occur.

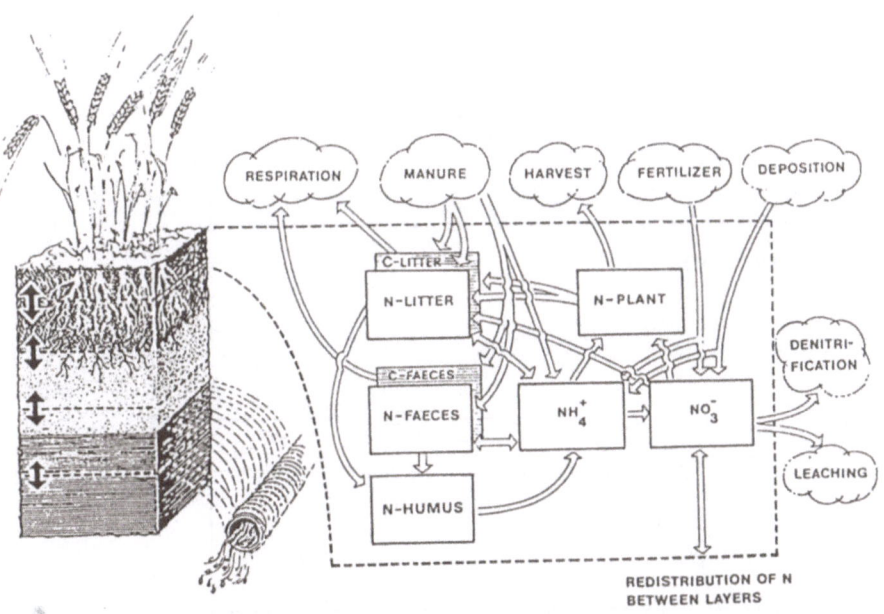

Figure 10. Structure of the nitrogen model (SOILN). Parts within the dotted line represent the uppermost soil layer. Subsurface layers have the same structure but do not receive any direct inputs from fertilizer and deposition (from Johnsson et al., 1987) (De Willigen et al., 1990).

310

Results of Simulation. To demonstrate the applicability of the model to different field data sets, two extreme applications concerning available information for validation of the various parts of the model are described. First, the model was applied to a field in central Sweden about which extremely detailed information was available. Data were collected on decomposition, denitrification, leaching, mineralization, mineral–N profiles, primary production, and soil abiotic conditions. Simulated and measured nitrate leaching and mineral–N profiles were compared in two treatments with barley (with and without N-fertilization) and a fertilized grass ley. The simulated periods were 3 and 4 years, respectively, and each plot size was 0.36 ha.

In the simulation, nitrate leaching depends on model output both of nitrate concentration and of drain–water flow. Time series of cumulative leaching calculated from measured nitrate concentrations and simulated drainage (i.e., "partly simulated") were compared to time series, where both nitrate concentrations and drainage flow are simulated, and to the observed time series of cumulative leaching (Figure 11). In this way deviations in leaching pattern caused by either poor predictions of drainage or of nitrate levels could be isolated. Figure 11 shows leaching from a barley field. Cumulative nitrate leaching was similar in all time series. The greatest differences between measurements and simulations occurred during the initial year (1981) and were largely due to underestimated drain–water flows in the simulations. For additional discussion see De Willigen et al. (1990).

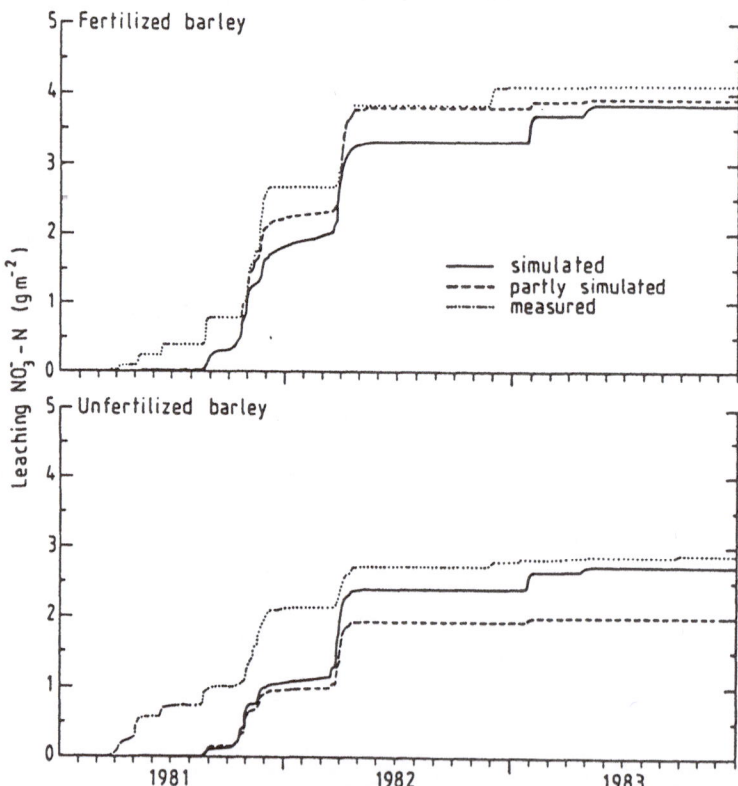

Figure 11. Simulated, "partly" simulated, and measured cumulative nitrate leaching from the barley crops at Kjettslinge (from Johnsson et al., 1987) (De Willigen et al., 1990).

In the second simulation, the model was tested on a whole watershed covering an area of 202 km^2 in Southern Sweden. Simulated values of nitrate leaching were compared with measurements for a 20–year period. More than 2/3 of the total flow of nitrogen in the river draining the basin originated from agricultural nonpoint sources. In the simulation a grain field was chosen to represent arable land in the area. Most of the parameter values in this application had to be estimated from the literature or from agricultural statistics for the region.

Data from the area showed that over the 20–year period N–fertilization rates in this area have increased from 90 kg Nha^{-1}yr^{-1} during the first 10 years to 120 kg Nha^{-1}yr^{-1} during the second 10–year period. The higher amounts of N–fertilizer used during the second 10–year period caused a distinct increase in simulated nitrate leaching, mainly when turnover of organic matter was assumed to be rapid (Figure 12). Observations showed a pattern similar

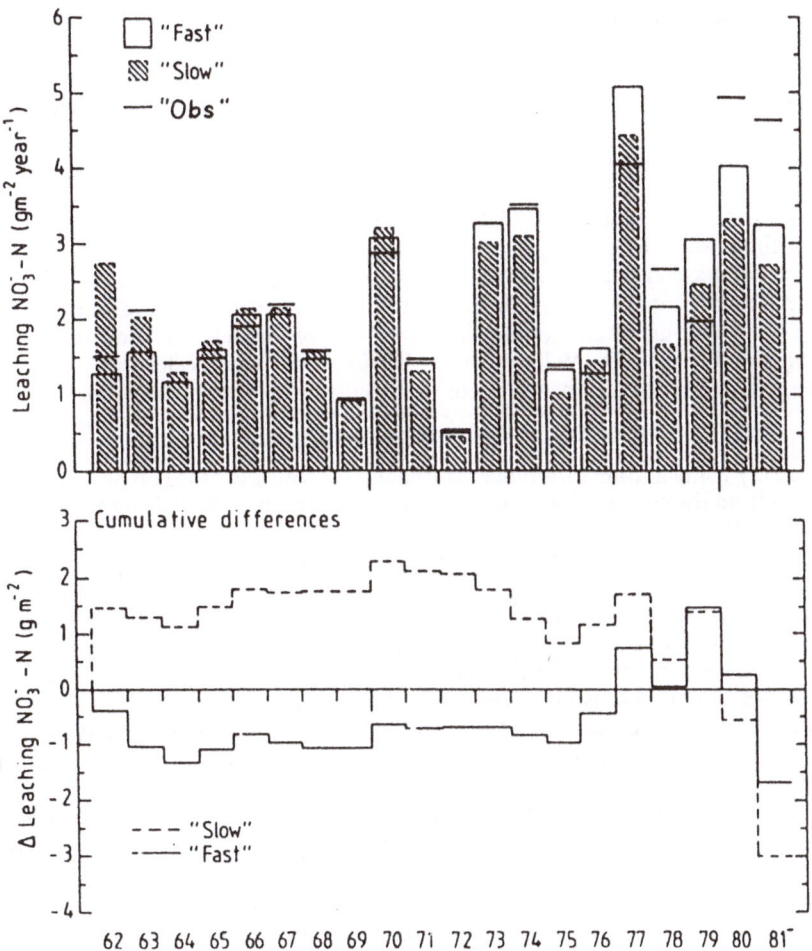

Figure 12. Annual amounts of simulated leaching with two different parameter sets and cumulative differences between simulated and measured leaching; observed values shown as a line (after De Willigen et al., 1990).

to that simulated by the fast turnover approach; while the simulation assuming a slow turn-over of organic matter led to an obvious underestimation of nitrate leaching, especially during the final years. See De Willigen et al. (1990) for more information.

Although there was a general lack of detailed information from the watershed, the results showed that both organic–N fractions in the soil contributed to the simulated change in leaching associated with increased N-fertilization. Discrepancies between measured and simulated response were mainly accounted for by uncertainties in hydrologic flow rather than uncertainties in biological processes. The discrepancies were attributed to macropore flow.

2.4.3. Leaching of Pesticides

The pesticides model described herein, Boesten (1986), is similar in many respects to several others: LEACHM (Leaching Estimation and Chemistry Model) by Wagenet and Hutson (1987); Opus by Ferreira and Smith (1990), RZWQM (Root Zone Water Quality Model) by DeCoursey and Rojas (1990), GLEAMS (Groundwater Loading Effects of Agricultural Management Systems) by Leonard et al. (1986) and PRZM (Pesticide Root Zone Model) by Carsel et al. (1984). One of the major differences between these models is the handling of water movement. The model described here by Boesten (1986), GLEAMS, and PRZM are all functional type models that approximate downward movement by assuming that water cascades from one layer to the next as water in the overlying layer exceeds field capacity; upward movement of water is not simulated. LEACHM, Opus, and RZWQM use the Richards' equation to redistribute water following infiltration and daily between storm events. Smith and Ferreira (1989) compare cascading and Richards' equation type models. Other differences between these models are described in Jones et al. (1990) and Wagenet et al. (1990).

Boesten (1986) presented an extensive and detailed study on models for simulating the behavior of herbicides in soil. De Willigen et al. (1990) report the model that Boesten developed has the possibilities of serving the needs of management. The model simulates the transformation and transport of herbicides, considering both instantaneous and time-dependent nonlinear sorption of the herbicide. Boesten solves the convective dispersion equation, Equation (2), using a finite difference approach. He considers the soil profile to be 40 cm deep, divided from the surface down, into 10 layers at 0.005 m each, 15 layers at 0.01 m each, and 10 layers at 0.02 m each.

Sorption of pesticides is accomplished by considering three classes of sorption sites. Sorption on Class 1 sites is considered so fast as to be considered instantaneous equilibrium. Sorption to, or desorption from, class 2 and 3 sites is slower with rate constants for Class 3 sites being about an order of magnitude slower than Class 2 sites. Equilibrium for class 1 and 2 sites is described by a Freundlich isotherm and that for the Class 3 sites is a linear isotherm. Decay, by all processes combined, is described as first order kinetics.

As mentioned previously, water movement when infiltration exceeds field capacity is by a functional model that allows water to move freely from one layer to the next as field capacity of each layer is exceeded.

Evaporation is calculated as a function of Penman evaporation and the difference between cumulative potential evaporation and rainfall. Evaporative water is taken from the soil profile as a function of the rooting pattern of the crop.

Transport of pesticides is by convection with the water, and by diffusion and dispersion as mentioned above. The diffusion coefficient is estimated from field experiments using bromide as a tracer.

De Willigen uses Figure 13 to demonstrate that the water movement part of the model performs satisfactorily. This model does not deal with upward movement of water because of the functional approach taken. However, the next model described is representative of the very few models that do handle upward movement as a function of potentiometric gradient.

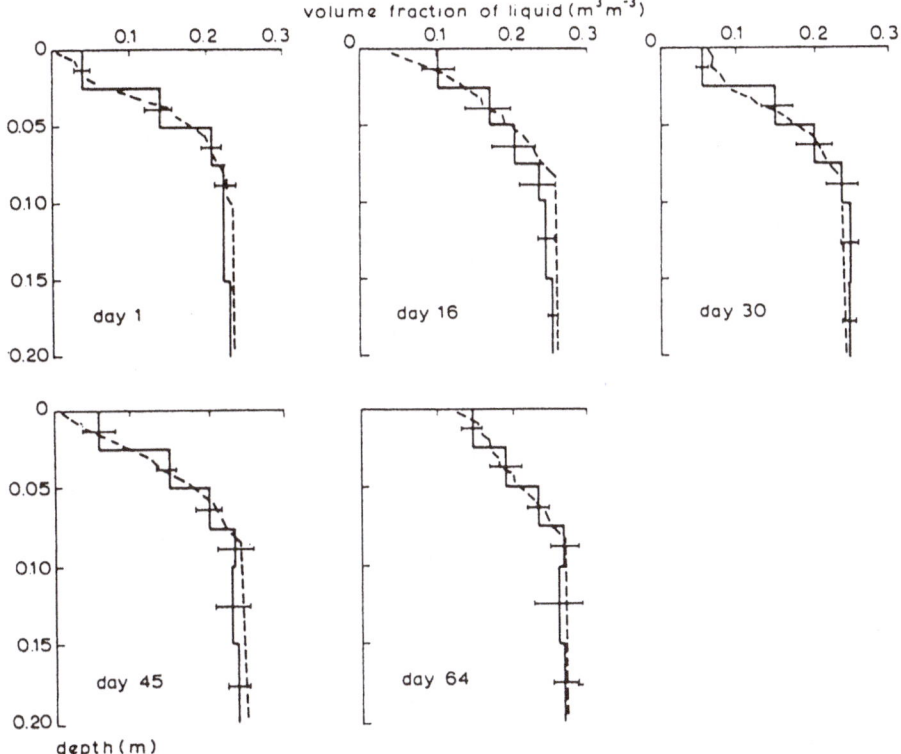

Figure 13. Profiles of water content on sampling dates in a field experiment (1981). Vertical solid line segments are averages of 5 measured profiles; horizontal bars are standard deviations; and dashed lines are simulated profiles (De Willigen et al., 1990).

Figure 13 shows moisture content profiles at five different dates. The simulated data falls well within the range of observation in all cases. Figure 14 shows how the model simulates the movement of the herbicide metribuzin. Simulated movement compares well with the observed data.

2.4.4. *Leaching Using a Research Level Model, Opus*

Opus (Ferreira and Smith, 1990) is a computer simulation model for runoff and pollutant transport in and from an elementary agricultural catchment area. Its purpose is to predict the effects of agricultural management practices on nonpoint-source pollution from field size areas. It is primarily a mechanistic model with water and solute movement defined by Equations (1) and (2). The model simulates the processes of sediment transport, nutrient cycling (nitrogen, phosphorus, and carbon), chemical transport, soil heat flow, crop growth, and residue decay. These processes are directly affected by choices made from a wide variety of agricultural practices that the model is capable of simulating. Management practices include crop rotations, all types of tillage and harvest operations, irrigation, fertilization, animal grazing, pesticide application, tile drainage, and hydrologic effects of local impoundments.

314

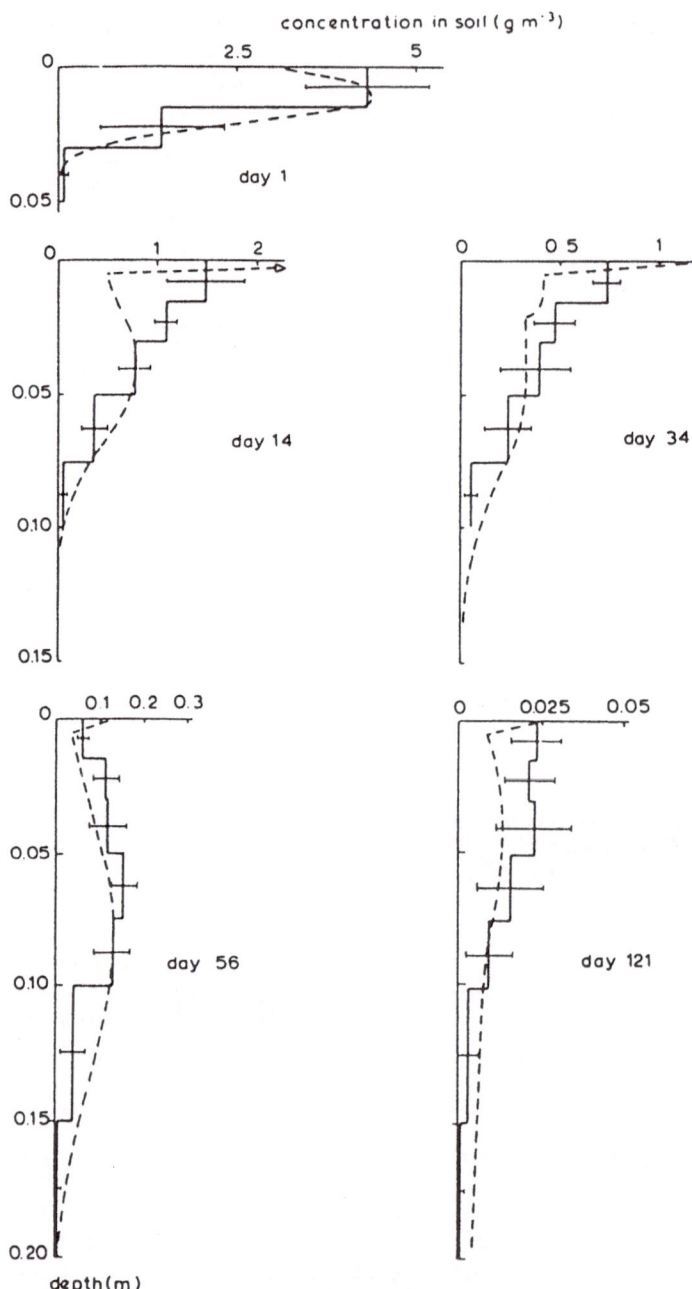

Figure 14. Comparison of calculated and measured profiles of the herbicide metribuzin. Vertical solid line segments are averages of measured concentrations; horizontal bars are standard deviations; dashed lines are calculated (De Willigen et al., 1990).

The model is limited to areas that can be characterized by a field with a single soil profile, whose hydrologic description does not require channel networks of order greater than two, and a single cropping system. The user has a choice of two levels of complexity or sophistication in the hydrologic component. One component uses or generates daily rainfall, the other component uses breakpoint rainfall data. Either daily net radiation (generated from monthly input statistics) or daily pan evaporation are used to estimate daily potential evapotranspiration.

Surface runoff volume from daily rainfall data is determined using the USDA–Soil Conservation Service runoff curve number method modified to give a continuous range of curve numbers that are functions of soil moisture content in the surface layers of the soil. Runoff from breakpoint rainfall is determined from infiltration rates based on solution of the Richards' equation as developed by Smith and Parlange (1978). A scheme based on surface soil moisture content allows for infiltration capacity recovery during interstorm hiatus. Redistribution of soil moisture during intrastorm hiatus is based on use of the Richards' equation. Flow through soil crusts that may develop as a function of rainfall energy is based on the hydraulic conductivity of the crust and underlying soil and the matric potential at the interface. The soil profile is assumed to consist of up to ten soil horizons that differ from one another in texture (and thus hydraulic characteristics) or in nutrient or organic matter contents. These horizons are divided up into as many as 20 computational layers by Opus.

Water movement across the soil surface is described using the kinematic or diffusive wave solutions of the Saint–Venant equations. The kinematic solution uses the Manning roughness relation as a rating between discharge and depth in the kinematic solution and the Newton–Raphson method of solving sets of equations for the diffusive wave solution. Leaching of plant nutrients and other soluble materials by rainfall and overland flow is assumed to take place as water flows through plant tissues, residues, and the surface layer of the soil. Erosion and sediment transport in overland flow is calculated by the modified universal soil loss equation (Williams and Berndt, 1977) for daily rainfall input and is calculated for breakpoint type input data using concepts of both sheet and rill flow erosion. The sheet and rill erosion, by particle size fraction, are functions of rainfall intensity, flow depth and shear stress, and transport capacity. Where water flow rates are significantly affected by ponded water conditions (ponds, terraces, etc.), both water and sediment loads are routed through the ponded water taking into consideration the flow velocities and settling rates of sediment particles.

Precipitation is accumulated as snow if air temperatures so indicate. Snowpack development or melt is a function of radiant and convective heat from the surface, evaporation, albedo, and soil heat input. Temperature gradients within the soil and heat flux are calculated using the general equation for heat diffusion which is applied, layer by layer, taking into consideration the soil density, moisture content, and energy advected into the soil by infiltrating water.

The crop growth submodel is a mechanistic approach to simulating plant growth. The incremental production of plant material is a multiplicative relation between a photosynthetic conversion coefficient for production of plant material from radiant energy, an energy conversion efficiency factor, plant age, plant size, radiation input, and a stress factor made up of water, nutrients, and temperature stresses. Each of the factors are crop dependent. Total mass of dry matter is partitioned between leaves and stems, roots, and fruit or grain. Roots move into the soil exponentially in time as a function of root mass and maximum potential depth. They can selectively garner water and nutrients as a function of availability.

Because nutrients, especially nitrate, are major pollutants of surface and subsurface waters, nutrient cycling and the effects of tillage are major features of the model. Organic matter is central to cycling of plant nutrients and their transformation (Parton et al., 1987), thus plant decomposition (along with a partitioning of phosphorus and nitrogen) is simulated. Microbial decomposition rates responsible for mineralization and immobilization of nutrients (N and P) are functions of organic matter, N and P levels, soil temperature, and soil moisture content.

As mentioned previously, water movement in the soil is simulated using the Richards' equation (dynamic mass balance coupled with the Darcy equation). Parameters relating hydraulic conductivity to moisture content are related to soil texture. Solute movement is assumed to be by convection only; diffusion is assumed insignificant. Adsorption is assumed to be governed by either a linear equilibrium partitioning coefficient or a first–order linear kinetic adsorption isotherm (see Equation [10]). Pesticide dissipation is assumed to be first–order kinetics (see Equation [3]), where the rate varies with water content and soil temperature.

The interesting features of the model discussed in this chapter relate to water and solute movement. Figure 15 shows water flux between the top two layers of the soil (top layer 1 cm thick). This simulation is for a corn field in the Georgia Piedmont. The soil is a sandy loam with a restricting layer at between 30 and 60 cm (see Smith and Ferreira, 1989, for more information). Figure 15 shows the cumulative flux as water infiltrates the soil and is then brought back to the surface by capillary action as the surface dries. A positive, upward, trend in the line on Figure 15 is the infiltration of rainfall between the two layers. The figure shows rapid evaporation from a wet surface then the decline in evaporation rate as the soil surface dries. It also shows very low rates of soil evaporation later in the season as the canopy cover retards surface evaporation. It's interesting to note the magnitude of evaporative flux (downward

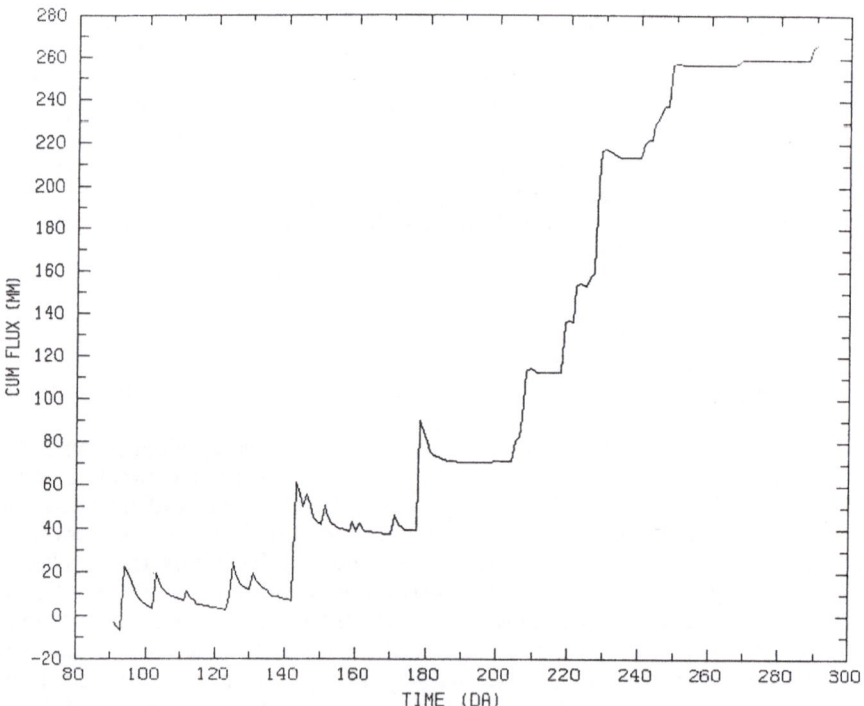

Figure 15. Simulation of the Opus model showing cumulative flux between layers 1 and 2 over a 200 day (Growing Season) simulation period.

trends) that will keep solutes in the upper layers of the soil. During the period from about the 90th day and until about the 210th day, the cumulative downward flux was about 260 mm and the evaporative flux was in excess of 150 mm. Thus it is obvious why models based on moisture tension that simulate upward movement should be utilized in leaching type studies.

Figure 16 shows the cumulative flux at the bottom of the rooting zone (60 cm). It shows a net downward movement throughout the period from the first of the simulation until about the 190th day. From that point until the end of the simulation, movement is upward into the root zone. During this period the crop uses all the infiltrated water and the additional demand pulls water from below the root zone.

Smith and Ferreira (1987) illustrate several features of adsorption kinetics. Figure 17 shows profile concentrations for two different pesticides, one much more mobile ($k_d = 20$) than the other ($k_d = 100$). The figure compares the two pesticides, assuming equilibrium and kinetic adsorption isotherms, 174 days after surface application. It shows a bulge in concentration at intermediate depths because the active plant root zone is a sink for soil water and thus an area of accumulation of the more mobile pesticide. The kinetic model exhibits much less bulge because its mobility is more limited. The less mobile more highly adsorbed pesticide ($k_d = 100$) remains near the surface. See Smith and Ferreira (1987) for more discussion of these features.

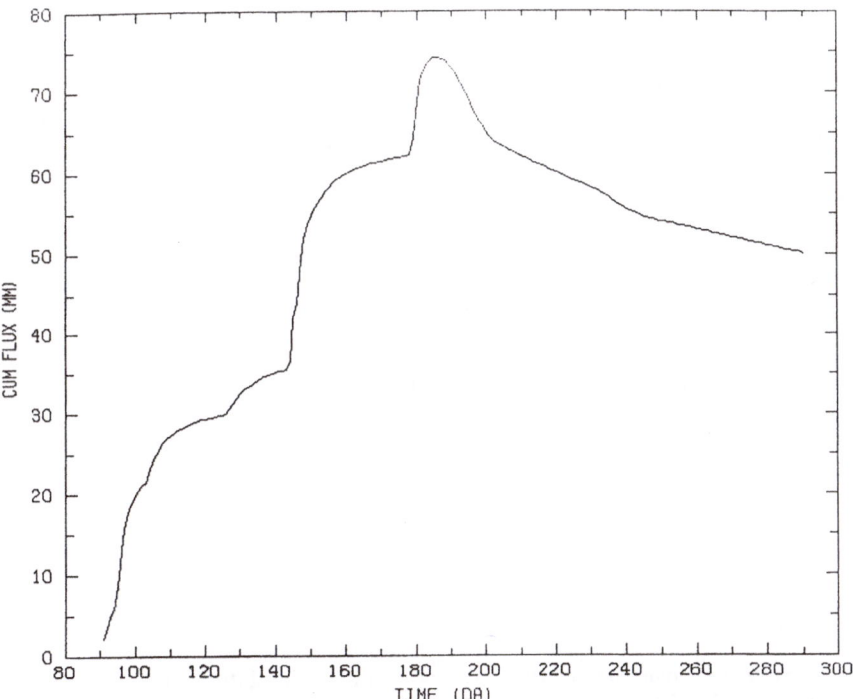

Figure 16. Simulation of the Opus model showing cumulative flux at the bottom of the root zone over a 200 day (Growing Season) simulation period.

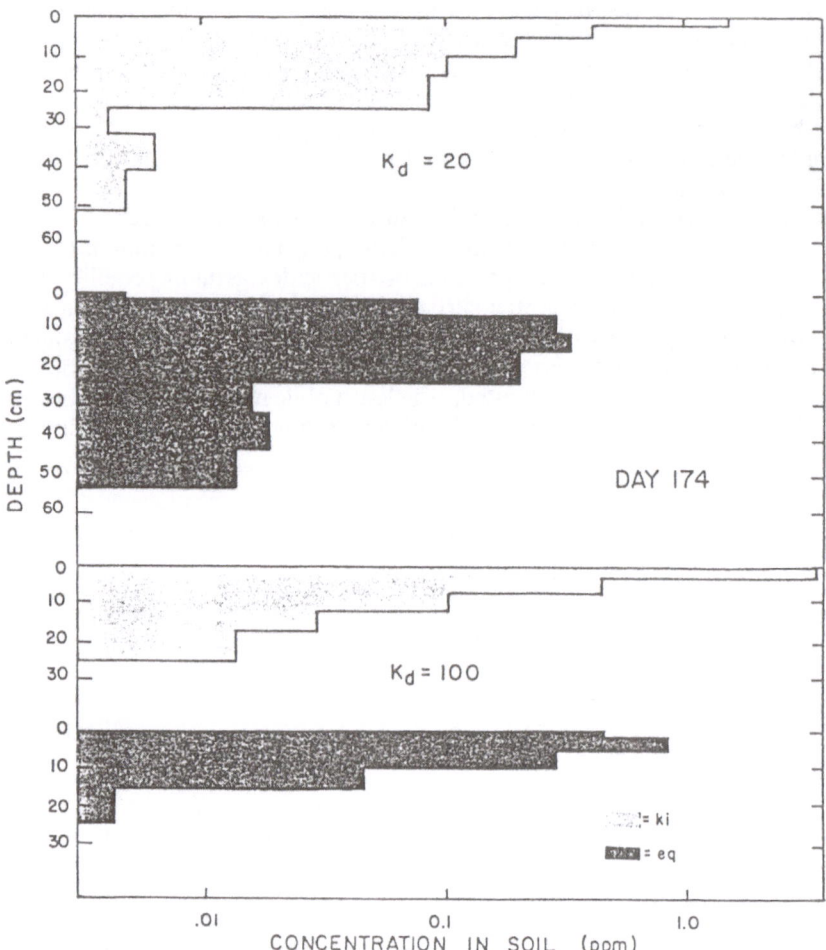

Figure 17. Simulations of the Opus model showing distribution of pesticides in a soil on day 174.

3. Conclusions

This has been a very brief overview of the variety of models simulating root zone leaching processes. A large body of literature on the subject is available. Space did not permit a description of all the types of models available. Hopefully, references included will provide an avenue for further investigation.

4. References

Addiscott, T. M.: 1982, 'Computer Assessment of the N Status During Winter and Early Spring,' in T. Batey et al. (eds.), *Assessment of the Nitrogen Status of the Soils*, University Press, Lueven, pp. 15–26.

Addiscott, T. M., and R. J. Wagenet: 1985, 'Concepts of Solute Leaching in Soils: A Review of Modeling Approaches,' *J. of Soil Sci.* **32**, 411–424.

Alonso, C. V., and D. G. DeCoursey: 1985, 'Small Watershed Model,' in D. G. DeCoursey (ed.), *Proceedings of the Natural Resources Modeling Symposium*, Pingree Park, CO, Oct. 16–21, 1983, USDA-ARS-30, pp. 40–46.

Aslyng, H. C., and S. Hansen: 1982, *Water Balance and Crop Production Simulation. Model WATCROS for Local and Regional Application,*' The Royal Vet. and Agric. Univ. Copenhagen, 200 p.

Baver, L. D., W. H. Gardner, and W. R. Gardner: 1972, *Soil Physics*, 4th ed., John Wiley and Sons, 498 p.

Berghuijs, J. T., P. E. Rijtema, and C. W. J. Roest: 1985, '*ANIMO, Agricultural Nitrogen Model,*' NOTA 1671, Inst. for Land and Water Management Research Wageningen.

Bergstrom, L., M. Brandt, and A. Gustafson: 1987, 'Simulation of Runoff and Nitrogen Leaching from Two Fields in Southern Sweden,' *Hydrol. Sci. J.* **32**, 191–205.

Boesten, J. J. T. J.: 1986, '*Behavior of Herbicides in Soil: Simulation of Experimental Assessment,*' Doct. Thesis, Inst. for Pest. Res. Wageningen, 263 p.

Bolt, G. H.: 1979, *Soil Chemistry, Physico–Chemical Models*, Elsevier Science, New York, 479 p.

Bresler, E., and G. Dagan: 1979, 'Solute Dispersion in Unsaturated Heterogeneous Soil at Field Scale. II. Applications,' *Soil Sci. Soc. Am. J.* **43**, 467–472.

Bresler, E., and G. Dagan: 1981, 'Convective and Pore Scale Dispersive Solute Transport in Unsaturated Flow Conditions,' *Water Resour. Res.* **17**(6), 1683–1693.

Bresler, E., and G. Dagan: 1983a, 'Unsaturated Flow in Spatially Variable Fields. 2. Application of Water Flow Models to Various Fields,' *Water Resour. Res.* **19**(2), 421–428.

Bresler, E., and G. Dagan: 1983b, 'Unsaturated Flow in Spatially Variable Fields. 3. Solute Transport Models and Their Application to Two Fields,' *Water Resour. Res.* **19**(2), 429–435.

Burns, I. G.: 1985, 'An Equation to Predict the Leaching of Surface–Applied Nitrate,' *J. Agric. Sci. Camb.* **85**, 443–454.

Carsel, R. F., L. A. Mulkey, M. N. Lorber, and L. B. Baskin: 1984, 'A Procedure for Evaluating Pesticide Leaching Threats to Groundwater,' *Ecological Modeling* **30**, 49–69.

Chardon, W. J.: 1984, '*Mobility of Cadmium in Soil,*' (in Dutch), Doctoral Thesis, Wageningen Agricultural University, Serie Bodembescherming No. 36, Staatsuitgeverij, 's–Gravenhage, 200 p.

Clothier, B. E.: 1990 'Root Zone Processes and Water Quality: The Impact of Management,' in D. G. DeCoursey (ed.), *Proceedings International Symposium on Water Quality Modeling of Agricultural Non–Point Sources*, June 19–23, 1988, Logan, UT, U.S. Dep. of Ag., Ag. Res. Serv., ARS-81, pp. 659–674.

Clothier, B. E., T. T. Sauer, and S. R. Green: 1988, 'The Movement of Ammonium Nitrate into Unsaturated Soil During Unsteady Absorption,' *Soil Sci. Soc. Am. J.* (in press).

Dagan, G., D. Russo, and E. Bresler: 1990, 'Effect of Spatial Variability Upon Subsurface Transport of Solutes from Nonpoint Sources,' in D. G. DeCoursey (ed.), *Proceedings International Symposium on Water Quality Modeling of Aricultural Non–Point Sources*, June 19–23, 1988, Logan, UT, U.S. Dep. of Ag., Ag. Res. Serv., ARS-81, pp. 523–548.

Dagan, G., and E. Bresler: 1983, 'Unsaturated Flow in Spatially Variable Fields. 1. Derivation of Models of Infiltration and Redistribution,' *Water Resour. Res.* **19**(2), 413–420.

Dautrebande, S., and J. P. Agneesens: 1985, 'Prevision du Transfert de Pesticides Dans Les Sols,' *Aspects de Standardisaation*, Ed. INRA no **31**, pp. 87–104.

DeCoursey, D. G., and K. W. Rojas: 1990, 'RZWQM – A Model for Simulating the Movement of Water and Solutes in the Root Zone,' in D. G. DeCoursey (ed.), *Proceedings International Symposium on Water Quality Modeling of Agricultural Non–Point Services*, June 19–23, 1988, Logan, UT, U.S. Dep. of Ag., Ag. Res. Serv., ARS-81, pp. 813–822.

De Smedt, F., P. J. Wierenga, and A. Van der Beken: 1981, 'Theoretical and Experimental Study of Solute Movement through Porous Media with Mobile and Immobile Water,' *VUB–Hydrology* **6**, 219.

De Willigen, P., L. Bergström, and R. G. Gerritse: 1990, 'Leaching Models of the Unsaturated Zone, Their Potential Use for Management and Planning,' in D. G. DeCoursey (ed.), *Proceedings International Symposium on Water Quality Modeling of Agricultural Non–Point Sources*, June 19–23, 1988, Logan, UT, U.S. Dep. of Ag., Ag. Res. Serv., ARS-81, pp. 105–128.

320

De Willigen, P., and J. J. Neeteson: 1985, 'Comparison of Six Simulation Models for the Nitrogen Cycle in the Soil,' *Fert. Res.* **8**, 157–171.

De Willigen, P., P. A. C. Raats, and R. G. Gerritse: 1982, 'Transport and Fixation of Phosphate in Acid Homogeneous Soils, II. Computer Simulation,' *Agric. Environ.* **7**, 161–174.

Enfield, C. G., R. F. Carsel, S. Z. Cohen, T. Phan, and D. M. Walters: 1982, 'Approximating Pollutant Transport to Ground Water,' *Ground Water* **20**, 711–722.

Ferrari, T. J., and J. L. Cuperus: 1973, 'Dynamic Simulation of Vertical Non–adsorbed Anion Transport,' *Plant Soil* **38**, 425–438.

Ferreira, V. A., and R. E. Smith: 1990, 'OPUS, an Advanced Simulation Model for Non–Point Source Pollution Transport at the Field Scale, An Overview,' in D. G. DeCoursey (ed), *Proceedings International Symposium on Water Quality Modeling of Agricultural Non–Point Sources*, June 19–23, 1988, Logan, UT, U.S. Dep. of Ag., Ag. Res. Serv., ARS–81, pp. 823–834.

Geng, Q. Z., G. Girard, Y. Soulard, and A. Blondel: 1987, '*Modelisation du Transfert de Nitrates sur le Bassin Versant de la Noe–Se'che–Bretange,*' Rapport CIG–ENSMP–LHM/RD/87:30 p.

Hansen, S., and H. C. Aslyng: 1984, '*Nitrogen Balance in Crop Production. Simulation Model NITCROS,*' The Royal Vet. and Agric. Univ. Copenhagen, 113 p.

Hebson, C. S., and D. G. DeCoursey: 1987, 'A Model for Assessing Agricultural Management Impact on Ground Water Quality,' *Jour. Env. Sci. and Health, Part B, Pesticides, Food Contaminants in Ag. Wastes.*

Holtan, H. N., C. B. England, G. P. Lawless, and G. A. Schumaker: 1968, '*Moisture–Tension Data for Selected Soils on Experimental Watersheds,*' USDA, ARS 41–44.

Huwe, B., and R. R. Van der Ploeg: 1987, '*Erfahrungen und Problem bie der Simulation des N–Haushalts Verschieden Genutzter Standorte Baden–Wurttembergs,*' Mitteilgn. Dtsch. Bodenkundl. Gesselsch. 55/I, pp. 181–187.

Jansson, P.-E., and S. Halldin: 1980, '*Soil Water and Heat Model, Technical Description,*' Swedish Coniferous Forest Project, Tech. Rep. 26, Swedish University of Ag. Sci., Uppsala, 81 p.

Johnsson, H., L. Berstrom, P.-E. Jansson, and K. Paustian: 1987, 'Simulated Nitrogen Dynamics in a Layered Agricultural Soil,' *Agric. Ecosystems Environ.* **18**, 333–356.

Jones, R. L., J. P. Gibb, and R. J. Hanks: 1990, 'Review of Unsaturated Zone Leaching Models from a User's Perspective,' in D. G. DeCoursey (ed.), *Proceedings International Symposium on Water Quality Modeling of Agricultural Non–Point Sources*, June 19–23, 1988, Logan, UT, U.S. Dep. of Ag., Ag. Res. Serv., ARS–81, pp. 129–140.

Jury, W. A.: 1982, 'Simulation of Solute Transport Using a Transfer Function Model,' *Water Resour. Res.* **18**(2), 363–368.

Jury, W. A., W. F. Spencer, and W. J. Farmer: 1983, 'Model for Assessing Behavior of Pesticides and Other Trace Organics Using Benchmark Properties. I. Description of Model,' *J. Environ. Qual.* **12**, 558–564.

Jury, W. A., G. Sposito, and R. E. White: 1986, 'A Transfer Function Model of Solute Transport Through Soil. 1. Fundamental Concepts,' *Water Resour. Res.* **22**(2), 243–247.

Knisel, W. G. (ed.): 1980, '*CREAMS: A Field–Scale Model for Chemicals, Runoff, and Erosion from Agricultural Management Systems,*' U.S. Dept. of Agric., Conservation Research Report No. 26, 640 p.

Kragt, J. F., and W. De Vries: 1987, '*Onderzoek Naar Effecten van Mestbeperking op de Nitraat Uitspoeling in Water–Wingebieden in Overysel. 1 – Beschrijving van RENLEM,*' Stiboka Rapport no. 1935, Wageningen.

Lafeber, D.: 1965, 'The Graphical Representation of Planer Pore Patterns in Soils,' *Aust. J. Soil Res.* **3**, 143–164.

Leonard, R. A., W. G. Knisel, and D. A. Still: 1986, '*Groundwater Loading Effects of Agricultural Management System,*' Trans. ASAE Paper #86–2511.

Marshall, T. J., and J. W. Holmes: 1979, *Soil Physics*, Cambridge University Press, 345 p.

Nicholls, P. H., A. Walker, and R. J. Walker: 1982, 'Measurement and Simulation of the Movement and Degradation of Atrazine and Metribuzin in a Fallow Soil,' *Pesticide Science* **13**, 484–494.

Nofziger, D. L., and A. G. Hornsby: 1985, 'A Microcomputer–Based Management Tool for Chemical Movement in Soil,' *Applied Agricultural Research* **1**, 50–56.

Office of Technology Assessment: 1984, '*Protecting the Nation's Groundwater from Contamination,*' OTA-O-233, Vol. I, 244 p.

Parton, W. J., D. S. Schimiel, C. V. Cole, and D. S. Ojima: 1987, 'Analysis of Factors Controlling Soil Organic Matter Levels in Great Plains Grasslands,' *Soil Science Soc.*

Raats, P. A. C.: 1978, 'Convective Transport of Solutes by Steady Flows. I. General Theory,' *Agric. Water Manag.* **1**, 201–218.

Reeder, J. D., and B. Sabey: 1987, 'Nitrogen,' in D. Williams, and G. E. Schuman (eds.), *Reclaiming Mine Soils and Overburdens in the Western United States: Analytic Parameters and Procedures*, Soil and Water Conservation Society, Ankeny, Iowa, pp. 155–184.

Richter, J., K. C. Kersebaum, and J. Utermann: 1988, 'Modeling of the Nitrogen Regime of Arable Filed Soils,' in D. S. Jenkinson, and K. A. Smith (eds.), *Nitrogen Efficiency in Agricultural Soils and the Efficient Use of Fertilizer Nitrogen*, CEC Seminar.

Rieu, M.: 1984, 'Simulation Nume'rique du Transfert GVertical des Solute's et des Equilibres d'e'change et de Pre'cipitation dans les sols Calcaires Gypseux,' *Sci. Geol. Bull.* **37**, 253–266.

Rose, C. W., F. W. Chichester, J. R. Williams, and J. T. Ritchie: 1982, 'Application of an Approximate Analytic Method of Computing Solute Profiles with Dispersion in Soils,' *J. of Environ. Quality* **11**, 151–155.

Shaffer, M. J., and W. E. Larson (eds.): 1987, *'NTRM, a Soil–Crop Simulation Model for Nitrogen, Tillage, and Crop–Residue Management,'* U.S. Department of Agriculture, Conservation Research Report 34-1, 103 p.

Shaffer, M. J., J. B. Swan, and M. R. Johnson: 1984, 'Coordinated Farm and Research Management (COFARM) Data System for Soils and Crops,' *J. Soil and Water Conser.* **39**, 320–324.

Smith, R. E., and V. A. Ferreira: 1987, 'Effect of Soil Adsorption Kinetics on Agricultural Pesticide Fate,' *Jour. Env. Sci. and Health, Part B, Pesticides, Food Contaminants in Ag. Wastes.*

Smith, R. E., and V. A. Ferreira: 1989, 'Comparative Evaluation of Unsaturated Flow Methods in Selected USDA Simulation Models,' in H. J. Morel-Seytoux (ed.), *Unsaturated Flow in Hydrologic Modeling, Theory and Practice*, NATO ASI Series, Kluwer Academic Publishers, Series C: Mathematical and Physical Sciences, Vol. **275**, pp. 391–412.

Smith, R. E., and J. Y. Parlange: 1978, 'A Parameter-Efficient Hydrologic Infiltration Model,' *Water Resour. Res.* **14**(3), 533–538.

Sposito, G., R. E. White, P. R. Darrah, and W. A. Jury: 1986, 'A Transfer Function Model of Solute Transport Through Soil. 3. The Convection–Dispersion Equation,' *Water Resour. Res.* **22**(2), 255–262.

Steenhuis, T. S., S. Pacenka, and K. S. Porter: 1987, 'MOUSE: A Management Model for Evaluating Groundwater Contamination from Diffuse Surface Sources Aided by Computer Graphics,' *Applied Agricultural Research*, in press.

Vachaud, G., M. Vauclin, and T. M. Addiscott: 1990, 'Solute Transport in the Vadose Zone: A Review of Models,' in D. G. DeCoursey (ed.), *Proceedings International Symposium on Water Quality Modeling of Agricultural Non–Point Sources*, June 19-23, 1988, Logan, UT, U.S. Dep. of Ag., Ag. Res. Serv., ARS-81, pp. 81–104.

Van der Zee, S. E. A. T. M., and W. H. Van Riemsdijk: 1986, 'Transport of Phosphate in a Heterogeneous Field,' *Transport in Porous Media* **1**, 339–359.

van Drecht, G.: 1985, *"Toetsing van model ONZAT m.b.v. de Meetgegevens Uit Het Hydrologisch Proefgebied 'Hupselve Beek',"* RIVM Report 847211001.

van Genuchten, M. T., S. M. Gorelick, and W. W.–G. Yeh: 1990, 'Application of Parameter Estimation Techniques to Solute Transport Studies,' in D. G. DeCoursey (ed.), *Proceedings International Symposium on Water Quality Modeling of Agricultural Non–Point Sources*, June 19-23, 1988, Logan, UT, U.S. Dep. of Ag., Ag. Res. Serv. ARS-81, pp. 731–754.

van Genuchten, M. T., and W. J. Alves: 1982, *'Analytical Solutions of the One-Dimensional Convective–Dispersive Solute Transport Equation,'* U.S. Dep. of Agric. Tech. Bull. 1661, 149 p.

van Genuchten, M. T., and P. J. Wierenga: 1976, 'Mass Transfer Studies in Sorbing Porous Media. I. Analytical Solutions,' *Soil Sci. Soc. Am. Proc.* **38**, 29–38.

Vinten, A., H. Frenkel, and J. Shalhevet: 1988, *'Validation of a Modified Steady State Model of Crop Response to Saline Water Irrigation under Transient Conditions,'* (submitted to Irrigation Sci.).

Wagenet, R. J., and J. L. Hutson: 1987, *'LEACHM, Leaching Estimation and Chemistry Model, A Process Based Model of Water and Solute Movement, Transformation, Plant Uptake and Chemical Reactions in the Unsaturated Zone,'* Continue, Vol. **2**, Water Resour. Inst., Cornell University, Ithica, NY.

Wagenet, R. J., M. J. Shaffer, and R. E. Green: 1990, 'Predictive Approaches for Leaching in the Unsaturated Zone,' in D. G. DeCoursey (ed.), *Proceedings International Symposium on Water Quality Modeling of Agricultural Non–Point Sources*, June 19-23, 1988, Logan, UT, U.S. Dep. of Ag., Ag. Res. Serv., ARS-81, pp. 63–80.

White, R. E., J. S. Dyson, R. A. Haigh, W. A. Jury, and G. Sposito: 1986, 'A Transfer Function Model of Solute Transport Through Soil. 2. Illustrative Applications,' *Water Resour. Res.* **22**(2), 248–254.

Williams, J. R., and H. D. Berndt: 1977, 'Sediment Yield Prediction Based on Watershed Hydrology,' *Trans. ASAE* **20**(6), 1100–1104.

Zandt, P. A., and P. De Willigen: 1981, '*Simulatie van de Stikstofverdeling in de Grond in Winter en Voorjaar,*' Inst. Bodemvruchtbaarheid, Haren (Gn), Rapp. 4–81.

Chapter 15

Integrated Quantity/Quality Modeling - Receiving Waters

Donn G. DeCoursey
USDA–ARS, Hydro–Ecosystem
243 Federal Building
301 South Howes, P. O. Box E
Fort Collins, Colorado 80522 U.S.A.

Abstract. In recent years point sources of pollution that have contaminated both surface and ground water supplies have been addressed and significant progress in cleaning up these sources has been made. As a result, emphasis has shifted to nonpoint sources of pollution that have become significant problems. This is emphasized by the fact that nitrates and pesticides are being increasingly discovered in ground and surface water supplies. This chapter describes the use of mathematical models used to simulate the movement of water and chemicals in receiving waters: ground water, surface runoff, and surface impoundments. As in the previous chapter comments will be confined to nonpoint source issues. The general features, structure, examples, and strengths and weaknesses of each are presented.

1. Introduction

The previous chapter of this two part presentation discussed the need for models as an aid in studying and solving the complex problems of nonpoint source pollution. Emphasis in that chapter was on root zone leaching processes. In this chapter emphasis will be placed on the receiving waters and the fate of contaminants as they move through these waters. Receiving waters are considered to be groundwater, surface runoff (and small streams), and surface impoundments (large rivers and lakes).

2. Discussion

2.1. Groundwater Quality Models

Saturated zone transport models for both point and nonpoint sources of contamination are similar with respect to modeling technique. The major difference between point and non-point sources is the clearly distinguishable plume from point sources of relatively small areal extent. In nonpoint source contamination, which is usually agricultural, the source area can cover relatively large land areas; however, the concentrations may be low and specific sources difficult to identify. Also in nonpoint sources, the area of contamination is much larger than the horizontal correlation scale of soil hydraulic properties; thus, the concentration properties of interest are the spatially averaged values and their distributions at a given

D. S. Bowles and P. E. O'Connell (eds.), Recent Advances in the Modeling of Hydrologic Systems, 323–353.

depth. Relating the properties of average concentration to average soil properties remains a problem (Dagan et al., 1990). In the following material the spatial properties, per se, are not investigated. Instead mass movement is the subject of investigation. The effect of spatially variable soil hydraulic properties on solute concentration was discussed in the previous chapter.

Processes associated with saturated zone contaminant transport include advection, dispersion, diffusion, adsorption, precipitation, and chemical transformation. Advection is the unsteady bulk movement of a contaminant with the mass of saturated water. Dispersion describes the deviations of flow about the mean, generally at a scale smaller than the advection; these deviations are frequently associated with hydrodynamic flow through pores of various sizes and the heterogeneity of the aquifer properties. Diffusion is the movement of solute ions in response to a concentration gradient and is generally at a scale smaller than dispersion. However, in very low gradient systems, where advection and dispersion are small terms, diffusion can be a very significant term. Both dispersion and diffusion spread the initial concentration of pollutant in both the longitudinal and transverse directions. Adsorption has a delaying influence on solute transport, as a fraction of the solute is adsorbed onto the matrix. When pH or a chemical reaction reduces the solubility levels of a solute to a point lower than the concentration of the existing solute solution, some of the material will precipitate (come out of solution). Chemical degradation and transformation leads to attenuation of the original concentration profiles or gradients and possibly to new species that may need to be tracked.

Duffy et al. (1990) in discussing mathematical models of nonpoint source transport, identify two simple approaches to looking at the transport problem. One approach is based on determining the transit time and age distribution of dissolved solutes using probability theory. In a steady state system the probability density functions of both transit time and age can be shown to be exponential functions of residence time. Residence time is defined as the mass of the aquifer divided by the steady state flux through the aquifer. In such well mixed systems the solute mass in the aquifer is proportional to the outflow mass flux or vice versa, the outflow mass flux is proportional to the mass in the aquifer.

The second approach described by Duffy et al. (1990) is the linear reservoir model for transient time. The linear reservoir concept is especially applicable to nonpoint source problems because the input is distributed over a large diffuse area and output is frequently at a point, i.e. a field drain, a well, or point of inflow to a stream. In this approach the mass of solute in the aquifer is volume of the aquifer times its spatial average concentration. For steady state conditions the concentration flowing out of the aquifer is equal to that in the aquifer, and the impulse response function for the aquifer is the same exponential function of time as that obtained by the probability approach. See Duffy et al. (1990) for more information on both modeling methods.

While the lumped approach may be a practical approach to solving the outflow problem where the solution is desired at a point, it does not provide any information about the spatial pattern within the system. Thus the rest of this presentation on groundwater quality models will deal with a more general solution.

All general transport models are dependent upon a saturated water flow model that describes average direction of pollutant movement, path lines, arrival times, isochrones and boundaries. If dispersion and diffusion are neglected, the model provides a full description of the advection. Since the emphasis of this chapter is on water quality features of models and other authors are concentrating on the hydraulics, the following discussion will deal mostly with the chemical and biological components. A very brief discussion of groundwater flow models will be presented to put the water quality features in perspective.

2.1.1. The Governing Equation

If one assumes that fluid density and viscosity are constant, both the medium and fluid are incompressible, and the medium is homogeneous and isotropic, the partial differential equation describing solute transport is:

$$R\frac{\partial C}{\partial t} = \frac{\partial}{\partial x_i}\left(D_{ij}\frac{\partial C}{\partial x_j}\right) - v_i\frac{\partial C}{\partial x_j} - \frac{C'W_*}{n} - \lambda CR \tag{1}$$

$$\quad\quad\text{(a)}\quad\quad\quad\quad\text{(b)}\quad\quad\quad\text{(c)}\quad\quad\text{(d)}\quad\quad\text{(e)}$$

in which C is the solute concentration, v_i is the seepage or average pore water velocity in the x_i direction, D_{ij} is the dispersion coefficient tensor (in this equation it is assumed to be the sum of the mechanical mixing and molecular diffusion), C' is the solute concentration in the source or sink fluid, W^* is the volume flow rate per unit volume of the source or sink, n is the effective porosity, R is the retardation factor associated with adsorption, λ is a half-life or decay constant, and x_i and x_j are Cartesian coordinates. The average pore water velocity, v_i, is defined as:

$$v_i = -\frac{K_{ij}}{n}\frac{\partial h}{\partial x_j} \tag{2}$$

in which K_{ij} is the hydraulic conductivity tensor and h is the hydraulic head. If the retardation factor can be considered to be an equilibrium-controlled ion exchange reaction (which is reasonable for groundwater situations), R can be defined as:

$$R = 1 + \frac{\rho_b K_d}{n} \tag{3}$$

in which ρ_b is the bulk density of the matrix, and K_d is a distribution coefficient.
Equation (1) assumes that the system is under steady-state conditions thus:

$$\frac{\partial^2 h}{\partial x_i^2} = 0 \tag{4}$$

Term (a) in Equation (1) is the solute concentration controlled by desorption, (b) is the dispersion component, (c) is the advective transport component, (d) is the source or sink term, and (e) is a dissipation or loss term. Duffy et al. divide Equation (1) into geochemistry and solute transport components and discuss solution with direct and indirect or two step coupling.
Boundary and Initial Conditions. Solution of Equation (1) requires appropriate boundary and initial conditions. Generally initial concentrations are assumed constant throughout the domain of interest, however in a more general case:

$$C(x, y, z, t,) = f(x, y, z) \quad at\ t = 0 \tag{5}$$

in which $f(x,y,z)$ is a known function.

Boundary conditions that can be specified, depending on physical constraints, include:
(1) Dirichlet boundary conditions that prescribe concentrations along a portion of the boundary:

$$C = C_o(x, y, z, t) \tag{6}$$

in which C_o is a given function of time and space along a particular portion of the boundary.
(2) Neumann boundary conditions that prescribe normal concentration gradients over a portion of the boundary:

$$\left[D_{ij} \frac{\partial C}{\partial x_j} \right] n_i = q(x, y, z, t) \tag{7}$$

in which q is a given function and n_i are directional cosines. The value of q is zero at impervious boundaries.
(3) Cauchy boundary conditions prescribe both the concentration and its gradient at the boundary:

$$\left(D_{ij} \frac{\partial C}{\partial x_j} - v_i C \right) n_i = g(x, y, z, t) \tag{8}$$

in which g is a known function. The first term in the left hand part of Equation (8) is dispersion flux and the second term is advection flux.

2.1.2. Analytical Solutions

Javandel et al. (1984) describes three methods of solving Equation (1) to predict the extent of subsurface contamination: analytical, semianalytical, and numerical. Even though the models and examples discussed in each of these classes of solution are oriented toward point sources, the material is applicable to nonpoint sources at the larger scale. This is an excellent reference and I have borrowed heavily from it in my introductory remarks to this section. Following are a few comments about each of the solution methods.

Analytical (closed form) solutions to Equation (1) require many simplifying assumptions and are generally limited to cases where flow is radial or uniform over the area of interest. Javandel et al. show derivation of a solution for a one–dimensional model of an infinitely long homogeneous isotropic porous medium with a steady uniform flow and a chemical, changing in time exponentially, injected at one end. Solutions are obtained at any time, t, and distance, x, from the boundary. They also obtain solutions for a constant input concentration, for no decay factor and constant input concentration, and with a retardation factor. Solutions to several two–dimensional models are also derived. One was flow from a seepage face through a homogeneous isotropic porous medium having unidirectional steady–state flow upon which is superimposed a line source of a pollutant normal to the flow field. Another derivation is dispersion in radial flow. Several examples of applying the solutions obtained to practical problems are illustrated.

Example of an Analytical Solution. To demonstrate an analytical solution, one of the examples is discussed. This problem is shown diagrammatically in Figure 1. It is a relatively thin shallow, homogeneous isotropic aquifer with a steady uniform seepage of 0.1 m/d. Liquid

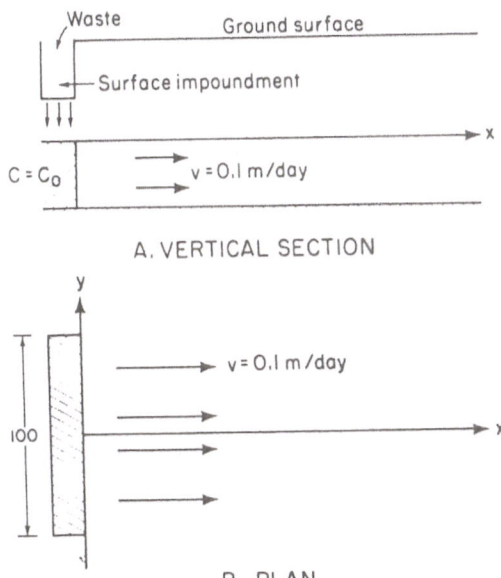

Figure 1. A schematic diagram showing (a) a vertical section of the aquifer and the surface impoundment, (b) a plan view of the flow and the source of contamination (Javandel et al., <u>Groundwater Transport: Handbook of Mathematical Models</u>, Water Resources Monograph 10, pg. 28, 1984, copyright by the American Geophysical Union).

waste from a factory is discharged into a surface impoundment 100 m long and 5 m wide perpendicular to the direction of groundwater flow. The discharge creates a constant concentration (no decay factor) in the groundwater of 1000 ppm in the area under the ditch. Transverse dispersivity, D_T, is 1/10 the longitudinal value, D_L, which was assumed, for comparison purposes to be 1 and 5 m²/day. The user is interested in (1) the concentrations at different distances downstream 1 and 5 years after application and (2) the area contaminated at concentrations in excess of 10 ppm 5 years from now. Figures 2 and 3 show the relative concentrations vs. y, the length of the injection field, at different distances downstream. The figures show the results for values of $D_L = 5$ and $D_T = 0.5$ m²/day. Figure 4 shows the area contaminated at concentrations in excess of 10 ppm, 5 years from the start of injection.

There are both advantages and disadvantages or limitations in using analytical solutions to the flow equations. Javandel et al. list the following advantages and limitations.

Advantages and Limitations of Analytical Methods. Analytical methods are advantageous (1) when data are sparce and uncertain, (2) when the physical system is simple and does not deviate far from the analytical representation, (3) because they do not require complex numerical codes and extensive data collection, (4) when only rough estimates are needed, and (5) because they are inexpensive to run and good for initial estimates.

Limitations in use of analytical methods are (1) requirements that they approach idealized conditions without complex boundary conditions, and (2) requirements that system properties be spatially and temporally constant.

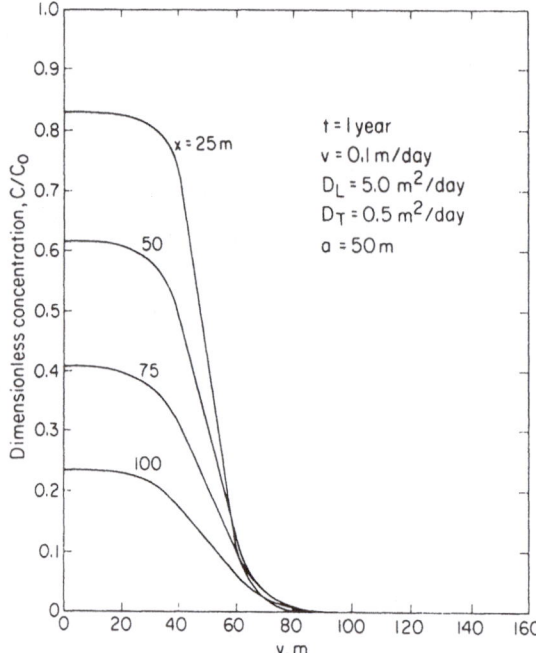

Figure 2. Dimensionless concentration C/C_0 versus distance from the x axis for values of x = 25, 50, 75, and 100 m downstream from the source and t = 1 year, v = 0, 1 m/d, D_L = 5 m^2/d, D_T = 0.5 m^2/d, a = 50 m (Javandel et al., <u>Groundwater Transport: Handbook of Mathematical Models</u>, Water Resources Monograph 10, pg. 30, 1984, copyright by the American Geophysical Union).

2.1.3. Semianalytical Solutions

Semianalytical methods apply the fluid mechanics concepts of the complex velocity potential using numerical tools and plotters to solve flow problems (Javandel et al., 1984). The methods are limited to simulations of steady state two–dimensional fluid flow through homogeneous media. The effects of transport by dispersion and diffusion are ignored; thus contaminants move by advection, but they can lag as a result of adsorption. Semianalytical methods are well suited to delineating the migration paths of natural flow systems such as groundwater flow under low–relief crop land.

Theory behind the semianalytical methods lies in use of the complex velocity potential, W:

$$W = \phi + i\psi \tag{9}$$

in which ϕ is hydraulic potential, generally defined as:

$$\phi = Kh + c \tag{10}$$

in which K is the hydraulic conductivity, h is the hydraulic gradient, and c is a constant. The stream function, ψ, is related to the hydraulic potential using the Cauchy–Riemann equations

$$\frac{\partial \phi}{\partial x} = \frac{\partial \psi}{\partial y} \tag{11}$$

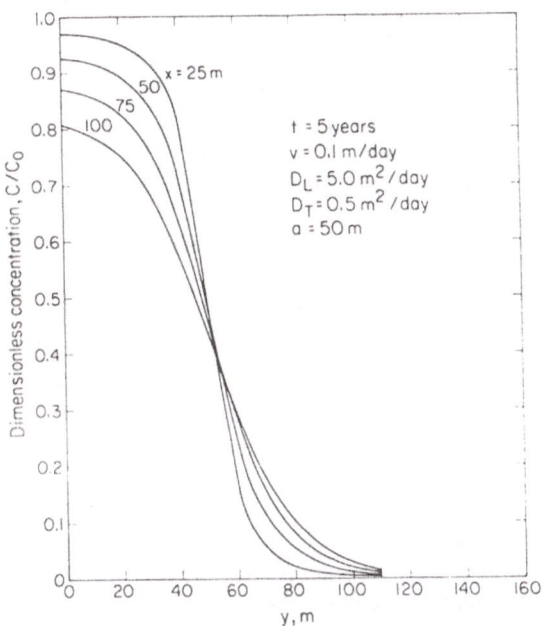

Figure 3. Dimensionless concentration C/C_o versus distance from the x axis for values of x = 25, 50, 75, 100 m downstream from the source and t = 5 years, v = 0.1 m/d, D_L = 5 m²/d, D_T = 0.5 m²/d, a 50 m (Javandel et al., <u>Groundwater Transport: Handbook of Mathematical Models</u>, Water Resources Monograph 10, pg. 31, 1984, copyright by the American Geophysical Union).

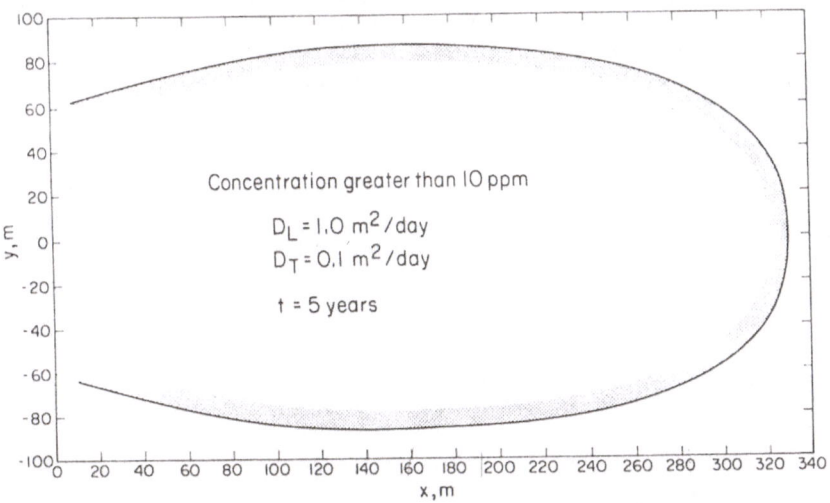

Figure 4. A map showing the zone of contamination with concentration greater than 10 ppm; v = 0.1 m/d, D_L = 1.0 m²/d, D_T = 0.1 m²/d, and a time of 5 years after the solute reached the aquifer (Javandel et al., <u>Groundwater Transport: Handbook of Mathematical Models</u>, Water Resources Monograph 10, pg. 32, 1984, copyright by the American Geophysical Union).

330

and

$$\frac{\partial \phi}{\partial y} = -\frac{\partial \psi}{\partial x} \tag{12}$$

Lines defined by ϕ = const. and ψ = const. describe the equipotential and streamflow lines, respectively, of the typical flow net. The governing equations for calculating the hydraulic potential $\phi = \phi(x,y)$ and the streamflow functions $\psi = \psi(x,y)$ are:

$$\frac{\partial}{\partial x}\left(K_{xx}\frac{\partial \phi}{\partial x}\right) + \frac{\partial}{\partial y}\left(K_{yy}\frac{\partial \phi}{\partial y}\right) = 0 \tag{13}$$

and:

$$\frac{\partial}{\partial x}\left(\frac{1}{K_{yy}}\frac{\partial \psi}{\partial x}\right) + \frac{\partial}{\partial y}\left(\frac{1}{K_{xx}}\frac{\partial \psi}{\partial y}\right) = 0 \tag{14}$$

in which K_{xx} and K_{yy} are principal components of the hydraulic conductivity tensor. Solution of Equations (13) and (14) with appropriate boundary conditions yields the typical flow net shown in Figure 5. For more information on these solution techniques see Javandel et al. (1984), Frind and Matanga (1985), Frind et al. (1985) and Bogardi et al. (1990).

An Example of a Semianalytical Solution. One of the examples of a semianalytical solution described by Javandel et al. (1984) pertains to a well problem superimposed on a flow field. The specific problem is shown in Figure 6. A well located at point A completely penetrates a 10 m homogeneous isotropic aquifer. Partially treated waste water is being recharged at a rate of 50 m³/hr through this well. At well B water is being withdrawn at the same rate. The aquifer has a flow rate of 50 m/yr as shown and has an effective porosity of 25 percent. Adsorption of the waste water is assumed to be negligible. Figure 7 shows progression of the pollutant front at 0.5, 2, and 4 years.

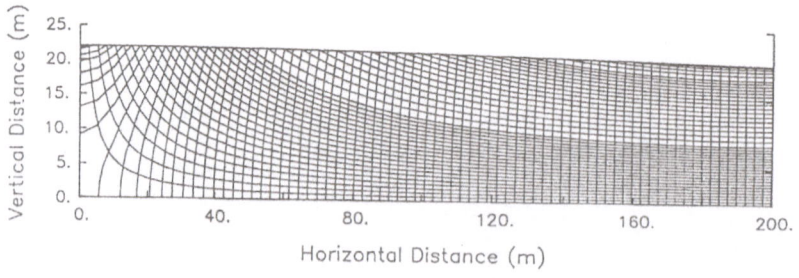

Figure 5. Typical flownet from a potential–streamfunction model (after Bogardi et al., 1990).

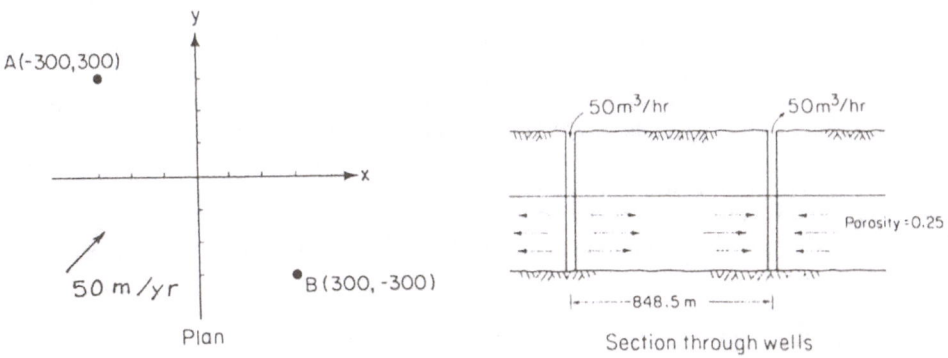

Figure 6. A schematic diagram showing the plan and vertical section of the recharge and discharge wells used in example 1 (adapted from Javandel et al., <u>Groundwater Transport: Handbook of Mathematical Models</u>, Water Resources Monograph 10, pg. 46, 1984, copyright by the American Geophysical Union).

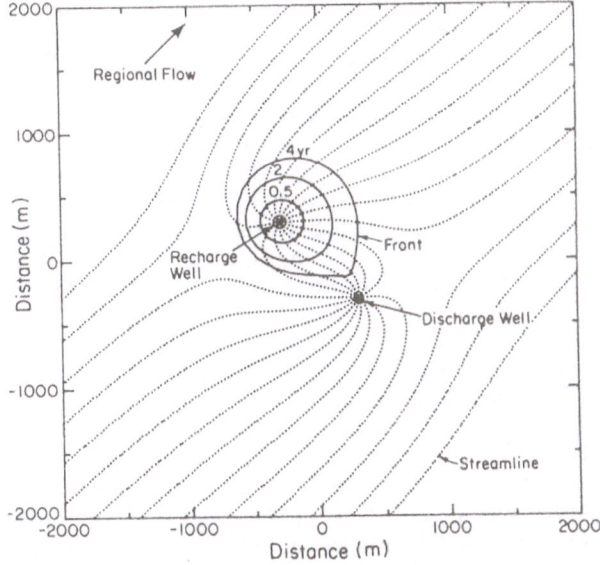

Figure 7. Streamlines and pollutant front 0.5, 2, and 4 years after recharge began. Streamlines coming from regional flow are included (Javandel et al., <u>Groundwater Transport: Handbook of Mathematical Models</u>, Water Resources Monograph 10, pg. 55, 1984, copyright by the American Geophysical Union).

The following advantages and limitations of semianalytical methods were presented by Javandel et al.

Advantages and Limitations of Semianalytical Methods. Semianalytical methods are advantageous to use when: (1) the flow field is simple but multiple sources and/or sinks (wells) preclude use of analytical methods; (2) order of magnitude results and limited time, data, or budget preclude other methods of estimating solute travel time; (3) computer availability limits fully numerical methods; and (4) only preliminary estimates are needed.

Use of semianalytical methods are limited because they (1) do not consider transport by dispersion and diffusion, (2) consider only two dimensional problems, (3) are not applicable to heterogeneous or anisotropic permeabilities and (4) generally only fit steady state problems.

2.1.4. Numerical Solutions

If the problem to be solved has irregular or unusual boundary conditions or lies in a heterogeneous or anisotropic formation, then analytical or semianalytical solutions may not be possible and numerical solutions may be necessary. These techniques approximate the governing equations by algebraic equations relating unknown variables at discrete nodal points and at different times. The accuracy and efficiency of the solution depend on (1) the numerical approximations used to evaluate the spatial gradients and time derivatives and (2) the solution scheme of the algebraic equations.

All such models have previously been classified as analytical, semianalytical, or numerical, and Equation (1) is the general governing equation. Bogardi et al. now describe an aquifer as consisting of mobile and immobile regions. Frequently aquifers are laced with clay or other relatively impervious regions with very low permeability that can interact with solutes in the more mobile fluid. This interaction causes a retarding effect on the front of a contaminant plume that cannot be obtained by an advection–dispersion model alone. Thus Bogardi et al. developed an alternative formulation that replaces terms (b) and (c) with advective–diffusive terms that enable the transfer of solute, by diffusion, between the mobile and stagnant regions. This advective–diffusive model holds considerable promise because of its more physically realistic representation of the transport mechanism. The model described has been solved both analytically and numerically, but probably holds greater promise in a numerical form because of the irregular nature of subsurface features.

Javandel et al. describe very briefly the general features of most numerical methods and different approaches used to determine the fluid flow field and solute concentration distribution. The equations solved are versions of Equation (1) that depend upon features of the specific flow field and the solutes. In general, it is not necessary to make simplifying assumptions regarding the flow field or solute features, but numerical dispersion can be a problem in some cases. Frind and Germain (1986) describe a procedure for the design of grids to minimize numerical dispersion using the Peclet and Courant criteria. The finite difference, finite element, and flow path network methods, explicit and implicit time stepping schemes, and iterative and direct equation solvers are all briefly described. Also the characteristics of a number of different numerical codes are summarized. Recent developments in spectral and in Eulerian–Lagrangian solutions have resulted in numerical schemes which eliminate numerical dispersion (Wheeler, 1988). These and other solution techniques are discussed in much more detail in other chapters of this book.

Example of a Numerical Solution. Examples of numerical solutions are numerous because most problems do not lend themselves to solution in any other way without tremendous physically unrealistic simplifications. This is evidenced by the large number of codes described by Javandel et al. (1984) and Duffy et al. (1990) who review several more recent codes oriented more specifically toward nonpoint sources. See their papers for a listing of the models and references. Figure 8 is an example of the solutions that can be obtained using a numerical scheme. It shows the evolution of a nonreactive contaminant in a sandy aquifer. The contaminant originates at a narrow nonpoint source as indicated. See Frind and Hokkanen (1987) for more information about the specific site. Figure 9 is an example of a vertical 2–D analysis of flow of nitrate along a stream line. The model shows path lines, travel times, and isochrones. For more information on this example see Kinzelbach et al. (1990).

Advantages and Limitation of Numerical Methods. Javandel et al. describe the advantages and limitations of numerical solutions; they are summarized here. Numerical methods are advantageous to use because: (1) spatial and temporal variation of system properties such as hydraulic conductivity, porosity, and dispersivity can be easily managed; (2) field problems

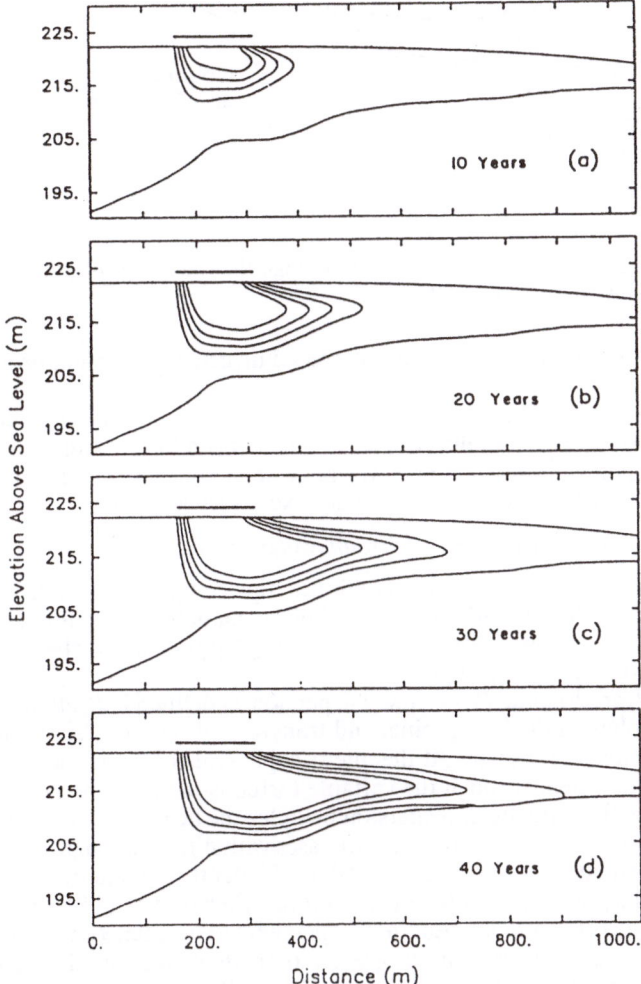

Figure 8. Evolution of the Borden plume to 40 years, constant source concentration; contour interval 0.2 relative concentration (Frind and Hokkanen, 1987) (after Bogardi et al., 1990).

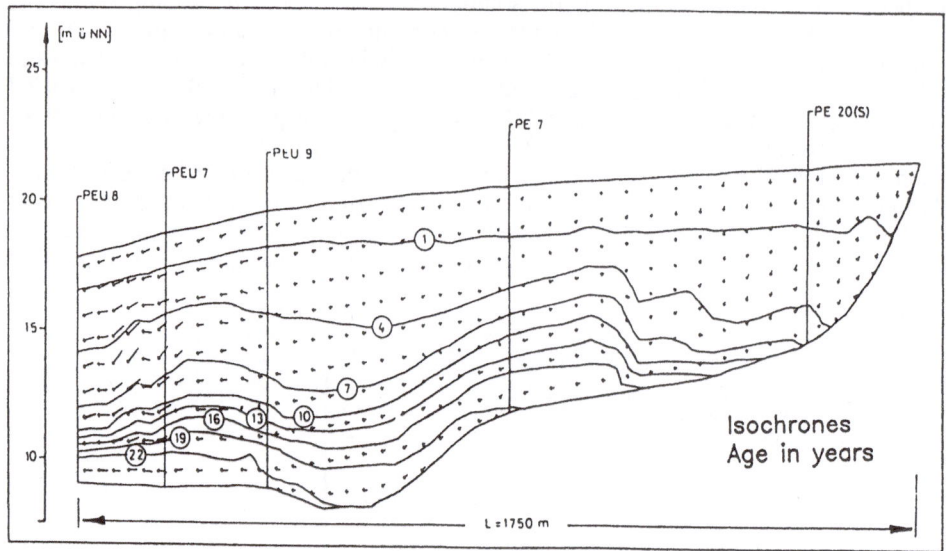

Figure 9. Isochrones and pathlines in a vertical cross–section through the catchment of Mussam water works (Obermann, 1981) (Kinzelbach et al., 1990).

with complex boundary conditions are tractable; and (3) three–dimensional transient problems can be solved.

Use of numerical methods is limited because (1) the programs are complex numerical codes that require a certain level of user familiarity that is time consuming to achieve, (2) errors due to numerical dispersion can overshadow physical dispersion in some codes, and (3) even for simple problems preparation of input data can take a very long time.

2.1.5. Special Geophysical and Chemical Features of Groundwater Models

Modeling agrochemical transport in groundwater combines the complexities of usual pollutant transfer in the aquifer with difficulties of a little–known spatially and temporally distributed source function. There are also several features of the geophysical and chemical state of the system that are particularly important.

Macrodispersion and Other Hydrodynamic Properties. Dispersion as defined in Section 2.1 represents the spreading of a solute in the longitudinal and transverse directions relative to the position predicted by advective movement. At the microscopic scale of the laboratory, dispersion is associated with movement through the variable tortuous paths in the porous medium. At the macroscopic scale of the field, dispersion includes the heterogeneity of the aquifer properties. In the laboratory Fickian diffusion processes with different longitudinal and transverse dispersivities adequately describe observed data. Under these conditions, the dispersivity is on the order of the grain size. In the field, however, Gelhar et al. (1985) compiled results from numerous tests that show longitudinal dispersivity to be a function of displacement and the values are orders of magnitude greater than those measured in the laboratory. Figure 10 shows values obtained by Gelhar et al. (1985). Reasons for the magnitudes observed by Gelhar et al. (1985) are not fully known; this is an area of much controversy. However, it is known that transverse dispersivity is an order of magnitude less than the longitudinal dispersivity. For additional discussion of the topic and references see Kinzelbach et al. (1990) and Gelhar et al. (1985).

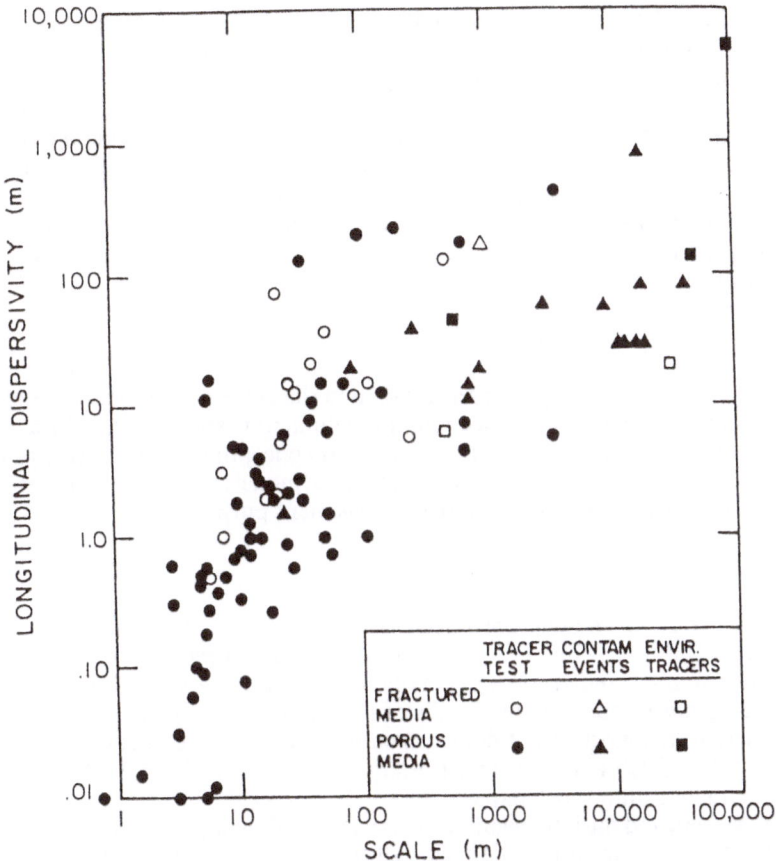

Figure 10. Scale of observation versus longitudinal dispersivity for the saturated zone (Gelhar et al., 1985) (Kinzelbach et al., 1990).

Anisotropy of hydraulic conductivity is confounded with vertical mixing and can be determined best near down–gradient boundaries where the effect of anisotropy on flow trajectories in the vertical plane is most pronounced. Plumes from point sources such as dairies, pesticide wash sites, etc., can provide valuable data for model testing if the source and plume are well defined.

Chemical Features. Some chemical processes have been studied extensively in the laboratory. These processes include equilibrium controlled ion–exchange, sorption, complexation and precipitation/dissolution. Equilibrium–controlled reactions apply when reaction rates are much faster than groundwater flow processes. Theoretical mathematical frameworks have been developed for many of these reactions and parameters entering into the equations have shown a consistent behavior, but most of these processes have not been validated with well documented field applications. Because of natural geological heterogeneity the laboratory determined rate relationships may not transfer to the field.

The state of modeling agricultural pollutants (nitrate and pesticides) is even less satisfactory than that for geochemicals (Kinzelbach et al., 1990). The fact that microorganisms are responsible for most of the nutrient cycling processes creates a common problem of lab–to–

field transferability of rate constants. Combined with a lack of knowledge about organic matter levels deep in the soil horizons and in aquifers, this leads to much uncertainty in modeling agricultural pollutants.

A major process of interest in groundwater modeling is denitrification. It requires anaerobic conditions. Denitrification occurs when microbes in the soil use organic carbon as the energy source and break down nitrate to get oxygen they need. In a second method of denitrification Thiobacillus denitrificans use pyrite as the energy source and nitrate for oxygen in a process similar to that of denitrification using organic carbon.

Adsorption parameters as needed in groundwater quality models are normally derived from batch adsorption and column tests; these values are thought to be satisfactory for field or regional scale (Kinzelbach et al., 1990).

2.2. Surface Runoff Water Quality Models

The hydrologic, chemical, and physical processes involved in the entrainment and transport of sediment, nutrients, and pesticides in surface runoff are numerous and complex. Models have been developed that provide a framework in which to conceptualize and investigate these processes and their interactions. Equally important these models have provided a set of tools to apply the resulting knowledge for resource management purposes.

2.2.1. Hydrologic Processes

A large number of surface runoff models have been developed; they include lumped- and distributed-parameter, continuous- and event-type models for use at plot, field, and small watershed scales. The number and diversity of models precludes extensive comparison. However, several recent state-of-the-art reviews of surface runoff water quality models provide excellent references to the most recent developments; see Leavesley et al. (1990), Rose et al. (1990), Oliver et al. (1990), and Seip and Botterweg (1990). Much of the following material has been abstracted from those papers.

The surface-runoff system and the interrelations of its water, sediment, nutrient, and pesticide components and processes are briefly defined. Of particular importance in the water quality aspects of surface runoff are the roles that vegetation and soil surface play. The vegetation canopy provides a surface for several processes. Rainfall is intercepted by leaf surfaces and evaporated back to the atmosphere. The leaves transpire water obtained from the soil profile, thus decreasing available soil moisture. Pesticide applications are intercepted by leaf surfaces where the chemicals reside for a period of time before being washed to the soil surface by rainfall. During residence on the surface of leaves or the soil, pesticides may be lost through photolysis, oxidation, volatilization, or degraded through biological and chemical processes.

The soil surface and profile provide major controls on the response of the surface-water system. During interstorm periods, nutrients and pesticides may be applied and undergo a variety of transformation and degradation processes affecting the total mass and chemical makeup of each constituent available for entrainment and transport. Land-use practices, such as tillage, may be applied to the soil affecting the infiltration, runoff, and erosion processes and the location of the chemicals.

Infiltration rates for rainfall and snowmelt are a function of soil properties and moisture content of the soil profile, especially the surface. When rainfall or snowmelt rates exceed infiltration rates, surface runoff occurs. With surface runoff comes the opportunity for the detachment and transport of sediment, nutrients, and pesticides. Nutrients and pesticides can be dissolved in surface runoff or adsorbed to sediment particles. The quantities available, the rates at which they are made available for transport, and their apportionment between the solution and sediment phases are functions of residence time, soil moisture and temperature, chemical and biological processes, and the physical characteristics of the individual nutrient and pesticide constituents and soils (Leavesley et al.).

The hydrologic aspects of surface runoff are described in great detail throughout this book; therefore, I will confine my comments to the erosion, sediment entrainment, nutrient and pesticides properties that are simulated and piggybacked on various hydrologic models. I will also confine my comments to more fundamentally based techniques.

2.2.2. Erosion and Sediment Entrainment Processes

Erosion and sedimentation involve the processes of detachment, transportation, and deposition of soil particles by raindrop impact and overland flow. All three processes are dominant under some circumstances in controlling the volume of sediment eroded from a given site as they respond to the meteorologic and hydrologic conditions prior to and during an event. Rainfall intensity, soil texture, soil–surface conditions, antecedent moisture, soil–tillage history, crop or plant cover and many other factors affect the movement and destination of soil particles. Foster (1982) reviews the various techniques used in modeling the erosion process. The following review is abstracted from his analysis.

Detachment/Entrainment Processes. Detachment of soil particles is caused by two different but related processes: raindrop impact and overland flow. Raindrop impact is assumed to be the dominant mechanism for removing particles from interrill areas and shear stress of flowing water the dominant mechanism for removing particles from rill areas. Detachment by raindrop impact, D_{ri} is generally expressed as:

$$D_{ri} = d_{ri} \, k_{ri} I^a \tag{15}$$

in which d_{ri} is an interrill detachment coefficient, k_{ri} is a rainfall soil erodibility factor, I is rainfall intensity, and a is an exponent. Field research has shown the exponent, a, to have a value of 2.0 (models Opus and ANSWERS use this value – see Table 1 and reference list). When Equation (15) is used as an emperical expression (HSPF AND PRMS – see Table 1 and reference list), "a" is a fitted parameter. When water becomes ponded on the surface to a depth greater than 3–4 drop diameters, it inhibits rainfall detachment. Foster presents more information on interrill erosion.

Detachment, more correctly defined as entrainment by shear stress, in rill areas is expressed in a variety of ways. When sediment transport is not capacity–limited, erosion by overland flow, D_{ro}, can be expressed as:

$$D_{ro} = d_{ro} \, k_{ro} \, (\tau - \tau_{cr})^a \tag{16}$$

in which d_{ro} is a rill detachment coefficient, k_{ro} is a runoff soil erodibility factor, τ and τ_{cr} are actual and critical shear stresses, respectively, and "a" is an exponent. Frequently τ_{cr} is assumed zero and "a" is given the value of 1.5. In some models τ_{cr} is calculated directly from flow and land surface characteristics (ANSWERS and CREAMS – see Table 1 and reference list). In other models rill detachment is a transport–capacity–deficit expression; PRMS relates rill detachment to the difference between transport capacity and current transport rate; Opus uses a relation of potential rill detachment to transport capacity. Rose and Hairsine (1988) relate entrainment to stream power, aggregate density, slope, and particle settling velocity, presenting an expression that describes how rilling increases the efficiency of sediment transport.

Transport and Deposition Processes. Sediment placed into flowing or surface ponded water may not be transported downstream if turbulence in the water is not capable of keeping the

Table 1. Partial list of agricultural nonpoint source pollution models.

Model Acronym and Name		Source	Reference
PTR	Pesticide, Transport and Runoff	Stanford/Hydrocomp	Donigian, 1982
ARM	Agricultural Runoff Model	"	Donigian et al., 1977
NPS	Nonpoint Simulation Model	"	Donigian & Crawford, 1976
HSPF	Hydrologic Simulation Program – Fortran	"	Johanson et al., 1984
ACTMO	Agricultural Chemical Transport Model	USDA/ARS	Frere et al., 1975
CREAMS	Chemicals, Runoff, and Erosion from Agricultural Management Systems	"	Knisel, 1980
SWAM	Small Watershed Model	"	Alonso & DeCoursey, 1985
AGNPS	Agricultural Nonpoint Pollution Model	"	Young et al., 1987
ANSWERS	Areal Nonpoint Source Watershed Environment Response Simulation	Purdue	Beasley et al., 1980
UTM-TOX	Unified Transport Model for Toxics	Oak Ridge	Patterson et al., 1983
LANDRUN	Overland Flow and Pollution Generation Model	Wisconsin	Novotny et al., 1979
GAMES/ GAMESP	Guelph Model for Evaluating the Effects of Agricultural Management Systems on Erosion and Sedimentation/Phosphorus	Guelph	Cook et al., 1985
SEDIMENT	Method for targeting of Agricultural Soil Erosion and Sediment Loading to Streams	Environment Canada	Snell, 1984
LOADING FUNCTION	Nonpoint source	EPA	McElroy et al., 1976
FESHM	Finite Element Surface Hydrology Model	Virginia	Ross et al., 1978
OPUS	Field Scale Water Quality Model	USDA/ARS	Smith and Ferreira, 1988

material in suspension. Figure 11 is a conceptual model of the delivery of interrill sediment showing that on very shallow slopes, water may not have the capacity to transport the material in suspension, but at steeper slopes it can carry more than may be contributed. Alonso et al. (1981) reviewed eight commonly utilized transport equations in studying sediment transport in rill areas. The equations are not all alike, four were bed–load equations and four were total load equations. Total–load equations of Laursen (1958) and Yang (1973) and the bed–load equation of Yalin (1963) were found to be more appropriate for erosion type models. The Yalin equation is used in the ANSWERS, CREAMS and FESHM models (see Table 1 and reference list). Transport and erosion of bed and bank materials of larger channels is simulated by particle size fraction, using fundamental processes that can simulate armor development. Alonso (1985) describes its use in SWAM (see Table 1 and reference list). Channel routing of sediment is used to calculate the "delivery ratio" as sediment loads are dropped or picked up by water moving down through the channel system.

Surface Leaching of Chemicals and Transport Downstream. As water ponds on the soil surface, soluble chemicals in the surface layers of the soil mix with the infiltrating and surface water. Mixing is induced by a pumping action of rainfall (advection) and diffusion (see Figure 12). The depth of mixing in the soil is a function of rainfall intensity, depth of water on the

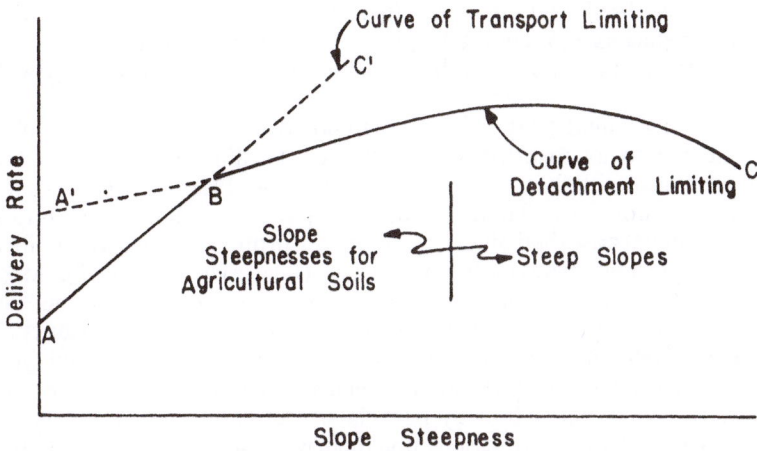

Figure 11. Conceptual model of the delivery rate of detached particles from interrill areas to rill flow (Foster, 1982, Hydrologic Modeling of Small Watersheds, ASAE Monograph 5).

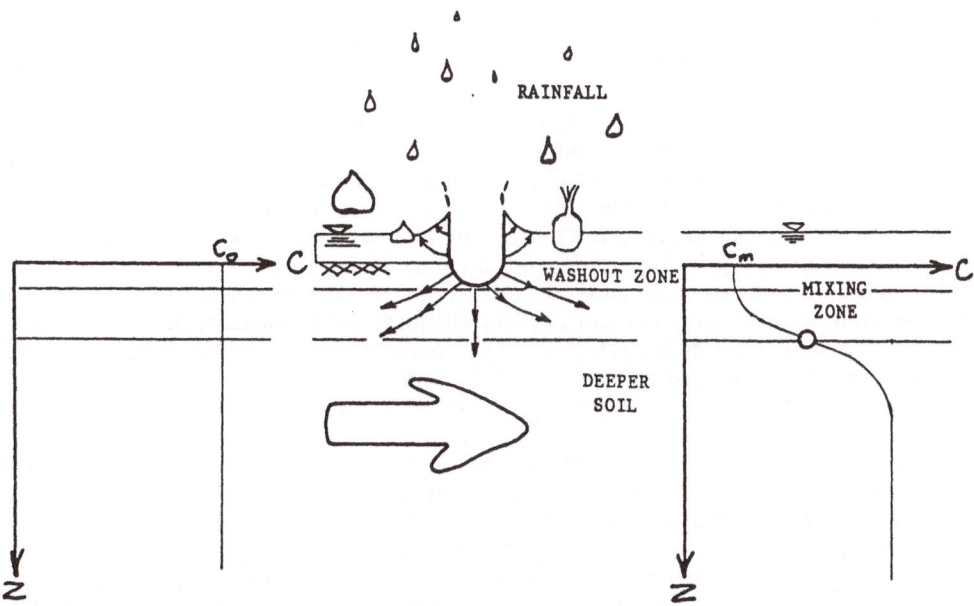

Figure 12. Cross section showing soil pore solution concentration profile before (left) and after a rainfall event (right) (Havis, 1986).

surface, and soil type. A soil depth of 1 cm or less is usually assumed, but in some cases diffusion from below the surface is allowed to take place. Ahuja (1986) assumes extraction can occur from as far down as 2 cm; however, the extraction decreases exponentially with depth.

Water and sediment leaving a field carry both soluble and adsorbed chemicals. Soluble chemicals such as nitrate that do not adsorb on organic matter or sediment can be treated as conservative elements and move with the water. For all practical purposes nitrate is the only plant nutrient that is lost in soluble form. The other nutrients (phosphorus and ammonia) are lost in both soluble and sediment adsorbed phases where the adsorption processes (equilibrium or kinetic) determine the ratios between the two phases (Leavesley et al., 1990). Most models do not simulate erosion and sediment transport by particle size fraction; thus, an enrichment ratio is calculated to specify the selective erosion of fine particles. Foster et al. (1985) and Rose and Dalal (1987) describe how the enrichment ratio depends on soil type, tillage, the erosion mechanism, and time. A few models such as Opus and SWAM route sediment through channels of concentrated flow by particle size fraction; thus, transport of adsorbed chemicals is calculated directly by associating them with the clay or organic matter portion of the sediment load. The processes of solution, sorption, and degradation of pesticides presented in Chapter 14 are equally applicable in stream flow routing to determine the distribution and concentrations of soluble and adsorbed phases as a function of time and distance (displacement) downstream. The calculation processes are mostly of a book-keeping nature if the channel erosion and deposition features (including particle size distribution) are known.

For a discussion of specific models and how the nutrient and pesticide processes are simulated, see Leavesley et al. (1990) and Rose et al. (1990). Table 1 lists some of the more popular models and their references. Nearly all of the processes discussed in Chapter 14 are also components of most complete surface water quality models. There is no need to repeat that discussion here.

2.3. Models of Surface Water Quality

Surface waters, as used in this chapter, consist of lakes, ponds, reservoirs, and larger river systems. Stefan et al. (1990) and Whitehead et al. (1990) recently completed excellent in-depth reviews of surface water quality models from the development and applications perspective, respectively. The following material is abstracted from those reports. Both papers present an excellent list of references and other current reviews of the literature.

Selection of an appropriate model depends upon the purpose, size, and shape of the water body and the time-scale of simulation needed. The time-scale is of particular importance because it influences the physical transport and chemical and biological transformation processes to be included. Expressions for energy transfer, chemical equilibrium or kinetics, and biological kinetics are integral components of the models. Models that have been developed simulate: sediment transport; temperature structure; organic wastes and dissolved oxygen concentration; and concentration of bacteria, nutrients, biomass, metals, synthetic organic chemicals, radioactive materials, and pesticides. Many of the lake and river water quality models discussed in the following materials are presented in Table 2.

The discussion of surface water quality models in the remainder of this chapter is divided into a description of hydrodynamic transport in standing and flowing waters, suspended sediment transport, organic waste and nutrient cycling and transport, and fate of toxic materials such as pesticides. The distinction between standing and flowing water is in thermo-hydro-dynamic differences. In rivers and streams, gravity is the main driving force and bed friction is the main resisting force leading to turbulent mixing. In meandering streams convective mixing caused by momentum induced secondary flows can overshadow turbulent mixing

Table 2. Some lake and river water quality models.

Model Name	Reference	Uses
WASP 4 (EUTRO4) and (TOX14)	Ambrose et al., 1987	All pollutants in most water bodies
EXAMS	Burns et al., 1982	Quasidynamic model for organics in lakes
RESQUAL II	Stefan et al., 1982	1-D model for sediments in lakes
CE-QUAL-R1	Env. Lab., 1986a	1-D model of sediment, organic wastes and nutrients in lakes and rivers
CE-QUAL-W2	Env. Lab., 1986b	2-D model of sediment organic wastes and nutrients in lakes and rivers
MINLAKE	Riley and Stefan, 1987	Organic wastes and nutrients in lakes
QUAL 2E	Brown and Barnwell, 1985	Intermediate level organic waste model for rivers
HSPF	Johanson et al., 1984	Sediments, organic wastes, nutrients, synthetic organics and metals in river systems
SERATRA	Onishi and Wise, 1982	2-D model for sediment, synthetic organics and metals transport in rivers
TODAM	Onishi et al., 1982	2-D model for sediment synthetic organics and metals transport in rivers
FETRA	Onishi, 1981	2-D model for sediment synthetic organics and metals transport in estuaries
UTM-TOX	Browman et al., 1983	A dynamic model for transport of synthetic organics and metals in rivers
AQUAMOD	Straskraba and Gnauck, 1985	Multi-layer, sediment, nutrient and organic matter transport model for river systems
Lake Balaton	Somlyody and Van Straten, 1986	2-D sediment, nutrient, organic matter transport model for large shallow lakes

(Fischer et al., 1979). The driving forces in lakes are wind, solar radiation, heat exchange, and advected energy from inflows and outflows. Because hydrodynamic forces in lake systems are weak, stratification, wind action, and other processes that are ignored in river systems are important. Another important distinction between lake and river systems lies in bed conditions. Lake beds tend to accumulate fine sediments and organic matter that support a continuous nutrient/biological exchange process that significantly influences lake water quality. River beds tend to remain clear and such activity is negligible.

2.3.1. Hydrodynamics of Lake Systems

Hydrodynamic models of lake systems can be divided into 0-, 1-, 2-, and 3-dimensional models (see Figure 13). The level 0 model is a surface-to-bed averaged process, the 1-D model is horizontally stratified and each layer is spatially averaged over the lake, the 2-D model is either vertically or horizontally stratified, and the 3-D model is as implied. Following is a discussion of each of these models.

0-D Models of Lake Water Quality. 0-D models contain no information on lake hydrodynamics. They are simple mass balance models, usually over long time intervals, of the form:

$$\frac{dC_i}{dt} = I_i - \frac{C_i}{\theta} \tag{17}$$

in which C_i is the concentration of material i, I_i is the sum of all material i inflows (overland flow, groundwater, atmosphere, internal sources, etc.) and θ is a weighted net dissipation rate

342

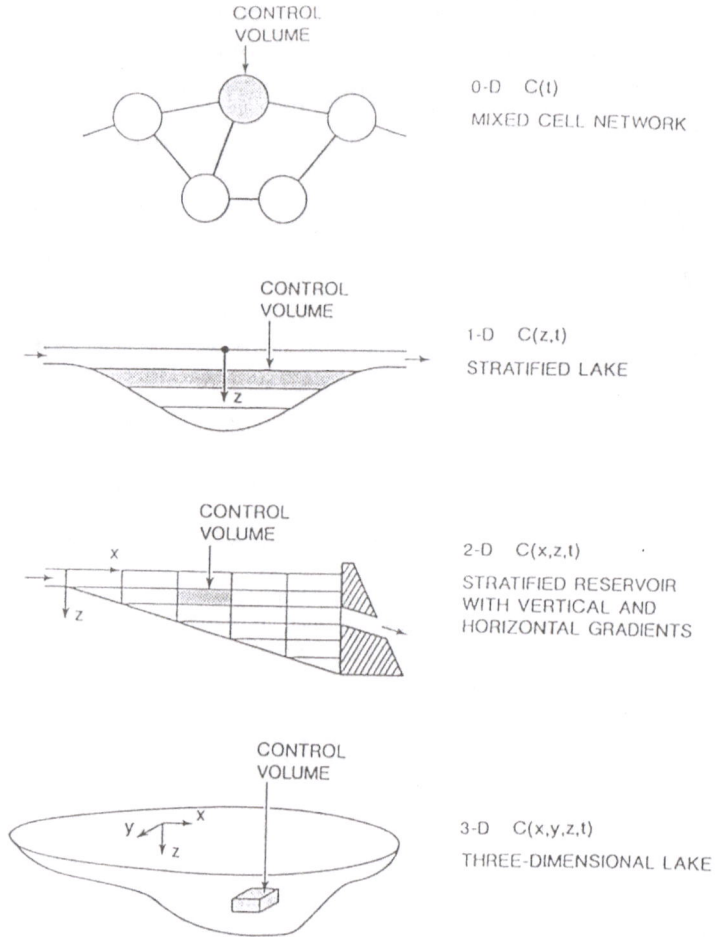

CONTROL
VOLUME

0-D   C(t)

MIXED CELL NETWORK

CONTROL
VOLUME

1-D   C(z,t)

STRATIFIED LAKE

CONTROL
VOLUME

2-D   C(x,z,t)

STRATIFIED RESERVOIR
WITH VERTICAL AND
HORIZONTAL GRADIENTS

CONTROL
VOLUME

3-D   C(x,y,z,t)

THREE-DIMENSIONAL LAKE

Figure 13. Dimensionality of lake/reservoir models (Stefan et al., 1990).

coefficient from all sources (outflow, biological degradation or chemical transformation, etc.). If components of the dissipation rate coefficient are appropriately scaled, then the model can be used in management applications such as lake flushing created by increased flows, reduction in pollutant load by watershed management, etc. Even though these models are relatively simple, they have proven to be very useful. An early spectacular application of a 0–D model was the restoration of Lake Washington (Sonzogni et al., 1976). It has also been used effectively in eutrophication control through phosphorus management, manifested in Phosphorus–ban detergents, Phosphorus–removal from municipal sewage, etc. (Chapra and Reckhow, 1983). Other studies have extended the concept to shorter time intervals and added features such as wind and gravity circulation (Demetracopoulos and Stefan, 1983).

1–D Models of Lake Water Quality. If continuous or semi–continuous change in water quality of lakes are to be predicted, internal hydrodynamics associated with temperature stratification must be considered in describing vertical water quality gradients that are developed. One–dimensional models developed to describe such conditions consist of a system of horizontal layers, each of which is well mixed. Vertical transport of heat or material compounds between layers is governed by a diffusion equation of the form:

$$A(z)\frac{\partial T(z,t)}{\partial t} + \frac{\partial (Q_v(z,t), T(z,t))}{\partial z} = \frac{\partial}{\partial z}\left(K(z)A(z)\frac{\partial T(z,t)}{\partial z} \right)$$
$$+ \frac{H(z,t)}{\rho c} + B(z)u_i T_i - B(z)u_o T(z,t) \tag{18}$$

in which $A(z)$ is the horizontal area of the lake at depth z, $T(z,t)$ is the water temperature at depth z and time t, $H(z,t)$ is the internal distribution of heat sources due to radiation absorption within the water column, ρ is the density of water, c is specific heat of water, $K(z)$ is the bulk vertical turbulent diffusion coefficient, $Q_v(z,t)$ is the vertical flow rate, $B(z)$ is the reservoir width, T_i is the inflow temperature, and u_i and u_o are horizontal inflow and outflow velocities, respectively. Heat fluxes due to solar radiation, atmospheric radiation, back radiation, evaporation, and convection are applied at the water surface of the first layer. See Stefan et al. (1990) for references on calculation of the radiation and water flux terms and solution techniques. McCormick and Scavia (1981) derived an expression for the turbulent diffusion coefficient as a function of wind velocity and von Karman constant.

The transport and transformation of dissolved substances in a 1–D vertically stratified water quality model is described by an equation similar to Equation (18) and is used to calculate movement of solutes, etc. (Orlob and Selna, 1970). Figure 14 is a schematic model of the various processes that are simulated in 1–D models.

A large number of models have been developed that simulate the 1–D processes described above [see Orlob (1983) and Stefan et al. (1990)]. One model in particular is worth mentioning: the Corps of Engineers CE–QUAL–R1 model (Environmental Laboratory, 1986a) simulates temperature and 34 other water quality parameters. It includes radiation inputs, wind induced mixing, advective inflows, density current inputs, selective withdrawal, and a variety of biochemical processes such as photosynthesis, respiration, organic matter decomposition, nitrification, phosphorus recycling, trophic relations for phytoplankton and macrophytes,

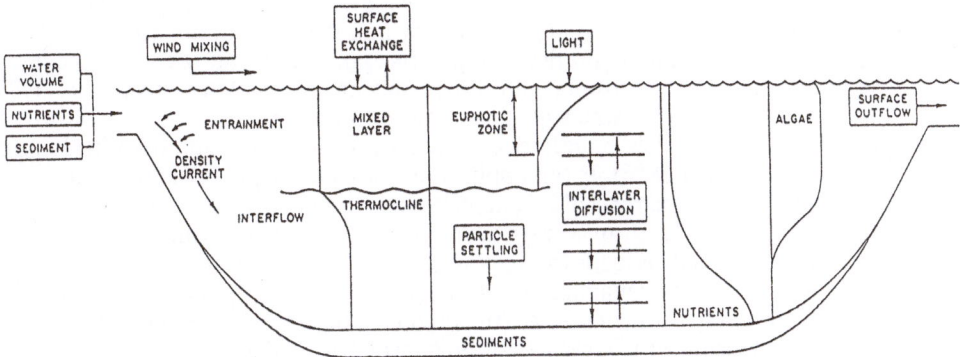

Figure 14. Schematic representation of the variables and processes simulated in the 1–D MINLAKE model (Stefan et al., 1990).

and a wide variety of other features. It is used to evaluate the water quality features of Corps of Engineers reservoirs.

2–D Models of Lake Water Quality. Two–dimensional water quality models were developed primarily for long deep lakes in which there are both vertical and horizontal water quality gradients. Lake Powell and Lake Mead on the Colorado River and the TVA lakes are examples of such lake systems. In general the 2–D models solve two–dimensional advection/ diffusion equations in a vertical longitudinal plane through a reservoir. An exception is the use of a 2–D model to simulate Lake Balaton in Hungary where the solution was on a horizontal plane. The following discussion of the solution mechanisms of 2–D models does not necessarily apply to the horizontal model.

Vertical 2–D models predict the spatial hydrodynamics and 2–D temperature structure of deep lakes throughout the annual stratification cycle. The hydrodynamic processes and temperature gradients provide the advective components of the 2–D models. 2–D models include varying numbers of water quality constituents. Temperature and velocity gradients are primarily in the vertical and longitudinal directions; lateral contributions are introduced through lateral integration. Turbulent processes occur at scales smaller than the model resolution and are included using empirical functions. Boundary conditions imposed on 2–D models are similar to those imposed in 1–D models. Implicit and explicit solutions of finite difference schemes are used to solve the equations in most models. Most recent development in 2–D reservoir modeling includes solution of the fundamental, fully convective primitive equations on supercomputers (Farrell and Stefan, 1986). See Stefan et al. and referenced papers for a more complete discussion of these processes.

There are far fewer 2–D water quality models than 1–D models. One is the LARM (Laterally Averaged Reservoir Model) (Edinger and Buchak, 1983) which has been modified by the Corps of Engineers to include 20 water quality constituents. A newer version of the model is referred to as CE–QUAL–W2 (Environmental Laboratory, 1986b). It is similar in many respects to CE–QUAL–R1, but its bottom sediment interaction is not as complete as in the 1–D version. See Stefan et al. for a summary of the differences. Figure 15 is an example of the output obtained from 2–D models (CE–QUAL–W2). It shows change in dissolved oxygen (DO) concentrations in both vertical and horizontal directions at different times of the year. Another model developed primarily for use in TVA lakes is COORS (Computation of Reservoir Stratification) by Waldrop et al. (1980), Harper and Waldrop (1980a,b), and TVA (1986).

An alternative approach to lake modeling is described by Brown and Barnwell (1985) in a model called Box Exchange Transport Temperature and Ecology of Reservoirs (BETTER). In this model the lake is arrayed into interacting elements or boxes with an upper surface and downstream conveyance. Layer spacing is arbitrary and dependent on vertical gradients. Flow patterns are based on inflow, outflow, temperature, and wind. Inflows are placed into a given layer based on density; horizontal and vertical mixing are simulated. Lam et al. (1983) used a box–type 2–D model for water quality studies of Lake Erie.

An application of 2–D models in the horizontal plane is the Lake Balaton study in Hungary (Somlyody and Van Straten, 1986). In this application, the lake is depth integrated rather than transverse integrated as in the vertical 2–D models. Lake Balaton is a very wide shallow lake (surface area about 600 km^2 with an average depth of only 3.2 m) that became eutrophic in the 1970's as a result of phosphorus loads from sewage and agricultural drainage. The 2–D model is being used to help determine management strategies for cleaning up the lake by 2005-2010 through removal of phosphorus in sewage (the predominant phosphorus source). A major contributor to lag in clean–up is recycling of the phosphorus in deposited lake sediments.

3–D Models of Lake Water Quality. Three–dimensional lake water quality models have not been used very extensively because of their computational expense. Also these large models suffer from computational problems in turbulence closing. An example of use of a 3–D model

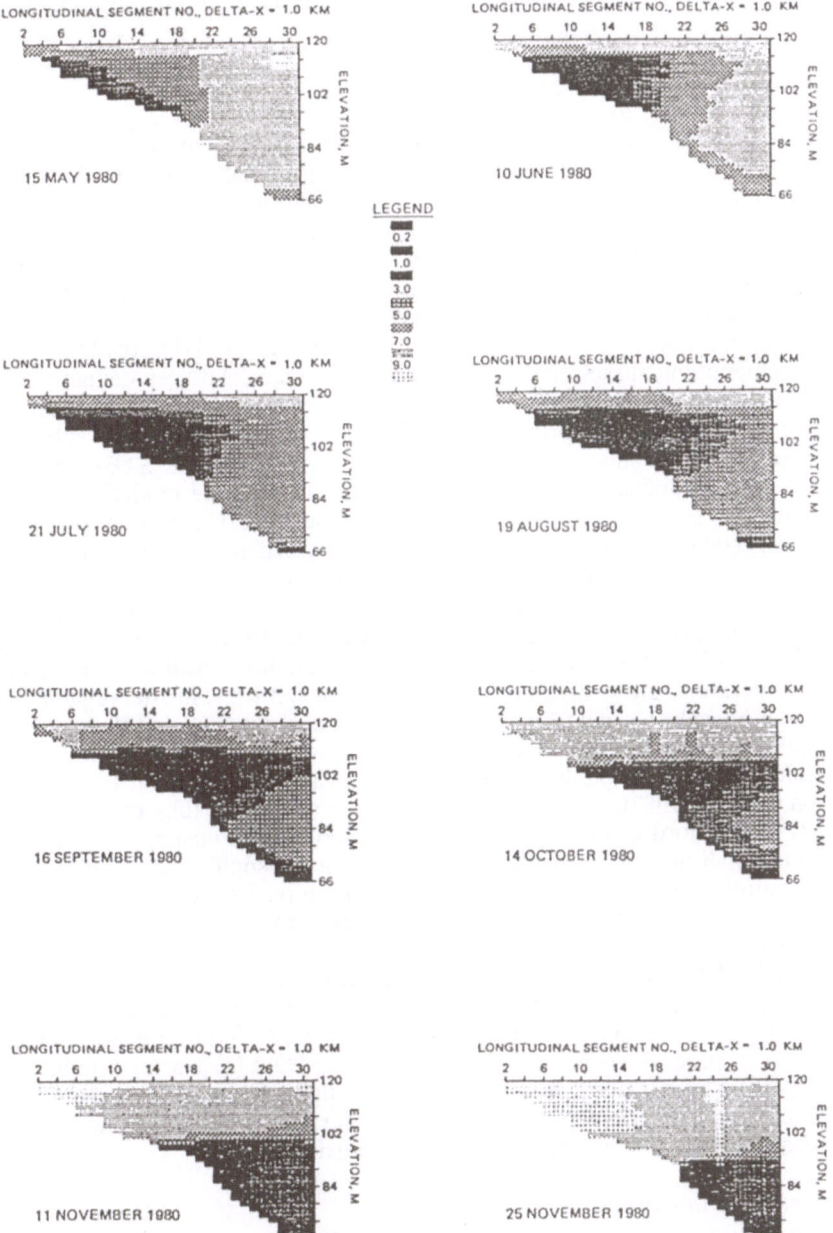

Figure 15. Shaded contours of predicted DO concentration using CE–QUAL–W2 (Martin, 1987) for the 1980 DeGray Lake verification simulations (concentrations milligrams per liter) (after Stefan et al., 1990).

is a study of pollutant transport in the Baltic Sea. It included seasonally and depth–varied eddy viscosities and eddy diffusivities (Stefan et al., 1990). 3–D models have also been used in modeling Lake Ontario (Thomann et al., 1975, 1979). More recently they are being developed in massive studies of the New York–New Jersey estuary and the Chesapeake Bay (Stefan et al., 1990).

2.3.2. Hydrodynamics of River Systems

Water quality models of river systems range from rather simple to very sophisticated unsteady flow models. Most models are 1–D, aligned longitudinally along the flow axis to coincide with dominant water quality gradients. In the following material a distinction is made between stream water quality and stream mixing models. Water quality models predict changes in water quality constituents due to transport, loading, and reactions. Stream mixing models typically predict downstream transport of a pollutant load but omit constituent interactions and reactions, i.e. the pollutant is considered conservative.

River Water Quality Models. The simplest water quality models are 1–D, steady state, first order decay–sedimentation, dissolved oxygen, DO, and biochemical oxygen demand (BOD) type models. Analytical versions of these models are easy to apply and have minimal input data requirements. They do not provide time varying response and require substantial simplification of stream channel geometry. The Corps of Engineers STEADY model (Martin, 1986) is an analytical example of this type of steady–state model; it simulates temperature, DO, and carbonaceous biochemical oxygen demand (CBOD). A similar U.S. Geological Survey (USGS) model simulates nitrogenous oxygen demand, CBOD, orthophosphate phosphorus, total fecal coliform bacteria, and three conservative substances.

Numerical versions of the above models are more flexible. An example of a model intermediate between fully dynamic and steady–state is QUAL2E (Brown and Barnwell, 1985). The 1–D version of QUAL2E simulates a total of 15 water quality constituents that include DO, CBOD, temperature, algae, organic nitrogen, ammonia, nitrate, nitrite, organic phosphorus, inorganic phosphorus, coliform, an arbitrary nonconservative, and three arbitrary conservative substances. It is an accepted standard in waste–load allocation studies.

Fully dynamic models are capable of simulating time varying constituent transport and transformation (Jobson, 1987). A Corps of Engineers' version of a fully dynamic model, CE–QUAL–RIV1 (Bedford et al., 1982), accurately resolves the transport of steep water quality gradients with little numerical diffusion. It incorporates sophisticated hydraulic and water quality routing components and simulates temperature, DO, CBOD, and nutrient kinetics. A reaeration feature takes into account stream reaeration, wind–driven reaeration, and control structure reaeration.

Stream Mixing Models. Stream mixing models are rather specialized point or nonpoint source models designed to determine the flow and concentration gradients in the vicinity of point source discharges. The models are very specialized hydrodynamic models as indicated by the types of mixing processes considered (see Stefan et al., 1990): (a) jets due to momentum of the discharge, (b) lateral displacement by effluent input, (c) downstream advection, (d) transverse mixing, (e) buoyant spreading due to temperature and sediment load density fields, (f) vertical turbulent mixing, and (g) navigation mixing.

2.3.3. Water/Sediment Interaction

Since many chemicals move in association with the fine sediment fraction (silts, clays, organic detritus, and plankton) and interact with sediments deposited in lake and stream beds, an important component of any lake/river model is sediment transport and sediment/water interaction. Through adsorption, biofilm, and other chemical/biochemical transformations, stream or lake sediments can become sinks or sources of materials including oxygen, toxic materials, and nutrients. Coarse sediments are also important in some streams where they bury contaminants, including radionuclides with long half lives. The fine sediments are characterized by size, shape, density, surface area, and surface physical and chemical properties, including electrical charges (Lal, 1977).

Physical and Chemical Processes. The physical and chemical processes related to sediment/ water quality of lakes are briefly described (see Stefan et al., 1990 for more detail and references). The fall velocity of particles is needed to characterize the movement of sorbed chemicals downward through the water column and estimate the probability of bed content at any given point. Enough is known about the drag coefficient of large mineral particles that their fall velocities can be calculated, but frequently the fall velocity of micron–size particles must be obtained experimentally. The possibility of flocculation which can increase fall velocity must be considered under certain conditions. Once deposited the material is subject to resuspension. If the material is granular and noncohesive the resuspension rate is a power function of excess shear stress (exponents as high as five have been observed). At some point deposition and resuspension are equal and the vertical sediment distribution will be uniform. The uniform distribution is expressed as a function of fall velocity and bed shear velocity. For cohesive materials the resuspension rate is a function of particle size, shear stress and cohesion or the degree of consolidation of surficial benthic deposits. Good estimates require site–specific calibration. Because the sizes of some of these deposited sediments is so small, perturbations caused by navigation of organisms can cause resuspension. When concentrations of suspended sediment in inflowing waters reach certain levels, the density of the water sediment mix becomes great enough that it causes the fluid to behave differently than surrounding water and density currents develop that can carry material over very long distances. Depositional features of these currents are still under investigation.

Sediment Transport in Lake Models. Models of sediment transport in most small to moderate size lakes are 1-D in the vertical. Advection in the horizontal direction is assumed rapid; thus, the water in horizontal layers is well mixed and only vertical gradients in suspended sediment concentrations are simulated. The governing equations for such models are:

$$A(z)\frac{\partial(C)}{\partial t} + \frac{\partial(V_s\, A(z)C)}{\partial z} = V_sC\frac{\partial A(z)}{(\partial z)} + \frac{\partial}{\partial z}\left(A(z)K(z)\frac{\partial C}{\partial z}\right) - R\frac{\partial A(z)}{\partial z} \qquad (19)$$

$$(1) \qquad\quad (2) \qquad\qquad (3) \qquad\qquad (4) \qquad\qquad (5)$$

in which C is the suspended sediment concentration, V_s is the fall velocity of suspended sediment in quiescent water, $A(z)$ is the horizontal area of the lake at depth z, $K(z)$ is the bulk vertical turbulent diffusion coefficient as a function of depth, and R is a resuspension coefficient. The first term is change in sediment content with time, the second term is rate of transfer by settling from one layer to the next, the third term is rate of deposition on the lake bed, the fourth term is the vertical mixing rate, and the fifth term is resuspension. RESQUAL II, previously described, is a good example of a 1–D model based on Equation (19).

HEC–6 (Hydrologic Engineering Center, 1977) is a frequently used model in which bed load and suspended sediment are combined. It is designed to analyze scour and deposition in rivers and reservoirs. It calculates transport of coarse sediments, sands, silts, and clays in both bed and suspended loads.

For larger lake or river systems where horizontal sediment gradients are significant, 2–D models have been developed. They exist as horizontal models in which the governing equations are depth integrated and vertical models in which the governing equations are width integrated. STUDH, a part of the TABS–2 flow model, (Thomas and McAnally, 1985) is a horizontal 2–D model designed for situations where flow and transport can be depth integrated, i.e. very shallow lakes. It calculates bed and suspended loads of silts and clays and can handle the wetting and drying of cells (the peripheral area of lakes) and consolidation. LAEMSED (Johnson et al., 1987), a vertical 2–D model designed generally for long deep reservoirs, is a modification of the LARM2 model, previously described, upon which CE–QUAL–W2 was based. LAEMSED handles only suspended sediment.

Several 3–D models have been developed. They are of both finite element and finite difference format with a free surface, time dependence, and allowance for stratification and complex geometry. A finite element version of the RMA series of models (King, 1982), computes water levels and velocities for density currents and salinity and temperature transport. CELC3D (Sheng, 1983) is a 3–D finite difference model for coastal, estuarine, and lake currents.

2.3.4. Organic Wastes and Nutrients

Lake water quality problems created by organic wastes and nutrients include depletion of dissolved oxygen and stimulation of nuisance aquatic growth. These organic wastes and nutrients come primarily from runoff from agricultural areas. In the previous chapter, nutrient cycling and related processes were discussed; therefore, this discussion will only be from the perspective of lake water quality modeling. Since zooplankton and fish impact phytoplankton, BOD, and DO, they are frequently included in lake water quality models along with simulation of temperature, nutrients, other chemical characteristics, detritus, bacteria and primary producers.

Nutrient fate processes (nutrient cycling), previously described, also take place at the surface of lake bed deposits. Lake models calculate these processes as part of the simulation because of their impact on water quality and interaction with phytoplankton and eutrophication processes. Phytoplankton kinetics affect nitrogen, phosphorus, and carbon cycles through uptake and death. The growth rate of a phytoplankton is a complicated function of the species present and their reactions to solar radiation, temperature, and nutrient availability. Data on growth kinetics of individual species are not sufficient to enable each species to be modeled individually; thus, they are simulated by classes such as greens, diatoms, blue–greens, and dinoflagellates. Their growth rate, G, is simulated as:

$$G = G_{\max} X_T X_L X_N \tag{20}$$

in which $G_{\max}$ is optimal growth rate; X_T, X_L, and X_N are temperature, light, and nutrient adjustment factors, respectively. See Stefan et al. (1990) for a description of the three adjustment factors. Overall phytoplankton death rates are normally expressed as the sum of first order endogenous respiration, death, and grazing rates. At times the grazing rate can be second order.

Organic carbon is present in water in various particulate and dissolved forms that oxidize and settle at different rates. Some models lump all organic carbon into a single state variable expressed in units of oxygen–carbonaceous biochemical oxygen demand (CBOD). Other models represent various fractions of organic carbon, with their separate oxidation and settling rates. Oxidation is generally modeled as a first order, temperature–corrected rate. Space does not permit a detailed discussion of these processes (see Stefan et al., 1990 and his references).

Dissolved oxygen is depleted by oxidation of organic carbon, nitrification, and respiration. Respiration may be combined or separated into components (bacterial, plankton, macrophyte, fish, etc.). The decomposition of organic material in benthic sediment can significantly affect the concentrations of oxygen and nutrients in overlying waters. Most models describe these benthic fluxes as spatially variable source and sink terms. Dissolved oxygen is replenished by phytoplankton growth (photosynthesis) and by reaeration.

The organic wastes, nutrient, and oxygen processes previously discussed have been incorporated in many models, but Stefan et al. (1990) describe a few that have been maintained for general application. To be satisfactory for general application to agricultural and organic waste simulations, a model should allow for unsteady transport and loading, phytoplankton kinetics, nitrogen and phosphorus cycles, and carbon oxygen balances. Several models previously described meet these criteria. They are CE-QUAL-R1, CE-QUAL-W2, CE-QUAL-RIV1.

2.3.5. Synthetic Organic Chemicals (Pesticides)

Synthetic organic chemicals present in lake and river waters are subject to processes of ionization, sorption, and other processes. Ionization is the formation of ionic species from neutral species and can occur by dissociation, protonation, and other reactions. This takes place rapidly and the equilibrium distribution between ionized and unionized species is pH controlled. These materials are also adsorbed onto suspended sediment, biological material, and dissolved or colloidal organic material. Sorption is important in controlling both the environmental fate and toxicity of these chemicals. It may cause chemicals to accumulate in bed sediments and bioconcentrate in fish. Sorption can also retard volatilization and base hydrolysis and enhance photolysis and acid–catalyzed hydrolysis. Sorption is usually rapid, and equilibrium conditions controlled by partitioning coefficients can be assumed. Even though hysteresis, in which desorption takes place at a different rate than sorption, exists, data do not provide enough information for it to be simulated.

Chemicals in lake and river water are lost over time as a result of physical and chemical reactions. These losses are caused by complexation, volatilization, hydrolysis, photolysis, bacterial degradation, and chemical oxidation and reduction as well as advection out of the system. Most of the processes are modeled as exponential decay.

Numerous simulation models have been developed for toxic chemicals in lake and river waters, but only a few include the important benthic and water column compartments, unsteady transport and loading, partitioning and transformation reactions necessary for adequate simulation. Stefan et al. (1990) discusses numerous models and their features.

2.3.6. Metals

Metals are found naturally in the Earth's crust. Irrigation return flows, evaporation, and other processes have brought these materials to the surface where they have been dissolved and transported into surface waters. These materials are subject to the processes described for synthetic organic chemicals; only some of the metals undergo processes of complexation. Complexation is the process of forming complexes with organic and inorganic liquids which precipitate or dissolve. Some models include these processes. See Stefan et al. (1990) for more information.

A summary of the various lake and river water quality models that have been discussed in this chapter are presented in Table 2.

3. Conclusions

Over the past several decades, efforts by the Environmental Protection Agency and environmental groups have reduced contamination of surface waters to the point that nonpoint sources of contamination are now the major contributors. Emphasis has also been placed on identifying and cleaning up point sources of groundwater contamination. Thus, efforts now are directed toward reducing nonpoint sources of contamination of both groundwater and surface water resources. In this chapter, mathematical models developed to aid in studying and solving the complex problems of nonpoint sources of pollution in receiving waters are briefly described. Receiving waters are assumed to be groundwater, surface water streams and large rivers and reservoirs. The volume of material describing models available to aid in studying receiving water quality and solving specific problems is very large. The numerous references identified in this chapter should lead the reader to appropriate models for solution to most problems.

4. References

Ahuja, L. R.: 1986, 'Characterization and Modeling of Chemical Transfer to Runoff,' *Advances Soil Sci.* **4**, 149–188.

Alonso, C. V.: 1985, 'Instream Sediment Routing in the SWAM Model,' in D. G. DeCoursey (ed.), *Proceedings of Natural Resources Modeling Symposium*, Pingree Park, CO, Oct. 16–21, 1983, pp. 490–494.

Alonso, C. V., and D. G. DeCoursey: 1985, 'Small Watershed Model,' in D. G. DeCoursey (ed.), *Proceedings of the Natural Resources Modeling Symposium*, Pingree Park, CO, Oct. 16–21, 1983, USDA–ARS–30, pp. 40–46.

Alonso, C. V., W. H. Neibling, and G. R. Foster: 1981, 'Estimating Sediment Transport Capacity in Watershed Modeling,' *Trans. ASAE* **24**(5), 1211–1220, 1226.

Ambrose, R. B. Jr., et al.: 1987, *WASP4, A General Water Quality Model for Toxic and Conventional Pollutants,'* U.S. Environmental Protection Agency, Athens, GA.

Beasley, D. B., L. F. Huggins, and E. J. Monke: 1980, 'ANSWERS: A Model for Watershed Planning,' *Transactions of the ASAE* **23**(4), 938–944.

Bedford, K. W., R. M. Sykes, and C. Libicki: 1982, *'A Dynamic, One–Dimensional, Riverine Water Quality Model,'* Draft Report prepared under Contract No. DALW39–82–M–3548 by Ohio State University for U.S. Army Engineers Waterways Experiment Station, Vicksburg, MS.

Bogardi, I., J. J. Fried, E. O. Frind, W. Kelly, and P. E. Rijtema: 1990, 'Groundwater Pollution Modeling for Agriculture Nonpoint Sources,' in D. G. DeCoursey (ed), *Proceedings International Symposium on Water Quality Modeling of Agricultural Non–Point Sources*, June 19–23, 1988, Logan, UT, U.S. Dep. of Ag., Ag. Res. Serv., ARS–81, pp. 227–252.

Browman, M. G., et al.: 1983, *'Physio–Chemical Processes in the Environment: Backgound Information for the ORNL Unified Transport Model for Toxicants (UTM–TOX),'* Oak Ridge National Laboratory, ORNL–5854.

Brown, L. C., and T. O. Barnwell: 1985, *'Computer Program Documentation for the Enhanced Stream Water Quality Model QUAL–2E,'* Report EPA–600/3– 85–065, USEPA, Athens, GA.

Burns, L. A., et al.: 1982, *'Exposure Analysis Modeling Systems (EXAMS): Users Manual and System Documentation,'* U.S. Environmental Protection Agency, Athens, GA, EPA–600/3–82–023.

Chapra, S. C., and K. H. Reckhow: 1983, *Engineering Approaches for Lake Management, Vol. 2: Mechanistic Modeling*, Butterworth Publishers, Woburn, WA.

Cook, D. J., W. T. Dickinson, and R. P. Rudra: 1985, *'GAMES – The Guelph Model for Evaluating the Effects of Agricultural Management Systems in Erosion and Sedimentation. Users Manual,'* Technical Report No. 126–71 School of Engineering, University of Guelph, Guelph, Ont.

Dagan, G., D. Russo, and E. Bresler: 1990, 'Effect of Spatial Variability Upon Subsurface Transport of Solutes from Nonpoint Sources,' in D. G. DeCoursey (ed.), *Proceedings International Symposium on Water Quality Modeling of Agriculture Non–Point Sources*, June 19–23, 1988, Logan, UT, U.S. Dep. of Ag., Ag. Res. Serv., ARS–81, pp. 523–548.

Demetracopoulos, A. C., and H. Stefan: 1983, 'Model of Mississippi River Impoundment,' *ASCE, Jour. Hyd. Div.* **109**(HY10), 1006–1029.

Donigian, A. S. Jr.: 1982, 'Water Quality Modeling in Relation to Watershed Hydrology,' in V. J. Singh (ed.), *Modeling Components of Hydrologic Cycle*, Water Resources Publications, Littleton, CO, pp. 343–382.

Donigian, A. S. Jr., and N. H. Crawford: 1976, *'Modeling Nonpoint Pollution from the Land Surface,'* EPA–600/3–76–083, July, 1976, 280 p.

Donigian, A. S. Jr., D. C. Beyerlein, A. H. Davis, and N. H. Crawford: 1977, *'Agricultural Runoff Management Model, Version II: Refinement and Testing,'* EPA-600/377-098, U.S. Environ. Prot. Agency, 293 p.

Duffy, C. J., C. Kincaid, and P. Huyakorn: 1990, 'A Review of Groundwater Models for Assessment and Prediction of Nonpoint-Source Pollution,' in D. G. DeCoursey (ed.), *Proceedings of International Symposium on Water Quality Modeling of Agricultural Non-point Sources,* June 19-23, 1988, Logan, UT, U.S. Dep. of Ag., Ag. Res. Serv., ARS-81, pp. 253-276.

Edinger, J. E., and E. M. Buchak: 1983, *'Developments in LARM2: A Longitudinal-Vertical, Time-Varying Hydrodynamic Reservoir Model,'* Technical Report E-83-1, U.S. Army Waterways Experiment Station, Vicksburg, MS.

Environmental Laboratory: 1986a, *'CE-QUAL-R1: A Numerical One-Dimensional Model of Reservoir Water Quality: Users Manual,'* Instruction Report E-82-1, Revised Edition, U.S. Army Engineer Waterways Experiment Station, Vicksburg, MS.

Environmental Laboratory: 1986b, *'CE-QUAL-W2: A Numerical Two-Dimensional Laterally Averaged Model for Hydrodynamics and Water Quality: User's Manual,'* Instruction Report E-86-5, U.S. Army Engineer Waterways Experiment Station, Vicksburg, MS.

Farrell, G. J., and H. G. Stefan: 1986, *'Buoyancy Induced Plunging Flow into Reservoirs and Coastal Regions,'* Project Report No. 241, St. Anthony Falls Hydraulic Laboratory, University of Minnesota, Minneapolis, MN, 251 p.

Fischer, H. B., et al.: 1979, *Mixing in Inland and Coastal Waters,* Academic Press, Orlando, FL, 483 p.

Foster, G. R.: 1982, 'Modeling the Erosion Process,' Chapter 8 in *Hydrologic Modeling of Small Watersheds,* Monograph 5, ASAE, St. Joseph, MI, pp. 295-380.

Foster, G. R., R. A. Young, and W. H. Neibling: 1985, 'Sediment Composition for Non-Point Source Pollution Analysis,' *Trans. ASAE* **28**, 133-139, 146.

Frere, M. H., C. A. Onstad, and H. N. Holton: 1975, *'ACTMO - An Agricultural Chemical Transport Model,'* U.S.D.A., A.R.S. Publication No. ARS-H-3, Hyattsville, MD.

Frind, E. O., and D. G. Germain: 1986, 'Simulation of Contaminant Plumes with Large Dispersive Contrasts: Evaluation of Alternating Direction Galerkin Models,' *Water Resour. Res.* **22** (13), 1857-1873.

Frind, E. O., and G. E. Hokkanen: 1987, 'Simulation of the Borden Plume Using the Alternating Direction Galerkin Technique,' *Water Resour. Res.* **23**(5), 918-930.

Frind, E. O., and G. B. Matanga: 1985, 'The Dual Formulation of Flow for Contaminant Transport Modeling: 1. Review of Theory and Accuracy Aspects,' *Water Resour. Res.* **21**(2), 159-169.

Frind, E. O., G. B. Matanga, and J. A. Cherry: 1985, 'The Dual Formulation of Flow for Contaminant Transport Modeling: 2. The Borden Aquifer,' *Water Resour. Res.* **21**(2), 170-182.

Gelhar, L. W., A. Mantoglow, C. Welty, and K. R. Rehfeldt: 1985, *'A Review of Field-Scale Physical Solute Transport Processes in Saturated and Unsaturated Porous Media,'* Report RP-2485-05, Electric Power Research Institute.

Harper, W. L., and W. R. Waldrop: 1980a, 'A Two-Dimensional, Laterally Averaged Hydrodynamic Model with Application to Cherokee Reservoir,' *Proc. of Symposium on Surface-Water Impoundments,* University of Minnesota, Minneapolis, MN.

Harper, W. L., and W. R. Waldrop: 1980b, 'Numerical Hydrodynamics of Reservoir Stratification and Density Currents,' IAHR, Trondheim, Norway, *Proc. Second Internat. Symp. on Stratified Flows.*

Havis, R. N.: 1986, *'Solute Transport from Soil to Overland Flow,'* Ph.D. Dissertation presented to Dept. of Civil Eng., Colorado State Univ., 123 p.

Hydrologic Engineering Center: 1977, *'Scour and Deposition in Rivers and Reservoirs; User's Manual,'* Report HEC-6, Davis, CA.

Javandel, I., C. Daughty, and C. F. Tsang: 1984, *Groundwater Transport: Handbook of Mathematical Models,* A.G.U. Water Resour. Mono. #10, 228 p.

Johanson, R. C., J. C. Imhoff, J. L. Kittle, Jr., and A. S. Donigian, Jr.: 1984, *'Hydrological Simulation Program - Fortran (HSPF): Users Manual for Release 8.0,'* EPA-600/3-84-066, June, 1984, 767 p.

Jobson, H. E.: 1987, 'Lagrangian Model of Nitrogen Kinetics in the Chattahoochee River,' *Jour. Env. Eng., ASCE* **113**(2), 223-242.

Johnson, B. H., M. J. Trawle, and P. E. Kee: 1987, *'A Numerical Model of the Effects of Channel Deepening on Shoaling and Salinity Intrusion in the Savannah Estuary,'* USAE Waterways Experiment Station, Vicksburg, MS.

King, I. P.: 1982, 'A Finite Element Model for Three-Dimensional Flow,' Resources Management Associates, Lafayette, CA.

Kinzelbach, W. K. H., P. Dillon, and K. H. Jensen: 1990, 'State of The Art of Existing Numerical Groundwater Quality Models of the Saturated Zone and Experience with Their Application in Agricultural Problems,' in D. G. DeCoursey (ed.), Proceedings of International Symposium on Water Quality Modeling of Agricultural Non-Point Sources, June 19–23, 1988, Logan, UT, U.S. Dep. of Ag., Ag. Res. Serv., ARS–81, pp. 307–326.

Knisel, W. G. (ed.): 1980, 'CREAMS: A Field–Scale Model for Chemicals, Runoff, and Erosion from Agricultural Management Systems,' U.S. Department of Agriculture, Conservation Research Report No. 26, 640 p.

Lal, D.: 1977, 'The Oceanic Microcosm of Particles,' Science 198(4321), 997–1009.

Lam, D. C. L., W. M. Schertzer, and A. S. Fraser: 1983, 'Simulation of Lake Erie Water Quality Responses to Loading and Weather Variations,' Scientific Series No. 134, Nat. Water. Res. Inst., Inland Water Directorate, Canada Ctr. for Inland Water, Burlington, Ontario.

Laursen, E.: 1958, 'The Total Sediment Load of Streams,' Jour. Hyd. Div., ASCE 84(HY1), Paper 1530, 36 p.

Leavesley, G. H., D. B. Beasley, H. B. Pionke, and R. A. Leonard: 1990, 'Modeling of Agricultural Nonpoint–Source Surface Runoff and Sediment Yield – A Review from the Modeler's Perspective,' in D. G. DeCoursey (ed.), Proceedings of International Symposium on Water Quality Modeling of Agricultural Non–Point Sources, June 19–23, 1988, Logan, UT, U.S. Dep. of Ag., Ag. Res. Serv., ARS–81, pp. 171–194.

Martin, J. L.: 1987, 'Application of a Two–Dimensional Model of Hydrodynamics and Water Quality (CE–QUAL–W2) to De Gray Lake, Arkansas,' Tech. Rept. E–87–1, U.S. Army Corps of Engineers Waterways Experiment Station, Vicksburg, MS.

McCormick, M. J., and D. Scavia: 1981, 'Calculation of Vertical Profiles of Lake–Averaged Temperature and Diffusivity in Lakes Ontario and Washington,' Water Resour. Res. 17(2), 305–310.

McElroy, A. D., S. Y. Chiu, J. W. Nebgen, A. Aleti, and F. W. Bennett: 1976, 'Loading Functions for Assessment of Water Pollution from Nonpoint Sources,' U.S.E.P.A. EPA–600/2–76–151.

Novotny, V., M. Chin, and H. V. Tran: 1979, 'LANDRUN – An Overland Flow Mathematical Model: Users Manual, Calibration, and Use,' International Joint Commission, Windsor, ONT.

Obermann, P.: 1981, 'Hydrochemische/Hydromechanische Untersuchungen zum Stoffgehalt von Grundwasser bei Landwirtscheftlicher Nutzung.,' Besond. Mitt., Dtsch. Gewässerkundl. Fahrk. 42.

Oliver, G., J. Burt, and R. Solomon: 1990, 'The Use of Surface Runoff Models for Water Quality Decisions – A User's Perspective,' in D. G. DeCoursey (ed.), Proceedings of International Symposium on Water Quality Modeling of Agricultural Non–Point Sources, June 19–23, 1988, Logan, UT, U.S. Dep. of Ag., Ag. Res. Serv., ARS–81, pp. 197–204.

Onishi, Y.: 1981, 'Sediment Contaminant Transport Model,' ASCE J. Hyd. Div. 107(HY9), 1089–1107.

Onishi, Y., G. Whelan, and R. L. Skaggs: 1982, 'Development of a Multimedia Radionuclide Exposure Model for Low–Level Management,' PNL–3370, Pacific Northwest Laboratory, Richland, WA.

Onishi, Y., and S. E. Wise: 1982, 'Mathematical Model, SERATRA, for Sediment–Contaminant Transport in Rivers and its Application to Pesticide Transport in Four Mile and Wolf Creeks in Iowa,' U.S. Environmental Protection Agency, Athens, GA, EPA 600/3–82–045.

Orlob, G. T., and L. G. Selna: 1970, 'Temperature Variations in Deep Reservoirs,' ASCE, Jour. Hyd. Div. 96(HY2), 391–410.

Orlob, G. T. (ed.): 1983, Mathematical Modeling of Water Quality in Streams, Lakes and Reservoirs, John Wiley and Sons, 518 p.

Patterson, M. R., T. J. Sworski, A. L. Sjoreen, M. G. Browman, C. C. Coutant, D. M. Hetrick, B. D. Murphy, and R. J. Raridon: 1983, 'A User's Manual for UTM–TOX, A Unified Transport Model,' Prepared by Oak Ridge National Laboratory, Oak Ridge, TN, for U.S. E.P.A. Office of Toxic Substances, Washington, D.C.

Riley, M. J., and H. G. Stefan: 1987, 'Dynamic Lake Water Quality Simulation Model MINLAKE,' St. Anthony Falls Hyd. Lab. Rept. 263, University of Minnesota, Minneapolis, MN, 140 p.

Rose, C. W., and R. D. Dalal: 1987, 'Erosion and Runoff of Nitrogen,' in J. R. Wilon (ed.), Advances in Nitrogen Cycling in Agricultural Ecosystems, Commonwealth Agricultural Bureau (CAB International).

Rose, C. W., W. T. Dickenson, S. E. Jorgensen, and H. Ghadiri: 1990, 'Agricultural NonPoint–Source Runoff and Sediment Yield Water Quality (NPSWQ) Models: Modeler's Perspective,' in D. G. DeCoursey (ed.), Proceedings of International Symposium on Water Quality Modeling of Agricultural

Non–Point Sources, June 19–23, 1988, Logan, UT, U.S. Dep. of Ag., Ag. Res. Serv., ARS–81, pp. 145–170.

Rose, C. W., and P. B. Hairsine: 1988, 'Processes of Water Erosion,' in W. L. Steffen, O. T. Denmead, and I. White (eds.), *Flow Transport in the Natural Environment*, Springer-Verlag (in press).

Ross, B. B., V. O. Shanholtz, D. N. Contractor, and J. C. Carr: 1978, '*A Model for Evaluating the Effect of Land Uses on Flood Flows,*' Bull. 85, Virginia Water Resour. Res. Ctr., Blacksburg, VA, 137 p.

Seip, K. L., and P. Botterweg: 1990, 'User's Experiences and the Predictive Power of Sediment Yield and Surface Runoff Models,' in D. G. DeCoursey (ed.), *Proceedings of International Symposium on Water Quality Modeling of Agricultural Non–Point Sources*, June 19–23, 1988, Logan, UT, U.S. Dep. of Ag., Ag. Res. Serv. ARS–81, pp. 205–224.

Sheng, Y. P.: 1983, '*Mathematical Modeling of Three–Dimensional Coastal Currents and Sediment Dispersion: Model Development and Application,*' TR CERC–83–2, USAE Waterways Experiment Station, Vicksburg, MS.

Smith, R. E., and V. A. Ferreira: 1988, '*OPUS, an Advanced Simulation Model for Non–Point Source Pollution Transport at the Field Scale,*' Model Documentation, Draft.

Snell, E. A.: 1984, '*A Manual for Regional Targeting of Agricultural Soil Erosion and Sediment Loading to Streams,*' Lands Directorate Working Paper No. 36, Environment Canada, Ottawa, ONT.

Somlyody, L., and G. Van Straten (eds.): 1986, *Modeling and Managing Shallow Lake Eutrophication With Application to Lake Balaton*, Springer-Verlag.

Sonzogni, W. C., et al.: 1976, 'A Phosphorus Residence Time Model: Theory and Application,' *Water Research* **10**, 429–435.

Stefan, H., R. Ambrose, and M. Dortch: 1990, 'Surface Water Quality Models: Modeler's Perspective,' in D. G. DeCoursey (ed.), *Proceedings International Symposium on Water Quality Modeling of Agricultural Non–Point Sources*, June 19–23, 1988, Logan, UT, U.S. Dep. of Ag., Ag. Res. Serv. ARS–81, pp. 329–380.

Stefan, H. G., J. J. Cardoni, and A. Y. Fu: 1982, '*RESQUAL II: A Dynamic Water Quality Simulation Program for a Stratified Shallow Lake or Reservoir: Application to Lake Chicot, Arkansas,*' Report 203, St. Anthony Falls Hyd. Lab., University of Minnesota, Minneapolis, MN.

Straskraba, M., and A. H. Gnauck: 1985, 'Freshwater Ecosystems. Modelling and Simulation,' Vol. **8**, *Development in Environmental Modelling*, Elsevier Science Publishers, 300 p.

Thomann, R. V., D. M. DiToro, R. P. Winfield, and D. J. O'Connor: 1975, '*Mathematical Modeling of Phytoplankton in Lake Ontario, 1. Model Development and Verification,*' Report No. EPA–660/3–75–005, USEPA.

Thomann, R. V., R. P. Winfield, and J. J. Segna: 1979, '*Verification Analysis of Lake Ontario and Rochester Embayment Three–Dimensional Eutrophication Models,*' Report No. EPA–600/3–79–094, USEAP.

Thomas, W. A., and W. H. McAnally: 1985, '*Users Manual for the Generalized Computer Program System: Open–Channel Flow and Sedimentation; TABS–2, Main Test,*' Instruction Report HL–85–1, USAE Waterways Experiment Station, Vicksburg, MS.

TVA: 1986, '*Hydrodynamic and Water Quality Models of the TVA Engineering Laboratory,*' Norris, TN, Issue No. 3.

Waldrop, W. R., C. D. Ungate, and W. L. Harper: 1980, '*Computer Simulation of Hydrodynamics and Temperatures of Tellico Reservoir,*' Report No. WR28–1–65–100, TVA Water Systems Development Branch, Norris, TN.

Wheeler, M. F.: 1988, 'Modeling in Highly Advective Flow Problems,' in M. A. Celia et al. (eds.), *Computational Methods in Water Resources*, Vol. **1**, Elsevier Science Publishing Co., New York, pp. 35–44.

Whitehead, P. G., L. Somlyody, and G. Van Straten: 1990, 'Surface Water Quality Models for Planning, Design and Operational Management,' in D. G. DeCoursey (ed.), *Proceedings International Symposium on Water Quality Modeling of Agricultural Non–Point Sources*, June 19–23, 1988, Logan, UT, U.S. Dep. of Ag., Ag. Res. Serv., ARS–81, pp. 423–446.

Yalin, Y. S.: 1963, 'An Expression for Bed–Load Transportation', *Jour. Hyd. Div., ASCE* **89**(HY3), 221–250.

Yang, C. T.: 1973, 'Incipient Motion and Sediment Transport,' *Jour. Hyd. Div., ASCE* **99**(HY10), 1679–1704.

Young, R. A., C. A. Onstad, D. B. Bosh, and W. P. Anderson: 1987, '*AGNPS, Agricultural Non–Point–Source Pollution Model,*' USDA Agricultural Research Service, Conservation Research Report 35, 80 pp.

Part IV – Current Modeling Issues

Chapter 16

Scale Considerations

Keith Beven
University of Lancaster
Centre for Research on Environmental Systems
Institute of Environmental and Biological Sciences
Lancaster, LA1 4YQ, United Kingdom

Abstract. This paper addresses the problem of changing hydrological processes with changing hydrological scale. The nature and effects of heterogenity in catchment topography, soils and rainfalls are discussed and the idea of the representative elementary area for runoff production is introduced. The joint problem of hillslope response and channel routing at different scales is considered, including the tendency to apparent linearity of response at larger scales. Finally a framework for considering the hydrological similarity of catchments with different characteristics and different scales is presented.

1. Discussion

1.1. Changing Scale and Changing Processes

Consider a zero–order catchment (Figure 1). This catchment is a collection of hillslopes draining to a stream head. The hydrological response of the zero–order basin is governed by the processes of hillslope hydrology; in most climatic regimes these are predominantly controlled by subsurface flow processes and their effect on the generation of surface runoff. At this scale we may often assume that rainfall inputs are constant in space, but soil and vegetation characteristics may well vary spatially. The time scale of individual storm response hydrographs will range from a few minutes to a few days after peak rainfall intensities, perhaps superimposed on a long–term pattern of subsurface outflows related to seasonal wetting and drying. Time delays associated with surface flows in such a small catchment will generally be small, although it must be recognized that for overland flow over densely vegetated surfaces, measured mean flow velocities may also be very small (see for example Newson and Harrison, 1978; Beven, 1978). There may be a surface flow network of ephemeral rills, pipes, and channels during storm periods (Weyman, 1973; Beven, 1978). The position of the stream head will vary dynamically within and between storms as a result of the interaction of surface and subsurface flow processes (see Blyth and Rodda, 1973; Gurnell, 1978), as outlined conceptually in the variable contributing area model of Hewlett (1974). Over longer periods of time the position of the stream head will also depend upon interactions between flow processes and sediment transport processes (see Dunne, 1980).

Consider now higher order catchments of increasing scale. As scale increases to that of a small catchment, each area includes a sample of hillslopes, subjected to rainfall patterns that may display distinct heterogeneity, particularly during convective rainstorms. Variability in

357

D. S. Bowles and P. E. O'Connell (eds.), Recent Advances in the Modeling of Hydrologic Systems, 357–371.
© 1991 *Kluwer Academic Publishers.*

358

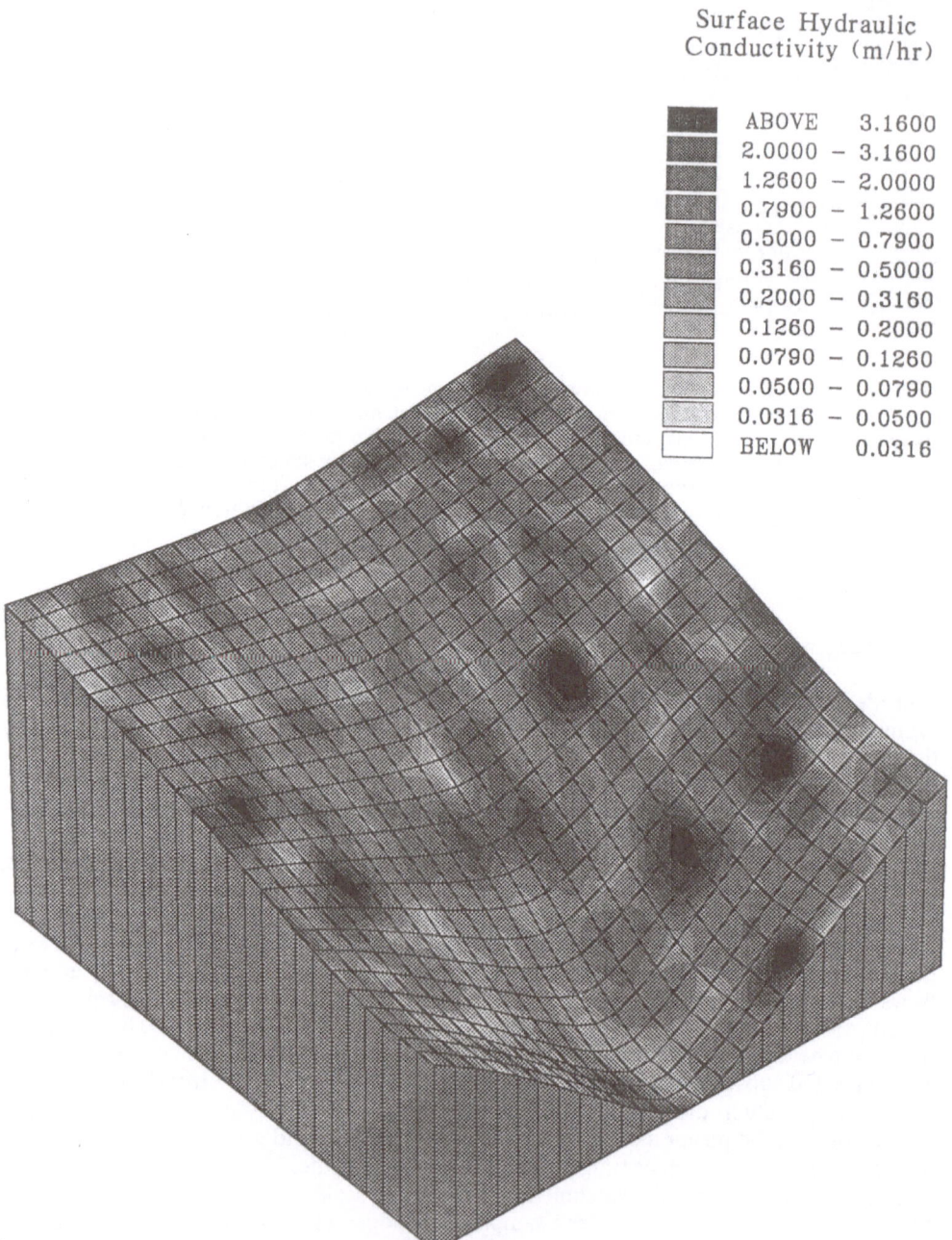

Surface Hydraulic
Conductivity (m/hr)

	ABOVE	3.1600
	2.0000 –	3.1600
	1.2600 –	2.0000
	0.7900 –	1.2600
	0.5000 –	0.7900
	0.3160 –	0.5000
	0.2000 –	0.3160
	0.1260 –	0.2000
	0.0790 –	0.1260
	0.0500 –	0.0790
	0.0316 –	0.0500
	BELOW	0.0316

Figure 1. Surface layer hydraulic conductivity distribution used in the three–dimensional simulation of a zero–order catchment or headwater hollow of dimensions 500 x 500 m (after Binley et al., 1989).

hillslope form, vegetation, and soil characteristics will increase, and the perennial channel network and channel flow processes may start to exert an important control over the shape of rainstorm response hydrographs. Kirkby (1976) notes that it is the relative time scales of hillslope and channel responses that are important in controlling hydrograph shape. The residence times associated with channel processes will increase with catchment scale, whereas within a particular physioclimatic regime the time scales of the hillslope responses may increase only slowly with catchment scale (unless there is a regional groundwater system).

Kirkby's work, and that of all other studies based on network topology, neglects any interactions between the geometry of the channel network and hillslope form that may affect the volume and timing of runoff production. The two are inextricably linked through the processes of morphological development of catchments. As scale increases so does the variety of hillslope forms. Average slope angles will tend to decrease, as recognized in the "law of slope angles" of Horton (1945). Hillslopes contributing laterally to interior network links will tend to make up a larger proportion of the catchment. Such hillslopes tend to be straight or divergent in plan, compared with the predominantly convergent slopes of the zero order basins contributing to and maintaining the heads of exterior links. Hydrological theory suggests that hillslope form may exert important controls on runoff production, with convergent hillslopes being more likely to maintain high antecedent moisture conditions at the base of slopes and consequently a greater likelihood of producing rapid subsurface or surface responses (Kirkby and Chorley, 1967; Beven, 1977, 1979a).

The relative contributions of runoff from headwater and sideslope areas will depend on hillslope form, soil characteristics and the prevailing rainfall regime. It would be expected that in many catchments the straight or divergent hillslopes may provide the greater mass of runoff, albeit at lower rates of production on a per unit area basis because the areas and lengths of streambank interface involved are so much greater. However, Burt and Arkle (1986) report on a contrary example in a catchment where short, steep sideslopes lead to higher rates of runoff production per unit area than convergent headwater areas, but the latter provide the greater mass of runoff because they comprise the greater area of the catchment.

Changes in geomorphology at larger scales may also lead to changes in the nature of the hydrological processes controlling the shape of the hydrograph. In particular, at a certain scale on any particular network, the channel morphology starts to develop a flood plain. For within bank flows the flood plain will tend to attenuate the *hillslope* hydrograph; for overbank flows the flood plain will lead to greater attenuation of the *channel* hydrograph peak as a result of the effects of flood plain storage.

1.2. Changing Scale and Storm Runoff Production

It is now increasingly recognized that there is a complete spectrum of processes by which a hillslope or catchment area responds to a storm rainfall, with the two end members being Hortonian infiltration excess surface runoff production and a purely subsurface stormflow response (Dunne, 1978). Between these lie various saturation excess surface runoff production, interflow, and throughflow mechanisms. All are recognized as producing runoff contributions to the stream from only part of the catchment area, the contributing area, that varies dynamically within and between storms. The heterogeneity of hillslope soil characteristics and topography, interacting with the received rainfall intensities, is important in controlling the contributing area dynamics. Position on a hillslope will affect antecedent moisture levels as a result of downslope flows during interstorm periods; surface runoff production will tend to occur first on areas of low conductivity spreading to areas of higher conductivity as a storm proceeds. Depending on antecedent conditions and rainfall intensities, infiltration excess, saturation excess, and subsurface stormflow runoff production may occur in the same catchment during the same storm or at the same location during different storms (eg. Freeze, 1980; Beven, 1986).

For the purposes of analyzing runoff production in heterogeneous catchments, we must consider the occurrence of soil and vegetation characteristics within the landscape in a probabilistic sense. There will be a certain joint probability of having a certain vegetation cover on soils of a given combination of conductivity and moisture characteristics at a certain position in the ensemble of hillslopes that make up a catchment area (where morphological position may affect antecedent moisture status prior to a storm and consequent hydrological response). Each point in the catchment may also be associated with a probability of receiving a certain storm rainfall volume or temporal pattern of rainfall intensities (that may depend upon the spatial correlation structures and movement of the rainstorm).

Let us look first at the case of a small catchment in which it may be possible to treat the variability in these hillslope, soil, vegetation, and rainfall characteristics as if drawn from stationary, if spatially correlated, statistical distributions. In a small area, however, only a small sample from the individual distributions will be included, particularly where the correlation scales of the individual variables is long relative to the scale of the area. The significance of this for analyzing and predicting runoff production lies in the fact that different samples drawn from the same distributions may have very different runoff production characteristics. The actual spatial *pattern* of characteristics may therefore be important. This has been shown, for example, by the simulations of Smith and Hebbert (1979) who compared the runoff produced by an infiltration excess mechanism on individual hillslopes with similar variations in soil hydraulic conductivities but either increasing or decreasing downslope. These simulations included the effects of the infiltration of surface "run–on" produced further upslope.

Within a physioclimatic region, therefore, we could envisage that there may be populations of small catchments that could be considered to be statistically similar in the sense that their patterns of hillslopes and other characteristics are drawn from the same underlying distributions. They may differ drastically, however, in terms of their actual patterns of properties and responses. The degree to which they differ will depend upon the nature of the variability of the individual properties. A familiar example of such differences occurs during convective rainstorms which have short spatial correlation scales. The probability of such a rainstorm providing flood producing rainfall intensities in a number of small adjacent catchments within a region may, over long periods of time, be considered to be identical. In an individual occurrence, however, a storm may be centered over one of the catchments, while the others may receive only minor amounts of rainfall. Newson (1980) for example reports the case of a storm in August 1977 which produced a 100 year recurrence interval flood peak on the 10 km^2 Severn catchment at Plynlimon in mid Wales, and only a minor hydrograph on the adjacent Wye catchment where there was zero rainfall at the stream gauge.

What happens now if we increase catchment scale? Consider the case first where the underlying distributions of catchment and rainfall characteristics stay the same. As scale increases so does the sample of properties sampled in each statistically similar catchment area. The result should be that the hydrological responses of the areas become more alike, even though the patterns of properties within the catchments differ. Above a certain scale, the sample may be sufficient that it may no longer be necessary to consider differences due to the pattern of characteristics within each area. Wood et al. (1988) have proposed the name Representative Elementary Area (REA) for this scale by analogy with the representative elementary volume of soil physics. Note, however, that in predicting runoff production at the scale of the REA it may still be necessary to consider the heterogeneity of the area in terms of the joint probability distributions of the individual characteristics, rather than representing the area in terms of uniform parameters.

Wood et al. (1988) used a version of the TOPMODEL structure (Beven, 1990) that predicts both saturation excess and infiltration excess runoff production on areas of heterogeneous soil, hillslope, and rainfall characteristics. They analyzed the predicted runoff production for catchment areas of different scales, in which the hillslope characteristics were sampled from one particular topography (Figure 2) and the soil conductivity and rainfall characteristics were sampled from distributions with specified mean and spatial covariance parameters. For the particular conditions considered they suggested that the REA for the volume of storm runoff production for these two mechanisms combined was of the order of 1 km^2 as determined by a minimum in the variance of predicted runoff volumes from areas of the same spatial scale (Figure 3). In this analysis, however, they did not consider the effects of run–on, or the effects of surface flow routing, or of subsurface runoff production on the volume and timing of runoff and variation in flood peak.

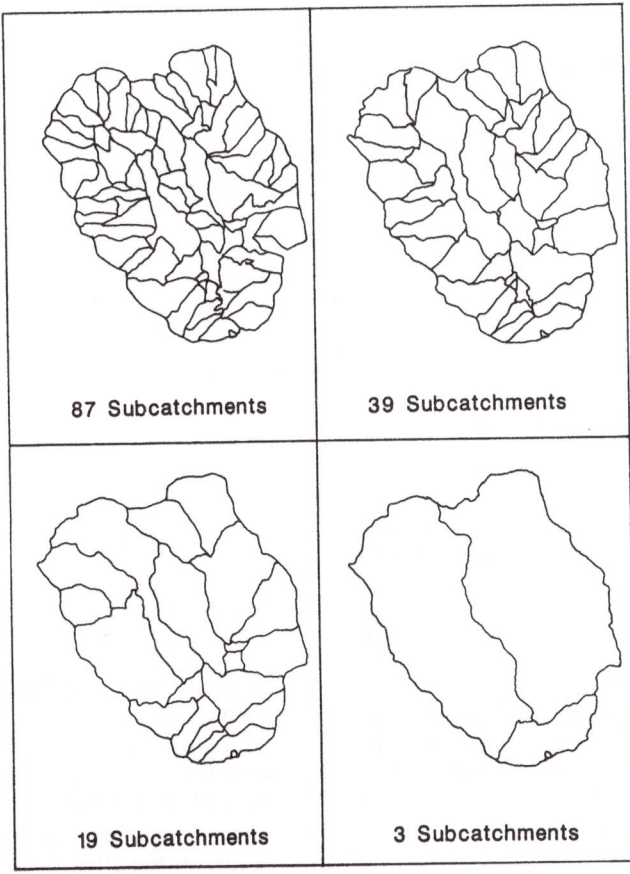

Figure 2. Subcatchment divisions based on digital terrain data used in the investigation of effects of catchment scale on runoff production (from Wood et al., 1988).

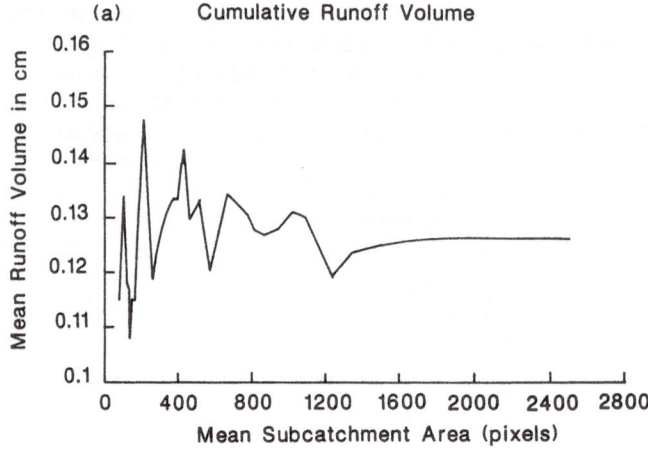

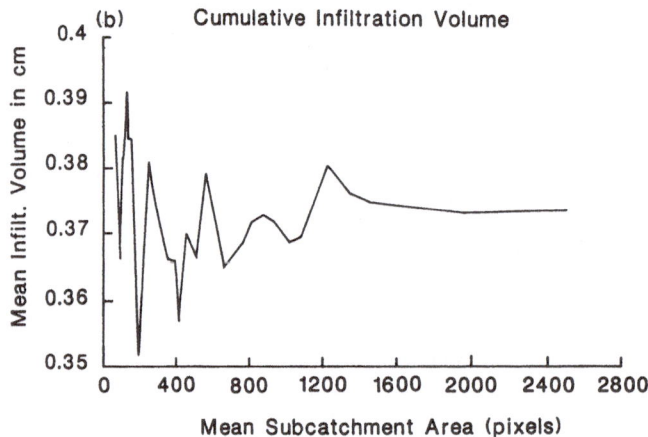

Figure 3. Variations in (a) mean runoff volume and (b) mean infiltration volume over areas of different scale; pixel size is 30 x 30 m (from Wood et al., 1988).

The concept of the REA is useful in clarifying the interaction of heterogeneity of catchment characteristics and catchment scale in runoff production but is clearly a great simplification of what occurs in reality in that it ignores the effects of channel routing, characteristics that have long or complex correlation structures relative to the catchment scale or else are non-stationary in space. Increasing catchment scale will generally result in different rock types, soil series, and hillslope forms being included in a catchment area, each with different distributions of hydrological parameters. Rainfall inputs may involve correlation structures at multiple scales, with the longer synoptic scales being very much greater than the catchment REA. Non-stationarity and extensive spatial correlations will lead to increased variability in catchment response beyond the scale of the REA, which might then represent the scale at which there is a minimum in the variability in catchment hydrological response.

1.3. Changing Scale and Channel Routing

We may apply a similar REA concept to the joint problem of runoff production and channel routing that results in the hydrograph seen at the catchment outlet. The variability arising from channel routing will be associated with the changing characteristics of channels of different orders, including flood plain initiation, and variations in the channel network geometry. The latter may have an important effect on hydrograph shape. This was first investigated by Surkan (1969) using a linear routing algorithm in randomly generated "topologically distinct channel networks." Later work by Kirkby (1976) suggests that, assuming spatially constant lateral inputs to the channel network, once the residence time of the channel network becomes significantly longer than the time–scale of the hillslope response, network geometry may affect hydrograph peaks by a factor of two and time to peak by a factor of four. Mesa and Mifflin (1986) suggest that channel network geometry may significantly influence the catchment hydrograph even at scales of a little over 1 km^2.

An important series of papers on the effect of changing catchment scale on the routing through the channel network was initiated by Rodriguez–Iturbe and Valdes (1979) who introduced the concept of the Geomorphological Unit Hydrograph (GUH). They made use of the generalized network parameters introduced by Horton (1945) and Strahler (1957) which relate the expected numbers, lengths, and areas to stream order. These parameters are:

the bifurcation ratio $\qquad R_B = (N_j)^{1/(\Omega-j)}$ $\qquad\qquad$ (1a)

the slope ratio $\qquad R_S = (S_j/S_\Omega)^{1/(\Omega-j)}$ $\qquad\qquad$ (1b)

the length ratio $\qquad R_L = (L_j/L_\Omega)$ $\qquad\qquad$ (1c)

and the area ratio $\qquad R_A = (A_j/A_\Omega)^{1/(\Omega-j)}$ $\qquad\qquad$ (1d)

where N is the number of channels of channel order j, Ω is the order of the whole network, L is expected channel length, S is expected channel slope, and A is the expected drainage area for subcatchments of order j.

The GUH is viewed as the frequency distribution of the times of arrival of individual water droplets at the catchment outlet, given an instantaneous uniform application of "excess" rainfall over the catchment. The mechanisms of runoff production and dynamics of contributing areas are not explicitly considered in GUH theory. The history of each individual drop is treated as a Markov process controlled by the probabilities of transitions from one stream order to a higher order. This assumption results in an exponential holding time distribution for the water drops in each stream link (equivalent to the impulse–response of a linear store). Neglecting any time delays associated with overland flow pathways, these probabilities are then related to the network characteristics to obtain the final probability density function of water droplets leaving the highest order stream, which is equivalent to the GUH. The peak flow and time to peak of the GUH are defined in terms of the network parameters of equations (l) by Rodriguez–Iturbe and Valdez (1979) as:

$$q_p = \frac{1.31\, R_L^{0.43} v}{L_\Omega} \qquad\qquad (2a)$$

$$t_p = \frac{0.44\, L_\Omega\, R_L^{-0.38}\, R_B^{0.55}}{v R_A^{0.55}} \qquad\qquad (2b)$$

where $L\Omega$ is the length (in km) of the highest order stream, and v is the peak streamflow velocity (in m s^{-1}) which is assumed constant throughout the network. The units of t_p and q_p are hours and per hour respectively. The assumption of an exponential holding time distribution for routing in the individual links is not a physically realistic one, although it has the advantage of involving only one parameter. Gupta et al. (1980) have generalized the GUH to deal with any arbitrary holding time distribution for the individual links and any combination of flow pathways, including overland flow pathways.

The GUH expresses the relationship between catchment scale (or order) and channel routing in a way that suppresses the unique structure of individual networks in favor of the expected characteristics of statistically similar networks, where that similarity is expressed in terms of the Horton parameters. As such it is a concept that should be applicable at a scale above the REA scale for the channel network, although it is not clear as yet what that scale should be. In fact, in any application to a particular network, under the same assumptions of a spatially constant wave velocity, it is relatively simple to use a routing algorithm using a unit hydrograph based directly on the actual network geometry, such as the link–frequency or width function algorithm used by Kirkby (1976) and Beven (1979b).

1.4. Catchment Scale and Linearity

Rodriguez–Iturbe et al. (1982) and Wang et al. (1981) have extended the theory of the GUH approach to take account of the nonlinear change of the network flow velocity with peak discharge, under the kinematic wave assumptions. The theory for including nonlinearity into the GUH framework depends rather crucially on the assumption that a velocity estimated by kinematic wave theory for shallow flow on a simple hillslope plane or first–order channel *at steady state* can be used as a routing velocity for the whole network. On the basis of this assumption Wang et al. (1981) can neglect the differences between mean flow velocities and kinematic wave velocities and suggest that the average network routing velocity should be proportional to the excess rainfall intensity raised to the power of $-\alpha$ where $\alpha = n/(1+n)$ and n is defined from the general shallow uniform flow relationship:

$$v \propto S_o^m h^n \tag{3}$$

For uniform laminar flow we would expect $m = 1, n = 2$; for the Manning equation $m = 1/2$, $n = 2/3$; and for the Chezy equation $m = 1/2, n = 1/2$. Wang et al. derive values for the parameter α for different sizes of catchment and suggest that it declines from the order of 0.66 at the scale of Izzard's (1946) runoff plot experiments (consistent with a laminar flow regime), through 0.6 for Minshall's (1960) classic nonlinear unit hydrograph data, towards zero for catchments of the order of several thousand square kilometers which would imply no dependence of the routing velocity on excess rainfall intensity and a purely linear routing algorithm.

It has been long accepted in hydrology that large basins are more nearly linear in their response than small basins (see Amorocho, 1963), but there is no good underlying theory to explain this behavior. The variation of α reported by Wang et al. provides empirical backing for increasing linearity with increasing catchment scale but has no explanatory power. It is, in fact, inconsistent with the uniform flow theory on which the dependence of velocities on intensity was based since it implies that in large catchments, velocities are independent of depths of flow. Beven (1979b) and Bates and Pilgrim (1982) show that there is field evidence to suggest that, at least in some catchments, the assumption of a constant wave velocity may be justified, even though mean velocities may be expected to change with discharge.

1.5. Changing Scale and Hydrological Similarity

The extension of hydrological knowledge from gauged to ungauged sites requires some theory or model of hydrological similarity. A traditional approach to this problem has been to treat all catchments in a certain physioclimatic region as similar in their response regardless of catchment scale and physical characteristics, allowing regional relationships to be established to predict the parameters of a unit hydrograph or flood frequency distribution for an ungauged catchment. Studies like the UK Flood Studies Report (NERC, 1975) have used this type of regional analysis to provide regression equations relating hydrological response to catchment characteristics. Pilgrim (1983) has pointed out some of the dangers of this approach arising from the variability of hydrological response within a region, while Acreman and Sinclair (1986) have shown that better predictive equations might be obtained by classifying basins within a region.

Surely hydrology should have advanced to the stage now when it should be possible to speculate about hydrological similarity and its relationship with catchment scale from a basis of physical reasoning. Some progress in this direction for the case of runoff routing has been made by the GUH studies described above, but despite a number of papers identifying and discussing the need for such studies (eg, Dooge, 1982; Klemec, 1983; Pilgrim, 1983) there have been very few studies that have tackled the far more difficult problem of similarity in runoff *production* in a realistic way. Such a study would have to incorporate the effects of heterogeneity of catchment and rainfall characteristics on the variability of runoff production processes. Yet, to be of practical use, it would have to involve only a small number of similarity parameters. It would need to be shown by empirical studies that those parameters did indeed encapsulate the essence of hydrological similarity across different scales and physioclimatic regions.

Some first steps towards this aim have been taken by Sivapalan et al. (1987, 1990) who have shown how a version of the TOPMODEL structure (Beven and Kirkby, 1979; Beven et al., 1984; Beven and Wood, 1983; Beven, 1986) can be expressed in dimensionless form. The version used predicts both infiltration excess and saturation excess runoff production on a heterogeneous catchment resulting from a single rainstorm with spatially variable, but constant in time, rainfall intensities. The form of the model used is based on a relationship, derived under steady flow assumptions, between a mean depth to water table for the catchment, z, the local depth to water table, z_i, and a soil transmissivity/topography index $ln\,(aT_e/T_o\,tan\beta)$, where "a" is the upslope area per unit contour length draining through any point in the catchment, $tan\beta$ is the local slope angle at that point, and T_o is the local transmissivity when the soil is just saturated to the surface and:

$$ln\,T_e = \frac{1}{A}\int_A ln\,T_o \tag{4}$$

This relationship takes the form for any point i:

$$z_i = \bar{z} - \frac{1}{f}[ln(a\,T_e/T_o\,tan\beta) - \lambda] \tag{5}$$

where $\quad \lambda = \frac{1}{A}\int_A ln(a/tan\beta)$

and f is a parameter of the assumed exponential decline of transmissivity with depth of saturation:

$$T(z_i) = T_o\,exp(-fz_i) \tag{6}$$

T_o is approximately equal to K_o /f where K_o is the saturated hydraulic conductivity of the soil at the surface. Equation (5) allows the saturated contributing area (where $z_i = 0$) in the catchment to be predicted if z, f and the pattern of the soil/topography index are known. The theory predicts that all points in the catchment with the same value of $ln(aT_e/T_o \tan\beta)$ will have an identical hydrological response. The soil/topography index is therefore an index of hydrological similarity at the point scale. Note that for the case of a homogeneous soil $T_e = T_o$ everywhere, and the index is dependent on topography alone.

It can be shown that under the assumptions of steady flow and exponentially declining transmissivity the subsurface outflow from the catchment should be governed by the equation:

$$Q_b = Q_o \exp(-f\bar{z})$$

$$\text{(7)}$$

where $\qquad Q_o = A\, T_e \exp(-\lambda)$

For single storm predictions, Equation (7) can be inverted to provide an estimate of $\bar{z}$ given an antecedent subsurface outflow from the catchment such that:

$$\bar{z} = -ln(Q_b(0)/Q_o)/f \qquad\qquad\qquad (8)$$

where $Q_b(0)$ is the antecedent subsurface flow at $t = 0$.

Runoff production will also depend upon antecedent conditions in the unsaturated zone. Infiltration excess runoff will depend primarily on the local surface hydraulic conductivity and moisture content; saturation excess runoff production will depend on the total storage deficit at a point. Sivapalan et al. (1987) assume that the unsaturated zone moisture profile is that of complete gravity drainage and use the soil moisture characteristic relationships of Brooks and Corey (1966), which involve the parameters Θ_s, the saturated moisture content, Θ_r, the residual moisture content, B, the Brooks–Corey parameter, and ψ_c the air entry potential or depth of the capillary fringe. It is assumed that these parameters are constant in the catchment even though the soil hydraulic conductivity is varying from point to point. The total storage deficit in any soil profile (where $z_i > \psi_c$) is then given by:

$$S_i = (\Theta_s - \Theta_r)\left[z_i - \psi_c - \frac{1}{(1-B)}\left\{\left(\frac{\psi_c}{z_i}\right)^B z_i - \psi_c\right\}\right] \qquad (9)$$

The initial saturated contributing area at the start of the storm is obtained by combining Equations (5) and (8) to determine the area for which $z_i < \psi_c$. If the redistribution of water within the saturated zone is ignored during the storm, the spread of the contributing area during the storm will be given by comparing the storage deficits given by Equation (10) with the cumulative volume of infiltrated water up to any time. That volume will depend upon any surface controls on infiltration and the production of infiltration excess runoff. This is predicted for a local rainfall intensity p_i, using a modified form of the Philip infiltration equation derived by Sivapalan and Wood (1986) in which the infiltration rate for any particular value of K_o is given by:

$$g(t) = p_i \qquad\qquad\qquad\qquad ;t < t_p$$

$$= C\,K_o + \tfrac{1}{2}\,S_r\,K_o^{\frac{1}{2}}\,(t-t_c)^{-\frac{1}{2}} \quad ;t > t_p \qquad (10)$$

where the values of t_c, t_p, C, and S_r depend upon the Brooks–Corey parameters, p_i, K_o, and the local value of z_i.

The nondimensional scaled version of the model is derived as follows. It is assumed that the distribution of $ln(a/\tan\beta)$ can be represented as a three parameter gamma (μ, φ, χ) distribution and the distribution of the logarithm of saturated hydraulic conductivity expressed in terms of the transmissivity ratio $ln(T_e/T_o)$ as a normal (O, σ) distribution. It can then be shown that the transformed variable:

$$v = \frac{ln(aT_e/T_o \tan\beta) - \mu}{\mu - \lambda} \tag{11}$$

follows a one parameter gamma distribution:

$$f_v(v) = \frac{\varphi^*}{\Gamma(\varphi^*)}(v\varphi^*)^{\varphi^*-1} \exp(-v\varphi^*) \tag{12}$$

with shape parameter φ^* scale parameter $1/\varphi^*$, mean 1 and variance $1/\varphi^*$. The variable v is scale–independent and φ^* is a dimensionless parameter that reflects only the soil and topographic distributions of the catchment, irrespective of scale. Derivation of Equation (13) assumes that there is no interdependence between the distributions of T_o and $a/\tan\beta$ and that the hydrological response does not depend on the *pattern* of soil parameters. These assumptions will only be appropriate at scales greater than the REA discussed above.

Sivapalan et al. (1987) further define the following dimensionless variables:

$$S^* = S_i/\psi_c(\Theta_s - \Theta_r)$$

$$z^* = fz_i/(\lambda - \mu) \tag{13}$$

$$t^* = t/t_d$$

where t_d is the duration of the storm and the dimensionless parameters:

$$Q^* = -ln(Q_b(0)/Q_o)/(\lambda - \mu)$$

$$\psi^* = \psi_c f/(\lambda - \mu)$$

$$K^* = K_o t_d/\psi_c(\Theta_s - \Theta_r) \tag{14}$$

$$p^* = p_i t_d/\psi_c(\Theta_s - \Theta_r)$$

The Brooks–Corey parameter B is already dimensionless. The assumed log normal distribution for T_o used in Equation (10) also implies a log normal distribution for K_o with coefficient of variation C_{vK}. Within this framework they show how both the infiltration excess and saturation excess runoff production calculations can be scaled for the case where the rainfall intensities are assumed to be gamma distributed in space, with mean p and coefficient of variation C_{vp}. The dimensionless model then involves: the catchment parameter φ^*; the soil parameters K^*, C_{vK}, ψ^*, and B; the rainfall parameters $p*$ and C_{vp}; and the antecedent condition parameter Q^*. The dimensionless theory predicts that catchments with identical values of these parameters will respond, in both infiltration and saturation excess runoff production, in a hydrologically similar way regardless of catchment scale or climatic regime. Some examples of the predicted variability in response as a result of varying some of these parameters one at a time from the values of Table 1 are shown in Figure 4.

This scaling theory has recently been extended by Sivaplan et al. (1990) to the long–term integration of storm responses to predict flood frequency characteristics for catchments with different similarity parameters. Their work combines the partial contributing area similarity theory for runoff production presented above with a generalized geomorphologic unit hydrograph routing model. Both infiltration excess and saturation excess runoff production are taken into account, and they show how the shape of the flood frequency curve for given catchment morphology, soil, and rainfall regime characteristics can be affected by the dominant runoff production process. One interesting implication of the results was that for catchments where saturation excess storm production dominates at low flood return periods and infiltration at high return periods, the resultant flood frequency curve may appear to be of extreme value type III, implying an approach towards a limiting flood. They suggest that this is not a correct interpretation and that in fact the curve is transitional to an infiltration excess dominated curve, where the transitional part of the curve is produced by long storms of medium intensity on initially wet catchments.

2. Conclusions

Clearly, there is much that is missing from this representation, notably the effects of channel routing, temporal variations of rainfall intensities, the effects of subsurface flows and runoff production, and the effects of evapotranspiration and redistribution of initial soil moisture deficits. It does, however, provide a useful theoretical framework for thinking about the factors involved in a definition of hydrological similarity and for exploration of the sensitivity of predicted runoff production mechanisms and volumes to the different parameters. Empirical verification of these (or future more complex) predictions would appear to be a vast undertaking in that the theory involves spatial distributions of parameters and state variables. This is, however, a direction in which theoretical hydrology, applied to appropriate field experimentation, must advance if we are ever to understand the relationship between hydrological response, hydrological similarity and catchment scale.

Table 1. Scaled parameter values used in the simulations of Sivapalan et al. (1987).

Soil parameters

K_o^*	=	0.657
C_{vK}	=	1.0
ψ^*	=	0.05
B	=	0.40

Catchment soil/topography distribution parameter

φ^*	=	2.338

Rainfall parameters

p^*	=	0.657
C_{vp}	=	1.0

Initial condition parameter

Q^*	=	0.555

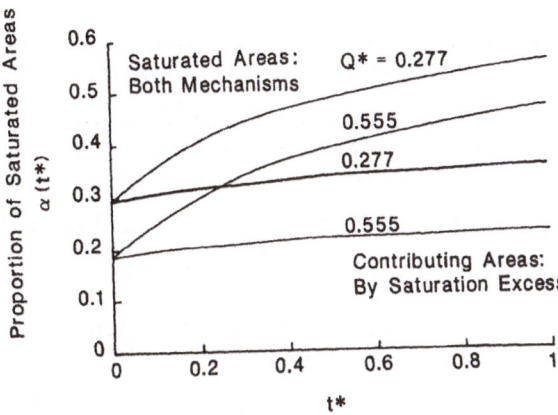

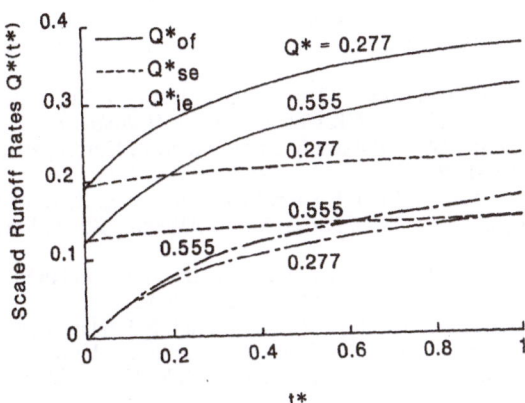

Figure 4. Sensitivity of (a) predicted saturated area and (b) scaled cumulative runoff volume to scaled antecedent moisture status Q^* (from Sivapalan et al., 1987).

3. Acknowledgements

Any presentation of ideas in this field must acknowledge the inspiration provided by Ignacio Rodriguez-Iturbe whose enthusiasm for the search for hydrological laws at the catchment scale has sparked many of those who have heard his presentations. Many of the ideas presented here have benefited from my long and fruitful collaboration with Eric Wood and Murugesu Sivapalan of Princeton University, supported in part by NASA grant NAG-5-491 and USGS grant 14-08-0001-G1138. I am grateful to Andrew Binley whose programs, developed with the support of NERC grant GR3/6264, produced Figure 1.

370

4. References

Acreman, M. C., and C. D. Sinclair: 1986, 'Classification of Drainage Basins According to Their Physical Characteristics: An Application for Flood Frequency Analysis in Scotland,' *J. Hydrology* **84**, 365.

Amorocho, J.: 1963, 'Measures of the Linearity of Hydrologic Systems,' *J. Geophys. Res.* **68**, 2237–2249.

Bates, B. C., and D. H. Pilgrim: 1982, 'Investigation of Storage–Discharge Relations for River Reaches and Runoff Routing Models,' in *Institution of Civil Engineers, Australia, Hydrology and Water Resources Symposium*, 120–126, Melbourne.

Beven, K. J.: 1977, 'Hillslope Hydrographs by the Finite Element Method,' *Earth Surface Processes* **2**, 13–28.

Beven, K. J.: 1978, 'The Hydrological Response of Headwater and Sideslope Areas,' *Hydrological Sciences Bulletin* **23**(4), 419–437.

Beven, K. J.: 1979a, 'Experiments with a Finite Element Model of Hillslope Hydrology – the Effect of Topography,' in *Surface and Subsurface Hydrology*, Proceedings of the Third International Hydrology Symposium, Water Resources Publications, Colorado.

Beven, K. J.: 1979b, 'On the Generalised Kinematic Routing Method,' *Water Resources Research* **15**(5), 1238–1242.

Beven, K. J.: 1986, 'Hillslope Runoff Processes and Flood Frequency Characteristics,' in A. D. Abrahams (ed.), *Hillslope Processes*, 187–202, Allen and Unwin, Boston.

Beven, K. J.: 1990, '*Spatially Distributed Modelling: Conceptual Approaches to Runoff Production,*' this volume.

Beven, K. J., and M. J. Kirkby: 1979, 'A Physically–based Variable Contributing Area Model of Basin Hydrology,' *Hydrological Sciences Bulletin* **24**(1), 43–69.

Beven, K. J., and E. F. Wood: 1983, 'Catchment Geomorphology and the Dynamics of Runoff Contributing Areas,' *J. Hydrology* **65**, 139–158.

Beven, K. J., M. J. Kirkby, N. Schoffield, and A. Tagg: 1984, 'Testing a Physically–Based Flood Forecasting Model (TOPMODEL) for Three UK Catchments,' *J. Hydrology* **69**, 119–143.

Binley, A. M., J. Elgy, and K. J. Beven: 1989, 'A Physically–Based Model of Heterogeneous Hillslopes,' *Water Resources Research* **25**, 1219–1226.

Blyth, K., and J. C. Rodda: 1973, 'A Stream Length Study,' *Water Resources Research* **9**, 1454–1461.

Brooks, R. H., and A. T. Corey: 1966, 'Properties of Porous Media Affecting Fluid Flow,' *J. Irrig. Drain. Div.. ASCE* **92**, 61–68.

Burt, T. P., and B. P. Arkle: 1986, 'Variable Source Areas of Stream Discharge and their Relationship to Point and Non–point Sources of Nitrate Pollution,' *IASH* **157**, 155–164.

Dooge, J. C. I.: 1982, 'Parameterization of Hydrologic Processes', in P. S. Eagleson (ed.), *Land Surface Processes in Atmospheric General Circulation Models*, 243–288, Cambridge University Press.

Dunne, T.: 1978, 'Field Studies of Hillslope Flow Processes,' in M. J. Kirkby (ed.), *Hillslope Hydrology*, 227–293, Wiley, Chichester.

Dunne, T.: 1980, 'Formation and Controls of Channel Networks,' *Progress in Physical Geography* **4**, 211–239.

Freeze, R. A.: 1980, 'A Stochastic–Conceptual Analysis of Rainfall–Runoff Processes on a Hillslope,' *Water Resources Research* **16**, 391–408.

Gupta, V. K., E. Waymire, and C. T. Wang: 1980, 'A Representation of the Instantaneous Unit Hydrograph from Geomorphology,' *Water Resources Research* **16**, 855–862.

Gurnell, A. M.: 1978, 'The Dynamics of a Drainage Network,' *Nordic Hydrology* **9**, 293–306.

Hewlett, J. D.: 1974, "Comments on Letters Relating to 'Role of Subsurface Flow in Generating Surface Runoff. 2. Upstream Source Areas' by R. Allen Freeze," *Water Resources Research* **10**, 605–607.

Horton, R. E.: 1945, 'Erosional Development of Streams and their Drainage Basins: Hydrophysical Approach to Quantitative Morphology,' *Bull. Geol. Soc. Amer.* **56**, 275–370.

Izzard, C. F.: 1946, 'Hydraulics of Runoff from Developed Surfaces,' *Proc. Highway Res. Bd.* **26**, 129–146.

Kirkby, M. J.: 1976, 'Tests of the Random Network Model and its Application to Basin Hydrology,' *Earth Surf. Process.* **1**, 197–212.

Kirkby, M. J., and R. J. Chorley: 1967, 'Throughflow, Overland Flow and Erosion,' *Bull. I.A.S.H.* **12**, 5–21.

Klemec, V.: 1983, 'Conceptualisation and Scale in Hydrology,' *J. Hydrology* **65**, 1–24.

Mesa, O. J., and E. R. Mifflin: 1986, 'On the Relative Role of Hillslope and Network Geometry in Hydrologic Response,' in V. K. Gupta, I. Rodriguez–Iturbe, and E. F. Wood (eds.), *Scale Problems in Hydrology*, 1–17, Reidel, Dordrecht.

Minshall, N. E.: 1960, 'Predicting Storm Runoff on Small Experimental Watersheds,' *J. Hydraul. Div.. ASCE* **86**(HY8), 17–38.

NERC: 1975, *The Flood Studies Report* (5 volumes), Wallingford.

Newson, M. D.: 1980, 'The Geomorphological Effectiveness of Floods – A Contribution Stimulated by Two Recent Events in Mid–Wales,' *Earth Surf. Process.* **5**, 1–16.

Newson, M. D., and J. G. Harrison: 1978, '*Channel Studies in the Plynlimon Experimental Catchments,*' Institute of Hydrology Report No 30, Wallingford.

Pilgrim, D. H.: 1983, 'Some Problems in Transferring Hydrological Relationships Between Small and Large Drainage Basins and Between Regions,' *J. Hydrology* **65**, 49–72.

Rodriguez–Iturbe, I., and J. B. Valdes: 1979, 'The Geomorphic Structure of Hydrologic Response', *Water Resources Research* **15**, 1409–1420.

Rodriguez–Iturbe, I., M. Gonzales–Sanabria, and R. L. Bras: 1982, 'A Geomorphoclimatic Theory of the Instantaneous Unit Hydrograph,' *Water Resources Research* **18**, 877–886.

Sivapalan, M., and E. F. Wood: 1986, 'Effect of Scale on the Infiltration Response in Spatially Variable Soils,' in V. K. Gupta, I. Rodriguez–Iturbe, and E. F. Wood (eds.), *Scale Problems in Hydrology*, Reidel.

Sivapalan, M., K. J. Beven, and E. F. Wood: 1987, 'On Hydrological Similarity: II. A Scaled Model of Storm Runoff Production,' *Water Resources Research* **23**(12), 2266–2278.

Sivapalan, N., E. F. Wood, and K. J. Beven: 1990, 'On Hydrological Similarity: III. A Dimensionless Flood Frequency Model using a Generalised Geomorphologic Unit Hydrograph and Partial Area Runoff Generation,' *Water Resources Research* **26**(1), 43–58.

Smith, R. E., and R. H. B. Hebbert: 1979, 'A Monte–Carlo Analysis of the Hydrologic Effects of Spatial Variability of Infiltration,' *Water Resources Research* **15**, 419–429.

Strahler, A. N.: 1957, 'Quantitative Analysis of Watershed Geomorphology,' *Trans. Amer. Geophys. Union* **38**, 913–920.

Surkan, A. J.: 1969, 'Synthetic Hydrographs: Effects of Network Geometry,' *Water Resources Research* **5**, 112–128.

Wang, C. T., V. K. Gupta, and E. C. Waymire: 1981, 'A Geomorphologic Synthesis of Non–Linearity in Surface–Runoff,' *Water Resources Research* **17**, 545–554.

Weyman, D. R.: 1973, 'Measurement of the Downslope Flow of Water in a Soil,' *J. Hydrology* **3**, 267–288.

Wood, E. F., M. Sivapalan, K. Beven, and L. Band: 1988, 'Effects of Spatial Variability and Scale with Implications to Hydrological Modelling,' *J. Hydrology* **102**, 29–47.

Chapter 17

Spatially Distributed Modeling: Conceptual Approach to Runoff Prediction

Keith Beven
University of Lancaster
Centre for Research on Environmental Systems
Institute of Environmental and Biological Sciences
Lancaster, LA1 4YQ, United Kingdom

"As scientists we are intrigued by the possibility of assembling our concepts and bits of knowledge into a neat package to show that we do, after all, understand our science and its complex interrelated phenomena." (W. M. Kohler, 1969)

Abstract. The relationship between perceptual models, conceptual models and physically based distributed models of hydrological processes is analyzed. It is shown that physically–based models must be considered as conceptual models at the scale at which they are used. A simplified distributed model (TOP MODEL) that can take account of heterogeneity in catchment topography and soils is introduced. A likelihood based procedure for estimating the uncertainties associated with the predictions of complex distributed models is described.

1. Discussion

1.1. On Models of Runoff Production

The processes of runoff production within a stream catchment are dependent upon transfers of energy and mass that are variable over space and time within a complex open system. In any particular field or modeling study of those processes, there are practical limits to the way in which we can observe or represent those processes. For example, we must apply some boundary to the system under study where both the position of that boundary and any boundary fluxes will be subject to considerable uncertainty, particularly for any groundwater and evapotranspiration fluxes. Such problems of observability will mean that our understanding of the operation of a catchment system must necessarily be limited and subject to simplification and generalization.

However, within every hydrologist there is a qualitative picture of the nature of hydrological processes within a catchment, a picture that reflects his or her preconceptions, training, and field experience (if any), together with those of teachers and colleagues. It may be highly individual to the hydrologist concerned, and some of the problems of hydrological prediction have arisen as a result of misconceived applications of an inappropriate set of processes

D. S. Bowles and P. E. O'Connell (eds.), Recent Advances in the Modeling of Hydrologic Systems, 373–387.
© 1991 *Kluwer Academic Publishers.*

descriptions to a new environment (see Beven, 1987a). It is worth noting in this context that despite the widespread application of the 'Hortonian' infiltration excess model of runoff production, Robert Horton may have observed widespread overland flow in his garden hydrology laboratory in Voorheesville, New York, only rarely.

This subjective and personal picture of catchment processes comprises what may be called a *perceptual model* of the catchment. It is of little use for prediction purposes. Many of the features of this picture may be highly complex and may not easily be described in a mathematical form suitable for quantitative prediction. It is therefore necessary to make simplifying assumptions, generalizing from the perceptual model to create a set of mathematical statements comprising a *conceptual model* of the catchment. Many of those assumptions may already be implicit in the perceptual model, others will be stated explicitly in the specification of the conceptual model. However, not all conceptual models are suitable for quantitative prediction. Hydrological processes are (perceived as being) nonlinear and (stochastically) variable in both space and time, so that it may not be possible to obtain an analytical solution to the mathematical description. In that case a further level of assumptions or approximation will be necessary to yield a *procedural* model that can actually be programmed for a computer to give quantitative predictions.

There are clearly many choices involved in moving from a personal perceptual model to a predictive procedural model. Individual hydrological modelers will make different choices, even though they may have the aim of producing a model of general applicability. This has resulted in a plethora of procedural models in hydrology, without the concomitant development of tools that would allow a rigorous choice between them. Simulating the 'loss' and 'delay' processes that shape the catchment hydrograph is not actually all that difficult if the parameters of the procedural model are calibrated against observed responses. All hydrological models work (to some extent); equally, all are incompatible with the hydrologist's perceptual model of catchment response (to some extent).

1.2. A Perceptual Model of Runoff Production

In order to place the following discussion of conceptual models of runoff production in perspective, it is worth considering a personal perceptual model of runoff production in a catchment that has not been subjected to major disturbance by man. The upper boundary condition will be taken as the surface of the vegetation canopy. Boundary fluxes during a storm rainfall may be highly variable in space and time resulting from the interaction of the wind field and storm characteristics with the structure of the vegetation canopy and catchment topography.

The effect of the vegetation cover will generally be to introduce greater variability in flux intensities at the soil surface as a result of drip and stemflow processes (Crabtree and Trudgill, 1985). Stemflow may account for as much as 15 percent of the incident rainfall, which will be concentrated into a limited area around the base of the plants. Such local intensity concentrations may be important in governing the flow of water into the soil surface or, in some cases, the production of surface runoff. Localized variability in soil properties may also be important. It is known that soil hydraulic characteristics can vary markedly in space, whilst in the vertical the presence of surface crusting, roots, cracks and other macropores, or an organic mat may also play an important role in governing infiltration. Infiltration into the larger structural voids of the soil may occur at application rates of the order of 1 mm/hr, even on quite permeable soils. The presence of structural voids, macropores, or channels will result in a wetting front that moves into the soil in a highly irregular manner (Beven and Clarke, 1986), while even in the matrix, instabilities in the wetting front may result in the development of "fingering" (see Hillel, 1988).

Locally, application rates to the ground surface may exceed the infiltration capacity of the soil. This will tend to occur first on parts of the catchment where the hydraulic conductivity of the soil surface is lowest or application rates are highest. This partial contributing area of infiltration excess overland flow will expand with increasing rainfall intensity or duration. The controls on the production of overland flow in this manner may involve only a thin layer at the surface of the soil, particularly where a surface crust has developed on a bare soil surface. On vegetated surfaces, "surface" flow may take place through a vegetation mat, and have quite low mean velocities (for example, see Beven, 1978). Only over very smooth surfaces will overland flow take place as a sheet flow; on most natural surfaces there will be concentrations of flow induced by the microtopography and vegetation (for example, see Emmett, 1978), and in some cases this will lead to rilling. Areas producing overland flow regularly will tend to be linked to the perennial channel network by ephemeral rills or channels allowing surface runoff to be shed efficiently (eg. Beven, 1978).

Within the soil there may be local breaks in the hydraulic conductivity profile, of the matrix or the structural porosity, that may lead to localized concentrations of moisture and perhaps to saturated pockets of soil, possibly perched above the base of the soil profile. These pockets will tend to expand during storm conditions. Downslope linkages of such saturated pockets may be important in controlling subsurface runoff responses. Anisotropy of the soil hydraulic conductivity can also lead to localized concentrations of moisture in certain parts of hillslopes as has been shown by Zaslavsky and Sinai (1981). The build up of pockets of saturated soil at the base of the profile will also be affected by the permeability and irregularity of the bedrock surface. Huff et al. (1982) have shown for example that subsurface outflows to a stream channel may be concentrated by bedrock steps coming closer to the surface.

Draining water in the soil will tend to flow downhill. The topographic form of hillslopes should, therefore, be expected to have an important effect on spatial patterns of soil moisture and soil saturation, particularly on soils overlying a relatively impermeable bedrock. Hillslopes that are convergent in plan will tend to lead to higher moisture levels at their base relative to straight or divergent slopes. Hillslopes of low angle will also tend to result in higher moisture contents because of lower hydraulic gradients (Kirkby, 1969; Beven and Kirkby, 1979; O'Loughlin, 1981,1986). High water tables can be expected to result in fast subsurface responses, particularly where there is a well developed capillary fringe in the matrix above the water table. Wave speeds associated with saturated subsurface flows are inversely proportional to the difference between the saturated and actual moisture content close to the water table (for example, see Beven, 1982). Within the capillary fringe this difference may be very small, so that the wave speeds for initial responses to infiltrated water reaching the capillary fringe may be very high. Rapid vertical fluxes within the structural voids, perhaps generated by localized concentration of inputs at the soil surface, may be important in providing water to the capillary fringe via preferential flow pathways.

Where the soil saturates to the surface, overland flow may be generated, both from further rainfall inputs and from subsurface "return flow" seeping back to the surface. There will be an interaction between this "saturation excess" surface runoff production, primarily controlled by subsurface flow processes, and the dynamic expansion of the channel network outlined in Hewlett's variable contributing area concept (Hewlett, 1974).

1.3. On Conceptual Models of Runoff Production

The perceptual model outlined above is one of highly complex, dynamic interacting flow pathways. As such it is a model that is beyond detailed physical analysis. The physics is too complex and the characteristics of individual flow pathways are too variable in space and time. The perceptual model must, therefore, be simplified to a practically useful conceptual model in which the processes of catchment response are represented at an appropriate scale.

There is some hope that this can be done in a physically realistic way, despite the complexity evident at small scales. The catchment is a naturally dissipative open system. Dissipative systems lead to a natural integration of complexity and heterogeneity (or disorder) resulting in order in their physical structure and relatively simple impulse–response functions (Stengers and Prigogine, 1984). We see this order in the forms of hillslopes and the structure of dendritic channel networks. We see the simplified response function in the similarity of hydrographs when scaled by area and lag time and in the success of the unit hydrograph concept (Dooge, 1977). The experience of two decades of hydrological modeling suggests that not more than four or five parameters can be justified in simulating a continuous discharge record from a catchment (see for example Kirkby, 1976a). Any more will result in the identifiability problems associated with overparameterization (see for example Blackie and Eeles, 1985), although it is clear that not all conceptual representations of catchment processes have adhered to such a limit (see for example Fleming, 1975).

The same of course applies to distributed models, although in this case there is the possibility of increasing the number of parameters if observations of distributed state variables are available. This is often true in the case of distributed groundwater models, and there is a considerable literature on the so–called inverse problem of using such data to calibrate groundwater models (see Yeh, 1986). It is rarely true in the case of distributed catchment models, albeit that because of the greater nonlinearity inherent in near–surface processes such additional data would be extremely valuable in modeling catchment response. Thus distributed models should also have only a few parameter values that require calibration.

1.4. Physically–based Distributed Models as Conceptual Models

The justification for using distributed catchment models cannot, therefore, be based on the availability of distributed state data (although in some rare cases semi–distributed models have been created to make use of distributed *input* data, see for example Wilson et al., 1979). The development of distributed models has usually been based on the argument that a model which more closely represents the physics involved in the processes of catchment response must be inherently superior to a model that is lumped at the catchment scale (Beven and O'Connell, 1982; Beven, 1985; Abbott et al., 1986a). The equations that describe these processes involve one or more space dimensions and will provide distributed predictions of say, the dynamics of contributing areas (see, for example, Freeze and Harlan, 1969; Freeze, 1972; Beven, 1977; Bathurst 1986a,b; Sharma et al., 1987; Binley et al., 1989a,b, Figure 1). The parameters of those equations have a physical interpretation and can, in principle, be measured in the field. Because the parameters are accepted as physically meaningful, we may have greater faith in modifying the values to simulate the effects of future land use changes which may occur only over part of the catchment, and for which no measurements will be available for calibration.

Most physically–based distributed models are based on the partial differential equations describing surface and subsurface flow processes. Commonly, as in the Système Hydrologique Européen (SHE) model (Abbott et al., 1986a,b) and Institute of Hydrology Distributed Model (IHDM) (Beven et al., 1987) these are the Richards equation for subsurface flows and a kinematic wave equation for surface flows. To obtain a procedural model with realistic boundary conditions, these equations must usually be solved by approximate numerical methods, with state variables predicted at a grid of points in space and time. Parameter values (and initial and boundary conditions) may be allowed to vary throughout this grid to reflect the spatial, and if necessary temporal, variability in catchment characteristics. This results in a very large number of parameter values, far more than is justified on identifiability grounds if those parameters are to be calibrated by comparing observed and predicted discharges. Beven (1989) has pointed out that resort to an argument that the

Water Table Depth (m)

ABOVE	1.1000
1.0000 –	1.1000
0.9000 –	1.0000
0.8000 –	0.9000
0.7000 –	0.8000
0.6000 –	0.7000
0.5000 –	0.6000
0.4000 –	0.5000
0.3000 –	0.4000
0.2000 –	0.3000
0.1000 –	0.2000
BELOW	0.1000

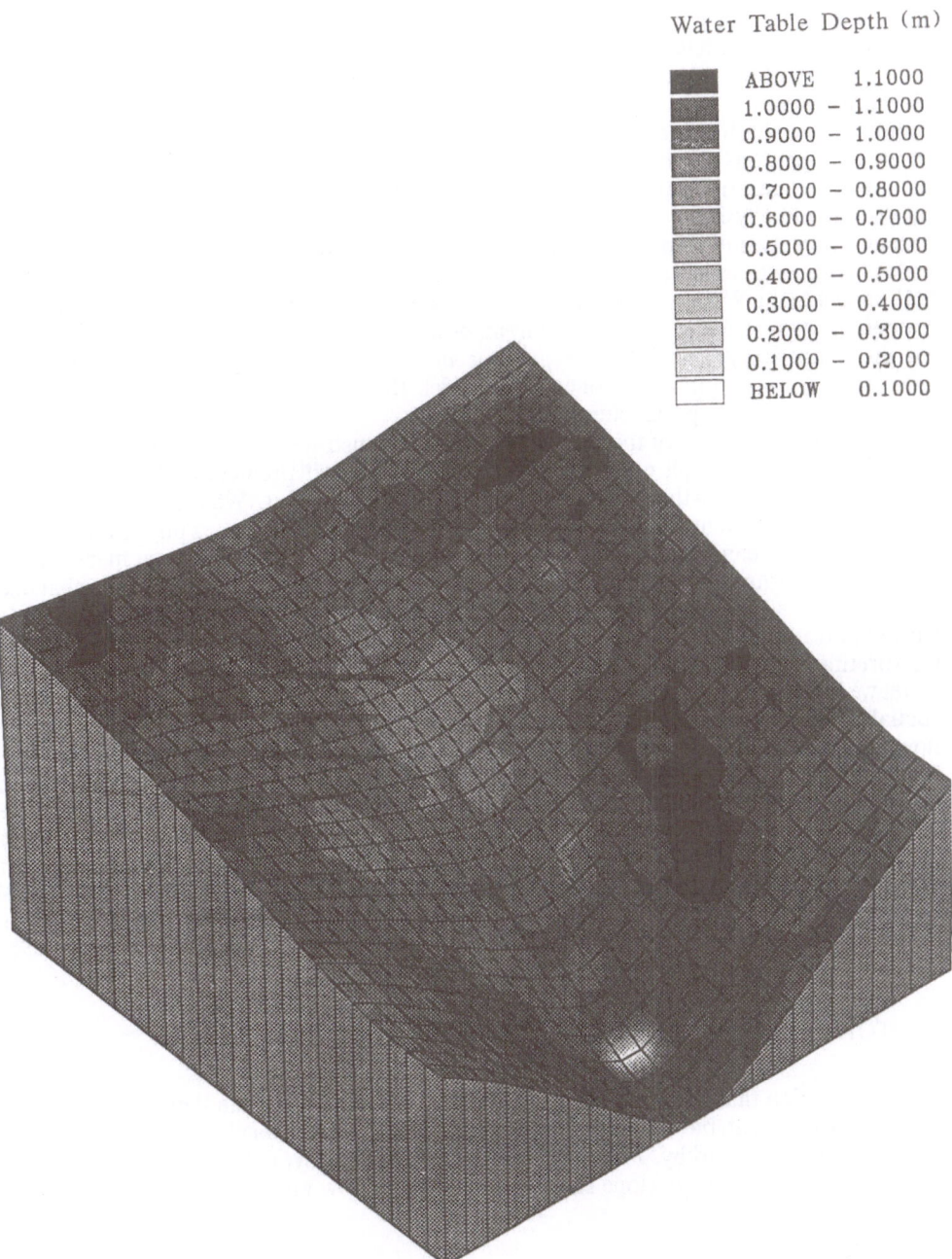

Figure 1. Predicted water table depths for a three–dimensional variably saturated heterogenous hillslope hollow. Calculations involved 12000 nodes in 12 vertical layers and were performed on a Cyber 205 supercomputer using the finite element code of Binley et al. (1989a,b).

physical nature of the parameters allows physical reasoning to be used in calibrating the parameters does not circumvent the problems of overparameterization unless additional observations of parameter values or distributed state variables are available *at an appropriate scale*. In this sense then, physically–based distributed models are no different from any other conceptual model.

Indeed, you will have realized that, despite the claims for such models being physically based, the equations on which these models are based are extreme simplifications of the procedural model described above. The equations are, in fact, descriptive equations of a different system; the physics on which they are based is the physics of small–scale homogeneous systems. In real applications of physically–based models we are forced to lump up the small scale physics to the model grid scale, for example the 250 x 250m used in the application of the SHE model to the Wye catchment (Bathurst, 1986a). There is no theoretical justification for assuming that the equations apply equally at the model grid scale, at which they are representing the lumped aggregate of heterogeneous nonlinear sub–grid processes. It is merely *assumed* that, in defining the conceptual model, the same small–scale equations can be applied with the same parameters.

The conceptual nature of these models can be demonstrated most clearly in terms of the variables used, for example capillary potential. The model will predict capillary potential at a grid point, representing the average value over some grid element. What does a grid element average capillary potential mean over an area of 250 by 250 m? It is not a physical variable in the sense that we can measure it. How can we compare it with a tensionmeter measurement within the grid square which may have a zone of influence of a few cm^3. Worse still, what does a grid square average capillary potential *gradient* mean when it is calculated from nodes 0.05m apart in the vertical for an area of $62500 \ m^2$. Certainly great care should be taken in interpreting such "physically–based" variables.

Yet we know from field evidence that the processes of catchment response show important spatial patterns, primarily arising from topographic variability and the tendency of water to flow downhill. Therefore, it should surely be possible to define a distributed conceptual model that can improve hydrological predictions by simulating such patterns, but in a parsimonious way with a limited number of parameters. There have been two important initiatives to try and develop such a model, both based on very similar quasi–steady state simplifications of "physically–based" equations. The first, TOPMODEL, has appeared in a variety of forms (see for example Beven and Kirkby, 1979; Beven and Wood, 1983; Beven et al., 1984; Hornberger et al., 1985; Beven, 1986a,b, 1987b; Sivapalan et al., 1987; Wood et al., 1988) and will be described in some detail below. The second is based on the work of O'Loughlin (1986) and is described in Moore et al. (1986).

1.5. TOPMODEL: A Distributed Procedural Model of Runoff Production

The presentation that follows is developed in terms of local and mean depths to water table within the catchment, but it can be equally developed in terms of local and mean storage deficits (see Beven and Kirkby, 1979 and other papers cited above). It is assumed that at any point i on a hillslope, downslope saturated subsurface flow rate q_i may be described by:

$$q_i = T_o \tan \beta \exp(-fz_i) \tag{1}$$

where T_o is profile transmissivity when the soil is just saturated to the surface; z_i is the local depth to water table; f is a parameter dependent upon the rate of change of transmissivity with depth; and $\tan\beta$ is the local slope angle and is taken as an approximate hydraulic gradient for downslope flow. Beven (1986a) has shown how Equation (1) can be related to an equivalent exponential decline in hydraulic conductivity with depth:

$$K_s = K_o \exp(-fz) \tag{2}$$

where K_o is hydraulic conductivity at the soil surface. Beven (1984) gives data to suggest that Equation (2) is a suitable form for a variety of soil types with a range of K_o from 0.01 to 100 m/hr and f from 1 to 12 m^{-1}. In soils for which Equation (2) holds and hydraulic conductivity at depth is small, $T_o \approx K_o/f$.

Under a steady and spatially homogeneous input to the water table, r:

$$q_i = a_i r \tag{3}$$

where a_i is the area of hillslope per unit contour length draining through point i. The value of a_i will be high for points at the base of convergent hillslopes and low for points near the divide and on straight and divergent hillslopes. Combining Equations (1) and (3):

$$z_i = -\frac{1}{f} \ln \frac{(a_i r)}{T_i \tan \beta} \tag{4}$$

Integrating over the hillslope or catchment area "A," we can obtain an expression for mean depth to water table, given r, as:

$$z = \frac{1}{A} \int_A -\frac{1}{f} \ln \frac{(a_i r)}{T_i \tan \beta} \tag{5}$$

Note that Equation (5) assumes that f is a constant and that the relationship Equation (1) holds for the case where water is ponded on the surface ($z_i < 0$). Given the exponential nature of Equation (1), and the relatively slow rates of surface flow to be expected for vegetated surfaces, this assumption may be acceptable in many catchments. Using Equations (4) in (5) to eliminate $ln(r)$ and rearranging:

$$\bar{z} = \frac{1}{f} [z_i + \gamma - \ln(a_i/T_o \tan \beta)] \tag{6}$$

where:

$$\gamma = \frac{1}{A} \int_A \ln(a_i/T_o \tan \beta) \tag{7}$$

Defining:

$$\ln(T_e) = \frac{1}{A} \int_A \ln(T_o) \tag{8}$$

Equation (6) can be rearranged into the form:

$$f(\bar{z} - z_i) = [\lambda - \ln(a_i/\tan \beta)] + [\ln(T_e) - \ln(T_o)] \tag{9}$$

where:

$$\lambda = \frac{1}{A} \int_A ln(a_i/\tan\beta) \qquad (10)$$

is a topographic constant for the catchment.

Equation (9) expresses, in dimensionless form, the deviation of the local depth to water table from the catchment mean, scaled by the parameter f, in terms of a deviation in the logarithm of transmissivity from its areal integral value, and a deviation in the local topographic index away from its areal integral value. For any given value of $\bar{z}$, Equation (9) can be used to predict the *pattern* of water table depths over the catchment. Points for which the right hand side deviations yield values of z_i less than the depth of a near saturated capillary fringe are of particular interest since they represent the predicted saturated contributing area, taking account of heterogeneity in both topography and soil transmissivity. Under assumptions of a homogeneous soil cover, the transmissivity deviation goes to zero ($T_o = T_e$ everywhere) and the pattern will depend on the topographic index alone. Figure 2 shows the patterns of soil saturation predicted from digital elevation data for the River Wye catchment at Plynlimon, mid-Wales.

In the procedural version of TOPMODEL the relationship (Equation [9]) (or a variant of it) is used at any time step to predict the saturated contributing area. Any rainfalls falling on this area, or in excess of that required top fill the storage deficit associated with small values of z, form the predicted saturation excess runoff production. Application of Equation (9), which is based on steady state assumptions, at successive time steps implies that a value of $\bar{z}$ is updated at each time step and that subsurface wave speeds are such that redistribution of inputs to the saturation zone is relatively rapid compared with the rate of change of $\bar{z}$. Updating of $\bar{z}$ requires continuous accounting of inflows and outflows for the saturated zone. An expression for subsurface outflows to the stream channel can be derived from Equation (1) as:

$$Q_b = Q_o \, exp(-f\bar{z}) \qquad (11)$$

where $Q_o = A\,T_e\,exp\{-\lambda\}$ (see Beven, 1986a,b; Sivapalan et al., 1987). Inflows to the saturated zone will depend upon infiltration rates and redistribution in the saturated zone, which in turn will depend on local values of K_o and z_i.

Infiltration rates will commonly be equal to the input flux rates at the ground surface, but for high input intensities and low local hydraulic conductivities there may be ponding at the surface and infiltration excess runoff production. Beven (1984) has provided an infiltration model based on Green–Ampt assumptions that is consistent with the assumption of an exponentially declining hydraulic conductivity. Sivapalan et al. (1987) and Wood et al.(1988) have also used a Philip equation infiltration model within the TOPMODEL structure.

It is very difficult to predict redistribution of infiltrated water in the unsaturated zone above the water table in a way that is both consistent with the perceptual model outlined above and reflects the pattern of soil properties and depths to water table in the catchment, but does not involve a large number of parameter values. Two forms, that add only one additional parameter, have been used in the TOPMODEL structure. In Beven and Wood (1983), which was formulated in terms of patterns of storage deficit, local vertical flux was predicted as:

$$q_v = S_{uz}/S_i \, t_d \qquad (12)$$

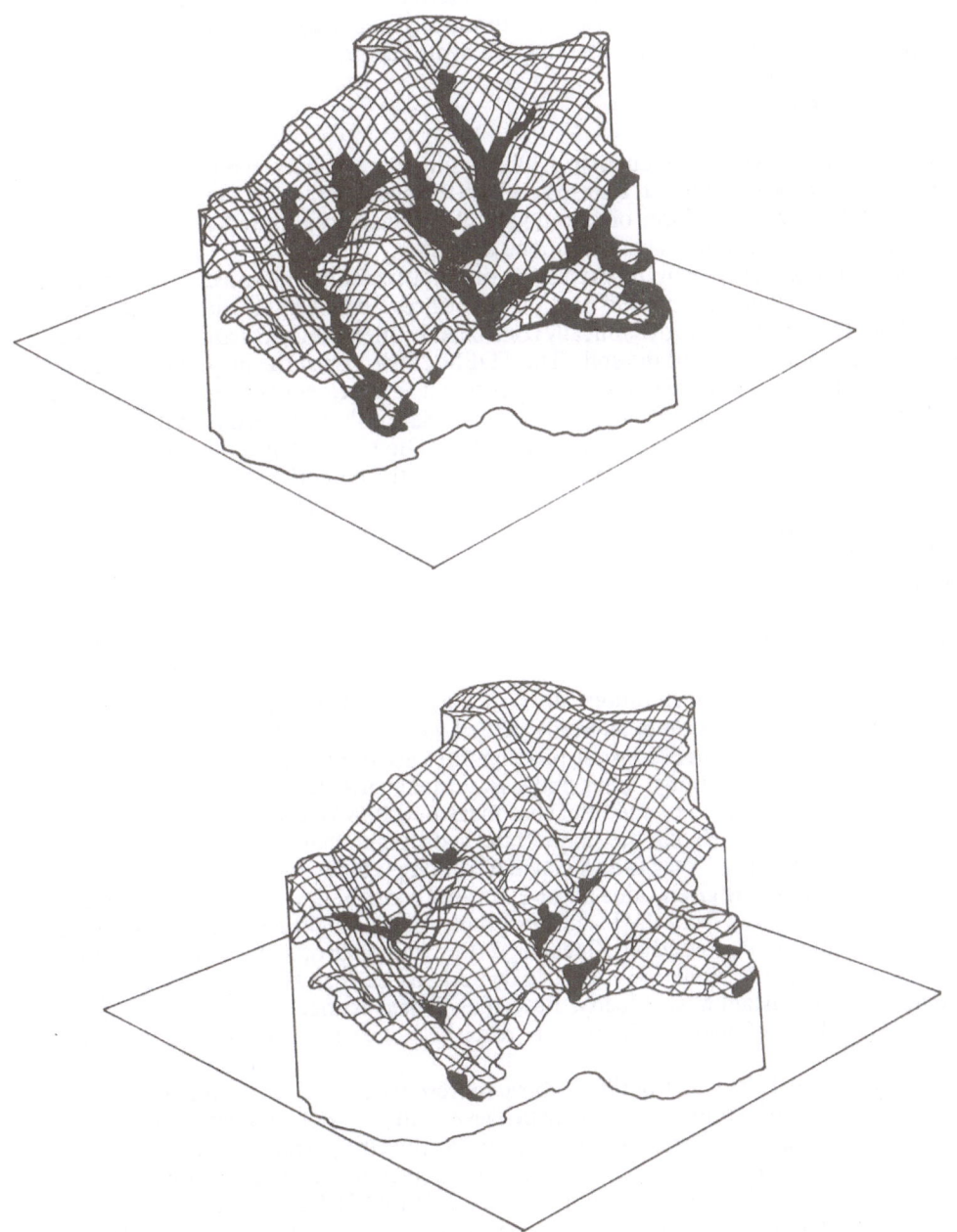

Figure 2. Predicted saturated areas at two different moisture deficits based on digital elevation data for the 10 km^2 Wye catchment Plynlimon.

where S_{uz} is local storage of mobile water in the unsaturated zone, S_i is local storage deficit, and t_d is a time delay per unit of storage deficit. Alternatively in Beven (1986a) the rate of flow into the saturated zone was set equal to the saturated hydraulic conductivity at the water table multiplied by a parameter, α, representing the hydraulic gradient ie.:

$$q_v = \alpha \, K_o \exp(-f z_i) \qquad (13)$$

In fact, Beven (1986a) sets α equal to 1, eliminating this as a parameter. Both forms allow the local delay in the unsaturated zone to vary with the depth of the water table. In the original Beven and Kirkby (1979) form of the model all the fast unsaturated zone drainage was assumed to reach the water table within one time step.

A similar problem arises in the prediction of evapotranspiration rates. Evapotranspiration takes place from both ground and plant surfaces and involves withdrawals of water throughout the rooting zone, dynamically controlled by meteorological and plant controls as well as the moisture status of the soil. The TOPMODEL structure, in common with many other conceptual hydrological models, assumes that if actual rates of evapotranspiration cannot be specified as a boundary condition, then the relationship between potential and actual rates is a function of root zone moisture status. The simplest form, involving only one parameter, allows evapotranspiration to take place at the potential rate wherever the soil is saturated to the surface or there is remaining storage of freely draining water in the unsaturated zone. When this storage is depleted, further evapotranspiration from the root zone is allowed at a rate determined by the ratio of a soil moisture deficit in the root zone to some maximum deficit, ie:

$$E_a/E_p = 1 - S_{rz}/S_{rmax} \qquad (14)$$

where E_a is actual evapotranspiration rate, E_p is the potential rate, S_{rz} is a root zone moisture deficit, and S_{rmax} is a maximum allowable deficit which is an additional parameter of the model. Areal differences in evapotranspiration rates are predicted by the model depending upon the predicted contributed areas and unsaturated zone storage.

Beven and Kirkby (1979) pointed out that measured flow velocities of surface runoff on vegetated surfaces are such that the shape of the hydrograph may be significantly influenced by overland flow routing. They used a constant routing velocity proportional to local gradient. This allows, for any given value of the saturated contributing area, a unique time delay histogram to be derived from the basin topography and a single velocity parameter.

Channel routing is also based on a linear algorithm (although other network algorithms can be easily implemented), similar to that used by Surkan (1969) and Kirkby (1976a), under the assumption of a constant wave velocity. Beven (1979) has justified the use of a constant wave velocity on the basis of field data for average reach velocities in the 10 km² Severn catchment in mid–Wales.

The TOPMODEL structure outlined above represents a simple distributed model with a minimum number of parameters. It may be viewed, in fact, not so much as a fixed model structure but as a set of concepts that can be tested against the perceptual model of the user in any application to a particular catchment. The distributed nature of TOPMODEL is important in this respect. Even at the very simplest level, TOPMODEL can predict, in space, where areas of saturated or near saturated soil will be found. A ready check can be made on such predictions in the field as a check of whether the concepts used are valid or need to be modified. Such comparisons between the predictions of a readily understood model and the hydrologist's *ideas* of how a catchment is working have been useful in a number of studies.

Perhaps the most important aspect of TOPMODEL and the Moore et al. (1986) model is the ability to take account of heterogeneity in soil and topographic characteristics.

Information about the spatial heterogeneity in soil characteristics is, of course, rarely available, but in the application of Equation (9) the deviations associated with the topographic variable are always likely to dominate the deviations associated with soil transmissivity. Therefore, in general little loss in accuracy would be expected in neglecting soil heterogeneity in the prediction of saturated contributing areas. Soil transmissivities are, however, more important in the determination of the drainage function, Equation (11), but here the estimate of the geometric mean transmissivity will be far more important than the estimate of the variability around that mean. Similar arguments cannot be used for catchments in which the prediction of infiltration excess overland flow is important, where the spread of the contributing area will be more directly related to the variability in hydraulic conductivities.

The model calculations may be carried out in either grid square or, since all points with the same value of $ln(a/T_o \tan\beta)$ are assumed to respond in a hydrological similar way, discrete distribution function forms. The distribution function form allows highly efficient computations, even though the predicted spatial patterns may still be mapped back onto the catchment topography if required. In either case, the catchment can be treated as a lumped unit for calculation of z and the contributing areas dependent on z, or where rainfall patterns are known to be highly varied in space, as a number of subcatchments, where each subcatchment may vary in its input, topographic or other characteristics. A spatially scaled version of the TOPMODEL structure has been developed by Sivapalan et al. (1987), while Beven (1986a,b, 1987b) has used versions of the model with a random rainstorm generator to predict flood frequency characteristics. Figure 3 shows the results of an application of the version of the model described here to predicting flows from the River Wye catchment at Plynlimon, mid–Wales. A more complex form of TOPMODEL, designed to provide hydrological

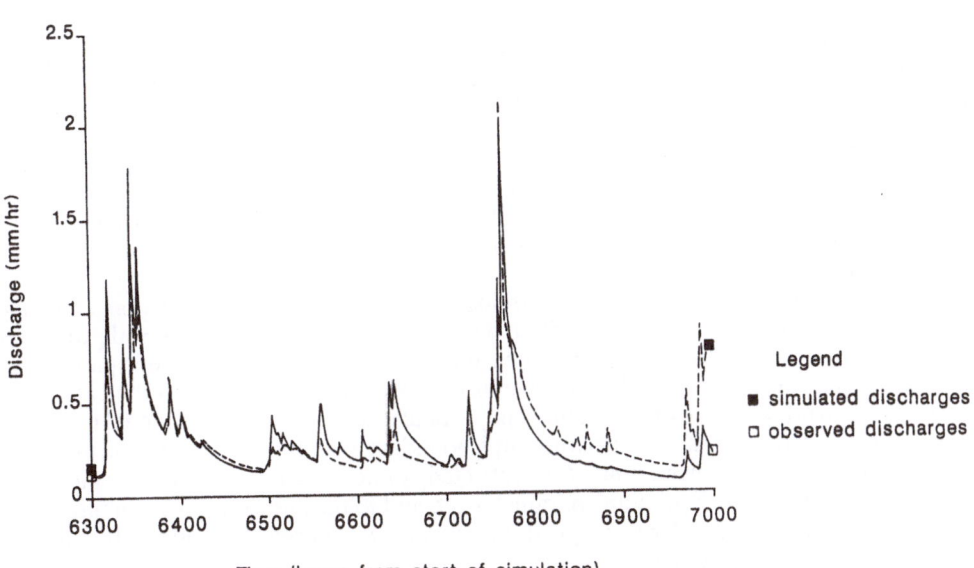

Figure 3. Validation run of TOPMODEL in a split record test for the River Wye catchment at Plynlimon.

predictions for use in multi–compartment geochemical calculations and involving 13 parameters, was evaluated by Hornberger et al. (1985) in an application to the White Oak Run catchment in Virginia, using the methods of regionalized sensitivity analysis proposed by Hornberger and Spear (1981). A number of different objective functions were evaluated in conjunction with a Rosenbrock automatic optimization routine to obtain an "optimum" set of parameter values for each objective function. The results indicate that all the objective functions were indifferent to all but a small number of the 13 parameters, but that the most sensitive parameters were different for different objective functions. Some interdependence between different parameter values was demonstrated, being evident as curved "valley" forms in the response surface. They also showed that it might be better, given such an over-parameterized model, to change the model structure to eliminate some of the parameters rather than simply fixing the values of insensitive parameters. They concluded:

"It is obvious that several of the parameters that were specifically included in the full model to account for processes that were observed in the field could not be estimated from the inflow–outflow data. The simpler model that we implemented gave almost an equally good fit to the data. We face the dilemma posed by Beck (1983): a 'small' (ie. rigorously calibrated) model may miss one or more key aspects of a system's behavior because the pertinent modes of behavior were not excited during the calibration period, but a "large" model, while theoretically capable of reproducing a wider range of behaviors can only be calibrated with a good deal of uncertainty." (p.1849)

A similar dilemma must inevitably face all attempts at spatially distributed modeling due to the number of parameters that arise in any distributed description of a hydrological system.

1.6. The Future of Distributed Modeling: Likelihood and Predictive Uncertainty

It is clear that there are inherent practical problems associated with distributed modeling in hydrology. We have a highly complex dynamic system that can be represented only crudely by the current generation of models. Even these, however, require the specification of large numbers of spatially distributed parameters. These parameters must inevitably be intercorrelated in their effect on the model simulations with the result that they will be very difficult to identify on the basis of comparisons between observed and simulated behavior. Measurements of internal state variables will help in this respect, but only to a limited extent since our measurement techniques tend to be at much smaller scales than is required for comparison with model variables.

It seems inevitable, in fact, that the hydrological systems that we are intent on simulating must remain unknowable in their distributed characteristics, particularly of soil properties. This "unknowability" must lead to predictive uncertainty, even where detailed spatial information on soils and their hydrological response is available for "points" in the catchment. There are also other causes of uncertainty, in particular unknown errors in model input data and unknown errors in model structure and boundary conditions. Studies of predictive uncertainty to date have been limited to restricted studies of uncertainty due to parameter variations around some "optimum" parameter set, and to the effects of errors in input data. It is argued here that a wider concept of uncertainty is required that reflects the ultimate "unknowability" of the system.

It follows from recognition of a concept of "unknowability," that there can be no ultimate validation of a model structure and set of parameters as a simulator of a catchment system (see Stephenson and Freeze, 1974, for an early discussion of the infeasibility of validation of hydrological models). In a sense, therefore, any model and set of parameters is in contention as an appropriate predictor for the real system. There are many reasons to consider simulation models in this light. For example, even under ideal hypothetical circumstances in which the correct model structure is known a priori, it can be shown that the introduction of errors into input data, or in sampling of spatial characteristics, can lead to a number of parameter sets giving close to "optimal" simulations. Also, Loague and Freeze (1985) have shown that in some circumstances there may be little to choose between alternative model structures in simulating catchment discharges. In addition, experience suggests that while a particular simulation may give poor results relative to others for one period, it may, over a longer period, prove to be a better predictor. Consequently, it is suggested that the idea of an "optimal" simulation model and set of parameter values be replaced by the concept of likelihood value, in which the value of every simulation of the behavior of the system under study is expressed as a likelihood index.

These likelihood values may be considered as an assessment of the probability of having a "correct" simulation. The likelihood function will have a value for each "set" of parameter values and as such will reflect the effects of any parameter interaction associated with the model structure. The likelihood values may be used in a number of ways; in particular, they may be used as probability weights to assess the predictive uncertainty associated with the model given the uncertainty in parameter values and input data, as conditioned by the available observations. Note that although the simulations may result from a nonlinear model or models, the combination of simulations in this way to calculate sensitivities and uncertainties is a linear process. The additional computations required to experimentally manipulate different likelihood functions should, therefore, be trivial.

Such a procedure also focuses attention on the value of observations in constraining uncertainty. Additional observations should impose additional constraints on the likelihood function and consequently on the uncertainty. This may be particularly important in the simulation of spatially distributed systems with observations of internal states, which may not be directly equivalent to simulated variables in the model. It should be noted that the definition of the likelihood function is necessarily subjectively chosen, particularly where different types of observations (for example discharges and water table levels) contribute, as some form of weighted sum, to the definition. Consequently, the predictive uncertainty resulting from the likelihood function will reflect this subjectivity of choice. It is an advantage of the proposed methodology that this dependence is made explicit.

2. Conclusions

I have argued elsewhere (Beven, 1987a, 1989) that the practical application of physically-based models must be accompanied by a realistic assessment of the uncertainty associated with the predictions. The generalized likelihood methodology proposed above provides a means for doing so in a readily understood way. In requiring multiple simulations to obtain values of the likelihood function it is a very computationally intensive procedure for many spatially distributed models. However, it is a procedure that would be readily implemented in parallel processing computers and may in the future become a valuable technique in the hydrological modeler's toolkit.

3. Acknowledgments

The original impetus for the development of the TOPMODEL structure was provided by Mike Kirkby of the University of Leeds. Much of the recent work on TOPMODEL has been carried out in collaboration with Eric Wood and Murugesu Sivapalan at Princeton University. The figures in this chapter have been prepared with the help of Andrew Binley and Paul Quinn of the University of Lancaster with the support of NERC grant GR3/6264.

4. References

Abbott, M. B., J. C. Bathurst, J. A. Cunge, P. E. O'Connell, and J. Rassmussen: 1986a, 'An Introduction to the European Hydrological System – Systeme Hydrologique Europeen, "SHE". 1. History and Philosophy of a Physically–Based, Distributed Modelling System,' *J. Hydrology* **87**, 45–59.

Abbott, M. B., J. C. Bathurst, J. A. Cunge, P. E. O'Connell, and J. Rassmussen: 1986a, 'An Introduction to the European Hydrological System – Systeme Hydrologique Europeen, "SHE". 2. Structure of a Physically–Based Distributed Modelling System,' *J. Hydrology* **87**, 61–77.

Bathurst, J. C.: 1986a, 'Physically–Based Distributed Modelling of an Upland Catchment using the Systeme Hydrologique Europeen,' *J. Hydrology* **87**, 79–102.

Bathurst, J. C.: 1986b, 'Sensitivity Analysis of the Systeme Hydrologique Europeen for an Upland Catchment,' *J. Hydrology* **87**, 103–123.

Beck, M. B.: 1983, 'Uncertainty, System Identification and the Prediction of Water Quality,' in M. B. Beck, and G. van Straten (eds.), *Uncertainty and Forecasting of Water Quality*, 3–68, Springer–Verlag, Berlin.

Beven, K. J.: 1977, 'Hillslope Hydrographs by the Finite Element Method,' *Earth Surface Processes* **2**, 13–28.

Beven, K. J.: 1978, 'The Hydrological Response of Headwater and Sideslope Areas,' *Hydrological Sciences Bulletin* **23**(4), 419–437.

Beven, K. J.: 1979, 'On the Generalised Kinematic Routing Method,' *Water Resources Research* **15**(5), 1238–1242.

Beven, K. J.: 1982, 'On Subsurface Stormflow: An Analysis of Response Times,' *Hydrological Sciences J.* **4**, 505–521.

Beven, K. J.: 1984, 'Infiltration into a Class of Vertically Non–uniform Soils,' *Hydrological Sciences J.* **29**, 425–434.

Beven, K. J.: 1985, 'Distributed Models,' in M. G. Anderson, and T. P. Burt (eds.), *Hydrological Forecasting*, 405–435, Wiley.

Beven, K. J.: 1986a, 'Hillslope Runoff Processes and Flood Frequency Characteristics,' in A. D. Abrahams (ed.), *Hillslope Processes*, 187–202, Allen and Unwin, Boston.

Beven, K. J.: 1986b, 'Runoff Production and Flood Frequency in Catchments of Order N: An Alternative Approach,' in V. K. Gupta, I. Rodriguez–Iturbe, and E. F. Wood (eds.), *Scale Problems in Hydrology*, 107–131, Reidel, Dordrecht.

Beven, K. J.: 1987a, 'Towards a New Paradigm in Hydrology, Water for the Future: Hydrology in Perspective,' *IASH* **164**, 393–403.

Beven, K. J.: 1987b, 'Towards the Use of Catchment Geomorphology in Flood Frequency Predictions,' *Earth Surf. Process. Landf.* **12**, 69–82.

Beven, K. J.: 1989, 'Changing Ideas in Hydrology: The Case of Physically–based Models,' *J. Hydrology* **195**, 157–172.

Beven, K. J., and M. J. Kirkby: 1979, 'A Physically–based Variable Contributing Area Model of Basin Hydrology,' *Hydrological Sciences Bulletin* **24**(1), 43–69.

Beven, K. J., and P. E. O'Connell: 1982, 'On the Role of Physically–based Models in Hydrology,' Institute of Hydrology Report No. 81, Wallingford.

Beven, K. J., and E. F. Wood: 1983, 'Catchment Geomorphology and the Dynamics of Runoff Contributing Areas,' *J. Hydrology* **65**, 139–158.

Beven, K. J., M. J. Kirkby, N. Schoffield, and A. Tagg: 1984, 'Testing a Physically–based Flood Forecasting Model (TOPMODEL) for Three UK Catchments,' *J. Hydrology* **69**, 119–143.

Beven, K. J., and R. T. Clarke: 1986, 'On the Variation of Infiltration into a Homogeneous Soil Matrix Containing Macropores,' *Water Resources Research* **22**, 383–388

Beven, K. J., A. Calver, and E. M. Morris: 1987, 'The Institute of Hydrology Distributed Model,' Institute of Hydrology Report No. 98, Wallingford.

Binley, A. M., J. Elgy, and K. J. Beven: 1989a, 'A Physically–based Model of Heterogeneous Hillslopes. I. Runoff Production,' *Water Resources Research* **25**(6), 1219–1226.

Binley, A. M., K. J. Beven, and J. Elgy: 1989b, 'A Physically–based Model of Heterogeneous Hillslopes. II. Effective Hydraulic Conductivities,' *Water Resources Research* **25**(6), 1227–1233.

Blackie, J. R., and C. W. O. Eeles: 1985, 'Lumped Catchment Models,' in M. G. Anderson, and T. P. Burt (eds.), *Hydrological Forecasting*, 311–346, Wiley.

Crabtree, R. W., and S. T. Trudgill: 1985, 'Hillslope Hydrochemistry and Stream Response on a Wood, Permeable Bedrock: The Role of Stemflow,' *J. Hydrology* **80**, 161–178.

Dooge, J. C. I.: 1977, 'Problems and Methods of Rainfall–Runoff Modelling,' in T. A. Ciriani, U. Maione, and J. R. Wallis (eds.), *Mathematical Models for Surface Water Hydrology*, 71–108, Wiley.

Emmett, W. W.: 1978, 'Overland Flow,' in M. J. Kirkby (ed.), *Hillslope Hydrology*, 145–176, Wiley.

Fleming, G.: 1975, 'Computer Simulation Techniques in Hydrology,' Elsevier.

Freeze, R. A.: 1972, 'Role of Subsurface Flow in Generating Surface Runoff. 2. Upstream Source Areas,' *Water Resources Research* **8**, 1272–1283.

Freeze, R. A., and R. L. Harlan: 1969, 'Blueprint for a Physically-based Digitally Simulated Hydrologic Response Model,' *J. Hydrology* **9**, 237–258.

Hewlett, J. D.: 1974, "Comments on Letters Relating to 'Role of Subsurface Flow in Generating Surface Runoff. 2. Upstream Source Areas' by R. Allen Freeze," *Water Resources Research* **10**, 605–607.

Hillel, D.: 1988, 'Unstable Flow in Layered Soils: A Review,' *Hydrological Processes* **1**, 143–147.

Hornberger, G. M., and R. C. Spear: 1981, 'An Approach to the Preliminary Analysis of Environmental Systems,' *J. Environ. Manage.* **12**, 7–18.

Hornberger, G. M., K. J. Beven, B. J. Cosby, and D. E. Sappington: 1985, 'Shenandoah Watershed Study: Calibration of a Topography–Based, Variable Contributing Area Hydrological Model to a Small Forested Catchment,' *Water Resources Research* **21**, 1841–1850.

Huff, D. D., R. V. O'Neill, W. R. Emanuel, J. W. Elwood, and J. D. Newbold: 1982, 'Flow Variability and Hillslope Hydrology,' *Earth Surf. Process. Landf.* **7**, 91–94.

Kirkby, M. J.: 1969, 'Infiltration, Throughflow and Overland Flow,' in R. J. Chorley (ed.), *Water Earth and Man*, 215–228, Methuen.

Kirkby, M. J.: 1976a, 'Tests of the Random Network Model and its Application to Basin Hydrology,' *Earth Surf. Process.* **1**, 197–212.

Kirkby, M. J.: 1976b, 'Hydrograph Modelling Strategies,' in R. Peel, M. Chisholm, and P. Haggett (eds.), *Processes in Physical and Human Geography*, Heinemann.

Loague, K. M., and R. A. Freeze: 1985, 'A Comparison of Rainfall–Runoff Modelling Techniques on Small Upland Catchments,' *Water Resources Research* **21**, 229–248.

Moore, I. D., S. M. Mackay, P. J. Wallbrink, G. J. Burch, and E. M. O'Loughlin: 1986, 'Hydrologic Characteristics and Modelling of a Small Forested Catchment in Southeastern New South Wales. Pre-logging Condition,' *J. Hydrology* **83**, 307–335.

O'Loughlin, E. M.: 1981, 'Saturation Regions in Catchments and their Relations to Soil and Topographic Properties,' *J. Hydrology* **53**, 229–246.

O'Loughlin, E. M.: 1986, 'Prediction of Surface Saturation Zones in Natural Catchments by Topographic Analysis,' *Water Resources Research* **22**, 794–804.

Sharma, M. L., R. J. Luxmoore, R. DeAngelis, R. C. Ward, and G. T. Yeh: 1987, 'Subsurface Water Flow Simulated for Hillslopes with Spatially Dependent Soil Hydraulic Characteristics,' *Water Resources Research* **28**, 1523–1530.

Sivapalan, M., K. Beven, and E. F. Wood: 1987, 'On Hydrological Similarity: II. A Scaled Model of Storm Runoff Production,' *Water Resources Research* **23**(12), 2266–2278.

Stengers, I., and I. Prigogine: 1984, *Order Out of Chaos*, Heinemann, London.

Stephenson, G. R., and R. A. Freeze: 1974, 'Mathematical Simulation of Subsurface Flow Contributions to Snowmelt Runoff, Reynolds Creek Watershed, Idaho,' *Water Resources Research* **10**, 284–298.

Surkan, A. J.: 1969, 'Synthetic Hydrographs: Effects of Network Geometry,' *Water Resources Research* **5**, 112–128.

Wilson, C. B., J. B. Valdes, and I. Rodriguez-Iturbe: 1979, 'On the Influence of the Spatial Distribution of Rainfall on Storm Runoff,' *Water Resources Research* **15**, 321.

Wood, E. F., M. Sivapalan, K. Beven, and L. Band: 1988, 'Effects of Spatial Variability and Scale with Implications to Hydrological Modelling,' *J. Hydrology* **102**, 29–47.

Yeh, W. W-G.: 1986, 'Review of Parameter Identification Procedures in Groundwater Hydrology: The Inverse Problem,' *Water Resources Research* **22**, 95–108.

Zavlavsky, D., and G. Sinai: 1981, 'Surface Hydrology (5 parts),' *J. Hydraul. Div., ASCE* **107**, 1–93.

Chapter 18

Hydraulic and Hydrologic Flood Routing Schemes

Ezio Todini
E. Todini & Partners, S. r. l.
Via Zanardi, 16
40131 Bologna, Italy

Abstract. The paper describes the different methods available for solving the problem of one-dimensional flood routing. Starting from the definition of the problem and the assumptions used to schematize the flood routing phenomenon, alternative methods are described in the light of different purposes for their use encountered in hydraulic or hydrological practices. Particular attention is given to simplified methods ranging from the so-called "hydrological models" to an intermediate model, known as PAB, which combines the simplicity of a hydrological model to the accuracy of a hydraulic model.

1. Introduction

The development of flood routing schemes has been tackled in the past mostly by hydraulic engineers who have mainly been concerned with the design of hydraulic structures and the analysis of their effectiveness. This has lead to the implementation of a number of computer programs, based on the numerical integration of the partial differential equations describing the flood routing phenomenon. These efforts have concentrated on the detailed description and accurate simulation of the propagation of a flood wave in a more or less complex channel reach. In general, computation time of the programs has not been considered as a limiting factor to their practical use.

Alternatively, hydrologists have proposed extremely simplified models, such as storage routing (Puls, 1928; Goodrich, 1931), Muskingum routing (McCarthy, 1938), Lag and K routing (Linsley et al., 1949), Kalinin and Miljukov routing (1958), linear reservoir and channel cascade routing (Maddaus, 1969), etc., which do not describe the propagation phenomenon but allow for the computation of discharges in a given downstream section as a function of discharges known at an upstream section. These models require virtually no running time but can only describe the overall effect of a channel or river reach, and they do not allow for the simulation of hydraulic structures inserted within the reach. They tend to give what Dooge calls the "external" description as, opposed to the "internal" description of the system, provided by the full dynamic models.

A comprehensive and detailed description of flood routing procedures can be found in "Channel Routing" by Fread (1985).

This chapter only tries to assess the use of the models in the light of the different objectives that can be encountered in the hydrological practices.

Finally, after a review of several methods proposed in the literature, a simplified hydraulic model is introduced, which allows for the "internal" description of a hydraulic system. The model which is virtually not affected by numerical stability allows for real–time simulations in a very limited amount of time and is, therefore, particularly suitable for real–time control purposes.

D. S. Bowles and P. E. O'Connell (eds.), Recent Advances in the Modeling of Hydrologic Systems, 389–405.
© 1991 *Kluwer Academic Publishers.*

2. Discussion

2.1. The Flood Routing Problem

In 1871 Barré De Saint Venant formulated the classical system of partial differential equations that allow for the study of one–dimensional, gradually varied unsteady flow. The two equations, describing mass and the momentum balance, can be written as follows:

$$\frac{\partial Q}{\partial x} + B\frac{\partial z}{\partial t} = 0$$

$$\frac{\partial Q}{\partial t} + \frac{\partial}{\partial x}\left(\frac{Q^2}{A}\right) + g\,A\left(\frac{\partial z}{\partial x} + J\right) = 0$$

(1)

where $Q(x,t)$ is the discharge, $z(x,t)$ is the water profile elevation above a horizontal datum, $A(z,x)$ is the wetted area, $B(z,x)$ is the surface width, J is the friction losses slope, and x,t is the space and time coordinates.

The first equation, known as continuity equation, represents the mass balance in a reach without lateral inflow or outflow. The second equation, known as momentum equation, represents the energy balance in the same reach.

The basic assumptions used to derive Equation (1) (Chow, 1959; Henderson, 1959) are:
- The flow is one–dimensional.
- The pressure has a hydrostatic distribution.
- The distributed friction losses can be evaluated using the usual uniform flow formula and must account for the concentrated head losses.
- The energy and momentum coefficients α and β can be taken equal to 1.
- There are no lateral inflows or outflows.

The solution of Equation (1), which constitutes a system of hyperbolic quasi–linear partial differential equations, has given rise to a number of numerical methods due to the lack of analytical solutions in the most general case of non–cylindrical channels. The solution methods are based either on the characteristic lines (Abbott, 1966; Evangelisti, 1961) or on time and space discretization using explicit or implicit finite differences schemes (Abbott, 1979; Lai, 1967; Price, 1973; Stoker, 1957; W.M.O., 1983; Zezulak, 1984). Applications of finite element schemes have also appeared (Cooley and Moin, 1976), although finite element techniques are not particularly recommended for the integration of hyperbolic systems and do not offer great advantages over the finite differences schemes for the solution of one dimensional problems.

The major problem in using the method of characteristics relates to the fact that the characteristic lines do not converge in a priori predictable time and space ordinates and, therefore, the solution may be found where the geometrical description of the reach is not available. This is particularly true in non–cylindrical channels (for example, all the natural river beds) where the geometrical description is provided through a number of cross sections, while no information is available between two successive sections defining a reach. In addition the characteristic lines slope tends to vanish at higher values of the discharge, thus becoming less accurate when the phenomenon becomes more interesting.

The methods based upon the finite differences discretization techniques are the most widely used (Ciriani et al., 1975; Cunge et al., 1980) and can be divided into explicit schemes and implicit schemes. The explicit schemes tend to have stability problems when the integration time step is greater than a specified upper limit, known as Courant–Friederichs–Lewy limitation (Barnett, 1976). This has given rise, in the last two decades, to the development of finite differences implicit schemes. In these schemes the solution of the flood routing phenomenon is found by recursive solutions of a system of linearized equations, of rank $2(n-1)$, n being the number of cross sections describing the problem geometry (Greco and Panattoni, 1975).

The time needed for solving the overall problem is, in general, a function of: the number of unknown; the complexity of the expressions needed for the coefficients; and, in the case of the implicit schemes, the rate of convergence. In any case, the computer time required for the solution of large systems is significant. This is the reason why many hydrologists have tried to simplify the equations in most practical applications.

2.2. The Simplified Models

If one disregards the so called "dynamic terms" in the second part of Equation (1), that is the terms containing the derivatives of Q with respect to x and t, the following parabolic equation can be derived:

$$\frac{\partial Q}{\partial t} = D\frac{\partial^2 Q}{\partial x^2} - C\frac{\partial Q}{\partial x} \tag{2}$$

where D and C are nonlinear functions of Q and z and are generally known as diffusivity and convectivity, respectively. The hypotheses made in order to perform this simplification are generally met in the real world practical applications. In fact, their validity is questionable only in special cases such as estuaries or reservoir routing or, in general, when both the friction losses slope and the bottom slope tend to vanish with respect to the "dynamic slope." Fortunately, this is not the case in most practical applications of flood routing, and the parabolic model can be successfully applied (Dooge, 1973; Hayami, 1951; Price, 1973; Tawatchai and Manhandar, 1985).

From a computational point of view the parabolic model shows more or less the same problems encountered in the integration of the full De Saint Venant equations (Price, 1973). Although the possibility of integrating the system by using the characteristic lines does not exist because the two lines are real but coincident (Todini and Bossi, 1986), the finite differences schemes, explicit and implicit, can be used with a reduction in size of the problem since now the unknown become the $(n-1)$ discharges at each time step.

An additional type of model is used when the bottom slope is large and the storage effect becomes small with respect to the time delay. This model is described by the equation:

$$\frac{\partial Q}{\partial t} = -C\frac{\partial Q}{\partial x} \tag{3}$$

and is known as the kinematic model. This model, simpler than the parabolic, can only be used for steep brooks with high bottom and friction slopes as a consequence of its basic assumptions (Ciriana et al., 1975).

Both the parabolic and the kinematic models may have an analytical solution if the coefficients D and C can be assumed constant, thus reducing the equations to ordinary partial differential equations. This will be the basis of the model proposed in this chapter and will be described in the next section.

2.3. The Hydrologic Models

Up to this point, only models based upon differential equations (which can be referred to as hydraulic models) with different degrees of simplification have been described.

Hydrologists have also proposed alternative models which represent a reach in lumped form as a function of one or two parameters to be estimated (Chow, 1964). The most widely known is the Muskingum–Cunge model (Cunge, 1969), where the volume of a reach is expressed as a function of a linear combination of inflow and outflow discharges, times a scale factor. Other models can be used to represent the overall effect of a reach, such as the Kalinin–Miljukov model (Kalinin and Miljukov, 1958) or the Nash cascade transfer function model (Dooge, 1973). Other possibilities include the Lag and K routing model or the more general class of ARMA models.

Threshold models are a category of models of growing interest in statistics (Tong, 1983). Examples include one developed by Becker and Glos (1970) or the Constrained Linear System (CLS) model (Natale and Todini, 1977) can also be used. In particular the CLS model has been successfully used as a piecewise linear representation of the overall nonlinear routing process for the development of operational real-time flood routing schemes in the Reno River in Italy, the Niger, and the Yellow River in China (Guanwu and Zhaohui, 1988).

To assess the domain of application of each one of these models, one must consider the fact that the hydrological models can provide only a lumped behavior of a certain system: the "external description" of the phenomenon. They are extremely useful when a forecast of discharge or of flow volume has to be made in one section as a function of upstream inflows, provided that no backwater effect is modifying the rating curve of the section.

If the rating curve is continuously modified by downstream level conditions, one has to resort to the full dynamic description. This is even more true if one needs an "internal description" of the reach. In particular, if the water profile is not modified by the backwater effect, one could use the Muskingum–Cunge model (Cunge, 1969) numerically integrated or, alternatively, models such as the parabolic or the kinematic, depending on the basic assumptions.

If a backwater effect is present, or if the knowledge of levels is an important requirement within the considered reach, one should use a full dynamic model. The purpose of this chapter is to introduce a model which can meet these requirements by successive linearization of a parabolic scheme about a backwater profile.

2.4. The Real–Time Flood Forecasting and Control Problem

In recent years the interest in real–time monitoring and control of hydraulic structures has grown with the availability of telemetering systems and the ability of microcomputers to provide advanced computational resources at low cost.

Objectives of real–time forecasting can be many and they include:
· Flood warning.
· Reservoir management.
· Water detention areas management.
· Diversion canals management.
· Irrigation and water distribution canals management.
· Sewer networks management.

The first two objectives can be met in general when future discharges can be forecasted in a given downstream section with a predefined lead time. Flood warning systems are in fact based upon the forecast of runoff, unless strong backwater effects modify the stage discharge law. Also, reservoir management requires discharge forecasting only since the reservoir releases can be estimated as a function of the dynamic behavior of the reservoir storage subject to the forecasted inflows. In these cases, hydrological models of different nature can be successfully used in practical applications.

When strong backwater effects are present in the canals and rivers and when dealing with most of the other real–time control problems one has to resort to hydraulic models which allow the forecasting not only of discharges but also of water levels at future times. In the latter case extremely efficient and unconditionally stable algorithms must be adopted to avoid failures of the forecasting and control system.

An extreme example is the case of real–time control systems developed for sewer networks where the available time for data collection, processing, and control varies between three and five minutes due to flash surges generated by storms in urbanized areas and the relatively small storage capacity of the system. One has to solve an unsteady flow problem in a tree or in a closed loop network of pipes within a very short time interval.

In other problems, such as water distribution or irrigation canals management or real–time control of diversion canals, the available time varies between 15 and 30 minutes. Even in these problems, however, the analysis of many management alternatives may prove impossible in such a short time interval, particularly when the available computer is not sufficiently powerful.

2.5. The Parabolic and Backwater (PAB) Approach

If one considers Equation (2) as the basic representation of the phenomenon and assumes that the parameters C and D remain constant during a time interval t, one can find the expression for the transfer function between two successive cross sections. In continuous time this expression is:

$$u(\Delta x, t) = \frac{\Delta x}{t \sqrt{4Dt}} e^{-\frac{(\Delta x - Ct)^2}{4Dt}} \tag{4}$$

for a reach of length Δx. This is a very well known approach (Hayami, 1951; Tawatchai and Manhandar, 1985). Unfortunately, the diffusivity and the convectivity change in time; therefore, as it stands this model is not able to cope with the time variation of the parameters. Additionally, the presence of backwater effects cannot be detected by the model.

The idea of the new model is to combine a discharge transfer model that is analytically integrated in time with a numerically integrated model in space of the backwater curve to recompute at each time step the two values of D and C (Franchini and Todini, 1986; Todini and Bossi, 1986). Therefore, at each time step, given the assumed linearity of the system, the solution in terms of discharge at a downstream section $Q(x + \Delta x, t)$ as a function of known inflow values $Q(x, t)$ can be obtained through a convolution integral:

$$Q(x + \Delta x, t) = \int_0^T Q(x, t - \tau)\, u(\Delta x, \tau)\, d\tau \tag{5}$$

where $u(\Delta x, t)$ is the impulse response function of the reach and T its finite "memory."

In practical applications, Equation (5) is used in discrete form. This means the integral is replaced with a summation, and the function $Q(x, t)$ becomes discrete with step Δt, that is with constant values between t and $t + \Delta t$. In addition, $u(\Delta x, t)$ is replaced by the discrete form of the unit graph relevant to a rectangular pulse of duration Δt, $U(\Delta x, i\Delta t)$. That is:

$$U(\Delta x, i\, \Delta t) = \frac{1}{\Delta t^2} \{ F\,[(i+1)\Delta t] - 2F\,[i\Delta t] + F\,[(i-1)\Delta t] \} \tag{6}$$

$$\forall\, i \in \aleph > 0$$

with:

$$F(i\Delta t) = \int_0^{i\Delta t} \int_{-\infty}^{\tau} u(\Delta x, \xi) \, d\xi \, d\tau =$$

$$\left[i\Delta t - \frac{\Delta x}{C} \right] N \left\{ -\frac{\Delta x - Ci\Delta t}{\sqrt{2Di\Delta t}} \right\} + e^{\frac{\Delta x C}{D}} \left[i\Delta t + \frac{\Delta x}{C} \right] N \left\{ -\frac{\Delta x + Ci\Delta t}{\sqrt{2Di\Delta t}} \right\}$$
(7)

where $N\{\bullet\}$ represents the standard normal probability distribution function:

$$N\{\cdot\} = \frac{1}{2} \int_{-\infty}^{\cdot} e^{-\frac{z^2}{2}} \, dz$$
(8)

In summary, Equation (5) becomes:

$$Q(x + \Delta x, j\Delta t) = \sum_{i=1}^{\tau} Q(x, (j - i + 1)\Delta t) \, U(\Delta x, i\Delta t)$$
(9)

As previously mentioned, this equation holds for D and C (or conversely $U(\Delta x, i \, \Delta t)$) invariant in time. If the unit graph does vary in time, this equation must be rewritten as:

$$Q(x + \Delta x, j\Delta t) = \sum_{i=1}^{\tau} Q(x, (j - i + 1)\Delta t) \, U_j \, (\Delta x, i\Delta t)$$
(10)

The basic idea of PAB can then be identified in the association of Equation (10) with the backwater curve equation which describes a subcritical downstream to upstream spatially variable backwater profile, given by:

$$\frac{\partial}{\partial x} \left(\frac{Q^2}{A} \right) + g \, A \left(\frac{\partial z}{\partial x} + J \right) = 0$$
(11)

This equation can be easily integrated in space using, for instance, the standard step method. At each time step Equation (2) can be linearized, and the parameters D and C can be recomputed for the levels derived using Equation (11). The levels then can be assumed constant for each of the channel sub-reaches and used to propagate discharges from upstream to downstream.

The PAB procedure can be summarized as follows:

(1) Enter the values of discharges for time $t = 0$ in all cross sections used to describe the geometry of the river bed.

(2) Impose the downstream level or compute it using either a rating curve or the continuity equation.

(3) Compute the backwater profile assuming Q variable with x.

(4) Evaluate the average value of parameters D and C for each sub-reach and compute the ordinates of the unit graphs.

(5) Propagate downstream the upstream discharges using the computed unit graph.

(6) Add a time increment and restart from 2.

The advantages of this procedure can be summarized as follows:

(1) Both the propagation of discharges from upstream to downstream and the propagation of levels from downstream to upstream are unconditionally stable—the first one being an analytical solution and the second one being only a steady state backwater profile computation.

(2) The propagation of levels and the propagation of discharges are carried out independently. This enables the definition of a coarser description of the geometry to propagate discharges, while the denser description may be reserved for a better analysis of the river levels.

(3) The computer time requirements are very small, being equivalent to a few convolutions and a backwater curve computation at each time step.

A number of experiments have been carried out to assess the validity of the method, and the results are described below.

2.6. Application Examples

To demonstrate the ability of the PAB scheme in simulating the dynamical behavior of a channel reach, these experiments have been carried out using the following geometrical data (Figure 1):

- trapezoidal cross sections
- length of channel (4500 m)
- distance between cross sections (500 m)
- bottom width (2.7 m)
- slope of banks (3h : 2v)
- bottom slope (.0002)
- Manning's roughness coefficient (.016)

The data refer to an existing irrigation canal run by La Société du Canal de Provence where a number of experiments were lead to derive a flood routing procedure (Linsley et al., 1949; Wignyosukarto, 1983) for the real-time operation of sluices.

The first two experiments compare the results obtained with PAB with those resulting from the integration of the full De Saint Venant equations derived by means of the characteristic lines method (Franchini and Todini, 1986). The two methods were compared by using the semi infinite channel assumption, which required the imposition of a downstream boundary condition very far from the upstream section (100,000 m) in the case of the method of characteristics. The integration time step for PAB was in both cases $\Delta t = 300$ s, and the initial condition was represented in both examples by $Q(\Delta x,0) = 3.65$ m^3/s with a backwater effect due to $z(4500,0) = 2$ m (the uniform depth being 1.14 m). The first example refers to the simulation of an increase of 2 m^3/s in 1,200 seconds, while the second one is relevant to an increase of 1 m^3/s in the same period of time.

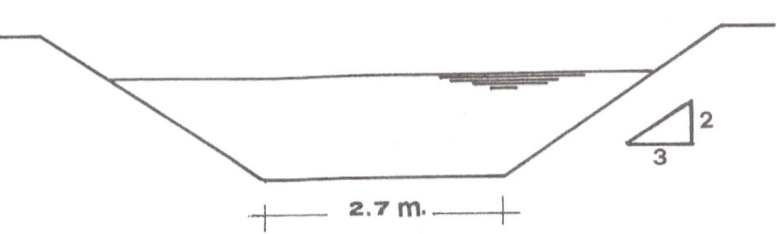

Figure 1. Cross sections shape and size.

Figures 2 and 3 show the pattern of discharges both in the upstream and in the downstream sections, while Figures 4 and 5 show the levels. Figures 2 and 3 show the small anticipation of PAB discharges, due to the parabolic approximation, which appear at the very beginning of the phenomenon. In terms of levels, though, the results of the two models lie within a range of 2 cm, an accuracy which is sufficient for practical applications.

A third example was also carried out by adding to the initial discharge $Q = 3.65$ m³/s a sudden increase of $\Delta Q = 2.95$ m³/s. The initial level conditions in this example were set to the uniform flow conditions of 1.14 m over the entire channel while the downstream condition was kept at critical depth throughout the experiment. The results of this application of PAB are shown in Figures 6, 7, 8, 9, 10, and 11 and are compared to the measured levels. In addition, discharges and levels are also compared to the results obtained by an implicit finite differences scheme (model Carima of Sogreah).

In Figures 6, 7, and 8 the results of PAB using a time step Δt of 100 seconds are compared to the experimental data and to the results of Carima obtained by integrating the full De Saint Venant equations with a $\Delta t = 30$ s for the first 18 minutes and a $\Delta t = 120$ s for the remaining time span.

Figures 9, 10, and 11 show the comparison of the results of PAB using a $\Delta t = 300$ s and $\Delta t = 600$ s with the results of Carima obtained integrating the full De Saint Venant equations with a time step $\Delta t = 300$ s.

It is worthwhile noting the initial stability of the PAB approach as opposed to the results of the integration of the full De Saint Venant equations. In addition, the results show a good agreement with the experimental data although they were obtained without calibration.

It must be noticed that PAB is properly working in this case although, theoretically speaking, the inertial effect of a sudden release should not be negligible, thus violating the parabolic hypothesis. A deeper insight of the phenomenon, though, reveals that the inertial term in Equation (1) is actually comparable with the bottom and friction slopes for a few seconds only, then dropping to almost one hundredth of its initial value.

The results show that the PAB approach results in good approximations, even when the discretization time intervals are by far larger than those generally required for the dynamic description of the routing phenomenon. This suggests that the PAB approach may be preferable to a model based upon the numerical integration of the full De Saint Venant equations when a stable and fast algorithm is required in the real-time control of hydraulic systems (Franchini and Todini, 1986).

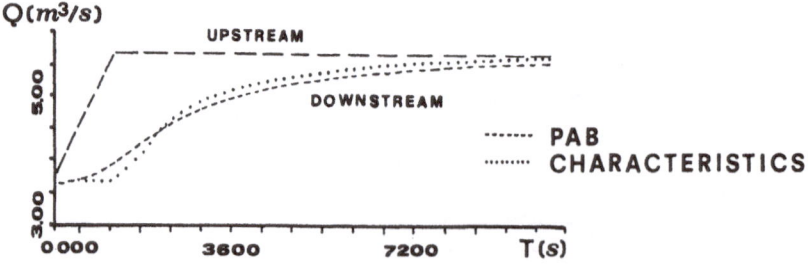

Figure 2.

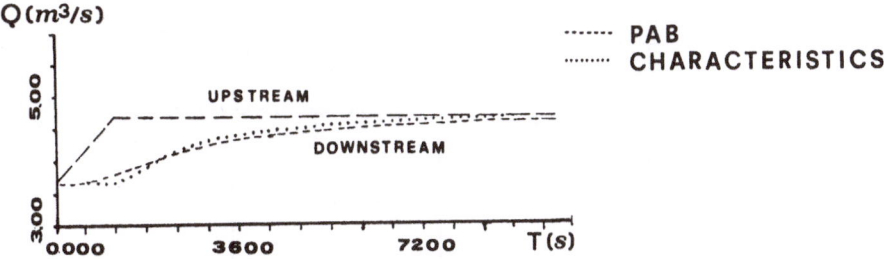

Figure 3.

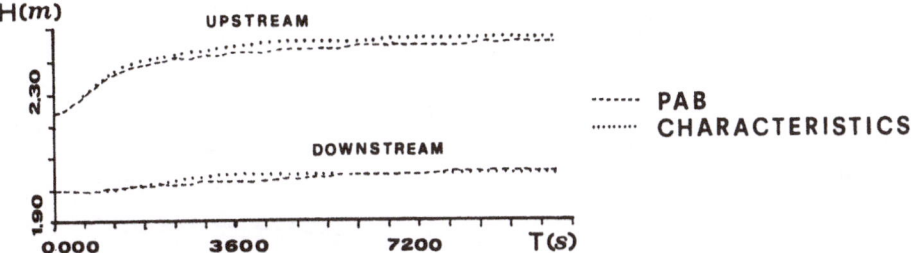

Figure 4.

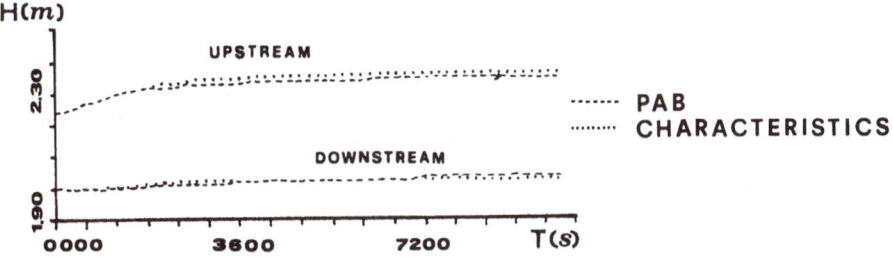

Figure 5.

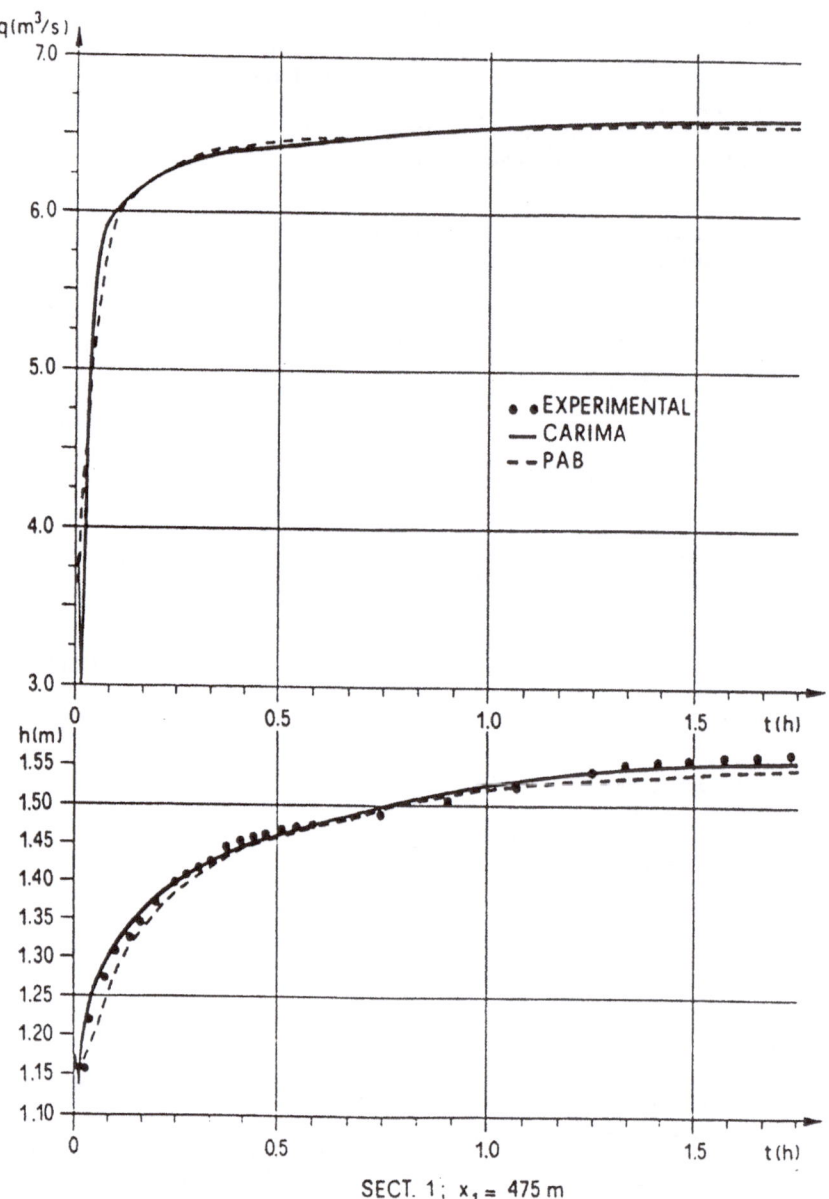

Figure 6.

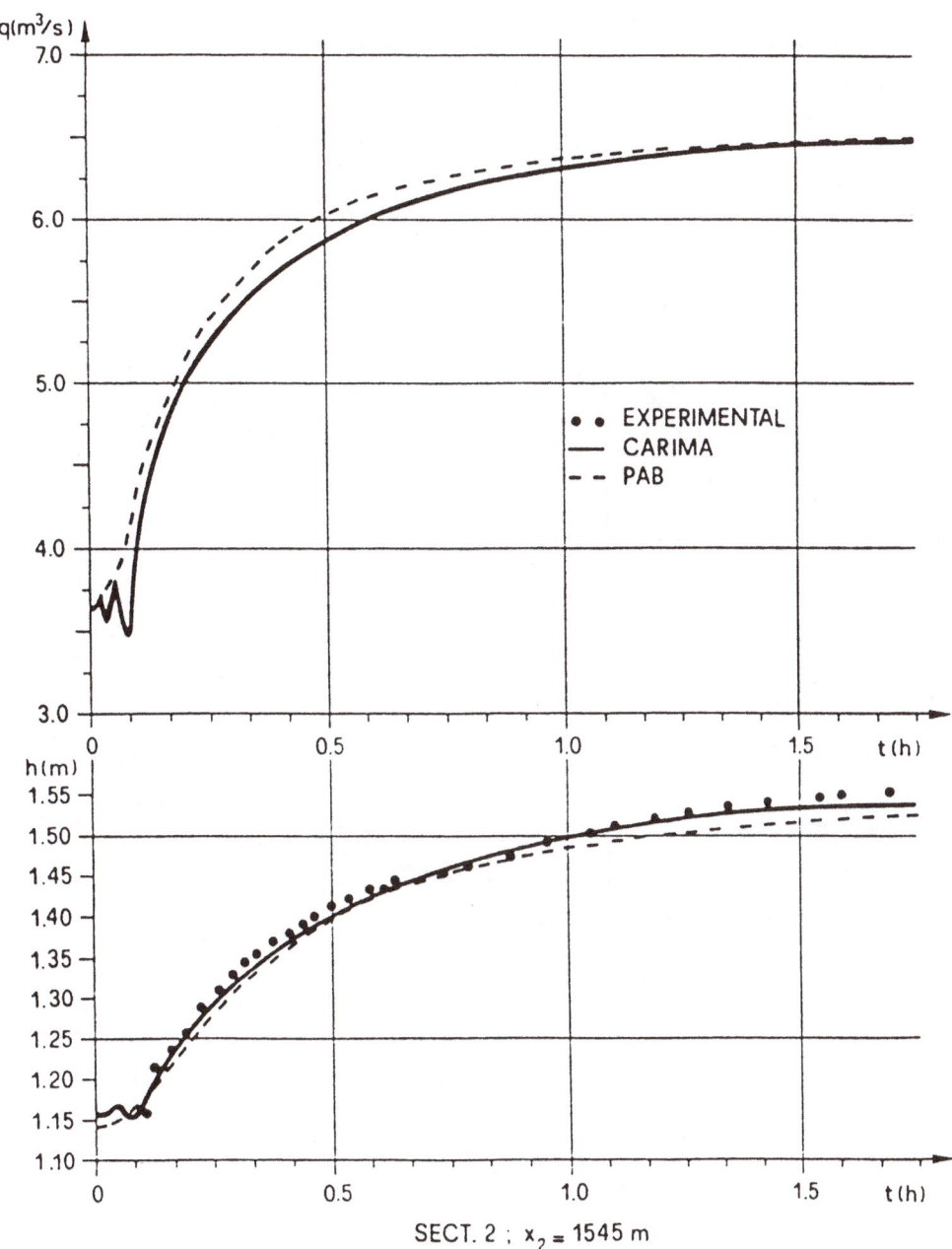

Figure 7.

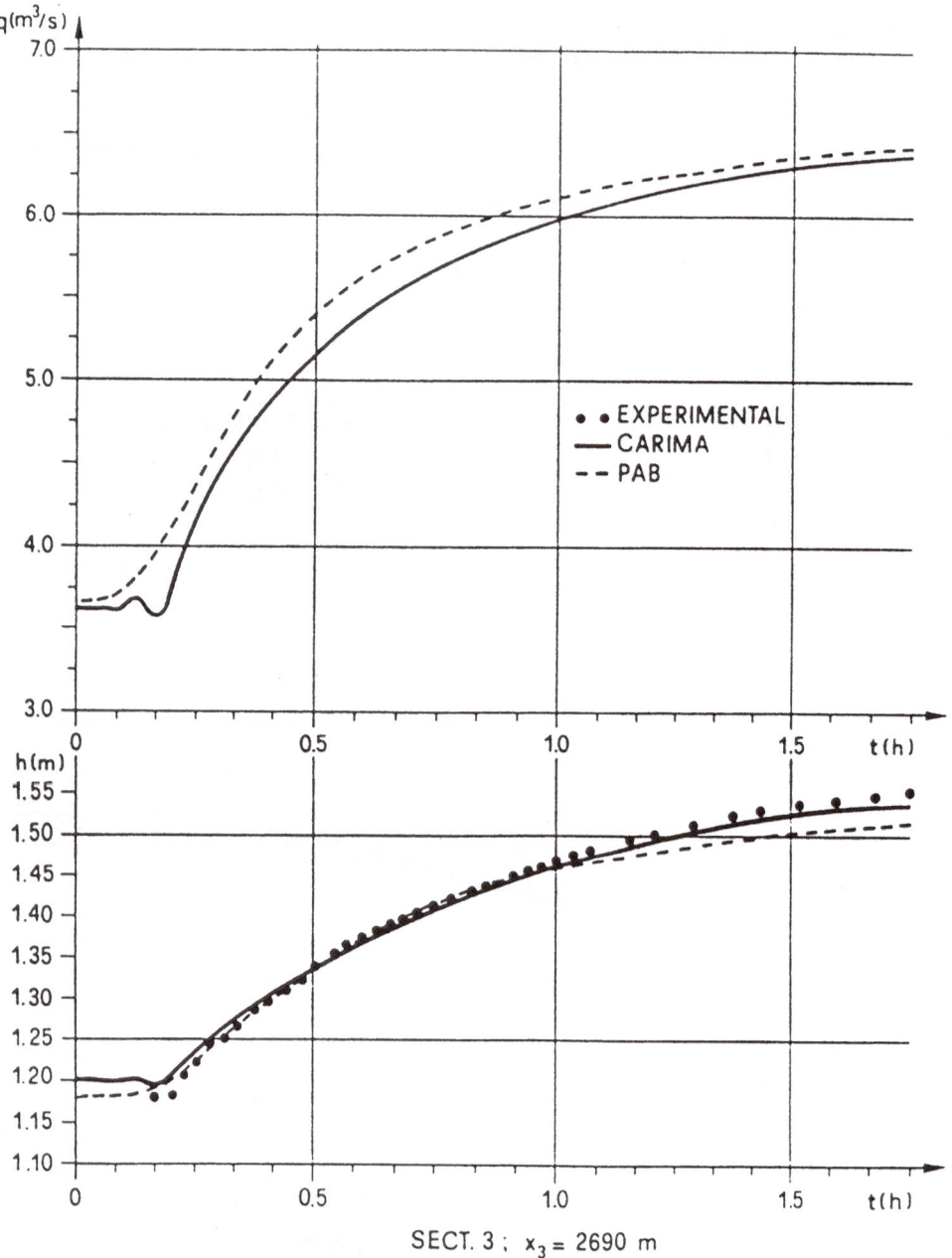

Figure 8.

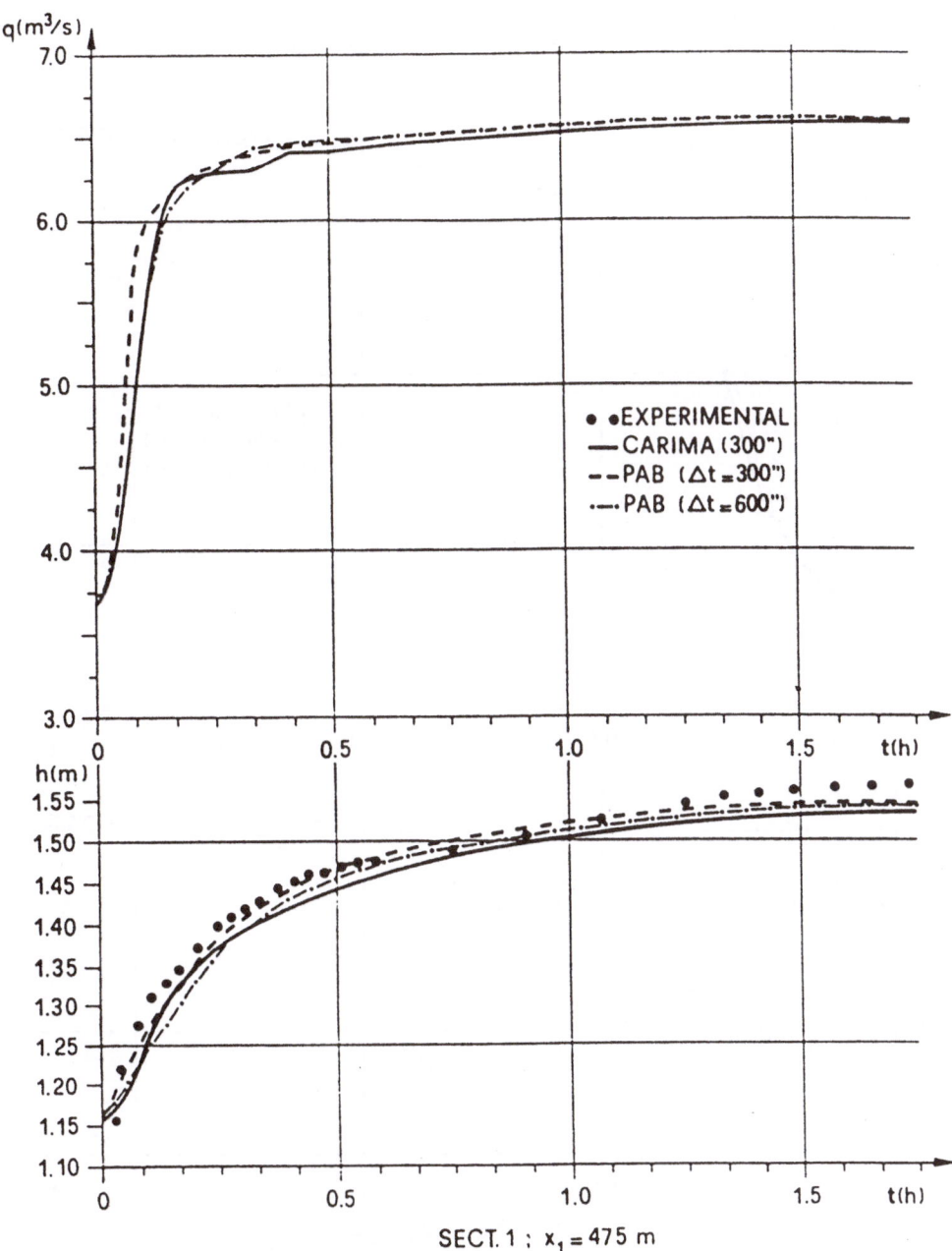

Figure 9.

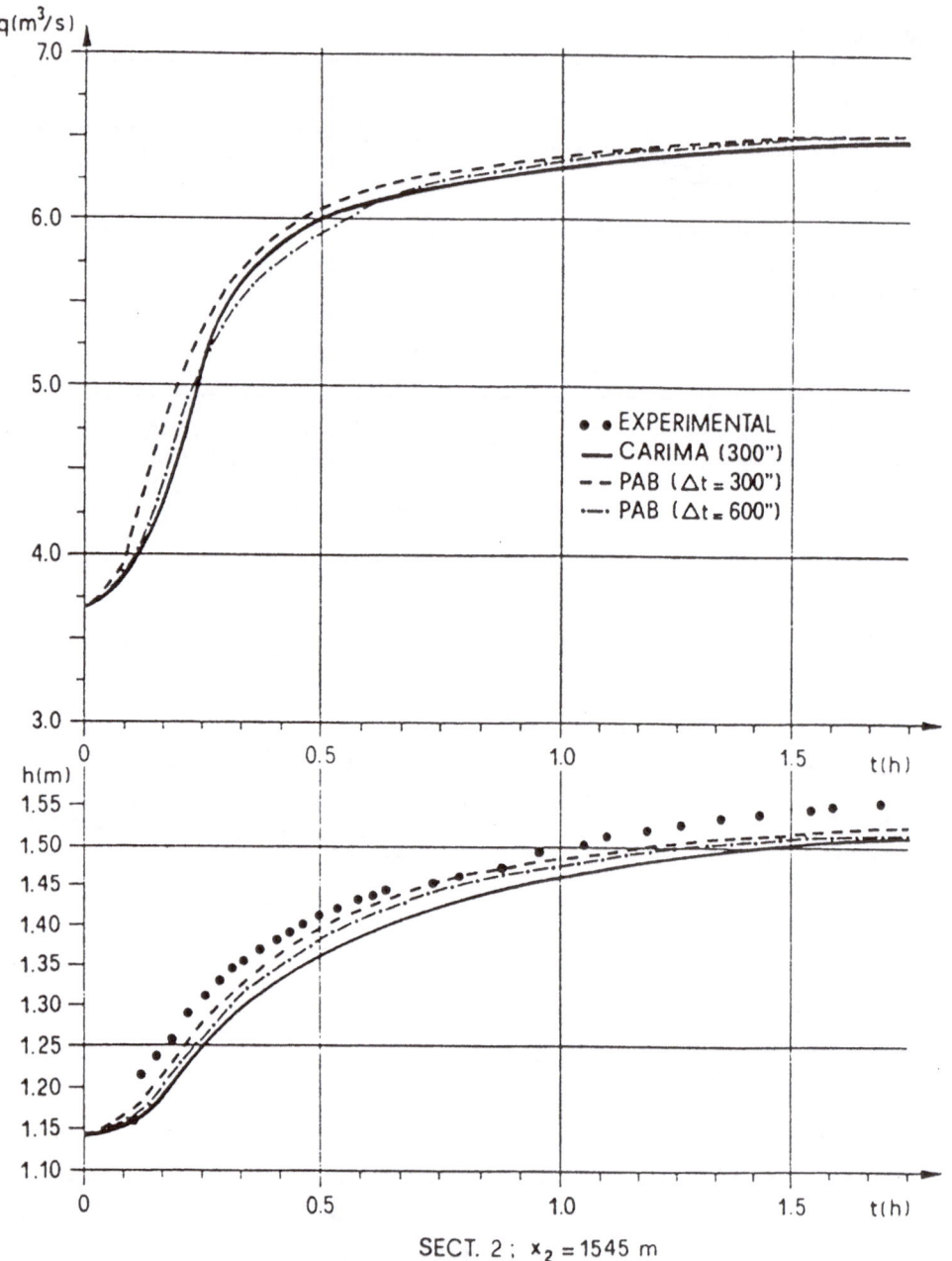

Figure 10.

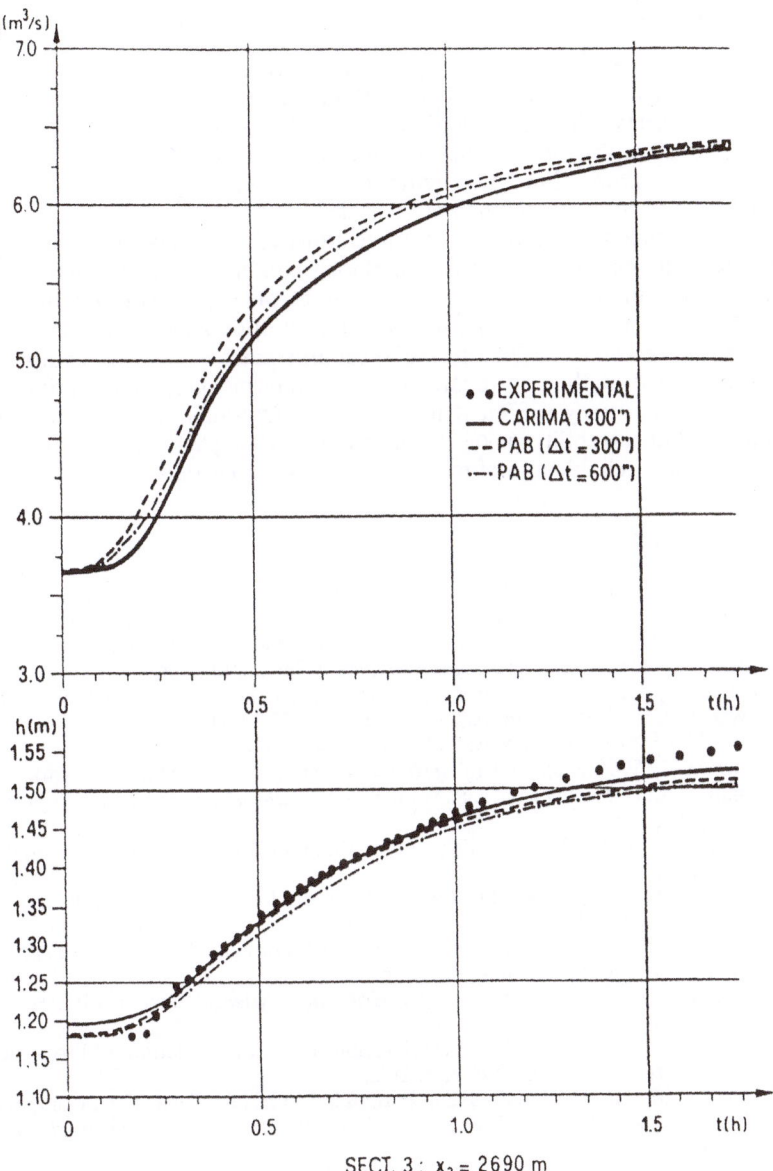

SECT. 3 ; $x_3 = 2690$ m

Figure 11.

3. Conclusions

In many real–time forecasting applications, where a hydraulic system must be regulated, a description of the levels associated with the propagation of a flood wave is generally needed.

The pure hydrologic models allow for an external description of the flood routing phenomenon, i.e. allow for the computation of discharges at a downstream section as a function of upstream inflows. Simplified hydraulic models may prove useful when an internal description of the routing phenomenon is needed, and in particular when the levels must be computed, provided that the downstream section of the reach is not affected by backwater effects. In the latter case only full dynamic models can be used, integrated numerically, to provide both levels and discharges at each time interval at the cost of relatively long computations.

An alternative method is proposed, based upon successive applications of a linearized parabolic scheme, based on analytical integration, and the computation of backwater profiles. The method has proven efficient in terms of 1) computational time requirements, and 2) in accuracy sufficient for a detailed hydraulic analysis. In addition, its inherent unconditional stability makes the proposed method particularly suited for real–time applications, where reliable and efficient algorithms are needed.

4. References

Abbott, M. B.: 1966, *An Introduction to the Method of Characteristics*, American Elsevier.

Abbott, M. B.: 1979, *Computational Hydraulics*, Pitman Advanced Publishing Program.

Barnett, A. G.: 1976, 'Numerical Stability in Unsteady Open Channel Flow Computations,' *Proc. Int. Symp. on "Unsteady Flow in Open Channels,"* Newcastle upon Tyne.

Becker, A., and E. Glos: 1970, 'Stifenmodell zur Hochwasserwellenberechnung in Ausufernden Wasserlaufen,' *Wasserwirtschaft–Wassertechnik*, 20 Jg, H.l., Berlin.

Chow, V. T.: 1959, *Open Channel Hydraulics*, McGraw Hill, New York.

Chow, V. T.: 1964, *Handbook of Applied Hydrology*, McGraw Hill, New York.

Ciriani, T. A., U. Maione, and J. R. Wallis: 1975, *Mathematical Models for Surface Water Hydrology*, John Wiley & Sons.

Cooley, R. L., and S. A. Moin: 1976, 'Finite Element Solution of Saint–Venant Equations,' *J. Hydraulic Div. ASCE* 102(HY6), 759–775.

Cunge, J. A.: 1969, 'On the Subject of Flood Propagation Computation Method (Muskingum Method),' *Journal of Hydraulic Research* 7, n. 2.

Cunge, J. A., F. M. Holly, and A. Verwey: 1980, *Practical Aspects of Computational River Hydraulics*, Pitman Advanced Publishing Program.

Dooge, J. C. I.: 1973, *'Linear Theory of Hydrologic Systems,'* Technical Bulletin n.1468, U.S. Dept. of Agriculture, Washington.

Evangelisti, G.: 1961, 'On the Numerical Solution of the Equations of Propagation by the Method of Characteristics,' *MECCANICA*, n. 1/2.

Franchini, M., and E. Todini: 1986, 'Parabolic Models for Real–Time Control of Irrigation Canals,' *Proc. Int. Workshop on "The Role of Forecasting in Water Resources Planning and Management"*, Bologna.

Fread, D. L.: 1985, 'Channel Routing,' in M. G. Anderson, and T. P. Burt (eds.), *Hydrological Forecasting*, John Wiley and Sons Ltd.

Goodrich, R. D.: 1931, 'Rapid Calculation of Reservoir Discharge,' *Civil Eng.* 1, 417–418.

Greco, F., and L. Panattoni: 1975, 'An Implicit Method to Solve Saint–Venant Equations,' *J. Hydraulic.*, n. 24, 171–185.

Guanwu, X., and C. Zhaohui: 1988, 'The System of Real–time Hydrological Information Processing and Flood Forecasting with PDP or VAX Series Computer,' *Hydrology*, n. 4. Beijing.

Hayami, S.: 1951, *'On the Propagation of Flood Waves,'* Disaster Prevention Research Institute, Kyoto University.

Henderson, F. M.: 1959, *Open Channel Flow*, McMillan Company, New York.

Kalinin, G. P., and P. I. Miljukov: 1958, *'Priblizenni Rascet Neustanovivsegosja Dvizenija Vodnych Mass,'* Trudy Centr. Inst. Progn., vyp. 55, Moskva.

Lai, C.: 1967, '*Computation of Transient Flows in Rivers and Estuaries by the Multiple–Reach Implicit Method,*' U.S. Geol. Survey Prof. Paper n. 575 D.

Linsley, R. K., M. A. Kohler, and J. L. H. Paulhus: 1949, *Applied Hydrology*, McGraw Hill Book Co., New York, pp. 502–530.

Maddaus, W. O.: 1969, '*A Distributed Representation of Surface Runoff,*' Hydrodynamics Lab. Report n. 115, MIT, Cambridge, Massachussetts.

McCarthy, G. T.: 1938, 'The Unit Hydrograph and Flood Routing,' *Conf. of the North Atlantic Div. US Corps of Engineers*, New London, Connecticut.

Natale, L. E., and E. Todini: 1977, 'A Constrained Parameter Estimation Technique for Linear Models in Hydrology,' in *Mathematical Models for Surface Water Hydrology*, John Wiley & S., U.K., Chichester.

Price, R. K.: 1973, '*Flood Routing Methods for British Rivers,*' Rep. III, Hydraulic Research Station, Wallingford.

Price, R. K.: 1973, '*Variable Parameter Diffusion Methods for Flood Routing,*' Rep. INT 115, Hydraulic Research Station, Wallingford.

Puls, L. G.: 1928, '*Construction of Flood Routing Curves,*' House Document 185, US 70th Congress, 1st Session, Washington, DC, pp. 46–52.

Societe Du Canal de Provence et D'amenagement de la Region Provencale: 1984, '*Ecoulements Transitoires à Surface Libre: Modèle d'Hayami,*' ref. FD/mp 40/84 – 240.

Stoker, J. J.: 1957, *Water Waves*, Interscience Publ., New York.

Tawatchai, T., and S. K. Manhandar: 1985, 'Analytical Diffusion Model for Flood Routing,' *Journal of Hydraulic Engineering* III, n. 3.

Todini, E., and A. Bossi: 1986, '*PAB (Parabolic and Backwater) an Unconditionally Stable Flood Routing Scheme Particularly Suited for Real–time Forecasting and Control,*' Institute of Hydraulic Construction, Pub. n. 1, Bologna.

Tong, H.: 1983, *Threshold Models in Non–linear Time Series Analysis; Lecture Notes in Statistics*, n. 21, Springer–Verlag.

Wignyosukarto, B.: 1983, '*Ecoulement Transitoire en Canal – Application à la Regulation,*' Rapport de D.E.A. de Mecanique des Fluides, Ecole Nationale Superieure de Grenoble.

W.M.O.: 1983, *Guide to Hydrological Practices*, Vol. 2, WMO–n. 168, Geneva.

Zezulak, J.: 1984, 'Deterministic Models II,' *International Postgraduate Training Courses in Hydrology*, Prague.

Chapter 19

OMEGA: Impact of Spatial Variability of Infiltration Parameters on Catchment Response

Francisco Nunes Correia
National Laboratory of Civil Engineering
Av. Brasil, 101
1799 Lisboa Codex, Portugal

and

Paulo Matias
High Institute of Agronomy
Tapada da Ajuda
1399 Lisboa Codex, Portugal

Abstract. OMEGA is a rainfall–runoff model for event or continuous simulation that incorporates to a large extent the physical conceptualization of the hydrological processes. This is principally achieved in the description of the highly nonlinear processes of infiltration, ponding, and redistribution of water in soil that play key roles as interfaces of surface and groundwater components of the hydrologic cycle. OMEGA is a distributed model with respect to the parameters and the inputs. Spatial variability is taken into consideration, and phenomena such as the movement of a storm over the watershed can be considered.

OMEGA is formulated in three consistent and compatible versions. The *simulation version* is the basic rainfall–runoff model and the core of the other versions. The *calibration version* has the purpose of allowing for a semi–automatic iterative calibration of the parameters of the model. The *real–time forecasting version* is a formulation that adds filtering capabilities to the model without sacrificing its physical and conceptual nature. In the filtering process physical feasibility is preferred to filter optimality; therefore, the filter results are constrained to merge values that are always physically acceptable.

The characteristics of the model make it very useful in studying the impact of spatial variability of parameters in catchment response. OMEGA was used to analyze the influence of the spatial variability of the saturated hydraulic conductivity in the processes of infiltration and runoff generation of small watersheds, under constant and variable rainfall intensities, and results were compared.

1. Introduction

Approximately fifteen years ago the World Meteorological Organization (WMO, 1975) conducted a study for the systematic comparison of ten conceptual hydrologic models over six sets of data. A somewhat surprising result from this study was that none of the models seemed to perform clearly better than the others, in spite of their significant differences in structure and conceptualization. Some other authors conducted similar studies and suggest that very different models may lead to similar results (Kite, 1975; Dawdy, 1982). These conclusions may lead some professionals to the erroneous idea that rainfall–runoff modeling is an overexplored field where nothing significantly better can be achieved. It should not be

D. S. Bowles and P. E. O'Connell (eds.), Recent Advances in the Modeling of Hydrologic Systems, 407–441.
© 1991 *Kluwer Academic Publishers.*

forgotten, however, that progress made in understanding the physics of the hydrological processes and in addressing issues such as problems of scale and spatial variability will necessarily have an impact in model performance and in its conceptual enrichment (Rodriguez–Iturbe, 1983; Dawdy, 1984).

Some of the most common shortcomings of available models are:

(1) Rainfall–runoff modeling is frequently dominated by a surface hydrology perspective (Morel–Seytoux et al., 1982; Morel–Seytoux, 1983; Dawdy, 1982; Pilgrim et al., 1982). The difficulties and "traumas" of unsaturated flow in porous media kept hydrologists away from treating carefully the interface processes such as preponding and postponding infiltration or soil moisture redistribution.

(2) Models are frequently developed in an inconsistent manner with respect to their internal structures (Fleming, 1976). A lot of attention is paid to the development and the perfecting of some components while some others are overlooked and treated in a crude fashion.

(3) Simple mathematical approaches and solutions are often adopted, neglecting physical concepts (Kitanidis and Bras, 1980a; O'Connell and Clarke, 1981) and frequently contradicting them. Parametric rather than conceptual hydrology is used with parameters being just fudge factors used tò get a specific and desired result instead of being physical variables with a physical meaning and a known range of values.

(4) Problems of scale, and principally time–scale, are often overlooked or neglected (Rodriguez–Iturbe, 1983). Different processes are described with the same time–scale and watersheds with very different sizes are described in the same way.

(5) Spatial variability is often ignored, and the watershed is considered as giving the same response throughout every point of its surface—like producing or not producing direct runoff—regardless of the fact that the area producing it may increase or decrease in time (Amorocho, 1979; Hawkins, 1982).

(6) Calibration is frequently treated as a mere optimization problem in which physical insight is lost (Nash and Sutcliffe, 1970; Fleming, 1976). Parameter values are chosen not for what they mean and how reasonable they are to explain the future behavior of the watershed, but for how they fit a historic and usually short record. In other words, a problem of inference is replaced by a problem of fitting (Correia, 1983).

(7) Models developed for off–line simulation are not used in real–time forecasting because they do not have built–in filtering capabilities (Rodriguez–Iturbe et al., 1978; Kitanidis and Bras, 1980a). Instead very crude hydrological models are usually associated with systems engineering techniques such as the Kalman filter. Very often the filter is not only a way of improving very short–term forecasting, but also a mask of the crudeness of these models (Kitanidis and Bras, 1980a,b).

(8) The fact that no process is "truly" linear in nature has been often used as a pretext to reject the potential of linear systems theory (Kalman, 1978). On the other hand the advantages of using linear models led to artificial and cumbersome linearization of highly nonlinear processes (Kitanidis and Bras, 1980b).

OMEGA was developed as a package of fully compatible and consistent computer watershed models designed for simulation or real time forecasting of river flows; it includes complete calibration procedures and represents an effort to tackle, and hopefully overcome to some extent, the problems and limitations previously mentioned. A more detailed description of the model can be found in Correia (1984) and Correia and Morel–Seytoux (1985). Some of its most important features are summarized below and are described in detail in this chapter:

(1) OMEGA attempts to incorporate to a large extent the physical conceptualization of the hydrological process. This is principally achieved in the description of the highly nonlinear processes of infiltration, ponding, and redistribution of water in the soil. The complexity of these processes make them unsuitable for description as simple black box systems or using

simple mass balance equations relating several storages, as is frequently done by most models. These processes, however, play a crucial role in the determination of the total amount and time distribution of excess rainfall, on one hand, and total amount and time distribution of aquifer recharge, on the other hand. Therefore, a lot of attention is dedicated to them and an attempt is made to describe them as physically as possible. No linearizations are considered because these processes are intrinsically highly nonlinear, and no simple storage balance equations are formulated although mass balance is obviously guaranteed.

(2) OMEGA is a distributed model with respect to both the parameters and inputs. In this way spatial variability can be taken into consideration and phenomena, such as the movement of a storm over the watershed, can be considered. This characteristic makes the model very adequate to study the impact of spatial variability, as it is presented in later sections of this chapter. Naturally, OMEGA can be also used as a lumped model if desired.

(3) OMEGA can be used as an event simulation model or as a continuous simulation model. Different time steps can be considered, and a changing time step can be accommodated, from one minute to a few hours. OMEGA can be used in an off–line simulation of the watershed behavior or in a real–time forecasting framework in which the filtering capabilities are activated but none of the physical conceptualizations are lost.

(4) OMEGA explicitly addresses the problem of the scale in time and space at which hydrological processes must be described for meaningful results. A much smaller time scale is used for the surface components as opposed to the groundwater components. The evolution of water content in the soil is described in more detail than surface water routing process which in turn is more detailed than the base flow. The degree of detail considered in each case depends on the intrinsic characteristics of the processes being studied and on the time scale at which they must be described.

(5) OMEGA explicitly addresses the problem of the hydrologic processes being linear or nonlinear. The processes of infiltration, ponding, and redistribution of water in the soil are highly nonlinear and are treated as such. On the contrary, the relationship between effective rainfall and surface runoff may be adequately described by a stationary linear model; therefore, the unit hydrograph approach is used. However, if the model is being used as a distributed one, the common lumped unit hydrograph is replaced by a more general and powerful methodology, the Larrieu's generalized unit hydrograph.

(6) OMEGA considers the filtering as an overhead procedure, neither impairing the physical meaning of the several components nor contradicting them. The filtering procedure allows for corrections in short–term forecasts. However when the lead time approaches the memory time of the watershed, the filtering procedure fades off and simulation results are more trusted than the filtering results.

(7) OMEGA was developed with very attractive calibration procedures. Automatic parameter optimization is combined with engineering judgement. All parameters have a physical meaning, their value ranges are known, and calibration proceeds as a multi–step interactive dialogue with the user. If few data are available, the physical basis of the model allows for a tentative estimation of parameters based on a field survey. This tentative is used as an *apriori* to be improved by the few data that can be used.

The fully distributed approach and the detailed description of the interface between surface and subsurface processes, accomplished by OMEGA, make it very attractive to use the model in the study of spatial variability of parameters and its impact on catchment response. In some circumstances, the model can be used as a limited but still useful surrogate for reality in this kind of study. In fact, it is not possible in nature to isolate one hydrologic process from the others and perform the analysis of its spatial variability. However, this can be done to some extent in a physically based and distributed model, enhancing modeling capabilities and still providing some knowledge about reality.

In this chapter OMEGA is used to assess the sensitivity of the characteristics of spatial variability of saturated hydraulic conductivity, $\bar{K}$, on catchment response under two different patterns of rainfall. The spatial variability of $\bar{K}$ is obtained, considering it as a random variable with a three parameter log–normal distribution; the influence of the coefficients of skewness and variation of $\bar{K}$ on the processes of infiltration and runoff is analyzed under constant and variable rainfall intensities. The results obtained with the distributed model are compared with those obtained with the lumped model to determine at what point only one single value of $\bar{K}$ can reflect the existing variability for the entire watershed. Two values are utilized in this analysis as single values of $\bar{K}$, namely the mean and the median values.

The name OMEGA suggests the ultimate perfection in watershed modeling. Naturally, this title must be taken as mere fun. The author is painfully aware that many improvements can be made in an endless process of incorporating theoretical knowledge in the engineering tool, which is ultimately what a watershed model is.

2. Discussion

2.1. Basic Characteristics of OMEGA

2.1.1. Structure

Physically sound infiltration theory, introduced by Morel–Seytoux (1978, 1981, 1982) may be considered as the core of the OMEGA simulation model. The computation of ponding time, infiltration rates (before and after ponding) and redistribution of water in soil when rainfall ceases are the main source of nonlinearity in the model behavior, as they are in nature. This is emphasized because it is considered to be an important characteristic of the model.

Another important characteristic of the model is its usability as a fully distributed model with respect to most input variables and parameters. In this way, spatial variability of watershed parameters and rainfall patterns can be accounted for. Table 1 presents a list of distributed and lumped variables and parameters. Examples of two approaches in subdividing the watershed and encompassing spatial variability are shown on Figure 1.

Table 1. List of lumped and distributed input variables and parameters in the watershed model OMEGA.

Lumped Input Variables and Parameters	Distributed Input Variables and Parameters
Pan evaporation	Rainfall hyetograph
Evaporation constant to adjust pan evaporation	Hydraulic conductivity at natural saturation
Surface water time step	Effective capillary drive
Groundwater time step	Exponent of normalized water content in
Discrete kernels of surface routing	relative permeability to water curve
Discrete kernels of base flow routing	Water content at natural saturation
	Residual water content
	Initial water content
	Retention storage depth
	Depth to the top of the past–root zone
	Depth to the average phreatic level
	Ultimate viscous correction factor

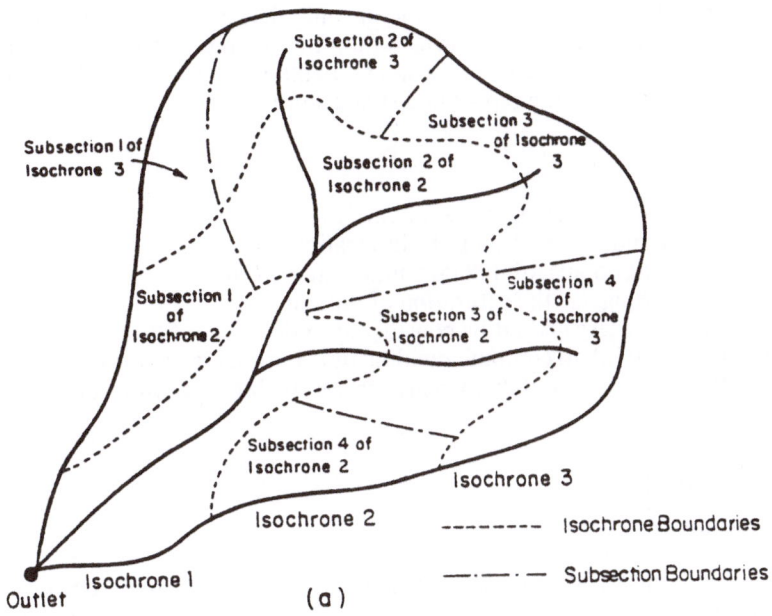

Subsection 2 of Isochrone 3

Subsection 3 of Isochrone 3

Subsection 1 of Isochrone 3

Subsection 2 of Isochrone 2

Subsection 1 of Isochrone 2

Subsection 4 of Isochrone 3

Subsection 3 of Isochrone 2

Subsection 4 of Isochrone 2

Isochrone 3

Isochrone 2

Outlet

Isochrone 1

------- Isochrone Boundaries

—·—·— Subsection Boundaries

(a)

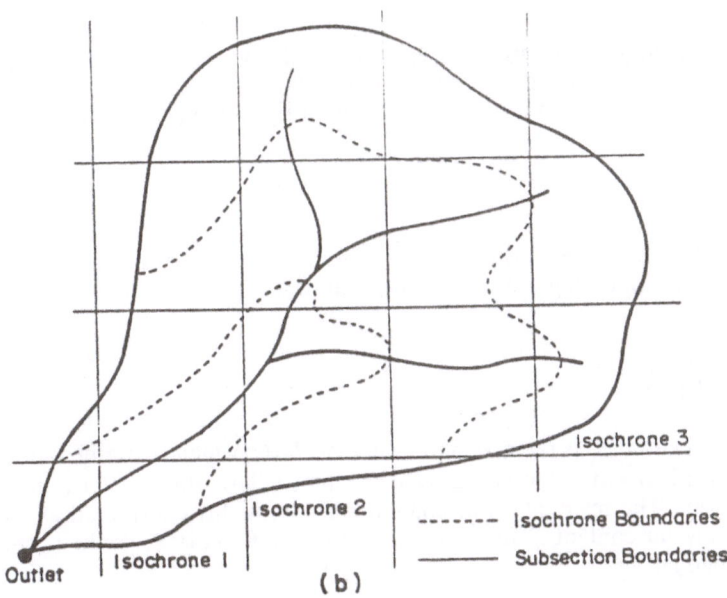

Isochrone 3

Isochrone 2

Outlet

Isochrone 1

------- Isochrone Boundaries

——— Subsection Boundaries

(b)

Figure 1. Using OMEGA as a distributed model: (a) Hydrologically based subdivision of the watershed; (b) Regular grid superimposed on isochrone subdivision.

The basic components of OMEGA are the ponding time determination, postponding infiltration calculation, retention storage and surface runoff initiation, surface routing, soil moisture accounting, calculation of excess infiltration and aquifer recharge, and base flow routing. The very basic structure of the model is displayed in Figure 2.

2.1.2. Ponding Time

Before ponding, infiltration rate equals precipitation rate and excess rainfall is equal to zero. After ponding, and while a ponded condition lasts, infiltration rate equals capacity infiltration rate and excess rainfall rate equals rainfall rate minus capacity infiltration rate. In other words, the occurrence of ponding, i.e. the saturation of the soil surface, acts as a threshold, determining which physical process should be considered to describe infiltration.

The expression used in OMEGA to compute ponding time is a generalization of the Mein and Larson equation for a variable rainfall intensity derived by Morel–Seytoux (1978):

$$t_p = t^o + \frac{1}{r}\left(\frac{S_f}{r^* - 1} - W^o \right) \tag{1}$$

where t_p is the time of ponding, t^o is the time at the beginning of the current time step, r is the rainfall rate during the current time step, S_f is the storage–suction factor, r^* is the normalized rainfall rate, and W^o is the cumulative infiltration depth since the beginning of the current continuous rainfall up to time t^o. The storage–suction factor is a composite factor affecting ponding and the infiltration process. Its value can be computed as:

$$S_f = (\bar{\theta} - \theta_i) \, H_c(\theta_i) \tag{2}$$

where θ is water content at natural saturation, θ_i is mean initial water content in an upper soil layer, and $H_c(\theta_i)$ is the effective capillary drive, a quantity which only slightly depends upon the initial water content. The normalized rainfall rate in Equation (1) is defined as:

$$r^* = \frac{r}{\bar{K}} \tag{3}$$

where r is the rainfall rate and $\bar{K}$ is the hydraulic conductivity at natural saturation. The effective capillary drive, appearing in Equation (2), can be estimated (Morel–Seytoux, 1978) by the formula:

$$H_c(\theta_i) = \left[1 - (\theta_i^*)^6 \right] H_c \tag{4}$$

for $\theta_i > \theta_r$. In this equation, H_c is the maximum value of $H_c(\theta_i)$ which is attained for $\theta_i = \theta_r$ with θ_r being the residual water content, a water content below which water is no longer mobile in liquid form. The residual water content can be roughly considered as equal to somewhat less than water content at field capacity. In Equation (4) θ_i^* is the normalized water content defined as:

$$\theta_i^* = \frac{\theta_i - \theta_r}{\bar{\theta} - \theta_r} \tag{5}$$

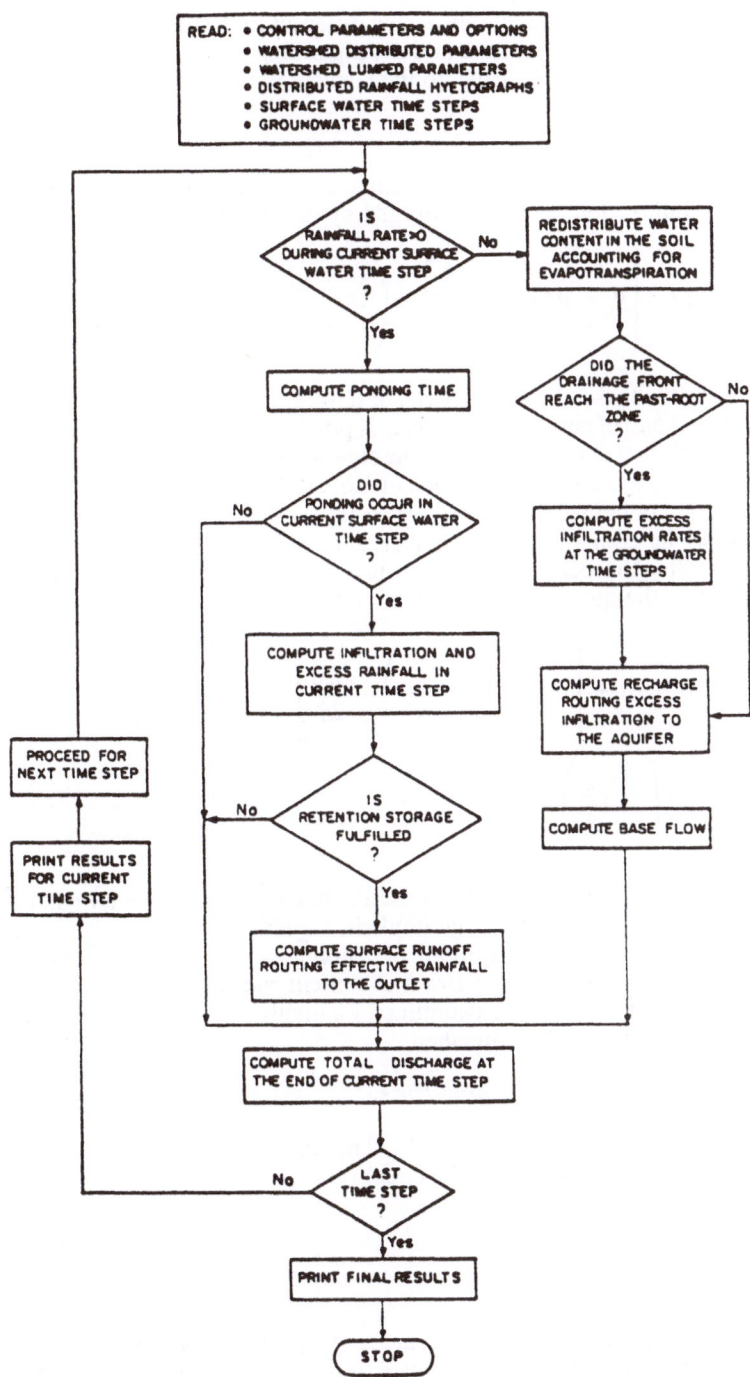

Figure 2. General flowchart of the simulation procedures of the watershed model OMEGA.

During rainfall periods the ponding time is computed for each subsection of the watershed. Ponding may occur in different subsections during different time steps. In this way the area contributing to surface runoff may change in time, expanding or contracting. This behavior leads, somehow implicitly but realistically, to the concept of variable source area.

2.1.3. Infiltration after Ponding

Under a ponded condition, infiltration rate is equal to the capacity infiltration rate. The capacity cumulative infiltration depth is computed by the equation:

$$\frac{\tilde{K}}{\beta}(t^v - t_p) = W^v - W_p - \left[S_f + W_p \left(1 - \frac{\beta_p}{\beta} \right) \right] \ln \left(\frac{S_f + W^v}{S_f + W_p} \right) \tag{6}$$

In this equation $\tilde{K}$ is the hydraulic conductivity at natural saturation, β is a viscous correction factor, t^v is time at end of current time step, t_p is ponding time, W^v is cumulative infiltration depth at the end of current time step, W_p is the cumulative infiltration depth at the time of ponding, S_f is the storage–suction factor defined by Equation (2), and β_p is the viscous correction factor at the time of ponding. Equation (6) gives time as a function of cumulative infiltration depth and cannot be explicitly inverted in an exact fashion. An approximate inverse can be obtained by expanding the logarithmic term in Taylor's series, truncating after the second–order term and solving the quadratic equation for $\Delta W = W - W^o$ in terms of $\Delta t = t - t^o$:

$$W^v = \left\{ \frac{2\tilde{K}}{\beta}(S_f + W^o) \, \Delta t + \left[W^o - W_p \left(1 - \frac{\beta_p}{\beta} \right) - \frac{\tilde{K}\Delta t}{2\beta} \right]^2 \right\}^{1/2}$$
$$+ \frac{\tilde{K}\Delta t}{2\beta} + W_p \left(1 - \frac{\beta_p}{\beta} \right) \tag{7}$$

The approximate but explicit Equation (7) is used to make better and better guesses of W^v that ultimately will verify the exact, although implicit, Equation (6). The same rainfall rate may produce different values of excess rainfall, depending on the parameters of the subsection or on its previous history, as described by water content and cumulative infiltration. This is a very realistic feature of OMEGA, accounting for a highly nonlinear, nonstationary, and spatially distributed behavior of the watershed.

2.1.4. Retention Storage and Runoff Starting Time

There will be no runoff as long as excess rainfall will not have satisfied the retention storage. The runoff starting time, t_e, is solution of the equation:

$$\int_{t_p}^{t_e} r_e(\tau) d\tau = R_M \tag{8}$$

where t_p is ponding time, $r_e(\tau)$ is excess rainfall rate, and R_M is retention storage depth. At the end of each ponded time step, OMEGA computes the cumulative excess rainfall depth and compares it with R_M. Until R_M is exceeded no runoff is generated. Once R_M is exceeded, retention storage is fulfilled and runoff starts.

2.1.5. Surface Routing

The excess rainfall resulting from the infiltration component is used to fulfill retention storage and to generate a runoff response determined by Larrieu's generalized unit hydrograph method. This method divides the watershed into isochrones, corresponding to areas centered about a line of constant travel time for water to reach the outlet. The overall runoff response for a general rainfall over the entire watershed is obtained by superposition in time and space. With this method, distributed rainfall is in some sense routed to the outlet by a lagging procedure and then superimposed in an artificial rainfall hyetograph convoluted with a regular unit hydrograph. When using this approach, discharge is given by the relation:

$$Q(m) = \sum_{\theta=0}^{\mu} f(\theta) \sum_{v=1}^{m-\theta} r_e(v, \theta) \; \delta(m - \theta - v + 1) \tag{9}$$

where $Q(m)$ is discharge at the end of the time step m, θ is the travel time corresponding to each isochrone, μ is the travel time of the most remote isochrone contributing to the outlet discharge at the end of time step m, $f(\theta)$ is the areal density associated with isochrone θ, $r_e(v, \theta)$ is the average effective rainfall rate during period v over isochrone θ, and the $\delta(.)$ are the unit hydrograph ordinates. If more than one subsection is considered in one isochrone, it is necessary to find the average effective rainfall rate over the isochrone—this value is used in Equation (9).

Larrieu's generalized unit hydrograph, used by OMEGA, is a method that allows for spatial variability of the excess rainfall inputs over the watershed. However, this method still keeps the excess rainfall–direct runoff relationship in the framework of the linear, time invariant systems. In fact, if the nonlinearities and nonstationarities are explicitly and correctly addressed in the ponding, infiltration, and retention components, it can be accepted that the excess rainfall–direct runoff process is described well enough by a time–invariant linear model.

2.1.6. Soil Moisture Accounting

Right at the cessation of precipitation the water content profile may resemble that shown in Figure 3. It seems reasonable to replace the actual profile at time zero (time being counted from the moment infiltration stopped) by the simpler rectangular profile with a uniform value of water content, θ_i. This value is water content at natural saturation, $\tilde{\theta}$ if ponding occurred or if the last rainfall rate, r, exceeds hydraulic conductivity at natural saturation $\tilde{K}$. If r is less than $\tilde{K}$, θ_i may be computed as (Morel–Seytoux et al., 1982):

$$\theta_i = (\tilde{\theta} - \theta_r) \left(\frac{r}{\tilde{K}}\right)^{1/n} + \theta_r \tag{10}$$

where r is the last rainfall intensity before cessation of the rainfall period, $\tilde{K}$ is hydraulic conductivity at natural saturation, $\tilde{\theta}$ is water content at natural saturation, θ_r is residual water content, and n is exponent in power law for relative permeability to water.

Naturally, the area under the rectangular approximation of the initial profile must be the same as under the true one. The initial position of the wetting front of the rectangular profile, z_f^0, is given by the mass balance equation:

$$z_f^0 = \frac{W}{(\theta_l - \theta_i)} \tag{11}$$

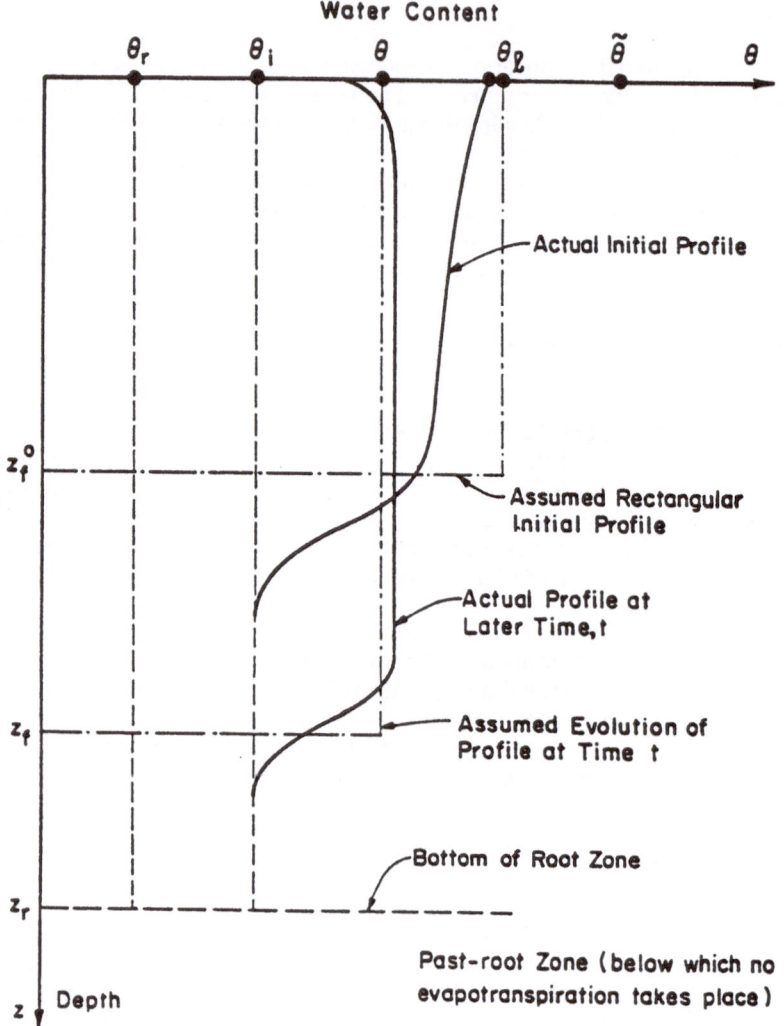

Figure 3. Actual redistribution of soil moisture and assumed rectangular redistribution.

where W is the total infiltration depth at the end of the rain and θ_i is mean initial water content when rain started. At a later time, t, when the assumed rectangular profile corresponds to a current water content, the position of wetting front, z_f, is given by the mass balance equation:

$$z_f = \frac{W - E - q_i t}{(\theta - \theta_i)} \qquad (12)$$

where E is the cumulative evapotranspiration depth since rainfall stopped and q_i is the water flux across a plane coinciding with the bottom of the root zone. The consideration of q_i expresses the mobility of the initial water content profile, defined by θ_i. From Figure 3 an expression can be deduced for the rate of change in water content:

$$\frac{d\theta^*}{dt} = \frac{-1}{z_f(\tilde{\theta} - \theta_r)} (q - q_i + e) \tag{13}$$

where θ^* is the normalized water content defined by Equation (5), t is time, z_f is the position of the wetting front, $\tilde{\theta}$ is water content at natural saturation, θ_r is residual water content, q is the water flux in a Darcy sense across the plane coinciding with the wetting front at time t, and e is the evapotranspiration rate.

Darcy's law for flow in unsaturated soil if capillary effects are neglected reduces to the form:

$$q = \tilde{K}k_{rw}(\theta) \tag{14}$$

where $k_{rw}(\theta)$ is the relative permeability to water that can be expressed as a power function of normalized water content of exponent n (Morel–Seytoux, 1979), $q = \tilde{K}(\theta^*)^n$. Naturally, this expression can also be used to compute the value of q_i as a function of θ_i^*. Substitution of expression for q and q_i in Equation (13) leads to a differential equation that can be integrated to provide time as a function of normalized water content:

$$t(\theta^*) = \int_{\theta^*}^{\theta_i^*} \frac{W - E - q_i t}{\left[e + \tilde{K}(u^n - \theta_i^{*n})\right](u - \theta_i^*)} du \tag{15}$$

This equation can be numerically integrated and inverted to obtain the value of the normalized water content at the end of each time step.

Any water flowing past the bottom of the root zone will become aquifer recharge with a certain delay, depending on the soil properties, the previous moisture conditions, and the depth to the aquifer.

The process governing the flow of water from the past root depth to the aquifer is shown to be a nonlinear process adequately represented by a nonstationary linear model. The theoretical approach used by OMEGA to represent this process was developed by Morel–Seytoux (1984) and Morel–Seytoux and Billica (1984).

During each groundwater time step the values of excess infiltration are added and an excess infiltration rate is computed for the entire groundwater time step. These values, obtained for each subsection, are used to perform a discrete convolution where the unit hydrograph ordinates no longer depend only on one argument, namely, the difference between response time m and excitation time v, i.e. $(m-v)$ but on two arguments, $(m-v)$ and v. It may happen that recharge occurs in one part of the watershed while in some other parts it is not occurring at all. However, the weighted recharge for the entire watershed is computed and used to generate baseflow. A more detailed description of these procedures can be found in Correia (1984).

The relationship between average aquifer recharge rates during the groundwater time steps and baseflow discharge at the end of the time steps, is assumed to be linear and stationary. Therefore, the recharge rates, $r_c(m)$, can be used to generate baseflow discharges with a simple discrete convolution equation. The values of baseflow discharge computed at the end of each groundwater time step are kept constant during the following groundwater time step and are added to the surface discharges computed at the end of each surface time step.

Surface time steps are smaller than groundwater time steps; therefore, surface discharges are allowed to change faster than baseflow discharges. This is a realistic and interesting feature of OMEGA because it uses two different time scales corresponding to two different time counters.

2.2. Calibration and Real–Time Forecast Versions of OMEGA

2.2.1. General Description of Model Calibration

The calibration procedures of OMEGA are based on an interactive dialogue of the user with the computer. This is documented in a separate report (Correia, 1984). Here, a summary description of the calibration methodology is presented. The most relevant steps of the calibration algorithm are displayed on Figure 4. This flowchart does not express in detail all of the computations performed by the calibration version of OMEGA. Its only objective is to provide some insight on the basic structure and sequence of the calibration procedures.

Three objectives are considered at different stages of the calibration process: minimize errors in mass balance, minimize errors in runoff starting time, and minimize the sum of the squares of discharge errors. The minimization of the errors in mass balance is considered the most important objective. This attitude is derived from the fact that the infiltration process is considered the core of the model, and this process is responsible for keeping mass balance. In fact, if good estimates are obtained for the parameters of the infiltration process, it is very likely that good estimates for runoff starting time and good matches between computed and observed surface discharges may also be obtained. The minimization of errors in mass balance is performed in an objective manner. In other words, given certain conditions a minimum in the mathematical sense is obtained.

The second objective, minimization of errors in runoff starting times, can be considered somewhat of a loose objective in the sense that no mathematical optimization is performed with respect to it. Instead, after mass balance is optimized the corresponding errors in runoff starting time are computed. If they are considered acceptable, calibration proceeds to the minimization of the third objective function. If they are not considered acceptable, some parameters are changed and a new optimization with respect to the mass balance error is performed. This procedure may be repeated until a good match between observed and computed runoff starting time is obtained and mass balance error is minimized. In the end, mass balance is optimized, conditioned on an acceptable error in runoff starting time.

Once the first two objectives have been achieved a time pattern of effective rainfall rates is available. The surface routing discrete kernels are then identified with the objective of minimizing the sum of the squares of discharge errors.

A good match in mass balance and in runoff starting time is usually associated with plausible time distribution of the effective rainfall pattern (Morel–Seytoux et al., 1981). This makes the third step of optimization easier to accomplish. The third objective function is mathematically optimized, conditioned on the results obtained in the previous steps.

In the calibration procedures not all parameters are treated alike. Some parameters are considered to have a driving role in the physical processes and are explicitly involved in the optimization procedures as decision variables. Some other parameters are considered useful only for fine tuning the model results and are not included in the optimization procedures. Instead, they are changed and adapted during the calibration procedures according to the experience and judgment of the person using the model. Finally, a third group of parameters are fixed and not changed throughout the calibration procedures. These parameters are considered to change very slightly from case to case or have a small impact on the model behavior. Naturally, they can be changed any time it is convenient; however, they are not explicitly considered in the calibration procedures.

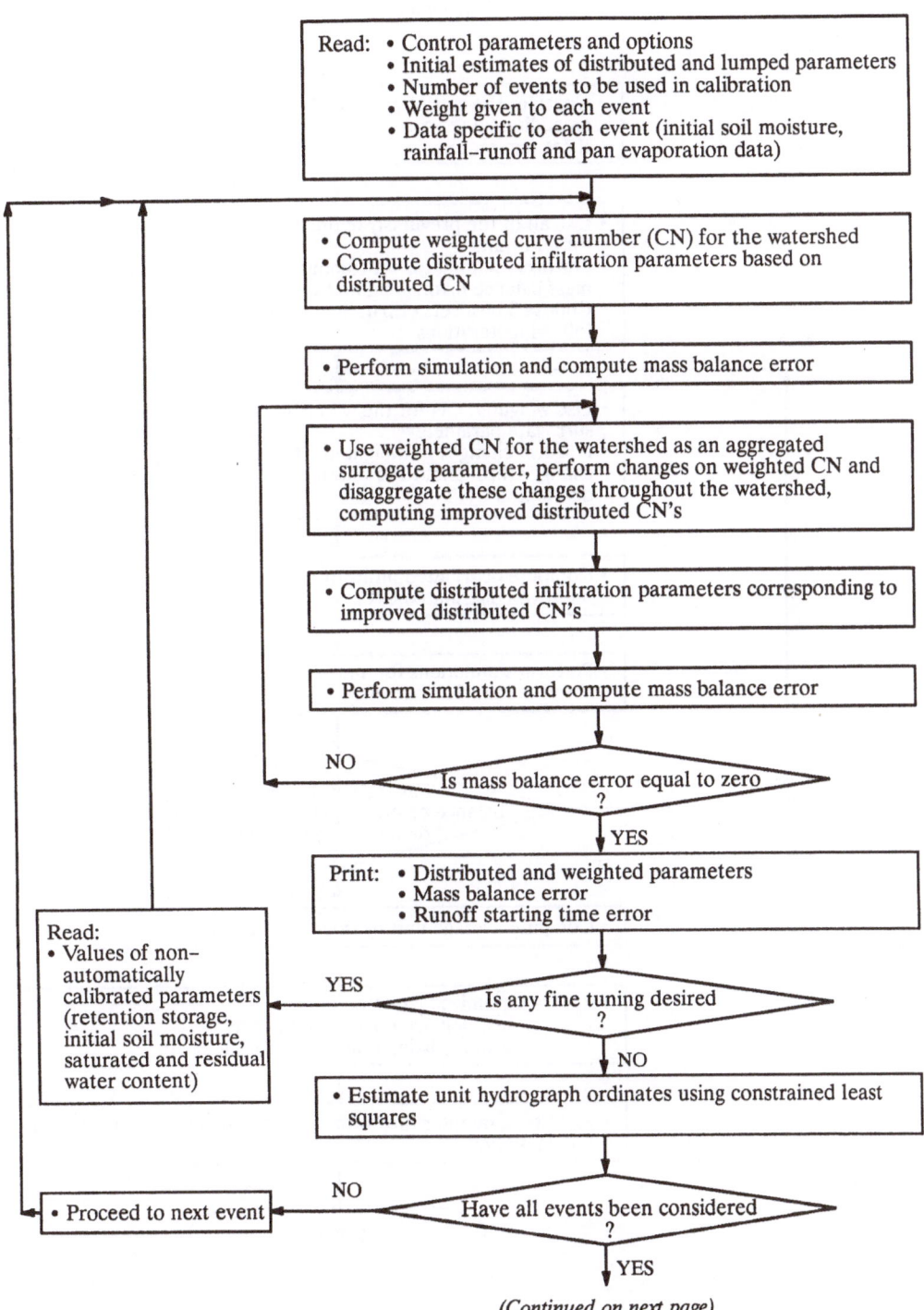

Read: • Control parameters and options
       • Initial estimates of distributed and lumped parameters
       • Number of events to be used in calibration
       • Weight given to each event
       • Data specific to each event (initial soil moisture,
         rainfall–runoff and pan evaporation data)

• Compute weighted curve number (CN) for the watershed
• Compute distributed infiltration parameters based on
  distributed CN

• Perform simulation and compute mass balance error

• Use weighted CN for the watershed as an aggregated
  surrogate parameter, perform changes on weighted CN and
  disaggregate these changes throughout the watershed,
  computing improved distributed CN's

• Compute distributed infiltration parameters corresponding to
  improved distributed CN's

• Perform simulation and compute mass balance error

NO   Is mass balance error equal to zero
          ?
          YES

Print: • Distributed and weighted parameters
        • Mass balance error
        • Runoff starting time error

Read:
• Values of non–
  automatically
  calibrated parameters
  (retention storage,
  initial soil moisture,
  saturated and residual
  water content)

YES   Is any fine tuning desired
          ?
          NO

• Estimate unit hydrograph ordinates using constrained least
  squares

NO   Have all events been considered
• Proceed to next event   ?
          YES

(Continued on next page)

Figure 4. General flowchart of the calibration procedures of the watershed model OMEGA.

420

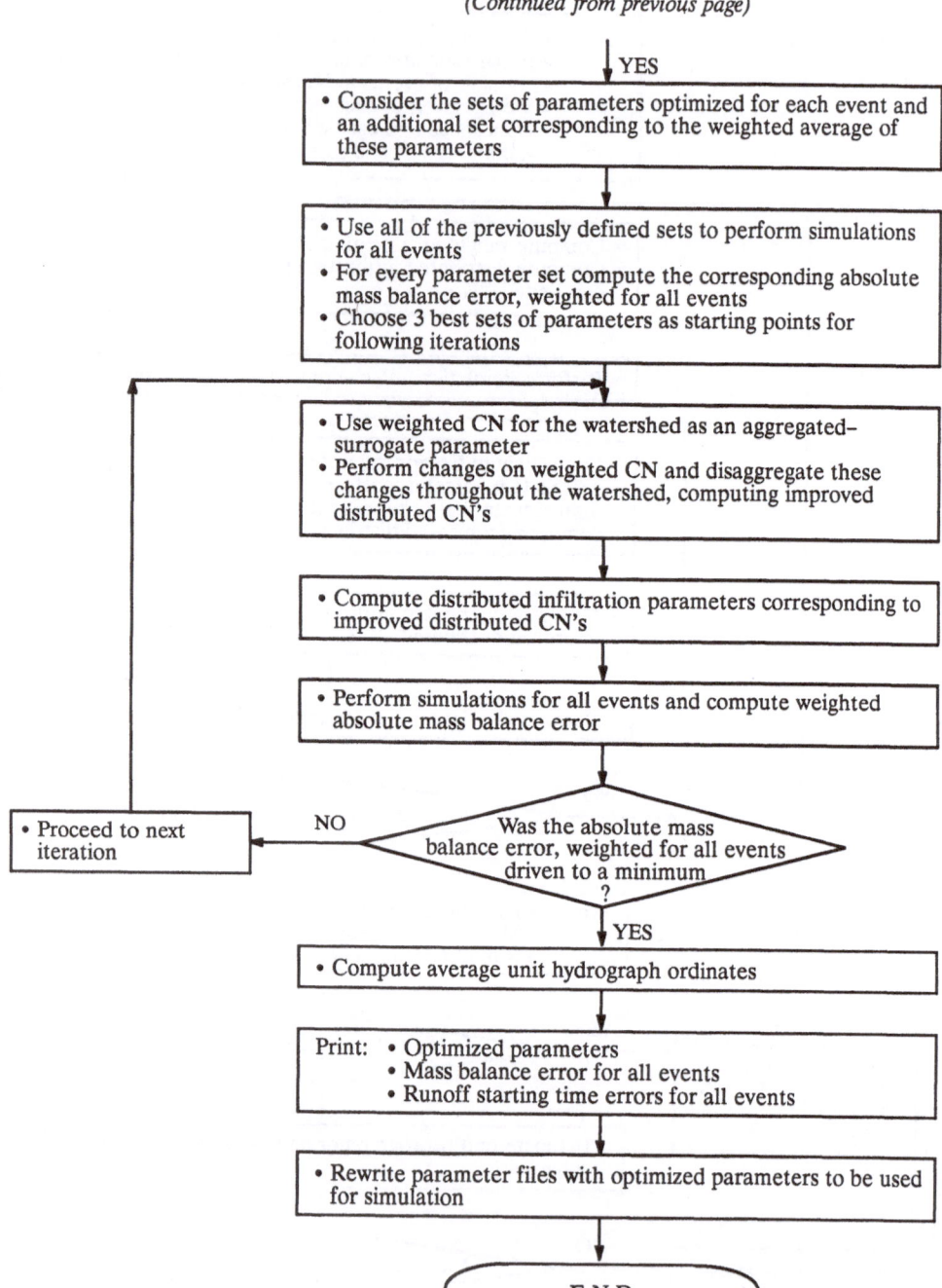

(Continued from previous page)

Figure 4. General flowchart of the calibration procedures of the watershed model OMEGA (cont'd).

This discrimination about the relevance of each parameter is based on experience and previous sensitivity studies. The parameters of the first type, i.e. explicitly optimized, are called "driving parameters." They are the hydraulic conductivity at natural saturation, $\tilde{K}$, the maximum effective capillary drive, H_c, and the surface routing unit hydrograph ordinates, δ (ν) for $\nu = 1,2...$, M where M is the memory time of the watershed. The parameters of the second type, i.e. those changed arbitrarily but judiciously by the user during the calibration procedures, are called "fine tuning parameters." These are the surface retention storage depth, R_M, water content at natural saturation, $\tilde{\theta}$, residual water content, θ_r, and initial water content at the beginning of every calibration event, θ_i. The fine tuning parameters are used mainly to get better results for runoff starting time. The driving parameters are mathematically optimized for a given set of fine tuning parameters. In other words if the fine tuning parameters are changed, even slightly, the driving parameters are recalibrated to give an optimum with respect to mass balance.

Finally, the third type of parameters will be called "implicit parameters" since they are taken for granted during the calibration procedures. Typical implicit parameters are: the ultimate viscous correction factor, β; the viscous correction factor at ponding time, β_p; the exponent of normalized water content in the relative permeability to water power law, n; and the evapotranspiration constant to adjust pan evaporation, C_E. Currently, the depth to the average phreatic level, D, the depth to the past–root zone, D_R, and the discrete kernels for base flow routing are also considered to be implicit parameters. Nevertheless, the depths to phreatic level and to the past root zone can be assessed fairly easily by a field survey. Some generic guidelines on how to estimate the discrete kernels for the base flow and how to decompose the watershed in isochrones and subsections are presented by Correia (1984).

A very important feature of OMEGA is that it uses parameters with physical meaning and has values that vary only in a limited domain. In this way not only very good first guesses can be made for the values of the parameters, but also some advantage can be taken from some degree of dependence among parameters, reducing somewhat the degrees of freedom of the optimization problems. For example, if a soil is a tight clay, it is likely to have a small hydraulic conductivity at natural saturation, a large maximum effective capillary drive, a small water content at natural saturation, and a relatively large storage capacity, measured as the difference between residual water content and wilting point. For a coarse sand these relationships will be reversed. The mathematical symbols, the numerical range, and the physical meaning of the parameters are presented in Table 2.

The dependence between two of the more important parameters of OMEGA, namely the hydraulic conductivity at natural saturation, $\tilde{K}$, and the maximum effective capillary drive, H_c, is used in the calibration procedures. A full description of these procedures, including the methodology that is used when more than one calibration event is considered can be found in Correia (1984). Also presented in that report, are very successful results of the application of the model in three Portuguese watersheds.

2.2.2. General Description of Real–Time Forecast Procedures

The most relevant steps of the real–time forecast procedures are displayed in Figure 5. This flowchart does not express in detail all the computations performed by this version of the model. An important objective of the formulation considered in OMEGA is that the filtering techniques are embedded in the conceptual and physically based watershed model without forcing it to behave linearly where it is clearly known that the watershed does not have such behavior. A real–time forecasting formulation is attempted in which the model still keeps all its physical background but has some possibility of learning from the forecast errors made in the recent past.

Table 2. Physical meaning and numerical range of OMEGA parameters.

Mathematical Symbol	Physical Meaning	Numerical Range	Corresponding Type of Soil
R	Hydraulic conductivity at natural saturation	0.0 - 5.0 cm/hr	Tight clay - coarse sand
H_e	Maximum effective capillary drive	5.0 - 30.0 cm	Coarse sand - tight clay
R_M	Surface retention storage	0.0 - 0.5 cm	Steep and flat surfaces - horizontal rugged surfaces
∂	Water content at natural saturation	0.25 - 0.35	Tight clay - coarse sand
$\partial - \theta_r$	Water content at natural saturation minus residual water content	0.07 - 0.18	Tight clay - coarse sand
θ_i	Water content at the beginning of the simulation period	$0.0 - \partial$	Completely dry - completely saturated soil
$\delta(\ )$	Ordinates of the surface unit hydrograph	$0.0 \geqq \delta(\ ) < 1.0$ $\Sigma \delta(\ ) = 1.0$	Usually monotonically increasing up to the peak and monotonically decreasing afterwards
$\delta_b(\ )$	Ordinates of the base flow unit hydrograph	IDEM	IDEM
β	Viscous correction factor	1.0 - 1.7	Silt loam-tight clay
n	Exponent of normalized water content in the relative permeability to water power law	4.0 - 12.0	Coarse sand - tight clay
C_E	Multiplicative factor to transfer pan evaporation into potential evapotranspiration	0.6 - 0.8	Depending on season and local conditions (0.7 is frequently considered)
D	Depth to the average phreatic level	0 - 5,000 cm	Depending on local conditions easily assessed
D_R	Depth of the top of the past-root zone	50 - 1,000 cm	Depending on season, type of soil and vegetation cover

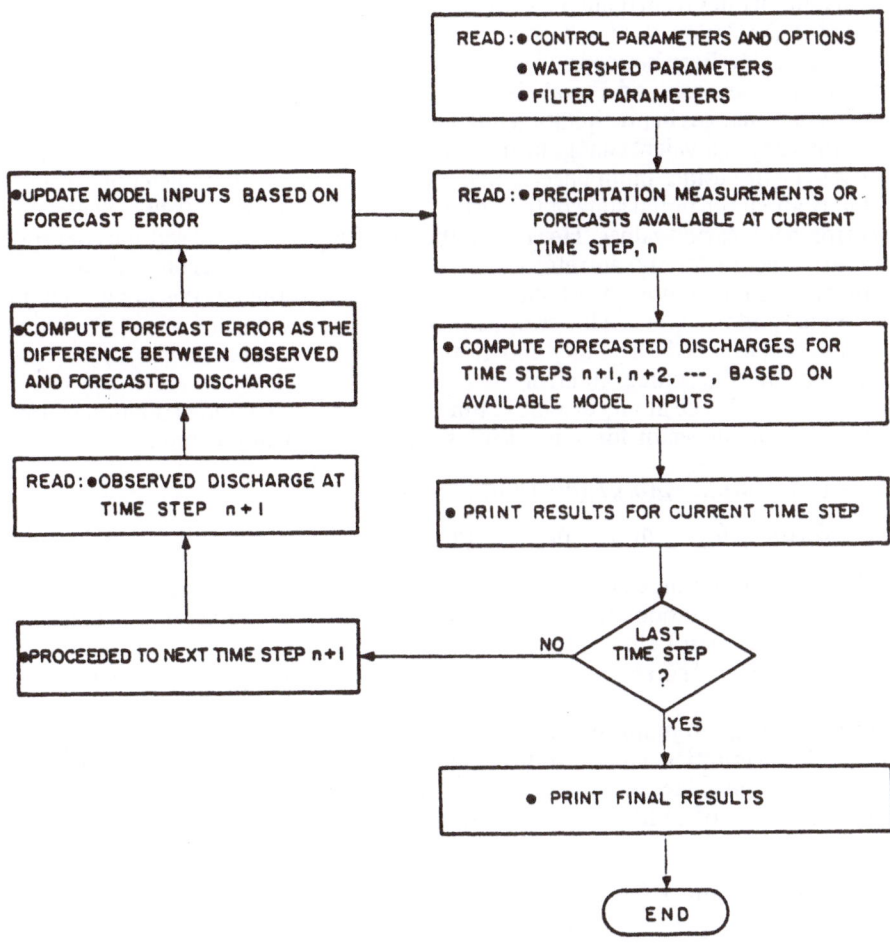

Figure 5. General flowchart of the real-time forecasting procedures of the watershed model OMEGA.

The Kalman filter methodology is used as the basic procedure for incorporating previous errors in the improvement of future forecasts. This procedure requires the formulation of a measurement equation and a state transition equation, as well as the definition of the states that have values which are going to be recurrently updated or filtered. If the model and its conceptual basis are trusted, one may believe that the main source of error in the forecasted discharges is the process of estimating precipitation and, eventually, errors in the computation of the infiltration rates.

This is an important assumption of OMEGA and a starting point for its formulation as a real-time forecasting model. The parameters of the model must be calibrated using the efficient procedures described in a previous section. When this is done the model hopefully adequately expresses the behavior of the watershed. Therefore, it is reasonable to assume that errors in forecasting the river discharges are mainly due to uncertainties in model inputs, namely the precipitation rates over the watershed. It is well known that the computation of average areal rainfall rates from point measurements is a process subject to significant errors. The sensitivity of simulation and forecasting models with respect to errors in evaluating the precipitation inputs has been noticed and emphasized by several hydrologists like O'Connell and Clarke (1981), Burnash and Ferral (1984), among others. According to Burnash and Ferral (1984), small changes in precipitation input may produce dramatic changes in forecast values, more than any equivalent change in any other input or parameter. Naturally, there are other sources of error besides input uncertainty. Model uncertainty, parameter uncertainty, and uncertainty about the initial state of the system (Kitanidis and Bras, 1980a) are certainly sources of error in forecasted values. However, if the model is physically based and conceptually structured it is very attractive and effective to consider precipitation as the main source of error. In doing so some minor uncertainties of the model and the parameters are unduly attributed to the rainfall inputs. However, as pointed out by Krzysztofowicz (1983), what matters to a forecast user is the accuracy and the timeliness of the forecast. The forecast user is not concerned with the sources of error.

To compute the discharge at any instant in time, a discrete convolution equation is used. This equation, written in vector form, is taken as the measurement equation:

$$
\begin{aligned}
Q(m) = &\ (r_1(m), r_2(m), \ldots, r_M(m), I_1(m), I_2(m), \ldots, I_M(m)) \\
&\cdot (\delta_1, \delta_2, \ldots, \delta_M, -\delta_1, \ldots, -\delta_M)^T + \emptyset(m)
\end{aligned}
\tag{16}
$$

where $r_1(m)$ denotes, at instant m, the current mean precipitation rate during period m, $r_2(m)$ the previous mean precipitation rate, etc. A similar notation is adopted for the antecedent infiltration rates $I_1(m)$, $I_2(m)$, etc. $\delta_1, \delta_2, \ldots \delta_M$, denote the discrete kernels or unit hydrograph ordinates. $\emptyset(m)$ is the measurement error and is assumed to be a random Gaussian process of zero mean.

The transition of state from instant m to instant $m+1$ will consist in discarding the most remote values of precipitation and infiltration rates, respectively $r_M(m)$ and $I_M(m)$, and considering the most recent values that were measured or computed as $r_1(m+1)$ and $I_1(m+1)$. Accordingly, all values of rainfall and infiltration rates will be shifted according to the expressions:

$$
r_j(m+1) = r_{j-1}(m) \text{ and } I_j(m+1) = I_{j-1}(m) \quad \text{for} \quad j = 2, \ldots, M
\tag{17}
$$

These expressions can be written in matrix form, leading to the transition equation. The variance covariance matrix of the transition equation is assumed to be diagonal and will play an important role in the updating of the state variables. It is reasonable that state variables contributing more to the observed discharge will be allowed to vary more. This can be reflected in the structure of the diagonal elements of this matrix. More details can be found in Correia (1984).

These procedures are easily adapted for a lead time larger than one. In this case not only the Mth most remote values of precipitation and infiltration, but all the L oldest values are discarded. Accordingly, the input vector is formed by the L new measures of forecasted values of precipitation rate and L new values of computed infiltration rate. In this case the state transition matrix has to be raised to the power L.

Briefly, the formulation of OMEGA as a real–time forecasting model makes possible the use of the Kalman filter algorithm and still relies on a very structured and physically based model. The nonlinear components of the model are not artificially linearized. When lead time increases, the importance of the filtering process decreases and the forecast model tends to the simulation model working in average conditions as it should. This model also accounts for the fact that large forecast errors are often caused by large errors in measuring or forecasting the precipitation inputs.

OMEGA adopts some procedures to prevent unfeasible values from happening. In the first place, the updated values of the state vector, i.e. precipitation and infiltration rates, are checked for non–negativity and set equal to zero if negative. As explained previously the precipitation and infiltration rates are considered the main sources of error and changed throughout the filtering process. However, these changes are not unbounded. If the Kalman filter generates an updated value more than a pre–defined percentage larger than the initial measured or computed value, this updated value is not accepted as such and is reset to a maximum value. If a generated value is smaller than a given fraction of the initial value, it is also reset to this lower bound. This criteria allows for larger absolute changes in the large values of precipitation and infiltration rates. This is consistent with the fact that these large values are more crucial for the model response and eventually subject to larger error. A third correction introduced in the filtering process relates to the non–negativity of effective rainfall rates. The infiltration rates appearing in Equations (16) and (17) are actual infiltration rates, not capacity infiltration rates. Therefore, their values cannot be larger than the corresponding precipitation rates. OMEGA checks these values; and if the infiltration rates are larger than the corresponding precipitation rates, the values are reset to obtain a null effective rainfall rate rather than a negative one.

With the adoption of these procedures the optimality of the Kalman filter procedures is naturally jeopardized. This optimality is unconstrained and, therefore, no bounds or any other limitations in the state space can be considered by the algorithm. However, the optimality itself is questionable in a physical system because reality may deviate from the assumptions of the filter. In these circumstances the Kalman filter is viewed simply as an effective device to guide the updating procedures rather than being entirely responsible for the updating algorithm. Physical feasibility is preferred to mathematical "optimality" and overreaction is avoided. These options are fully consistent with the very structured and physically based model that OMEGA is (Correia, 1984; Correia and Morel–Seytoux, 1985).

2.3. Impact of Spatial Variability on Catchment Response

2.3.1. Characteristics of the Experiments

The structure of OMEGA makes it very suitable to perform studies on the impact of spatial variability of hydrologic parameters on catchment response. Some work that was done in this topic is reported in this chapter.

In hydrologic modeling of the rainfall–runoff processes, the need for dealing with space and time variability of the hydrologic variables and parameters is often recognized. Although it is not possible, in nature, to isolate one hydrologic process from others that interact with it, the analysis of the impact of the variability of one process while ignoring the others is still a valuable and interesting contribution.

Smith and Hebbert (1979) studied the influence of spatial variability of the saturated hydraulic conductivity, $\bar{K}$, in the infiltration process under constant rainfall intensities. Freeze (1980) investigated the spatial stochastic properties of some hillslope parameters on the statistical properties of the runoff events under stochastic rainfall patterns. In both of these studies the infiltration model developed by Smith and Parlange (1978) and the two parameter log normal distribution for the infiltration parameter was assumed. Vauclin (1983) summarized the distribution functions usually tested for samples of the saturated hydraulic conductivity values, $\bar{K}$, namely the normal and the two parameter log normal distributions, suggesting that a positively skewed distribution was adequate.

The experiments made with OMEGA were based on the consideration of the saturated hydraulic conductivity as a random variable with a three–parameter log normal distribution. The impact of the variation of the coefficient of skewness and the coefficient of variation of

$\bar{K}$, on the processes of infiltration and runoff, were analyzed under constant and variable rainfall intensities.

The results obtained with the distributed model, considering one isochrone subdivided in one hundred subsections, were compared with those obtained with a corresponding lumped model. Based on these experiments it is possible to assess if one single value of $\bar{K}$ for the all watershed can reflect the existing variability, be it the mean or the median of the random $\bar{K}$.

This was done for six watersheds, considered as homogeneous and uniform, except for the the saturated hydraulic conductivity, $\bar{K}$, randomly distributed over the 100 subsections. The characteristics of the watersheds are listed in Table 3. The spatial variability of $\bar{K}$ was obtained by considering it as a random variable not spatially autocorrelated. The probability distribution function considered for $\bar{K}$ was the three–parameter log normal distribution, $LN3$. The parameters of that distribution are the location parameter $\bar{K}_o$, the scale parameter μ_y, and the shape parameter T_y. $\bar{K}_o$ is defined in the domain of $\bar{K}$ as the lower bound of the distribution, μ_y and T_y are defined in the log domain of $\bar{K}$ as the mean value and standard deviation of the normal random variable Y, defined as $Y = ln(\bar{K} - \bar{K}_o)$.

Six sets of parameters $\bar{K}_o$, μ_y and T_y were considered, defining six $LN3$ distributions. One hundred values of $\bar{K}$ were generated for each case, determining a random variability of the infiltration parameter. In Table 4 the theoretical values and the values estimated from the generated samples are presented. It can be seen that all the watersheds have the same mean value $\bar{K} = 1$ cm/h and that there are two groups corresponding to two different values of standard deviations: 0.5 cm/h for the watersheds A, B, and C; and 1 cm/h for the watersheds D, E, and F—that is, a coefficient of variation $CV_K (CV_K = \bar{K}/s)$ of 50 percent and 100

Table 3. Characteristics of the six watersheds considered in the experiments.

Input Parameters	Symbol	Unities	Values
Area of the watersheds	A	Km^2	5
Maximum effective capillary drive	H$_c$	cm	13.4
Water content at natural saturation	$\bar{\theta}$	cm^3/cm^3	0.3
Residual water content	θ_r	cm^3/cm^3	0.05
Initial mean water content	θ_i	cm^3/cm^3	0.05
Viscous correction factor	β	–	1
Viscour correction factor at the time of ponding	β_p	–	1
Time step	Δt	min	5
Memory time of the watersheds	M	–	11
Ordinate of the direct runoff unit hydrograph (discrete step kernels)	h (.)	–	0.1227
			0.3301
			0.2023
			0.1246
			0.0831
			0.0559
			0.0362
			0.0238
			0.0128
			0.0063
			0.0022

Table 4. Theoretical and estimated parameters of the populations of the saturated hydraulic conductivity, $\tilde{K}$, of the six watersheds simulated. $\tilde{K}$ is a random variable with a three–parameter log normal probability distribution.

Watershed	Mean value	Standard deviation	Skew coefficient		Location parameter
	cm/h $\mu_{\tilde{K}} = \bar{K}$	cm/h $T_{\tilde{K}} = s$	$\tilde{K}$	$g_{\tilde{K}}$	cm/h $\tilde{K}_o = \hat{K}_o$
A	1	0.5	1.62500	1.62525	0.00000
B	1	0.5	2.43750	2.43678	0.28446
C	1	0.5	3.25000	3.24975	0.42295
D	1	1	4.00000	4.00451	0.00000
E	1	1	6.00000	5.99933	0.22355
F	1	1	8.00000	8.03029	0.33895

Watershed	Scale parameter		Shape parameter		Median value	
	ln (cm/h)		ln (cm/h)		cm/h	
	μ_y	$\bar{y}$	T_y	s_y	μ_{md}	$\hat{\mu}_{md}$
A	-0.11157	-0.11173	0.47238	0.47244	0.89443	0.89429
B	-0.53352	-0.53322	0.63057	0.63045	0.87100	0.87117
C	-0.82987	-0.82979	0.74838	0.74835	0.85906	0.85909
D	0.34657	-0.34770	0.83255	0.83301	0.70711	0.70631
E	0.74194	-0.74184	0.98886	0.98882	0.69974	0.69979
F	1.00913	-1.01256	0.09106	1.09235	0.70349	0.70224

428

percent, respectively. These values are within the range presented by Freeze (1975) and Vauclin (1983).

In a first set of simulations three constant rainfall rates were considered ($r_1 = 1.5$ cm/h, $r_2 = 2.5$ cm/h and $r_3 = 3.5$ cm/h) with a four hour duration. These storm events correspond to return periods of 10, 200 and 6000 years, respectively, for the region of Lisbon.

In a second set of simulations three variable rainfall rate events were considered with the same duration and corresponding to the same total rainfall depths of the first set. The time distribution of the rainfall depths was based on the rainfall patterns given by the second Huff quartile time distribution with 50 percent of exceedence probability (Huff, 1967). These rainfall events are displayed in Figure 6.

2.3.2. Results for Constant Rainfall Rates

Mean infiltration rate as function of time for the six watersheds and for the three rainfall rates are displayed in Figure 7. The impact of the coefficient of variation of the saturated hydraulic

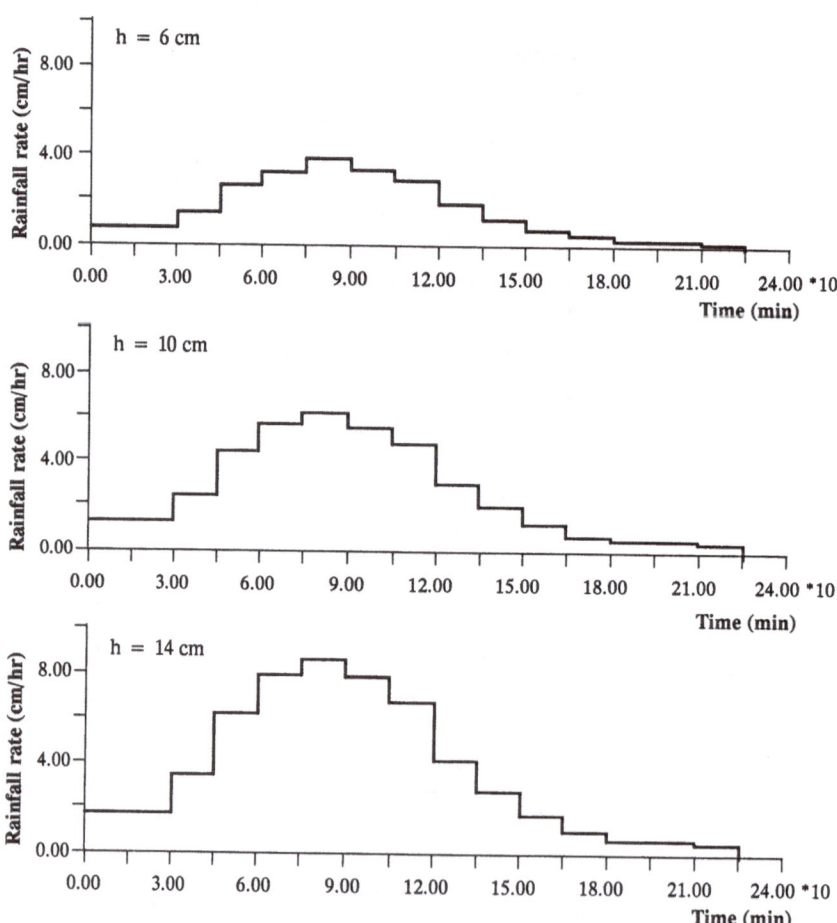

Figure 6. Hyetographs obtained with the second Huff quartile, for a 50 percent exceedence probability, total depth h, and four hour duration.

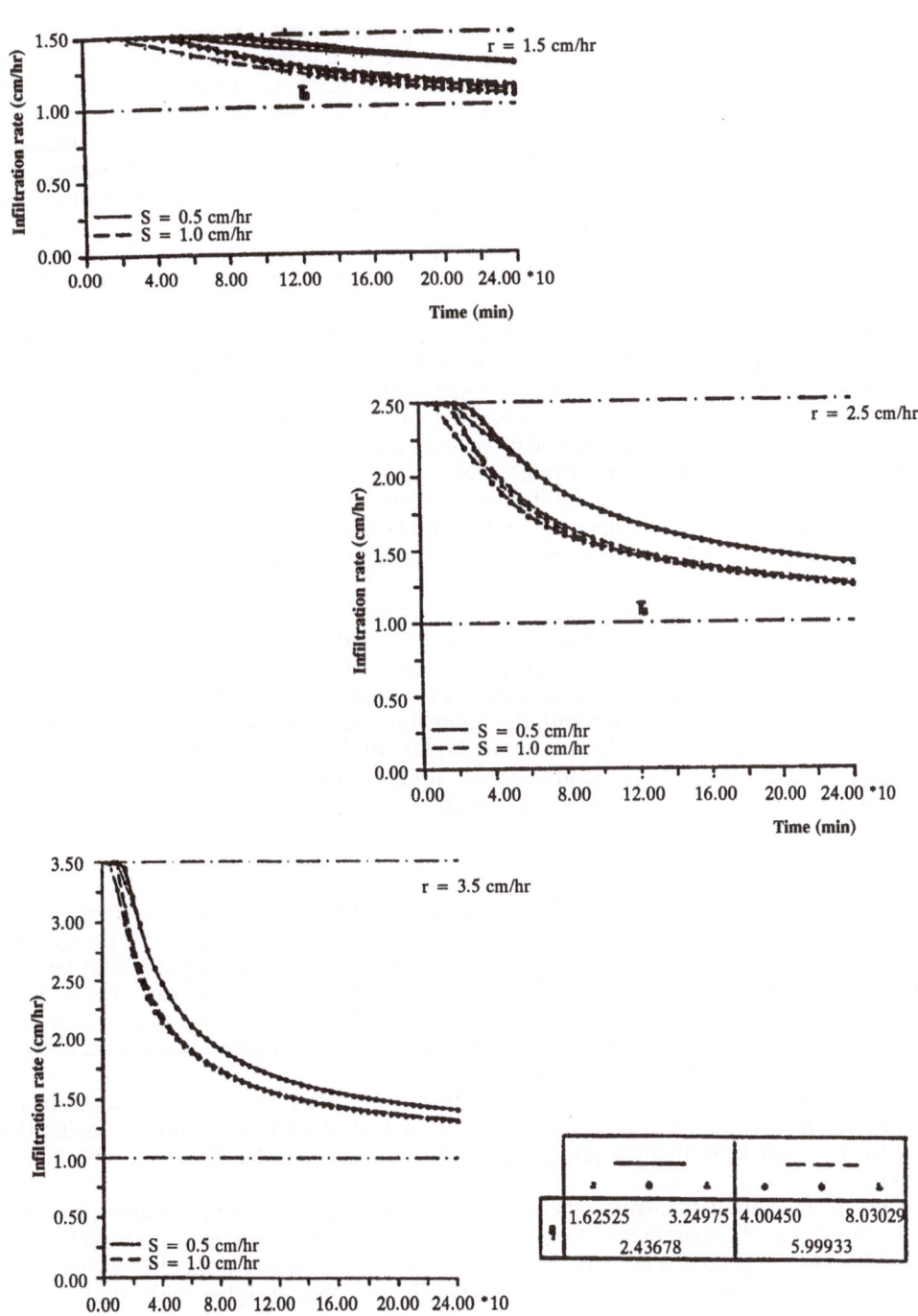

Figure 7. Mean infiltration rate curves for the six watersheds and constant rainfall rates, r.

conductivity, $CV_{\bar{K}}$, on the mean infiltration rate pattern is much more significant than the impact of the skew coefficient, $G_{\bar{K}}$. An increase of $CV_{\bar{K}}$ has the same qualitative effect as a decrease of $G_{\bar{K}}$, leading to an earlier ponding time in some of the subsections of the watershed and, consequently, a larger portion of the basin contributing to runoff.

This effect can also be seen in Figure 8 which shows the evolution in time of the cumulative frequencies of ponding time for the six watersheds and the three rainfall rate events. The ordinates in these drawings can be read as the percentage of the watershed area contributing to runoff. Table 5 shows the mean ponding time, t_p, coefficient of variation, CV_{tp}, and skew coefficient, g_{tp}, obtained in the simulations for the six basins and the three rainfall events. It is clear from Figures 7 and 8 that the impact of the spatial variability of $\bar{K}$ is strongly attenuated with an increase of rainfall rates.

The sample probability density function of the current infiltration rates tends very slowly to the probability density function of $\bar{K}$, even with the higher rainfall rate, at the end of the fourth hour of precipitation. This result is exemplified for watershed B in Figure 9.

Table 5 also presents the total runoff obtained for each watershed and for the three rainfall events, as well the total runoff obtained when using the lumped model with homogeneous and uniform value of $\bar{K}$. Two lumped values were considered: the mean value and the median value. Figure 10 shows the mean infiltration rate pattern of watershed B and the infiltration rate curves for the uniform values of $\bar{K}$, and Figure 11 shows the hydrographs produced in the same conditions. The last two columns of Table 5 display the relative errors when a lumped instead of distributed model is used.

For the smaller rainfall rate the use of the lumped model gives poor results when compared with the use of the distributed model. For the intermediate rainfall rate the lumped model approaches the solution obtained with the distributed model when the single value of $\bar{K}$ is considered to be equal to the median value of the distributed parameters. For the largest rainfall rate the results obtained with the lumped model are close to the solution obtained with the distributed model, either for lumped values of $\bar{K}$ equal to the mean or the median. The total runoff is always overestimated when using the median value in the lumped model and underestimated when using the mean value.

2.3.3. Results for Variable Rainfall Rates

For this set of simulations only the results concerning the discharge hydrographs are presented. Figure 12 displays the hydrographs for the six watersheds and for the three rainfall depths. The impact of spatial variability of $\bar{K}$ is strongly attenuated compared with what was observed for constant rainfall rates. Only the effect of the coefficient of variation is visible, especially for the lower rainfall depth. Figure 13 compares the hydrographs for the three rainfall events, for watershed B, using distributed and lumped models. The results obtained for all basins are summarized in Table 6.

Time to peak is well estimated with the lumped model with both the mean and the median values for $\bar{K}$. Relative errors in the estimation of peak discharge are always smaller than relative errors in the estimation of total runoff when a mean value for $\bar{K}$ is used in the lumped model.

For smaller rainfall intensities, the use of a median value in the lumped model gives quite good results in the computation of peak discharge and total runoff, with absolute errors smaller than 14 percent and 7 percent, respectively. For larger rainfall intensities both lumped solutions provide results close to the distributed solution. When the median value for $\bar{K}$ is considered in the lumped model, peak discharge and total runoff are always overestimated. This is the best lumped solution for the estimation of total runoff. When the mean value of $\bar{K}$ is considered, peak discharge and total runoff are always underestimated; this is the best lumped solution for the estimation of peak discharge.

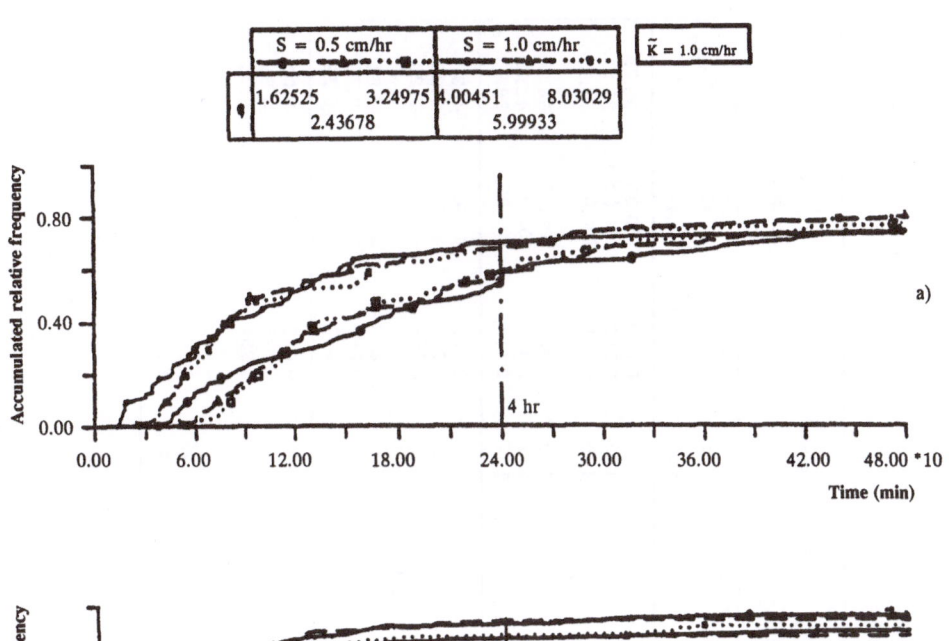

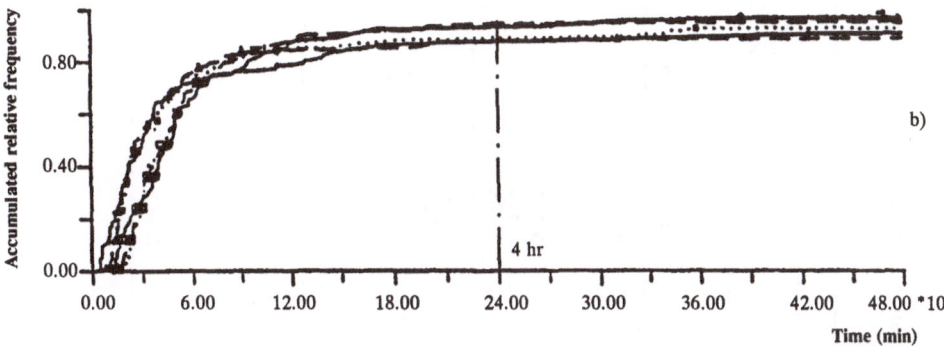

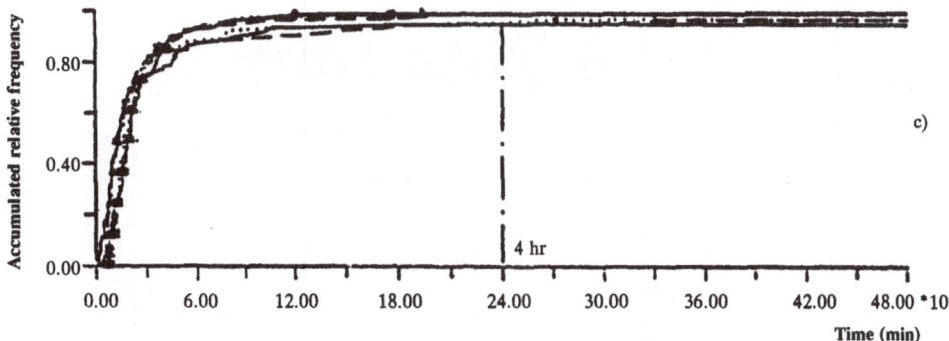

Figure 8. Time evolution of ponding time cumulative frequencies for the six watersheds and constant rainfall rates, r.

Table 5. Results of numerical simulations for constant rainfall rates – values for ponding time, T_p, and total runoff, Q.

r cm/h	watershed	Distributed model T_p min	CV_{t_p}	g_{t_p}	Q cm	Lumped model $\tilde{K}=\bar{K}$ t_p min	Q cm	$\tilde{K}=\mu_{md}$ t_p min	Q cm	Relative errors in computing Q $E_{\bar{K}}$ %	$E_{\mu_{md}}$ %
15	A	1618.93	4.85	7.27	0.420	∞	0.000	197	0.039	100.00	90.75
	B	379.72	2.44	7.33	0.343			185	0.068	100.00	80.31
	C	287.26	1.30	4.53	0.318			179	0.085	100.00	73.31
	D	227.67	2.99	5.88	0.996			119	0.455	100.00	54.30
	E	207.91	2.52	7.66	0.835			117	0.481	100.00	42.33
	F	184.39	1.38	4.40	0.775			117	0.474	100.00	38.93
2.5	A	84.92	1.99	6.72	2.828	53	2.405	44	2.895	14.95	-2.37
	B	65.22	1.38	6.61	2.822			43	3.006	14.79	-6.25
	C	65.09	1.11	5.65	2.806			42	3.064	14.28	-9.20
	D	73.64	1.92	4.30	3.676			31	3.846	34.58	-4.61
	E	136.57	5.67	9.26	3.600			31	3.880	33.20	-7.78
	F	68.03	2.55	7.84	3.509			31	3.867	31.46	-10.22
3.5	A	31.93	2.22	9.14	6.354	22	6.115	19	6.672	3.75	-5.01
	B	32.83	2.16	8.24	6.351			19	6.796	3.72	-6.99
	C	24.98	0.89	8.37	6.350			18	6.861	3.70	-8.03
	D	22.57	1.23	4.46	7.071			14	7.714	13.51	-9.11
	E	30.03	1.81	4.48	6.995			14	7.752	12.57	-10.83
	F	26.01	1.59	4.96	6.961			14	7.738	12.15	-11.15

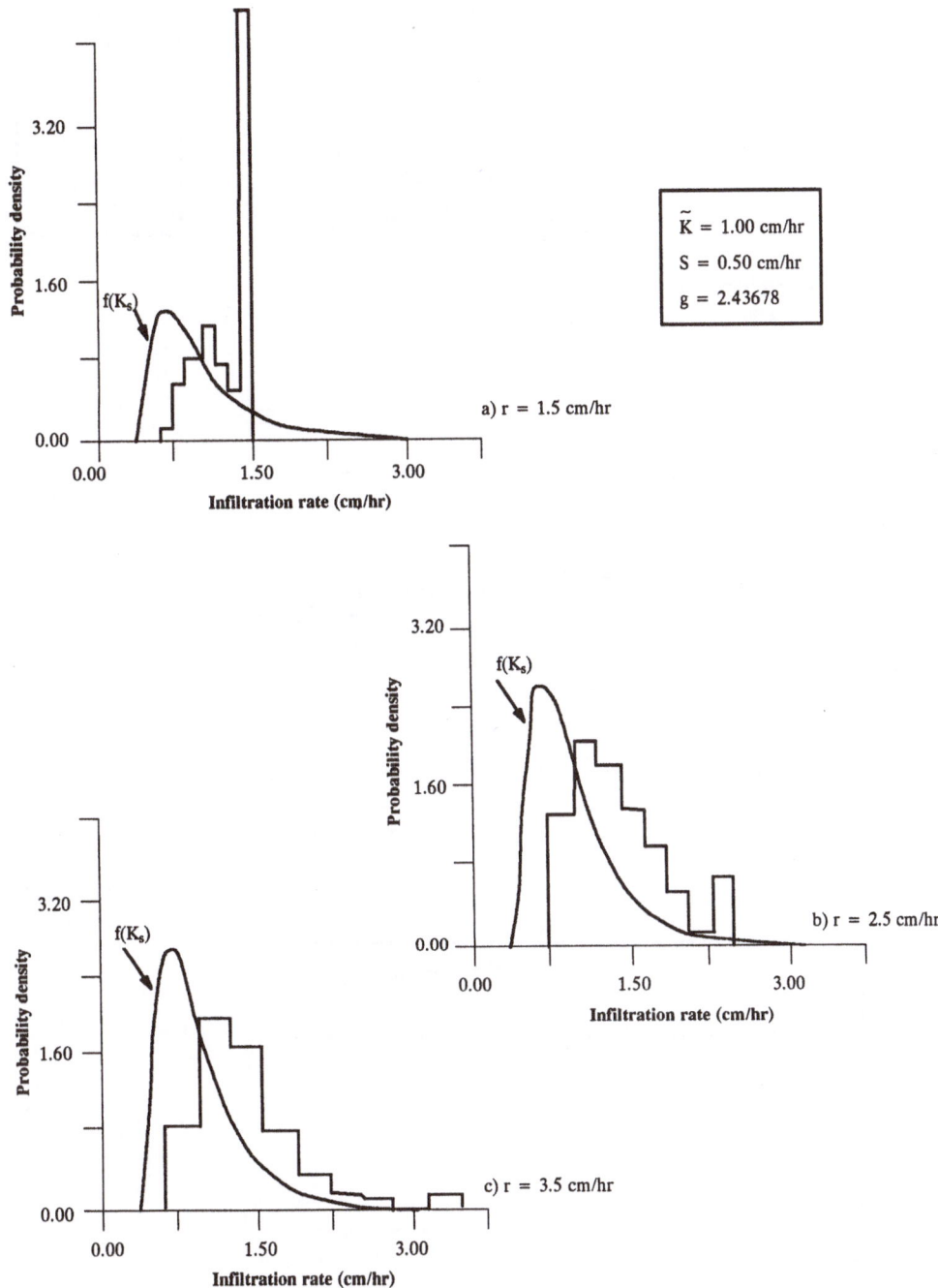

Figure 9. Sample probability density function of infiltration rate at the end of the fourth hour of precipitation with constant rainfall rate, r, and probability density function of $\tilde{K}$, $f(\tilde{K})$, for watershed B.

434

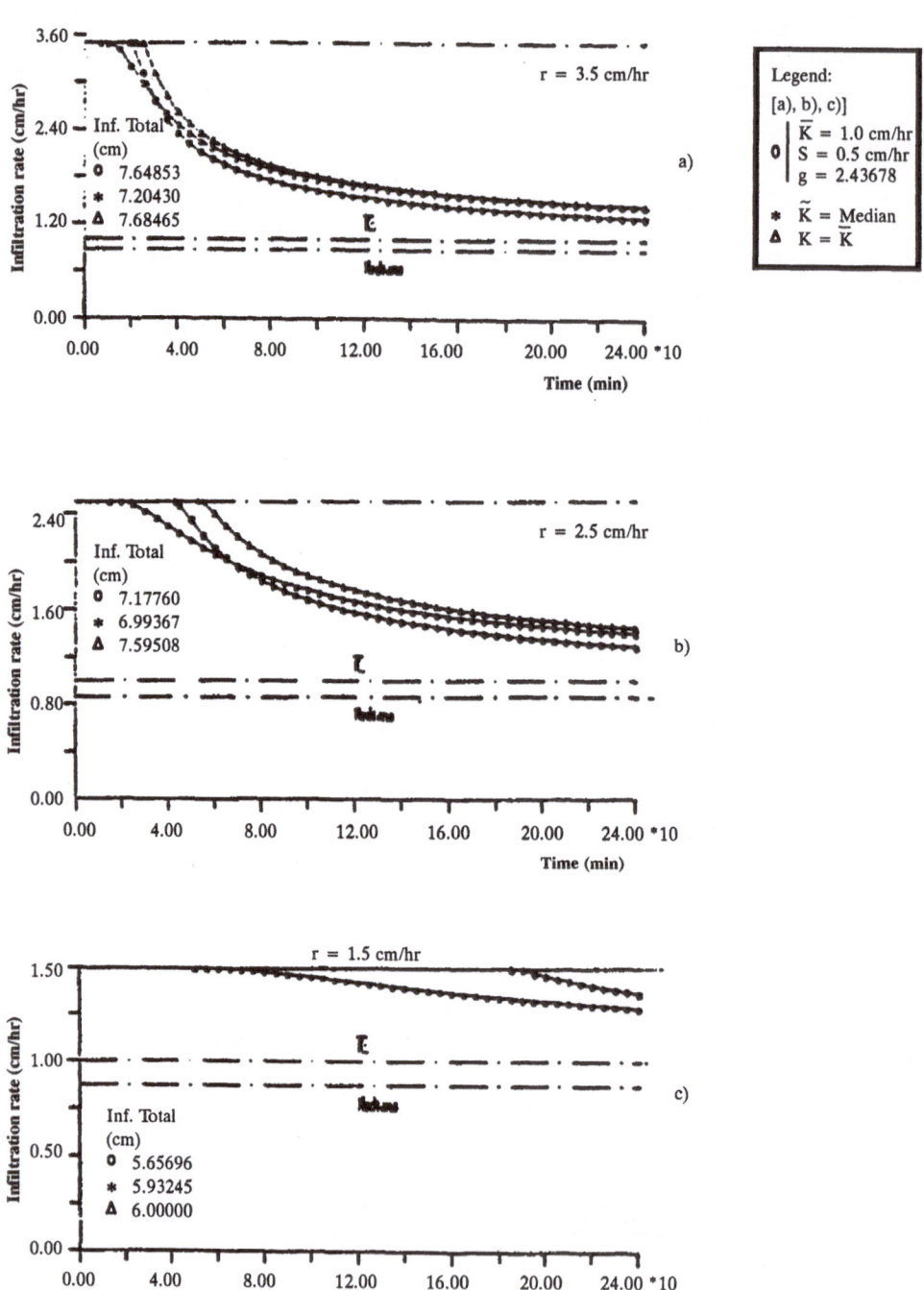

Figure 10. Mean infiltration rate curves for watershed B and constant rainfall rates, r. Distributed and lumped models, with $\tilde{K}$ equal to the mean and median values.

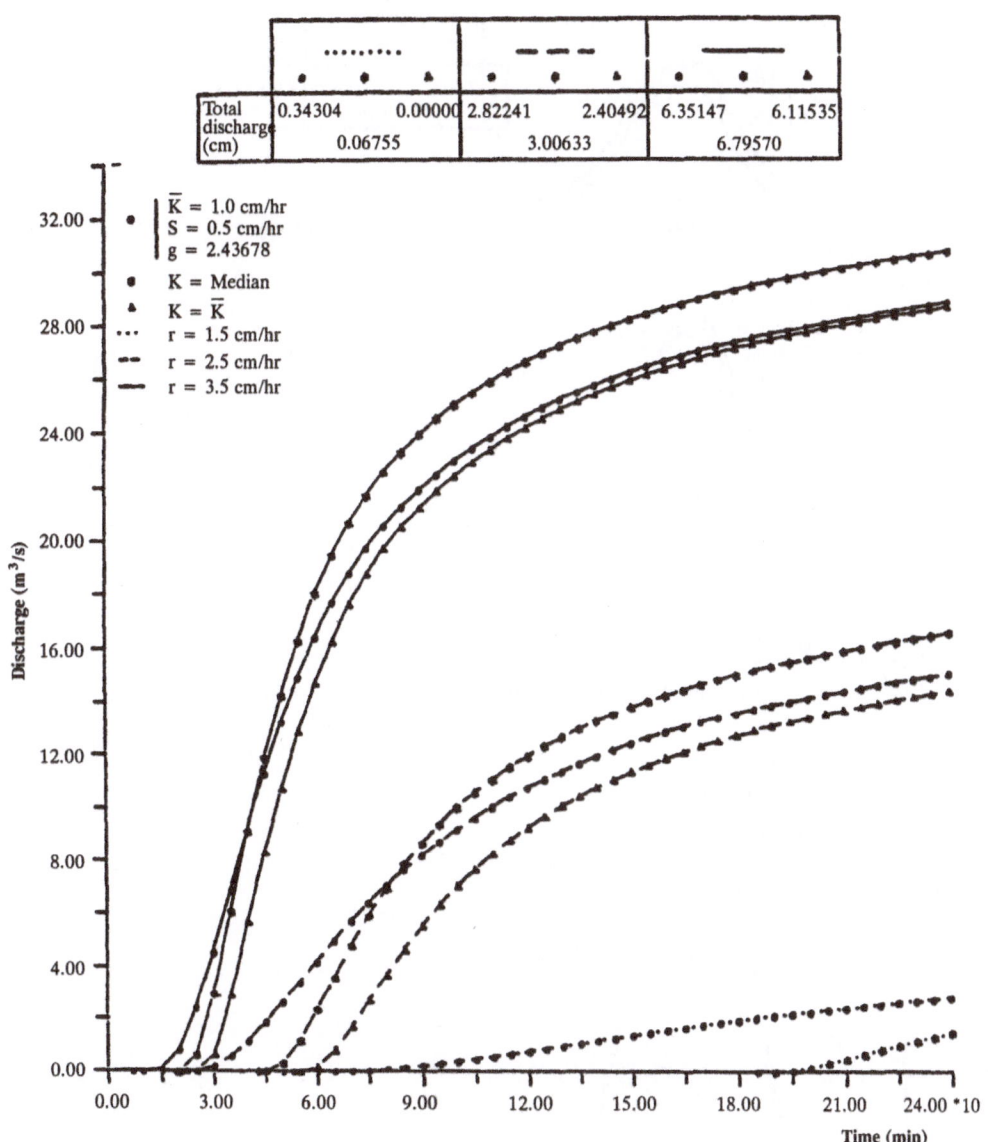

Total discharge (cm)	0.34304	0.00000	2.82241	2.40492	6.35147	6.11535
		0.06755		3.00633		6.79570

$\overline{K}$ = 1.0 cm/hr
S = 0.5 cm/hr
g = 2.43678

K = Median
K = $\overline{K}$
r = 1.5 cm/hr
r = 2.5 cm/hr
r = 3.5 cm/hr

Figure 11. Hydrographs for watershed B and constant rainfall rates r. Distributed and lumped models with $\widetilde{K}$ equal to mean and median values.

436

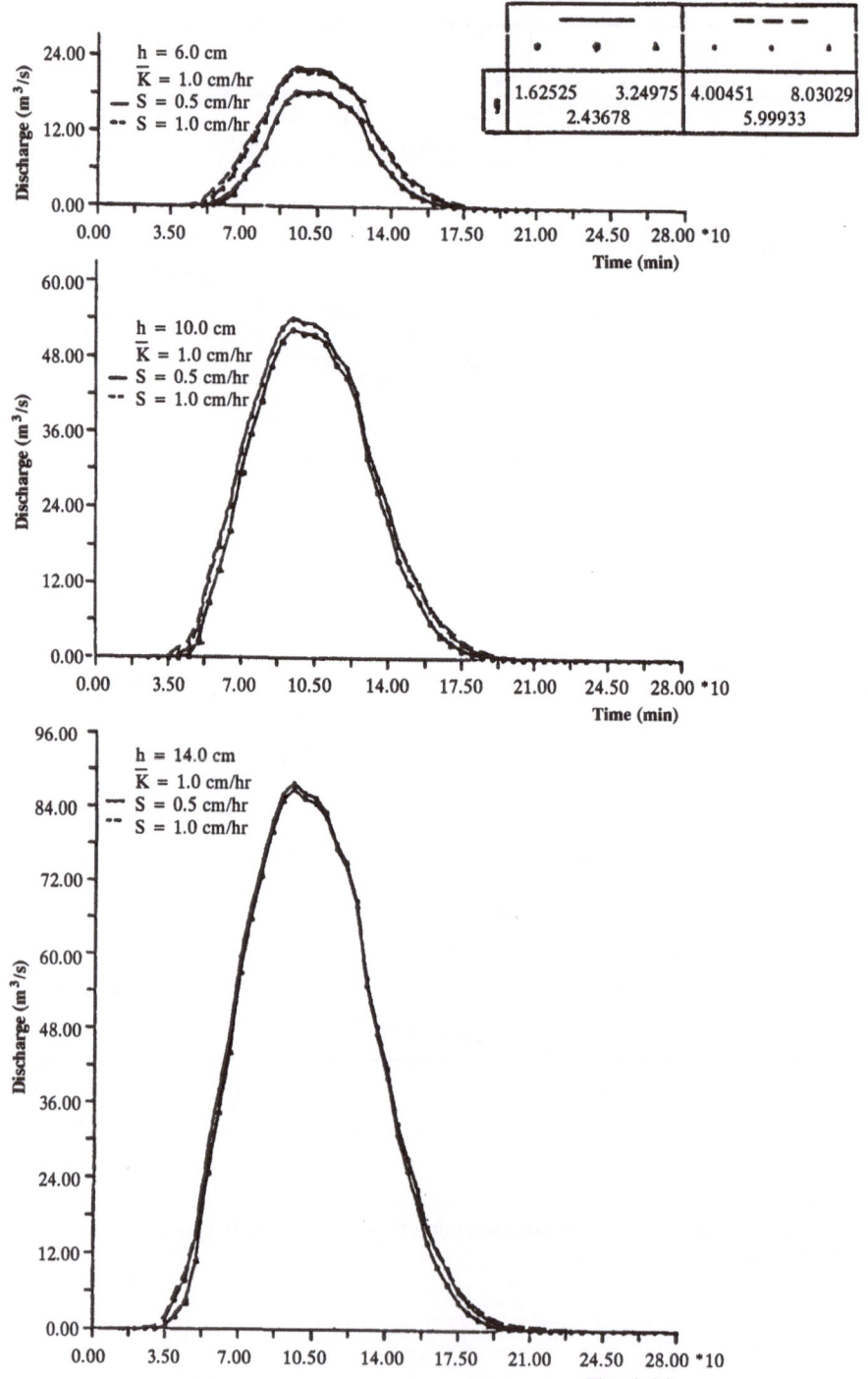

Figure 12. Hydrographs for the six watersheds, variable rainfall rates, and total rainfall depth, h.

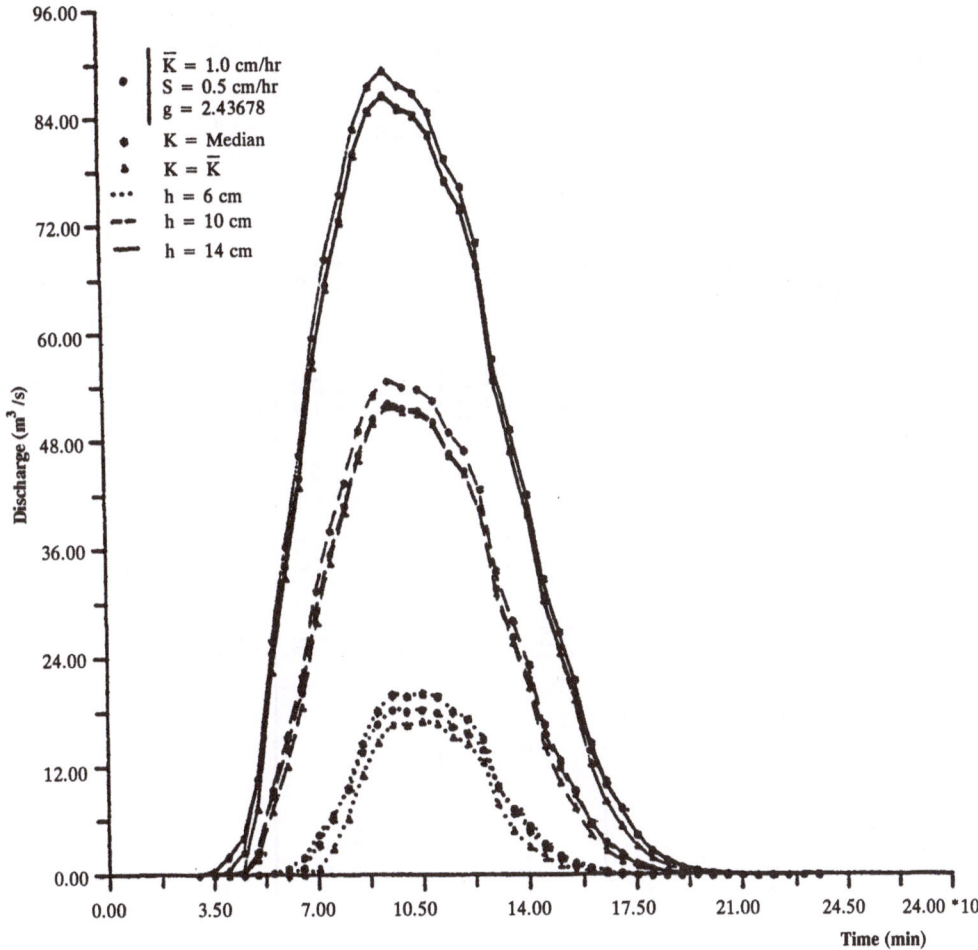

Figure 13. Hydrographs for watershed, B, variable rainfall rates and total rainfall depth, h. Distributed and lumped model, with mean and median values for $\widetilde{K}$.

Table 6. Results of numerical simulations for variable rainfall rates – values for time to peak T_{Q_p}, peak discharge, Q_p, and total runoff, Q.

h cm	Watershed	Distributed model t_{Q_p} min	Q_p m³/s	Q cm	Lumped model $\tilde{K}=\tilde{K}$ t_{Q_p} min	Q_p m³/s	Q cm	Lumped model $\tilde{K}=\mu md$ t_{Q_p} min	Q_p m³/s	Q cm	Rel. errors Qp $E_{\tilde{K}}$ %	$E_{\mu md}$ %	Rel. errors Q $E_{\tilde{K}}$ %	$E_{\mu md}$ %
6	A	95	18.24	1.245	105	17.03	0.989	105	19.60	1.219	6.61	-7.48	20.54	2.10
	B	105	18.38	1.226				105	20.18	1.272	7.33	-9.79	19.34	-3.73
	C	105	18.30	1.214				105	20.48	1.300	6.93	-11.92	18.54	-7.08
	D	95	22.26	1.701				95	24.55	1.664	23.48	-10.30	41.87	2.21
	E	95	22.15	1.632				95	24.74	1.680	23.11	-11.66	39.41	-2.92
	F	95	21.72	1.584				95	24.67	1.674	21.59	-13.55	37.55	-5.68
10	A	95	52.17	4.314	95	51.64	4.145	95	54.14	4.458	1.01	-3.77	3.93	-3.33
	B	95	52.14	4.297				95	54.69	4.528	0.94	-4.40	3.54	-5.37
	C	95	52.15	4.293				95	54.98	4.564	0.97	-5.43	3.46	-6.31
	D	95	54.02	4.753				95	58.79	5.039	4.40	-8.81	12.80	-6.01
	E	95	52.91	4.683				95	58.95	5.060	4.21	-9.34	11.50	-8.03
	F	95	54.00	4.660				95	58.89	5.052	4.36	-9.05	11.06	-8.40
14	A	95	86.62	7.797	95	86.33	7.596	95	88.71	7.956	0.34	-2.41	2.58	-2.03
	B	95	86.58	7.779				95	89.19	9.044	0.29	-3.00	2.35	-3.41
	C	95	86.57	7.773				95	89.46	8.087	0.27	-3.33	2.28	-4.04
	D	95	87.59	8.216				95	92.88	8.660	1.40	-6.08	7.54	-5.41
	E	95	87.53	8.136				95	93.03	8.685	1.37	-6.28	6.63	-6.75
	F	95	87.53	8.113				95	92.97	8.676	1.37	-6.22	6.37	-6.94

3. Conclusions

OMEGA can be used for event or continuous simulation of river discharges and can be utilized in a lumped or distributed fashion. When used as a distributed model the spatial variability of the input and the watershed response may be fully considered. OMEGA incorporates, to a large extent, the physical conceptualization of the hydrological processes. The description of the complex and highly nonlinear processes of infiltration, ponding, and redistribution of water in soil is addressed with special attention. The modern theory of infiltration developed tools that can be adequately incorporated in rainfall–runoff models, enhancing the possibilities of these models and improving their results. Three totally compatible versions of the model were implemented, namely: the basic simulation version, the parameter calibration version, and the real–time forecast version. This triple formulation adds a large degree of versatility. The model showed a remarkable ability to adapt to different hydrologic conditions, going from a mountainous and wet region to a semi–arid and flat one, and provided consistently good results in practical applications reported by Correia (1984).

The fully distributed formulation of OMEGA and the careful description of the physical processes make this model very suitable for studying the sensitivity of catchment response to the variability patterns of the hydrologic parameters. In this chapter the model was used with the purpose of analyzing the influence of the spatial variability of the saturated hydraulic conductivity in the processes of infiltration and runoff, under constant and variable rainfall rates. Saturated hydraulic conductivity was defined as a random variable with a three–parameter log normal probability distribution function. Six sets of parameters were used for the generation of one hundred values for the infiltration parameter. One value for the population mean, two different values for the coefficient of variation, and six values for the skewness coefficient were considered. The coefficient of skewness and the coefficient of variation of the distributed infiltration parameter proved to have a decreasing impact on the characteristics of the infiltration rate curves and on the discharge hydrographs and an increasing impact on rainfall intensities. That attenuation of this impact is still more visible for variable rainfall rates. The coefficient of variation of the distributed parameter is by far larger and much more important than the skewness coefficient. It is reasonable to admit that a two parameter probability distribution describes with sufficient accuracy the random variability of the hydraulic conductivity, as far as it is relevant for the hydrologic process of infiltration and runoff generation.

The same rainfall events were considered with the lumped version of the rainfall–runoff model. Two lumped values of $\bar{K}$ were utilized, namely the mean and the median of the distributed values. It was found that for rainfall depths with a low return period, serious errors occur for lumping the infiltration parameter; therefore, it is preferable to use the distributed model. When variable rainfall rates are considered, the performance of the lumped model increases and the use of the median value of hydraulic conductivity proves better than the use of the mean value. For intense rainfall rates, events with large return periods, the use of the lumped model gives results that approach those obtained with the distributed model. With variable rainfall rates the consideration of the mean value in the lumped model gives good estimates of peak discharge (although underestimated), and the use of the median value provides reasonable estimates of total runoff (although overestimated).

Clearly, the question of using distributed infiltration parameters is not very relevant in a flood assessment context when dealing with intense rainfall rates of large return periods. But it may be crucial for continuous simulation of hydrologic systems.

The situation of ittermitent rainfall was not addressed in this chapter, but it will be considered in future work. It can be expected that an ittermitent rainfall pattern seriously affects the discrepancy between results obtained with the lumped and distributed versions of the model. The impact of the variability of other parameters and the cross effects of the parameter variabilities will be addressed in future work.

When presenting these results, there is an underlying supposition that the distributed model is closer to reality. Therefore, the errors caused by using a lumped model are computed with respect to the distributed model. Of course, there are good reasons to admit that the behavior of a distributed model is closer to the behavior of a real watershed; it makes sense to see how much we sacrifice when using lumped representations of hydrologic systems. Although significant, the results that were presented are based on "fictitious" watersheds, in which the complex distributed version of the model is somewhat considered the "truth" and the simple (lumped) version of the model is considered to be "erroneous."

A general recommendation for using a distributed model instead of a lumped model should not be deduced from the results presented in this chapter. Experience shows that in many circumstances a simple model may perform better and be more robust than a complex one. Therefore, this research effort should be seen as an investigation on space-scale problems, trying to assess how much results change when a lumped instead of a distributed approach is used, rather than an attempt to describe nature in the "best" way.

4. References

Amorocho, J.: 1979, 'Spatially Distributed Variables in Hydrologic Modeling,' in E. H. Loyd, T. O'Donnell, J. C. Wilkinson (eds.), *The Mathematics of Hydrology and Water Resources*, Academic Press, London.

Burnash, R., and R. Ferral: 1984, 'Hydrological Forecasting – Advantages and Limitations of Continuous Analysis Systems,' *American Water Resources Association Symposium on a Critical Assessment of Forecasting in Western Water Resources Management*, Seattle, WA.

Correia, F. N.: 1983, 'Methods for the Analysis and the Estimation of Flood Discharges,' Laboratório Nacional de Engenharia Civil, Lisboa, 389 pp. (in Portuguese).

Correia, F. N.: 1984, 'OMEGA, A Watershed Model for Simulation, Parameter Calibration and Real-Time Forecast of River Flows,' Department of Civil Engineering, Colorado State University, Fort Collins, Published also as Memória Nº 643, Laboratório Nacional de Engenharia Civil, Lisboa, 146 pp.

Correia, F. N., and H. J. Morel-Seytoux: 1985, 'User's Manual for Omega. A Package of FORTRAN 77 Programs for Simulation, Parameter Calibration and Real-time Forecast of River Flows,' ITH 18, Laboratório Nacional de Engenharia Civil, Lisboa, 119 pp.

Dawdy, D. R.: 1982, 'A Review of Deterministic Surface Water Routing in Rainfall-Runoff Models,' *Proceedings of the International Symposium on Rainfall-Runoff Modeling*, Mississippi State University, Water Resources Publications, Littleton, CO, pp. 23-36.

Dawdy, D. R.: 1984, 'State of the Art in Hydrologic Forecasting: What Next,' *American Water Resources Association Symposium on a Critical Assessment of Forecasting in Western Water Resources Management*, Seattle, WA.

Fleming, G.: 1976, *Computer Simulation Techniques in Water Resources*, Elsevier Publishing Company, New York, NY.

Freeze, R. A.: 1975, 'A Stochastic-Conceptual Analysis of One-Dimensional Groundwater Flow in Nonuniform Homogeneous Media,' *Water Resources Research* 11(5), 725-741.

Freeze, R. A.: 1980, 'A Stochastic-Conceptual Analysis of Rainfall-Runoff Processes on a Hillslope,' *Water Resources Research* 16(2), 391-408.

Hawkins, R. H.: 1982, 'Interpretations of Source Area Variability in Rainfall-Runoff Relations,' *Proceedings of the International Symposium on Rainfall-Runoff Modeling*, Mississippi State University, Water Resources Publications, Littleton, CO, pp. 303-324.

Huff, F. A.: 1967, 'Time Distribution of Rainfall in Heavy Storms,' *Water Resources Research* 3(4), American Geophysical Union, Washington, DC.

Kalman, R. E.: 1978, 'A Retrospective after Twenty Years: From the Pure to the Applied,' in C. L. Chiu (ed.), *Applications of Kalman Filter to Hydrology, Hydraulics and Water Resources*, University of Pittsburgh, Pittsburgh, PA.

Kitanidis, P. K., and R. L. Bras: 1980a, 'Real-Time Forecasting with a Conceptual Hydrological Model; 1. Analysis of Uncertainty,' *Water Resources Research* 16(6), 1025-1033.

Kitanidis, P. K., and R. L. Bras: 1980b, 'Real–Time Forecasting with a Conceptual Hydrological Model; 2. Applications and Results,' *Water Resources Research* **16**(6), 1034–1044.

Kite, G. W.: 1975, 'Performance of Two Deterministic Hydrologic Models,' *Symposium on the Application of Mathematical Models in Hydrology and Water Resources*, Bratislava, International Association of Hydrologic Sciences, 7 p.

Krzystofowicz, R.: 1983, 'A Baysian Markov Model of the Flood Forecast Process,' *Water Resources Research* **19**(6), 1455–1465.

Morel-Seytoux, H. J.: 1978, 'Derivation of Equations for Variable Rainfall Infiltration,' *Water Resources Research* **14**(4), 561–568.

Morel-Seytoux, H. J.: 1979, 'Infiltration', in R. W. Fairbridge, and C. W. Finki, Jr. (eds.), *Encyclopedia of Soil Science, Part I*, Dowden, Hutchinson and Ross, Inc. Publishers, Stroudsburg, PA, pp. 223–237.

Morel-Seytoux, H. J.: 1981, 'Applications of Infiltration Theory for the Determination of Excess Rainfall Hyetograph,' *Water Resources Bulletin, AWRA* **17**(6), 1012–1022.

Morel-Seytoux, H. J., L. A. Lindell, and F. N. Correia: 1981, 'Runoff Model Based on Larrieu's Generalized Unit Hydrograph Theory and Modern Infiltration Theory,' *Proceedings of International Symposium on Rainfall–Runoff Modelling*, Water Resources Publications, Littleton, CO, pp. 49–68.

Morel-Seytoux, H. J.: 1982, 'Analytical Results for Prediction of Variable Rainfall Infiltration,' *Journal of Hydrology* **59**, 209–213.

Morel-Seytoux, H. J., F. N. Correia, J. H. Hyre, and L. A. Lindell: 1982, '*Rainfall–Runoff Modeling, Contribution to Frontiers in Hydrology,*' Volume Published in Honor of the Late V. T. Chow, Hydrowar Program Report, Department of Civil Engineering, Colorado State University, Fort Collins, CO, 20 pp.

Morel-Seytoux, H. J.: 1983, '*Elementary Physical Modeling of Infiltration and Excess Rainfall for Small Watersheds in Dry Climates,*' Hydrowar Program Report, Department of Civil Engineering, Colorado State University, Fort Collins, CO, 29 pp.

Morel-Seytoux, H. J.: 1984, 'Elementary Physical Modeling of Infiltration for Small Watersheds in Dry Climates,' *Proceedings of the 4th Annual AGU Front Range Branch Hydrology Days*, Colorado State University, Fort Collins, CO, pp. 50–78.

Morel-Seytoux, H. J., and J. Billica: 1984, '*A Two–Phase Numerical Model for Prediction of Infiltration: Applications to a Semi–Infinite Soil Column,*' Hydrowar Program Report, Department of Civil Engineering, Colorado State University, Fort Collins, CO.

Nash, J. E., and J. V. Sutcliffe: 1970, 'River Flow Forecasting Through Conceptual Models, Part I – A Discussion of Principles,' *Journal of Hydrology* **10**, 282–290.

O'Connell, P. E., and R. T. Clarke: 1981, 'Adaptive Hydrological Forecasting – A Review,' *Hydrological Sciences Bulletin* **26**(2), 179–205.

Pilgrim, D. H., I. Cordery, and M. J. Boyd: 1982, 'Australian Developments in Flood Hydrograph Modeling,' *Proceedings of the International Symposium on Rainfall–Runoff Modeling*, Mississippi State University, Water Resources Publications Littleton, CO, pp. 37–48.

Rodriguez-Iturbe, I.: 1983, *Workshop on Scale Problems in Hydrology*, International Association of Hydrologic Sciences, Hamburg.

Rodriguez-Iturbe, I., J. B. Valdes, and J. M. Velazquez: 1978, 'Applications of Kalman Filter in Rainfall–Runoff Studies,' in C. L. Chiu (ed.), *Applications of Kalman Filter to Hydrology, Hydraulics and Water Resources*, University of Pittsburgh, Pittsburgh, PA.

Smith, R. E., and J.-Y. Parlange: 1978, 'A Parameter-Efficient Hydrologic Infiltration Model,' *Water Resources Research* **14**(3), 533–538.

Smith, R. E., and R. H. B. Hebbert: 1979, 'A Monte Carlo Analysis of the Hydrologic Effects of Spatial Variability of Infiltration,' *Water Resources Research* **15**(2), 419–429.

Vauclin, M.: 1983, 'Méthodes d'Étude de la Variabilité Spatiale des Propriétés d'un Sol,' *Editions INRA, Les Colloques de l'INRA* **15**, 9–43.

WMO: 1975, '*Intercomparison of Conceptual Models Used in Operational Hydrological Forecasting. Operational Hydrology,*' World Meteorological Organization, Report No. 7, No. 429, Geneva, Switzerland, 172 pp.

Chapter 20

Parameter Estimation, Model Identification, and Model Validation: Conceptual-Type Models

Soroosh Sorooshian
University of Arizona
Department of Hydrology and Water Resources
Tucson, Arizona 85721 U.S.A.

Abstract. A variety of modeling approaches have been adopted to represent the hydrologic response (runoff) of catchments to meteorologic input (precipitation). The available models vary greatly in their degree of complexity and realism, and they range from simple empirical relationships and time–series models to conceptual–type and physically–based models. The purpose of this lecture is to focus on the conceptual rainfall–runoff (CRR) models. A summary of the work reported since 1983 is presented. As it will be demonstrated, CRR models contain structural peculiarities which make the application of model identification and parameter estimation methods difficult. Recent developments which have been intended to overcome some of these difficulties will be discussed.

1. Introduction

A variety of modeling approaches have been adopted to represent the hydrologic response (runoff) of catchments to meteorologic input (precipitation). The available models vary greatly in their degree of complexity and realism, and they range from simple empirical relationships and time–series models to conceptual–type and physically based models. The purpose of this lecture is to focus on the conceptual rainfall–runoff (CRR) models.

The CRR models are intended to incorporate within their structure the general physical mechanisms which govern the hydrologic cycle. The internal description of the various sub-processes is modeled according to empirically–determined functions which are linked in their conceptualized logical order. The development and popularity of CRR models would have not taken place without the advent of digital computers to meet their computational needs.

Historically, Crawford and Linsley (1966) are credited with the development of the first CRR model, known as the Stanford Watershed Model (SWM). The earliest known reference on the application of an automatic technique for evaluation of CRR model parameters is the work of Dawdy and O'Donnell (1965). Since the mid–1960s, many other CRR models have been developed. The most prominent among them is the version of the Sacramento model (Burnash et al., 1973) adopted by the U.S. National Weather Service known as the Soil Moisture Accounting (SMA) model of the National Weather Service River Forecast Service (NWSRFS). The primary purpose of this model is forecasting of river flows. During the same period, many investigators have addressed various problems related to CRR models (e.g., calibration issues and model application, identification, and verification, among others). For details of the reported work through 1983, the interested reader may review Sorooshian

443

D. S. Bowles and P. E. O'Connell (eds.), Recent Advances in the Modeling of Hydrologic Systems, 443–467.
© 1991 *Kluwer Academic Publishers.*

(1983). A summary of the work reported since 1983 is presented in the following section. As it will be demonstrated, CRR models contain structural peculiarities which make the application of model identification and parameter estimation methods difficult. Recent developments which have been intended to overcome some of these difficulties will be discussed.

2. Discussion

2.1. CRR Model Structure

Most CRR models available today recognize the presence of different vertically stratified zones of soil in the ground. The soil mass is most often conceived to have two layers or "zones," the upper and the lower, connected through vertical percolation. Each zone is commonly modeled using reservoirs (usually linear). Channel inflow consists of baseflow, interflow, and overland flow, and consideration is made for evapotranspiration depletion. The set of operations performed by most conceptual models can be summarized as follows: (1) interception of rainfall, (2) direct runoff from impervious areas, (3) infiltration of water from surface to upper zone, (4) surface runoff of water not accepted by the upper zone, (5) percolation of water from the upper to the lower zone, (6) interflow (lateral subsurface flow) from the upper zone, (7) baseflow (lateral groundwater outflow into river), (8) evapotranspiration from all storages to the atmosphere, and (9) routing of river inflow to the downstream gaging point. CRR models usually accept lumped precipitation and potential evapotranspiration data input, at fixed intervals of time, which represent equivalent uniform depths of precipitation and potential evapotranspiration over the catchment area. The model output is usually the stream stage or flow volume but may also consist of evaporation water loss data and/or volume of subsurface water storage. Two important features of the models are: (a) they keep track of the present state of moisture conditions of the watershed; and (b) they simulate some of the dominant nonlinearities of the system, such as those associated with saturation of the soil mass. Figure 1 displays a schematic representation of the SMA–NWSRFS model, and the description of model parameters is presented in Table 1. This model has been the subject of numerous investigations in recent years (Kitanidis and Bras, 1980; Brazil and Hudlow, 1981; Sorooshian et al., 1983; Sorooshian and Gupta, 1983; Gupta and Sorooshian, 1983; Hendrickson et al., 1988, Gan, 1987; Brazil, 1988; among others).

2.2. Parameter Estimation Requirements

CRR models are general and must be calibrated for specific applications. The calibration process refers to the parameter identification phase of catchment modeling wherein a given CRR model is made specific to a given site. Some of the model parameters can often be estimated by the study of the watershed characteristics, but others relating to the internal subprocesses of the catchment, which cannot be determined in this manner, must be indirectly estimated based on available information, such as historical precipitation and streamflow observation. Implementation of the so–called "automatic" approach entails the following:

(1) Specification of a measure of "closeness" (called estimation criterion) between the model and the catchment. This is usually defined in terms of the differences between the model and catchment outputs when both are subjected to the same inputs.

(2) Selection of a method for identifying those parameter values which "optimize" (minimize or maximize, as appropriate) the chosen estimation criterion. Because the estimation criterion is usually nonlinear in the parameters, this usually involves implementation of an iterative optimization algorithm.

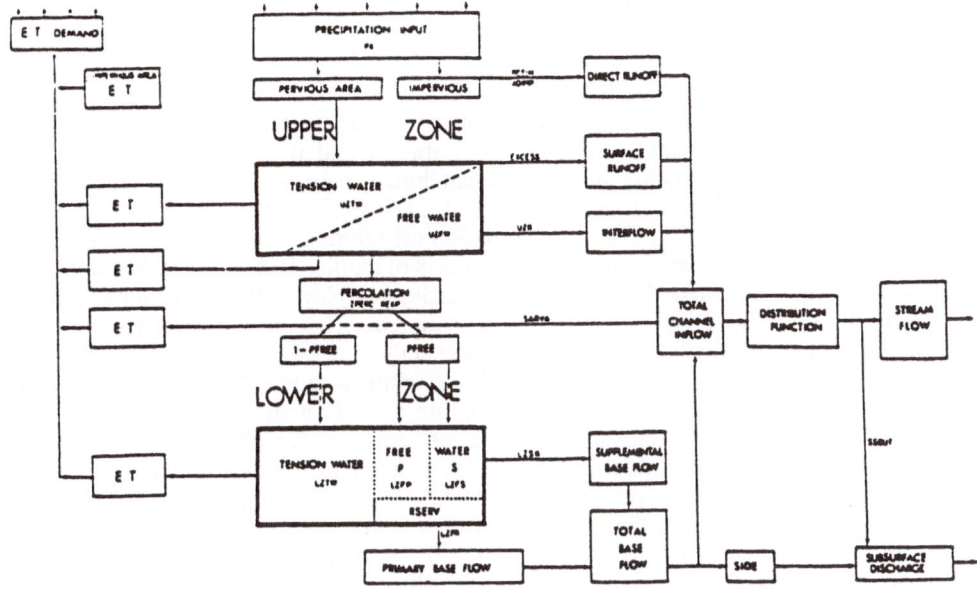

Figure 1. SMA–NWSRFS catchment model (Source: Brazil and Hudlow, 1981).

Table 1. Parameters of the soil moisture accounting model.

Parameter	Explanation
UZTWM	maximum capacity of upper–zone tension storage. mm
UZFWM	maximum capacity of upper–zone free storage. mm
LZTWM	maximum capacity of lower–zone tension storage. mm
LZFPM	maximum capacity of lower–zone primary free storage. mm
LZFSM	maximum capacity of lower–zone secondary free storage. mm
ADIMP	fraction of basin which becomes impervious as all tension storage is met
UZK	lateral drainage rate of upper–zone free storage. fraction/day
LZPK	lateral drainage rate of lower–zone primary storage. fraction/day
LZSK	lateral drainage rate of lower–zone secondary storage. fraction/day
ZPERC	percolation parameter which indicates. when used with other parameters the maximum possible percolation rate. dimensionless
REXP	percolation parameter. an exponent. determining the rate of change of the percolation rate as the lower–zone moisture varies from full to dry. dimensionless
PCTIM	fraction of basin which is impervious and contiguous with stream channels
RIVA	fraction of basin covered by streams, lakes, and riparian vegetation
PFREE	fraction of percolation water entering free storages, regardless of tension water deficiency
SIDE	ratio of groundwater flow entering channel to that bypassing channel
SAVED	fraction of lower–zone free water unavailable for evapotranspiration

The mathematical representation of the above technique may be given as follows:

optimize $g(E) = g[Q-F(I,\theta)]$

where $Q^T = [q_1, ... , q_n]$ is the vector of single output variables (streamflows), I is the multi-variate vector of input variables (usually precipitation and potential evapotranspiration records, $F(-)$ is the deterministic model of the catchment response, θ is the vector of model parameters (the values of which must be estimated), E is the stochastic time series of the additive errors, and $g(-)$ is the selected estimation criterion.

It is interesting to note that in many of the reports of studies involving catchment models, aside from a brief mention of the calibration procedure employed, there is rarely any discussion of the merits of the calibration results. The main reason for this omission is probably the lack of agreement or understanding as to what exactly is to be achieved in the calibration stage (in view of the assumptions accompanying catchment models). It is important to ask whether the purpose of calibration is (1) to obtain a "unique and conceptually realistic" parameter set which closely represents our understanding of the physical system, or (2) to obtain a (any) parameter set which gives the best possible fit between the model simulated and the observed hydrograph (i.e., for the calibration period). From the point of view of physical modeling, it is clear that a method which emphasizes both aspects is desirable. Unfortunately, as Sorooshian and Dracup (1980) pointed out, most of the commonly employed calibration methods tend to emphasize the latter aspect. As they discussed, the danger in this case is that the resulting parameter set, though producing a close reproduction of the observed hydrograph during calibration, may result in poor performance when used for forecasting. This conjecture was strongly supported by the results reported by Sorooshian et al. (1983) and recently by Gan (1987) that stated "CRR models, which have been used extensively for simulating catchment streamflow, have limitations when operated beyond the range of calibration and validation experience; there is no guarantee that they will predict runoff accurately."

Certainly, much of the blame for poor forecasting ability can be placed on the inability to obtain unique and conceptually realistic parameter sets. In the context of the calibration methodology discussed above, some researchers have looked into the causes of our inability to find accurate parameter estimates. Ibbitt and O'Donnell (1971) and Johnston and Pilgrim (1976) listed the following features of catchment models and their calibration methodology as the primary reasons:

(1) Interdependence between model parameters.
(2) Indifference of the objective function to the values of "inactive" (threshold–type parameters).
(3) Discontinuities of the response surface.
(4) Presence of local optima due to the nonconvexity of the response surface.

2.3. Model Identification

Over the last decade, the author and his students have examined these problems further and have been able to explain some of the underlying reasons for what causes the four attributes identified by Ibbitt and O'Donnell and Johnston and Pilgrim. Our approach to studying these issues has been to examine CRR models in the framework of model identification. In this framework, the relationships between the calibration approach, the model structure, and calibration data are examined, the idea being that any degree of incompatibility between these three elements is the most common cause of an inability to find accurate parameter estimates. In brief, the search for the "best" parameter set takes place in an m–dimensional (m is the number of parameters) space. Therefore, whatever influences the shape of the parameter space (e.g., criterion, calibration data, model structure, etc.) plays a significant role in our ability (or lack of) to recover parameter values.

2.4. Response Surface Study

The inability to visually examine an m–dimensional space leaves us with the less–than–desirable option of two–dimensional response surface analysis. In this approach, only pairs of parameters are examined at a time to look for parameter interactions, local optima, discontinuities, as well as other peculiarities. Of course, the ideal situation is that the response surface be convex (in the case of minimization), having a single extremum, and be influenced little by parameter interactions, data, and selected estimation criterion. In the following sections, response surface studies will serve as a means of demonstrating the findings to date. Furthermore, in many of the cases presented below, a simple two–parameter model depicted in Figure 2 will be used. This simple discrete–time linear reservoir model is a common element in many catchment models where upper and lower zone storages are modeled. The two parameters are M and K, where M (unit: length L) represents the threshold parameter limiting the depth of the reservoir, and K (unit: time inverse T^{-1}) is the recession rate coefficient. Notice that surface runoff, R_t (unit L), is equal to zero as long as storage level, X_t (unit L), does not exceed the value of the threshold parameter M. On the other hand, depletion (S_t) of the reservoir from below occurs at a rate, K, proportional to its contents.

2.5. Choice of Estimation Criterion

In most reported cases, as well as the operational models used by most US agencies (e.g., NWS, HEC) where automatic techniques are used, the selection of the estimation criterion has been rather subjective. The common choice has been the Simple Least Squares (SLS) or the Root Mean Squares (RMS) which is [SLS/number of observations]$^{1/2}$. Response surfaces for our two–parameter model using three different estimation criteria are presented in Figures 3, 4, and 5. The criteria were RMS, the sum of the absolute value of the differences between simulated and observed (AMS), and the Root Mean Squares of the log–transformed flows (LOG). All conditions were kept identical for comparison purposes. For each criterion, the response surface was generated by perturbing the parameters away from their assumed true values. A hypothetical input sequence was used each time to generate the outputs.

As noticed, the choice of estimation criteria definitely impacts the shape of the response surface. The gradient of the criterion is greatest for RMS, intermediate for ABS, and least for the LOG function.

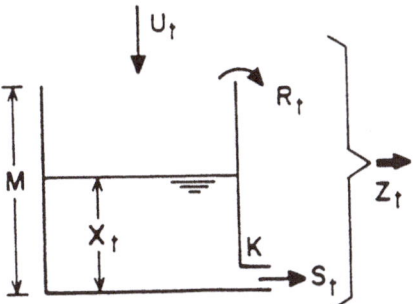

Figure 2. Two–parameter linear reservoir model.

448

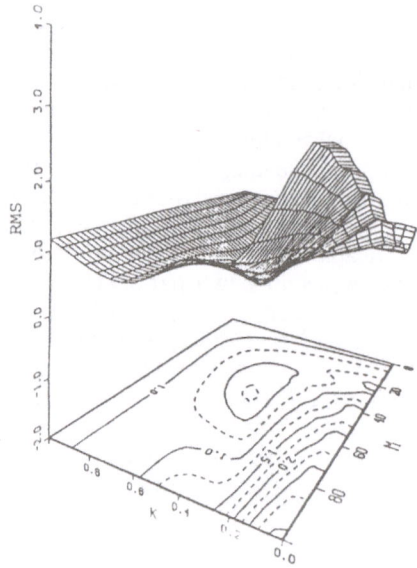

Figure 3. Response surface contours of the two–parameter
model for Root Mean Square (RMS) criterion.

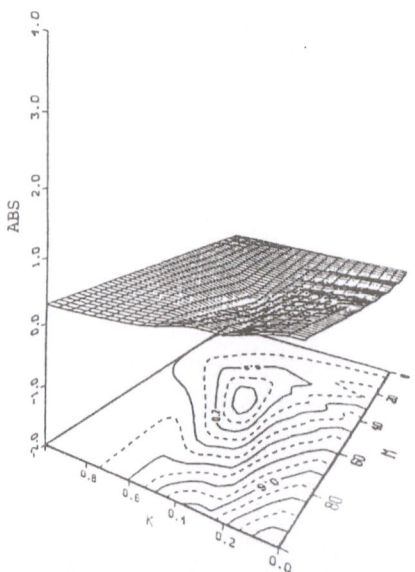

Figure 4. Response surface contours of the two–parameter
model for Absolute Value (ABS) criterion.

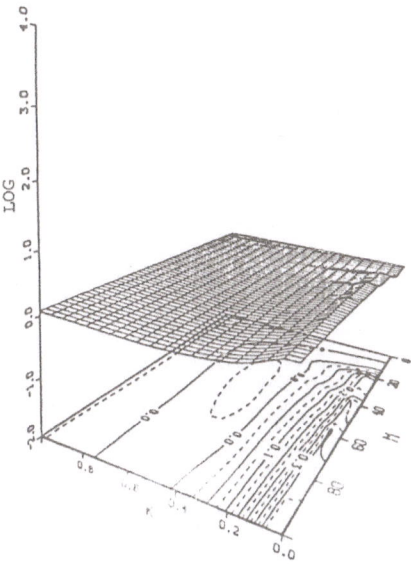

Figure 5. Response surface contours of the two–parameter
model for RMS of Log–transformed (LOG) flows.

To study the influence of measurement errors on the response surface, synthetically generated errors were added to both inflows and outflows. Figures 6 and 7 depict the RMS response surface with identical conditions as those of the previous study except for the following:

(1) Ten percent uniformly–distributed noise added only to the input data for the case presented in Figure 6.

(2) Five percent uniformly–distributed noise added only to the output data for the case presented in Figure 7.

Comparison of Figures 3, 4, 5, 6, and 7 shows that measurement errors also influence the shape of the response surface, with output errors producing a very rough surface with numerous local optima. In this case, the ability of an optimization algorithm (particularly Newton-type) to recover the global optimum is left to the judgement of the reader.

Sorooshian (1978) discussed the importance of measurement errors and pointed out that the selected estimation criterion must take into consideration the stochastic nature of the errors present in the measurement data. For this purpose, the author (Sorooshian, 1978; Sorooshian and Dracup, 1980) proposed the use of Maximum Likelihood (ML) theory as an appropriate framework for parameter estimation. In particular, estimation criteria for two kinds of error structures commonly believed to exist in hydrologic data—correlated (systemic) and heteroscedastic (nonconstant variance)—errors were proposed. The ML methods have proved to be successful in practice. Our work with the U.S. National Weather Service's model (SMA–NWSRFS), which was reported by Sorooshian et al. (1983), indicated that the ML approach is superior to the SLS method. In particular, it was observed that the effects of heteroscedasticity in the data were more severe than those of autocorrelation, at least for daily streamflow measurements. The ML estimator for the heteroscedastic error case (HMLE) provided parameter estimates that (a) were more conceptually realistic, and

450

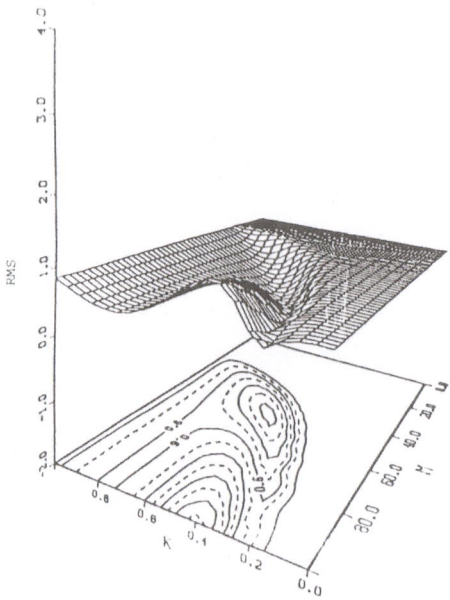

Figure 6. Response surface contours of the two–parameter model in presence of noise on the input and output data. Ten percent uniformly–distributed noise added only to input data.

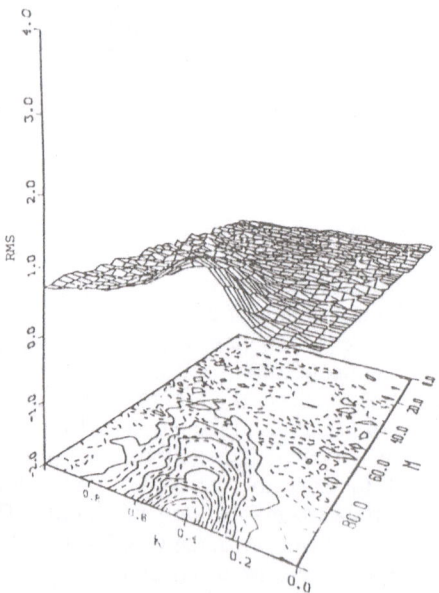

Figure 7. Response surface contours of the two–parameter model in presence of noise on the input and output data. Five percent uniformly–distributed noise added only to output data.

(b) provided consistently superior forecasts to those obtained using either the SLS or the ML autocorrelated error procedure (AMLE). The HMLE was also more efficient in terms of its ability to extract relevant information from the data set. HMLE parameter estimates seemed to be remarkably less affected by the nature and quality of data used for the calibration. Further, good parameter estimates were obtained using just one year of data; use of longer periods such as two and three years served only to marginally improve the estimates. Some of these conclusions have recently been confirmed by independent researchers: Lemmer and Rao (1983), Ibbitt and Hutchinson (1984), and Delleur et al. (1984). Willgoose (1987) reported that, for the cases he studied, the AMLE produced superior results, and he provided some explanation for this behavior.

Reasons for improved parameter precision with the HMLE criterion were investigated through response surface analysis. Figures 8 and 9 present the contour plots for two of the parameters of the SMA–NWSRFS model, namely ZPERC versus REXP0 (Sorooshian and Gupta, 1983). Figure 8 shows the resulting response surface for the Simple Least Squares (SLS) criterion, and Figure 9 displays the resulting response surface for the HMLE criterion. Attention is drawn towards two important observations:

(1) The location of the optimum is significantly different for the two criteria.

(2) The response surface is better shaped for the HMLE estimator (i.e., elliptically–shaped with no evidence of local optima and a smoother surface than the SLS contours).

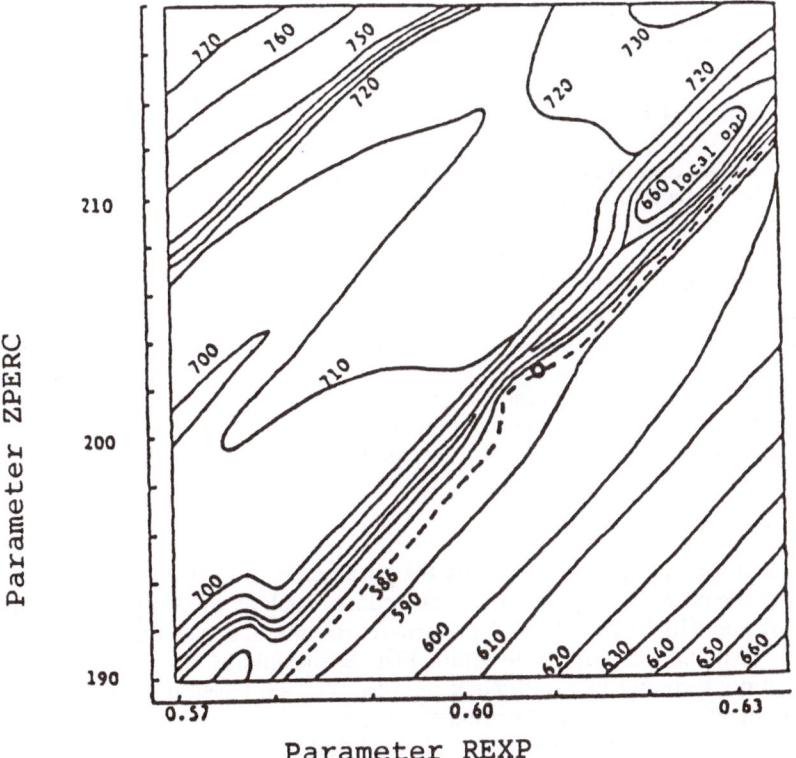

Figure 8. Response surface contour map in region of SLS calibration best parameter set (ZPERC versus REXP). Source: Sorooshian and Gupta, 1983.

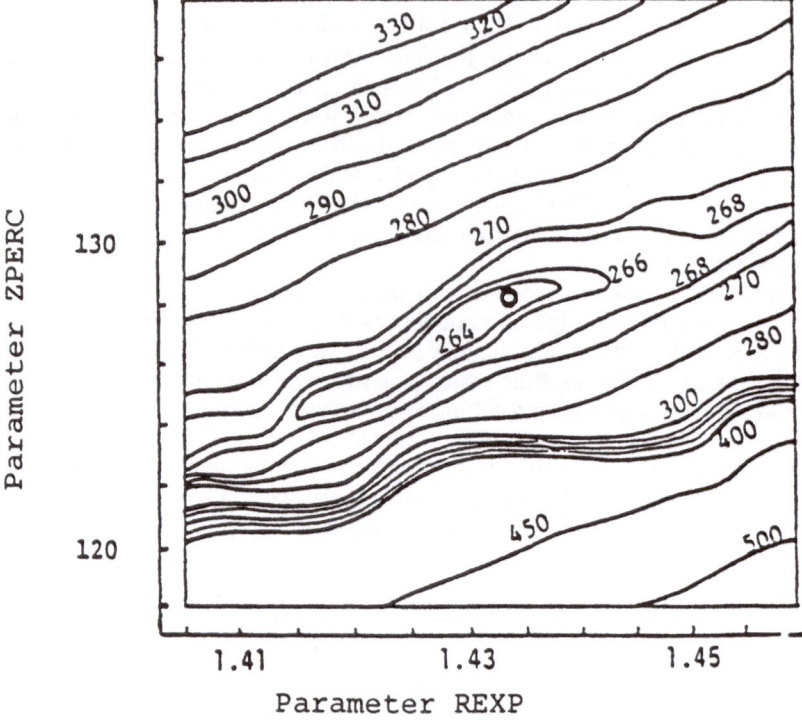

Figure 9. Response surface contour map in region of HMLE calibration best
parameter set (ZPERC versus REXP). Source: Sorooshian and Gupta, 1983.

These observations clearly support the fact that estimation criteria which properly account
for the stochastic properties of measurement errors produce better–shaped response surfaces, making it easier for a selected optimization method to search for the best parameter
values.

Recently, Duan et al. (1988) extended the ML approach and developed a criterion suitable
for model calibration using data which are recorded at unequal time intervals and which contain autocorrelated errors. These ML estimators are presented in the appendix section of this
chapter.

2.6. Optimization Algorithm

The algorithms most frequently used to identify the parameters of CRR models have been
those belonging to the class of "direct–search" procedures such as the simplex method of
Nelder and Mead and Hooke and Jeeves' pattern-search method (see e.g., Dawdy and
O'Donnell, 1965; Nash and Sutcliffe, 1970; Ibbitt, 1970; Sorooshian et al., 1983). As is well
known, such techniques are not highly efficient because they are based on trial-and-error
testing schemes to determine a feasible direction of movement. Derivative–based techniques
have seldom been used, the reason being a general belief that the values of the derivatives of
the model equation with respect to its parameters cannot explicitly be obtained due to the
presence of threshold–type parameters (e.g., Johnston and Pilgrim, 1976; Moore and Clarke,
1981). Some researchers have compared the performance of direct-search algorithms with

those wherein the gradients are approximated using finite–difference techniques (e.g., Ibbitt, 1970; Johnston and Pilgrim, 1976; among others). In most cases, the direct–search procedures were reported to be superior; Johnston and Pilgrim (1976) suggested that this was probably due to inaccuracies arising in the numerical gradient approximation procedures.

The problem of explicitly computing exact values for the derivatives of models containing threshold parameters has recently been addressed by Gupta (1984). He proposed a method based on a state–space analysis of the behavior modalities of such models which does not require replacing the threshold structures by smoothing functions. As an example of how this procedure works, let us consider the two–parameter model shown in Figure 2. Let:

x_{t-1} = state of model at beginning of time interval t,

u_t = precipitation input during time interval t,

and

$$x_t = x_{t-1} + u_t \tag{1}$$

x_t represents the intermediate model state, i.e., the state of the model increased by the amount of input into the reservoir and prior to the outputs being computed and depleted from the reservoir. When $x_t \leq M$, the model operates in mode 1, and the equations are:

$$x_t = (1-K)(x_{t-1} + u_t) \tag{2}$$

$$z_t = S_t = K(x_{t-1} + u_t) \tag{3}$$

When $x_t > M$, the model operates in mode 2, and the equations are:

$$x_t = (1-K)\, M \tag{4}$$

$$z_t = S_t + R_t = x_{t-1} + u_t - (1-K)\, M \tag{5}$$

The source of this modality of behavior is the "threshold parameter" M. The model is clearly linear in each mode ($z_t \propto u_t$), but the proportionality constant is equal to K in model 1, while it is equal to 1 in mode 2. Mode 2 is interpreted as the behavior of the system under conditions of "saturation." Using Gupta's method, the equation of the above model can be written as:

$$x_t = T_{1t}[(1-K)(x_{t-1} + u_t)] + T_{2t}[(1-K)M] \tag{6}$$

$$z_t = T_{1t}[K(x_{t-1} + u_t)] + T_{2t}[x_{t-1} + u_t - (1-K)M] \tag{7}$$

where the threshold functions, T_{1t} and T_{2t}, are defined as:

$$T_{1t} = \begin{bmatrix} 1 & \text{for } x_{t-1} \leq M - u_t \\ 0 & \text{for } x_{t-1} > M - u_t \end{bmatrix} \tag{8}$$

$$T_{2t} = \begin{bmatrix} 0 & \text{for} & x_{t-1} \le M - u_t \\ 1 & \text{for} & x_{t-1} > M - u_t \end{bmatrix} \tag{9}$$

Using the chain rule of calculus, it can be shown that the required differential equations are:

$$\frac{\partial x_t}{\partial K} = T_{1t}\left[-(x_{t-1} + u_t) + (1-K)\frac{\partial x_{t-1}}{\partial K}\right] + T_{2t}[-M] \tag{10}$$

$$\frac{\partial x_t}{\partial M} = T_{1t}\left[(1-K)\frac{\partial x_{t-1}}{\partial M}\right] + T_{2t}[(1-K)] \tag{11}$$

and

$$\frac{\partial z_t}{\partial K} = T_{1t}\left[(x_{t-1} + u_t) + K\frac{\partial x_{t-1}}{\partial K}\right] + T_{2t}\left[\frac{\partial x_{t-1}}{\partial k} + M\right] \tag{12}$$

$$\frac{\partial z_t}{\partial M} = T_{1t}\left[K\cdot\frac{\partial x_{t-1}}{\partial M}\right] + T_{2t}\left[\frac{\partial x_{t-1}}{\partial M} - (1-K)\right] \tag{13}$$

where the initial conditions to start the recursions in Equations (10) through (13) are:

$x_0 =$ some initial assumed/known constant value ($0 \le x_0 \le M$)

and hence:

$$\frac{\partial x_0}{\partial K} = 0$$
$$\frac{\partial x_0}{\partial M} = 0 \tag{14}$$

If necessary, the second derivatives can be computed in a similar fashion.

Gupta (1984) tested the usefulness of the method by comparing the performance of a Gauss/Newton–type (derivative–based) and the simplex (direct–search) optimization algorithms on a six–parameter CRR model. His results indicated that the derivative–based method is more efficient and uses less computer time, especially when the number of parameters to be optimized is large.

Recently, Hendrickson (Hendrickson et al., 1988) extended Gupta's approach and applied it to the SMA–NWSRFS model. The performance of a Newton–type optimization algorithm was compared with that of the pattern–search of Hooke and Jeeves. The results indicated that the direct–search algorithm is the more robust of the two because the Newton–type algorithm is more susceptible to poor conditioning of the response surface. It should be mentioned that the Newton algorithm used several times less computer run time, and when it worked, it worked very well. But when it didn't work, it tended to do very poorly.

The reasons for the behavior of the Newton algorithm were further explored through graphical studies. As an example, Figure 10 represents the cross–section and derivatives of an SLS criterion along the axis of parameter "upper zone tension water maximum" (UZTWM), with the arrow marking the location of the discontinuity shown at a finer scale in Figure 11. The following conclusions emerged as a result of the graphical studies:

(1) The first and second derivatives of the criterion (with respect to parameters) were more roughly textured and contained more numerous and more severe discontinuities than the criterion itself, a factor which contributes to the relatively poor robustness of the Newton–type algorithm. Derivatives were found to exist on many scales.

(2) The discontinuous estimation criterion and its derivatives were found to be piecewise continuous and everywhere finite. It is to be expected that the discontinuities will have some adverse impact on Newton–type algorithms. They will not, however, prevent their use.

(3) The major cause of discontinuities in a CRR response surface was found to be the existence of thresholds in the model, in conjunction with a discrete time step. For example, the worst discontinuities were associated with the tension reservoir which must fill before surface runoff can occur.

(4) The performance of Newton–type algorithms is additionally hampered by the method's reliance on estimation criterion characteristics at a single point (higher likelihood of a perturbation or bump at that single point) as compared to multiple points in a direct–search algorithm.

The most recent work on the application of optimization algorithms to CRR models is that of Brazil (1988). In his dissertation, he proposed a three–level optimization strategy of parameter estimation of SMA–NWSRFS. In the second level of his approach, two random–search techniques were used. The first procedure, a uniform random–search technique,

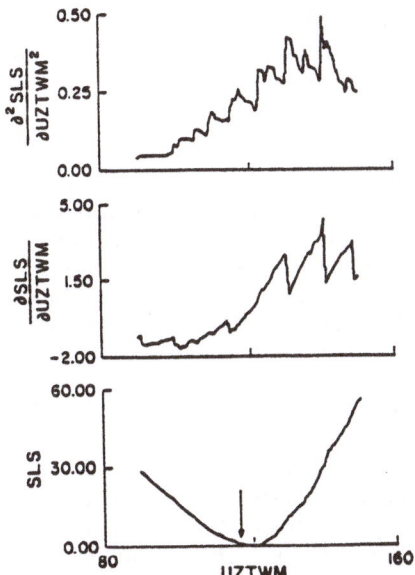

Figure 10. Cross section and derivatives of SLS function along axis of parameter UZTWM. Arrow marks the location of the discontinuity shown at a smaller scale in Figure 11.

456

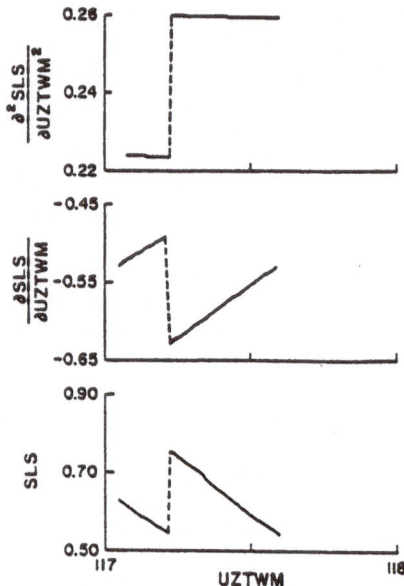

Figure 11. Example of a discontinuity in a SLS function and its derivatives. This figure is a close–up of a very small discontinuity shown at full scale in Figure 10.

randomly selects parameter values and generates an output data set of various statistics. The second random–search procedure converges as the simulation statistics improve. Brazil reported relative satisfaction with the random–search procedures.

2.7. Issues Related to Model–Structure Identifiability

As pointed out in the previous sections, proper choice of estimation criterion can result in improvements in the calibration phase. However, major problems associated with determination of unique and consistent parameter estimates still remain. Sorooshian and Gupta (1983, 1985) and Gupta and Sorooshian (1983) examined these problems in some detail, and the conclusion is that a major source of the problem is poor identifiability of the subprocess equations which make up the structure of CRR models. An example of this type of problem was discussed by Gupta and Sorooshian (1983) in relation to the percolation equation of SMA–NWSRFS.

Since our research work in 1983, which clearly demonstrated the seriousness of structural identifiability, we have been exploring ways to address this problem.

A major issue to be tackled here is how to detect structural nonidentifiability. Sorooshian and Gupta (1985) recently attempted to establish a mathematical basis for such a study. They define "a model M parameterized by θ to be globally observable if, and only if, different parameter values of M give rise to different model output vectors." In a similar manner, "a model structure is [defined to be] locally identifiable at θ^* if, and only if, there exists an open neighborhood of θ^* in which it is identifiable." On the basis of these definitions, a mathematical framework for studying local structural identifiability was proposed. In particular, a measure called the sensitivity ratio was developed and was shown to be effective in determining poorly–identifiable parameter combinations in multiparameter vector spaces. In brief, the region in the parameter space around θ for which the model output sequences are considered to be indistinguishable is approximated using a hyperellipsoid. The sensitivity ratio, η_i, for the ith parameter is then computed as:

$$\eta_i = \frac{PS_i(\theta)}{CPS_i(\theta)} \tag{15}$$

where $PS_i(\theta)$ is parameter sensitivity index for η_i, and $CPS_i(\theta)$ is conditional parameter sensitivity index for η_i.

A two–dimensional example is presented in Figures 12 and 13. As can be seen, $PS_i(\theta)$ represents the maximum that the parameter θ can vary (allowing other parameters to vary freely) while remaining within the hyperellipsoid. Similarly, if all other parameters are assumed fixed at their chosen values, $CPS_i(\theta)$ represents the amount that θ_i can vary. By taking the ratio of $PS_i(\theta)$ to $CPS_i(\theta)$, we get a non–dimensional measure η_i of the amount by which the other model parameters ($\theta_j \neq \theta_i$) compensate for the changes in model output caused by perturbations in the parameter θ_i. Note that, when $\eta i = 1$, as in Figure 13, there is no compensation for the effects of the parameter θ_i on the model output by the other parameters. As η_i gets larger, this indicates poorer and poorer identifiability of θ_i in relation to other model parameters. The mathematical details can be found in Sorooshian and Gupta (1985).

At the present time, the practical utility of global identifiability techniques is limited due to the high degree of nonconvexity and nonlinearity of the problem. The uncertainty about the location to which the optimization algorithm converges casts serious doubt on the results of global identifiability studies. Clearly, much more work needs to be done on these issues.

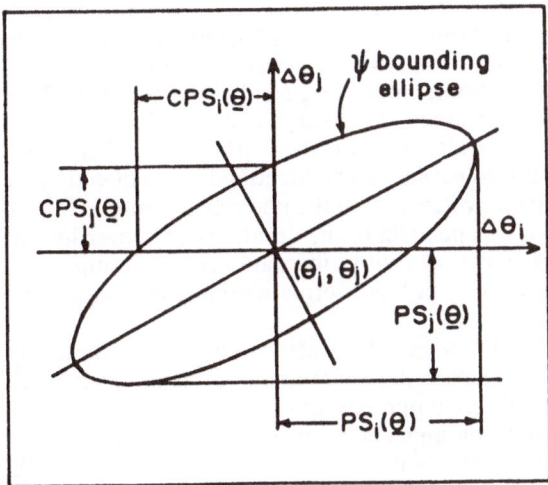

Figure 12. Two–parameter example of an indifference region, with some parameter interaction.

458

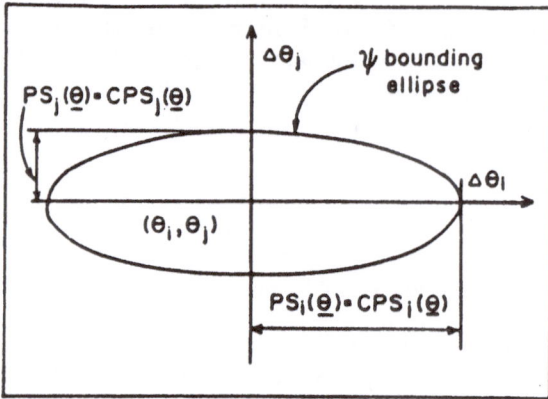

Figure 13. Two–parameter example of an indifference region, with no parameter interaction.

2.8. Choice of Calibration Data

The type and quality of the data also influence the shape of the response surface and, hence, the observability of the model parameters. It has often been suggested or implied in the literature that the data used should be "representative" of the various phenomena experienced by the catchment. Many researchers have attempted to satisfy this requirement by using as large a data set as possible without demonstrating superior results. Sorooshian et al. (1983) pointed out that, rather than length, it is the quality of information contained in the data which is important. They also stated that the data sequences which contain greater "hydrologic variability" are more likely to activate the various operational modes of the model sufficiently to result in reliable parameter estimates. The issue of how to measure "hydrologic variability" has recently been presented by Gupta and Sorooshian (1985a,b). A theoretical investigation into the relationship between the data, the model structure, and the precision of the parameter estimates was conducted. Some interesting results, particularly related to threshold parameter identifiability, were obtained. For example, it was shown (Gupta and Sorooshian, 1985a) that the precision to which the threshold parameter M of the two–parameter linear reservoir model mentioned earlier can be identified is: (1) directly proportional to the number of times the reservoir switches from the nonoverflow mode ($R_t = 0$) to the overflow mode ($R_t \neq 0$), and (2) is independent of the duration of the overflow mode. This has interesting implications in the choice of calibration data because it implies that a data set containing a few large storm events may be less informative than a data set containing many storms of moderate size.

We have numerically confirmed the theoretical results by using simulation studies. Figures 14, 15, and 16 depict the RMS response surface in the K–M parameter space for three different data sets of the same length but with different numbers of runoff–generating storms. The figures clearly show that the precision of the parameters improves directly as a function of the number of reservoir spills (overflow events). Notice that, with no spill (Figure 14), the response surface shows a horizontal extended valley in the direction of parameter M. This insensitivity results in the nonidentifiability of parameter M when no spills occur. The

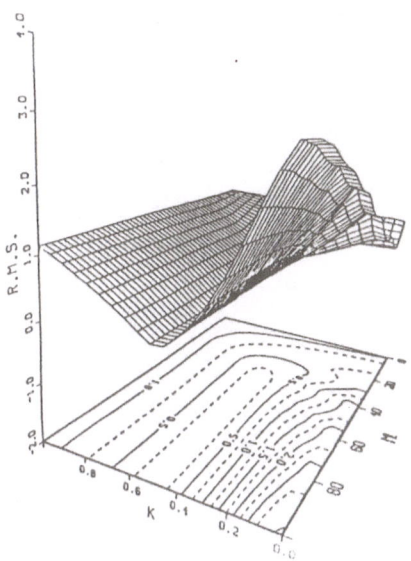

Figure 14. RMS response surface contour plots for the two–parameter model for no overflow.

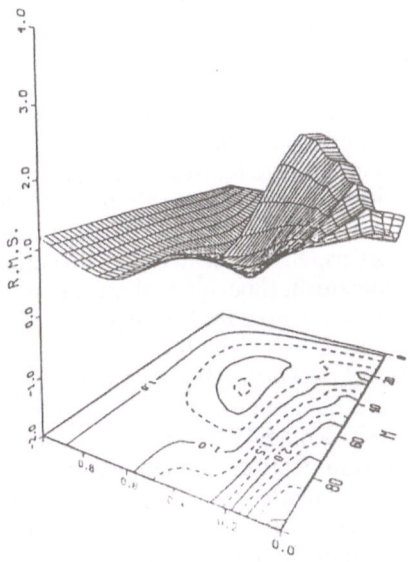

Figure 15. RMS response surface contour plots for the two–parameter model for one overflow.

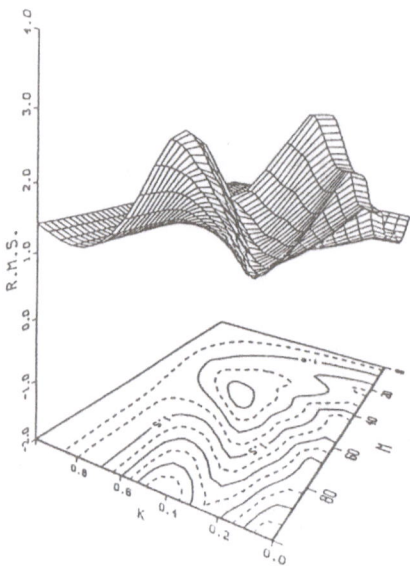

Figure 16. RMS response surface contour plots for
the two–parameter model for multiple overflow.

identifiability improves as the number of spills increases. This is evident by the degree of improvement and the concentricity and steepness of the response surface. With improved concentricity and steepness of the response surface in a multiple–spill case, our parameters are estimated with improved precision.

This simple example relates to a very basic two–mode switching structure, and the result in this case indicates that identifiability is mainly a function of one characteristic of the data, namely the switching frequency. Typical CRR models have more complex structures and several different modes of operation. Consideration of the relationship between different vertical storages and lateral flow components of a more complex operational model will require more detailed analysis. In these cases, it seems reasonable to expect that other characteristics of the data, such as storm duration, interstorm time–interval, variation in storm intensity, etc., will become relevant. This is an important area for future research which we are currently trying to address. The thrust of our work is to see whether it is possible to determine, for a given model, the region of the combined parameter–input data vector space in which the calibration procedure can be expected to be more successful. The feasibility of such an analysis depends on our ability to characterize a model in terms of all possible modes of operation and, within each mode, determine certain features of the input data that best trigger all (or some) of the parameters involved in that mode. Ideally, it is desirable to find similar identifiability "clues" for all model parameters as was found for the threshold parameter of the two–parameter model (discussed above). Indeed, once such a scheme is fully developed, one may find that certain parameters are nonobservable by any possible and imaginable type of data. Such cases may well result in restructuring the model to eliminate or replace such parameters.

2.9. Model Validation

Once a CRR model is calibrated, it must first be validated before applying it for its intended forecasting purpose. The main objective of the validation process is to ensure that the estimated parameter values satisfy the following two conditions: (1) the parameter values are conceptually realistic, and (2) confidence is high that, using the estimated parameter values, the model will be able to forecast. Reported literature on CRR models seldom addresses the validation issue, except for possibly presenting the criterion value as a measure of goodness-of-fit for the calibration phase.

At a minimum, one should expect the model's forecast to be significantly better than the "Zero–Order Forecast" (ZOF). For instance, in the case of daily forecast, the ZOF for tomorrow's flow is the value of the flow observed today. This is the "best" forecast with no model. On the other hand, a model should be able to do much better because it incorporates a lot of information about the physical processes and the dynamics of a catchment. In fact, the objective of the validation should not be to check a model's prediction accuracy under average or slowly changing conditions, but to test a model's credibility under extreme or rapidly-varying conditions. A model should be able to predict runoff, with a reasonable degree of accuracy, beyond the range of calibration and with input data exhibiting different properties than those used for calibration.

Unfortunately, there are no established and standardized guidelines to be followed by a would-be modeler to check the validity of his/her models, except for applying some of the measures proposed in statistical literature. In our work, we found the following approach to work reasonably well.

First, the estimated parameter values should be screened for unrealistic and unreasonable values. If the selected model is one that is in common use (e.g., SMA–NWSRFS), then the parameter values should be checked against the reasonable range suggested by NWS.

Second, the estimated parameter values should be used to generate a streamflow hydrograph for a selected forecast period, and the model output should be compared to the measured hydrograph. (Preference should be given to data sets exhibiting different properties than the sequence used for calibration.) Because visual comparison is a highly subjective method, use of a statistical measure(s) should be considered seriously. We have used SLS and/or RMS [or Daily RMS (DRMS)] of the difference between observed and simulated hydrograph ordinates. A lower SLS (or RMS, DRMS) would indicate a closer reproduction of the observed flows.

Another useful measure is the percent bias (PBIAS) statistic of the residuals by flow groups. This test is useful for detecting deficiencies in the reproduction of various aspects of the hydrograph. The PBIAS statistic is defined as follows:

For each flow group i, PBIAS$(i) = [\overline{Q}_{obs,t} - \overline{Q}_{sim,t}]/\overline{Q}_{obs,t}$, where $\overline{Q}_{sim,t}$ and $\overline{Q}_{obs,t}$ are the simulated and observed mean flows at time t, and t is such that $Q_{obs,t}$ falls in the flow group i (a flow group is defined as all the measured flows falling within a prescribed flow range). The breakdown of flow groups will depend on the range of data (lowest flow to the largest storm on that record). The result of PBIAS analysis would reveal inconsistencies in the forecast performance. An accurate and well calibrated model should result in values of PBIAS consistently close to zero in all flow groups. Additional statistical measures can be found in Aitken (1973).

2.10. The Issue of Model Choice

The problem facing many practitioners in the field of hydrology is the selection of an appropriate model for a particular watershed. "Appropriate" means that the model is capable of adequately reproducing the various aspects of the output hydrographs which are of interest to the hydrologist.

At the present time there are two general classes of models from which a selection might be made. One may select a CRR model which purports to be physically based, or one may select a simpler model which is either empirically based or is derived based on linear systems theory and/or time–series analysis. As was highlighted in this chapter, CRR models which are believed to be inherently more accurate in their representation of watershed behavior are plagued with parameter identifiability problems which limit their usefulness. On the other hand, the two simpler models (e.g., time series, linear state–space models, etc.) are usually easier to construct and calibrate, but have been strongly criticized as employing unrealistic assumptions about the nature of the physical system (e.g, ignoring the nonlinear dynamics of the watershed process). It has sometimes been argued that, from an engineering point of view, the usefulness of a watershed model need not depend on its conceptual realism so much as on its capability to reproduce input–output behavior. Some researchers have attempted to compare the performance of CRR models with simpler models from this point of view. More often than not, the published reports have supported the simpler models. However, Kitanidis and Bras (1980) found, for example, that under rapidly–changing hydrologic conditions, the conceptual SMA–NWSRFS model performed significantly better than an autoregressive moving average with exogenous inputs (ARMAX) linear stochastic model with on–line, adaptively estimated parameters and states. Although the ARMAX model was found to forecast satisfactorily in the recession limb, the conceptual model was found to be more reliable in forecasting the most important features of the hydrograph, such as the beginning of the rising limb, the time to and height of peak, and the total water volume. Similar results have been noted by Todini and Wallis (1978), Andjelic and Szollosi–Nagy (1980), and O'Connell and Clarke (1981).

The problem of whether to employ a simple or CRR model is really just a smaller facet of a larger problem, i.e., how to decide on a level of complexity of model structure appropriate to the modeling of a given watershed. The heart of this problem is the issue of system scale which is often ignored. In fact, CRR models (as well as, say, state–space and time–series models) are often applied without any concern about the temporal and spatial scale of the problem. For example, the SMA–NWSRFS model is used to model catchments as small as tens of square miles to those as large as tens of thousands of square miles in area and to provide forecasts with lead times ranging from minutes to days. In recent years, the importance of scale in modeling hydrologic processes has received a great deal of attention. Introducing a *Journal of Hydrology* issue devoted to hydrologic scale (August 1983), Rodriguez–Iturbe and Gupta stated, "The understanding of the collective type of behavior which takes place in the basin (at basin scale) is one of the most challenging and crucial problems in hydrology." Several papers on hydrologic scale also appeared in a special issue of *Water Resources Research* entitled, "Trends and Directions in Hydrology" (August 1986). In particular, Dooge (1986) stated that the alternative to complex physically based distributed models is to discover "hydrologic laws at the catchment scale that represent more than mere data fitting."

In 1984 a conference devoted to scale problems in hydrology (Gupta et al., 1986) was held to examine progress made and to put forth new ideas. From the results presented, it is clear that to understand hydrologic processes at the basin scale, we need more information on the interaction of various hydrologic subprocesses (infiltration, routing, etc.) at different scales. An initial framework for this was presented by Goodrich (1986), along with evidence for the existence of subprocess dominance and scale thresholds. This approach intimately links scale to the issues of model complexity and the spatial and temporal variability of hydrologic components.

Variability in the hydrologic output of a watershed system is influenced by two major factors: (1) variability in the inputs, and (2) properties associated with the physical structure of the watershed (e.g., expansion and compression of time scales, damping and attenuation, nonlinearities, input–dependent system modality, losses, etc.). It seems entirely possible that, in certain watersheds, most of the variability in the outputs can be fairly simply related to variability in the input. That is, the influence of input variability on output behavior predominates over that of system structure so much so that the latter influences are difficult to separate out. An extreme example is a small, impervious, steeply–sloped parking lot in which the input variability is only slightly dampened or attenuated by the parking lot. In fact, Goodrich et al. (1988) showed that the relative dominance of overland and channel flow processes in semi–arid watersheds is intimately tied to the scale of the rainfall event. Large storms overwhelm the terrestrial variability in watershed processes, making it possible to model the highly–complex system with a very simple representation. This, no doubt, helps explain why unit hydrograph theory (Sherman, 1932) works relatively well in flood estimation. At other scales, however, the spatial and temporal variability of climatic and basin factors will need to be considered.

3. Conclusions

A major objective of hydrologic modeling is to accurately forecast the effects of climatic variations on watershed behavior. While much research has been devoted to this issue, most of the myriad hydrologic models that have been developed suffer from one or both of the following two problems:

(1) The model assumptions usually require that the dynamics of the watershed behavior be represented adequately using a lumped approach and point–based flux data.

(2) Most models were developed for specific types of basins and climatic regimes and suffer great inaccuracies when applied to situations where the space–time scales are significantly different. In this regard, few if any guidelines exist to help the hydrologist objectively select a suitable model and the proper space–time scales over which data sampling and model parameterization are to take place.

There are several reasons why existing hydrologic models suffer from the aforementioned inadequacies. Probably the most pervasive is that historically it has been either too difficult or too expensive to obtain information regarding the distributed nature of the watershed parameters, states, and fluxes. This lack of adequate data has served to bias the modeling approach towards utilization of easily available information. Furthermore, it has retarded research aimed at uncovering the appropriate hydrologic principles and physical laws which operate at the different basin space–time scales.

The approaching era of remote sensing and sophisticated ground–truth measurement devices (such as Next Generation Radar [NEXRAD]) will no doubt revolutionize the science of watershed hydrology. For the first time it becomes possible to utilize a completely distributed approach to hydrologic modeling and, hence, make it practical to consider a different modeling approach other than CRR modeling. This will enable the development of a theoretical framework that identifies the factors which control and dominate the hydrologic cycle as a function of basins and climatic scale. As a result we should be able to predict watershed–specific space–time thresholds and dominant subprocesses, thereby greatly enhancing our conceptual understanding and modeling ability.

4. Acknowledgments

The material presented in this chapter is based on several research projects and the research work of many former and current students. For support, the author acknowledges the National Science Foundation for grant ECE86-10584 and NOAA for cooperative grant NA85AA-H-HY088. The assistance of Vijai K. Gupta, Jené Michaud, Qingyun Duan, and James Washburne is greatly appreciated.

464

5. References

Aitken, A. P.: 1973, 'Assessing Systematic Errors in Rainfall–Runoff Models,' *Journal of Hydrology* **20**(2), 121–136.

Andjelic, M., and A. Szollosi–Nagy: 1980, 'On the Use of Stochastic Structural Models for Real–time Forecasting of River Flow on the River Danube,' in *Hydrological Forecasting* (Proc. Oxford Symp. April 1980), IAHS Publication No. 129, pp. 371–380.

Brazil, L. E.: 1988, *'Multi–level Calibration Strategy for Complex Hydrologic Simulation Models,'* Ph.D. Dissertation, Department of Civil Engineering, Colorado State University, Fort Collins, CO.

Brazil, L. E., and M. D. Hudlow: 1981, 'Calibration Procedures Used With the National Weather Service River Forecast System,' in Y. Y. Haimes, and J. Kindler (eds.), *Water and Related Land Resources Systems*, pp. 457–466, Pergamon Press, New York.

Burnash, R. J. C., K. L. Ferral, and R. A. McGuire: 1973, *'A Generalized Streamflow System: Conceptual Modeling for Digital Computers,'* Report, Joint Federal/State River Forecast Center, U.S. National Weather Service, and California Department of Water Resources, Sacramento, CA.

Crawford, N. H., and R. K. Linsley: 1966, *'Digital Simulation in Hydrology: Stanford Watershed Model Mark IV,'* Dept. of Civil Engineering, Tech. Rep. 39, Stanford University, Stanford, CA.

Dawdy, D. R., and T. O'Donnell: 1965, 'Mathematical Models of Catchment Behavior,' *Journal of Hydraulics*, ASCE, **91**(HY4), 113–137.

Delleur, J. W., J. M. Bell, M. S. Blumberg, M. H. Houck, H. R. Lemmer, L. E. Ormsbee, H. R. Potter, A. R. Rao, and H. Schweer: 1984, *'Problem–oriented Evaluation of Institutional Decision Making and Improvement of Models Used in Regional Urban Runoff Management: Application to Indiana,'* Water Resources Research Center, Technical Report #164, Purdue University, West Lafayette, IN.

Dooge, J. C. I.: 1986, 'Looking for Hydrologic Laws,' *Water Resources Research* **22**(9), 46S–58S.

Duan, Q., S. Sorooshian, and R. P. Ibbitt: 1988, 'A Maximum Likelihood Criterion for Use With Data Collected at Unequal Time Intervals,' *Water Resources Research* **24**(7), 1163–1173.

Gan, T. Y.: 1987, *'Applications of Scientific Modeling of Hydrologic Responses From Hypothetical Small Catchments to Assess a Complex Conceptual Rainfall–Runoff Model,'* Ph.D. Dissertation, Department of Civil Engineering, University of Washington, Seattle, WA.

Goodrich, D. C.: 1986, *'Utilization of Time–Space Scale Dependencies in Hydrologic Modeling,'* AGU, Hydrology Section Horton Research Grant Application, February 25.

Goodrich, D. C., D. A. Woolhiser, and S. Sorooshian: 1988, 'Model Complexity Required to Maintain Hydrologic Response,' *Proceedings ASCE National Conference on Hydraulic Engineering*, Colorado Springs, CO, August 6–12.

Gupta, V. K.: 1984, *'The Identification of Conceptual Watershed Models,'* Ph.D. Dissertation, Department of Systems Engineering, Case Western Reserve University, Cleveland, OH.

Gupta, V. K., and S. Sorooshian: 1983, 'Uniqueness and Observability of Conceptual Rainfall–Runoff Model Parameters: The Percolation Process Examined,' *Water Resources Research* **19**(1), 269–276.

Gupta, V. K., and S. Sorooshian: 1985a, 'The Automatic Calibration of Conceptual Watershed Models Using Derivative–based Optimization Algorithms,' *Water Resources Research* **21**(4), 473–485.

Gupta, V. K., and S. Sorooshian: 1985b, 'The Relationship Between Data and the Precision of Parameter Estimates of Hydrologic Models,' *Journal of Hydrology* **81**, 57–77.

Gupta, V. K., I. Rodriguez–Iturbe, and E. F. Wood (eds.): 1986, *Scale Problems in Hydrology*, D. Reidel Publishing Company, Boston, Massachusetts.

Hendrickson, J. D., S. Sorooshian, and L. E. Brazil: 1988, 'Comparison of Newton–type and Direct–Search Algorithms for Calibration of Conceptual Rainfall–Runoff Models,' *Water Resources Research* **24**(5), 691–700.

Ibbitt, R. P.: 1970, *'Systematic Parameter Fitting for Conceptual Models of Catchment Hydrology,'* Ph.D. Dissertation, University of London.

Ibbitt, R. P., and P. D. Hutchinson: 1984, 'Model Parameter Consistency and Fitting Criteria,' in *Proceedings of International Federation of Automatic Control (IFAC) 9th World Congress*, Budapest, Hungary, July 2–6, 1984, Volume IV, pp. 153–157.

Ibbitt, R. P., and T. O'Donnell: 1974, *'Designing Conceptual Catchment Models for Automatic Fitting Methods,'* International Association of Hydrologists, Publication 101, pp. 461–475.

Ibbitt, R. P., and T. O'Donnell: 1971, 'Fitting Methods for Conceptual Catchment Models,' *J. Hydraulics Division*, ASCE, **97**(HY9), 1331–1342.

Johnston, P. R., and D. Pilgrim: 1976, 'Parameter Optimization for Watershed Models,' *Water Resources Research* **12**(3), 477–486.

Kitanidis, P. R., and R. L. Bras: 1980, 'Real–time Forecasting With a Conceptual Hydrologic Model, 2: Applications and Results,' *Water Resources Research* **16**(6), 1034–1044.

Lemmer, H. R., and A. R. Rao: 1983, '*Critical Duration Analysis and Parameter Estimation in ILLUDAS,*' Technical Report #153, Water Resources Research Center, Purdue University, West Lafayette, Indiana.

Moore, R. J., and R. T. Clarke: 1981, 'A Distribution Function Approach to Rainfall–Runoff Modeling,' *Water Resources Research* **17**(5), 1367–1381.

Nash, J. E., and J. V. Sutcliffe: 1970, 'River Flow Forecasting Through Conceptual Models, Part I—A Discussion of Principles,' *Journal of Hydrology* **10**(3), 282–290.

O'Connell, P. E., and R. T. Clarke: 1981, 'Adaptive Hydrological Forecasting: A Review,' *Hydrological Sciences Bulletin* **26**(2), 179–205.

Sherman, L. K.: 1932, 'Streamflow From Rainfall by the Unit Graph Method,' *Eng. News Record* **108**, 501–505.

Sorooshian, S.: 1978, '*Considerations of Stochastic Properties in Parameter Estimation of Hydrologic Rainfall–Runoff Models,*' Ph.D. Dissertation, University of California, Los Angeles.

Sorooshian, S.: 1983, 'Surface Water Hydrology: On–line Estimation,' *Rev. Geophys. Space Phys.* **21**(3), 706–721.

Sorooshian, S., and J. A. Dracup: 1980, 'Stochastic Parameter Estimation Procedures for Hydrologic Rainfall–Runoff Models: Correlated and Heteroscedastic Error Cases,' *Water Resources Research* **16**(2), 430–442.

Sorooshian, S., and V. K. Gupta: 1983, 'Automatic Calibration of Conceptual Rainfall–Runoff Models: The Question of Parameter Observability and Uniqueness,' *Water Resources Research* **19**(1), 260–268.

Sorooshian, S., and V. K. Gupta: 1985, 'The Analysis of Structural Identifiability: Theory and Application to Conceptual Rainfall–Runoff Models,' *Water Resources Research* **21**(4), 487–495.

Sorooshian, S., V. K. Gupta, and J. L. Fulton: 1983, 'Evaluation of Maximum Likelihood Parameter Estimation Techniques for Conceptual Rainfall–Runoff Models—Influence of Calibration Data Variability and Length on Model Credibility,' *Water Resources Research* **19**(1), 251–259.

Todini, E., and J. R. Wallis: 1978, 'A Real–time Rainfall–Runoff Model for an On–line Flood Warning System,' in C.-L. Chiu (ed.), *AGU Conference on Application of Kalman Filtering Theory to Hydrology, Hydraulics, and Water Resources*, Dept. of Civil Engineering, University of Pittsburgh, Pittsburgh, PA, pp. 355–368.

Willgoose, G. R.: 1987, '*Automatic Calibration Strategies for Conceptual Rainfall–Runoff Model,*' M.S. Thesis, Department of Civil Engineering, Massachusetts Institute of Technology, Cambridge, Massachusetts.

Appendix—Description of Maximum–Likelihood Estimators

1. The Maximum–Likelihood Estimator for the Autocorrelated–Error Case (AMLE)

We have:

$$\min_{\theta, \rho, \sigma_v^2} AMLE = \frac{n}{2} ln(2\pi) + \frac{1}{2} ln \frac{\sigma_v^{2n}}{1-\rho^2} - \frac{1}{2} \rho^2 \sigma^2_v \epsilon_1^2 + \frac{1}{2\sigma_v^2} \sum_{t=2}^{n} [\epsilon_t - \rho\epsilon_{t-1}]^2 \tag{16}$$

where:

$$\sigma_v^2 = \frac{1}{n} \left[-\rho^2 \epsilon_1^2 + \sum_{t=2}^{n} (\epsilon_t - \epsilon_{t-1})^2 \right] \tag{17}$$

and ρ is estimated from the implicit equation:

$$\epsilon_1{}^2 - \sum_{t=2}^{n} \epsilon_{t-1}^2 \rho^3 + \left(\sum_{t=2}^{n} \epsilon_t \, \epsilon_{t-1} \right) \rho^2 +$$

$$\left(\sigma_v^2 - \epsilon_1{}^2 + \sum_{t=2}^{n} \epsilon_{t-1}^2 \right) \rho - \sum_{t=2}^{n} \epsilon_t \, \epsilon_{t-1} = 0 \tag{18}$$

where $e_t = q_{t,obs} - q_{t,sim} =$ (residual at time t) with $q_{t,obs}$, $q_{t,sim}$ the measured and the simulated flows at time t, q is the set of parameters to be estimated, n is the number of data points, s_v^2 is a constant variance term, and r is the first–lag autocorrelation coefficient. The above estimator is developed based on the assumption that the output errors are Gaussian with a constant variance and correlated according to a first–lag autoregressive scheme (for details see Sorooshian and Dracup, 1980). Note that in the case $r = 0$, the estimates obtained using AMLE are equivalent to those of the simple least–squares (SLS) criterion:

$$\min_{\theta} \; SLS = \sum_{t=1}^{n} \epsilon_t^2 \tag{19}$$

2. The Maximum–Likelihood Estimator for the Heteroscedastic–Error Case (HMLE)

We have:

$$\min_{\theta, \lambda} \; HMLE = \left[\sum_{t=1}^{n} \omega_t \, \epsilon_t^2 \right] \left[n \left(\prod_{t=1}^{n} \omega_t \right)^{\frac{1}{n}} \right]^{-1} \tag{20}$$

where ω_t is the weight at time t, computed by:

$$\omega_t = f_t^{2(\lambda-1)} \tag{21}$$

where f_t is the expectation of $q_{t,true}$ (either $q_{t,obs}$ or $q_{t,sim}$) and λ is the unknown transformation parameter which stabilizes the variance. The implicit expression to estimate λ is:

$$\left(\sum_{t=1}^{n} ln(f_t) \right) \left(\sum_{t=1}^{n} \omega_t \, \epsilon_t^2 \right) - n \left(\sum_{t=1}^{n} \omega_t \, ln(f_t) \, \epsilon_t^2 \right) = 0 \tag{22}$$

Briefly, the HMLE estimator is derived based on the assumption that the errors are Gaussian with mean zero and covariance matrix V, where $V_{t,t} = \sigma_t^2$ and $V_{t,t+s} = 0$ for $s \neq 0$.

Stabilization of variance is attempted through the use of the Box–Cox (Box and Cox, 1964) power transformation which relates the variance of each error to its associated output value (see Sorooshian and Dracup, 1980). In this study, $f_t = q_{t,obs}$ was used in the computation of weights (the original procedure reported in the aforementioned papers used $f_t = q_{t,sim}$).

Fulton (1982) has shown that this results in a more stable estimation scheme. It is interesting to note that if the variances of the additive errors are homogeneous (independent of the time or magnitude of the associated flows), then the procedure will automatically select the value of $\lambda = 1.0$. This results in $\omega_t = 1$ for all t, and the estimation criterion reduces to the SLS. If, however, the variance of the errors is proportional to a power function of the magnitude of the flows, then the procedure will select a value of $\lambda = 1.0$. Pertinent to our problem is the case where the error variance increases as the flow values get larger [Sorooshian and Dracup (1980) have discussed the underlying reasons at great length]. In this situation, the ML estimate of λ will be less than unity, and this ensures that, in the estimation criterion, the errors associated with lower flows (which contain more reliable information) are weighted more heavily.

3. The Maximum–Likelihood Estimator for the Autocorrelated–Error Case for Unequal Time Interval Data (UMLE)

The maximum–likelihood estimator for the autocorrelated–error case for unequal time interval data is:

$$\min_{\theta, \alpha, \sigma_v^2} UMLE = \frac{n}{2}ln(2\pi) - \frac{1}{2}ln(\sigma_v{}^{2n}) - \frac{1}{2\sigma_v^2}\sum_{i=1}^{n}\left\{\epsilon_{t_i} - \exp[-\alpha(t_i - t_{i-1})]\cdot\epsilon_{t_{i-1}}\right\}^2 \qquad (23)$$

where:

$$\sigma_v^2 = \frac{1}{n}\sum_{i=1}^{n}\left\{\epsilon_{t_i} - \exp[-\alpha(t_i - t_{i=1})]\epsilon_{t_{i-1}}\right\}^2 \qquad (24)$$

and α is estimated from the following implicit equation:

$$\sum_{i=1}^{n}\left\{\epsilon_{t_i} - \exp[-\alpha(t_i - t_{i-1})]\cdot\epsilon_{t_{i-1}}\right\}\left\{\epsilon_{t_{i-1}}\cdot(t_i - t_{i-1})\cdot\exp[-\alpha(t_i - t_{i-1})]\right\} = 0 \qquad (25)$$

where e_{ti} is the residual at time t_i with $e_{t0} = 0$, and α is a constant of the time–variant auto-correlation coefficient, $r(t_i,t_{i-1}) = \exp[-\alpha(t_i - t_{i-1})]$. For fixed α, the degree of correlation decreases as the time interval, $t_i - t_{i-1}$, increases. The degree of correlation for a fixed time interval is determined by the magnitude of α. If a is small, widely–spread measurements may be significantly correlated, while if it is large, even closely–spaced measurements may be uncorrelated. If $\alpha = 1$, then the correlation among measurement errors is zero. In this case, the UMLE estimator is equivalent to the simple least–squares (SLS) criterion. It is interesting to note that, if $t_i - t_{i-1}$ is constant for all time intervals, the UMLE estimator is essentially the same as the AMLE estimator.

Part V – Current Technological Trends

Chapter 21

Recent Advances and Future Implications of Remote Sensing for Hydrologic Modeling

Edwin T. Engman
NASA Goddard Space Flight Center
Hydrologic Science Branch
Code 974
Greenbelt, Maryland 20771 U.S.A.

and

Robert J. Gurney
University of Reading
Reading, United Kingdom

Abstract. Remote sensing has begun to have an impact on applied hydrology as well as hydrologic research. Because remote sensing is basically spatial information it is providing a new form of data that hydrologists are not accustomed to working with. Remote sensing can also provide new types of information, such as the microwave measurements of soil moisture, that are not available via other means. This chapter reviews applications of remote sensing to hydrology and current research thrusts, and looks forward to new sensors and concepts that may be available in the near future.

1. Introduction

Remote sensing involves measurements of the electromagnetic spectrum that can be used to characterize the landscape, or infer properties of it, or in some cases, actually measure hydrologic state variables. Photography in the visible wavelengths was one of the first remote sensing techniques to be used. Over the years remote sensing techniques have expanded to the point that now they include most of the electromagnetic spectrum. Different sensors can provide unique information about properties of the surface or shallow layers of the Earth. For example, measurements of the reflected solar radiation give information on albedo, thermal sensors measure surface temperature, and microwave sensors measure the dielectric properties of surface soil or snow. Remote sensing and its continued development have added new techniques that hydrologist can use in a large number of applications.

It is perhaps easiest to examine the role of remote sensing in hydrology by examining the individual fluxes and storages; that is, precipitation, evaporation, runoff, and soil moisture storage. Each of these have different potentials for using remote sensing and each have been developed to differing degrees. This chapter will attempt to summarize the state of the art as well as recent advances in applications of remote sensing to hydrology. It will also discuss current modeling trends and needs as well as future programs of interest to hydrologists.

D. S. Bowles and P. E. O'Connell (eds.), Recent Advances in the Modeling of Hydrologic Systems, 471–495.
© 1991 *Kluwer Academic Publishers.*

2. Discussion

2.1. Precipitation

Recognizing the practical limitations of raingauges, hydrologists have increasingly turned to remote sensing as a possible means for quantifying the precipitation input to the globe. Because the fundamental approach to measuring rainfall and snow are different with respect to remote sensing, snow will be discussed separately.

2.1.1. General Approach

Direct measurement of rain from satellites for operational purposes has not generally been feasible because the opacity of clouds prevents direct observation of precipitation with visible, near infrared, and thermal infrared sensors. However, improved analysis of rainfall can be achieved using both satellite and conventional ground–based data. Satellite data are most useful in providing information on the spatial distribution of potential rain producing clouds, and gage data are most useful for accurate point measurements. Ground–based radar has also proven useful for locating regions of heavy rain and for estimating rainfall rates. Useful data can be derived from satellites used primarily for meteorological purposes, including polar orbiters such as NOAA–N and DMSP, and geostationary satellites such as GOES, GMS and Meteosat, but their visible and infrared images can only provide information about the cloud tops.

However, since these satellites do provide frequent observations, even at night with the thermal sensors, the characteristics of potentially precipitating clouds and the rates of changes in cloud area and shape can be observed. From these observations, estimates of rainfall can be made which relate cloud characteristics to instantaneous rain rates and/or rain totals over time.

The general approach used in making quantitative estimates of rainfall varies greatly depending on the type of remote sensing instrument used, type of precipitation, and region of the globe.

2.1.2. Visible and Infrared Techniques

The availability of meteorological and Landsat satellite data has produced a number of techniques for inferring precipitation from the visible and/or infrared (VIS/IR) imagery of clouds (Barrett and Martin, 1981; D'Souza and Barrett, 1988). These techniques have led to the development of three dominant approaches: a cloud indexing approach, the thresholding approach, and the life–history approach. Cloud indexing, which is time–independent, identifies different types of rain clouds and estimates the rainfall from the number and duration of clouds or their area. Thresholding techniques consider that all clouds with low upper surface temperatures are likely to be rain clouds. Life–history methods are time–dependent and consider the rates of change in individual convective clouds or in clusters of convective clouds (Scofield and Oliver, 1977). All such methods are essentially empirical in that they use statistical coefficients based on historical cloud and ground measured rainfall.

2.1.3. Microwave Radiometry

Microwave techniques offer a great potential for measuring precipitation because at some microwave frequencies clouds are essentially transparent, and the measured microwave radiation is directly related to the rain drops themselves.

Microwave radiometry or passive microwave techniques react to the rain in two fundamental ways: by emission/absorption, and by scattering. With the emission/absorption approach, rainfall is observed through the emission of thermal energy by the raindrops themselves. With the scattering approach the rain attenuates upwelling radiation from the Earth's surface and scatters or reflects cold, cosmic background radiation to the radiometer antenna. Considerable effort is now being spent on the development of passive microwave rainfall algorithms using dual or multifrequency principles, or polarizations at a single frequency. The results are promising (Spencer et al., 1988) although for operational applications in the near to mid–term future it appears that the most successful operational methods for general rainfall inventory will likely use passive microwave data in conjunction with visible and infrared (Barrett and Kidd, 1987).

2.1.4. Ground–Based Radar

Ground–based radar is conceptually similar to the spaceborne radar, except that the radar is stationary and its area of measurement is limited to a circle with a radius up to about 100 km (Schultz, 1989). A recording rainguage is desirable for calibration. Thus, the rainfall rate, R, can be estimated according to:

$$Z = aR^b \tag{1}$$

where Z is the measured radar reflectivity, and a and b are calibration parameters.

Ground–based radars have been used in several ways in operational hydrology. Osano et al. (1988) have used the radar with geostationary meteorological satellite data to estimate rainfall intensity. The Japan Meteorological Agency is combing digital radar data with rain gage data to improve the estimates of areal precipitation (Takemura et al., 1984). Klatt and Schultz (1983) have reported on the use of ground–based radar measurements being extended in time by a probabilistic model and used as input to a runoff model for flood forecasting. Creutin et al. (1988) have used a cokriging approach to obtain a linear estimator of ground-level rainfall from gage and radar data.

2.1.5. Future Considerations

It appears likely that considerable progress will continue to be made on improving our ability to measure rainfall and rain rates using remote sensing instruments. It is also expected that our capabilities for doing this from space will develop rather rapidly, although no single approach or wavelength will be the answer. It is increasingly obvious that progress will be made using multifrequencies and multipolarizations, as well as a host of non–spaceborne instrumentation such as atmospheric soundings, ground data, etc. It also seems obvious that owing to the inexact nature of the science, the use of interactive computers and expert systems will be incorporated to assist analysis in the more subjective aspects of estimating rainfall.

Additionally, it is likely that hierarchical operational approaches will be developed and perfected. Such approaches may use an objective VIS/IR method for estimating large-area, long–term (5 days, 25 or more) rain totals. Within this large framework, interactive techniques such as BIAS (Barrett et al., 1986) with or without passive microwave data would be used to analyze significant events and frontal storms. Embedded further in the system might be an interactive severe storm analysis procedure for predicting high–intensity rain rates and making flash flood forecasts. Such a hierarchical system could be developed for relatively inexpensive personal computers (PCs) so that the entire system could be located in the field where the information is needed.

There are also a number of precipitation–related satellite programs on the drawing board. One of these is Tropical Rainfall Measuring Mission (TRMM) (Simpson et al., 1988). TRMM is a proposed, experimental rainfall–measuring satellite designed for a minimum three year mission to measure the distribution and variability of rainfall and latent heat release over tropical and subtropical regions of the globe. The purpose of TRMM will be to improve short–term climatological models, general circulation models, and the global hydrologic balance—particularly as it is affected by tropical oceanic rainfall. TRMM is expected to resolve problems in the VIS/IR/PMW data interpretation areas and should, therefore, make possible improvements in the existing types of satellite techniques discussed above.

Another future program that will undoubtedly have great potential for estimating precipitation will be the Earth Observing System (EOS) being proposed by NASA (NASA, 1984). EOS will be a multiplatform system of sensors dedicated both to monitoring environmental conditions and to measuring earth resources. Several instruments that have been selected will be valuable for estimating rainfall volumes and rates.

2.2. Snow Hydrology

Snow is a form of precipitation; however, in hydrology it is treated somewhat differently because of the lag between when it falls and when it produces runoff, groundwater recharge, or is involved in other hydrologic processes. Remote sensing offers a new and valuable tool for obtaining snow data for predicting snowmelt runoff.

Just about all regions of the electromagnetic spectrum can provide some useful information about the snowpack and its condition. Ideally, one would like to know the areal extent of the snow, its water equivalent, and the "condition" or grain size, density, and presence of liquid water. Although no one region of the spectrum can provide all these properties, certain regions of the spectrum can be used to measure individual properties.

2.2.1. Gamma Radiation

The water content of some snowpacks can be measured with low elevation aircraft carrying sensitive gamma radiation detectors. This method takes advantage of the natural emission of low level gamma radiation from the soil. The aircraft passes over the same flight line before and during snow cover to measure the attenuation resulting from the snow layer which is empirically related to an average snow water equivalent for that site (Carroll and Vadnais, 1980). This approach is limited to low aircraft altitudes (approximately 150 m) because the atmosphere attenuates a significant portion of the radiant energy. In the NOAA operational airborne gamma radiation snow water mapping program, procedures for correcting for the soil moisture are included in the system (Carroll and Carroll, 1989). Currently, this operational program flies over 1400 flight lines in the United States and Canada (Carroll and Carroll, 1989).

2.2.2. Visible/Near Infrared

Albedo of the snow surface is the property most easily measured by remote sensing. Typically, new snow will have an albedo of 90 percent or greater, whereas older snow that has been weathered and has accumulated dust and litter can have an albedo as low as 40 percent (Foster et al., 1987). The reflectivity of new snow decreases as it ages in both the visible and infrared regions of the spectrum; however, the decrease is more pronounced in the infrared region.

Snow can be readily identified and mapped with the visible bands of satellite imagery because of its high reflectance in comparison to non–snow areas. Generally this means selecting the NOAA VHRR visible channel, Landsat MSS channels 4 or 5, SPOT, or Landsat TM channels 2 and 4. Although snow can be detected at longer wavelengths, i.e., in the near infrared, the contrast between snow and non–snow area is considerably lower than with the visible region of the spectrum. However, the contrast between clouds and snow is greater in Landsat TM Band 5 (1.57 – 1.78 μm), and this serves as a useful discriminator between clouds and snow (Dozier, 1984).

Use of satellite data for snow mapping has become operational in several regions of the world. Currently, NOAA develops daily snow–cover maps for 56 river basins in the US for use in streamflow forecasting (Carroll and Allen, 1988). NOAA also produces maps of mean monthly snow cover for the northern hemisphere (Wiesnet and Matson, 1975; Matson and Wiesnet, 1981) using the lower resolution (4–8 k_m) visible data. A handbook to assist potential users of satellite data in the mapping of snow covered areas has been published by Bowley et al. (1981).

2.2.3. Thermal Infrared

Thermal data are perhaps the least useful of the common remote sensing products for measuring snow and its properties; however, thermal data can be useful for helping identify snow/non–snow boundaries and discriminating between clouds and snow with AVHRR data.

2.2.4. Microwave

Microwave remote sensing offers great promise for future applications to snow hydrology. This is because microwave data can provide information on the snowpack properties of most interest to hydrologists; i.e., snow cover area, snow water equivalent (or depth), and the presence of liquid water in the snowpack which signals the onset of melt (Kunzi et al., 1982). Unfortunately, operational use of microwave technology is still some time in the future.

2.2.5. Snowmelt Runoff

Predicting snowmelt runoff for either water supply or flood prediction is a major task for hydrologists in many parts of the world. In some areas such as the western United States, snowmelt runoff provides nearly all the water for industry, agriculture, and domestic use. Accurate and timely prediction of snowmelt runoff is necessary for efficient reservoir management and planning the distribution of the water. Also, certain parts of the world (e.g. Bangledesh) are habitually plagued by flooding from rapidly melting snow.

Snowmelt runoff procedures using remote sensing have followed two distinct paths: empirical approaches, and modeling based. The choice of approach depends somewhat on the available data and to a great extent on the detail in output desired.

The most direct approach is to relate the satellite snow cover area on a given date to the seasonal runoff. Using data from several snow seasons enables one to develop an empirical curve (see Figure 1) such as that developed by Rango et al. (1975) from Landsat data for some basins in the Wind River Mountains in Wyoming (USA).

Snowcover depletion curves are another empirical approach that uses estimates of snow covered area obtained from satellite imagery. The snowcover depletion curves show a melt sequence that is repeatable from one year to the next (Hall and Martinec, 1985). The displacement of the curves from one year to the next is related to the snowpack water content. Although these curves are developed for one basin, they may possibly be transposed to other nearby basins as an estimate of the melt process in a nearby basin.

2.2.6. Modeling Snowmelt

A number of models for predicting and simulating snowmelt runoff have been developed or have been modified from existing models to incorporate satellite–derived snow cover data.

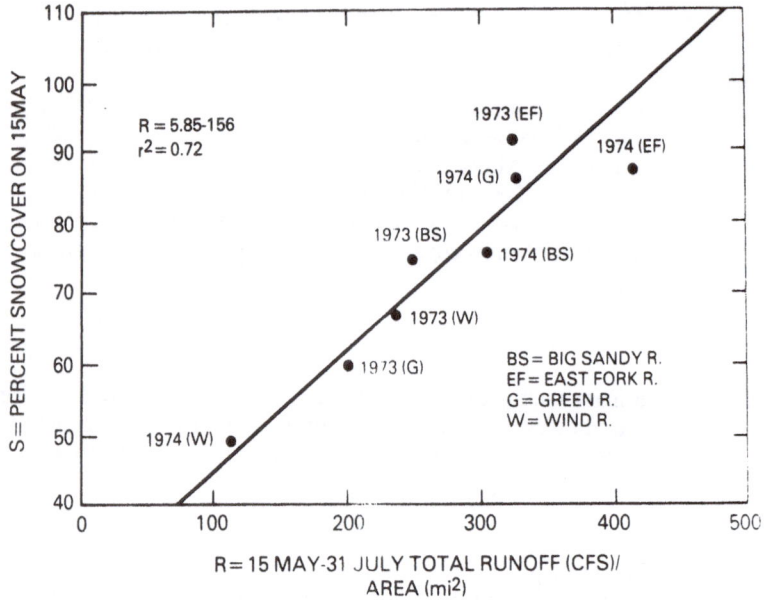

Figure 1. An example of the relationship between Landsat derived snowcover and measured annual runoff for four watersheds in the Wind River Mountains, Wyoming (after Rango et al., 1975).

The California Department of Water Resources has taken an existing hydrologic model for the Kings River Basin, removed the original snowmelt component, and replaced it with a procedure based on the snowcover area (Hannaford and Hall, 1980). In the Pacific Northwestern United States, satellite snow cover data are being used operationally in the Streamflow Synthesis and Reservoir Regulation (SSARR) model (Dillard and Orwig, 1979).

The snowmelt runoff model (SRM) (Martinec, 1975) for simulating snowmelt is one of the better known and most thoroughly tested models available. SRM has been developed to simulate and forecast daily streamflow in mountain basins where snowmelt is a major component of the annual water balance. SRM is a degree–day model that uses the percentage of the basin or elevation zone covered by snow as the primary input. Using Landsat data, SRM has been successfully run for various sized basins in Europe and the US (Rango and Martinec, 1979; Rango, 1980; Rango, 1983; Martinec and Rango, 1986).

Examples of simulations are shown for two basins in Figures 2 and 3. It can be seen from these that the model has done quite a good job of simulating the measured flows.

2.2.7. Future Considerations

There are several areas in which remote sensing will have an impact on snow hydrology in the future. These advances can be looked at from two perspectives: one will be from the development of new sensors, especially microwave sensors; and the second will be the modification of models and the utilization of new types of data and databases.

New sensors and methods of analysis will have a major impact on remote sensing of snow and snowmelt hydrology. The combination of visible, near–infrared, and microwave measurements at various spatial scales as planned for the EOS mission will provide much of

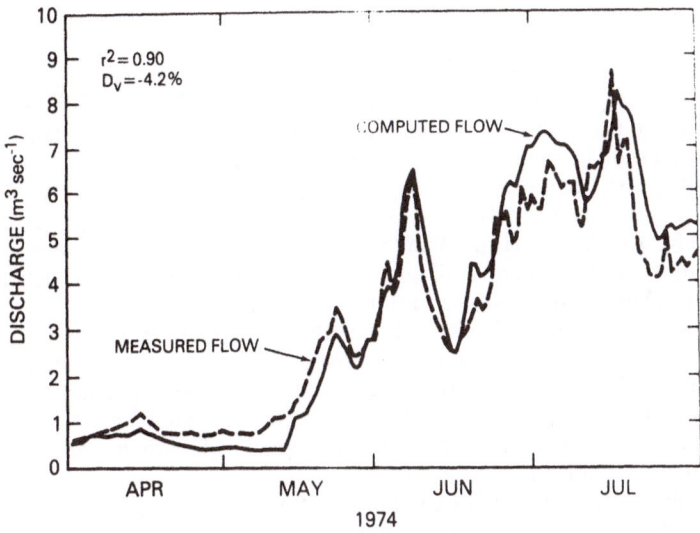

Figure 2. An illustration of discharge simulation for the Dischma Basin (43.3 km²) Switzerland using SRM (after Rango, 1985).

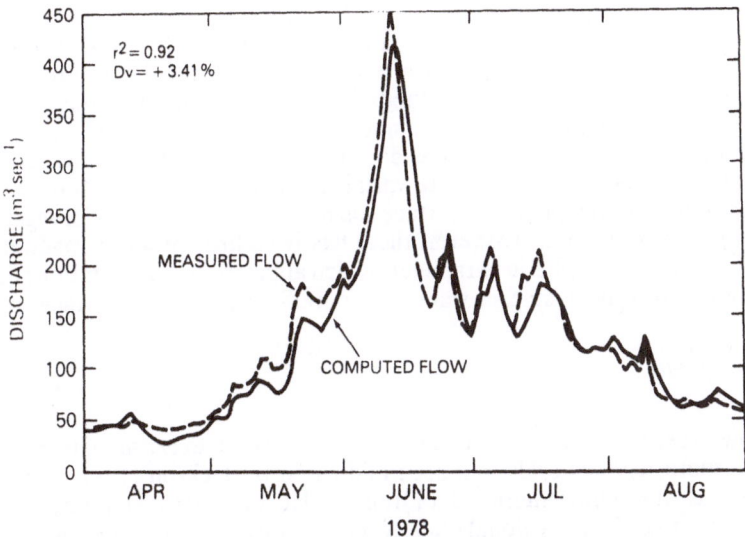

Figure 3. An illustration of discharge simulation for the Durance River Basin (2170 km²), France, using SRM (after Rango, 1985).

the necessary data for sophisticated snow measurements and analysis of the snowmelt process.

New instruments, together with innovative methods of analysis, should provide new insight to the solution of snow hydrology problems. For example, Danes and Danes (1988) demonstrated how the difference in polarization between the SMMR 37 GHz and 19 GHz passive microwave measurements could be related to snow–water equivalent. Their research showed good results for areas of sparse vegetation in the upper Colorado River Basin. New models that are developed to use remote sensing data will also improve snowmelt runoff predictions. SRM is being modified for forecasting (Rango and van Katwijk, 1990). To do so, a series of snow depletion curves must be developed (Hall and Martinec, 1985). These snow depletion curves usually can be developed from historic Landsat, Spot, or AVHRR data.

Another model that is perhaps more physically based than SRM has been recently modified to use remote sensing data of the snowpack. The Precipitation Runoff Modeling System (PRMS) (Leavesley et al., 1987) is a modular, distributed parameter watershed model which treats the snowpack as a two–layer system. Heat and mass (rain or snow) are transferred across a 3–5 cm surface layer to the main body of the snowpack, which is maintained and modified both as a water reservoir and a heat reservoir.

The merging of remote sensing data with digital elevation modeling (DEM) and geographical information systems (GIS) enables different types of data to be combined objectively and systematically. DEM's are used to normalize imagery by using the elevation of the sun and the slope, aspect and elevation of the terrain (Baumgartner, 1988; Miller et al., 1982). GIS are helpful for combining vegetation masks with satellite imagery (Keller, 1987).

Snow hydrology was one of the first areas in water resources to make effective use of remote sensing data, and it appears to be able to take advantage of future developments as they become available. The parallel developments of understanding sensor response and modeling are major areas of research in the present state of the field.

2.3. Evapotranspiration

Remote sensing techniques cannot measure evaporation or evapotranspiration (ET) directly. However, remote sensing does have two potentially very important roles in estimating ET. First, remotely sensed measurements offer methods for extending empirical relationships, such as the Thornthwaite (1948), Penman (1948), and Jensen and Haise (1963) methods, to much larger areas including those areas where data may be sparse. Secondly, remotely sensed measurements may be used to measure variables in energy and moisture balance models. These include surface temperature, surface soil moisture, surface albedo, vegetative cover, and incoming solar radiation. However, there has been little progress made in the direct remote sensing of the atmospheric parameters which affect evapotranspiration such as near–surface air temperature, near–surface water vapor gradients, and near–surface winds.

2.3.1. Present Applications

The question of how to use the spatial nature of remote sensing data to extrapolate point ET measurements to a more regional scale has been addressed by Jackson (1985) and Gash (1987). Price (1982) has shown how thermal data from the Heat Capacity Mapping Mission (HCMM) could be used to estimate regional scale evapotranspiration rates which were comparable to pan evaporation data. Figure 4 illustrates the sensitivity of thermal data for delineating surface conditions, especially the irrigated agricultural areas that show up as cool in 4b.

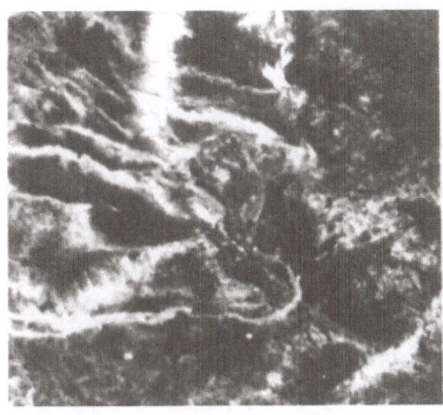 

2 30 LT (Early Morning)   13:30 LT (Early Afternoon)

7 - 16°C   33 - 55°C

Day-Night Temperature Contrast
Washington State
July 22, 1978

Figure 4. Early morning (a) and early afternoon (b) thermal imagery of eastern Washington State, USA. Dark areas are cold and light areas are warm. The irrigated areas are identifiable in the afternoon scene (b) by their darker tone (reproduced by permission of John C. Price).

The incoming solar radiation can be estimated from satellite observations of cloud cover, primarily from geosynchronous orbits (Brakke and Kanemasu, 1981; Tarpley, 1979; Gautier et al., 1980). For clear sky conditions, the surface albedo may be estimated by measurements covering the entire visible and near infrared wave band, while measurements at narrow spectral bands can be used to determine vegetative cover (Jackson, 1985; Brest and Goward, 1987).

The surface temperature can be estimated from measurements in thermal infrared wavelengths of the emitted radiant flux, i.e. the 10.5 and 12.5 micron waveband, and from some estimate of the surface emissivity (for natural surfaces, the emissivity is usually close to unity). Using the temperature sounders on the meteorological satellites in a linear regression model, Davis and Tarpley (1983) have estimated shelter temperatures with an error of about 2°K for clear or partly cloudy conditions. Seguin et al. (1989) estimated ET for the Sahelian region using the relationship proposed by Jackson et al. (1977) and studied by Seguin and Itier (1983) where:

$$ET = Rn + A - B(T_s - T_a) \tag{2}$$

Here ET is a daily value of evapotranspiration, Rn is the net radiation, T_s is the surface temperature measured by satellite, and T_a is the air temperature measured by a ground network.

One formulation of potential evapotranspiration that lends itself to remote sensing inputs is that developed by Priestley and Taylor (1972):

$$LE = \alpha[\Delta/(\Delta + \gamma)](R_n - G) \tag{3}$$

where Δ is the slope of vapor pressure versus temperature curve, and α is an empirical evaporation constant which they determined to be 1.26. Barton (1978) and Davies and Allen (1973) have modified the result for an unsaturated surface by treating α as a function of the surface layer soil moisture. Barton used airborne microwave radiometers to sense soil moisture remotely in his study of evaporation from bare soils and grasslands. Equation (3) is essentially the basis for the model which Kanemasu et al. (1977) have used for estimating ET with satellite data. Kotoda et al. (1983) applied Equation (3) to estimate evapotranspiration using surface temperatures estimated from an airborne radiometer.

Estimates of the net radiation from geostationary satellite data are used by Heilman et al. (1977) in a Priestley–Taylor type of equation to estimate evapotranspiration. The resulting moisture flux is then used to drive a water balance model for predicting crop yields. They also used Landsat multispectral data to obtain vegetation indices.

Reginato et al. (1985) combined reflected solar radiation and surface temperatures gathered from remote sensing with ground–based meteorological data (incoming solar radiation, air temperature, wind speed, and vapor pressure) to estimate the net radiation and the sensible heat flux. The soil heat flux was estimated to be a fraction of the net radiation and the height of the plant canopy through air empirical equation.

2.3.2. Future Considerations

The major problems with all remote sensing methods of evapotranspiration are: (1) the process of transpiration is still not well understood and parameterized either for structured crops such as cereals or for complex vegetation such as trees; (2) in the presence of vegetation, the surface temperature estimated by a thermal infrared sensor is at an unknown level within the vegetation; and (3) the most appropriate use of microwave observations of surface soil moisture in the presence of vegetation needs to be determined. For the future, we expect that the most practical method will probably use a multispectral approach including repetitive observations at the visible, near and thermal infrared, and microwave wavelengths. This will afford the possibility of estimating solar insolation, surface vegetative cover and/or albedo, surface temperature, and surface soil moisture from remotely sensed data and incorporating them into models of the type described here.

Fortunately some of the space initiatives that are currently being planned should help provide the amount and type of remote sensing data needed for further advances in estimating the global ET fluxes. Modelers have also been developing land–phase models at a scale compatible with the needs of general circulation models (Sellers et al., 1986) and capable of being coupled with them. Of special interest is the EOS being planned by NASA, ESA, and Japan with major contributions being made from a number of other countries. With respect to the energy budget, the surface temperatures will be measured by several planned instruments. The multichannel capability of these instruments will also be able to make independent measurements of surface emissivity. In addition, the distribution of water vapor will be measured in varying degrees by atmospheric sounders.

Components for determining the sensible heat flux can also be measured by the EOS instruments. As mentioned above, the temperature of the atmospheric boundary layer can be measured. Although convection and turbulence within the boundary layer cannot be measured directly, this information can be inferred from boundary layer height measurements with Lidar sounder. Also, estimates of wind speed at the top of the boundary layer can be made from cloud motions observed from geosynchronous satellites. One other component to the sensible heat flux, the surface roughness, may be able to be estimated with synthetic aperture radar of the land surface and with scatterometers and altimeters over the sea.

The latent heat flux cannot be measured directly, but EOS instrumentation will provide some sampling capability. Looking at the land–phase of ET, it can be seen that the capabilities for measuring solar radiation from albedo and surface temperatures will provide valuable information for such models as the Priestly-Taylor relationships. Plant conditions themselves may be able to be monitored through thermal data, changes in reflectance caused by stress and changes in geometry and dielectric constant (moisture content) may be measured with radar. Various spaceborne radars will also be operating in the same time period and will give additional useful information on soils and vegetation.

In summary, future programs, such as EOS and the multispectral capabilities of the instrumentation should provide unusual and exciting new data for evaluating evaporation and transpiration on local, regional and global scales.

2.4. Soil Moisture

Remote sensing of soil moisture can be accomplished to some degree or other by all regions of the electromagnetic spectrum. Successful measurement of soil moisture by remote sensing techniques depends upon the type of reflected or emitted radiation. Table 1 summarizes the advantages and disadvantages of each approach. A comprehensive summary of remote sensing approaches for measuring soil moisture has been represented by Schmugge et al. (1980). However, only the microwave region offers the potential for truly quantitative measurement from a spaceborne instrument.

2.4.1. Microwave Techniques

Microwave techniques for measuring soil moisture include both the passive and active microwave approaches with each having distinct advantages. The theoretical basis for measuring soil moisture by microwave techniques is based on the large contrast between the dielectric properties of liquid water and dry soil. Thus, as the soil moisture increases, the dielectric constant can increase to a value of 20 or greater (Schmugge, 1983). Figure 5 illustrates the change in dielectric constant for three representative models. For passive microwave remote sensing, this change in dielectric constant would result in a decrease of emissivity from about 0.95 to 0.5 or lower, and for the active case, the measured radar backscatter would increase by about 10 db or more.

2.4.2. Microwave Applications

The fact that microwave instrumentation can be used to measure soil moisture and has been demonstrated with numerous experiments should be of great interest to hydrologic modelers. Unfortunately, to date there are not many data sets for researchers to experiment with. However, there has been one experiment using the spatial and temporal soil moisture data developed from remote sensing (Engman et al., 1989).

In this recent aircraft experiment, the four–beam pushbroom microwave radiometer (L–Band) was used to map the spatial distribution of soil moisture (Wang et al., 1989). By using overlapping flight lines for several flights during a drying period, the spatial patterns of soil moisture within a small 37.7 ha watershed were mapped, and their changes during drying were also mapped. In addition to the microwave maps of soil moisture, the small watershed

Table 1. Summary of remote sensing techniques for measuring soil moisture (modified after Engman, 1982).

Wavelength Region	Property Observed	Advantages	Disadvantages
Gamma Radiation	Attenuation of naturally emitted	Existing airborne program	Limited spatial. Limited to low elevation flights. Averages over a line, calibration necessary.
Reflected Solar	Albedo; index of refraction	Data available	No unique relationship between spectral reflectance and soil moisture; thin surface layer; cloud interface.
Thermal Infrared	Surface Temperature (measured diurnal range of surface temperature or crop canopy temperature)	High spatial resolution, large swatch. Relationship between temperature and soil water pressure is independent of soil type.	Bare soil only; cloud interference; surface topography and local meteorological conditions can cause noise; surface layer only
Active microwave (1–100 cm)	Backscatter coefficient; dielectric constant.	All weather high resolution; limited swath width.	Surface roughness; vegetation; topography, interference from communications at large wavelengths.
Passive microwave (1–100 cm)	Brightness temperature (microwave emission); dielectric constant; soil temperature	All weather; penetrates some vegetation; large areal coverage hot as sensitive to roughness and vegetation as active microwave.	Limited spatial resolution; (1–soil temperature; interference from communication.

was instrumented to measure the water balance components—precipitation, ET, and stream-flow. The microwave data were converted to volumetric soil moisture (Wang et al., 1989) and plotted as contours of equal soil moisture over the 37.7 ha watershed (Figure 6). These patterns illustrate the drying that we expected and also give some hints of the hydrologic processes involved which were downslope drainage to form channel base flow.

2.4.3. Future Considerations

Future applications of soil moisture to hydrologic questions and applications are bound to become more common as more experimental data become available. Hydrologists are going to have to realize that temporally frequent spatial measurements of soil moisture will some-day be available on a routine basis and they are going to have to learn how to use these new data. This may involve a considerable learning process because none of our existing models use soil moisture as an input nor do they predict it as output (Peck et al., 1981). In general, existing models have represented soil moisture conceptually to make the model work but have not considered the possibility of independent determination of soil moisture or soil

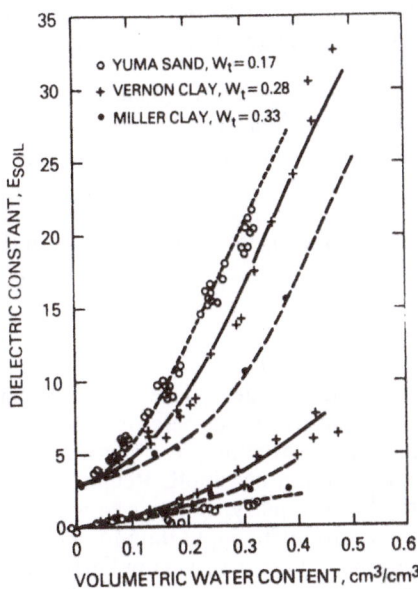

Figure 5. A comparison of laboratory measurements of the real and imaginary parts of the dielectric constant and model predictions (smooth curves) for three representative soils as a function of volumetric water content, 21 cm wavelength (after Wang and Schmugge, 1980).

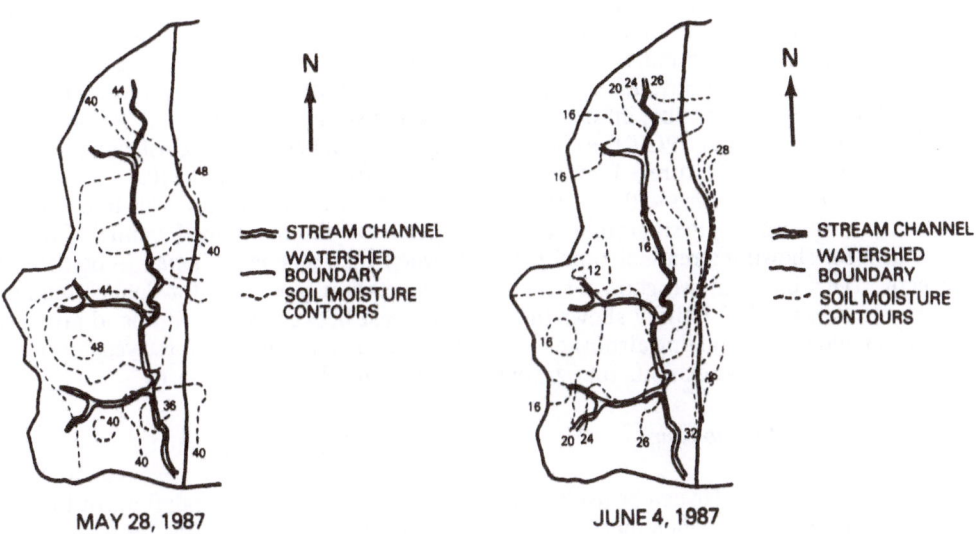

Figure 6. Contours of equal volumetric soil moisture illustrating the spatial and temporal changes during drying (after Wang et al., 1989).

parameters. For the most part this approach has been justified because soil moisture data have not been generally available and hydrologists have not been able to deal with the spatial variability of soil moisture and soil properties.

It also appears that operational remote sensing will involve more than one sensor. Bernard et al. (1986) have demonstrated how a future space system built around visible, thermal infrared and microwave sensors could provide operational information of the soil/vegetation/atmosphere system. In another look to the future, Jackson (1988) reviewed previous work that would lead to an operational passive microwave system for soil moisture.

The use of remotely sensed soil moisture in hydrologic modeling is discussed further at the end of the next section, Runoff.

2.5. Runoff

Runoff cannot be directly measured by remote sensing techniques. The role of remote sensing in runoff calculations is generally to provide a source of input data or as an aid to estimating equation coefficients and model parameters. There are three general areas where remote sensing currently has been used as input data for computing runoff. Remote sensing data are very useful for obtaining information on watershed geometry, drainage network, and other map–type of information. The second is based on producing input data for a class of empirical flood peak, annual runoff, or low flow equations. These first two applications use satellite imagery in the same way as aerial photography is used. In the third approach, runoff models that are based on a land–use component have been modified to use digital analysis or image interpretation of multispectral data to delineate land use classes.

2.5.1. Watershed Geometry

Remote sensing data can be used to obtain almost any information that is typically obtained from maps. In many regions of the world, remotely sensed data, and particularly Landsat or SPOT data, may be the only source of good cartographic information. Drainage basin area and the drainage network are easily obtained from good imagery, even in remote regions. There have also been a number of studies to extract quantitative geomorphic information from Landsat imagery (Haralick et al., 1985). Selection of imagery is important if one is to obtain the maximum possible information. Vegetation state is an important consideration. In the case of Landsat or SPOT, the choice of imagery with a low sun angle will enhance topographic and drainage features. Landsat MSS bands 5 (0.6 – 0.7 μm) and 7 (0.8 – 1.1 μm) and TM bands 3 (.63 – 0.69 μm), 4 (0.76 – 0.9 μm) and 5 (1.55 – 1.75 μm) have proven to be the best choices for discerning physiographic features. The visible red band (MSS band 5, TM band 3) is best for showing stream channel networks when their size is too small to be detected directly. This band is also good for separating vegetation types and for delineating non-vegetated areas. MSS Band 7 shows the most contrast between water and land areas. In tropical regions, side looking airborne radar (SLAR) can penetrate the dense vegetation and produce an image that exhibits topographic features and drainage patterns.

2.5.2. Empirical Relationships

Empirical flood formulae can be useful for making quick estimates of peak flows or for making peak flow estimates when there is very little other information available. A user must be cautioned, however, to be fully aware of their limitations. Generally these equations are restricted to a size range of the basin and the climatic/hydrologic regions of the world in which they were developed.

Landsat data can be used to improve empirical equations of various runoff characteristics. Regression equations relating runoff to basin characteristics have been proposed by a number of hydrologists, including Thomas and Benson (1970). Recent work by Allord and Scarpace (1979) has shown how the addition of Landsat-derived land cover data can improve regression equations based on topographic maps alone.

2.5.3. Runoff Models

To date, most uses of satellite remote sensing data have consisted of relatively straightforward extensions of photogrammetry; however, there have been a number of successful applications where Landsat data have been used to determine both urban and rural land use for estimating runoff coefficients. Land use is an important characteristic of the runoff process that affects infiltration, erosion, and evapotranspiration. Thus, almost any physically based hydrologic model uses some form of land use data or parameters. Distributed models, in particular, need specific data on land use and its location within the basin. However, most of the work on adapting remote sensing to hydrologic modeling has been with the Soil Conservation Service (SCS) runoff curve number model (RCN) (US Department of Agriculture, 1972).

The empirical SCS models have widespread appeal because the major input parameters are defined in terms of land use and soil type and do not require hydrologic data for calibration. The model parameters can be chosen from handbook tables and maps. A desirable feature of the SCS models is the ability for the hydrologist to simulate various design alternatives and compare the results. The parameter defined by land use allows the user to experiment with alternative forms of land development and management and to assess the impact of the proposed changes.

The general approach is to use remote sensing data as a substitute for land cover maps obtained by conventional means. Figure 7 is an example of how the satellite data would fit into the curve number process.

The results using satellite data have been very good. For example, in a study of the Upper Anacostia River Basin in Maryland, Ragan and Jackson (1980) demonstrated that Landsat-derived land use data were used for calculating synthetic flood frequency relationships. The Landsat-derived results were compared to predicted values developed by conventional procedures that used low level aerial photography to determine the land use. Table 2 is a comparison of these two procedures.

2.5.4. Flood Monitoring

The area inundated by floods and floodplains can be mapped effectively with remotely sensed data. Satellite data, such as those from Landsat, can be used to define coverage of an entire river basin but may have some limitations on small basins because of the spatial resolution. Black and white photography, infrared photography, thermal infrared data, multispectral scanner data, and radar have all been successfully used to map the areal extent of flooding. For most approaches using remotely sensed data, determining areas of inundation depends upon measuring reductions in reflectivity caused by standing or flowing water, high soil moisture, moisture-stressed vegetation, and temperature changes. These effects may be detected for up to two weeks or longer after the passage of a flood; thus, the need to obtain data exactly during the flood peak may not be necessary. A number of studies using Landsat data and infrared photography have been reported by a series of papers directed to this subject in the American Water Resources Association (1974).

Cloud cover is frequently a problem in mapping floods with Landsat data because the 18-day coverage may not provide a clear image during or shortly after the flood. The NOAA satellites have an advantage of more frequent coverage over the target area (twice daily). In

486

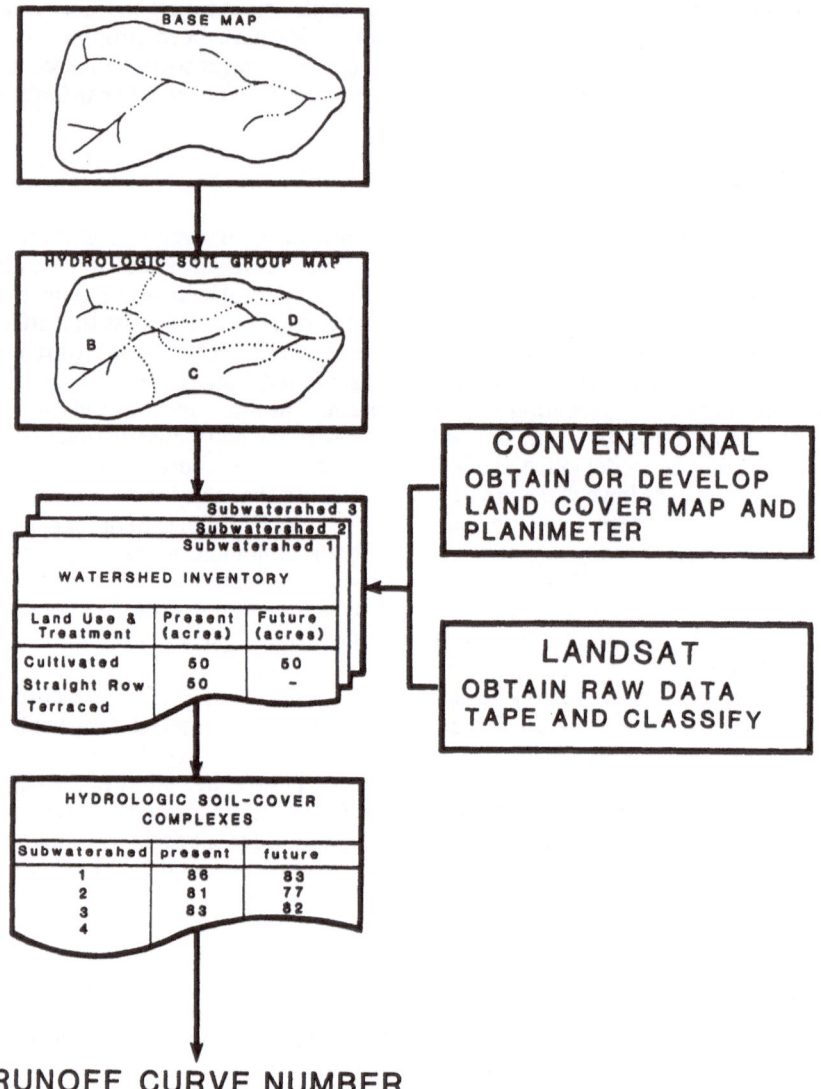

RUNOFF CURVE NUMBER

Figure 7. A schematic diagram showing the role of Landsat in the curve number procedure (after Welle and Jackson, 1982).

spite of the coarser spatial resolution (approximately 1100 m vs. 80 m for Landsat MSS and 30 m for the Thematic Mapper), the NOAA satellite thermal infrared sensor has proved effective in measuring areas of flood inundation (Berg et al., 1979; Berg et al., 1981; Wiesnet et al., 1974).

All weather flood mapping is possible with microwave sensors. Radar systems are capable of higher spatial resolution than passive systems under similar situations and should be well suited for this task. Lowry et al. (1981) demonstrated that airborne synthetic aperture radar

Table 2. Comparison of aerial photo origin and Landsat–derived discharges computed with the SCS runoff curve number model (from Ragan and Jackson, 1980).

Return Period (Years)	Precipitation (Inches)	Discharge in Cubic Feet per Second	
		Aerial Photographic	Landsat
2	3.0	3850	3490
5	3.3	6064	5140
10	5.4	7580	6900
25	5.8	9300	8759
50	6.7	10,400	9900
100	7.3	11,806	11,100

(SAR) could provide all weather flood area delineation. The Shuttle Imaging Radar (SIR–B) was used to map flood boundaries and assess damage over areas of Bangladesh. The digitally processed and geometrically rectified radar data were compared to earlier Landsat data to estimate the areas of innundation subsidence (Imhoff et al., 1987).

2.5.5. Floodplain Mapping

Floodplains have been delineated using remotely sensed data to infer the extent of the floodplain from vegetation changes, soils, or certain anthropogenic features commonly associated with floodplains. Rango and Anderson (1974) have developed indicators that can be used to infer floodplains from Landsat data.

Sollers et al. (1978) describe how flood and floodprone maps have been produced at 1:24,000 and 1:62,500 scales using digital Landsat data. One can reasonably expect better results and more accurate delineation of floodprone areas with the TM and SPOT data, although for many legal requirements it is necessary to map floodprone areas from high resolution aerial photography.

Results of another study (Jackson et al., 1977) indicate that for planning studies the Landsat approach is highly cost effective. The authors estimated that the cost benefits were on the order of 2.5 to 1 or 6 to 1 in favor of the Landsat approach, depending on the experience of the analysts and the availability of data and background information. These benefits would increase even more for larger basins or for multiple basins in the same general hydrological area.

Strubing and Schultz (1983) have developed a runoff regression model that is based on Barretts (1970) indexing technique. The cloud area and temperature are the satellite variables used to develop a temperature–weighted cloud cover index. This index is then transformed linearly to mean monthly runoff. Their results, from the Boise River Basin, are very promising.

2.5.6. Future Considerations

A number of hydrologists have speculated on how remote sensing will have an impact on hydrology and help to improve our data and models. A general consensus is that remotely sensed data will be extremely valuable but current techniques must be modified or new procedures developed to optimally use these new data forms.

Peck et al. (1981) conducted a detailed study on the suitability of seven hydrologic forecasting and simulation models to assimilate remotely sensed data. In general, they concluded that remote sensing had limited usefulness for those models in their form at that time. Their study identified model variables in addition to land use and snow cover area that could be provided by remote sensing. These included soil moisture, frozen soils, and snow–water equivalent.

These three characteristics are currently not used as input data to any of the models discussed by Peck et al. (1981). However, because of their importance in determining runoff rates and volumes, it appears that demonstrable improvements in forecast accuracy may be achieved if these quantities could be measured for input data. For example, Wiesnet (1976) stated that hydrologists in the NOAA River Forecast Center in Kansas City, Kansas, believed soil moisture to be the most uncertain parameter affecting their forecasts.

Since existing conceptual hydrologic models have been designed and calibrated for point measurements and treat soil moisture as a "fitting parameter," measurements of areal soil moisture have been of no value. The state of moisture conditions conceptually indicated by existing models probably does not relate to the actual field conditions. Thus, with present models, improved measurement of actual conditions may result in very little improvement in the accuracy or timeliness of hydrologic forecasts or simulations. To use measured soil moisture for forecasting, the models will have to be modified or new models developed to use remotely measured soil moisture as input data or as feedback to check on predictions and to update system states.

For example, the possibility of using soil moisture as input data would allow the hydrologists to redesign the models so that the soil zone, subsurface flow to groundwater, interflow, or base flow can be conceptualized in a more physically realistic manner. The surface runoff could be simulated by an infiltration model with parameters that could be either determined independently or calibrated. Likewise, the subsurface water transport could be modeled using procedures based on our knowledge of flow in porous media.

Soil moisture could also be used in a feedback loop to update the model state. Most models are based on a mass balance, taking rainfall (or snowmelt) as input and routing it to streamflow. Usually a portion is stored temporarily as soil moisture. The soil moisture defines the state of the system and, as such, controls the rate of sequential processes (i.e. infiltration or deep seepage). If there are errors in the initial state of the model, errors in the predicted output may increase with time since each successive computation is based on the previous state of the system. The prediction accuracy of models may be improved significantly if the system state could be checked periodically and updated or corrected as necessary to reduce the propagation of initialization errors. Repetitive measures of soil moisture used as feedback to the model could function in this way. This is one of the many possible future applications of remote sensing to hydrology.

3. Conclusions

Many techniques for using remotely sensed data have been described in this chapter. Most of them have been proven by field use and are either in use by an operational agency or under development for such agencies. However, the interest of the scientific community in global change has resulted in an increased awareness of hydrologic science as opposed to water resources and has promulgated several new initiatives that are likely to change remote sensing and hydrology in the future. While it is hard to predict the scientific developments that will occur, it behoves us to look at the developments which are likely to take place over the next few years.

Several factors are changing the comparative lack of attention to the studies of scientific hydrology. First, the observed changes of atmospheric carbon dioxide and methane concentrations have led to pressure for a more quantitative understanding of the Earth's system, which requires an understanding of the hydrologic cycle. Second, the computational hardware and software tools needed to handle these complex global problems are being developed to the point where they can carry out the calculations involved. Third, remote sensing instruments and conventional data communications and archiving are developing to the point where the data needed for initializing and updating global earth system models are available in a calibrated form that allow comparisons to be made over time. These are allowing great changes to be made in the type of hydrologic questions that can be addressed (e.g. Eagleson, 1990) and allowing a much greater emphasis on hydrology studies than previously. The advances in each of these three areas are described below.

3.1. New Sensors and Platforms

Several new satellites are planned for launch over the next decade which will carry payloads and make measurements relevant to the land part of the hydrologic cycle. There are several satellites, such as ERS-1 to be launched by the European Space Agency, the J-ERS-1 to be launched by the Japanese, and RADARSAT to be launched by the Canadians, which have primarily oceanographic applications but which will also be used for acquiring data over land. All will carry single-polarization single-wavelength synthetic aperture radars, plus radiometers at various wavelengths which may also give useful data over land. Continuing high spatial resolution data from the Landsat and SPOT satellites, passive microwave data from the Special Sensor/Microwave Imager (SS/MI) on the DMSP satellite series, and continuing meteorological satellite coverage from the NOAA, GOES, GMS and Meteosat series all mean that the remotely sensed techniques described in this chapter can continue to be employed and expanded upon. However, there are many other sensors and satellites being planned that will have considerable hydrologic interest.

The Tropical Rainfall Measurement Mission (TRMM) and EOS are two of the planned systems that will be of great interest to hydrologists. The Tropical Rainfall Measurement Mission (Simpson et al., 1988), to be launched into an orbit inclined at about 30°, will have on board a rainfall radar that will be capable of estimating rainfall rates over land. The orbit is such that any point within the tropics will be sampled twice each month at every hour.

EOS (Butler et al., 1988) and its counterpart European and Japanese platforms will lead to considerable advances in the understanding of all earth sciences, including hydrology. Although the draft manifest of instruments is impressive, the organization of the data and of other earth science in an information system where time series of all the data will be easily available is of at least equal importance. The data system will also allow many types of data to be used simultaneously to calibrate or be assimilated into numerical models. This advance is considered next.

3.2. Advances in Numerical Modeling

Most hydrologic models have been calibrated or verified by a single type or only a few types of data. This is because many of the estimation techniques used are inverse techniques, where there are fewer measurements than unknowns, and these can be easily mathematically intractable. People are only just starting to understand how to use several types of data together in numerical models (e.g. Hollingsworth et al., 1986). Computers have also become larger and are more capable of handling complex problems.

While standard hydrologic models continue to improve and become more physically based, there is a whole new class of model that is now receiving attention from hydrologists (Eagleson, 1990). These are global models that include the whole water budget. While it may be recalled that almost all the water at the surface of the earth is in the oceans, locked in ice sheets, or in deep groundwater, the cycling of the remaining water between the ocean and land surfaces and the atmosphere regulates the Earth's weather and climate, as well as its hydrology.

Early global models were developments of atmospheric general circulation models for weather forecasting and included only the simplest representations of atmospheric hydrologic processes, such as condensation, and only minimal descriptions of the land and ocean surfaces. However, as these models were refined for climatological studies, ocean circulation and the air and sea interaction had to be included. The land–surface hydrology has also begun to be represented in global models, with models varying from the very simple (Manabe, 1969) representation of the land surface as a "bucket" to the very complicated (Dickinson et al., 1986; Sellers et al., 1986).

One feature of the large–scale hydrologic models that is not well described anywhere, is the spatial variability of the land surface. Some advances are being made (for instance, see Entekhabi and Eagleson, 1989) but many models just average measured variables by taking the arithmetic mean for all quantities over large areas. Remote sensing can help deal with the spatial variability of the land surface through observations, but experiments are needed to understand what these remotely sensed data are related to and how they can be used. Some early observational results are described in the next section.

3.3. Observational Advances

There have been several recent experiments to try to understand the spatial variability of hydrologic fluxes at the land surface and to relate these to remotely sensed data. Two in particular will be noted here because of their size. The hydrologic–atmospheric pilot experiment, or HAPEX (Andre et al., 1988), took place in south–western France in 1986. A 100 x 100 km^2 area was covered with radiosondes to estimate the atmospheric component of the water budget, runoff and soil moisture samples, and some surface evaporation and sensible heat flux measurement sites. Aircraft were also flown over the site to estimate sensible and latent heat fluxes and to make remote sensing measurements. Early measurements have shown that there are considerable variations in the radiant surface temperature in the site (Schmugge and Goutorbe, 1989), which is half agriculture and half forested, and that sites which contain different types of land use in a limited area (e.g. trees and grass) behave differently from areas that are of uniform land cover (e.g. all trees or all grass) (Shuttleworth, 1988). Early modeling studies have shown, however, that the evaporation and water balance can be modeled as a whole using a coupled mesoscale atmospheric and surface model (Andre et al., 1988), leading one to hope that the data requirements for routine monitoring will not be prohibitive. Clearly, more work needs to be done, though, on how to couple surface and mesoscale atmospheric models and then calibrate them with remotely sensed data.

Under the international satellite land surface climatology project (ISLSCP), the first ISLSCP field experiment was carried out on a natural prairie grassland in Kansas in 1987 and 1989 (Sellers et al., 1988). It is similar in motivation to HAPEX, but with more intensive surface and remote sensing measurements in a more limited area (15 x 15 km^2). The experimental team was large (29 principal investigators) and a wide variety of useful results should be forthcoming, but it has already been shown (Shuttleworth et al., 1989) that the site appeared to behave very uniformly spatially in 1987, when the vegetation was not lacking water, and that the changes in remotely sensed data, particularly reflectance changes, tracked the seasonal changes in the site rather than any shorter term changes.

3.4. Future Considerations

There are many new and exciting observations of the hydrologic cycle that are going to be available from new satellite systems and, concurrently, new models to allow the data to be analyzed which will address heretofore intractable problems. As we become more proficient at dealing with continental–scale problems, new uses will also be found for existing data sets.

Many experiments will be needed to tie these observations together with conventional observations and validate them, both through field experiments such as FIFE (Sellers et al., 1988) and through global experiments such as the Global Energy and Water cycle Experiment (GEWEX) (WMO, 1988) of the World Climate Research Program. Remote sensing can provide many of the necessary data to supplement conventional data, thereby expanding hydrology in new and exciting directions and also providing entirely new data types and forms that will help hydrologists tackle previously unsolvable questions.

4. References

Allord, G. J., and F. L. Scarpace: 1979, 'Improving Streamflow Estimates Through Use of Landsat,' in *Satellite Hydrology, 5th Annual William T. Pecora Memorial Symposium on Remote Sensing*, Sioux Falls, SD, pp. 284–291.

American Water Resources Assoc.: 1974, 'Satellite Analysis of the 1973 Mississippi River Floods,' *Water Resour. Bull.* **10**, 1023–1096.

Andre, J. C., et al.: 1988, 'Evaporation Over Land Surfaces: First Results from HAPEX–MOBILHY Special Observing Period,' *Annales Geophys.* **6**, 477–492.

Barrett, E. C.: 1970, 'The Estimation of Monthly Rainfall From Satellite Data,' *Monthly Weather Rev.* **98**, 322–327.

Barrett, E. C., M. J. Beaumont, A. Harrison, and T. S. Richards: 1986, '*BIAS I²S Users Guide,*' Final Report to U.S. Dept. of Commerce, Washington, DC, Remote Sensing Unit, University of Bristol, UK, 72 pp.

Barrett, E. C., and C. Kidd: 1987, 'The Use of SMMR Data in Support of a VIR/IR Satellite Rainfall Monitoring Technique in Highly–Contrasting Climatic Environments,' in J. C. Fischer (ed.), *Passive Microwave Observing From Environmental Satellites, a Status Report*, NOAA Tech. Report NESDIS 35, Washington, DC, pp. 109–123.

Barrett, E. C., and D. W. Martin: 1981, *The Use of Satellite Data in Rainfall Monitoring*, Academic Press, London, 340 pp.

Barton, I. J.: 1978, 'A Case Study Comparison of Microwave Radiometer Measurements Over Bare and Vegetated Surfaces,' *J. Geophy. Res.* **83**, 3513–3517.

Baumgartner, M. F.: 1988, '*Snowmelt Runoff Simulation Based on Snow Cover Mapping Using Digital Landsat–MSS and NOAA/AVHRR Data,*' USDA–ARS, Hydrology Laboratory Technical Report.

Berg, C. P., D. F. McGinnis, and D. G. Forsyth: 1980, '*Mapping the 1978 Kentucky River Flood from NOAA–5 Satellite Thermal Infrared Data,*' Tech. Pap., ACSM–ASP Convention, American Society of Photogrammetry, St. Louis, MO, pp. 106–111.

Berg, C. P., M. Matson, and D. R. Wiesnet: 1981, 'Assessing the Red River of the North 1978 Flooding from NOAA Satellite Data,' *Satellite Hydrology*, American Water Resources Association, Minneapolis, MN, pp. 309–315.

Bernard, R., O. Taconet, and D. Vidal–Madjar: 1986, 'Toward a Satellite System to Monitor the Spatial and Temporal Behavior of the Soil Water Content,' *Proc. IGARSS'86 Symposium*, Zurich, **110**, 751–753.

Bowley, C. J., J. C. Barnes, and A. Rango: 1981, '*Applications Systems Verification and Transfer Project. Vol. VIII: Satellite Snow Mapping and Runoff Prediction Handbook,*' NASA Tech. Paper 1829, Goddard Space Flight Center, Greenbelt, MD, 97 pp.

Brakke, T. W., and E. T. Kanemasu: 1981, 'Insolation Estimation From Satellite Measurements of Reflected Radiation,' *Rem. Sens. of Environ.* **11**, 157–167.

Brest, C. L., and S. N. Goward: 1987, 'Deriving Surface Albedo Measurements From Narrow Band Satellite Data,' *International Journal of Remote Sensing* **8**, 351–367.

Butler, D., et al.: 1988, '*From Pattern to Process: The Strategy of the Earth Observing System,*' NASA, Washington, DC.

Carroll, S. S., and T. R. Carroll: 1989, 'Effect of Forest Biomass on Airborne Snow Water Equivalent Estimates Obtained by Measuring Terrestia Gamma Radiation,' *Remote Sensing Environment* **7**, 313–320.

Carroll, T. R., and M. Allen: 1988, '*Airborne Gamma Radiation Snow Water Equivalent and Soil Moisture Measurements and Satellite Areal Extent of Snow Cover Measurements. A User's Guide,*' Version 3.0, National Weather Service, NOAA, Minneapolis, MN, 54 pp.

Carroll, T. R., and K. G. Vadnais: 1980, 'Operational Airborne Measurement of Snow Water Equivalent Using Natural Terrestrial Gamma Radiation,' in *Proc. 48th Annual Western Snow Conf.*, Laramie, WY, pp. 97–106.

Creutin, J. D., G. Delrieu, and T. Level: 1988, 'Rain Measurement by Raingage – Radar Combination: A Geostatistical Approach,' *J. Atmospheric and Ocean Technology* **5**(1), 102–115.

Danes, Z. F., and P. L. R. Danes: 1988, 'Polarization of Passive Microwave Signals as Indicators of Snow Water Equivalent,' *Proc. IGARSS'88 Symposium*, Edinburgh, Scotland, ESA SP-284:441–442.

Davies, J. A., and C. D. Allen: 1973, 'Equilibrium, Potential and Actual Evaporation From Cropped Surfaces in Southern Ontario,' *J. Appl. Meteor.* **12**, 649–657.

Davis, P. A., and J. D. Tarpley: 1983, 'Estimation of Shelter Temperatures From Operational Satellite Sounder Data,' *J. of Climate and Appl. Meteor.* **22**, 369–376.

Dickinson, R. E., A. Henderson–Sellers, P. J. Kennedy, and M. F. Wilson: 1986, *'Biosphere–Atmosphere Transfer Scheme (BATS) for the NCAR Community Climate Model,'* NCAR, Boulder, CO, Tech. Note TN 275 + STR.

Dillard, J. P., and C. E. Orwig: 1979, 'Use of Satellite Data in Runoff Forecasting in the Heavily Forested, Cloud Covered Pacific Northwest,' *Proc. Workshop on Operational Applications of Satellite Snow Cover Observations*, NASA Conf. Publ. **2116**, 127–150, Sparks, NV.

Dozier, J.: 1984, 'Snow Reflectance From Landsat–4 Thematic Mapper,' *IEEE Trans. Geosci. and Rem. Sens.* **GE–22**(3), 323–328.

D'Souza, G., and E. C. Barrett: 1988, *'A Comparitive Study of Candidate Techniques for U.S. Heavy Rainfall Monitoring Operations Using Meteorological Satellite Data,'* Final Report to U.S. Department of Commerce: Cooperative Agreement No. NA86AA-H–RA001, Amendment No. 3, University of Bristol, pp. 39.

Eagleson, P. S.: 1990, *Bull. Amer. Meteor. Soc.,* in press.

Engman, E. T.: 1982, 'Remote Sensing Applications in Watershed Modeling,' in *Applied Modeling in Catchment Hydrology*, Water Resources Publications, Littleton, CO, pp. 473–494.

Engman, E. T., G. Angus, and W. P. Kustas: 1989, 'Relationship Between the Hydrologic Balance of a Small Watershed and Remotely Sensed Soil Moisture,' *Proc. IAHS Third Intl. Assembly*, Baltimore, IAHS Publ. **186**, 75–84.

Entekhabi, D., and P. S. Eagleson: 1989, 'Land Surface Hydrology Parameterization for Atmospheric General Circulation Models Including Subgrid Scale Spatial Variability,' *J. Climate* **2**, 816–831.

Foster, J. L., D. K. Hall, and A. T. C. Chang: 1987, 'Remote Sensing of Snow,' *EOS* **68**(32), 681–684.

Gash, J. H. C.: 1987, 'An Analytical Framework For Extrapolating Evaporation Measurements by Remote Sensing Surface Temperature,' *Int. J. Rem. Sens.* **8**(8), 1245–1249.

Gautier, C., G. Diak, and S. Masse: 1980, 'A Simple Model to Estimate Incident Solar Radiation at the Surface from GOES Satellite Data,' *J. Appl. Meteorol.* **19**, 1005–11.

Hall, D. K., and J. Martinec: 1985, *Remote Sensing of Ice and Snow*, Chapman and Hall, London, x + 189.

Hannaford, J. F., and R. L. Hall: 1980, 'Application of Satellite Imagery to Hydrologic Modeling Snowmelt Runoff in the Southern Sierra Nevada,' in *Operational Applications of Satellite Snowcover Observations*, NASA CP-2116, pp. 201–222.

Haralick, R. M., S. Wang, L. G. Shapiro, and J. B. Campbell: 1985, 'Extraction of Drainage Networks by Using a Consistent Labeling Technique,' *Remote Sensing of Environment* **18**, 163–175.

Heilman, J. L., E. T. Kanemasu, J. O. Bagley, and V. P. Rasmussen: 1977, 'Evaluating Soil Moisture and Yield of Winter Wheat in the Great Plains Using Landsat Data,' *Rem. Sens. of Environment* **6**, 315–326.

Hollingsworth, A., D. B. Shaw, P. Lowenberg, L. Illari, K. Arpi, and A. J. Simmons: 1986, 'Monitoring of Analysis and Observation Quality by a Data Assimilation System,' *Mon. Weather Rev.* **114**, 861–879.

Imhoff, M. L., C. Vermillion, M. H. Story, A. M. Choudhury, A. Gafoor, and F. Polycyn: 1987, 'Monsoon Flood Boundary Delineation and Damage Assessment Using Space Borne Imaging Radar and Landsat Data,' *Photogramm. Eng. Remote Sens.* **53**, 405–413.

Jackson, T. J.: 1988, 'Research Toward an Operational Passive Microwave Remote Sensing System for Soil Moisture,' *J. Hydrology* **102**, 95–112.

Jackson, R. D.: 1985, 'Evaluating Evapotranspiration at Local and Regional Scales,' *IEEE Trans. Geosci. Remote Sensing* **GE–73**, 1086–95.

Jackson, T. J., R. M. Ragan, and W. N. Fitch: 1977, 'Test of Landsat–Based Urban Hydrologic Modeling,' *J. Water Resources Planning and Management Div., ASCE* **103**(WR1), 141–158, Proc. Papers 12950.

Jackson, R. D., R. J. Reginato, and S. B. Idso: 1977, 'Wheat Canopy Temperature: A Practical Tool for Evaluating Water Requirements,' *Water Resources Research* **13**, 651–656.

Jensen, M. E., and H. R. Haise: 1963, 'Estimating Evapotranspiration From Solar Radiation,' *Proc. American Society of Civil Engineering, J. Irrigation and Drainage Division* **89**, 15–41.

Kanemasu, E. T., L. R. Stone, and W. L. Powers: 1977, 'Evapotranspiration Model Tested for Soybeans and Sorghum,' *Agronomy Journal* **68**, 569–572.

Keller, M.: 1987, *'Ausaperungskartierung mit Landsat–MSS daten zur Erfassung Oekologische Einflussgroessen im Gebirge,'* Ph.D. Thesis, Remote Sensing Series, 10 Dept. of Geography, Univ. of Zurich.

Klatt, P., and G. A. Schultz: 1983, 'Flood Forcasting on the Basis of Radar Rainfall Measurement and Rainfall Forcasting,' *Proc. Hamburg Symposium*, IAHS Publ. **145**, 307–315.

Kotoda, K., S. Nakagawa, K. Kai, M. M. Yoshino, K. Takeda, and K. Seki: 1983, 'Application of Equilibrium Evaporation Model to Estimate Evapotranspiration by Remote Sensing Technique in Remote Sensing of Snow and Evapotranspiration, Part II,' *Proc. Second U.S. Japan Snow and Evapotranspiration*, Honolulu, Hawaii, NASA Conf. Pub. **2363**, 115–123.

Kunzi, K. F., S. Patil, and H. Rott: 1982, 'Snow–covered Parameters Retrieved From NIMBUS–7 SMMR Data,' *IEEE Trans. Geosci. Rem. Sens.* **GE–20**, 452–467.

Leavesley, G. H., A. M. Lumb, and L. G. Saindon: 1987, 'A Micro–computer Based Watershed–Modeling and Data–Management System,' *Proc. 55th Annual Western Snow Conference*, Vancouver, B.C., pp. 108–117.

Lowry, R. T., E. J. Langham, and N. Mudry: 1981, *'A Preliminary Analysis of SAR Mapping of the Manitoba Flood, May 1979,'* Satellite Hydrology, American Water Res. Assoc., Sioux Falls, pp. 316–323.

Manabe, S.: 1969, 'Climate and Ocean Circulation: 1. The Atmospheric Circulation and the Hydrology of the Earth's Surface,' *Mon. Weather Rev.* **97**, 739–774.

Martinec, J.: 1975, 'Snowmelt–Runoff Model for Streamflow Forecasts,' *Nordic Hydrology* **6**(3), 145–154.

Martinec, J., and A. Rango: 1986, 'Parameter Values for Snowmelt Runoff Modeling,' *J. Hydrology* **84**, 197–219.

Matson, M., and D. R. Wiesnet: 1981, 'New Data Base for Climate Studies,' *Nature* **289**(5795), 451–456.

Miller, W. A., M. B. Shasby, W. G. Rhode, and G. R. Johnson: 1982, 'Developing In–place Data Bases by Incorporating Digital Terrain Data into the Landsat Classification Process,' in *Place Resource Inventories: Principles and Practices, Proc. Workshop*, August 9–14, 1981, Univ. of Maine, Orono. Spon American Society of Photogrammetry.

NASA: 1984, *'Earth Observing System, Volumes I and II,'* NASA TM–86129, NASA Goddard Space Flight Center, Greenbelt, MD.

Osano, S., T. Motoki, and K. Suzuki: 1988, 'An Estimation of Precipitation Intensity From GMS Data Using Digital Radar Echo Intensity Data,' in J. S. Theon, and N. Fugone (eds.), *Tropical Rainfall Measurements*, pp. 459–461.

Peck, E. L., T. N. Keefer, and E. R. Johnson: 1981, *'Strategies for Using Remotely Sensed Data in Hydrologic Models,'* NASA Report No. CR–66729, Goddard Space Flight Center, Greenbelt, MD, 52 pp.

Penman, H. L.: 1948, 'Natural Evaporation From Open Water, Bare Soil and Grass,' *Proc. Roy. Soc. London* **A193**, 129–145.

Price, J. C.: 1982, 'Estimation of Regional Scale Evapotranspiration Through Analysis of Satellite Thermal–Infrared Data,' *IEEE Trans. on Geosci. and Rem. Sensing* **GE–20**, 286–292.

Priestley, C. H. B., and R. J. Taylor: 1972, 'On the Assessment of Surface Heat Flux and Evaporation Using Large–Scale Parameters,' *Mon. Weather Rev.* **100**, 81–92.

Ragan, R. M., and T. J. Jackson: 1980, 'Runoff Synthesis Using Landsat and SCS Model,' *J. Hydraul. Div., ASCE* **106**(HY5), 667–78.

Rango, A.: 1980, 'Operational Applications of Satellite Snow Cover Observations,' *Water Res. Bull.* **16**(6), 1066–1073.

Rango, A.: 1983, 'Application of a Simple Snowmelt–Runoff Model to Large River Basins,' *Proc. Western Snow Conference*, pp. 89–99.

494

Rango, A.: 1985, 'Results of the Snowmelt–Runoff Model in an International Test,' *Proc. 53rd Western Snow Conf.*, Boulder, CO, Western Snow Conference, pp. 99–106.

Rango, A., and A. T. Anderson: 1974, 'Flood Hazard Studies in the Mississippi River Basin Using Remote Sensing,' *Water Resour. Bull.* **10**, 1060–1081.

Rango, A., and J. Martinec: 1979, 'Application of a Snowmelt–Runoff Model Using Landsat Data,' *Nordic Hydrology* **10**(4), 225–238.

Rango, A., V. V. Salomonson, and J. L. Foster: 1975, 'Employment of Satellite Snow Cover Observations for Improving Seasonal Runoff Estimate,' in *Operational Applications of Satellite Snow Cover Observations*, NASA SP-391, Washington, DC, pp. 157–174.

Rango, A., and V. van Katwijk: 1990, 'Developing and Testing of a Snowmelt–Runoff Forecasting Techniques,' *Water Resour. Bull.* **26**(1), 135–144.

Reginato, R. J., R. D. Jackson, and P. J. Pinter, Jr.: 1985, 'Evapotranspiration Calculated From Remote Multispectral and Ground Station Meteorological Data,' *Rem. Sens. of Environ.* **18**, 75–89.

Schmugge, T. J.: 1983, 'Remote Sensing of Soil Moisture: Recent Advances,' *IEEE Trans. on Geosci. and Remote Sensing* **GE-21**(3), 336–344.

Schmugge, T. J., and J. P. Goutorbe: 1989, 'Remotely Sensed Surface Temperature Observations in HAPEX,' *Proc. IGARSS'89, IEEE*, 89CH2768-0, 2127–2129.

Schmugge, T. J., T. J. Jackson, and H. L. McKim: 1980, 'Survey of Methods for Soil Moisture Determination,' *Water Resour. Res.* **16**(6), 961–979.

Schultz, G. A.: 1989, 'Remote Sensing of Watershed Characteristics and Rainfall Input,' in H. J. Morel–Seytoux (ed.), *Unsaturated Flow in Hydrologic Modeling–Theory and Practice, NATO ASI Ser., Ser. C.*, vol. 275, Kluwer, Dordrecht, pp. 301–23.

Scofield, R. A., and V. J. Oliver: 1977, '*A Scheme for Estimating Convective Rain from Satellite Imagery,*' NOAA/NESS Tech. Memo. 86, U.S. Dept of Commerce, NOAA, Washington, DC, 47 pp.

Seguin, B., E. Assad, J. P. Feteaud, J. Imbernon, Y. Kerr, and J. P. Lagouarde: 1989, 'Use of Meteorological Satellites for Water Balance Monitoring in Sahelian Regions,' *International Journal of Rem. Sens.* **10**, 1101–1117.

Seguin, B., and B. Itier: 1983, 'Using Midday Surface Temperature to Estimate Daily Evapotranspiration From Satellite Thermal IR Data,' *International Journal of Rem. Sens.* **4**, 371–383.

Sellers, P. J., F. G. Hall, G. Asrar, D. E. Strebel, and R. E. Murphy: 1988, 'The First ISLSCP Field Experiment,' *Bull. Amer. Meteor. Soc.* **69**, 22–27.

Sellers, P. J., Y. Mintz, Y. C. Sud, and A. Dalcher: 1986, 'A Simple Biosphere Model for Use Within General Circulation Models,' *J. Atmos. Sci.* **43**, 505–531.

Shuttleworth, W. J.: 1988, 'Macrohydrology: The New Challenge for Process Hydrology,' *J. Hydrol.* **100**, 31–56.

Shuttleworth, W. J., R. J. Gurney, A. Y. Hsu, and J. P. Ormsby: 1989, 'FIFE: The Variation in Energy Partition at Surface Flux Sites,' in A. Rango (ed.), *Remote Sensing and Large–Scale Global Processes*, IAHS Pub. **186**, Wallingford, UK, pp. 67–74.

Simpson, J. R., R. F. Adler, and G. R. North: 1988, 'A Proposed Tropical Rainfall Measuring Mission (TRMM) Satellite,' *Bull. Amer. Meteor. Soc.* **69**, 278–295.

Spencer, R. W., H. M. Goodman, and R. E. Wood: 1988, 'Precipitation Retrieval Over Land and Ocean with the SSM/I, Part 1: Identification and Characteristics of the Scattering Signal,' *J. Atmos. and Oceanic Tech.*

Strubing, G., and A. Schultz: 1983, 'Estimation of Monthly River Runoff Data on the Basis of Satellite Imagery,' *Proc. Hamburg Symposium*, IAHS Publ. **145**, 491–498.

Sollers, S. C., A. Rango, and D. L. Henniger: 1978, 'Selecting Reconnaissance for Flood Plain Surveys,' *Water Resourc. Bull.* **14**, 359–373.

Takemura, Y., K. Takase, and Y. Makihara: 1984, 'Operational Precipitation Observation System Using Digital Radar and Rain Gauges,' *Proc. Nowcasting II*, Norrkopping Sweeden, ESA SP-208, pp. 411–416.

Tarpley, J. D.: 1979, 'Estimating Incident Solar Radiation at the Surface From Geostationary Satellite Data,' *J. Appl. Meteor.* **18**, 1172–1181.

Thomas, D. M., and M. A. Benson: 1970, '*Generalized Streamflow Characteristics from Drainage Basin Characteristics,*' USGS Water Supply Paper, 1955, Washington, DC, 55 pp.

Thornthwaite, C. W.: 1948, 'An Approach Toward a Rational Classification of Climates,' *Geophys. Rev.* **38**, 55–94.

U. S. Department of Agriculture: 1972, *Soil Conservation Service, National Engineering Handbook, Section 4. HYDROLOGY,'* U. S. Govt. Printing Office, Washington, DC, 544 pp.

Wang, J. R., and T. J. Schmugge: 1980, 'An Empirical Model for the Complex Dielectric Permittivity of Soils as a Function of Water Content,' *IEEE Trans. Geosci. and Remote Sensing* **GE-18**, 288–295.

Wang, J. R., J. C. Shiue, T. J. Schmugge, and E. T. Engman: 1989, 'Mapping Soil Moisture With L–Band Radiometric Measurements,' *Remote Sens. Environ.* **27**, 305–312.

Welle, P. I., and T. J. Jackson: 1982, *'Application of Landsat Data and Computer Data Bases for Runoff Curver Number Estimation,'* Am. Soc. Agric. Engr. Paper No. 82-2097, 1982 Summer Mtg, Madison, Wisconsin, 14 pp.

Wiesnet, D. R.: 1976, 'Remote Sensing and Its Application to Hydrology,' in J. C. Rodda (ed.), *Facets of Hydrology*, John Wiley & Sons, London, 368 pp.

Wiesnet, D. R., and M. Matson: 1975, *'Monthly Winter Snowline Variation in the Northern Hemisphere From Satellite Records,'* NOAA Tech. Memo NESS 74, National Environmental Satellite Service, Washington, DC, 21 pp.

Wiesnet, D. R., D. F. McGinnis, and J. A. Pritchard: 1974, 'Mapping of the 1973 Mississippi River Floods by the NOAA-2 Satellite,' *Water Resour. Bull.* **10**, 1040–1049.

WMO: 1988, *'Concept of the Global Energy and Water Cycle Experiment,'* World Meteor. Organ. WCR P-5, WMO/TD 215.

Chapter 22

The Impact of ESPRIT Projects upon the Modeling of Hydrologic Systems

M. B. Abbott
International Institute for Hydraulic
  and Environmental Engineering
Oude Delft 95
2601 DA Delft, The Netherlands

and

J. B. Nielsen
Danish Hydraulic Institute
Agern Allé 5
DK–2970 Hørsholm, Denmark

Abstract. If only by virtue of its sheer size and many–sidedness, the European Strategic Programme for Research and Development in Information Technology (ESPRIT) must have a considerable impact upon hydrologic modeling. This impact is naturally conditioned by traditions already established in this field, not all of which are positive from the point of view of practical applications. After a brief review of the tradition, some more obviously relevant ESPRIT projects are described and their possible impacts are outlined. The notion of the "cost of capture" of such technology is introduced and the economic viability of capture of ESPRIT technology by hydrologic modeling is investigated.

1. Introduction

The European Strategic Programme for Research and Development in Information Technology (ESPRIT) constitutes one of the most ambitious scientific and technological programs of our times. Even when expressed in quantitative terms alone, the project is clearly far–reaching: all together some 6.6 billion dollars are being spent over a five year period. One result of this investment will be the introduction of a considerable range of new information-technology tools, many of which are of value in the modeling of hydrologic systems. In this chapter an attempt will be made to place the ESPRIT program within the perspective of both current practice and the ongoing development of a modern practically–orientated hydrology.

2. Discussion

2.1. Historical Perspective

Throughout the 1960's and 70's, and indeed during the elaboration of the tools that are currently predominant in hydrologic practice, questioning voices were raised concerning the

D. S. Bowles and P. E. O'Connell (eds.), Recent Advances in the Modeling of Hydrologic Systems, 497–507.
© 1991 *Kluwer Academic Publishers.*

real–world value of these tools for engineering and water–resources management applications. We shall here make extensive recourse to two review papers of Klemes, who himself recalled some of the questions posed at this relatively early stage. Although Klemes' earlier paper (1982) is of great value for its explanations of the manner in which many hydrological models appear to work quite well, but for the wrong reasons, the second (1986) contribution is of primary interest here for its systematic criticism of the misuse of a whole range of current tools taken over from mathematics as practiced in other fields. In particular:

"Hydrology, having no solid foundation of its own and moving clumsily along on an assortment of crutches borrowed from different disciplines, has always been an easy victim of this practice. Every mathematical tool has left behind a legacy of misconceptions invariably heralded as scientific breakthroughs. The Fourier analysis, as was pointed out by Yevjevich (in 1968), had seduced the older generation of hydrologists into decomposing hydrologic records into innumerable harmonics in the vain hope that their reconstruction would facilitate prediction of future hydrologic fluctuations (fortunately few computers were available at the time so that the Fourier fever did not become an epidemic); various statistical methods developed for evaluation of differences in repeatable experiments have been misused to create an illusion of a scientific analysis of unrepeatable hydrologic events; linear algebra has served to transform the idea of a unit hydrograph from a crude but useful approximation of a soundly based concept into a pretentious masquerade of spurious rigor now exercised in the modeling of flood events; time series analysis has been used to remake inadequate 20–year streamflow records into 'adequate' 1000–year records, or even more adequate 10,000–year records, and the theory of pattern recognition is now being courted in the vain hope that it will lend scientific legitimacy to the unscientific concept of mindless fitting that dominates contemporary hydrologic modeling. In all these cases, mathematics has been used to redefine a hydrologic problem rather than to solve it."

For his own part, the present author concluded in a UNESCO report of 1972 that there was little of substance in the modeling tools that were then available that would make them useful in the teaching of hydrology. Instead, already at that stage, it was pointed out that a worthwhile teaching aid would need to take the form of an integrated, physics–based, distributed, deterministic modeling system of a kind that did not at that time exist. The study that supported this conclusion (Abbott, 1972) in fact provided a first specification for the Systèm Hydrologique Européen (SHE), although of course the general practicabilities of such an approach had earlier been investigated elsewhere. (See, for example, the references given in Abbott et al., 1986.)

The point about the SHE was that it not only brought to bear upon hydrology a considerable body of modeling technique and experience accumulated in hydraulics, but it did this in such a way that the 'non–hydraulic' components (evapotranspiration, soil moisture balance, snow thermodynamics, etc.) were not suppressed by the weight of the hydraulics. The SHE was "less than fifty percent hydraulics", and indeed it was a product of a team composed of individuals with that wide range of educational backgrounds which is such a characteristic feature of hydrology generally.

The manner in which modeling system development leads to a need for expert systems and similar products of artificial intelligence has been outlined in Chapter 23 of this volume. Another view of this matter is provided by Bugliarello (1987), who extends the discussion to encompass also the question of adapting information–technology developments to problems of hydrologic data collection, enhancement, and interpretation. In general it may be supposed that hydrological modeling will be increasingly applied to environmental problems, often in combination with advanced biochemical, biological, and ecological modeling. It will also find its way into both off–line and on–line control, safety and management systems, components for measurement and control ("robotics" applications), CAD systems, local and more global data and data interpretation systems, and many other products of a future water resources engineering. It is in all events clear that immense new opportunities are opening up in this field – as in so many others. The question that has to be posed, however, is this: are these new techniques to be used as a means of coming to terms with real-world problems in an authentic way; or will they only be used to extend the number of ways "to redefine a hydrological problem rather than to solve it"?

The ESPRIT clearly provides exceptional opportunities for both kinds of development.

2.2. The First Phase of ESPRIT

The European Strategic Programme for Research and Development in Information Technology constitutes the principal effort to combine resources in this field within the European Community. In the main, ESPRIT complements the various national programs of the Member States. The methodologies and tools developed within ESPRIT are clearly of interest to a wide range of end users, of which the hydrology and water resources communities are particular representatives.

Under the general heading of 'Advanced Information Processing' the ESPRIT work program addresses the following key areas:

(1) Knowledge Engineering—which covers the change from processing of numeric data to processing of information and knowledge. Particularly important is the development of knowledge acquisition, representation, and manipulation techniques, which are also vital in improving human access to conventional databases.

(2) External Interfaces—through which the computing system communicates with its environment. This area deals with the recognition, understanding, and synthesis of signals. These signals include those exchanged with humans.

(3) Information and Knowledge Storage—dealing with the developments in data and knowledge bases as well as with techniques of access to these bases and with new hardware devices.

(4) Computer Architectures—dealing with the organization of the processing elements used by the computing system. The development of novel computer architectures and their associated programming environments is a major part of the work in this area.

This first phase is seen to be directed towards projects that are of an almost purely Informational Technological (IT) nature, with very little attention being given to "building bridges" between the possible results of these projects (their "deliverables") and those real-world needs of manufacturing and service industries and agriculture with which engineering hydrology is associated. A first consequence of this specialized direction of resources has been that the deliverables of ESPRIT projects have appeared to be inaccessible to exploitation by many scientific and technological communities, including the hydrologic community. At first sight it may then appear that the impact of the first phase of ESPRIT on hydrology in general and hydrologic modeling in particular will be entirely negligible; the ESPRIT results will, so to speak, "pass way over the heads" of the hydrologic community. If this were to happen, such benefits as could be realized would only be attained through indirect channels, such as from developments being accelerated in other areas (e.g. remote sensing).

In any application of high technology, however, there is always an element of *capture* of the technology. By this is meant that a scientific or technological community must exert itself to understand what is appearing or "surfacing" in its environing high-technology areas, and, having identified promising developments, it must find ways to transform these new developments into procedures and products that are useful to itself. Without capture there is usually no impact. Unfortunately, capture is a very demanding and often expensive undertaking that can only be done in a systematic way by organizations with considerable resources. These organizations must, of course, still weigh the potential benefits accruing from a successful capture of high–technology against the costs of achieving such a capture. The basic problem with the first phase of ESPRIT is that the benefits may well appear too distant and too uncertain relative to the costs of their capture for most organizations working in the area of hydrologic modeling.

By way of exemplification of the above comments, the following first phase ESPRIT projects may be mentioned as being of direct interest to hydrologic modeling (see also "ESPRIT 87", in the list of references: the project, P, numbers refer to this publication):

2.3. Knowledge Engineering Tools

P 96, The Expert System Builder (ESB)

The ESPRIT project (P 96) has a capacity of approximately 110 man years. The partners are PLESSEY Defense Systems, U.K., Compagnie d'Informatique Militaire Spatiale et Aeronautique (CIMSA), France, Centre Studi e Laboratori Telecommunicationi (CSELT), Italy, and Soeren T. Lyngsoe A/S, Denmark. Five more firms and universities participate as subcontractors.

The main objective of P 96 is to provide a universal frame for the development of ES's with the ultimate goal of reducing the production cost of ES's and making available Advanced Information Processing (AIP) aspects to non experts. A layered model has been used for the planning of the project. A particular emphasis is placed from the user's point of view on the structure of the domain layer and the product layer which surround the AI tools layer and the Common Base Programming Environment, the later being the kernel of the layered model.

A task analysis is carried out in the product and the domain layer to select the proper Knowledge System (KS) and expert system shell. More specifically, each of the three parts of a KS, i.e. the Rule Base (RB), the Reasoning Engine (RE), and the World Model (WM) are to be selected in a manner most appropriate to the problem. In a prototype ESB shell, the user can build up his own expert system incrementally by picking KS's and ES's out of libraries by adding, modifying, or deleting rules in a product expert session and updating the individual ES's and KS's in a domain expert session.

The expert system shell used for the implementation of the Prototype Expert System Builder, PESB, is the POKUS shell (Powerful Object Oriented Knowledge Utilizing System). In POKUS the expert systems are objects themselves having their own problem space and operator base. Operators are either simple rules or reasoning systems. The rules produce simple state transformations. POKUS is written in Common Lisp and an interface to Lisp is provided. The ESB is implemented on a Symbolics 3600 machine.

Four expert systems developed independently by each partner of P 96 are currently candidates for the testing of the ESB. The test expert system provided by STL is able to trace malfunctions in a power plant. The second expert system is concerned with test pattern generation for digital boards by splitting the board into macro components. The third is a diagnostic system on a M68000–based microprocessor environment. The last is a troubleshooting aid in electronic equipment using both deep and shallow knowledge.

One promise of the ESB deliverable of P 96 is that of obtaining a much more transparent overview of control flow in any ES so constructed (Abbott, 1988). This should be of considerable value in the elaboration of human–knowledge–intensive modules, especially those which embody decisions and decisiveness conditions. It should accelerate the construction, or at the very least the prototyping, of water resources control and management systems, so long as the problem of coordinating and interaction of these systems with simulations systems can be overcome.

P 820, Design and Experimentation of a KBS Architecture and Tool Kit Real–Time Process Control Applications

This constitutes another approach to KBS building, using a "tool kit", specifically aimed at a class of applications in process control and supervision. The tool kit comprises a general systems architecture, a set of special–purpose building modules, a set of support tools for the construction and testing of the knowledge base, aids to construction of interfaces to traditional software packages and sensors, and a KBS analysis and design methodology.

This toolkit, which is to be available in March 1990, will be used initially in the development of process control applications. The demonstrators chosen for this project are (1) a system for the control of a power plant, (2) a support system for operators in the control room of a geostationary satellite for communications, and (3) a system for the control of a chemical plant. P 820 is a British–Danish–French–Italian cooperation, rated at 59 man years.

This development appears to be relevant not only to process control and supervision in water resources problems but also, to some extent, to the construction of software for the control and supervision of the numerical model simulations themselves.

P 440, Message Passing Architectures and Description Systems: The Omega System

The Omega system has been developed partly with support from ESPRIT project P 440, "Message Passing Architectures and Description Systems". It is an integrated environment for the development of knowledge–based systems in the sense that it does not so much consist of several components (rule base, data base, inference mechanism) but offers more an object description facility, combined with a deductive system.

The basic construct in Omega is the description denoting a class of objects in terms of their attributes and their relations to other descriptions. The basic relation for prediction assignment is the *is* relation. The descriptions in Omega form a boolean lattice, where each node corresponds to a description and links to other descriptions proceed through the *is* relationship. The bottom of the lattice is the description *Nothing*, which plays the role of the null entity and the top of the lattice is the description *Something*, which has the capacity to encompass all the descriptions. The insertion of a new description is accomplished in a consistent way, such that soundness and completeness properties are satisfied.

The deduction process in Omega (taxonomic reasoning) is based on a traversal of the lattice of descriptions. The inference mechanism can be tailored to the problem. In fact the user can provide his own reasoning strategy to the system. Strategies are programmable in an object oriented language and a standard library of strategies is provided, with the strategies being handled by a strategy interpreter at the current implementation level. A compiler for more efficient performance is envisaged as a future development.

Omega has been tested in office system environments. It could be used as a fault diagnosis system. In this case only the description of the normal operational mode of the system is required. The system is modeled as a lattice as just described. The possible malfunction states are detected through a pattern matching mechanism after issuing a query to Omega as to whether a certain state of the system can actually exist.

The current research work of the Omega group is directed towards the applicability of the Omega system in a message passing environment. In this connection a concurrent strategy execution is envisaged. This can be accomplished through parallel search in the lattice of descriptions.

The released Omega version includes the classification mechanism for the descriptions, the taxonomic reasoning system, and the strategy interpreter. A graphic interface (Omega Presenter) is also provided. The Omega Presenter offers the possibility to the user to interact graphically with the knowledge base. Portions of the description base can be displayed and the user can add or remove descriptions in an interactive way. The search through the lattice can be made visible at run time but the user is not authorized to redirect the routing path during the search. Another tool, the Omega Designer, can be used to draw icons of objects representing their structures and interconnections with other objects. Descriptions can be generated automatically from these icons.

Omega is implemented in Common Lisp and runs on the Symbolics 3600 Series, on the Texas Instruments Explorer, and the LMI Lambda family. It is to be made available on SUN workstations and the micro VAX–II of DEC. An interface to Common Lisp exists but there is at present no interface to Prolog or Fortran. (By way of indication it may be mentioned that the acquisition price is approximately 5000 U.S. dollars for universities and companies participating in ESPRIT, provided that Omega will be used within the framework of an ESPRIT project. The service costs amount to 500–1000 U.S. dollars per annum.)

It is anticipated that the deliverables from P440 will be applicable to a wide range of problems of land use, and especially those of agriculture and forestry, placed within a more general ecological frame. Malfunctioning states could then be associated with the arising of conditions that are unacceptable within the framework of the overall ecological objectives. In such an application, a "message" is to be understood in the general information–theoretic sense.

2.4. User Interfaces

P 857, The GRADIENT Project

The GRADIENT (Graphical Dialogue Environment for Process Supervision and Control) project has emerged from the fusion of ESPRIT project P 857 (Graphics and Knowledge–Based Dialogue for Dynamic Systems) with CRI (Computer Resources International A/S), Copenhagen, Denmark as the Prime Contractor and a smaller project with K.U. Leuven as the Prime Contractor. It has an allocated manpower amounting to 125 man years. The main object of the work carried out at K.U. Leuven is to construct expert systems dealing with corrosion prevention and control. A metrication study of expert systems in an industrial environment is a further objective of this project.

An expert system for corrosion prevention and proper material selection, PRIME (Process Industries Material Expert) is currently under construction. For this purpose, an object oriented language, KEE, is used. KEE has evolved to an existing standard in Knowledge Engineering. The latest version of KEE (version 3) is installed on the SUN workstation at the Chemical Engineering Department of the K.U. Leuven.

The team at K.U. Leuven is also working on another expert system, SECOND, which is to be used for multisensor signal processing. The goal of this system is to monitor on–line the quality of feed or cooling water in an industrial plant, but it can be easily adapted for other applications. Use is made of an expert shell named KAPPA. KAPPA consists of a number of object definitions with declarative value attachments, which may be facts or rules of any complexity. The KAPPA compiler transforms the facts and rules into an internal format, performs a number of checks on the represented knowledge, and stores the knowledge base in a file. Contradictions are excluded in the compiling phase. The file produced by the KAPPA compiler is an ASCII file and therefore it is portable to any computer. The KAPPA inference engine works in a monotonic forward chaining manner. No backtracking is possible. After all deductions have been made, and if necessary after some consultation with the user, all possible conclusions are presented to the user. The whole system is implemented in Pascal on an IBM–AT. The KAPPA system is currently refined to include trace options and a graphics interface. It has been developed by a subcontractor, the Expert Software Systems (ESS), Ghent, Belgium.

Other partners in the P 857 group are concerned with the development of a Quick Response Expert System (QRES) destined for operator support during system failure (CRI, Denmark) and with a Graphical Expert System (GES) to serve as a man–machine interaction interface

(University of Strathclyde, Turing Institute, Glasgow, U.K. and University of Kassel, FB 15, FRG). A team at BBC (Brown Boveri Corporation) participating in P 857 serves as a knowledge–engineering team.

Process supervision and control is expected to be more widely applied in water resources projects, and especially as a means of economizing water use. The widespread introduction of such means is indeed essential if severe conflict situations are to be avoided in the next decade (e.g. Starr and Stoll, 1987).

P 280, EUROHELP

A necessary development during the next century will be that of providing expert advice on a more uniform basis for all manner of information processing systems (IPS's). For such purposes, user–friendly "advice–serving" instruments will be required that can be connected, through a suitable network, to 'banks' of data, knowledge, and expertise. Such a system of advice servers and their network might be constructed along the lines proposed in this project.

In general, the function of a help system is to provide information about the use of some IPS when needed by any type of user. The need can be expressed by the user or inferred by the help system itself. A help system should play the role of a human coach, who looks over the shoulder of the user to interpret performance, who interrupts when the user makes an inadmissible step or when there is an opportunity to extend the repertoire of the user, and who is able to answer questions in the context of the current use of the IPS.

Monitoring on–line the performance of the user entails many conceptual and computational problems, but these are not qualitatively different from those in intelligent teaching systems in general. Because user needs may either be identified by the system or by the user, EUROHELP contains a "Question Interpreter" and a "Performance interpreter," or a passive and an active side. Both communicate with a "Diagnoser" which tries to identify the actual or "local" need of the user. This local need can always be stated in terms of a specific lack of knowledge or misconception. This lack of knowledge may not only refer to concepts, but also to current states of the system. Many local needs are identified by the fact that the user makes an error.

The tracing of errors in using an IPS is simplified to some extent by the fact that many erroneous actions of the user are not executable. However, the variety of causes of executable errors is as complicated as in any "natural" domain, because users may acquire all kinds of misconceptions, ranging from wrong models of the underlying "virtual machine" to not knowing about a side effect of some command. The identification, or diagnosis, of problems of users is not only complicated by the fact that lack of knowledge may lead to executable errors, but particularly by the fact the system has to infer what the intentions of the user are. The system can look over the shoulder of the user, but it would become a nuisance if it asked what the user intended too often. In intelligent teaching systems this is a minor problem because the system itself prescribes what goals the user should accomplish. Such a system presents specific problems to the student–user and makes safe assumptions about at least the global goals. This means that in a help system there is a great emphasis on plan recognition in interpreting the performance of the user. Because the *skill* (not the insight) in handling an IPS consists mainly of planning actions, a planner may work in close cooperation with the plan recognizer to provide top–down constraints on the interpretation of user intentions but may also work independently. For instance, this is required when the user makes an executable error and an unintended state is created. Help does not only imply an ability to tell the user about what a correct action would have been, but also to explain how to retrieve a situation starting from a specific undesirable current state.

The P 280 project should provide a number of tools, if not a complete methodology not only for advice serving for individual water–resources systems and their data and knowledge bases but also for coordinating systems and networks of bases.

2.5. External Interfaces

P 322, CAD–I

A great deal now has to be done in order to interface CAD systems with numerical models, functioning as "Knowledge encapsulators." It should then be possible to analyze "the language of the designer" and the relation of this to the mode of functioning of the numerical modeling components on the basis of deliverables of this project.

The objective of this project is to develop a family of consistent, compatible standardized interfaces for CAD, allowing:

- Representation of 2–D and 3–D geometrical models of CAD design objects
- Archiving and retrieving these models by various CAD/CAM systems
- Exchange of such models over networks
- Storage of parametrized part libraries in data bases connected to networks
- Access to such libraries from various CAD/CAM systems
- Use of advanced modeling techniques for model generation
- Standardized application of different finite element model analysis programs
- Comparison of experimental and analytical dynamic analysis results
- Dynamic model optimization resulting from experimental and analytical analysis.

The availability of a standard interface will facilitate the free flow of geometrical design data between different CAD systems without expensive conversion and data restructuring. It will also permit CAD systems to be interfaced with Computer–Aided Engineering systems.

Although this project is geared to the use of finite element codes of the type that have become dominant in problems of static analysis and in the dynamic analysis of linear systems, it should be possible to modify the tool sets so developed in order that these can work with the faster, more accurate, but considerably more complicated codes that will be necessary for most hydrologic applications.

It should be observed that this project appears, at least at first sight, to overlap with the more recent P 1062, ACCORD.

To the above list may be added, for independent reference in (ESPRIT, 1987):

- P 112, "Knowledge Integration and Management System";
- P 387, "Knowledge Representation and Inference Techniques in Industrial Control (KRITIC)";
- P 316, "An Architecture for Interactive Problem Solving by Cooperative Data and Knowledge Bases";
- P 117, "Next Generation Integrated Data Base and Knowledge–Based Management Systems (KIWI)".

It should be added that the first phase of ESPRIT has led to considerable developments on the hardware and firmware side, such as an industrial approach to custom–built electronic components. As the "art of capture" appears at the moment to be so high in these cases, they have not been introduced here.

A particular interest accrues here to one other ESPRIT project, if only because the present author is its proposer and an active participant in it. This is P 1074, "Shipboard Installation of Knowledge–Based Systems, KBSSHIP". This project, which itself draws upon results from P 96 and P 387 through cross partnerships, necessitates the combination of a large amount of human intervention (legislation, general conditions of tender and contract, charter conditions, standard procedures, etc.) interacting at several levels with voyage–fixed "natural" conditions (ship properties, cargo properties, etc.) and voyage–variable "natural" conditions (weather, sea state, hazards, etc.). It is quite simple to make analogies with water resources systems where the same basic elements come into play. P 1074 has already provided a valuable proving ground for already formulated notions, and an inspiration for new ones again, while it is expected that its deliverables will throw more light on the problems and potentialities of the management, operations, and control systems of the 1990's and beyond.

2.6. The Second Stage of ESPRIT

The aim here is more that of "building bridges" between the results of the first phase and the needs of main–line industries and agriculture. A complete description is given in the "Draft ESPRIT Workprogramme" of the European Commission. The first round of contracts under this phase will be concluded during 1988. It must suffice here only to list, in Table 1, the areas that are of a more obvious and immediate interest to hydrological modeling, leaving their descriptions to the reference. The areas listed are primarily those named as so–called "A–type projects", the terms of which are formulated by the Commission. There are also "B–type projects", which are much more loosely formulated, as "topics" being more innovative or longer term.

2.7. Capture Strategies for ESPRIT Deliverables

Few organizations, and certainly none working in hydrology and water resources, can capture the totality of ESPRIT–driven technology. A selective approach has to be adopted, where the selection has to be made on the basis of the perceived strengths of the existing organization

Table 1.

II INFORMATION PROCESSING SYSTEMS

II.1 SYSTEM DESIGN: Project ASEE (Advanced System Engineering Environment; Techniques for Requirements Capture, Analysis and Animation; Automatic Support of Host–target Aspects of System Development; Application of Knowledge–based Techniques to System Design; Automatic Generation of System Components; Reusable System Components; Methods for Supporting the Continued Evolution/Enhancement of Systems; System Maintenance; Evaluation of Current Methods and Tools; Metrics for Application Projects; Support of Effective Management of Systems.

II.2 KNOWLEDGE ENGINEERING: Real–time Knowledge–Based Systems; Cooperating Expert Systems; Enhancement of Systems by IKBS; Knowledge Based Systems Life Cycle; Performance Evaluation of Knowledge Based Systems; Man Machine Interaction; Large Knowledge Bases; Explanation Facilities; Time Dependent Reasoning; Machine Learning; Knowledge Elicitation and Acquisition.

III.2 INTEGRATED INFORMATION SYSTEMS

III.2.1 ARCHITECTURES AND SOFTWARE TOOLS: Integrated Application Support System; Office Document Architecture; Decision Oriented Integrated Tools; User Controlled Operating Environment; B–Type Topics.

III.2.2 APPLICATION SYSTEMS ENGINEERING: Support Environment for Organizational Systems; System Management Support.

III.3.1 HUMAN FACTORS AND HUMAN MACHINE INTERFACES: Human–System Interaction in Integrated Systems; Visual Interface Design and Integration.

---

and the perceived developments of its market. It is a common practice to represent such selection processes diagramatically, as follows.

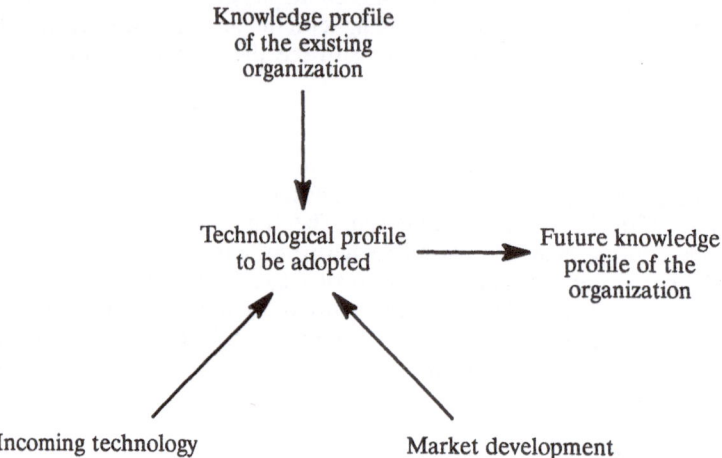

The profile of an organization is just as commonly represented by a "disc diagram," in which a set of concentric circles carrying scores running from 0 to 100, as one proceeds outwards from the center to the circumference, are divided up into "market segments". Often the areas of the segments are set proportionally to the market share that they represent, or are expected to come to represent. The perceived strength of the organization relative to its competitors, as measured subjectively, are then marked by a point on each segment. The line joining the set of such points around the disc provides a graphical representation of the knowledge profile of the organization. Such diagrams commonly provide the starting points of strategy gaming, in which the impact of incoming technology is tested against market expectations and this impact is compared with the expected cost of the technology "capture" or "acquisition."

When this standard "business–school" approach is followed in the case of hydrologic modeling relative to ESPRIT impacts, the results indicate an increase in activity that transcends what is generally credible. Either hydrologic modeling is going to increase in the volume of its turnover by a factor of between one hundred and one thousand or there is something very wrong in the standard approach. On the one side, the situation appears to resemble that of the computer industry itself when the Apple Corporation had just introduced the first commercially viable "personal" machine. On the other side, there are a number of negative influences which could inhibit this development, and the most dominant of these is the entrenched, established tradition of inauthentic modeling in this field, as described by Klemes. In so far as a physically–sound and practically–orientated hydrologic modeling can be pursued to its logical conclusion, so an immense market of farmers, foresters, industrial organizations, local authorities and other end users can be mobilized and satisfied. On the other hand, the further extension of the established tradition of modeling, which can itself be promoted by ESPRIT developments, may seriously reduce the penetration of the more serious approach.

3. Conclusions

The potential impact of ESPRIT projects upon hydrologic modeling is immense, but this potential will almost surely be dissipated to a considerable extent by the tradition already established in such modeling, with its emphasis on curve fitting, "mindless extrapolation," "black-box" modeling and the other non-physical approaches described so vividly by Klemes. On the one hand the market for hydrologic modeling could be transformed out of all proportion by ESPRIT impacts, but on the other hand this transformation could be so retarded by the established tradition that the financial returns accruing from it would not be sufficient to recompense the costs of capturing the required technology. Although all such outcome estimates must of necessity be highly subjective, it is the author's estimate that it will take at least ten years to overcome the established habits of thought and behavior in this area, and so ten years before a major ESPRIT impact can be realized. This is roughly the time span between the first specification of the SHE and its first large-scale application. Experience has indeed suggested that a ten year gestation time is quite normal in the case of any such strategic initiative.

It follows from this assessment that the capture of ESPRIT technology is scarcely viable in hydrology in any mundane actuarial sense, but only in a long-term, strategic, "industry building" sense. In order to satisfy the exigencies of the mundane and average world of financial viability, it is necessary to place ESPRIT impacts within a wider frame, which includes application areas that are not so constrained and inhibited by inauthentic traditions. This analysis suggests that ESPRIT technology capture should be directed more to other areas than hydrologic modeling in the shorter term – and in particular to the offshore industry, with its much more sophisticated range of end-users. It would then appear to make more sense in hydrology to concentrate on third-generation modeling, with some "spill-over" to fourth generation-modeling, at least over the years up to about 1992, while concentrating ESPRIT captures and general fifth-generation modeling development upon other areas. As these areas begin to be consolidated, so the knowledge accumulated there can be redirected towards the (much more extensive) areas of hydrology and water resources.

The revolution in hydrology, in which ESPRIT has a central part to play, is already upon us, but it still has a very long way to go.

4. Acknowledgement

The author acknowledges the inputs of other participants in P 1074 who have analyzed ESPRIT projects, and especially the contribution of Dr. L. Bardis, Technical University of Athens.

5. References

: 1987, *ESPRIT 87, Achievements and Impact*, Parts I and II, North Holland, Amsterdam.

Abbott, M. B.: 1972, '*The Use of Digital Computers in Hydrology,*' Teaching Aids in Hydrology, Tech. Paper in Hydrology No. 11, UNESCO, Paris, pp. 29–32.

Abbott, M. B., J. C. Bathurst, J. A. Cunge, P. E. O'Connell, and J. Rassmussen: 1986, 'An Introduction to the European Hydrologic System – Système Hydrologique Européen, SHE,' *J. Hydrology* 87, 45–59 and 61–77.

Bugliarello, G.: 1987, 'Computers and Water Resources Education: A Projection,' ASCE, *J. Water Res. Planning and Management* 113(4), 498–511.

Klemes, V.: 1986, 'Dilettantism in Hydrology: Transition or Destiny,' *Water Resources Research* 22(9), 177S–188S.

Klemes, V.: 1982, 'Empirical and Causal Models in Hydrology,' in *Scientific Basis of Water Resources Management*, Nat. Acad. Press, Washington, D.C.

Starr, J., and D. Stoll: 1987, *U.S. Foreign Policy on Water Resources in the Middle East*, Centre for Strategic Studies, Washington, D.C.

Yevjevich, V.: 1968, 'Misconceptions in Hydrology and their Consequences,' *Water Resources Research* 4(2), 225–232.

Chapter 23

Contributions of Computational Hydraulics to the Foundation of a Computational Hydrology

M.B. Abbott
International Institute for Hydraulic
  and Environmental Engineering
Oude Delft 95
2601 DA Delft, The Netherlands

Abstract. Experience acquired in the development of computational hydraulics is applied to computational hydrology. Both the deductional and computational aspects are considered. The special features of hydrology, as a "scientific conglomerate," are described.

1. Introduction

1.1. The Pre–Computer Era

In view of the precedent already established in this workshop, of going back in time to the origins of the subject, we can as well begin this survey by referring back to the era:

B.C.

which, in the present context, stands for *Before Computers*. As is well known, civilization was already at that time quite considerably advanced, which is to say that people had already acquired a considerable ability to *make problems*. In this first period, which precedes and prepares the way for the period proper to the present survey, our ability to make problems was still restricted by the comparative simplicity of our relations to our outer world, which could be schematized as follows:

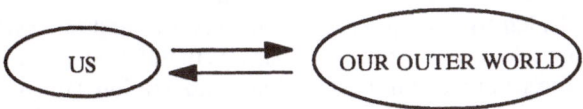

As the most civilized people of their times, the Greeks were preeminent in making problems, and it was in this preeminence that they developed mathematics in terms of arithmetic, as one distinct part, and geometry, as another distinct part.

Arithmetic had to do with numbers, but these numbers had two very widely different significances. On the one side, numbers have in all civilizations prior to our own, including that of the ancient Greeks, been regarded as "holy." For the Hindus, the numbers sprang forth in the very creation dance of Siva, and the first ten natural numbers were always endowed with

509

D. S. Bowles and P. E. O'Connell (eds.), Recent Advances in the Modeling of Hydrologic Systems, 509–537.
© 1991 *Kluwer Academic Publishers.*

spiritual or (to express the matter more precisely) *numinous* properties (e.g. Jung, 1944). We really have here to do with an *allegorical use of numbers*, which in this case takes the form of a natural–number symbolism. Through the resurgence of interest in Neo–platonism in the European Renaissance, the number symbolisms of the Greeks were reintroduced into the beginnings of modern science, strongly influencing the investigations of Kepler, for example, and leaving their mark as well on the works of Leonardo da Vinci. With the current reawakening of interest in Leonardian fluid dynamics (e.g. Macagno, 1988) – a reawakening that does not really lie so far away from the interests of computational fluid dynamics – this number symbolism still leaves a trace in our current societies. Normally, however, number symbolism only enters into modern science surreptitiously, "through the back door" so to speak, as in Quantum Logic (e.g. Manin, 1977, pp. 82–94), in Fractal Theory (e.g. Mandelbrot, 1981) and, in a really subtle way, in the so–called Arrow Impossibility Theorems that originated in Economic Welfare Theory but have very much wider fields of applications (e.g. Kelley, 1978; Abbott and Basco, 1989). Indeed, as we shall subsequently argue, it has been almost entirely succeeded in modern science by another kind of allegorical description in terms of numbers, which we call *numerical modeling*.

These same (for us, nowadays) natural numbers were also used, of course, in commerce, where they were indispensable in every market place. We may suppose that it was the market place, in the first instance, that gave rise to the introduction of the *rational* numbers. However the Greeks had also a more philosophical, if not theosophical interest in geometry, as a "model" of the continuum, and it appears that it was this interest that led them to invent the *irrational* numbers. After all, given the universal truth of Pythagoras' theorem and the existence of a triangle with two sides of exactly unit length at right angles to one another, it seemed reasonable to enquire into the *length* of the hypothenuse of this triangle. Essentially, they tried to introduce the concept of number into the continuum of space. The "discovery" that this length of the hypothenuse could not be expressed as a rational number shocked the ancient Greeks, but for a very long time now we have managed duly to remember the results while forgetting the shock. We have got quite accustomed to using the word "irrational," while forgetting its deeper connotation, as something that is *opposed* to reason. This was something that the old Greeks, due to their understanding of the generality of the word "ratio" itself, could not forget. Nowadays we re–experience the shock as our computers, working with rational numbers, fail to reproduce the apparently equivalent continuum result, as happens in the event of numerical instability, and as we are driven in this case of instability to realize that the closer we come to "modeling" the continuum results, as by refining our numerical grids, the worse the discrepancies become.

In fact, the Greeks left us a great legacy in this respect, and one with an apparently unlimited potential for making new problems. This was that the set of points on any segment of the "real" line and the set of real numbers (rational and irrational) were equivalent. We shall speak of this as the *standard model for one–dimensional worlds*, applying as it does to the field of *standard real numbers* (Rosinger, 1988).

The problems that arise from the use of this paradigm were already experienced during the controversy over whether the physical world should be treated as a discrete system or as a continuous one, and the nature of the relations existing between these systems – a controversy that lasted for well over two hundred years (Hankins, 1967). Galileo posed one such problem when discussing the free falling of bodies, such as occurred in the already socially important science of ballistics. He observed that the distance travelled in free fall under gravity (so a fall that neglects resistance effects) is proportional to the square of the elapsed time from the initiation of the fall. We may schematize this as shown in Figure 1.

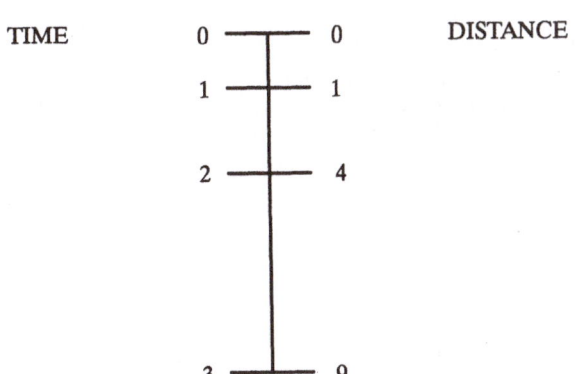

TIME   DISTANCE

Figure 1. Schematization of the Galilean equivalence of time and distance sets.

Using normalized units we then have the following set of equivalence relations:

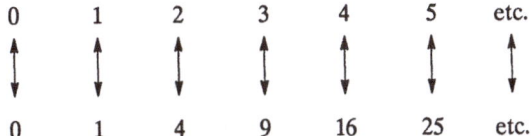

The continuum *motion* is then conceived as the limit of such a set as the number of its elements becomes infinite. The problem then arises, however, that the set associated with distances is a proper subset of the set associated with times, and indeed that the first of these sets contains "infinitely fewer" elements than the second set as the number of elements of both become infinite. How, in that case, could these two sets remain equivalent? Galileo, concluded — one imagines with a figurative shrug of the shoulders — that "there must be different kinds of infinities."

Newton, with an understanding that extended from the religious mysticism of Jakob Böhme to the logic of modern science, could probably accept the implied paradox with equanimity. In particular, in his *Principae* he first described and defined all his subject matter in discrete terms, and only after doing this did he proceed to any kinds of "limits," to introduce his "fluxions." Thus, for example he introduced the second law of motion as follows:

"The change in motion is proportional to the motive force impressed and is made in the right line in which that force is impressed."

The form of the second law that we all know today, and which we call "Newton's second law of motion," was in fact introduced only by Euler in 1750, and it was that time seen as a completely new law, with its origins situated more in the thought–world of Galileo than in that of Newton (as expressed by Maupertuis: see Hankins, 1967).

The total dominance of the continuum paradigm in science, as established by Euler (a dominance especially well described in the pseudonymous "Bourbaki" *Elements d'histoire etc.* of 1960) had its clear mathematical consequences. As Heidegger (1949/1977), among others, has emphasized, modern science is in its essence already mathematical before it is ever cast into a mathematical form. However, the continuum paradigm established by the

Greeks cast it into one particular form. This is the form of a statement that is true for any point P in a given domain, so that it becomes either a differential statement or one that applies to a corresponding perfectly–integrable world, as an integral statement.

In the case of only one variable this descriptive paradigm can be represented by the dual:

$$\frac{df}{dt} = -af \quad \text{and} \quad f = f_o \exp(-at) \tag{1}$$

while in the case of two variables it can be represented by the dual:

$$\frac{\delta f}{\delta t} + u\frac{\delta f}{\delta x} = 0 \quad \text{and} \quad f(x, t_2) = f\left[x - \int_{t_1}^{t_2} u\,dt, t_1\right] \tag{2}$$

These two examples are schematized in Figure 2 (See also Abbott, 1966).

The application of the above continuum paradigm in fluid mechanics led to the subject called *hydrodynamics*, and for over two hundred years this paradigm almost totally dominated the scientific description of fluid motion. Its influence still persists in the use of differential descriptions, such as those of the equations named after Darcy and Richards.

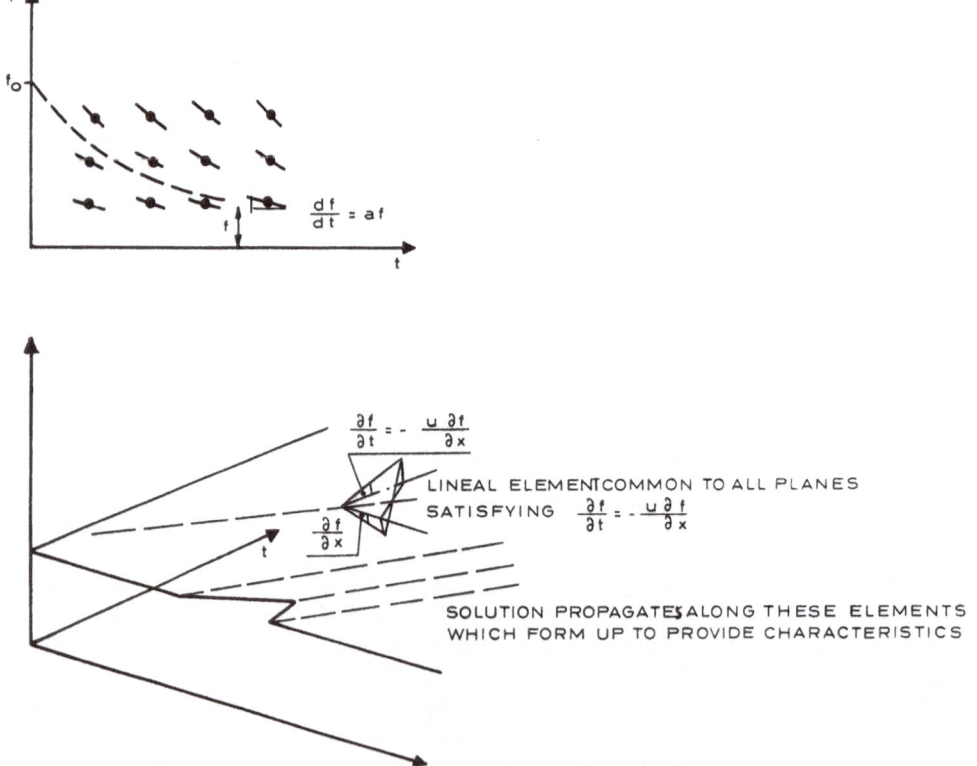

Figure 2. Schematizations of differential forms and their solutions.

The continuum paradigm, with its *standard analysis* based upon the paradigm of the field of standard real numbers, still left room to make new problems, and by far the most significant of all problems of the last century was that proposed by Cantor when he showed that:

"The set of points on the interval [0, 1] is not enumerable."

This is to say that the set of standard real numbers cannot be paired–off in a one–to–one way with the set of natural numbers. This already contains the paradox that even as every irrational point can be said to *be* the limit of a sequence of rational points, the number of elements in the set of irrational numbers is "infinitely greater" than the number of elements in the set of rational numbers. (This is *not* the "Cantor Paradox" as it is usually stated, but it can be seen as a consequence of that paradox; it is taken from Lauwerier, 1987, p. 32.)

At about the same time, starting with Frege, towards the end of the last century (see van Heijenoort, 1967), *logic* itself became steadily more arithmetized; which is to say that logical operations were reduced to arithmetic operations and the "ways of logic" and the "ways of arithmetic," or the processes of *deduction* and the processes of *computation*, were found to be increasingly congruent. One representation of this congruence is the formalization of essentially enumerable logical statements in one or the other form of a *first order language of set theory*, L_1Set, and the formalization of essentially enumerable arithmetic statements in one or the other form of a *first order language of arithmetic*, L_1Ar. It is a common practice nowadays to use the Zermelo–Fraenkel language of set theory as a standard for L_1Set and the Peano language of arithmetic as a standard for L_1Ar.

In computational fluid dynamics and computational hydraulics – and this will presumably hold also in any future computational hydrology – an important role is played by two problems, which, as they follow upon certain conjectures, are often expressed in the form of propositions. Altogether, David Hilbert in 1900 proposed twenty-three such problems or propositions that he thought would keep mathematicians busy throughout the twentieth century. The two that currently most concern us are the *thirteenth* and the *second propositions*.

The thirteenth proposition was originally stated (for all practical purposes) in the form that not all continuous functions of three variables can be represented as superpositions of continuous functions of two variables. This proposition was definitively refuted by Kolmogorov and Arnold in 1957. These authors showed even more than this; however, in effect they showed that it was possible to represent a wide range of continuous functions of several variables by superpositions (that is, functions of functions) and sums of functions of one variable. In effect – and at a certain level of abstraction – a wide range of problems are always susceptible to a *dimensional decomposition* into superpositions and sums of solutions of problems of lower dimension. The most obvious realizations of this possibility is that of the collection of techniques that are called *Boundary–Element Methods* (BEM). However, such techniques are just as widely used in Finite–Difference Methods (FDM: see Abbott et al., 1985), and are increasingly used in Finite–Element Methods (FEM). Such decompositions are sure to be just as essential to computational hydrology as they have already proved to be to computational hydraulics.

The second proposition (which Hilbert subsequently extended to embrace a whole program) has to do with the possibility that a formal logical system might be found through which the entirety of all mathematical proofs could be generated. In equivalent arithmetic terms it has to do with the possibility of finding a formal arithmetic system that could realize any computation whatsoever. The possibility of executing such a program, even on an ideal (infinite memory, infinite speed, infinite accuracy) machine, was shown to be illusory in a sequence of epochal papers by, principally, Gödel, Church, and Turing in the 1930's. In the course of this work the arithmetization of logic came to a certain B.C. completion, through the introduction of "Gödel numbers" to represent logical propositions, while the B.C.–formal expression of

514

propositions came to a certain completion in the λ–calculus of Church, which in turn formed a foundation for all machine–formalized symbolic–logical expressions in the LISP programming language. Similarly, the B.C.–conceptual mechanization of the process reached a certain conclusion in the Turing machine, as an ideal conceptual representation of a digital computer.

Correspondingly, and almost simultaneously, one branch of logic led into quantum logic (e.g. Manin, 1977, pp. 82–94), the problem of calculating trajectories in ballistics—the problem of Galileo—led to improved aids to such calculations, and the Turing machine contained the germ for Turing's military decoding machines. These mathematic problems then acquired a new social significance at Las Alamos, at the Aberdeen ordnance proving grounds in the United States, and in the British decoding centers. Their common need and invention was, of course, the *digital computer*.

1.2. The New Era

This brings us to the years:

A.D.

since each such year is now an *Anno Digiti*. The possibilities to make problems now becomes even more expanded due to a further complication in human existence, which can be represented as follows:

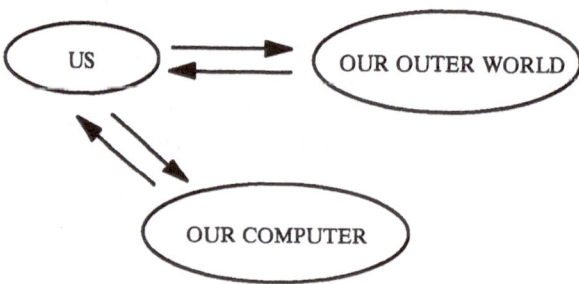

The computer that then appeared, and with which most of us still coexist, is a so–called "von–Neumann machine," in which logical operational elements and numerical operational data are indistinguishable at the level of the electronic processes, but only become differentiated through the intercession of the operating system of the machine. Thus, for example

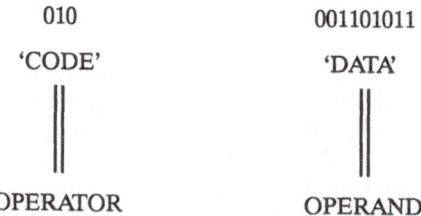

might be placed in a register in order to add two numbers represented as "data." In any event, the logical and the numerical have, at some level, an identical representation.

2. Discussion

2.1. The Generations of Modeling and the Rise of Computational Hydraulics

The notion of "generations" of numerical models derives largely from applications outside hydrology, primarily in the companion areas of hydraulics, coastal, and offshore engineering. On the other hand, the application of later–generation systems is expected to be particularly widespread in hydrology and general water resources management.

Indeed, it appears that in the case of hydrology, as in many other areas, applications only become viable with the introduction of the later–generation systems, so that the earlier generations are, so to say, "sprung over." In order to introduce the notions of these later–generation systems, it is first necessary to review the development of modeling in the companion areas.

Many of the first users of digital computers employed their machines only to calculate numerical values from formulae, in much the same way as they had earlier used slide rules and other analogue devices. This use of the digital machines as "super slide rules" is nowadays called "first–generation modeling." For example, a solution such as $f = f_0 \exp(-at)$ was expressed as a convergent polynomial expansion, as this was amenable to arithmetic, at least for some region of convergence. The process of solution was, so to say, kept as long as possible in the continuum domain, only being translated into arithmetic "at the last moment."

Others of the first users went beyond this stage, however, to use methods which, though impractical for non–machine computation, were well suited to the sequential, repetitive, and recursive mode of operation of the digital machine. These particularly "machine–friendly" methods were especially well adapted to building models that describe variations of properties in space and in time. The most common were the so–called "finite–difference" methods, in which a differential–equation description of the behavior of a system could be approximated, or "modeled" by a difference–equation description through the use of Taylor's series expansions (e.g. Abbott and Basco, 1989). Over the first decade of modeling, which extended roughly from 1960 to 1970, each model of a particular physical area was constructed individually, so that the model so constructed applied only to that specific area. This approach, of one–off or customized modeling, is referred to as second–generation modeling.

For example, the primitive expression $df/dt = -af$ might be translated directly into:

$$\frac{f^{n+1} - f^n}{\Delta t} = -a \left(\Theta f^{n+1} + (1 - \Theta) f^n \right), \quad 0 \le \Theta \le 1, \tag{3}$$

so as to provide the recursion in time of :

$$f^{n+1} = \left(\frac{1 - a(1 - \Theta)\Delta t}{1 + a\Theta\Delta t} \right) f^n, 0 \le \Theta \le 1, \tag{4}$$

This is a more efficient procedure than that of solving a sequence of polynomial expansions. However, this process, of replacing a problem originally posed in terms of the continuum directly by one posed in terms of arithmetic, introduced new problems again. For example, if we set $\Theta = 0$ in the above expression we have:

$$f^{n+1} = (1 - a\Delta t) f^n \tag{5}$$

Thus, for $\Delta t > 2/a$ we have:

$$| f^{n+1} | > | f^n \qquad (6)$$

which is not only unrealistic but also propels f towards infinity as n is increased indefinitely. We thus have a situation where the behavior of the arithmetic "model" differs *essentially* from the behavior of its continuum "prototype."

This kind of difference becomes more pronounced when we have to do with numerical "models" of partial differential equations (PDE's). Thus, to take the simplest PDE and one of its simplest numerical models, we use:

$$\frac{\delta f}{\delta t} + u \frac{\delta f}{\delta x} = 0 \quad \textit{modeled by} \quad \frac{(f^{n+1} - f^n)}{\Delta t} j + u \frac{(f_j - f_{j-1}^n)}{\Delta x} = 0 \qquad (7)$$

By the expression "modeled by" we mean in this case that when the arithmetic statement on the right is expanded in Taylor's series, it provides a statement which becomes the statement on the left as $\Delta t \to 0$. We then say that the arithmetic statement *is consistent with* the continuum statement. The difference between the Taylor's series expansion of the arithmetic statement, on the one hand, and the continuum statement with which it is consistent, on the other, is then called *the truncation error*. The truncation error is non–zero for almost all combinations of finite Δt and Δx.

The above arithmetic "model" can be written as a recursion in space and time as:

$$f_j^{n+1} = (1-a) f_j^n + a f_{j-1}^n \quad ; \quad a = \frac{u\Delta t}{\Delta x}, \qquad (8)$$

This solution procedure, or "scheme," can be represented geometrically as shown in Figure 3.

It is seen to correspond to the choice of a first–order interpolating polynomial, i.e. an interpolation that is linear in x, at least locally. This is experienced in terms of the Taylor's series expansion of the scheme as a truncation error that contains terms of order Δx and Δt.

This linear–interpolation procedure makes sense so long as we have to do with a true interpolation: in this case we are asking the scheme to do something while providing the information necessary to do it.

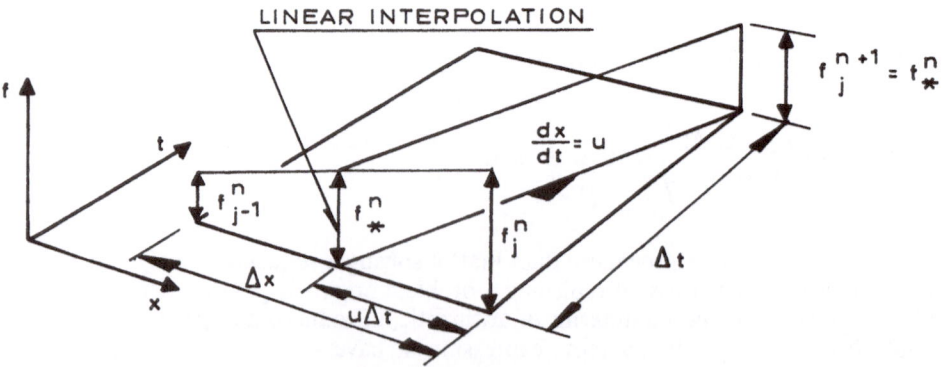

Figure 3. Schematization of the arithmetic model.

If, however, we use a Δt so great that $a > 1$, we are actually extrapolating linearly, and this is quite unreasonable since we do not know anything about what happens outside of the interval $j\Delta x \leq x \leq (j{-}1)\Delta x$.

We cannot then be too surprised if the scheme does something strange when $a > 1$. Indeed, we see that setting $a = 2$, just as an example, we have:

$$F_j^{n+1} = -f_j^n + 2f_{j-1}^n \tag{9}$$

and if we then let:

$f_o^n = 1$ for every n, $n = 1, 2, \ldots$, and $f_j^o = 0$ for every j we have

$f_1^1 = 2$

$f_2^2 = 4$

$f_3^3 = 8$

.

.

.

$f_i^i = 2^i$

.

.

.

etc.

which is, again, not only unrealistic but which propels f to infinity as n tends to infinity. In this case, however, we are dealing with an α which is the ratio $u\Delta x/\Delta t$, so that we can conceive of a sequence of solutions of even smaller Δt and Δx in which the ratio $u\Delta t/\Delta x$ is kept constant at, say, $a = 2$. Then for any finite total prototype time of the computation, $T, = nn\Delta t$, where nn is the maximum value attained by n, we can always find some nn such that $|f|$ exceeds that of any number which we can conceive. In this case we say, of course, that the solution is *unstable*.

We may summarize this situation in the schematization of Figure 4.

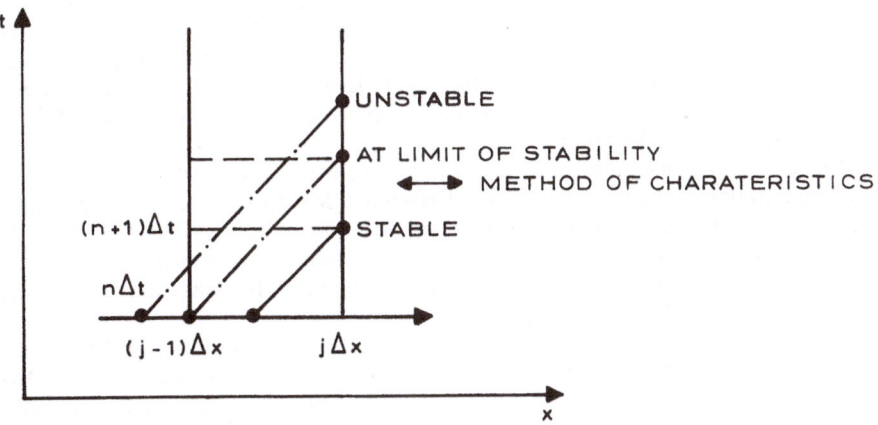

Figure 4. Schematization of the stability condition.

At the limit of stability, at $a = 1$, we have a special, or singular situation, for here the exact solution:

$$f(x, t_2) = f\left(x - \int_{t_1}^{t_2} u\,dt, t_1\right) \qquad (10)$$

can be written as:

$$f\left[j\Delta x, (n + 1)\Delta t\right] = f\left[(j - 1)\Delta t, n\Delta t\right] \qquad (11)$$

which is exactly the same statement as that given by the numerical model when $a = 1$:

$$f_j^{n+1} = f_{j-1}^n \qquad (12)$$

In this special or singular situation, therefore, the continuum statement and the discrete, numerical statement say exactly the same thing.

In terms of the truncation error, we find that its terms cancel out, one against the other, when $a = 1$, so that it disappears completely.

We then have a method for making exact solutions (at least for $u \geq 0$), which, for historical reasons, is called *The Method of Characteristics*.

If we now try to describe what we have done in this process we see that it amounts to a process of *translating between two languages*. The first of these is the language of arithmetic, which we have already denoted by L_1Ar, augmented by operational descriptions expressible in the language of set theory, already denoted by L_1Set, to form a working code.

The denotation of "L_1" then stands for a first–order language, which can be thought of, in the era AD, as any language that can be realized on an ideal digital machine (not necessarily a Turing machine) which is limited to definite (explicitly definable, operational) statements, but which has no round–off, or indeed any other error.

The second of these kinds of languages, is, generally, *the language of the continuum*, which we represent by the letter C. It may be shown (for all practical purposes through Skolem's paradox: see Manin, 1977, p. 64) that there is no unique way of expressing this language that is universally satisfactory. The most economical way is through a language on the set of real numbers, as the standard number paradigm of the continuum. This language is called a "second–order" language, commonly denoted by L_2Real because its elements can be obtained by setting up fundamental sequences (which correspond to power sets in set theory) on elements of "first–order" languages.

The process of translation can be carried out in a large number of ways, such as are commonly subsumed under the headings of Finite–Difference Methods (FDM), Finite–Element Methods (FEM), Boundary–Element Methods (BEM), Fast–Fourier–Transform Methods (FTT), and so on. The most common "dictionaries" used for making translations are those of Taylor's series, polynomial and more general rational function representations, and Fourier series.

It may be shown that in almost every case a statement of finite length in L_2Real translates into a statement of infinite length in L_1Ar and, similarly, a statement of finite length in L_1Ar translates into a statement of infinite length in L_2Real. As already mentioned, in the exceptional case in which a statement of finite length in one language translates into a statement of finite length in the other, the statements together constitute a "method of characteristics."

The prejudice still persists that the language of the continuum is more "rich and sophisticated" than is the language of arithmetic, but in fact we find that exactly the converse can be maintained, and that indeed L_1Ar and L_1Set are intrinsically richer languages than is L_2Real (see, for example, Manin, 1977, p. 112). We can now introduce the manner in which this greater richness is experienced and exploited by considering further the phenomenon of instability.

We may first observe that the PDE, which is essentially a statement about the continuous variations occurring in a process *at a point* in space and time, is contained within the numerical statement; indeed, we only have to go to the limit with the discretion intervals tending to zero to obtain the PDE from the numerical scheme. However, the numerical statement also says something *more* than the continuum statement: it also says something about how the described process varies *over* space and time. The numerical statement cannot, however, proclaim the any–point–in–the–continuum statement without, at the same time, proclaiming the nature of the variations over space and time (unless it be a method of characteristics). Instability occurs when these two proclamations fall into contradiction. *Instability is the numbers' way of telling us that our numerical scheme contains contradictory statements.*

We next observe that the existence of instability implies the following paradox, which, following Richtmyer (1957), we shall call *the stability paradox*:

If we refine the numerical grid, but in such a way as not to satisfy the conditions for stability, the equation in L_1Ar comes closer and closer to satisfying the solution of the equation with which it is consistent in L_2Real, but the solution of the equation in L_1Ar departs further and further away from the solution of the equation in L_2Real.

There is an unfortunate tendency to regard instability in a numerical model as a mere calamity, whereas, in fact, whenever we encounter instability, we enter into a situation in which the model is "trying to tell us something." Instability always arises when our model is subject to an "internal conflict," i.e. when we are "asking the scheme to do the impossible." For this reason we often compare numerical instability with a "nervous breakdown."

In either case, whether of numerical instability or psychic instability, the proper response is not to "dampen–out" the instability – such as by introducing a quite foreign and unphysical diffusion in the one case or prescribing tranquilizing drugs to much the same effect in the other case. The proper response is to get down to the root of the problem, and so to get down to the basic contradiction and to treat that.

The main point that we need to introduce here, however, is that the behavior of the numerical model (i.e. a set of numbers and set of operations on these numbers) *symbolizes* something, and it most commonly does this in a manner which is, scientifically seen, paradoxical. We have to do, in effect, with another kind of number symbolism, which is another kind of *allegory*.

Numerical models are the "number allegories" of our time and the successful modeler is then one who can "read meanings" into sets of numbers and their associated operations. The successful numerical modeler is a person with a feeling for this form of number allegory. Such a person may read into one set of numbers the motions of the waters of the northern polar basin, and into another set of numbers the diffusion of an antibiotic in the human eye. These pretensions of the modern modeler must, of course, appear every bit as fantastic to the "uninitiated" as do, nowadays, the number symbolisms of earlier times!

Modeling, then, is on the one side a scientific activity and on the other side it is an art – and indeed for some it clearly looks more like a "black art" (see, for example, Leendertse, 1981, and the references given there). We may try to identify the source of this impression by observing, more generally, that the model teaches us things of which we were not previously aware, even though the model is itself entirely a product of our own minds. At first sight it may seem that this function of the model can be explained by our not seeing all the consequences of the laws that we have programmed into the model, but we are thereby soon led into intractable problems about "how the numbers get to know" that these laws are operating. Indeed, the characterization of instability as "the numbers way of telling us, etc." is not scientifically justified, but is only a conclusion drawn from a very large number of individual experiences, so that it is a *dogmatic statement*. Of course we can in no way exclude the possibility that a scientific relation may be found between the way of behaving of the numbers and "the logic behind the numbers"—indeed the topological methods used in modern analysis point in just this direction (e.g. Rosinger, 1987) – but certainly this notion did not arise from out of science itself, but out of the practice of modeling.

In order to proceed to the next stage in this introduction along the way of second-generation modeling, let us take as an example the following problem:

How do we describe the effects on the numbers that we do have of the numbers that we do not have?

In the case of fluid dynamics, this problem is the central problem of the *theory of turbulence*. In this case historical prejudice continues to hold sway for the most part, and the problem is reposed in terms of the language of the continuum, and indeed for all practical purposes in L_2Real. We introduce a notion of averaging in such a way as to move, as quickly as possible, from the discrete into the continuous. For example, we may start out from the ideal–gas model of a fluid, as one composed of point masses, or molecules, travelling with a completely randomly direction but with their velocities distributed normally about a finite mean. We average (conceptually) the masses and momenta of all molecules over all areal elements passing through a point. We then divide the mean momentum by the mean mass to obtain the mean velocity vector for each and every area, and we choose the vector of greatest magnitude from the resulting set of mean velocities to obtain "the velocity of the fluid" as a vector at the given point.

By going through this process, however, we are obliged to "loose" certain features that were inherent in the original discrete model. For example, we loose the feature that molecules situated in areas with a larger fluid velocity will move into areas with a smaller fluid velocity, and vice versa, due to the directions of motion being completely random about the mean. This is to say that we loose the feature that the larger fluid velocities will tend to be reduced by the increasing presence in time of molecules arriving from the smaller–fluid–velocity region and the smaller fluid velocities will tend to be increased by the increasing presence in time of molecules arriving from the larger–fluid–velocity region. At that macroscopic scale at which we define the fluid velocity we can only correct for this lost influence by introducing a further macroscopic effect, which is that of "the viscosity." Similarly, if we average over the *energies* of the molecules, we will have to introduce a further macroscopic effect, which is that of "heat conduction." Even at this level, we establish the paradigm whereby averaging causes us to loose something and we try to put this something back by introducing a sort of "de-averaging" process. Within the frame of the mechanical description, the concepts of "fluid velocity" and "viscosity," for example, appear as "duals": the one cannot be properly thought without the other within that frame.

On each occasion that we proceed to an increasing distance or time scale of averaging, we introduce a further set of de–averaging variables.

In the "classical" theory of turbulence, as already introduced, we keep within L_2Real, and within this language we conventionally describe the averaging process by a process of convolution, denoted by *, of the better resolved velocity field with a filter function, G, in order to provide the less–well–resolved velocity field. Thus, if u is the fluid velocity resolved at the smallest macroscopic scale, we define a mean velocity $\bar{u}$ by

$$\bar{u} = (u * G) \tag{13}$$

and a perturbation velocity u' by $u' = u - \bar{u}$. The principle de–averaging variable is then most commonly introduced as an "eddy viscosity." In free–surface–flow hydraulics we commonly integrate further over the fluid depth, from the free surface to the bed, and the de–averaging term is usually called "the bed resistance." We can proceed further to integrate also over the width of a water course, to obtain such conceptual representations of the de–averaging process as that of the "resistance radius."

When we come now to processes necessarily described in L_1Ar on a computer, we commonly continue using this same, continuum paradigm by reformulating the resolution problem that there arises: that we possess no description beyond certain scales $2\Delta x_i$, where Δx_i is the length of the grid interval in x_i, and Δt. We do this in space by introducing a filter function G with effective widths of $2\Delta x_i$, so that as much as possible of the flow information contained at scales smaller than these are eliminated.

Now when we apply this process to differential equations, which describe conditions at a point in space and time, we find that linear terms retain their form under filtering, or averaging, while nonlinear terms give rise to additional terms under this process. The form, and even in some cases the very existence of such terms depends not only on the nature of the instability, but also on the form of the filter function G (e.g. Leonard, 1974; Abbott and Basco, 1989, Ch.9). In particular, some terms remain defined entirely in terms of the filtered variable $\bar{u}$, so that they are representable at this scale. In the case of the terms of the Navier–Strokes equations these "redistribution terms" take the form of stresses and are commonly called *Leonard Stresses* (Leonard, 1974: see again Abbott and Basco, 1989, Ch.9). Such terms then appear as *logical absurdities*, in that they are statements in the language of the continuum (which can resolve down to any scale) of the consequences of our being intrinsically incapable of resolving our world below certain finite scales.

As hydrology comes more to terms with scale effects, such as cannot be avoided in deterministic distributed modeling, so it will have to confront problems of this same kind, which problems are difficult conceptually.

Further averaging processes occur when we plot our numerical results in a plane so as to obtain the illusion of a two–dimensional continuous flow, but in this case we still do not have a specific de–averaging procedure (see Abbott and Larsen, 1986).

All of this provides a workable paradigm, but one that is still essentially restricted to the "world" of L_2Real. The way of the numbers, however, can be more devious than this—and a lot more interesting so long as we care to listen. In order to introduce this yet more devious way of the numbers, let us consider the case of the formation of a hydraulic jump in free–surface–flow hydraulics. (The hydrologist may prefer to think of a wetting front in the unsaturated soil zone, even though the thermodynamics are very different.) As we successively refine the grid, keeping the ratio of Δx to Δt constant (or calculating a steady state so that Δt becomes irrelevant), we obtain a succession of *self–similar* profiles in the water elevation (Figure 5).

In this case, *the truncation error does not disappear as Δx, $\Delta t \longrightarrow 0$*, since the derivatives all increase at exactly the same rate as Δx (or Δt) decrease—since we are dealing with self–similar curves. Similarly, and equivalently, the space integrals of the derivatives remain constant.

In the limit each element of the infinite set of derivatives that appears in the truncation error tends toward an entity which, although it is not itself a function in the sense of standard analysis, can be regarded as a limit, in some sense, of a sequence of functions recognizable by standard analysis. Since the time of Laurent Schwarz (see Rosinger, 1987), it has been usual

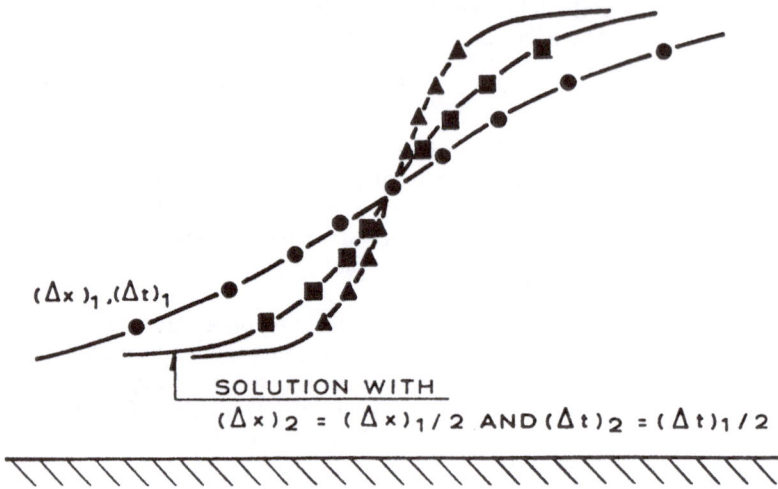

Figure 5. Self–similar solutions with $\Delta x/\Delta t$ constant.

to describe these entities as *distributions*. Thus, the first derivative, the "δ–distribution," is defined by what it does to any test function $w(x)$ which tends rapidly to zero with x:

$$\int_{-\infty}^{\infty} \delta(x)\, w(x)\, dx = w(o) \tag{14}$$

However, this way of defining such "generalized functions" through the scalar product that they provide with "standard functions," was soon found (see Rosinger, 1987) to have a serious disadvantage, in that the product of two such generalized functions could not be uniquely defined—while discontinuities form–up precisely in the case of *non–linear* equations, where the limit necessarily involves products of generalized functions.

The procedure thus held particularly true to Bertrand Russell's dictum that "mathematics is the subject in which we do not know what we are talking about"!

The practitioners of fluid dynamics found a way around this problem by applying the Theory of Weak Solutions, in which recourse was made to a sort of local linearization that can be attained when the PDE's are written in *conservation form* (see e.g. Abbott, 1979, Ch.5). This introduces the infinite set of duals of "levels" and "flux densities," of which only the first five have ever actually been used in practice. Unfortunately—yet again—many of the most interesting cases could not be described in conservation form, so that the theory had distinct limitations. It is in fact only during the last decade that a general way of describing the processes involved has been invented – but then using a mathematical apparatus that is more associated with the realms of pure mathematics than with those of "classical" applied mathematics – not to speak of "engineering mathematics" (see Rosinger, 1987 and Colombeau and Le Roux 1987, with references given in these works).

This example may serve to prepare the way for the next example, for which we have as yet no description in terms of standard analysis – and possibly not even a satisfactory paradigm in the standard model of one–dimensional worlds. We consider the following sequence of flows occurring in the plane that are schematized in Figure 6.

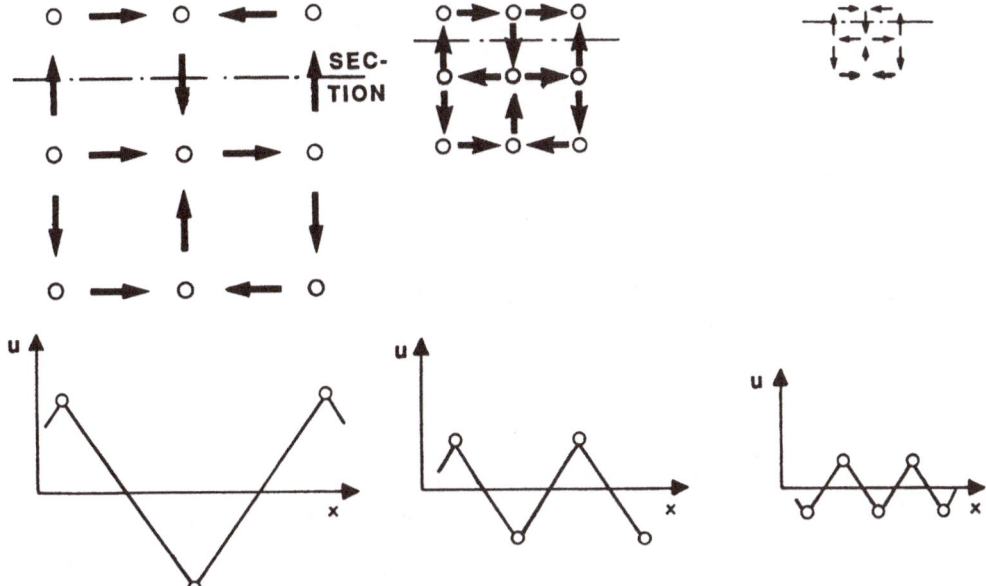

Figure 6. Schematization of plane flows.

Consider now the case of a first–order truncation error that is translated into L_1Ar:

$$\frac{\delta^2 v}{\delta s^2} \Delta s + H.O.T. = \frac{v_{j+1} - 2v_j + v_{j-1}}{\Delta s} + H.O.T. \tag{15}$$

where *H.O.T.* stands for "higher order terms." It is clear that this leading term, at least, can remain constant, even as $\Delta s \longrightarrow 0$ so that, again, *the truncations error does not disappear.*

We may already observe that whenever the truncation error remains constant as $\Delta t \longrightarrow 0$ and the requisite $\Delta x_i \longrightarrow 0$, we have always to do with a sequence of representations that remain essentially self similar as we proceed through this same limiting process. This is to say, however, that in all cases of a constant truncation error, and so in all situations in which the solution in L_1Ar comes to differ essentially from the solution in L_2Real, we have to do with the *theory of self–similar representations*, which is to say, nowadays, that we have to do with the *Theory of Fractals* (e.g. Mandelbrot, 1981). Similarly, the inner and essential reason for our casting PDE's into conservation form in numerical work is that we thereby assure ourselves of the self–similar properties of their solutions (see Abbott and Basco, 1989).

(Let us also observe at this place that, also in the case of instability, it is the truncation-error part of the translation of the scheme in *C* which moves off towards infinity, as at least some of the derivatives in the individual terms of the truncation error increase more rapidly than the products of Δt, and the requisite Δx_i appearing with them in these terms decrease during the limiting process).

There are many other such computational hydraulics' results that must have equivalents in any future computational hydrology. There is space here for only one of these. Consider the solution of $\partial f/\partial t + u \partial f/\partial x = 0$ with the well-worn implicit scheme:

$$\frac{1}{2}\left[\frac{(f^{n+1}-f^n)_j}{\Delta t} + \frac{(f^{n+1}-f^n)_{j-1}}{\Delta t}\right] + \frac{u}{2}\left[\frac{(f_j-f_{j-1})^{n+1}}{\Delta x} + \frac{(f_j-f_{j-1})^n}{\Delta x}\right] = 0 \qquad (16)$$

solved with a single left–to–right sweep:

$$f_j^{n+1} = \frac{(-1+r)\, f_{j-1}^{n+1} + (1-r)\, f_j^n + (1+r)\, f_{j-1}^n}{(1+r)} \qquad (17)$$

Now let $u = 0$, so that $r = 0$, $f_j^o = 0$ for every j and $f_0^1 = 1$.
Then we have the solution:

$$f_j^{n+1} = (-1)^j \qquad (18)$$

If we now proceed to the limit, $\Delta x \longrightarrow 0$, we have a solution which we can only describe as one that "alternates between $+1$ and -1 as we proceed from the one rational point to the next." But how does one write this solution in any standard language––and what then *is* the truncation error?

We can conclude from this, apparently rather long digression into computational fluid dynamics, that in modeling we have to do not only with standard–scientific knowledge on the one side and the nonscientific aspects of modeling on the other, but also with nonstandard scientific endeavors.

It was this complex of different kinds of understanding that developed already within the second–generation modeling groups, even as it rose to it full stature in third–generation modeling. However, the accumulation of this understanding was limited in second–generation modeling to the experience and knowledge of the individual modeler within the environment of the modeling group, and it did not get "fixed," so to say, "at the level of the code." The resulting product was the one–off or customized model of low portability. Even so there emerged, even at this stage, an *aesthetic* of modeling, which we can describe as an experience of a congruence, or a consonance, between the right way of numbers and the observed way of nature (see, more generally, Dilthey, 1976, and Husserl, 1900/1970).

2.2. Third–generation Modeling

The needs of engineering practice were such that the second–generation approach proved inadequate: the lead times for model construction and testing were too uncertain and the corresponding costs were too high for most of the applications arising from hydraulic, coastal, and ocean engineering. Accordingly, from about 1970 onwards, a third–generation approach was developed, which consisted of constructing *design systems* or, as they are nowadays most commonly called, *modeling systems*. These are software packages that construct and run any model of a given, wide class when the description of the model is provided to the package in a definite, specified format. The modeling system is, in effect, "a factory for building models." Whereas the second–generation approach necessitated the writing and testing of special codes for each model, the third–generation approach necessitated only the coding and testing of the system; thereafter, the models could be constructed and run without recourse to further coding. The introduction of the third–generation approach reduced lead times and costs by at least one order of magnitude and made possible a considerable increase in the range and power of numerical modeling methods in practice. In other areas, such modeling systems are commonly called *shells*. Whereas second–generation modeling established the scientific paradigm for modeling in this field, as computational hydraulics, third–generation modeling established a new technological paradigm concerning how this science could best be applied within society at large.

The third–generation approach to modeling thus had a number of other practical consequences. In the first place, it became possible to invest much more in a modeling system than in almost any single one–off model, since the initial investment could be amortized over many tens, if not hundreds of individual models. Similarly, the modeling system could be constructed "off–line," without those constraints of a tight time schedule to which a one–off model would normally be subject. This made it possible to use numerical methods and model–organizational procedures that were far more complicated than those that could be justified for almost any one–off model. These more "refined" methods in turn opened the door to more accurate models and to models that were more flexible and more robust. All of this made numerical models more attractive to end users and increased the throughput of modeling systems further again.

In the second place, the modeling system could be made "to learn from experience," by modifying it in the light of practical experience in its use. Improvements could be introduced into the code, new capabilities introduced, the range of applications extended, and the overall utility of the system enhanced by incorporating changes at the level of the code, rather than at the level of the memory and the notes of the individual modeler.

In the third place a standardization could be achieved in the software that supported the modeling system: charts could be digitized from a plotting table using standard software, distributions of variables could be displayed on screens in standardized displays, and results could be plotted directly in a manner that was easily understandable to the client. Third–generation modeling thus accelerated the movement towards a computer–integrated environment in the centers where such modeling was practiced.

In the fourth place, considerably larger teams of individuals could be built up to practice modeling at this scale, and this in turn allowed a greater differentiation of educations, interests, and talents. The modeling environment that was so created became correspondingly richer and more interesting.

As an example of a third–generation system, constructed on the basis of experience in the companion areas, reference is made to the European Hydrologic System – Système Hydrologique Européen (SHE) (e.g. Abbott et al., 1986).

Through the introduction of third–generation modeling, the following "virtuous circle" could be set into motion:

This circle could be best maintained when the construction of systems and their application occurred in one and the same group of individuals. It was for this reason, for example, that the Danish Hydraulic Institute set up its "Computational Hydraulic Centre" (CHC) both to construct systems and to apply them in practice.

The advance of the third–generation approach was contemporaneous with the more widespread introduction of time–sharing facilities together with reliable compilers for using higher–level languages. It led, however, to a situation where a considerable part of the world's modeling work was concentrated in only a few centers, where highly–differentiated teams could be built up and where powerful computing facilities were available. A considerable gap thereby opened up between the working environment of those providing modeling services and the environments of most end users. Moreover, these end users constituted only a small group of "connoisseurs," and then usually only of model results and not of the models themselves.

The third–generation systems themselves reflected the "expert" environment of their builders and users, in that investment was channeled into improving the range of application and, closely associated with this, the accuracy of the system–built model, with only a limited attention being given to "user–friendliness." The result was that these systems were often quite notoriously delicate and difficult to handle. The delicacy and the difficulty naturally increase as the dimensionality of the model increases since a two–dimensional model has already proceeded through one degree of filtering or averaging, and a one–dimensional model through one degree again, and the effect of each such averaging is always to introduce de–averaging stabilizing terms.

In a comparatively new subject, such as numerical modeling, it is often necessary to seek for analogies, and for this purpose one needs analogies that are exhaustively researched and well documented. In terms of the thousands of pages of popular text expended on them, some of the best analogies are those of motor cars and military aircraft. In terms of those analogies, we might compare a typical third generation system with a Type 35 Bugatti of the 1920's, still a jewel of a machine for the enthusiast and priced accordingly, but surely a horror for the average motorist; or a Messerschmitt Bf 109 of the Second World War, which was for ten years almost unbeatable in the hands of an "ace," but which was also more "self–destructive" in the hands of novice pilots than was almost any other aircraft.

Thus, the effectiveness of the third–generation system depends strongly upon the skill and experience of its users. Important variations may, however, occur also in the degree of "transparency" of the code and its documentation. In an organization like CHC, in which a considerable number of engineers with widely varying backgrounds work upon the same system, a high priority is placed upon transparency, while it is observed that in some other environments this aspect is not considered as so important.

During the first years of operation of a third–generation system, only its users can run it, but after some years of experience and development—and a corresponding investment—such a system can be transferred to another environment. Even then, however, a very extensive training component is required, and this in turn usually presupposes that the end–users are already trained and experienced in modeling generally.

A further disadvantage of the third–generation system is that once such a system is in operation, constant attention must be given to its further development and refinement. Once such a system is "launched upon the market," its further development is commercially unavoidable. We can then say of such a system that "its investment levers are locked on." It is then only too easy, however, to invest one's way out of business! This has the result that third–generation modeling, by itself, can only be at best marginally profitable in the convention actuarial sense. Like the 19th century entrepreneurs described by Rosa Luxemburg and Maynard Keynes, those who construct these systems are constantly investing, but never showing a profit!

2.3. Fourth–generation Modeling

The third–generation of modeling proceeded on a "hard" foundation of time–sharing, main-frame computer systems, as initiated by the IBM 360 series in the 1960's. During the 1980's, however, Personal Computers (PC's) appeared as serious professional tools, accessible to a much wider range of users than were the main frames of the larger modeling centers. This development naturally led to the desire to make modeling services available in a compact user–friendly form on PC's. Moreover, by this time the modeling teams were in place that could satisfy this need.

In view of the fact that modeling services now had to be usable outside of the environment of the larger centers, a considerable effort had to be put into providing efficient menus and menu drives, help–facilities, and other means to provide an independent operating environment. At the same time, a considerable effort had also to be put into determining default values of parameters that would allow these systems to continue functioning in inexperienced hands while still providing results of adequate accuracy. Furthermore, means had to be found to make such systems "fault tolerant," so that even if they were incorrectly used at a first attempt, at least some part of the work done in setting–up a model should be reusable in further attempts. ("If all else fails, read the users–guide"!) It was soon found that these requirements necessitated a considerable rethinking of the whole mode of operation of the corresponding third–generation modeling systems. The resulting products—menu driven, micro–based, default–adaptive, fault–tolerant, data–base–integrated and help–menu–supported systems—then provided the "tools" of the fourth generation of modeling. This then constituted a further step in the "social dimensions" of hydraulics.

This change in the type of modeling brings with it far–reaching changes in the organization and way of working of the modeling group, while it alters the commercial viability of such a group completely. In effect, if some play of alliteration be permitted, the group passes from *a performer of projects to a producer of products*.

To take the commercial aspect first, we observe that the performer of projects sells in the first place the time of its staff on a direct hourly basis. A minute of genius costs the same as a minute of mediocrity. The producer of products, on the other hand, sells primarily that complex of standard and nonstandard scientific understanding, phenomenological insight, hermeneutic understanding and, in the last analysis, good taste, that is incorporated in the group as a whole. If we look around for analogies here we come across that of the business of the couturier, as this has developed during the present century. Earlier in this century the principals in this business served a small clientele of highly demanding and generally very sophisticated users. Each garment was tailored individually, although, to keep to the third–generation analogy, the same design and even the same pattern might be used several times.

The cost of each garment was then largely taken up in the hourly cost of labor used to manufacture it. Later, these same principals (Chanel, Dior, Yves St. Laurent, etc.) moved over to more of a "mass market," of "ready–to–wear" products, whereby the income came much more from what these organizations really always sold—"good taste"—than it did from some overhead on the hourly–paid work done to produce the products. Indeed, today, these businesses (which have become very big) rarely have any real production facilities at all, but are producers of designs—and labels to sew into the finished products!

There is undoubtedly something comical in the prospect of rather coarse engineers going over histories of very refined couturiers, and even of visiting exhibitions to try to "enter into the spirit" of the symbiotic competition that pervaded their rivalries—such as existed between Schiaparelli and Chanel already in the 1930's. However, such analogies are almost always instructive. One of the lessons to be learned from this analogy is that, once such a shift from projects to products is made, the range of products can open up to quite an extraordinary degree. Thus, the couturiers that started by making garments first moved into perfumes, then into shoes, then in custom jewelry and finally, through various stages, they arrived at such items as hotel lobbies, car and airliner interiors and any number of other "aesthetic exercises." At the same time, these organizations had to move away from the world of their select and highly selective clientele and come to terms with the taste of the "big world," of the "mass market."

This analogy shows that a future computational hydrology may have to "come to appearance in the world," or "surface into the world" in a large number of forms, all of which *encapsulate* that with which this computational hydrology has really always been concerned, which is its complex of standard and nonstandard scientific understanding, phenomenological insight, hermeneutic understanding and so on, leading, at the end of the day, simply to: good taste. The analogy shows further, however, that this development can only be realized to the extent to which the organizations concerned learn to satisfy the needs of their "mass market." They have to move from the "small world" of satisfying a limited clientele of sophisticated end-users to the "big world"—that which Heidegger described as "the mundane and average world."

If we then refer to Heidegger's general description of this world (as in his "Being and Time," 1927/1962, e.g. pp. 212-213), we see that what we must come to terms with is the problem of expressing ourselves with "the average intelligibility of the discourse" of this world. The way of discourse of the computational center is quite incomprehensible within the mundane and average environment. Such experiences as those of the transcendental glory of Cantorian Set Theory and the aesthetic joys of the Theory of Fractals simply cannot be transmitted within this environment. It is terrible pity, but that is how it is! It is this transition in the manner of its discourse that really divides the way of third–generation modeling from the ways of fourth and—as we shall shortly see—fifth- generation modeling. Very little happens here, at least in the first place, scientifically; almost everything happens at the level of the intelligibility of the discourse.

To the extent that computational hydrology succeeds in expressing itself with this average intelligibility of discourse, so it can extend its domain into any number of areas which are so far quite untouched by hydrologic knowledge and understanding. We may mention the development of "the intelligent greenhouse," whereby modeling techniques may be encapsulated in more advanced control systems so as to optimize further on energy consumption and (which is much the same thing) water consumption, as well as other inputs. Similarly, the increasingly critical water-supply situation that arises in many parts of the world will in time necessitate the construction of water–economizing field–enclosing structures and soil treatments. We may expect to see an increasing number of *encapsulations of hydrologic modeling expertise*, whereby this expertise is made available on an everyday basis to many millions of people.

By the same token, of course, it will be seen that the fourth and fifth generations of modeling do not replace the third generation. The computation–hydrologic expert is not going to be at all impressed by the "Mickey-Mouse" of the "averaged" products and is certainly not going to waste his or her time in running backwards and forwards through menus. The third generation system will remain for some time still the instrument for the computational expert, and it will develop accordingly in the direction of the corresponding expertise.

To return to an earlier analogy, the relation between the fourth and fifth–generation systems and the third–generation systems is much the same as the relation between a more or less electronically–automated family saloon car, on the one hand, and a Formula–One racing car, on the other.

2.4. Fifth–generation Modeling

Almost parallel with the development of systems for fourth–generation modeling, was that of the elaboration of working expert systems and other products of earlier research in artificial intelligence. It was soon found that no matter how well a fourth–generation system was designed, a permanent "trouble–shooting" service had to be maintained in order to deal with the flow of questions that came back to the manufacturer from end–users. The resulting service center then provided a source of expertise and the next natural step was to incorporate this advice into the modeling systems themselves. As of 1988, this incorporation had proceeded only so far as the first provisions of expert–system "front–ends" on existing fourth–generation systems. (The resulting product is called, with a certain levity, a "four and one–half" generation system.) However, a considerable research effort is currently being concentrated on the future provision of expert–system–integrated modeling facilities, probably incorporating a learning capability, which will then provide the fifth generation of modeling services. This research is backed within the European environment by a wide range of projects sponsored under the ESPRIT program of the Commission for the European Communities (see Chapter 22 in this volume).

We observe that another possible way of naming a fifth–generation modeling system is as an "intelligent modeling system" (which may, however, sound rather disparaging, by inference, towards third– and fourth–generation systems).

2.5. Commercialism and Professionalism

As presented here, the development of computational hydraulics may appear as inseparable from a market–led approach to modeling which in turn depends upon a commercial attitude. However, this mode of development is distasteful, if not actually repugnant, to many hydrologists, who resent the intrusion of "commercial interests" into their science. It has been the experience of computational hydraulics, however, that only through a commercial approach has it been possible to mobilize a sufficient level of professionalism in the sense of finding realistic solutions to real–world problems. In fact, it is the experience of computational hydraulics (at least in the European environment) that modeling groups with sufficient breadth and depth can only be built up within a professional–commercial environment. Certainly, it is essential that the appropriate institutional structures are set in place when making an advance of this kind. A computational hydrology cannot avoid these same imperatives.

2.6. The Problem of Applying Expert Systems in Hydrology

As presented above, the way to the introduction of expert systems into hydrologic practice appears at present to proceed through the further development of fourth–generation modeling systems, if only because the application of these instruments is inhibited by the lack of expertise of the majority of their potential users. On the other hand, advance knowledge-based systems themselves require a simulation "package" with which to represent their "world" of domain knowledge, even if only as a means of organizing inputs from on–line data streams. This package, however, corresponds, by and large, to the "fourth–generation" system. Thus, on the one hand we observe simulation packages looking for expert systems and on the other hand we observe expert systems looking for simulation packages. These two are now meeting up, and one outcome of their union will be fifth–generation modeling systems, incorporating a large range of artificial–intelligent facilities.

Thus, on the one side the application of expert systems will provide entirely new potentialities for applications of hydrologic modeling while, on the other side, hydrologic modeling will provide entirely new potentialities for the use of expert systems. The question concerning the application of expert systems has to be seen from both of these sides. When so seen, however, it becomes clear that a new approach—which can be called a new technology—has to be developed. We have to do with a new kind of *Knowledge Product* (KP) for which indeed, as of 1988, we did not even have a name (see Abbott, 1991). When seen in terms of present products, it appears as a combination of a classical (or, rather, academic) AI logic code, complete with rule bases, inference engines and so on, and a classical (applied) algorithmic code, complete with topological interfaces, algorithmic structures, system frames and all other such elements. When seen from the deeper point of view of mathematical logic, we have to do with the "demonstrational" and "computational" sides of the subject, and thus unavoidably with certain equivalences (homomorphisms) that have long been established between these sides. When seen from a decision–theoretic point of view, we have to do, on the one hand, with the setting of decisiveness conditions and the possibilities and impossibilities thereof, and, on the other hand, with the simulated consequences of imposing these conditions. We shall return to these approaches shortly.

In classical or academic Expert Systems (ES's), it is usual to distinguish between domain expertise (roughly speaking, the "facts of the case") and AI expertise (even more roughly speaking, "how the facts are interpreted in a given situation"). In the classical or academic approach, however, these two classes of expertise are *separated*, by what really constitutes a *massive simplification*. This simplification consists in the assumption that domain knowledge can be expressed in terms of a finite sequence of simple statements, all of which have definite unequivocal meanings. Even when some "fuzziness" is allowed, this is still "decomposed," so that the "fuzzy" problem can be represented in the same kind of definite and unequivocal way (e.g. Turner, 1984; Genesereth and Nilsson, 1987). In order that a computer can be used at all, of course, the statements must be representable within a first–order language, but in the academic approach the "first orderedness" is introduced already at the level of the individual indivisible symbols (e.g. words of natural language) of the expressions. Thus, to take an extreme example, a code constructed in these terms may "understand" statements like "it is raining" and "it is not raining" and possibly something like "it is raining a little," but it cannot comprehend something like "the net precipitation rate varies as shown in the accompanying graph" (unless, of course, it breaks this graph down into a sequence of statements of the type just exemplified). Indeed, probably the best that current academic ES's can at present achieve in this respect is to provide sophisticated search routines, with the sophistication residing essentially in the search control. This facility in turn necessitates a particularly carefully formalized control specification, which in turn appears to require the explicit introduction of some kind of "metaknowledge" (e.g. Davis, 1980; Aiello, 1984). This in its turn leads to the difficulty that most existing ES Shells (ESS), such as would be used, by preference, for the constructing of an ES (e.g. ART, KEE, Knowledge Craft, and NEXPERT), although otherwise allowing for a wide range of knowledge formalisms and providing powerful reasoning techniques, do not provide any great flexibility in control specification, or at least their flexibility in this respect cannot be assessed by their users. In particular, the constraints that are naturally introduced from a decision–theoretic point of view (such as through the Arrow impossibility theorems) cannot be explicitly introduced into the major existing ESS's.

These considerations alone would appear to invalidate the project of using more sophisticated search routines in ES's so as to represent numerical functional data and all that is done with these data in numerical modeling.

2.7. The Problem of Combining Logic Coding and Numeric Coding

As already observed, many works on Mathematical Logic are divided into two parts, the one of which has to do with *demonstration*, and the other with *computation*. As has already been explained, these two sides are often intimately connected, and indeed we have already observed that it has become a common practice to "model" a theory in the one part by a theory in the other part, a strategy that was pursued progressively in the era between Frege and Gödel through the "arithmetization of logic" (Van Heijenoort, 1967).

The task of elaborating a system for fifth-generation modeling, can be described in rather similar terms. In effect, there is a logical demonstration component which reflects the *structure* of decisiveness conditions, as imposed by the "outer world" of, primarily, human decisions and concomitant interventions, and a "computational component" that describes, at least by implication, the structure of the "outer world" of physical system behavior. The problem of combining logic coding and numeric coding can be posed as one of associating these two structures through means that are best suited to their respective coding languages. In effect a mapping has to be defined that can be interpreted formally as a homomorphism.

One approach to this problem proceeds through a system of "types" which can be thought of as *identifiers*, with interpretations in both the logic and numeric contexts. These identifiers must then themselves have a structure which is homomorphic to the structures of the logic and numeric components: the identifiers must themselves "model" both classes of structure.

As an example of such a combined logic and numeric systems, a schematization is given in Figures 7 and 8 of one possible architecture for use on board a ship for the design of an optimal routing (see Abbott et al., 1988). On the left are the logical constraints, which can be expressed as facts and rules if this should turn out to be productive, and on the right are the specifically numerically expressed inputs. In such a system the facts are "merged", using the rules to create more global fact or constraints sets: we have to do with a sort of "constraint propagation" proceeding in Figure 7 from the left towards the right. Similarly, numerical data is combined into numerical constraint sets, proceedings from the right towards the left in Figure 8. In this architecture, the fact-and-rule part is called the "office" part and the numerical part is called the "factory" part, thus playing along with the prejudice that the performance of logical operations has a "higher function" than the performance of numerical operations. The reader will easily translate this schematization into one for optimal water utilization in a hydrologic catchment.

2.8. The Question Concerning Applications

Already, at the time of the SHE market survey of 1971, it became clear that traditional methods of hydrologic modeling, such as rainfall–runoff modeling, could cover only a small part of the potential demand for hydrologic modeling services. It was against this background that the SHE was developed, as a third-generation system, for practical engineering applications. However, operational experience with the SHE has since shown that the level of background knowledge, training, and experience required to use this tool efficiently exceeds by a considerable margin that which is usually available. In analogical terms, the difference in difficulty between using a conventional rainfall–runoff model and using the SHE is the difference between riding a bicycle and flying a commercial airliner! The step required to proceed from simple lumped models to physics–based distributed and dynamic models constitutes a quantum leap which is difficult to realize. It then appears that further progress in making hydrologic modeling useful to water resources management necessitates the provision of ES facilities that are fully integrated with the numerical facilities. These tools must then incorporate intelligent default options, interrogation facilities, and all the usual means for making the system user-friendly and operationally transparent: they must constitute fifth generation systems. The question concerning applications then is this: are such systems commercially feasible?

This question of the commercial, or micro–economic feasibility of fifth-generation modeling may be placed within context of the generic problem of elaboration software tools, and

532

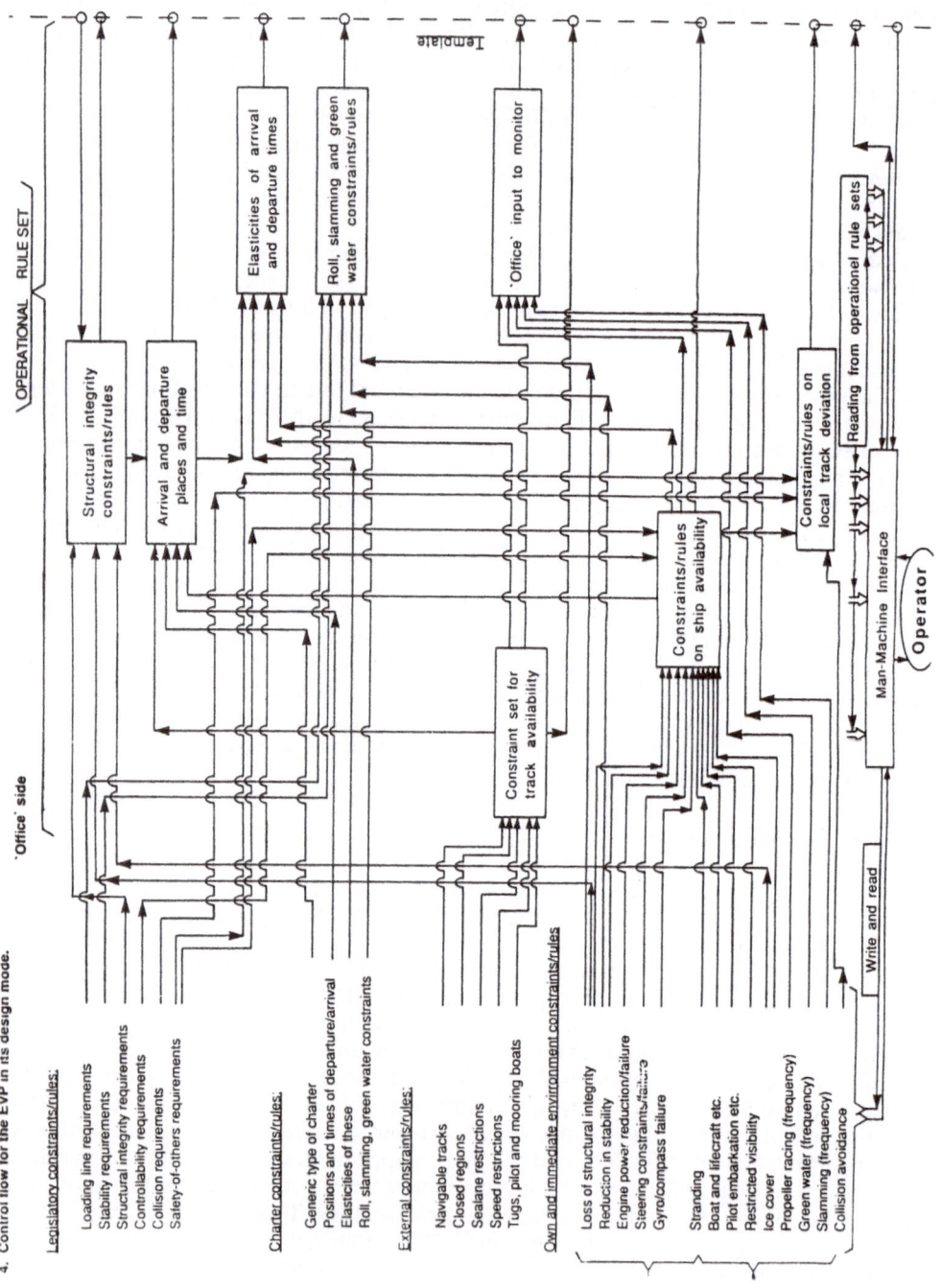

Figure 7.

4. Control flow for the EVP in its design mode.

Legislation constraints/rules:

Loading line requirements
Stability requirements
Structural integrity requirements
Controllability requirements
Collision requirements
Safety-of-others requirements

Charter constraints/rules:

Generic type of charter
Positions and times of departure/arrival
Elasticities of these
Roll, slamming, green water constraints

External constraints/rules:

Navigable tracks
Closed regions
Sealane restrictions
Speed restrictions
Tugs, pilot and mooring boats

Own and immediate environment constraints/rules:

Loss of structural integrity
Reduction in stability
Engine power reduction/failure
Steering constraints/failure
Gyro/compass failure

Stranding
Boat and lifecraft etc.
Pilot embarkation etc.
Restricted visibility
Ice cover
Propeller racing (frequency)
Green water (frequency)
Slamming (frequency)
Collision avoidance

OPERATIONAL RULE SET

'Office' side

Template

Structural integrity constraints/rules

Arrival and departure places and time

Elasticities of arrival and departure times

Roll, slamming and green water constraints/rules

'Office' input to monitor

Constraint set for track availability

Constraints/rules on ship availability

Constraints/rules on local track deviation

Reading from operational rule sets

Man-Machine Interface

Operator

Write and read

533

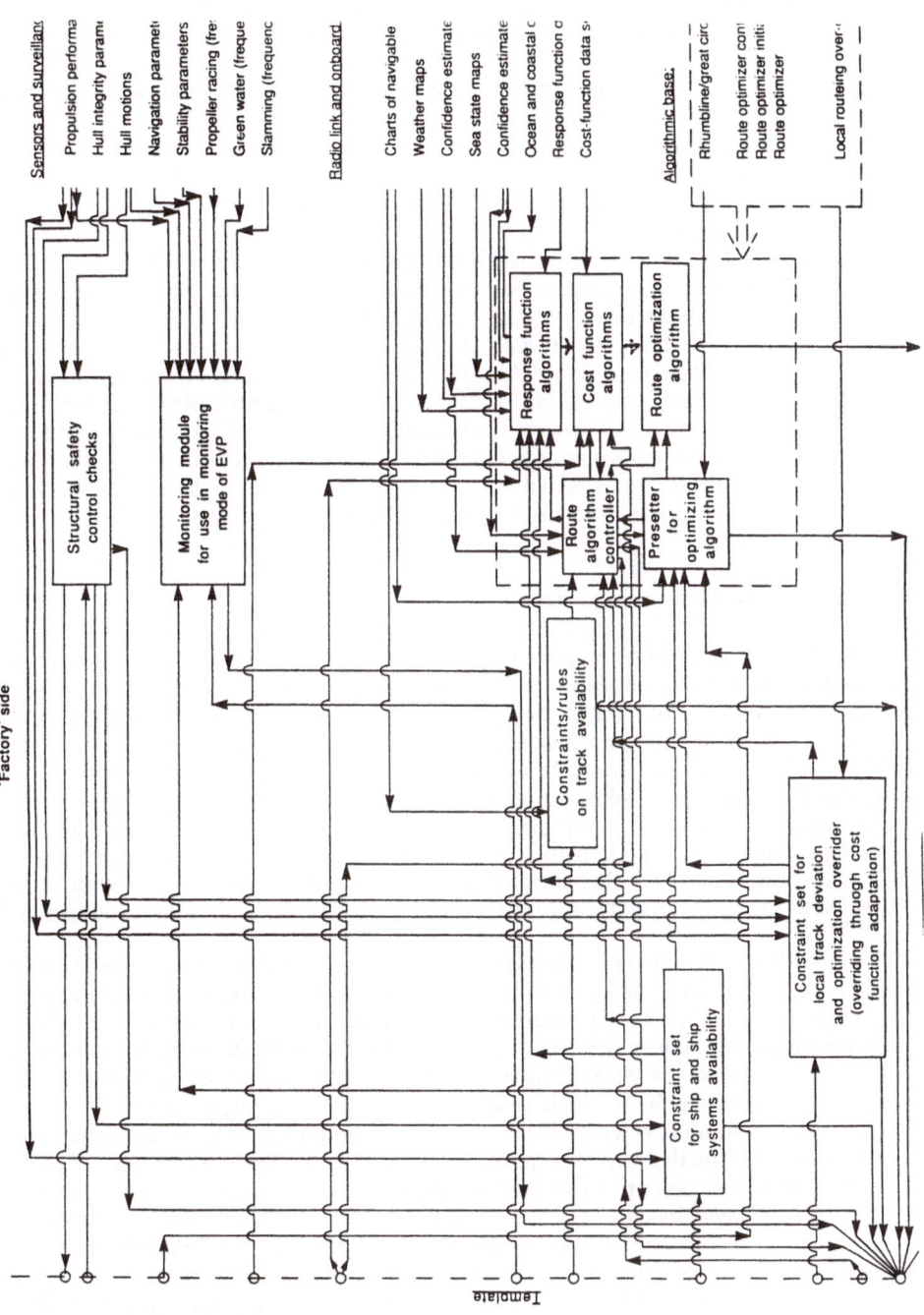

Figure 8.

then tracing this problem back to the most basic of such tools. Within the European context, this can be related to the problem of exploiting the results of the six–billion–dollar ESPRIT program.

ESPRIT, the European Strategic Programme for Research in Information Technology, can be divided into four main branches:

(1) Knowledge engineering
(2) External interfaces
(3) Information and knowledge storage
(4) Computer architectures

The first three of these are of obvious relevance to the subject of this lecture, and indeed, as ESPRIT has progressed, it has appeared to be more and more concerned with tools that will make expertise available within a mundane and average environment. The problem of "capturing" the ESPRIT technology, however, remains a formidable one. The relevance of the ESPRIT and the problematics of capturing its technology are discussed in Chapter 22. To the conclusions of that chapter should be added that the cost of capture of any such information technology depends upon the level of authenticity of the domain knowledge and level of authenticity of the basic information technology tools. In the event that both are questionable, the cost may easily rise to such a level as to make the capture commercially uninteresting. In any event, quite a new institutional arrangement is necessary to effect such a technology capture.

2.9. The Question Concerning Education

The question that now has to be posed is how the introduction of such systems as have just been outlined will change education and research in the area that has traditionally been the domain of hydrology. From the immediate and practical point of view, simulation packages are already available that far transcend the possibilities of even the most elaborate and sophisticated laboratory equipment, at the very least in the range of their applications but often also in the prototype physical realism of their results. For example, the student can already have, at his or her fingertips, the complete sewer network of a large town, or the complete canal system of a major irrigation system. How is education to proceed in the presence of these and any number of other such simulation packages that are already being prepared, and which must increasingly intrude into engineering hydrology and water resources applications?

This question raises at once the matter of the extent to which education is to remain "application–area driven" or is to become more "methodology driven." It can be argued that a new emphasis must be placed on methodologies that are common to a variety of application areas, for the student has now to come to terms with the methodologies behind simulation packages, CAD packages, work–station operating systems, and other items of information processing equipment if he or she is to use such facilities in a responsible way. The actual execution of designs and projects will clearly be much simplified by the use of expert systems, but the understanding that will be required to justify the use of expert systems will be at least of another type, and perhaps of another order, than that required to execute designs and projects by traditional means. The "production process" becomes generally simplified, but this is only possible because the tools that make this production possible become more complicated. The tendency, then, appears to be one of a moving away from the details of the production and a moving towards the *software tools* that make this production possible, and the most significant of these software tools must be the expert systems themselves. The knowledge required is less that of doing the job and more that of understanding how the job is done. It follows, in passing, that a range of expert systems, together with expert system shells and builders, must be made available in fully documented and possibly even source–code form to Universities. This may well lead to the dismantling of existing departmental divisions, not so much through "interspecialization" (which has been found only to lead to new specializations), but by the formation of a new kind of generalism.

2.10. The Question Concerning Research

The problem concerning the application of expert systems in hydraulics, hydrology, and water resources research appears to lead to quite other solutions than those that apply to teaching. Again keeping to only a rather superficial level of analysis, it is clear that the bulk of research will be directed to applications of software tools, which requires that an extensive range of software tool sets is available to the researcher. While the drive to generalism speaks for the renewal of the university, the needs of research speak for other environments, in which powerful and correspondingly expensive software and other equipment are available. In this respect it should be remembered that research spending on software exceeds spending on hardware in most major hydraulic and water resources institutes already, and this tendency can only be expected to strengthen in the future. Already there seems to be little prospect of equipping universities with software facilities even resembling those available in the major research institutes (let alone of maintaining such equipment!). Thus, by way of an example, the major European institutes working in the computational hydraulics field each have between 30 and 50 engineers (with M.Sc. and Ph.D. degrees) working just on research development and maintenance. It is, moreover, anticipated that this number will double within five to ten years. Thus, it is difficult to see how universities can continue with their research activities except at a certain level of fundamental or abstract investigations or in relation to a few selected systems. These considerations speak for a more generalized education at the university level and ongoing "specialization-by-specialization" education at research organizations, except in the case where the university can lay claim to a unique advantage in a particular area of system development and application.

It then appears, at least at this rather superficial level of analysis, that a clear distinction has to be drawn between applications of hydrologic-modeling systems in education and their applications in research. The line appears to divide two very different environments.

3. Conclusions

Numerical modeling in hydrology and water resources has come into something of an impasse, caused by a mismatch between, on the one hand, the social requirements placed upon modeling services and the concomitant complexity of the models, and, on the other hand, the training and experience of practitioners in these fields. It is a reflection of this state of impasse, that a veritable "crisis" is descried in this area, and indeed the future of hydrology as a science is itself called into question (e.g. Klemes, 1986).

From the analyses so far made, the following features of this crisis and its resolution can be discerned:

(1) Hydrology is not a science in the same way as are its constituent parts: meteorology, soil physics, plant physiology, agronomy, etc. It is essentially for this reason that there is no undergraduate education in hydrology. Hydrology is a kind of "scientific conglomerate," grouping together a large number of otherwise rather separate sciences, extracting material from these sciences and combining this material according to objectives which are usually quite other than the objectives of the individual sciences themselves.

(2) This feature of hydrology is antipathetical to the main thrust of modern science, which is directed to producing increasingly narrowly specialize work, with the objective of providing increasingly fast returns on investment. Accordingly, hydrology does not attract funding to

the same extent as, even, certain of its sub–domains do. In this respect, hydrology suffers in the same way as other "scientific conglomerates," such as geography.

(3) From the *scientific* point of view this situation can only be redressed by allowing one of the constituent sciences of the "conglomerate" to attain to a predominant position, so that it is able, so to say, to "dictate" to the other sciences as to their functions in the conglomerate. To some extent this is happening in the case of meteorology, where the drive towards Global Circulation Models (GCM's) can be seen as a "scientific unifier." The funding of GCM's is in turn driven by concerns that are highly community– and national– political, so that very considerable resources will continue to flow into this area over the next decade.

(4) However, this approach cannot come to terms at all directly with such real–world problems as deforestation, soil erosion, flooding, pollution, and other features which proceed at quite other scales than those used in current or foreseeable GCM's, and which problems are themselves closely linked with geopolitical, legal, agricultural– precedent, and other such constraints. From the *technological* point of view, therefore, quite other scenarios present themselves for the future of hydrology.

(5) These scenarios indicate that hydrology can only come to terms with real–world technological problems, and thus break out of its present impasse, to the extent that it can make a quantum leap to the levels of fourth– and fifth–generation modeling. This, however, is increasingly a modeling that has both deductional and computation components, unified into a number of homogeneous–modeling systems and domain–knowledge encapsulators.

(6) It follows that computational hydrology can never fully realize itself without the elaboration of a "deductional hydrology." The computer–integrated development of hydrology must take account of hydrology's "scientific–conglomerate" nature, and so cannot be modeled directly on computer–integrated developments in other fields, such as hydraulics.

(7) The development of software tools in this area has to conform, on the whole, to this form of development. This is to say that although an instrument like the SHE must be developed as far as possible scientifically, as a third–generation system, it will never realize anything like its full potential until it is itself incorporated into, and indeed integrated with, deductional–computational devices, along the lines of the KBSSHIP devices introduced above.

(8) Beyond this again, the design and construction of the tools necessary for this technological realization of the potential of hydrology call for new institutional arrangements, bringing together scientific and industrial partners in new and as yet untried combinations. This in turn poses new problems of financing, for which again innovative solutions will be required.

(9) Corresponding to this analysis, there is no other way to go technologically that makes any sense in economic terms. Third–generation modeling in hydrology, with the SHE or any other such instrument, is only an expedient: a way of acquiring operational experience, understanding and a "higher aesthetic," and so a way of investing in a future capability. The realization of this capability, and the breakthrough to commercial viability and thus to a long–term, capital–intensive development in hydrology and water resources engineering, necessitates the jump to fourth– and fifth–generation modeling.

Over and above these conclusions it must be observed that this leap must have far–reaching consequences for education and research activities, including the organization of these activities.

4. References

Abbott, M. B.: 1966, *An Introduction to the Method of Characteristics*, Thames and Hudson, London, and American Elseviers, New York.
Abbott, M. B.: 1979, *Computational Hydraulics*, Pitman, London.
Abbott, M. B.: 1991, *Hydroinformatics: Information Technology and the Aquatic Environment*.

Abbott, M. B., and D. R. Basco: 1989, *Computational Fluid Dynamics*, Longman, London, and Wiley, New York.

Abbott, M. B., L. Bardis, C. P. W. Hornsby, P. S. Katsoulakos, M. Lind, and T. Wittig: 1988, 'An Architecture for a Shipboard Knowledge–Based System,' *ESPRIT '88, Putting the Technology to Use, Part 1*, North Holland, Amsterdam, pp. 780–795.

Abbott, M. B., J. C. Bathurst, J. A. Cunge, P. E. O'Connell, and J. Rassmussen: 1986, 'An Introduction to the European Hydrologic System – Système Hydrologique Européen, SHE,' *J. Hydrology* **87**, 45–59 and 61–77.

Abbott, M. B., J. Larson, and Jinhua–Tao: 1985, 'Modelling Circulation in Depth–Averaged Flows,' *J. Hyd. Res.* **23**, 309–326; 397–420.

Aiello, L.: 1984, 'The Uses of Metaknowledge in AI Systems,' *Proc. ECAI 84, Advances in AI*, Elseviers, Amsterdam and New York.

Bourbaki, N.: 1960, *Elements d'Histoire des Mathematiques*, Herman, Paris.

Colombeau, J. F., and A. Y. LeRoux: 1987, 'Numerical Methods for Hyperbolic Systems in Non–Conservative Form Using Products of Distributions,' in R. Vichnevetsky, and R. S. Stepleman (eds.), *Advances in Computer Methods for Partial Differential Equatorial*, IMACS.

Davis, R.: 1980, 'Meta–Rules: Reasoning about Control,' *Artificial Intelligence* **15**.

Dilthey, W.: 1976, *Selected Writings*, J. Mcquerrie, and E. Robinson (Trans.), Blackwell, London.

Genesereth, M. R., and N. J. Nilsson: 1987, *Logical Foundations of Artificial Intelligence*, Morgan Kaufmann, Los Altos, California.

Hankins, T. H.: 1967, 'The Reception of Newton's Second Law of Motion in the Eighteenth Century,' *J. History of Ideas*, 43–65.

Heidegger, M.: 1927/1962, *Being and Time*, J. Mcquerrie, and E. Robinson (Trans.), Blackwell, London.

Heidegger, M.: 1949/1977, *The Question Concerning Technology and Other Essays*, W. Lovitt (Trans.), Harper and Row, New York.

Van Heijenoort, J.: 1967, *From Frege to Gödel: A Source Book in Mathematical Logic, 1879–1931*, Harvard Univ. Press, Cambridge, Mass.

Husserl, E.: 1900/1970, *Logical Investigations*, J. N. Findley (Trans.), Routledge and Kegan Paul, London.

Jung, C. G.: 1944/1953, *Psychology and Alchemy*, Routledge and Kegan Paul, London.

Kelly, J. S.: 1978, *Arrow Impossibility Theorems*, Academic Press, New York.

Klemes, V.: 1986, 'Dilettantism in Hydrology: Tradition or Destiny,' *Water Resources Research* **22**(9), 177–188.

Lauwerier, H.: 1987, *Fractals: Meetkundige Figuren in Eindloze Herhaling*, Arimith, Amsterdam (in Dutch).

Leendertse, J. J.: 1989, 'Discussion,' in H. B. Fischer (ed.), *Transient Models for Inland and Coastal Waters*, Academic, New York.

Leonard, A.: 1974, 'Energy Cascade in Large–Eddy Simulations of Turbulent Fluid Flows,' *Advances in Geophysics* **184**, 237–248.

Macagno, E.: 1988, '*Leonardian Fluid Mechanics: What Remains to be Investigated in the Codex Hammer. A Critical Study and a Challenge*,' Iowa Inst. of Hydraulic Research, Iowa City.

Mandelbrot, B. B.: 1981, *The Fractal Geometry of Nature*, Freeman, New York.

Manin, Y. I.: 1977, *Mathematical Logic*, Trans. Koblitz, N, Springer, New York.

Richtmyer, R. D.: 1957, *Difference Methods for Initial Value Problems*, John Wiley, New York.

Rosinger, E. E.: 1987, '*Generalised Solutions of Nonlinear Partial Differential Equations*,' Notas de Matemática 146, North Holland, Amsterdam.

Rosinger, E. E.: 1988, '*One Dimensional Cyclic Dynamics*,' Dept. of Mathematics, Univ. of Pretoria.

Turner, R.: 1984, *Logics for Artificial Intelligence*, Ellis Harwood, Chichester, U.K.

Part VI – Synthesis of Hydrologic Models for Different Physical Settings and Various Applications

Chapter 24

Problems of Runoff Modeling Which are Particular to the Area or Climate Being Modeled

David R. Dawdy
3055 23rd Avenue
San Francisco, California 94132 U.S.A.

Abstract. Models of hydrologic systems are influenced by local climate and topography. The choice of models should consider particular characteristics and problems as related to various climates and physiographic regions. Arctic hydrology, snow–melt hydrology, temperate as opposed to arid–zone hydrology, and hydrology of very flat areas, problems involved in modeling in particular areas or climates, what is being done to overcome those problems, and areas of needed research are discussed.

1. Introduction

Problems involved in the modeling of hydrologic systems are influenced by the local climate and topography. The statement is a truism, but it is often forgotten in the research and application of hydrologic models. Frequently a hydrologic problem is stated, then the modeler looks at his favorite model and asks how that model can solve the problem. Often it can't. A recent Workshop of the International Association of Scientific Hydrology (IAHS, 1987), held at the IUGG General Assembly in Vancouver, addressed the choice of models in terms of the particular characteristics and problems of models as related to the various climates and physiographic regions. In particular, sessions were developed around Arctic hydrology, snow–melt hydrology, temperate as opposed to arid–zone hydrology, and the hydrology of very flat areas. The emphasis was on problems involved in modeling in particular areas or climates, what is being done to overcome those problems, and areas of research still needed for those particular problems. This chapter is a summary of the contents of some of the keynote papers from that workshop.

2. Discussion

2.1. General Overview of Modeling Problems

Most model development is for the temperate zones. That is where most people live and most research into modeling of the hydrologic cycle takes place. Because most models are developed for application to temperate regions, modeling for those regions is relatively better developed than elsewhere. The level of understanding or the lack of understanding in such models should give some insight into where models for other regions and their problems are heading. "In order to obtain useful results...a variety of water balance models is needed, developed in different scales. In general, there is a trend to physically based models with spatially distributed parameters and an adequate temporal resolution." (Dyck, 1987)

D. S. Bowles and P. E. O'Connell (eds.), Recent Advances in the Modeling of Hydrologic Systems, 541–547.
© 1991 *Kluwer Academic Publishers.*

The problem of scale in basin modeling and the desire for physically based models work against each other to an extent. The smaller the area being modeled, the more physically meaningful the parameters of a model can be. As area increases more averaging takes place, and the more the parameters and the variables, such as infiltration or soil moisture, become averages; then, at larger scales, the better the indices which maintain the structure of the physical representation of the system can simulate the system's large–scale response. This is partly overcome by the use of distributed parameter models, so that the physically based concept is applied at an appropriate scale and results aggregated to obtain the output from the basin. However, data requirements grow with the need to define parameters at a small enough scale. Computation time also increases with the more detailed subdivision of a basin, so that for large enough basins scale averaging intrudes even for distributed parameter models. The Mississippi River or the Rhine at their mouths cannot be modeled in enough detail to maintain a set of "true" physical parameters.

Part of the drive toward physically based models results from the need to predict changes in the hydrology of a basin which result from man–made changes. This need is greater for smaller basins, where man–made changes may affect a major part of the basin and cause an easily visible impact on the hydrologic response. As the basins that are modeled become larger, man's impact is more masked, and, once again, the problem of scale requires a different means to assess those impacts.

However, there is some hope. Eagleson (1982) states that "water–limited natural vegetation systems are in stable equilibrium with their climatic and pedologic environments when the canopy density and species act to minimize average water demand stress...Using these hypotheses, only the soil effective porosity need be known to determine the optimum soil and vegetation parameters in a given climate." He further states, with proper qualification, "...it appears possible to make rational estimates of soil and vegetation properties from a minimum of prior information, a capability which should be of great use to hydrologists and hydroclimatologists." The conditions for the stable equilibrium of Eagleson appear to hold best for temperate regions, however. Therefore, this gain for modeling in temperate regions may partly explain why models for such regions are not directly transferrable to other, different, regions. Regions of desert or of permafrost, for example, may have a different condition of stable equilibrium, or none at all. In an area with frozen ground, the infiltration may be almost zero. Upon thawing, the soil may have a very high infiltration rate until the energy of the raindrops compacts the surface soil and it returns to its "normal" infiltration rate.

Thus, modeling for temperate regions is moving in two contradictory directions. The first is toward physically based, distributed–parameter models, which leads to more and more detail in the model. The second is toward an understanding of the influence of scale in modeling, which leads toward simpler, more averaging and averaged models as the modeled area grows larger. These need not be contradictory. Perhaps the dominant process modeled should change with scale, even in physically based models. On top of both of these trends is the continuing interest in error analysis of hydrologic models and in the underlying stochastic nature of any predicted output (Troutman, 1985).

2.2. Runoff Modeling in Areas of Low Relief

In regions of very low relief, natural drainage occurs mainly as infiltration, evaporation, and, only very slowly, drainage. Groundwater rises easily to the surface, soils become saturated, and there is waterlogging. A major determinant of systems modeling in such regions depends upon whether the major direction of water movement is mainly vertical or horizontal (Kovacs, 1984). For gravity flow and drainage, the driving force often must be provided through human effort. In that case, the hydrologic response is not a natural catchment characteristic. Kienitz (1987) terms this *influenced runoff* to differentiate it from the natural basin response. Thus, the modeling of the hydrology of these areas is to produce tools to facilitate human interference with the natural system. The components of any model must include the human structures and the parameters must include those of the structures.

Planning in regions of low relief involves a choice of the hydrologic system to be imposed on the natural system. The first step is to assess the relative importance of various components of the hydrologic model, in particular the storages. Kienitz categorizes the storages and inputs in terms of dominance:

A. Dominant storage space
    A1. Groundwater aquifer
    A2. Saturated upper soil layers
B. Dominant input
    B1. Precipitation
    B2. Surface inflow from outside
    B3. Groundwater intrusion

Based on this preliminary assessment, the hydrologist can begin modeling the hydrologic cycle. However, that planning is done to achieve a goal, and the goal or objective influences the resulting hydrology imposed on the area. Thus, the objective can be to keep the area as nearly as possible in its natural state, such as a park, or it can be to optimize growing forests, pasture, or crops, or it can be to develop the area for human habitation. The objective is more of a determiner of the resulting hydrologic system than is the natural environment. Thus, modeling in areas of low relief is more concerned with testing alternative treatments of the watershed than in determining the natural hydrology, although the natural hydrology may constrain the alternatives which can be chosen.

2.3. Modeling Runoff in Arid and Semi-arid Regions

Arid regions, on the other hand, are regions where the climate controls the hydrology (Pilgrim and Chapman, 1987). What is peculiar to arid-zone hydrology? First, rainfall tends to be both less and more variable in time and space than in humid regions. Pilgrim and Chapman cite a site in northwestern Australia where, for two of the four years of record the total rainfall was less than 60 mm, whereas for the other two years each had single days which recorded over 500 mm. Data on the rainfall in arid regions tends to be sparse. This, combined with the greater natural variability of the rainfall, introduces a large element of uncertainty into the estimates of rainfall characteristics.

Base flow is generally absent and channel losses are of critical importance, both from infiltration into the channel bottom and from growth of phreatophytes. Although the water table usually is disconnected from the surface drainage system, a temporary saturated hydraulic connection may occur during flood events. More interestingly for the hydrologist, the whole nature of the hydrology and values of model parameters may be changed by a prolonged wet or dry sequence. However, there are some compensating factors. Evaporation and transpiration in arid regions commonly account for 95 percent of the runoff, but this aids modeling to an extent. The problem of the estimation of antecedent moisture conditions is simplified. They usually are dry, and all storages are empty; many modelers automatically assume this in their models. In addition, most action concerning soil moisture takes place in a shallow surface layer, so that modeling of the soil zone can be simpler in models for arid regions. However, an intense storm on a previously wet basin can cause particularly extreme runoff events.

Water losses in alluvial channels are widely reported, but not often modeled in basin models. Pilgrim and Chapman (1987) cite studies of channel losses in Arizona, southern Spain, and Australia. About half the volume of flow can be lost through seepage into the channel bottom. This can have a greater effect than the effect of flood routing, which usually is included in the models.

Are channel losses Hortonian or non–Hortonian? In arid and semi–arid regions, the predominant runoff mechanism is Hortonian. When runoff occurs, the rate of rainfall exceeds the potential rate of infiltration. Times to ponding in arid regions can be short, with less than 10 minutes reported. Small amounts of rainfall can cause runoff, even with the dry antecedent conditions. However, variable source area concepts have been shown to occur in some cases as a result of variations in infiltration capacity. (The analog of source–area modeling in the Stanford Watershed Model was explained as the modeling of a distribution of potential infiltration over the basin.)

The channel transmission losses affect the shape of the outflow hydrograph by causing the time of rise from little or no flow to peak flow to decrease, because the antecedent lower runoff amounts infiltrate. The intense rainfalls plus the transmission losses and overriding translatory waves cause the wall of water at the front of the flood wave often encountered in streams in arid basins. The steep rising limb of the hydrograph in desert streams is not modeled by most of the usual algorithms used in models developed for temperate regions (for examples, see French, 1987).

The emphasis in arid zone modeling has been on high–intensity thunderstorm events because of the predominance of work done on basins in the arid Southwest, USA. Pilgrim and Chapman stress that the predominance of high intensity thunderstorms in arid regions is not universal. Some areas in Australia, in particular, are subject to long, low–intensity rainfalls equally likely throughout the year. The prevailing rainfall pattern must be taken into account in the design of a rainfall–runoff model.

Similarly, the alluvial fans found in the Southwest are not indicative of alluvial fans everywhere, and local conditions must be considered. In particular, the southwestern USA is an area of high relief, and the alluvial fans tend to be steep, with heights of tens to hundreds of meters. In Australia, in areas of low relief, the alluvial fans are on the order of meters. Similarly, alluvial fans in Pakistan appear to be flatter (modal slope .02) and longer than those in the USA (modal slope .04) (French, 1987).

2.4. Modeling Runoff in the Arctic

The Arctic is semi–arid over much of its extent, but its hydrologic distinctiveness arises not only from its relative dryness as much as from its cold, in particular its permafrost, and the seasonality of its energy supply (Woo, 1987). Summers have 24 hours of sun, winters have none. Thus, most of the hydrology concerns snow and ice, and most of the runoff occurs in a rush during the short period of thaw. There are hydrologic and thermal processes unique to the Arctic which are not represented in rainfall–runoff models designed for more temperate climes. For example, thawing of the active layer in the permafrost region deepens continuously from spring to late summer, causing a change in the storage capacity for soil moisture. Snow meltwater often refreezes in the cold snow cover and, therefore, can delay runoff for days after surface melt has begun. Even when melt starts to run off, the channels are choked with snow and ice, so that the timing of the flows cannot be predicted in a simple manner. Thus, a hydrologic model for the Arctic should encompass the major physical processes of a permafrost environment, yet should demand a minimum amount of parameter estimation and meteorologic data. Solar energy may be more easily determined than can soil properties and may be more important.

During the winter, up to the time the surface of the ground thaws, hydrology is simple. Any runoff occurs as overland flow over the frozen substrate. During the thawed season, the soil column receives water from meltwater, rainfall, and inflow from upslope and loses water to evapotranspiration and outflow. The water table occurs in the thawed zone and varies with the thawing front. Throughout summer, when large rainfall raises the water table above the ground surface to exceed the depression storage, saturation overland flow occurs; according to Woo (1987), that is the major mechanism through which water is delivered to the streams. During the fall, surface cooling causes a descent of the freezing front. This may squeeze some water from the saturated zone out as streamflow, but usually it causes icing. In the unsaturated zone there is a redistribution of soil moisture storage in the profile.

Thus, the problem of modeling in the Arctic is one of simulating the energy fluxes and their impact on the location of the moisture in storage. The moisture in storage as snow on the surface is subject to spatial distribution. Much of the Arctic is of relatively low relief, and snowfall is redistributed through drifting. Thus, location of amount of snow should be related to areas of preferential drifting rather than to point of snowfall. This preferential location of snowpack leads to wet and dry areas in a basin, and can lead to a phenomenon similar to the source–area concept used in temperate climates. Ice jams are common in the northern temperate zones and extend into the southern fringe of the Arctic. Snow jams are more common in the Arctic. Both must be included in any model of basin response in the important early breakup period.

2.5. Snowmelt-Runoff Modeling

Snowmelt–runoff modeling would appear to have much in common with Arctic hydrology, but they are distinct in their approaches. Snowmelt–runoff modeling in temperate regions is concerned with modeling of snowfall in areas of high relief, whereas much of the Arctic is of low relief. Therefore, distributed parameter models with snow occurrence definition is important. Three general approaches are taken to dividing basins: elevation zones, basin characteristics, and grids. Model complexity and data requirements vary with the model approach.

Snowmelt models are divided into temperature–index and energy–balance approaches. The first is more empirical; the second more physically based. The temperature–index models have a base temperature and a melt factor as parameters. They must be defined for each sub–basin. Some models have the parameters vary through the season. The physically based models require even more data to calibrate. Many of the problems in simulating snowmelt runoff are related to data availability, data quality, and extrapolation of point to areal values. In fact, a World Meteorological Organization report (1986) concluded that the determination of the spatial distribution of precipitation was more important than the selection of a model. As reported by Woo (1987), the redistribution of snow on the ground is a major problem, drifting occurs other than in the Arctic. Remote sensing is suggested as a solution to this problem.

The spatial distribution of air temperature is needed for distributed area snowmelt models. Elevation and slope–aspect affect the distribution. An inversion can reverse the lapse rate relation and complicates snowmelt prediction. Only more on–site temperature data can overcome this uncertainty. Temperature–driven empirical models contain a critical temperature above which rain rather than snow occurs. Storms which have temperatures above and below the critical must be simulated at a finer time scale in order to simulate the complex nature of the storm. If only daily maximum and minimum temperatures are available, such storms cannot be properly simulated, and another source of uncertainty is introduced. Once again, more data are needed to overcome this uncertainty.

The more physically based energy balance methods must estimate albedo of the snow. Small errors in the estimation of albedo can have substantial effects on simulated snowmelt runoff. Remote sensing — more data — is suggested as a solution. Regions with sustained periods of below–freezing temperatures and relatively shallow snowpacks are susceptible to the development of frozen soils. Thus, the discussion of Woo (1987) and Leavesley (1987) converge, and the problems and solutions presented by each of them is applicable to the other. Similarly, Leavesley (1987) discusses the problems of glacial melt, which Woo (1987) mentioned only in passing as an intractible problem to date. Both quote Fountain and Tangborn (1985) as the state–of–the–art for modeling of glaciers.

Updating as a tool in forecasting is discussed by Leavesley, but it concerns all forecast models, not just snowmelt modeling (Dawdy, 1985). Model parameters are more fitted to the data for forecast models because data are available for assessment and improvement. In addition, the constant fitting during forecasting allows the errors in data to enter the model parameters and improve immediate forecasts. For prediction, however, credibility of results must be based on measures of model performance. The need for physically based parameters thus may assume more importance for prediction models than for forecasting models. Parameter estimation is affected by scale. As one moves from small plots to large basins, different sets of physical laws dominate (Klemes, 1983). At larger scales the parameters might be expected to represent some sort of averages or integrals of those at smaller scales. Parameters for small basins, thus, may not be representative of those for larger basins. Short- and long–term time simulation has similar effects on meaning of parameters. These, once again, apply to all models, as mentioned earlier, not just to snowmelt modeling.

3. Conclusion

Leavesley suggests the use of modular–designed simulation models with a master library of compatible subroutines which simulate most of the components of the hydrologic cycle. Alternative procedures can then be compared directly for choice of better methods. "The modular concept provides a powerful operational and research modeling tool." However, it also can lead to a cook–book approach, such as the HEC models, where quite good models are mixed with quite bad ones in the same model package, and the choice is left to the bias of the modeler with no value judgment as to goodness or badness.

Woo (1987) quotes Popov (1968) on the criteria for modeling a hydrologic system. They hold today as then, for elsewhere as well as the Arctic; they are: 1) to have physically correct ideas about the major processes; 2) to have a model with a minimum number of variables and parameters for a complete description of the processes, the principle of parsimony; 3) the need to have a clear physical meaning of the parameters and how to determine their values; 4) to have a minimum number of assumptions limiting the use of the model; and 5) to have the possibility to improve the model on the basis of new information.

Leavesley concludes "A number of advances in snowmelt–runoff simulation have been made during the past few decades. These advances resulted from an improved understanding of the physical processes of snowmelt and basin runoff, and the development of new technologies in the areas of data collection and computer technology." That conclusion can apply to each of the areas of runoff modeling, or to the field as a whole. It is a fitting ending to the overall discussion of problems of runoff modeling in various climatic and physiographic regions.

4. References

Dawdy, D. R.: 1985, 'State of the Art of Hydrologic Forecasting: What Next?,' *A Critical Assessment of Forecasting in Western Water Resources Management, AWRA Symposium*, Seattle.
Dyck, S.: 1987, *'Problems of Rainfall–Runoff Modeling in Temperate Regions,'* Keynote Paper Presented at IAHS General Assembly, Vancouver, B.C., Canada, Aug. 11–12, 1987.
Eagleson, P. S.: 1982, 'Ecological Optimality in Water–Limited Natural Soil-Vegetation Systems 1. Theory and Hypothesis,' *Water Res. Res.* 18(2), 325–340.
Fountain, A. G., and W. Tangborn: 1985, 'Overview of Contemporary Techniques,' in *Techniques for Prediction of Runoff from Glacierized Areas*, IAHS Publ. 149, pp. 27–41.

French, R. H.: 1987, 'Hydraulic Processes on Alluvial Fans, Vol. 3,' in *Developments in Water Science*, Elsevier, Amsterdam.

IAHS: 1987, *'Methods of Runoff and Streamflow Simulation Applied to Various Physiographic and Climate Conditions,'* Workshop No. 1, IUGG General Assembly, Vancouver, British Columbia.

Kienitz, G.: 1987, *'Simulation of Influenced Runoff Phenomena in Regions of Low Relief,'* Keynote Paper Presented at IAHS General Assembly, Vancouver, B.C., Canada, Aug. 11–12, 1987.

Klemes, V.: 1983, 'Conceptualization and Scale in Hydrology,' *J. Hydr.* **65**, 1–23.

Kovacs, G.: 1984, 'General Principles of Flat–Land Hydrology,' *Proc. Symp. Hydrology of Flatlands*, Olivarria, Argentina.

Leavesley, G.: 1987, *'Snowmelt–Runoff Modeling in Selected Physiographic and Climatic Regions,'* Keynote Paper Presented at IAHS General Assembly, Vancouver, B.C., Canada, Aug. 11–12, 1987.

Pilgrim, D., and T. Chapman: 1987, *'Problems of Rainfall–Runoff Modeling in Arid and Semi–arid Regions,'* Keynote Paper Presented at IAHS General Assembly, Vancouver, B.C., Canada, August 11–12, 1987.

Popov, E. G.: 1968, 'Principles of Modelling and Prospects of Using Electronic Computers in Hydrological Forecasting,' *Intern. Assn., Sci. Hydr. Publ.* **81**, 448–455.

Troutman, B. M.: 1985, 'Errors and Parameter Estimation in Precipitation–Runoff Modeling, 1. Theory, 2. Case Study,' *Water Res. Res.* **21**(8).

WMO: 1986, *'Intercomparison of Models of Snowmelt Runoff,'* WMO Operational Hydrology Rept. No. 23, WMO Publ. 646, Geneva.

Woo, M.: 1987, *'Arctic Hydrology,'* Keynote Paper Presented at IAHS Gen. Assembly, Vancouver, B.C., Canada, August 11–12, 1987.

Chapter 25

Comparison of USGS and HEC-1 Kinematic Wave Runoff Models

David R. Dawdy
3055 23rd Avenue
San Francisco, California 94132 U.S.A.

Abstract. Distributed parameter rainfall–runoff models are a recent development for use in basin modeling. They are useful for modeling the flow to the channel system, combining channel flows, and routing flows to the basin outlet. Two models that are widely available in the United States are models of the Geological Survey and of the Corps of Engineers. These models are particularly appropriate for use in urban systems modeling for master planning because of their capability to model changes in basin physical characteristics. The effects of proposed changes can be predicted, ultimate buildout conditions can be modeled, and downstream effects can be determined in a consistent, overall impact assessment. In addition, for urban basins the relative timing of the various channels determines the response of the basin.

The parameters for the kinematic wave method included in these models are based on physical descriptions of the channels and overland flow areas, which can be determined from maps and physical inspection of the basins. Kinematic wave models require little data to calibrate runoff characteristics. Despite their availability on personal computers, the models are not widely used, partly because they are new and different. To help overcome resistance to their use, a comparison of the two models, their similarities, and their differences is described.

1. Introduction

Distributed parameter rainfall–runoff models which use kinematic wave solutions for the surface runoff routing are a recent development for use in basin modeling. The two most widely available in the United States are the models of the US Geological Survey (Dawdy et al., 1978; Alley and Smith, 1982; Leavesley et al., 1983) and of the Corps of Engineers (HEC, 1981). These models are particularly appropriate for use in urban systems modeling for master planning because of their capability to model changes in basin physical characteristics. Thus, the effects of proposed changes can be predicted, ultimate buildout conditions can be modeled easily, and downstream effects can be determined in a consistent, overall impact assessment. In addition, the kinematic wave approximation included in these models seems appropriate because of the small size of urban basins, their generally simple channel configurations, and the fact that the major determinant of peak characteristics is the timing of flows from the various tributaries in the urban environment. For larger basins for which dynamic channel storage becomes important, kinematic wave models may be inappropriate.

The kinematic wave parameters are determined from the physical descriptions of the channels and overland flow areas, which can be determined from maps and physical inspection of the basins. Therefore, kinematic wave models require little data to calibrate runoff characteristics. The major need for calibration is to determine the volume of runoff which results from a volume of rainfall.

D. S. Bowles and P. E. O'Connell (eds.), Recent Advances in the Modeling of Hydrologic Systems, 549–557.
© 1991 *Kluwer Academic Publishers.*

Despite their easy availability for use on personal computers, the USGS and HEC–1 models are not widely used. There is considerable controversy concerning the applicability of kinematic wave approximations, particularly for overland flow. There is no better solution available at this time. The two models are much preferable to the widely used unit hydrograph procedures, which are difficult to adjust to simulate the effects of man–made changes in the urban environment. In addition, the two models are not widely used because they require an extra degree of knowledge. That which is new and different is not freely accepted or used. In order to help overcome this natural resistance on the part of practicing engineers to the new and exotic, a comparison of the two models, their similarities, their differences, their strengths, and their weaknesses is needed.

2. Discussion

2.1. Direct Comparison of USGS and HEC–1 Models

A direct comparison can be made between the USGS and HEC–1 kinematic wave models if a hypothetical basin is constructed which can be modeled identically by the two models. A USGS model for such a basin is shown in Table 1. The basin is 1200 feet by 5000 feet, with 300 feet of overland flow length to the channel system. Excess precipitation is used as rainfall input with a fully impervious surface for all the examples in this demonstration chapter so that the surface water response of the various models which depends upon the kinematic wave routing can be compared. The discharge from the input shown with a time step of two minutes is 64.7 cfs. The parameters for the model are as shown in Table 2.

The description of the input and of the various parameters for the USGS model are given in Dawdy et al. (1978). This version of the USGS model is chosen because it is in "vanilla Fortran" and easily implemented.

The output discharge for the overland flow element (OF02) is 0.003 cfs. That discharge can be used to compute the celerity of the kinematic wave, which by the Courant condition

Table 1. USGS input for hypothetical model. Parameters for kinematic routing are length, slope, Manning's n, and bottom width.

```
LIST10
          1PROBLEM COMPARISON                                   0.215 (DA)
          2IUH DAILY RAINFALL
          3IUH DAILY EVAPORATION
                         85  1  1    85  1  3 (DATES FOR PERIOD MODELED)
          4IUH UNIT RAINFALL                                 10.0
          485 1 210 1     0    25    0     0...0    0     0     0     0    0 1

          285 11    0-100    0    0    0    0....0    0    0    0    0    0    0 3
          385 11    0  00    0    0    0    0....0    0    0    0    0    0    0 4

      7   1   1     0.06 (CONTROL VARIABLES FOR OPTIMIZATION)
          0.000      -1.000     15.000      )THESE
          0.001      -1.000      1.000      )SEVEN
         10.000       5.000     20.000      )CARDS
          5.000       1.000     15.000      )DEFINE
          0.700       0.500      1.000      )INFILTRATION
          0.900       0.500      1.000      )PARAMETERS
          0.800       0.100      1.000      )
      5
        2  2.0  1.0    1  .05 (CONTROL PARAMETERS FOR ROUTING)
   OF02 (OVERLAND FLOW PARAMETERS)  5 110 300.  .020 .100  1.0   0.0 1  1.0
   CH21              OF02OF02OF02OF02 3 1 65000.  .010 .035  2.0  (CHANNEL)
      1 1
      1  15 0.250
```

Table 2. Kinematic wave parameters for the sample USGS model.

Computation Sequence		Kinematic Channel Parameters	
Index	Segment	Alpha	M
1	OF02	2.11	1.670
2	CH21	3.20	1.330

determines the relation of distance steps to time steps for the kinematic wave equation to obtain a correct solution. The Courant condition requires:

$$\Delta x / \Delta t = c = amA^{(m-1)/m} \qquad (1)$$

The discharge combined with the kinematic wave parameters shown in Table 2 gives a celerity from 0.24 to 0.27 fps, for an optimal N^*, the number of subdivisions of the overflow length, of 9 or 10 for the overland flow plane. For the channel N^* is 4.6 for a discharge of 64.7 cfs. $\Delta x / \Delta t$ should equal the celerity. However, the main channel is 5000 feet long, and the celerity is computed at the outlet only. If we divide the channel in two, the outflow for that model is 64.7 cfs. In fact, the USGS model is quite robust in its output in the vicinity of the proper $\Delta x / \Delta t$. This is shown in the solutions for the "two–division" model as compared with the "one–division" model:

Table 3. Outflow peak discharge for different subdivisions for the USGS model for the sample basin.

Division	N	Q	N	Q	N	Q	N	Q	N	Q	N	Q
1	4	32.8	4	32.8	4	32.8	3	32.7	5	32.8		
2	4	64.3	3	64.7	2	64.4	3	64.7	3	64.4		
Undivided			7	64.6			6	64.7	8	64.3	5	64.4

The USGS model can be mimiced with the HEC–1 kinematic wave option model. The description of the parameters and of the input are given in the HEC–1 Users Manual (1981). The listing for a "one–division" model is shown in Table 4. The standard HEC–1 kinematic wave option presently available has a built–in algorithm for the determination of the

Table 4. The HEC–1 kinematic wave model for the sample basin.

```
ID SAMPLE PROBLEM FOR MODEL COMPARISON
ID KINEMATIC WAVE ROUTING --- EXPONENTIAL LOSS FUNCTION
IT      2 01JAN01        0001    50
IN      2 01JAN01        0001
IO      1
PG      1
PI 0.050    0.050    0.050    0.050    0.050
KK   SAM1
KM      RUNOFF FROM SAMPLE SUBAREA
KO      1
PR      1
PW    1.0
BA   .215
LE   0.20       0        9        0      100
UK    300    0.02      0.1      100
RK   5000   0.010    0.035           TRAP    0        1
ZZ
```

subdivision of the channel, unlike the USGS model for which it is necessary to specify the number of subdivisions as an input. The "one–division" model for HEC–1 divides the overland flow into six parts and the main channel into six parts, and the resulting peak discharge is 62.6 cfs. By dividing the HEC–1 model into two successive reaches of 2500 feet of the main channel, the HEC–1 program keeps six subdivisions for the overland flow, but divides the main channel into four and three subdivisions, to obtain a peak of 64.63 cfs. If the overland flow subdivision is specified as 10, the main channel is subdivided into five and four parts, and the peak increases to 64.68, which is the USGS solution.

Table 5. Resulting peak discharge for different subdivisions for sample problem with standard HEC–1 kinematic wave model compared with USGS solution, both with two–minute solution time increment.

Subdivisions	NXOF	NXCH	Q
1	6	6	62.61
1	10	9	63.59
2	6	7	64.63
	10	9	64.68
USGS	6	10	64.7

Results so far have used two minutes as the time increment for solution of the kinematic wave equation. By use of one minute for HEC–1 the various subdivisions yield:

Table 6. Resulting peak discharge for different subdivisions for sample problem with HEC–1 kinematic wave model with one–minute solution time increment compared with USGS solution for two minutes.

Subdivisions	NXOF	NXCH	Q
1	12	11	64.43
2	13	11	64.96
2	20	11	64.88
USGS	6	10	64.7

Thus, with a one–minute time increment the HEC–1 model reproduces the results of the USGS model with a two–minute time increment. A similar comparison for the USGS model is shown in Table 7. For the USGS model, reducing the time increment *reduces* the computed peak discharge. This results because the USGS model is constrained to allow only 10 subdivisions at most of an element, and the overland flow plane, OF02, is at its limit. There is little change at 2500 feet and only 2.6 percent decrease at 5000 feet. Thus, the USGS solution algorithm is robust to the time increment and to the number of distance subdivisions.

As noted, the USGS solution algorithm is quite robust. This is shown in Table 7, where different subdivisions are used for the USGS model. The different configurations for the one–minute increment differ among themselves by but 0.5 percent. To run the USGS model with 14 subdivisions for one hour with a one–minute time increment took four seconds on a 286. The identical model configuration with HEC–1 took 27 seconds on a 286 and seven seconds on a 386. Both had detailed output suppressed.

The conclusion drawn from this first sample problem is that the USGS model is faster than the HEC–1 model. The two models agree in their solution and are quite robust in terms of choice of distance subdivisions and solution time increment. The sample problem is for a

Table 7. Comparative results for USGS model of sample basin with one- and two–minute time increments.

Time	NXOF	NXCH	Length	DX/DT	Q	% Diff.
2	10	3	2500	6.9	32.733	
	10	3	2500	6.9	64.738	
1	10	6	2500	6.9	32.502	–0.7
	10	6	2500	6.9	63.026	–2.6
1	10	8	3333			
	10	4	1667	6.9	63.027	–2.6
1	10	3	1000	5.6		
	10	4	2000	8.3		
	10	4	2000	8.3	63.273	–2.3
1	10	3	990	5.5		
	10	3	1330	7.4		
	10	4	2135	8.9		
	10	1	545	9.1	63.355	–2.1

main channel length of 5000 feet, which is longer than one normally should use in urban modeling. The HEC–1 and USGS results should be even closer for channels for which the main channel length is shorter than 5000 feet.

A recently revised version (1990) of the HEC–1 model was used for these analyses. It is a great improvement over previous versions in its solution of the kinematic wave equations, and it includes a Muskingum–Cunge routing for main channel flows which will be useful for over-bank conditions. Much of the variation in the revised version of HEC–1 seems to be in its rounding for time and distance, not in its solution algorithm. The time steps are never an integral part of the output time, and thus each step must be rounded to obtain the input for the next step. The discharges listed in this report are prior to rounding, however, so that they include numerical dispersion only. The new HEC–1 is a major improvement. All previous versions should be replaced.

2.2. Modeling of Collector Channels Using USGS Model

One advantage of HEC–1 in its representation of a basin is its use of collector channels and its input of the drainage area. The basin configuration for the USGS model is used to compute the drainage area and to compare that area with the input area. If they differ by more than one percent, the USGS program aborts. The definition of overland flow, thus, is constrained in the USGS model as compared to the HEC–1 model. Collector channels simulate a series of similar channels, such as storm drains, and are entered as a line source. They have a similar use to the repetitive elements of the USGS model in that they need to be modeled but once. For demonstration, let us expand our sample basin to include collector channels for the HEC–1 model, as shown in Table 8.

The output peak for the HEC–1 model with the collector channel is shown in Table 9 for time increments of one and two minutes. Similar input for a USGS model is shown in Table 10. The USGS model seems more complex in order to mimic the HEC–1 construct. However, CH21, the collector channel, is computed once, then used 16 times. HEC–1 computes the collector channel as a discrete source but treats it as a line source to the main channel; whereas, the USGS model must treat the collector channel as discrete inputs, here distributed uniformly along the channel. The resulting USGS peak outflow is shown in Table 11.

Table 8. The HEC–1 kinematic wave model for a sample basin with collector channel.

```
ID SAMPLE PROBLEM FOR MODEL COMPARISON --- KINEMATIC WAVE ROUTING
IT      1 01JAN01    0001      50
IN     10 01JAN01    0001
IO      1
PG      1
PI 0.250
KK   SAM1
KM        RUNOFF FROM SAMPLE SUBAREA
PR      1
PW    1.0
BA    .400
LE   0.20        0        9        0      100
UK 312.5      0.02      0.1      100
RK   1115     0.015     0.045    .025     TRAP      0        1
RK   5000     0.010     0.035             TRAP      0        1
```

Table 9. Resulting peak discharge for sample problem for HEC–1 kinematic wave model with collector channel using one– and two–minute time steps for solution.

Subdivisions	IT	NXOF	NXCCH	NXMCH	Q
1	1	12	5	11	114.71
1	2	5	3	5	112.21

Table 10. USGS equivalent model for HEC–1 model with collector channel.

```
    10  1.0   1.0      1   .05
OF02                          5 110312.5 .020 .100 1.0  0.0 1  1.0
CH21              OF02OF02    3 1 41115.  .015 .045 2.0
CH22CH21CH21                 3 1 1 625.  .010 .035 2.0
CH23CH22CH21CH21             3 1 1 625.  .010 .035 2.0
CH24CH23CH21CH21             3 1 1 625.  .010 .035 2.0
CH25CH24CH21CH21             3 1 1 625.  .010 .035 2.0
CH26CH25CH21CH21             3 1 1 625.  .010 .035 2.0
CH27CH26CH21CH21             3 1 1 625.  .010 .035 2.0
CH28CH27CH21CH21             3 1 1 625.  .010 .035 2.0
CH29CH28CH21CH21             3 1 1 625.  .010 .035 2.0
```

Table 11. Resulting peak discharge for different subdivisions for sample problem with USGS model of HEC–1 kinematic wave model with collector channel using one–minute time steps for solution.

Subdivisions	NXOF	NXCCH	NXMCH	Q
8	10	4	8	113
8	10	2	8	111

The HEC–1 peak for one minute is highest, the best USGS solution is one or two percent less, and all four solutions are within three or four percent of the maximum result. The celerity for the collector channel is 5.0 fps for the USGS model, which compares with $\Delta x/\Delta t$ of 4.65; whereas, the HEC–1 imposed $\Delta x/\Delta t$ is 3.8 for one–minute and 4.0 for two–minute time increments. The subdivision of the main channel for the USGS model is dictated by the configuration of the basin and the need to distribute the collector channels along the main channel. Therefore, channel reach lengths are not optimal. However, the robustness of the USGS solution algorithm yields a peak discharge close to the HEC–1 "optimal" subdivision. The comparative run times are five seconds for the USGS model and 38 seconds for HEC–1, both on a 286. Once again, the two models are in close agreement.

2.3. Modeling a One Mile Square Basin with the Two Models

A one mile square basin with a central main channel requires a longer collector channel system. The USGS model for the basin is shown in Table 12. Each collector channel is a half mile long. The overland flow distance is 330 feet to obtain the correct drainage area. The output peak discharge for the USGS model for the basin is shown in Table 13 for various time and length increments.

The two–minute time increment is best, in that it yields the highest discharge, which implies the least numerical dispersion. With eight subdivisions optimal for the 2.5 minute run for the overland flow, 10 should be correct for 2.0 minutes. For 1.0 minute the overland flow constraint of 10 subdivisions reduces the peak by 1.8 percent, and the constraint of channel length for the main channel reduces the peak for 2.5 minutes by almost the same amount.

Now let us see how HEC–1 would handle this by use of a collector channel, now 2640 feet long, and by the equivalent of the USGS construct. First, let us revise the HEC–1 model from Table 8 so that the lengths are 330, 2640, and 5280 feet, as in the USGS model in Table 12. The output is shown in Table 14.

The predicted peak outflow for the HEC–1 model is lower than that for the USGS model. The closest agreement with the USGS model is at one minute, but the highest result is USGS

Table 12. One mile square basin modeled with USGS model with collector channels.

```
      10   2.0   1.0     1   .05
    OFO2                              5 110330.0  .050 .150 1.0  0.0 1  1.0
    CH21              OFO2OFO2        3 1 42640.  .012 .040 2.0
    CH22CH21CH21                      3 1 1 660.  .010 .035 2.0
    CH23CH22CH21CH21                  3 1 1 660.  .010 .035 2.0
    CH24CH23CH21CH21                  3 1 1 660.  .010 .035 2.0
    CH25CH24CH21CH21                  3 1 1 660.  .010 .035 2.0
    CH26CH25CH21CH21                  3 1 1 660.  .010 .035 2.0
    CH27CH26CH21CH21                  3 1 1 660.  .010 .035 2.0
    CH28CH27CH21CH21                  3 1 1 660.  .010 .035 2.0
    CH29CH28CH21CH21                  3 1 1 660.  .010 .035 2.0
```

Table 13. Peak discharge for one square mile basin using the USGS model with different time increments and subdivisions.

DT	NOF	NCCH	NMC	Q	% diff
1.0	10	9	1X8	278.84	–1.8
2.0	10	4	1X8	284.00	0
2.5	8	3	1X8	278.94	–1.8

Table 14. HEC-1 model output for longer collector channel basin.

DIVISIONS	DT	NOF	NCCH	NMCH	Q
1	1	12	9	9	270
1	2	6	5	5	264
1	5	5	3	3	258

for two minutes. Once again, the USGS model is constrained to 10 subdivisions. Also, because of the structure of the USGS model, the main channel had to be segmented. In this case, that is an advantage because it reduces numerical dispersion. The two "best" solutions differ by only five percent.

2.4. Numerical Instability

A phenomenon known as "numerical instability" sometimes occurs when a high–intensity rainfall is applied to a kinematic wave model. Theoretically, it has to do with the solution scheme used to solve the kinematic wave equations. However, in practice it occurred more often in the earlier versions of the HEC-1 model than in the USGS model. By "unstable" is meant a spurious peak, which usually occurs after the main peak has passed. The spurious peak can be many times larger than the true peak and, thus, is obvious. However it can be on about the same order of magnitude as the true peak and can, thus, cause confusion and produce erroneous results, particularly for detention basin design where the output discharges are routed through a basin to attenuate the peak. The USGS model does not produce the instability because its solution algorithm is supposedly unconditionally stable. The HEC-1 model has adopted the same solution algorithm as the USGS model, and instabilities which previously occurred with HEC-1 no longer occur. The conditions which create the instability usually occur when modeling extreme events, say the 100–year flood, and, if they occur, can be avoided by minor adjustment of the rainfall input.

2.5. Recent Revisions to HEC-1 Model

The Hydrologic Engineering Center has revised HEC-1 to overcome some of the problems the users of the program have experienced in the past. Preliminary versions of the revised model were made available for this comparison. An understanding of the internal workings of the kinematic models is still desirable, but they may not be as necessary in the future in order to use the models. First, HEC has used the USGS solution algorithm, which is unconditionally stable. That should eliminate the problems of unstable solutions for major floods. Second, they have reprogrammed HEC-1 to choose Δx and Δt based on the discharge occurring, and then to interpolate to a common time base step by step. Thus, the Courant condition should be met during each time step. This, combined with the robustness of the USGS algorithm, should make the HEC-1 solutions contain little or no numerical dispersion, although it will introduce some rounding error, as mentioned earlier. However, this leads to longer solution times, even though their new algorithm seems to be faster than for earlier versions. For the model of a complex basin with many subdivisions, time for solution may be increased considerably. Thus, the present advantage which the USGS model has in solution time may remain. Now that HEC-1 is revised, any basins modeled with HEC-1 kinematic wave option in the past should be rerun with the revised program and a decision made as to whether the different answer, if any, is sufficiently different to cause a revision of decisions made with the earlier version.

3. Conclusions

Recently developed distributed parameter rainfall–runoff models which use kinematic wave solutions for the surface runoff routing are useful tools for basin modeling. These models are particularly appropriate for use in urban systems modeling for master planning because of their capability to model changes in basin physical characteristics. Thus, the effects of proposed changes can be predicted, ultimate buildout conditions can be modeled easily, and downstream effects can be determined in a consistent, overall impact assessment. In addition, the kinematic wave approximation included in these models seems appropriate because of the small size of urban basins, their generally simple channel configurations, and the fact that the major determinant of peak characteristics is the timing of flows from the various tributaries in the urban environment.

Two of the most widely available models (Dawdy et al., 1978; HEC, 1981) are shown to be equivalent, if used properly. However, in order to be used the two models require an extra degree of knowledge. In particular, the choice of time step and of basin subdivision and choice of length step are critical for correct use of the models. This process is described explicitly in the documentation for the USGS model, and the modeler must then choose those parameters for the model solution. The documentation for the HEC–1 discusses the problem, and the program chooses the length step properly to combine with the time step to produce a correct answer. The procedures outlined in the example problems presented show the process for the proper selection of time and length increments for both the USGS model and for HEC–1 to obtain correct solutions for the kinematic wave equations so that the programs produce the solutions which they are intended to produce.

Revisions to the HEC–1 program now solve the kinematic wave equations more accurately. A choice of a small time step and subdivision of the basin into short lengths of main channel reaches increase the accuracy of the HEC–1 solution. The HEC–1 and USGS models will give similar results if the two model a basin in the same manner.

4. References

Alley, W. M., and P. E. Smith: 1982, *'Distributed Routing Rainfall–Runoff Model—Version II User's Manual,'* U.S.G.S. Open–File Report 82–344.

Dawdy, D. R., J. C. Schaake, Jr., and W. M. Alley: 1978, *'Users Guide for Distributed Routing Rainfall–Runoff Model,'* U.S.G.S. Water Resources Investigations 78–90.

HEC: 1990, *'HEC–1 Flood Hydrograph Package Users Manual,'* Hydrologic Engineering Center, U.S. Army Corps of Engineers, September 1981. (Updated version February 1990).

Leavesley, G. H., R. W. Lichty, B. M. Troutman, and L. G. Saindon: 1983, *'Precipitation–Runoff Modeling System: User's Manual,'* U.S.G.S. Water Resources Investigations 83–4238.

Chapter 26

Hydrologic Modeling for Water Resources Planning

L. Douglas James
Utah State University
Utah Water Research Laboratory
Logan, Utah 84322–8200 U.S.A.

Abstract. Technology is the motor of human progress, but it moves slowly. Recent advances in hydrology may not be adopted in practice. We need to assess why and what can be done. A model is a set of equations organized for rapid computation in linked systems of spatial and linear transforms. Model assessments can help decision makers in facility, operations, and nonstructural planning; but planner response is limited by organizational structure, laws and policies, politics, and pressure groups. Thus, hydrologic modeling is an institutional process. Acceptance requires good credentials, understandability, and validity. Models are expected to convey facts, answer doubts, and be readily updated. They should free time for professional reflection. Modelers should state explicit goals to reduce model misuse and track the many interactions among large numbers of alternatives. Then institutions would accept them more readily.

1. Introduction

Irrigation and civilization were born at the dawn of history, and they grew together over the centuries. Ancient civilizations advanced water technology to the point where Bonnin (1988) has asked whether modern urban water supply systems are really better than those that Rome designed 2000 years ago to deliver a cubic meter of water per capita per day.

Mesthene (1970) sees technology as the motor of human progress. Water resources technology achieved the construction of multiple billions of dollars worth of infrastructure for irrigation, flood control, navigation, urban water supply, waste treatment, and hydropower. Early engineers made rough estimates of design quantities, and their methods became the ancestors of our hydrologic models. With increasing experience, hydrologists examined available data, defined relationships, and made thoughtful judgments for engineering designs. Slowly, they improved the hydrologic estimates and thereby the reliability of the designs. In Rome a more dependable water supply is now distributed more widely throughout the city and with greater protection against pollution.

The accuracy of hydrologic estimation made a quantum leap forward in the early 20th century, after the installation of streamflow and climatological gauging networks, as the data were used to derive more sophisticated quantitative relationships. The last few years have brought a new round of advances in instrumentation, catchment characterization (Gupta et al., 1986), data storage and retrieval, physical–mathematical modeling, computer simulation, and the use of tracers in tracking water movements (Sklash et al., 1988). With these means to make another quantum leap forward (Burges, 1986), we must do our best to use them to gain hydrologic understanding and to apply what we learn to improve water resources planning.

However, it may not be easy. Rogers and Fiering (1986) note that developing countries are more apt than the advanced countries to use the latest methods. History contains many

D. S. Bowles and P. E. O'Connell (eds.), Recent Advances in the Modeling of Hydrologic Systems, 559–570.
© 1991 *Kluwer Academic Publishers.*

examples where technical leadership shifted to new countries as older institutions proved unable to modernize. Mesthene (1970) gets to the root of the problem in describing how institutional sluggishness blocks progress. This chapter analyzes those roots as they relate to hydrologic modeling.

Hydrologic modeling is invaluable in simulating information for water resources planning and management, but hydrologic models drift into misuse when they replace thinking (Linsley, 1986; Klemes, 1986). The temptation is exacerbated by the way institutions judge water development projects. An engineer who explains how he sat down at his desk and thoughtfully compared the alternatives will lose out every time to an engineer who presents an elaborate computer program, displays graphic output, and draws a red box around the bottom line.

If we are to make hydrologic models effective servants in water resources planning, we need guidelines for model use in education and practice. They should cover components, linkages, uses, and support systems. They should show how to organize information more effectively and how to organize information more logically in succinct communications to decision makers in government and to property managers in the private sector.

2. Discussion

2.1. Concept of Model Role

A water resources planner or engineer should begin with a sound concept of hydrologic modeling. In essence, today's model is an expression of algebraic equations found useful in former years. In the days of manual computation, an engineer estimated variables, performed calculations, and applied results. He scrutinized every step for reasonableness. Today, an engineer uses computer algorithms to estimate parameters in equations evolved through generations of experience calculates and applies the results. The computer speeds the calculations and delivers much more information for comparing many more alternatives.

Ideally, the time saved should be devoted to critical review of the additional information and lead to better judgments. However, hydrologic modeling has not achieved this result. People become so intrigued with the execution of their models that they short-change the interpretation of their output. Pressures crowd schedules and rob people of the blocks of time required for thoughtful reflection. Furthermore, computer programs separate computational details from human scrutiny as their output is too bulky for easy review and their complex algorithms are difficult to follow. Finally, the greater prestige of office work (Pilgrim, 1986) separates modelers from observing the processes modeled.

Hydrologists should not forget that the fundamental objective of hydrologic modeling, as was that of its ancestors, is to provide reliable information for managing water resources to increase human welfare (Howe, 1971, p. 3). For this to happen, the people responsible for water resources planning must begin to think interactively with the computer models provided by hydrologists. In conclusion, the primary role of hydrologic modeling is providing information for management application, and the primary constraints to fulfilling that role are inefficiencies at the human–computer interface.

2.2. Process Kernels

The problems at this interface begin when model users expect to get too much with too little thought. A user can review published guidelines (Renard et al., 1982; Office of Technology Assessment, 1982) and select and apply a model. However, effective human–computer interfacing is not going to be achieved by taking a model from the shelf and making a quick application; nor at the other extreme is it practical for planners "to be intimately involved with model development, solution, and analysis" (Loucks et al., 1981). Neither way is reasonable. The first gives bad answers, and the second takes too much time.

We must teach the art of thinking through interactive interfacing with computer algorithms developed by others. A promising middle course is to develop participatory interfacing by subdividing the models into components that we will call process kernels. A planner in need of hydrologic information can then select appropriate kernels, link them, and interact with the assemblage. The advantages are psychological as well as practical. The planner avoids becoming mired in programing details but acquires the needed general understanding of the assumptions and limitations of his tool at a larger scale.

Many kinds of kernels should be tried and used. In order of increasing complexity, their outputs might be volumes in storage, rates of movement, spatial distributions of the volumes and rates, and the changes with time of all three (volumes, rates, and spatial distributions). For example, Rodriquez–Iturbe (1986) discusses modeling rainfall fields with temporal rainfall at a point, areal rainfall over a storm, and space–time rainfall representations from cell lives and intensities. The kernels can be used in scales from laboratory to global (Eagleson, 1986).

In terms of approach, kernels at the theoretical end of a spectrum simulate hydrologic processes from "white box" models that apply mathematical transforms to express known physical laws (Brutsaert, 1986). Differential equations are used with physiographic and geomorphic characterizations of catchment media to quantify the rates at which precipitation, radiation, and gravity produce hydrologic storages and flows at times and locations of interest. At the other extreme, "black box" (operational or empirical) models correlate output with input from historical associations without pretense of expressing causal ties.

Between these two extremes are the "gray box" (conceptual) models. Their approach subdivides a hydrologic transformation (rainfall–runoff) into component processes (interception, infiltration, interflow, etc.). Each process is then represented by a kernel at a scale that compromises between the small time–space grids used in physical–mathematical models and the system-wide associations expressed by statistical models. The modeler calibrates the lumped parameters for the kernels so that simulations of the linked assembledge match historical records. Principles of similarity can then be applied in catchment classification and model extrapolation (Dooge, 1986).

Whether black, white, or gray, models can be applied deterministically to give single estimates or stochastically to generate probability distributions. A vast literature can be consulted in searching for suitable relationships. One can even try the new perspectives of the physics of chaos (Gleick, 1987) or the mathematics of fractals (Mandelbrot, 1983). However, following our theme of modeling at an aggregate level, this chapter will not pursue the details of formulating process kernels but rather will move on to linking them for use in water resources planning and management.

2.3. Linked Systems

The linking concept can be presented by means of illustration. We will consider planning water resources development above a dry but fertile plain. A snow–fed river flowing through the plain can be diverted to irrigate fields (Figure 1), and the farmers will attain a more satisfying life style. In this situation, mountain snowfall makes happy farmers, and a statistical analysis of a time series of annual data would probably reveal a high correlation, but application of such an empirical equation to other locations would be conceptually and practically unsatisfactory for water resources planning.

A water resources planner should instead apply the gray-box approach of subdividing the "snow–to–happiness" transform into steps, choose kernels for each step, and link them. In

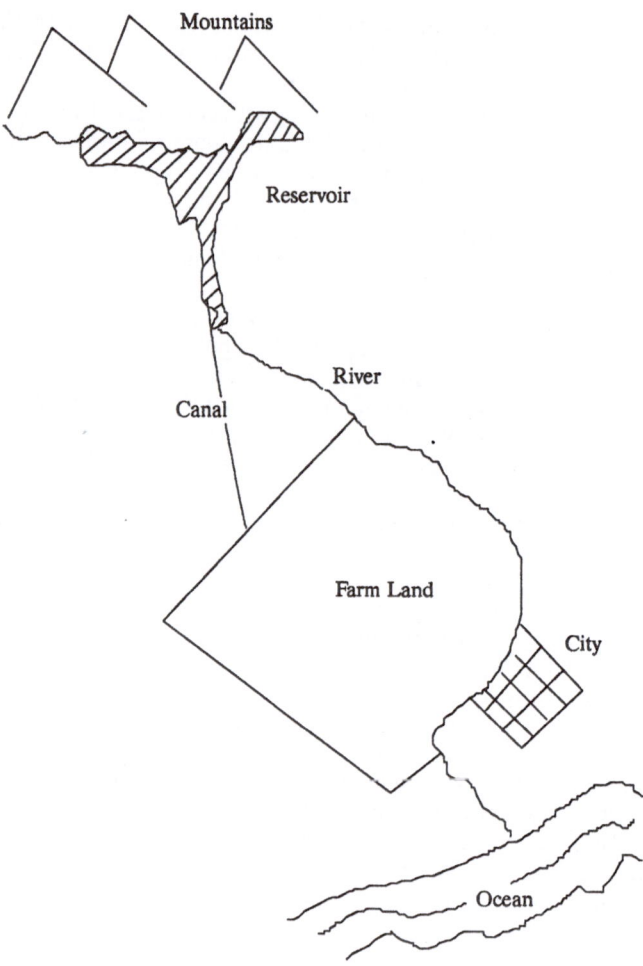

Figure 1. Schematic diagram of river flows for water resources planning.

Figure 2, the flows are designated by number, and convolution kernels for computing a succeeding from a preceding flow are designated by letter. Flows can be movements in space or time of water, transported salts, crops, income, etc. The linking kernels have four forms:

Type 1. Linear to linear (Figure 2: C). Flows can be routed down a river or canal based on channel characteristics. The use of linear kernels assumes that flow in the transverse dimension (expansions, contractions, bends, and eddies for kernel C) is not an important influence on the transform (channel routing). The user of a linear kernel must choose a balanced combination of time units and the incremental distances to use in the computations, and this choice of scale should depend on the application (flood peaks, reservoir yield, environmental impacts, low flows for waste dilution, etc.).

Type 2. Spatially distributed–to–spatially distributed (Figure 2: A, E, F, G). In many cases, flow distribution in the transverse dimension does have a significant impact on the transform. The familiar example of estimation of runoff volume from spatially distributed

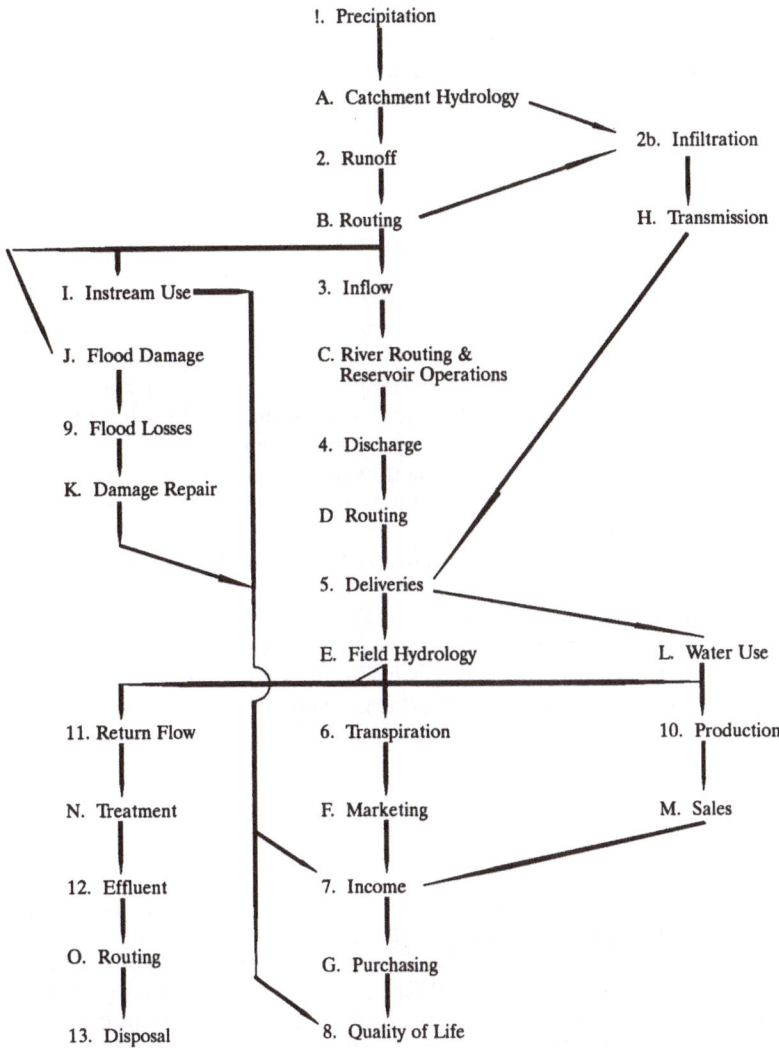

Figure 2. General chart of flows (numbers) and convolutions (letters) for water resources planning.

rainfall (Figure 2: A, the flow of water from the sky) on a spatially varied catchment illustrates this type.

In reality, the Type 1 kernel is a special case of the Type 2. Nature is three–dimensional, and all physical processes are spatially distributed. However, the modeling is greatly simplified when the process can be represented as point transforms. The modeler must determine whether use of a transverse distribution improves the accuracy of a given simulation, given the input data availability and the desired resolution for the application at hand. For spatially distributed kernels, the modeler must determine how finely to divide the space units and select matching time units. Examples requiring different resolutions are total river flows, local runoff totals into many farm ponds, and spatially distributed ground water recharge.

Type 3. Spatially distributed–to–linear (Figure 2: B). If some kernels are to be linear and others spatially distributed, some linkages must go from one type of flow to the other. For example, Kernel B aggregates runoff from spatially distributed sources by routing through collection channels to a common point. Time–area histograms, linear reservoirs, or channel maps and cross sections can be used.

Type 4. Linear–to–spatially distributed (Figure 2: D). Some flows must be delivered from a common source to spatially distributed users. The routing is an inverse of the Type 3 process, but the algorithm is more complicated because most flow divisions are humanly controlled. The modeling of these human choices must incorporate the policies prescribed for systems operations and the reliability of the operators in executing them.

Structural projects can sometimes be designed from Type 1 convolutions. However, the varied local impacts of watershed management and nonstructural measures over a basin must generally be analyzed with Type 2 kernels. Whether the modeler is investigating differences in water availability at various locations to develop an operation for trading water between areas of surplus and areas of deficiency or just distributing the benefits to water users over a river basin, Type 2 algorithms pose a major research challenge.

The modeler links the hydrologic kernels to represent water flows and may add external linkages to portray impacts on the larger socio–economic system. The model parameters can be tied functionally to the characteristics of structural and nonstructural measures, estimated for specific installations and varied to compare alternatives. With suitably integrated kernels, a planner can compute system responses to exogenous events.

In the end, a planner who learns to interface with kernel linkages is in a stronger position to incorporate new hydrologic advances and additional combinations of processes in simulations to serve multiple needs. A user who links kernels in simulation has developed the often missing skills of critical thinking at the computer interface. Each discipline can concentrate on its own kernels, and interdisciplinary teams can work on the linkages and on understanding of the total system.

2.4. Planning Issues

Once planners gain confidence in using and linking hydrologic kernels, they can begin to use their computers thoughtfully in organizing information for assessing the four principal water resources planning issues: budgeting, facilities, operations, and water use. The four inter-related assessments should be interactive (Maass, 1962). For example, the planning of water supply facilities should be coordinated with program budgeting, setting system operating policy, and designating nonstructural efforts to promote water conservation. Taken as a whole, the information obtained by modeling should help management identify structural, operational, and water management components in a plan for securing the necessary funds. The assessments by issue require the following sorts of hydrologic information:

(1) *Budgeting*: Cost estimation requires hydrologic information for projecting future water needs (water supply, water treatment, flood control, hydropower, etc.) and quantifying the supplies that can be made available (from various sources, with various facilities, by trades with other jurisdictions, and through operating plans and water conservation programs) to meet those needs. Water use varies with culture and depends on management policy (pricing, regulation, water education, etc.), and supplies can be increased with greater effort in resource development and management. Finally, the imbalances between supplies and demands are used to project funding requirements for budgeting purposes.

(2) *Facilities*: The planning of the structural measures used in water resources development can be integrated by nesting plans at smaller scales into plans at larger scales. This can be accomplished for two widely used levels:

(a) At the basin scale, information on hydrologic interactions among projects should be used to design a cost effective (net benefit maximizing) system of facilities to serve a river basin as a whole.

(b) At the smaller project scale, information on the hydrologic impacts of alternate configurations should be used to design a cost effective facility for a particular site to accomplish fixed goals within the basin plan.

(3) *Operations*: For any facility set, the operating policies must be integrated at three levels:

(a) Planning: Information on the probabilities of hydrologic events of various sizes can be used for computing expected values of water stored in reservoirs and storage space held empty for flood control by month of the year. These expectations can then be used to derive rule curves or, with present technology, computerized guidance systems that achieve the optimal balance in making use of the storage.

(b) Real Time: Information projecting the future course of floods and droughts while they are in progress should be used for applying the rule curve or computerized guidance system to decide what to do during actual events.

(c) Emergency: Floods and droughts may worsen. Information is needed for estimating the probabilities of the situations intensifying to the point that the overriding of normal operating procedures is justified and that emergency procedures are followed to reduce losses.

(4) *Use (nonstructural)*: Information on the severity and spatial distributions of water shortages, flood hazard, and pollution risks and their probabilities should be prepared. The governmental agencies managing these programs can then help people more efficiently use (achieve greater benefits from) their land and water resources. Separate assessments are needed for programs for:

(a) Water conservation.
(b) Flood plain management.
(c) Waste disposal management.

The nonstructural programs also have downstream impacts. The impacts of upstream programs on these assessments are also important to regulatory agencies that are trying to protect people downstream.

The resolutions of these four issues should proceed with progressively greater refinement through a sequence of reconnaissance screening, system and component optimization, nonstructural program development, and cost analysis. Sensitivity studies can add tradeoff values that can be used to compare plans at the margin.

2.5. Decision Systems

For the people who actually make the decisions, hydrologic models can supply facts about these four planning issues and their interrelationships. The process should iterate until the studies provide the information that the decision makers consider important. Grigg (1985, p. 35) describes decision makers in water resources management as influenced by organizational structure, laws and policies, politics and pressure groups, leadership, economic and time constraints, background and training, professional experience, cognitive orientation, preferences for types of information, and patterns of information use. Models should be used not only for greater accuracy in representing nature but also for greater clarity in communicating the hydrologic facts that the decision makers want.

Here, we can return once more to the human–computer interface and the potential that it provides for establishing interactive communications so that models address the concerns of the decision makers (Rogers and Fiering, 1986). By interaction, decision makers can clarify what they want to know, and hydrologists can clarify what their estimates say. Hydrologic models add information for this interactive exchange because additional simulation runs can be conducted easily and quickly.

Good communications at the human–computer interface can serve decision makers of diverse experience in many roles and at levels from the top officials in government to small water users. Some decisions are made by lay individuals and others are made by professional consultants who represent clients. Some decisions are made directly on water issues (national water policy, river basin development and system operation, state water policy, project development, and personal water use), others are made on related issues (land use, economic development, budgetary policy, etc.) that indirectly impact water.

To a large degree, the information needs of these many decision makers are met by presenting factual descriptions of the same natural phenomenon in different forms to match the resolution, precision, and reliability required by different users. Hydrologic models can be applied on different space–time grids and connected to different tabular and graphic displays.

Hydrologic models can also assist decision makers by making it easier for them to interact with the experts in clarifying and supplementing information. Different groups participating in the decision making can propose policies, plans, projects, programs, and regulations. The experts can run a reconnaissance simulation for whatever the decision makers have chosen. Where analysis is needed in greater depth, more thorough studies can be conducted. Afterwards, studies can be revisited as conditions change and new issues arise. The quick delivery of reliable information gives decision makers confidence that they can react effectively to changing conditions which may be the greatest single contribution of hydrologic modeling.

2.6. Support Systems for Hydrologic Studies

Modeling should be considered as a tool in a larger strategy for hydrologic data gathering and information delivery. The hydrologists doing the modeling must be careful to ensure that their studies are using correct and full information and then communicating this information fully to the decision makers. In this process, the hydrologists need support from:

(1) *Data Gathering.* The models need measurements of: the climatological processes driving hydrology, the catchment characteristics controlling hydrologic response, and the stream flows and aquifer conditions that are providing the data for model calibration and verification. In each case, the precision, timing, and space–grid for data collection should support and anticipate model requirements. The interface between modeling and data gathering also involves procedures for approximating missing data, screening provided data for reasonableness, making needed adjustments, extending records, converting measurements under changing conditions to a homogeneous base (often virgin flow conditions), and estimating how the data would be altered by various human changes to the land and water system. Additions or revisions to existing data acquisition programs may be necessary where they cannot service model needs.

(2) *Process Research.* Many older models mix outdated with state–of–the–art components. Hydrologic modeling should be updated to utilize hydrologic research. The process will be facilitated as research finds more effective ways to move from publishing results in equations and theories to communicating applications (computer programs can help here). An important interfacing goal is to standardize the linkages so that new relationships can be more easily placed in updated kernels and used in simulation. Efforts are underway to develop the needed overall model structure, but the lead organizations are understandably reluctant to share their algorithms and have them misused.

(3) *Systems Research.* The thrust in hydrologic research is moving from the laboratory-scale to the field– and meso–scales (Gupta et al., 1986), and the results will have to be incorporated into hydrologic modeling at these larger scales. Major improvements in the physical

concepts and mathematical representations used in linking the process kernels can be expected. At some point, a quantum change may be needed in the structure of the larger system.

(4) *Institutional Facilitation*. The results of hydrologic studies are largely used by water planning and management institutions. Even though little research has been done on how to communicate information efficiently to institutions for decision making (James and Caswell, under preparation), recent research is contributing valuable insights. The use of research in communications to transfer the findings of hydrologic modeling more efficiently to decision makers should facilitate institutional support for model development.

2.7. Engineering Soundness

Each hydrologist bears ultimate responsibility for the results of his application. Criteria that help improve accuracy and, thus, should be used to guide the hydrologic study are:

(1) *Site-specific measurements*. A hydrologic model works best with site–specific data; this means that a model should be programmed to receive measurements in the form that they are locally being taken. In some situations, cost effective methods may have to be implemented for taking new measurements. For example, no major water project should be built without some gauged streamflow record, and the time frame of project planning permits this. An approximate relationship using local measurements often gives better results than does a theoretically more sound relationship using data imported from afar. As an example here, planners may obtain better results by estimating evapotranspiration from local temperature data than by using the Penman equation with data from a distant site (McVicker, 1982).

(2) *Causative constructs*. The ideal hydrologic model uses "white box" kernels to the extent practical in order to have more reliable estimation in a wide variety of conditions. However, this approach is limited by both theory and data. Where "white box" theory has not been developed, engineers must use their experience to develop "gray box' representations. Where "white box" data are not obtainable for an acceptable cost, hydrologists must estimate quantities by engineering judgments and from correlations based on general data and experience. "Black box" approaches should generally be resisted as inappropriate for hydrologic simulation.

(3) *Consistency*. The model should produce compatible results for similar situations in different times and different places.

(4) *Sensitivity*. The model should give different results when the input describes different conditions that in fact alter catchment responses. An indirect test of sensitivity is that modeling errors should be explainable from physical causes.

(5) *Economy*. Engineers must consider costs as well as results. The modeling effort should set and pursue specific goals in data preparation, computation, and interpretation. It is very easy to devote a great deal of work to tasks that bring little hydrologic gain.

2.8. Hydrologic Modeling as an Institutional Process

A major effort is required to develop and maintain a hydrologic modeling system (where multiple kernels may be used or dropped according to need). Still greater effort is needed in coordination with support and user systems. Successful efforts must be formally institutionalized and supported with adequate budget and expertise. For this to happen, the models must fill the goals of many sponsoring users and research and academic entities. Management agencies have different goals than scholars (Grigg, 1985, p. 36), and all institutions want to further their missions and protect their credibility, as I discussed in Chapter 2. It is particularly important to recognize the vigor with which agencies protect their credibility against political pressures growing out of changes in the economic, social, and environmental goals of the public (Hamilton, 1974).

Given this need for institutional support, a modeling effort should:

(1) Provide the hydrologic quantities considered important by people who will be given the information. Most people want to know how they will be impacted by floods and droughts and need information on their local conditions. Others want to know about economic and environmental impacts at a larger scale.

(2) Identify the hydrologic processes that people think should be incorporated into the modeling to estimate the desired quantities. Users frequently ask whether or not some factor has been considered. They insist that the models seem reasonable in the light of their experiences and be explainable to satisfy their legal constraints. Background explanatory materials may have to be prepared to overcome doubts.

(3) Select equations and algorithms that represent the processes in a manner consistent with established manuals and policies. Institutional objections to controversial estimation methods may constrain the adoption of new approaches. Conflicts between agency practice and sound hydrology should be explicitly identified and resolved.

(4) Synthesize kernels simulating the processes into a computational framework that provides an opportunity for institutions to review the framework and make suggestions. This opportunity is very important for building user confidence at the time a new method is adopted. Afterwards, periodic reviews and assessments continue to be beneficial.

2.9. Institutional Tests

An important principle to remember is that scientists and engineers seek accuracy; institutions seek closure (Grigg, 1985, p. 36). Hydrologists as scientists want to keep refining their work; institutions want to find answers and then move on to something else. In evaluating the proposed closure, hydrologic orientation is important. For example, India places priority on water supply (National Water Policy, 1987), and Japan on floods (Sambongi, 1988).

The requirements listed in the last section are rather technical, and institutions are generally more comfortable in indirect criteria for model acceptance, as I noted in Chapter 2. When they examine hydrologic models, institutions place their emphasis on such criteria as:

(1) *Credentials.* The model must be programmed and applied by capable people who properly represented the context and used good judgment in interpreting the results. They must obtain reliable data, use accepted methods, and be trained in using the hardware and software.

(2) *Peer Acceptance.* The study must receive widespread peer acceptance (lack significant peer criticism). Institutions that are not adept at technical evaluations are troubled by criticism from experts.

(3) *Understandability.* The model must be understandable at all levels of decision making and through all communication channels dealing with the problem addressed. The principles used in the model must be explained in simple enough terms for the users to understand.

(4) *Face Validity.* The model results must be reasonable in the experience of the decision makers in the desired applications. The results should seem reasonable to the decision system dealing with the particular issue and to experienced people who may not be familiar with the particular model.

2.10. Communication

The information obtained by hydrologic modeling is presented to engineers, generalists planners, the public at large, and government officials—some of these are in decision making capacities and others have professional roles, all must be involved to make water planning fully effective (Fox, 1975). The presentations should convey the model results and use thoughtful assessments in their interpretation. Good information delivery:

(1) Conveys the facts established by the study in a manner that builds correct understanding of the situation (different presentations should be prepared for different audiences).

(2) Communicates the doubts that were not resolved and the uncertainties that remain (water planning institutions should disclose the full situations in their public presentations).

(3) Updates the presentations periodically as new information or techniques become available or as conditions change.

(4) Takes advantage of such tools as:

(a) The distribution of information through electronic communication systems with emphasis on computer graphics.

(b) The placement of output summaries and readings on the concepts in public libraries. This would help people interpret the information and, for example, use the results in implementing nonstructural measures.

3. Conclusions

Generations of hydrologists have learned to apply their understanding of physical systems thoughtfully and interactively with equations and formulas in supplying information for water resources planning. Recently, we have been able to automate the numerical work in hydrologic models. We must now concentrate on improving our approach in our professional evaluation of the results, and we must now link the much larger amount of information we can supply to the many types and levels of decision making in water resources planning and management.

We have advocated placing the technical details of the computations into process kernels for evaluation and refinement by experts in hydrologic science, and we have recommended linking these kernels in the larger models for the required thoughtful assessment at the planning scale. The assembled models should provide the information needed to address the basic planning issues with respect to issues that decision makers consider important. Here, hydrologic models make the important contribution of providing a means to respond rapidly to the myriad questions people ask when confronted with choices on their future.

To make this approach to hydrologic modeling work effectively, hydrologists must institutionalize supporting data gathering and research, find cost–effective ways to produce accurate results, structure their models to satisfy institutional goals, and document their procedures to pass institutional tests.

In reflecting on this overall situation, the hydrologic modeler must remember that technological innovations require support from multiple institutions within the social and political system for their benefits to be realized (Mesthene, 1970, p. viii). A hydrologic model is an invaluable tool for organizing extensive information for critical thinking and public display. Modeling is also a dangerous tool because the time committed to model development, the aura of sophisticated technology, and the over–zealous claims of proponents can make a model a master rather than a servant.

A summary suggestion for reducing model misuse is for modelers to write down their modeling goals, modeling assumptions, and procedures for interpreting model results. Users should be given this documentation when making their evaluations. The very exercise would be a continuing reminder of how far modeling falls short of full understanding.

Each hydrologic study should begin from specific questions in an explicit context. The relevant process kernels should be linked to answer those questions, and the modeler should interact with the water planners in sifting through multiple simulations for formulating logical management plans.

Ultimately, the role of hydrologic modeling is as support for human understanding in thinking through large numbers of alternatives with widely scattered ramifications. A model can be used to track many more factors than any one expert can ever consider simultaneously. Both agencies and the public at large need the information from hydrologic models for decision making in water resources planning.

4. References

Bonnin, J.: 1988, 'Were Urban Water Systems Improved Over the Last 20 Centuries?,' *Water International* **13**(1), 13–16.

Burges, S. J.: 1986, 'Trends and Directions in Hydrology,' *Water Resources Research* **22**(9), 1S–5S.

Brutsaert, W.: 1986, 'Catchment–Scale Evaporation and the Atmospheric Boundary Layer,' *Water Resources Research* **22**(9), 39S–45S.

Dooge, J. C. I.: 1986, 'Looking for Hydrologic Laws,' *Water Resources Research* **22**(9), 46S–58S.

Eagleson, P. S.: 1986, 'The Emergence of Global–Scale Hydrology,' *Water Resources Research* **22**(9), 6S–14S.

Fox, I. K.: 1975, *'Strategic Considerations in Attaining Water Resources Planning Goals,'* Selected Works in Water Resources, International Water Association, pp. 231–239.

Gleick, J.: 1987, *Chaos: Making a New Science*, New York: Viking.

Grigg, N. S.: 1985, *Water Resources Planning*, McGraw–Hill Book Company.

Gupta, V. K., I. Rodriguez–Iturbe, and E. F. Wood: 1986, *Scale Problems in Hydrology*.

Hamilton, A. L.: 1974, 'Organizational Impediments to Effective Environmental Impact Research,' in *Agassiz Center for Water Studies, The Allocative Conflicts in Water–Resource Management*, Winnipeg, pp. 391–403.

Howe, C. W.: 1971, *'Benefit–Cost Analysis for Water System Planning,'* American Geophysical Union, Washington, D.C.

James, L. D., and P. Caswell: Under preparation, *'From Construction to Resource Management: The Evolution of Irrigation Institutions.'*

Klemes, V.: 1986, 'Dilettantism in Hydrology: Transition or Destiny?,' *Water Resources Research* **22**(9), 177S–188S, August 1986.

Linsley, R. K.: 1986, 'Flood Estimates: How Good Are They?,' *Water Resources Research* **22**(9), 159S–164S.

Loucks, D. P., J. R. Stedinger, and D. A. Haith: 1981, *Water Resources Systems Planning and Analysis*.

Maass, A., et al.: 1962, *Design of Water–Resources Systems*, Cambridge, Mass.: Harvard University Press.

Mandelbrot, B. B.: 1983, *The Practical Geometry of Nature*, W.W. Freman and Co., N.Y.

McVicker, R. R.: 1982, *'The Effects of Model Complexity on the Predictive Accuracy of Soil Moisture Accounting Models,'* Utah State University, Logan, Utah.

Mesthene, E. G.: 1970, *Technological Change: Its Impact on Man and Society*, Harvard University Press, Cambridge, Mass.

National Water Policy: 1987, Government of India, Central Water Commission, New Delhi.

Office of Technology Assessment: 1982, *Use of Models for Water Resources Management Planning and Policy*, Washington, DC.

Pilgrim, D. H.: 1986, 'Bridging the Gap Between Flood Research and Design Practice,' *Water Resources Research* **22**(9), 165S–176S.

Renard, K. G., W. J. Rawls, and M. M. Fogel: 1982, *Currently Available Models. Hydrologic Modeling of Small Watersheds.*

Rodriguez–Iturbe, I.: 1986, 'Scale of Fluctuation of Rainfall Models,' *Water Resources Research* **22**(9), 15S–37S.

Rogers, P. P., and M. B. Fiering: 1986, 'Use of Systems Analysis in Water Management,' *Water Resources Research* **22**(9), 146S–158S.

Sambongi, K.: 1988, 'Characteristics and Functions of Water Institutions,' *Water Resources Development* **4**(1), 53–57.

Sklash, M. G., I. D. Moore, and G. J. Burch: 1988, *Environmental Isotope Tracer Studies of Catchment Processes: Tools for Verifying Integrated Water Quality Models.*

Chapter 27

Learning by Parametric Modeling: Hydrologic Investigations of Utah Mountain Catchments

L. Douglas James
Utah State University
Utah Water Research Laboratory
Logan, Utah 84322–8200 U.S.A.

Abstract. As a hydrologic study finds answers to problems, the learning experience builds human capital. This "learning by doing" is illustrated in studies of flood risks below Utah canyons. A version of the Stanford Watershed Model (SWM) with five principal parameters was applied to several heterogeneous catchments. The Wasatch Front rises 1600 meters above the Great Salt Lake and is drained by many parallel mountain canyons whose side slopes average about 35 percent. Since most precipitation is snow, the modeling covered snow accumulation, solar radiation, melt runoff, soil moisture, landslides, debris flows, routing, and sediment transport. The poor explanation of certain physical processes suggested use of a stochastic parametric model that can be segmented into progressively smaller units to find a scale of best fit. Critical thinking can then examine the identified problem conditions with a physical model at that scale.

1. Introduction

Every hydrologist should do his best to provide his agency or client with full and accurate information. Part of doing one's best is to make each study a learning experience that adds new capabilities. Over time, the learning should be building human capital; and, indeed, this Advanced Study Institute is largely organized for the exchange of practical ideas for achieving this goal.

This chapter presents an example joining these short– (doing) and long– (learning) term objectives of hydrologic investigations. The doing part describes hydrologic studies to assess the risk of snowmelt flooding in the urban areas below Utah mountain canyons. The risk assessment used a parametric model for continuous simulation of spatially distributed snow accumulation, ground temperatures, snowmelt, runoff, soil moisture, landslides, debris flows, and sediment transport in a cold, arid climate. The modeling revealed a great deal about the hydrology of these catchments and gave valuable clues for better understanding of the hydrologic implications of their extreme spatial heterogeneity and temporal changes. The success of this study and the costliness of such fine–scale measurements suggests that the added effort of using a physically based model should not be undertaken without first taking maximum advantage of the information that can be learned by parametric modeling from large–scale data. Science and judgement should be brought together in linking fine–scale (soil properties) and large–scale (gauged flows) phenomena in advancing hydrologic understanding and making better hydrologic estimates for engineering design.

D. S. Bowles and P. E. O'Connell (eds.), Recent Advances in the Modeling of Hydrologic Systems, 571–587.
© 1991 *Kluwer Academic Publishers.*

Accordingly, this presentation of "lessons learned from doing" begins from an historical perspective and leads to ideas for a strategy for hydrologic learning. The original proponents of continuous hydrologic simulation sought to use longer precipitation records to extend shorter stream flow records to make better probabilistic estimates of flood peaks and volumes, sediment movements, long–term runoff volumes, low flows, and other quantities useful in water resources management. Many of them envisioned extending this estimating capability to ungauged catchments. They charged ahead with enthusiasm in applying the parametric methods that were the only practical means of simulation with the computers and data available in their day.

However, these modelers were not able to fulfill these dreams. One of the primary reasons was that they never established objective ways to estimate model parameters from measure-ments of the physical characteristics of a catchment. The underlying difficulties are both conceptual and numerical. Conceptually, the model parameters are indices rather than physical properties, and they indicate aggregate rather than local conditions. Mathematical-ly, the parameters are unknowns in equations placed in conceptual models and are estimated by statistical methods to match observed hydrologic response.

We now have the alternative of going to a smaller scale and applying distributed models that use physical laws in differential equations that have unknown physical properties. With this approach, the modeler selects a scale, takes measurements, and finds a satisfactory way to represent the spatial distributions and correlations of these properties. However, without an objective basis for selecting the scale, that decision becomes arbitrary. Without a concep-tual model of how catchments respond at larger scales, physically based modeling improves the ties to measured catchment characteristics at the expense of separating from the ties to measured hydrologic responses at the catchment scale. Distributed, physically based model-ing often finds that assumptions based on limited data do not aggregate to reliable results at a larger scale (Bevan, 1989).

The many years of doing parametric modeling have taught us many important lessons on how catchment–scale hydrologic systems work. The following ideas are reflections based on applying the Stanford Watershed Model to catchments in Utah. To set the stage, we will first give a streamlined presentation of the model and its parameters. The results of the simulation can then be better understood in drawing lessons in selecting a scale for analysis, identifying processes to incorporate in modeling at that scale, and measuring catchment characteristics.

2. Discussion

2.1. Understanding the Parametric Model

A good hydrologist understands his tools. Parametric models, like other statistical tech-niques, can be made to reproduce numbers without portraying real hydrologic processes. Consequently, our first step was to assess the principal features of the selected tool as a framework for interpreting the results. The resulting interpretation provides valuable back-ground for considering applications to meet different needs.

2.1.1. The Heart of the Stanford Watershed Model

The Stanford Watershed Model (Crawford and Linsley, 1966) pioneered lumped, parametric hydrologic simulation and became the ancestor of many models that are widely used today. After peeling away outer layers of inputs, outputs, runoff routing, and water–balance track-ing, the inner kernel of the model routes precipitation through upper and lower zone storages (Figure 1) to its ultimate disposition as either evapotranspiration or streamflow. Water

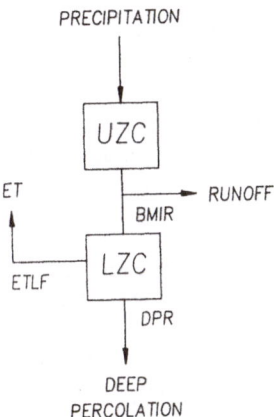

Figure 1. Simplified SWM schematic.

temporarily held in the upper zone is exposed to evaporation and infiltration. A larger upper-zone storage capacity absorbs smaller storms and delays runoff from larger ones. Subsequent evaporation can then dry the soil surface after summer showers. The process reduces direct runoff and adds infiltration and base flow.

A larger lower–zone storage absorbs more rainfall, discharges more precipitation back to the atmosphere as transpiration, and reduces both high flow and low flows. The lower zone loses water by transpiration and by deep percolation, with both loss rates diminishing as the soil dries. The equations have four principal parameters:

(1) An Upper Zone Capacity (UZC) that indexes the volume of precipitation that depression storage on the surface of a catchment can hold. Amounts increase with the roughness and decrease with the slope of the soil surface.

(2) A Lower Zone Capacity (LZC) that indexes the volume of infiltration that the soil in a catchment can hold. Amounts increase with soil depth and soil porosity from wilting point to field capacity in the hydrologically active layers.

(3) BMIR, an infiltration index that represents the maximum rate (volume per unit time) at which water can enter the surface of the catchment. Actual rates, except as limited by precipitation amounts, are assumed to vary linearly from zero to this maximum; and the maximum at any time varies with soil wetness indexed to current lower zone storage as a fraction of LZC. BMIR also controls the rate at which water passes through the hydrologically active zone and percolates to the water table.

(4) ETLF, a transpiration index that represents the maximum rate at which vegetation can transpire water from the hydrologic active zone to the atmosphere. Actual rates, except as limited by the potential evaporation, vary linearly from zero to this maximum.
The continuous functions used in spatial representation to model infiltration and evapotranspiration provide explicit recognition of spatial variability that is an important model feature.

2.1.2. Linking SWM Parameters with Measured Catchment Characteristics

Most of the Utah mountain catchments studied were ungauged. The conventional approach for parametric simulation of flows from ungauged catchments is to estimate the parameter values from model calibrations for similar gauged watersheds. Calibrations done by trial-

and–error work poorly because of the same sort of subjective error found in comparing lines of best fit drawn by eye. A computerized optimization routine called OPSET (Liou, 1970) is available to reduce this subjective noise in the estimates.

Over time, a large number of catchments with diverse climates, topographies, soils, and geologies have been calibrated. Past studies have sought to correlate these parameter values with catchment characteristics estimated from maps, aerial photography, and field observations (Ross, 1970; James, 1972). Some of the correlations proved significant, but they were not sufficiently convincing to be widely adopted, largely because the methodology faces important obstacles in measuring catchment characteristics.

The available mapping was performed over many years by different groups using different standards, and their quantitative interpretations were not consistent. Hydrologic interpretation of aerial photography also contains considerable subjective noise.

Nowadays, we have alternative. Catchment characteristics may be measured by remote sensing for pixel elements about 30 meters square. Statistics can be computed from pixels aggregated in various combinations in a search for the best correlation with the parameter values estimated by model calibration. However, translation of sensor images to physical characteristics influencing catchment hydrology is still in its infancy.

Other problems are caused by the model structure. First, because the SWM was programmed to simulate flows rather than to estimate catchment characteristics, it was streamlined to fewer parameters to facilitate calibration. Some conceptual variables were taken as constant ratios of other parameters. When several concepts are mixed in the indexing by a single parameter, calibrations that are interpreted from a primary process (e.g., surface infiltration) actually incorporate the other processes (e.g., deep percolation). The ratio assumed between these rates for model streamlining may not match the field data at other locations. This difficulty could be overcome by restructuring the SWM to separate concepts by using different parameters.

The SWM takes precipitation as uniformly distributed over area. The two–dimensional catchment surface is represented by a one–dimensional distribution using fraction of the area as the abscissa. Infiltration and evapotranspiration rates are represented by triangular distributions along this abscissa. Volumes calculated from these triangles are then taken as spatially uniform inputs to the next process in the simulation. The major weakness of this approach is that the simulated flow from one process to the next does not consider sequential correlations when in fact, for example, high transpiration rates may or may not be spatially correlated with high infiltration rates. In fact, each flow in the hydrologic cycle is spatially distributed, interacts with media differently distributed, and yields an outflow following a third distribution. Simulation that uses media but not inflow variability undoubtedly biases the estimates of the parameter values.

Distributed process linking might be approached by either deterministic or stochastic modeling. One such method would be to devise deterministic representations of the spatial variability and their sequential correlations; however, reliable calibration of such parameters may be difficult. A second approach would be to take all the distributions as uniform and keep subdividing the catchment into smaller subareas until their internal distributions no longer matter. During a simulation, different precipitation amounts and parameter values could then be taken stochastically for each subarea. Such a stochastic model would preserve the sequencing of moisture conditions on each subarea. The difficulty would be in calibrating the variabilities and correlations among parameters.

2.2. Applying the Model to Understand Western Mountain Catchments

The above assessment indicated that it would be no simple matter to model the heterogeneous Utah catchments. However, by using a parametric model for conceptual thinking, a research hydrologist can refine and expand the model in a living documentation of ideas that improve hydrologic simulation. Since the details and results of the mountain catchment modeling in Utah are published in the cited theses and dissertations, the following presentation is limited to interpretation, modeling issues, and suggestions for future investigations.

2.2.1. Catchment Characteristics

The Wasatch Front rises steeply above the eastern side of the Great Salt Lake from plains at an elevation of about 1400 meters above sea level to about 3000 meters at the ridge crest which is about eight kilometers further east. The north–south mountain range is drained by creeks flowing westward in the bottoms of dozens of deep (up to 1000 meters), parallel canyons. A typical creek flows on a gradient of about 20 percent in a canyon bottom between sunny south-facing slopes of 30 percent and shaded north-facing slopes of 40 percent.

The map in Figure 2 shows Steed Creek Canyon as an example. The elevation contours are marked in feet (0.31 meters), and the grid is in square furlongs (201 meters). These units may seem archaic, but they are a product of a surveying heritage in which streets are eight per mile and blocks are square furlongs of ten acres (4.05 hectares). A point here is that people perceive hydrologic processes from spatial relationships sensed from observable landmarks.

Soil depths vary from none on the ridges and cliffs to more than 5 meters in some swales. Precipitation falls primarily as snow from September to May in storms where a strong orographic influence causes the amounts to reach maximum average annual water contents of about 1200 mm about two thirds the way up the slope. The remaining precipitation falls as summer rain in convective storms that vary little in amount with elevation.

The years 1983 and 1984 experienced record precipitation, primarily in winter storms, after decades of drier weather (James and Bowles, 1986). During the spring snowmelt of both years, urban areas below the canyon mouths were flooded, damaged by sediment deposition, and subjected to debris flows (Pack, 1985).

The U.S. Forest Service once maintained a number of short–term gaging stations on the side streams in one of the larger canyons. A typical instrumented catchment drained about 200 hectares on a north–or south–facing slope. The periods of record varied as the gauges were moved from one site to another, and the last gauge was discontinued in the 1970s. Unfortunately, we do not have gauged flows during the extremely wet years in the 1980s.

2.2.2. Snow Accumulation Modeling

The employed version of the SWM contained a snow accumulation and melt subroutine (Anderson and Crawford, 1964), and the simulation was checked against measured snow depths. The greatest modeling difficulty was in differentiating rain from snow during autumn and spring storms that bring rain at lower elevations and snow at higher ones. The application of lapse rates (differentiated by time of year and whether precipitation was occurring) to temperatures measured at the base of the mountains worked poorly in modeling the diurnal movement of the snowline up and down the mountainside because actual lapse rates vary considerably during storm events. Also, the lapse rates in cold climates frequently exhibit strong inversions during midwinter, but failure to simulate correct temperatures during this period has little hydrologic impact because conditions are continuously subfreezing at all elevations.

2.2.3. Solar Radiation Modeling

The snow begins melting on the south-facing slopes during the first sunny days of spring. The shaded north-facing slopes hold the snow longer, and the snow there melts more quickly

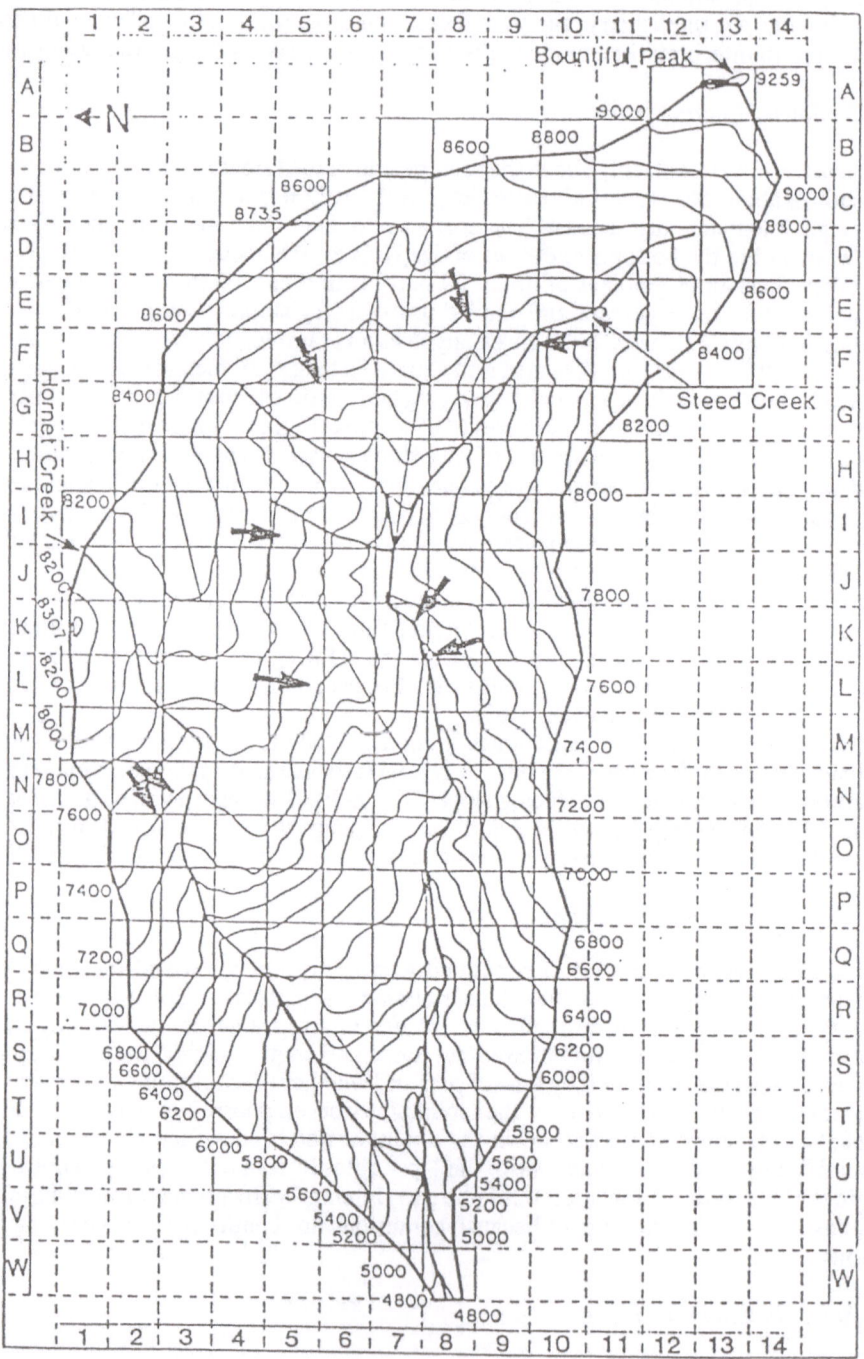

Figure 2. Steed Canyon: location of 1983/1984 landslips.

when air temperatures are warmer. This period of more rapid late–spring melt causes higher stream flows.

Spring weather with prolonged cloudiness reduces incoming radiation and delays the melt on the south–facing slopes. With sufficient delay, both slopes can be subjected to simultaneous rapid melt when sunshine and warm weather finally arrive, and this condition causes the largest snowmelt floods. Because the more intense radiation on the south–facing slopes causes greater diurnal temperature variations, weathering and soil formation occur more rapidly, and the greater accumulations of soil in the swales favor landslides.

The solar radiation reaching the ground surface was modeled based on field tests at Logan, Utah, where Aurasteh (1983) showed that the energy reaching surfaces in shadows averages only about 16 percent of that reaching surfaces in sunshine. From information on the topography of the mountains and resultant shading, on angles of the sun in the sky over the day and year, and on the type and extent of cloud cover, Lambson (1988) used this ratio to simulate hourly and daily totals of solar radiation reaching the ground surface in mountain areas. The simulation matched ground radiation and temperature measurements during sunny days but gave poor results during overcast periods because the recorded data on cloud cover poorly depict the variability among clouds of different types and thicknesses. The impacts of cloud cover on subsequent snowmelt should be given greater attention in quantifying flood flows and soil instability as long periods of cloudy weather suggest that there will be problems as spring progresses.

2.2.4. Runoff Modeling

For simulating runoff from Steed Canyon, Pitcher (1986) modified the conceptual model to cover the thin soil mantle perched on the bedrock slopes, as shown on Figure 3. Groundwater storage was divided between water trapped in the more porous soil lenses that were behind the clay layers above the bedrock surface (GWS) and deep groundwater in the bedrock (DGWS).

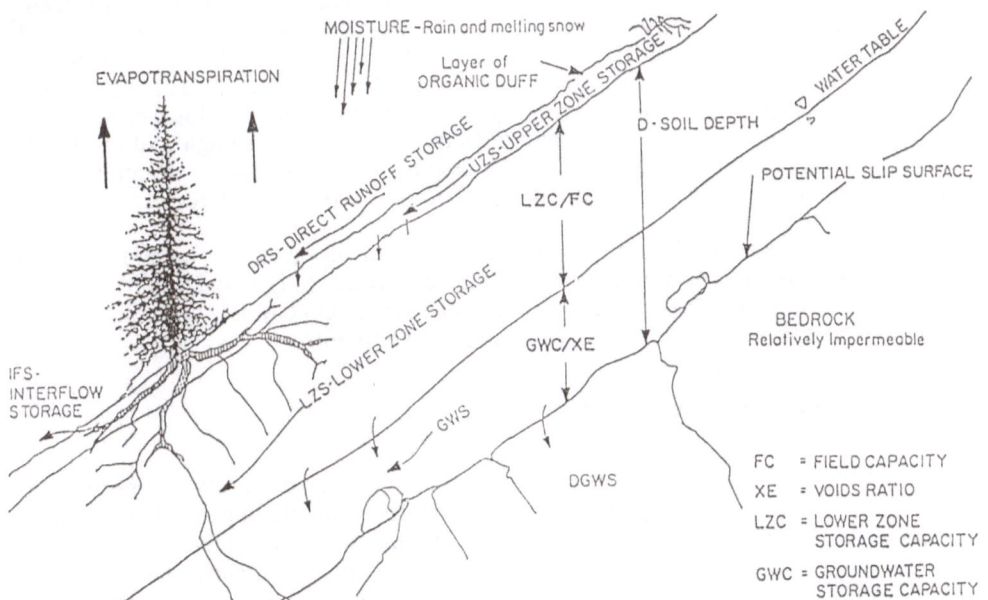

Figure 3. Groundwater and soil moisture zones.

Runoff varies considerably with the elevation and orientation of these mountain slopes. Pitcher calibrated several gauged catchments with different elevations and orientations against 1956–59 data. For the 1983 and 1984 applications, three zonal classifications were used. The parameter values for six cases in two classifications (north v. south facing slopes and three elevation ranges) were calibrated directly. The elevation ranges were above, within, and below the zone of primary landslide activity as explained below. The parameter values for the third classification (natural surfaces v. surfaces scoured by slides) were estimated by adjusting the calibrated values for natural surfaces to represent the observed field conditions in slide areas. Zonal hydrographs were summed to estimate total daily runoff volumes from Steed Creek. The findings were generally confirmed by field observations during the events, and the parametric modeling was particularly useful in establishing relative orders of magnitude, thus pointing our priority hydrologic processes for more detailed investigation.

Even in the two wet years of 1983 and 1984, the model simulated no direct runoff from the natural catchment surfaces. The simulation indicated that about 24 percent of the runoff percolated into the bedrock and discharged into the stream at springs located near the main channel. This flow source dominates the runoff during drier years when snow melt rates seldom exceed the permeability of the underlying geologic formations. This discharge caused no downstream hazards.

During the two wet years about 75 percent of the simulated flow came from melt rates faster than the underlying formations could absorb and percolated down–slope through the soil mantle tributary gullies or the canyon bottom. This interflow–type runoff causes most of the stream sediment loading because it reduces the stability of the soil in the swales, erodes gully heads, and causes stream banks to slough (generally at the base of north–facing slopes, as shown on Figure 2). This flow dominates the runoff during wet years. In drier years, it supplies little flow, and catchment sediment production is relatively small. Simulations indicating that this flow is more than about half of the total suggest pending trouble.

Land slides form when the hydrostatic pressure behind the clay lenses becomes excessive and causes the soil in the swales (generally perched at higher elevations on south–facing slopes as shown on Figure 2) to give way. The mass moves down the steep hillsides generally leaving a track of exposed bedrock. In 1983 and 1984, the remaining 1 percent of the flow simulated from Steed Canyon came from the much smaller fraction of the land surface that has experienced these slides. Their margins become primary source areas for sediment as the water reaching them through the soil mantle causes the banks to slough and carries the loosened material down the hillside. During extremely large snowmelt events (perhaps 1000–year) these areas of exposed bedrock could become a primary source of flood runoff and sediment production.

These areas are likely to generate much more than their proportional share of runoff for years. The hydrologic recovery of these areas deserves further study, and data will have to be collected over a period of years to make this possible. Unfortunately, it is difficult to sustain funding for this sort of research after conditions return to normal.

2.2.5. Soil Moisture Modeling

The runoff simulation uses an index of water accumulation in the soil to vary infiltration rates. Good replication of runoff suggests that soil moisture accumulations are being indexed reasonably. From Figure 3 one can see that the accumulation of water in the thin soil mantles of these mountain slopes depends not only on infiltration at the catchment surface but also on the rate at which deep percolation passes from the soil to the underlying formations. In addition, clay layers can restrict flow downslope through the soil mantle, add to soil water accumulations, and increase the hydrostatic pressures in sand lenses.

It would be extremely rare, in this climate, for enough water to infiltrate during the short summer convective summer storms to cause significant hydrostatic pressures. The calibrated model could be used to simulate soil moisture conditions during major storms of record for specific information.

One strength of a general parametric model is that the intermediate variables can be checked against field observations. Four such checks were used in this study. A previous study showed that the simulated values of LZS matched field estimates of soil moisture (James, 1968). Some soil moisture data were obtained from studies on the mountain slopes during the two wet years. Second, the soil water contents simulated for days when landslides occurred can be compared with critical amounts determined by slope stability analyses. Third, the timing of the maximum simulated soil water accumulations can be compared with the timing of the observed landslides. Finally, the calibrated value for LZC can be compared with the soil depth required to generate the slides. Use of multiple checks drives the calibration much closer toward a unique solution.

2.2.6. Landslide Modeling

Pack (1985) reported 90 landslides in Davis County during the period of snowmelt in 1983 and 41 more in 1984. Six sites are shown in Figure 2. He characterized landforms throughout the area and found that the hazard of slide formation was greatest in concave swales that were perched on canyon walls above sharp breaks in the slope down gradient where the soil mantle was more than three feet thick. Most of the slides originated in a relatively narrow band of mountainside areas between 2050 and 2250 meters in elevation, the location of the snowline where rapid melt is sustained for the longest periods. Snow does not accumulate to as great a depth at lower elevations, and the melt at higher elevations is slowed by prevailing cooler temperatures. Pack (1985) succeeded in defining locations of high landslide risk, but his multivariate predictor lacked a dynamic element for determining when the water content of catchment soils was reaching threatening proportions.

This dynamic assessment turned to the above hydrologic models. The simulated hourly values of GWS gave an index that Weston (1987) used as a driving force in modeling landslide occurrences by summing forces in an infinite–slope analysis. In the freebody diagram, the weight of the water and the soil mass are resisted at the sloped–rock surface by friction that can be quantified from soil cohesion (C_s), the angle of internal friction (O), and root cohesion (C_r). Water content causes landslides by adding weight, buoyancy (reducing the normal force for internal friction), and hydrostatic pressure and by reducing soil resistance (C_s and O). The added forces can be estimated from the principles of fluid mechanics and the reduced resistance from studies of soil properties.

Data were collected on slope and vegetative cover on the 10–acre (squares approximately 200 meters on a side) grid. Data were not available for mapping soil characteristics. Field measurements showed that within the swales, pore pressures in the sand lenses rise and ebb diurnally over wide ranges (Jeppson and Farmer, 1985). Rises occur when the melting snow infiltrates faster than the water can percolate through the clay layers in the heterogeneous soil, and pressures decline during the night as melt stops but percolation continues. Failures occur when the pore pressures reach critical levels. The simulation (Weston, 1987) linked times and zones with maximum values of GWS (by hydrologic zone) to the observed occurrence of landslides.

Water accumulates as the swales intercept interflow from three uphill directions. Examination of the force equilibrium when slides began suggested that the ratio of maximum water content (in swales) to average water content (over the zone) is about three. This ratio suggests a factor of safety to apply when assessing slide hazard. A local physical model would have to be applied to individual swales to model specific events.

2.2.7. Debris Flow Modeling

A landslide may either halt immediately or turn into a debris flow that reaches the bottom of the canyon. The primary indices to watch in determining whether a landslide becomes a debris flow are soil depth (a measure of mass and momentum) and water content (a measure of fluidity and thus friction resisting liquid flows). Debris flows that are traveling perpendicularly to the stream when they reach the bottom of the canyons are likely to block the flows. In cases where the debris flow is traveling parallel to the stream, it may continue scouring and growing all the way to the mouth of the canyon. An initial assessment of the debris flow hazard can be based on the landslide potential (Pack, 1985) from mountainside areas where the contours indicate that debris flows would be generally parallel to the stream.

In the moving mass, the trailing material travels faster, overrides the nose, and forms a churning roller at the leading edge that thoroughly mixes the material (Rodriguez, 1982). The resulting grinding action totally destroys any property encountered as the debris flow emerges from the canyon mouth and continues in a straight line through the urban areas on flatter ground. Finally, the moving mass of soil stops abruptly.

DeLeon (1982) derived equations for routing these debris flows. Modeling the stopping points is more difficult and extremely important for assessing the hazard at various distances from the canyon mouth. The primary factors causing the sudden halt are the loss in momentum per unit width of flow (the depth decreases as material spreads laterally) and the loss in water content with infiltration into the underlying soil.

2.2.8. Flow Routing and Sediment Transport

In order to translate runoff and sediment production into hazards in the urban areas, it was necessary to route the water and sediment down the steep mountain streams through alternating pools and cascades of super critical flow. The process becomes much more complicated than simple applications of standard flow routing and sediment transport equations when one considers the physical phenomena involved.

In normal years, most runoff travels through the underlying formations and reaches the creeks in springs. This flow contributes little sediment. The baseflow hydrographs rise and fall so slowly, compared to the travel time, that routing has little impact on downstream peaks and even if it did the flows would be too small to cause a significant hazard.

The problems emerge as more of the runoff comes as interflow and land slides begin to occur. Some of the debris flows dam the channel. When they are overtopped, a moving wall of water and mud rushes down the stream and spreads out over the lowlands.

Where landslides are not involved, water comes from baseflow, interflow, and direct runoff from landslide–denuded areas. Sediment comes from bank sloughing caused by intercepted interflow, from debris deposited in the creek channels, and from direct runoff from landslide-disturbed areas. The routing poses problems because the underlying phenomena of pool and ripple flow are poorly understood. The exploratory modeling used a parametric approach based on kinematic routing (Paxman, 1987).

The hydraulic routing suggested that most flows were much faster than super critical; however, super critical flow is seldom encountered in natural streams (Jarrett, 1984). Instead, nature dissipates energy by bank cutting and channel reshaping, gauging energy dissipation pools, and moving sediment. Steeper slopes generate more energy and thus cause greater irregularities, larger pools, and the movement of more and bigger rocks (Sannareddy, 1988).

This problem could be overcome by adjusting the value of Manning's upward to that required for critical flow and repeating the kinematic routing. However, these results correspond poorly with observed flows where changes in size, shape, flow directions, and velocity profiles are so drastic over short distances that the dynamic terms dominate the kinematic term in St Venant's equation.

The problem becomes more evident in trying to quantify sediment transport. The standard equations (Meyer–Peter and Mueller, 1948; Smart, 1984) seemed to severely underestimate the increase in sediment transport with flow (Paxman, 1987). Significant research in mountain stream hydraulics is needed to understand the feedback between energy dissipation and sediment transport in these steep streams. This situation is one in which lumped parameter models seem to give reasonable flows but the physical processes must be examined at a smaller scale to quantify erosion and deposition.

2.2.9. Flood Severity Modeling

The study goal was to route the water and sediment outflows from the canyon through the urban development in order to map hazard areas by degrees of risk (Heefner, 1983). After the stream passes the canyon mouth, the flows gradually infiltrate or overflow the banks. The reduced transport capacity causes deposition, and the channels become smaller downstream. The long periods between floods usually bring bank sloughing and human encroachment that further reduce the channel capacity. The water leaving the channels spreads laterally and infiltrates into the alluvium. In a surprisingly short distance, the flood flows from the smaller canyons disappear entirely. The situation is in sharp contrast with more humid climates where flows continue to increase all the way to the sea. To work with nature, flood control measures must emphasize flow dissipation rather than conveyance. The hazard mapping must estimate rates of dissipation, and the most effective method of flood control is to increase the dissipation rate.

The deposition of sediments below the canyon mouths slowly forms alluvial fans where elevations drop at approximately the critical hydraulic slope (French, 1987). These fans, the places where man's actions, either in the stream or on the ground, have flattened the slopes become targeted by nature for sediment deposition. Engineering design and city planning can take advantage of these locations to collect and remove the material.

Heefner (1983) and Hall (1985) developed a model for delineating these flood risks. However, the flood hazards in these communities are still poorly defined, and the prospects for additional studies are dim as the wet years become memories.

2.2.10. An Idea for Catchment Characterization

Modelers should remember that soils are largely formed by hydrologic processes, and modeling these processes offers possibilities for improved catchment characterization. On steep mountain slopes, the weathering of rocks midway up the mountain slopes is the principal sediment source. Our understanding of the soil formation processes may be advanced through analogies with studies of snowpack evolution. Soil patterns are similar to snowpack distributions, except that the soil is deeper on the sunny slopes and shallower on shaded slopes. Landslides are similar to avalanches, as Morris discussed in Chapter 5.

2.2.11. Research in Process Modeling

The modeling identified ten specific research topics for gaining the research understanding of snowmelt flooding required to estimate the magnitudes and consequences of major events, a subject that has not received its deserved attention. The indication is that the ratio of rare-to-common event magnitudes is much larger for snowmelt than for rainfall flooding. The comparative ratios become even larger when one considers how much faster snowmelt magnifies sediment transport and damages. Rainfall events can bring larger flows causing less damage.

Some research topics that should be emphasized in trying to better understand snowmelt flooding are:

(1) The movements of the line separating rain from snow up and down a mountain during storm periods and how they impact depths of snow accumulation.

(2) The temporal (cloud related) and spatial (shading related) ground radiation patterns in mountain topography and how they impact maximum rates of snow melt. Measurements of incoming solar radiation (light intensity) should be used more widely to characterize cloud cover and assess its impact on snowmelt flooding.

(3) Soil formation in mountain swales, the dynamics of geomorphological processes that intersperse clay layers with sand lenses, and practical methods for draining swales where adverse conditions are identified (National Research Council, 1987). Soil formation is particularly important in quantifying the hydrologic recovery of swales after landslides and of debris flow paths.

(4) Determination, from assessments of precipitation cloud–cover and temperature records, of the conditions necessary for increasing interflow runoff to the point of causing significant bank sloughing and land slides, and determination of the probability of their occurrence. The definitions of the conditions can help identify impending hazards, and the probabilities can guide the level of effort that should be spent in preparation.

(5) The situations that stop debris flows within mountain canyons and a methodology for estimating the directions that these flows may take when leaving the canyons and how far they will travel with various probabilities.

(6) The recovery of the channel geometry after being scoured by debris flows, and the changing characterization of the channel for flood and sediment transport routing during the recovery.

(7) The formation and failure of landslide dams, and the routing of dambreak floods and sediment transport.

(8) The interaction between energy dissipation and sediment transport in streams where hydraulic analysis suggests super critical flow. The dynamics of how temporary situations with super critical flow establish an energy balance in nature.

(9) Mapping the declining hazard (depth, duration, velocity, and sediment content) down slope on alluvial fans where nature disposes of water by infiltration and sediment by deposition. How to blend urban land use into this setting with minimal disturbance to the environment and risk of damage.

(10) Measurement of extreme events is important for risk assessment, and the opportunity to obtain descriptive information on the rare events of the 1980s was lost when the gauging network was discontinued. Consistent data gathering is particularly difficult in an arid or semi–arid climate because of the institutional difficulties in preserving funding over the long periods between large events.

2.2.12. A Larger Issue

The Steed Creek catchment has vast spatial and temporal variability. The hydrologic processes change with wetter catchment conditions. New processes (land slides, debris flows, scour, and the formation and overtopping of debris dams) emerge and magnify downstream hazards much more than one would estimate by statistical extrapolation. Parametric modeling is useful in identifying these conditions, and physical modeling is needed for detailed assessment.

Most catchments have more variability than we think. Past hydrologic studies have sought to characterize catchments with representative parameter values, but we should also measure their variability. Variability measurement offers a challenge for modern surveillance techniques, but another possibility is described in the last section of this chapter.

2.3. Modeling Hydrologic Heterogeneity

Past parametric modeling has concentrated on explaining recorded flows. Beck and Young (1976) led the way in trying to extract additional information from the differences between recorded and simulated flows. Some differences may be explained by catchment heterogeneity in that some rains happen to concentrate on more and others on less permeable areas. A stochastic parametric model may be able to try to capture this effect and quantify the heterogeneity. Other differences may be explained by the emergence of processes that were not considered in the conceptual model, often because they rarely occur. In order to examine the utility of this possibility, it is useful to consider the issues in scale selection, how they are assessed by distributed models, what lumped models have to offer, and ideas for coordinating between them.

2.3.1. Issues in Model Scaling

The parametric modeling approach used for Steed Creek subdivided the catchment into subareas and conceptualized the transformation of rainfall to runoff by defining hydrologic storages (boxes in Figure 1). The flows from storage-to-storage (arrows in Figure 1) were quantified with parametric equations that were calibrated against recorded flows. Catchment runoff was estimated by routing and summing outflows from the subareas.

The process of routing only the streamflows assumes subareas that discharge entirely by streamflow, but this method breaks down for smaller subareas with substantial subsurface flows. Figure 4 plots a conceptual relationship of error versus scale and gives a U-shaped curve. Parametric simulation deteriorates at smaller scales where water flows laterally among the conceptual storages (to and from areas with recent landslides). It deteriorates at larger scales because the heterogeneity within the storages is not adequately represented.

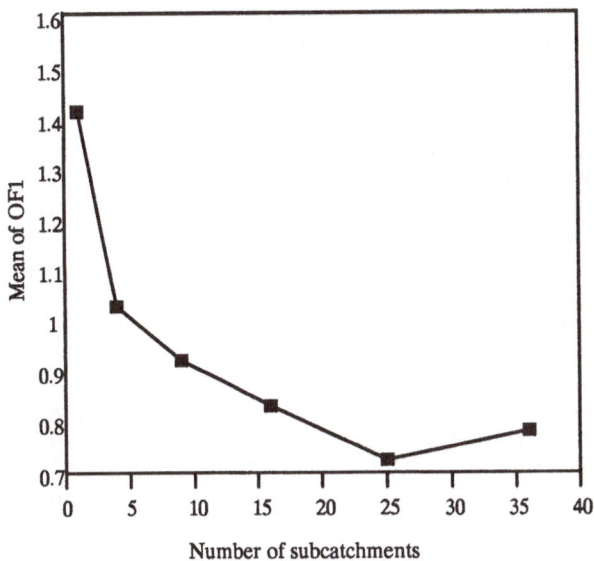

Figure 4. A conceptual relationship of error versus scale.

584

Two major issues must be addressed: how to determine a reasonable scale (avoid scales that work poorly), and how to represent catchment heterogeneity at the chosen scale. Perhaps, a stochastic version of the SWM could be calibrated to find a scale of best–fit and then misses between simulated and recorded flows could be used to measure hydrologically significant heterogeneity at that scale. Extreme or persistent cases could then be probed by physically based modeling.

2.3.2. *The Distributed Model Approach*

The physically based method for simulating the hydrologic response of a heterogeneous catchment is to map the important physical characteristics which can then be discretized and used in a spatially distributed model, as discussed by Bathurst and Wicks in Chapter 13. The catchments are subdivided into elements, and the equations are selected to quantify processes at the chosen elemental scale. However, such a model will not capture interactions at larger scales unless appropriate additional expressions for interactions among groups of elements can be defined and incorporated—we must resolved the two issues raised in the last paragraph.

As we learned long ago when mathematical models in hydraulics never eliminated the need for hydraulic testing, the grouping of elements for defining interactions at multiple scales is too complex to address from a purely theoretical approach. We cannot build practical physical models of catchments nor wait for storms to occur on real catchments, but we can use parametric modeling in sensitivity analysis. The scales of interaction found to be associated with the greatest hydrologic sensitivity would provide promising sizes for grouping elements in interactive distributed models. The scales selected for their sensitivity would suggest what catchment properties should be measured and how to take advantage of evolving sensing capabilities.

2.3.3. *Potential Contributions from Lumped Parametric Models*

In order to evaluate this strategy, we should review how lumped–parameter models work in simulating streamflow from precipitation. First, they separate the hydrologic processes (infiltration, interflow, transpiration, etc.). Then, they use parameters to index the influence of catchment physical characteristics (surface permeability, soil depth, vegetation density, etc.) on each process. Finally, they sum the results by following a conceptual flow chart.

When these models are calibrated to estimate parameter values that minimize differences between simulated and recorded flows, the resulting parameter values cannot represent hydrologic characteristics any better than the separability of the represented processes. Inseparable processes are lumped, and their characteristics are aggregated. For example, calibration of an infiltration parameter lumps a characterization of the infiltration process at the scale of the process that dominates the infiltration process at the scale used in model calibration. Systematic analysis of many calibrations may reveal differences in the way infiltration should be characterized for computational efficiency for different scales. The primary difference is in the characterization of catchment heterogeneity.

Since heterogeneity affects runoff (Wood et al., 1988) differences between simulated and recorded flows, if the impact of spatial variation in the precipitation can be represented, it may be used to infer the amount of spatial variance in the physical characteristics of a catchment. This can be shown at the scale that has the greatest influence on the hydrologic simulation. If this approach can be made to work, the differences between simulated and recorded flows will provide raw data to use in calibrating the variance in parameter values over a catchment. Data on parameter variance can be used to guide the selection of a grid size for a physical model.

Such an approach is fraught with practical difficulties. Noise from modeling the spatial variability in precipitation, the influence of spatially variable evapotranspiration, and interactions among the hydrologic processes may mask the influence of spatial variability of catchment characteristics. Our experiences with deterministic parametric modeling suggest that several years of data are required to damp out noise (James and Burges, 1982). Stochastic modeling may take an even longer record to calibrate parameters of variance.

2.3.4. Some Issues in Implementation

Other factors that must be considered when attempting to use stochastic–parametric hydrologic modeling to quantify catchment heterogeneity are:

(1) *Precipitation Patterns.* A stochastic hydrologic model must be linked with a stochastic precipitation model to generate a large number of equally probable rainfall distributions over a catchment. These simulated storms should reasonably represent the mean, variance, and spatial correlation of rain that might fall on each subarea.

(2) *Temporal Parameter Variations.* Seasonal parameter changes, presently programmed within the SWM, are used to account for differences in temperature, between wet and dry seasons, in agricultural activity, etc. Present state adjustments make infiltration decline and transpiration increase as soils grow wetter. Calibrated spatial variability must be considered along with the temporal variability.

(3) *Spatial Parameter Variations.* Spatial variability can be modeled by taking parameter values for subareas from representative distributions (the log normal is mentioned most often).

(4) *Parameter–Precipitation Juxtaposition.* Rainfall amounts may not be randomly distributed with respect to catchment characteristics. For example, rainfall is obviously correlated with vegetation. An alternative is to take the element properties as randomly distributed over rainfall zones.

(5) *Correlations among Parameters.* Published sets of catchment parameter values (Ross, 1970) show that some parameters are highly intercorrelated and that others vary independently. Preservation of these relationships would require a multivariate model. The parameter values assigned to a segment might have a component which: characterizes the catchment as a whole; correlates with measurable element properties, such as slope or land use; represents the way a group of parameters vary collectively, and a component representing the way a given parameter varies individually. Some of these may not be significant in a given application. Since much of the runoff is expected to be simulated from cases with favorable combinations of parameter values, research defining source areas should contribute valuable information on intercorrelations on the tails of the parameter distributions.

2.3.5. A Recapitulation of the Needed Model

In summary, a stochastic–parametric model can be segmented into elements that can be varied in size. The model should simulate runoff from multiple elements that are randomly distributed within rainfall fields which sequentially represent precipitation in the catchment over a year. The parameters can be calibrated to estimate the means and the variances of their distributions over the catchment. The calibration should use repeated simulations to develop distributions of flows summed catchment wide for each trial set of parameter values. It should also iterate toward mean values that will reproduce the recorded flow volumes and variances that will generate flows that encompass the departures of recorded from simulated flows.

Calibration can proceed over a range of numbers of elements in searching for the size that best represents the catchment. The calibration for each element size will conclude with some minimum achievable sum of squares of the errors, and the series of calibrations is expected to identify some element size that minimizes this sum. Large elements will have too few points to represent the distribution tails, and small elements will give catchment characteristics on a scale too fine to impact catchment runoff hydrology. The optimal number of elements and the associated parameter values can then be recorded. As studies are repeated over catchments classified by ranges of area, the growing data base promises significant information for assessing the hydrologic processes operating at various scales. If this sort of stochastic-parametric modeling proves practical, flow–based calibration to index the spatial variability in catchment characteristics may reduce the cost of gathering remote sensing and field data for research in hydrologic scaling.

3. Conclusions

This chapter on the hydrology of mountain catchments has been an attempt to demonstrate how lumped parametric hydrologic modeling can be applied with critical thinking to enhance hydrologic understanding. In a sense, parametric hydrologic models are a growing expert system that incorporates the thoughtful assessments of leading hydrologists over the years. We should use their thinking and then improve the programs to record ours.

4. References

Anderson, E. A., and N. H. Crawford: 1964, *'The Synthesis of Continuous Snowmelt Hydrographs on a Digital Computer,'* Technical Report No. 36, Stanford University, Department of Civil Engineering, Stanford, California.

Aurasteh, M. R.: 1983, *'A Model for Estimating Lawn Grass Water Requirement, Considering Deficit Irrigation, Shading and Application Efficiency.'*

Beck, M. B., and P. C. Young: 1976, 'Systematic Identification of DOBOD Model Structure,' *Proceedings of the American Society of Civil Engineers, Journal of Environmental Engineering Division* **102**(EE5), 902–927.

Bevan, K.: 1989, 'Changing Ideas in Hydrology – The Case of Physical Based Models,' *Journal of Hydrology* **105**, 157–172

Crawford, N. H., and R. K. Linsley: 1966, *'Digital Simulation in Hydrology: Stanford Watershed Model IV,'* Technical Report No. 39, Department of Civil Engineering, Stanford University, Stanford, California.

DeLeon, A. A.: 1982, *'Hydraulics of Debris Flow,'* Ph.D Dissertation, Dept. of Civil and Environmental Engineering, Utah State University, Logan, Utah.

French, R. H.: 1987, *Hydrologic Processes on Alluvial Fans*, Elsevier, Amsterdam.

Hall, B. R.: 1985, *'Aggravation of Snowmelt Floods by Sediment Deposition at Culverts,'* M.S. Thesis, Dept. of Civil and Environmental Engineering, Utah State University, Logan, Utah.

Heefner, S.: 1983, *'Analysis of Flood Characteristics on an Urbanized Alluvial Fan: Mill Creek Drainage Basin, Bountiful, Utah,'* M.S. Thesis, Dept. of Civil and Environmental Engineering, Utah State University, Logan, Utah.

James, L. D.: 1968, *'Economic Analysis of Alternative Flood Control Measures,'* Research Report No. 16, University of Kentucky Water Resources Institute, Lexington, Kentucky.

James, L. D.: 1972, 'Hydrologic Modeling, Parameter Estimation, and Watershed Characteristics,' *Journal of Hydrology* **17**.

James, L. D., and D. S. Bowles: 1986, 'Assessment of Control Alternatives for the Great Salt Lake,' in *Improving the Effectiveness of Floodplain Management in Arid and Semi–Arid Regions*.

James, L. D., and S. Burges: 1982, *'Selection, Calibration, and Testing of Hydrologic Models,'* Hydrologic Modeling of Small Watersheds, American Society of Agricultural Engineers.

Jarrett, R. D.: 1984, 'Hydraulics of High–Gradient Streams,' *American Society of Civil Engineers, Journal of the Hydraulics Division* **110**(HY11), 1519–1539.

Jeppson, R. W., and E. E. Farmer: 1985, '*Mechanics Associated with Utah's 1983 Slides and Debris Flows,*' Ogden, Utah: Intermountain Forest and Range Experiment Station, Forest Service, USDA.

Lambson, C. H.: 1988, '*Simulated Radiation Patterns on Mountain Slopes, and their Effect on Snowmelt Runoff,*' M.S. thesis, Dept. of Civil and Environmental Engineering, Utah State University, Logan, Utah.

Liou, E. Y.: 1970, '*OPSET: Program for Computerized Selection of Watershed Parameter Values for the Stanford Watershed Model,*' Research Report No. 34, Water Resources Institute, University of Kentucky, Lexington, Kentucky.

Meyer–Peter, E., and R. Mueller: 1948, 'Formulas for Bed Load Transport,' in *International Association of Hydraulic Research*, Second Congress, Stockholm, Sweden, pp. 39–64.

National Research Council, U.S. National Academy of Sciences: 1987, '*Confronting Natural Disasters: An International Decade for Natural Hazard Reduction,*' Washington, DC.

Pack, R. T.: 1985, '*Multivariate Analysis of Relative Landslides Susceptibility in Davis County, Utah,*' Ph.D. CEE.

Paxman, S. W.: 1987, '*The Routing of Water and Sediment Down Steep Natural Channels,*' M.S. Thesis, Dept. of Civil and Environmental Engineering, Utah State University, Logan, UT.

Pitcher, D. O.: 1986, '*Parametric Surface Erosion Modeling of Mountain Watersheds in Utah,*' M.S. Thesis, Dept. of Civil and Environmental Engineering, Utah State University, Logan, UT.

Rodriguez, S. A. G.: 1982, '*Simulation of Unsteady Debris Flow in Open Channels,*' M.S. thesis, Dept. of Civil and Environmental Engineering, Utah State University, Logan, UT.

Ross, G. A.: 1970, '*The Stanford Watershed Model: The Correlation of Parameter Values Selected by a Computerized Procedure With Measurable Physical Characteristics of the Watershed,*' Research Report 35, Water Resources Institute, University of Kentucky, Lexington.

Sannareddy, R. B.: 1988, '*Energy Losses and Sediment Transport in the Complex Hydraulic Geometry of a Steep Mountain Stream,*' M.S. Thesis, Dept. of Civil and Environmental Engineering, Utah State University, Logan, UT.

Smart, G. M.: 1984, 'Sediment Transport Formula for Steep Channels,' *American Society of Civil Engineers, Journal of the Hydraulics Division* 110(HY3), 267–276.

Weston, A.: 1987, '*Sediment Yield From Landslips in Steep Mountain Watersheds in Utah,*' M.S. Thesis, Dept. of Civil and Environmental Engineering, Utah State University, Logan, UT.

Wood, E. F., M. Sivapalan, K. Beven, and L. Band: 1988, 'Effects of Spatial Variability and Scale with Implication to Hydrological Modeling,' *Journal of Hydrology* 102, 29–47.

Part VII – Relationship of Experimental Studies and Hydrologic Modeling

Chapter 28

Mathematical Models: Research Tools for Experimental Watersheds

Donn G. DeCoursey
USDA–ARS, Hydro–Ecosystem Research Unit
243 Federal Building
301 South Howes, P.O. Box E
Fort Collins, Colorado 80522 U.S.A.

Abstract. Concurrent development of hydrologic models and research watersheds is necessary for adequate development of both. The models developed and objectives used in setting up a research watershed must be centered on the needs of the decision maker. The impact of human activity on hydrologic response has increased to the point that long–term global impacts are being felt. Thus the tools that the decision maker needs to make adequate decisions are different than they were a few years ago. Causal models that include probability distributions of input parameters are needed to respond to the problems we now face. Models that are developed concurrently with field experiments will be used to study the research site, select appropriate parameters, evaluate the accuracy and precision required of variables and parameters, evaluate the significance of spatial and temporal representativeness, and illustrate the concepts of time–space trade off.

1. Introduction

The title of this chapter may be somewhat misleading. It implies that mathematical models are subservient to experimental watersheds when in fact they should be parallel in their development. It is impossible to properly develop an experimental watershed without a concept of the hydrologic system and what must be measured. In almost all cases, excepting only the very simple, this concept takes the form of an analytical expression—a model. Thus, prior to development of an experimental watershed, a statement should be prepared describing: what the experimental watershed is to accomplish; what mission, objective, or purpose it has; and how the challenge is to be met.

In the following presentation the role of the decision maker or natural resource manager in defining the structure of the mathematical model and experimental watershed is described. Also discussed is the influence of man on the hydrologic cycle and how such activities affect our approach to the modeling activity. These introductory comments are provided to put perspective on the use of experimental watersheds. Because such watersheds are very expensive to manage, optimal use should be made of them. Thus, the really important global hydrologic problems that exist should be given consideration at a watersheds' inception and in developing a model or models to guide its development. Several examples illustrate the synergism between mathematical models and use or development of experimental watersheds.

2. Discussion

2.1. The Role of the Decision Maker in Watershed Research

Before discussing the interaction of mathematical models and experimental watersheds, it is appropriate to consider the role of mathematical models in decision making. Hewlett (1981),

D. S. Bowles and P. E. O'Connell (eds.), Recent Advances in the Modeling of Hydrologic Systems, 591–612.

in a keynote address at a Workshop on the Effect of Rural Land Use and Catchment Management on Water Resources, divides hydrologic models into three broad categories: stochastic, deterministic, and decision models. He considers them primarily stochastic if the output is expected to vary because some element of the input has a known or predictable distribution. Deterministic models simulate the hydrologic processes mechanistically. Decision models are designed to yield useful answers to economic or managerial type questions.

Normally when we are classifying models we do not consider the term "decision" as one of the classifications. Whether we use the term or not is not important as long as we recognize that the need for such models exists. Notwithstanding the need for research to further the science of hydrology, the ultimate end product of hydrologic research (watershed science and modeling included) is the development of tools to improve the manager or decision maker's ability to make sound decisions. Hewlett (1981) described this very effectively:

"Here then lies one clear role for research: What models, descriptive relationships or facts are needed to advance the art of decisionmaking? What decision models are used now? Do they properly embrace the goals of management? Do they recognize all feasible alternatives? Do they include the essential constraints and facts? Are they broadly normative," (normative models consist of components that describe the state of the hydrologic system) "as in multiple use, or scientifically normative, as in linear programming, with strategies, states of nature, and feed–back mechanisms arrayed against a hierarchy of specified goals? The need for research emerges clearly when the decision model in use fails to distinguish between the efficiency of several possible management strategies, or fails to predict with useful accuracy the outcomes from different strategies."

Use of water supplies world wide continues to grow, diminishing excess supplies; thus it becomes increasingly important that we manage existing supplies with greatest possible efficiency. Until recently, the time scale and aerial extent of hydrologic projects have been such that our knowledge of and ability to forecast watershed response have been satisfactory. However, the effects of human activity are now large enough to measurably affect the hydrologic cycle in both quality and quantity (Matalas et al., 1982).

2.1.1. Global Problems and the Decision Maker

The National Research Council (NRC) (1982) identified several long–term and large–scale water management problems in the United States that are also applicable world wide; these are: (1) groundwater contamination due to toxic and nuclear-waste disposal; (2) nonpoint sources of pollution on stream systems; (3) impacts of change in both flow and water quality on the aquatic ecosystem; (4) the frequency, duration, and impacts of drought, including long–term trends toward desertification; (5) long–term hydrologic budgets for assessing the adequacy of regional or national water resources; (6) global geochemical cycles such as the fate of nitrogen and sulfur (e.g. acid rain); and (7) protection of engineered systems (e.g. nuclear power plants) against hydrologic extrema.

Thus with dwindling supplies and increasing human influence on the system, projecting the impact of water related projects becomes more important. What is going to be the long term impact? Is it likely that the collective effect of a series of projects could have a global effect? We are intensifying the use of the hydrosphere for storage, detoxification, and assimilation of wastes; energy production, intensive agricultural development, and urbanization have the potential of producing global hydrologic effects; and possibilities of climatic change have shifted much of the emphasis in prediction of long–term hydrologic phenomena. Except for superficial evaluations, existing methodologies are not designed to answer these questions.

Yet these large–scale, long–term problems involve large investments and the health and well–being of much of the world's population. Thus these problems demand increasingly precise and accurate predictive statements expressed in a way that will enable decision makers to make knowledgeable choices in an increasingly "tight" economy.

2.1.2. The Role of Mathematical Models

With respect to the decision maker, the Congress of the United States recognizes the impact that mathematical models have had on the successful management and planning of water resources. They instructed the Office of Technology Assessment (OTA) to (1) assess the Nation's ability to use models efficiently and effectively in analyzing and solving water resource problems and (2) make recommendations for improving the use of available technologies. The OTA Report (1982) presented the United States Congress with an assessment that is applicable worldwide. The over–riding concerns are similar and the needs are the same in many countries.

The OTA Report (1982) found that:

"Mathematical models have significantly expanded the Nation's ability to understand and manage its water resources" ... "models have increased the accuracy of estimates of future events" ... "They have made possible, analyses that could not be performed empirically or without computer assistance" ... "Models are often the best available alternative for analyzing complex resource problems. While many of the economic and social factors in water resource decisions cannot be fully enumerated, models can be used to integrate the available data, and provide estimates of future effects and activities." "Models have the potential to provide even greater benefits for water resource decision–making in the future." ... "they will be able to increase the efficiency of water resource management and encourage cost–effective decision–making."

The Office of Technology Assessment also found that many of the analytical tools needed to aid the decision–maker were not available and that there was no overall strategy for developing, using, disseminating, and maintaining these tools. Thus, strategy for model development remains a very high priority need of water resource management and decision–makers.

Since mathematical models have become an acceptable tool of the hydrologist, water resource manager, and decision–maker, I will describe some of the ways that models contribute to furthering watershed science (France and Thornley, 1984).

(1) Hypotheses expressed in mathematical terminology can provide a quantitative description and understanding of chemical, biological, and hydrologic processes.

(2) Complete mathematical models can provide a conceptual framework which may help pinpoint areas where knowledge is lacking, and might stimulate new ideas and experimental approaches.

(3) Mathematical models can be a good way of providing a recipe by which research knowledge is made available in an easy–to–use form for the user.

(4) The economic benefits of methods suggested by research can often be investigated and highlighted by a model, thus stimulating the adoption of improved methods of prediction.

(5) Modeling may lead to less ad hock experimentation, as models sometimes make it easier to design experiments to answer particular questions, or to discriminate between alternative mechanisms.

(6) In a system with several components, a model provides a way of bringing together knowledge about the parts, thereby giving a coherent view of the behavior of the whole system.

(7) Modeling can help provide strategic and tactical support to a research program, motivating scientists and encouraging collaboration.

(8) A model may provide a powerful means of summarizing data, and also a method for interpolation and cautious extrapolation.

(9) Data are becoming increasingly precise but also more expensive to obtain; a model can sometimes make more complete use of data.

(10) The predictive power of a successful model may be used in many ways: playing the "what if games" in assigning priorities in research and development, management, and planning.

(11) Models validated by data from experimental watersheds provide a mechanism to transfer data from study areas to other areas where less data may be available.

2.2. Mathematical Models and Important Hydrologic Issues

The literature is repleted with descriptions of many types of hydrologic models. For the purpose of this presentation, it is not necessary to categorize and define the variety of models that exist, except as necessary to relate hydrologic modeling to the development of experimental watersheds. In this section of the chapter I have quoted extensively from a book, "Scientific Basis of Water–Resource Management," by the NRC (1982). The purpose of the study, presented in this book, was to provide assessments, from the scientific community, that would aid policymakers in water related societal decisions. The study was motivated by the need for substantial improvements in the hydrologic sciences to support the decisions. The book contains a series of 11 essays on all aspects of the hydrologic cycle written by a panel of well known scientists. The report also contains an "Overview and Recommendations" that summarized the highlights of the 11 essays and formulated conclusions and recommendations. The entire report is pertinent to the objectives of this Institute and should be read by anyone considering the development of an experimental watershed. It very aptly describes the gaps in knowledge and data base development that must be filled to develop management tools.

The "overview" of the 11 papers identified the need for new material by describing problems and differences in approaches with emphasis on gaps in our scientific understanding. The overview emphasized the development of causal models but also recognized the need for empirical models. It presented a summary of its findings and arguments for further understanding the elements of hydrology and for dealing with problems of hydrologic scale. Because these are so pertinent to research watershed development the following is abstracted from the NRC report.

2.2.1. Differences in Approach Used in Modeling

The National Research Council identified seven long–term large–scale water management problems which were presented in a previous section of this chapter. They involve large investments and the health and well–being of much of the worlds population; thus, they set the stage for the need to develop new and improved analytical procedures. These problems also emphasize issues and differences in approaches that are currently in use. Three significant examples are:

(1) The difference in viewpoint and practice between empirical and causal models. Empirical models, frequently used with today's problems, require continuous updating and fitting to on–site data; they express a relation of a dependent variable upon selected independent variables based on observations of both. They are very difficult to extrapolate beyond the range of the data or to transfer from one time or place to another. Causal models, based on physical principles which are assumed to be invariant in time and space, can, in theory,

incorporate the effects of change in the environment. In principle they are also transferable in space and time; frequently, this does not require additional data because of a priori evaluation of the coefficients. From an operational point of view hydrologic determinism incorporates both causal relations and statistical relations. The statistical relations provide information and a means of incorporating uncertainties.

(2) The difference between models for prediction and understanding. Generally prediction is directed toward immediate needs and may use whatever is practical. Understanding is oriented toward improved ability to explain observed phenomena in both space and time. These should be combined, because progress in predicting the performance of complex systems requires both.

(3) The solution of most problems is focused at the local spatial and temporal scale with little thought given to the global hydrologic scale of which they are a part.

These three controversies are important in both mathematical model and research watershed development and focus attention on the difference between a short–run and long–run evaluation. Even though short–term forecasts are sensitive to many factors that are not important in long–range prediction, the reverse is more important; i.e., short–term forecasts are insensitive to many imperfections that are intolerable in long–range schemes. Because the investment in major water resources projects is so great, it becomes almost impossible to discard a project, after construction begins, in favor of a significantly different alternative. Thus accurate assessment of risk is becoming increasingly important and management decision making critical. Because of the possible irreversible, unanticipated effects—perhaps on a global scale, such as on climate, water chemistry, and land erosion—the long–term "before and after" effects of major projects on the societal and hydrologic environment must be accurately assessed. These types of projects, the complexity of the hydrologic system, and the long time period over which they must be evaluated require that we continue to develop and improve causal models of the entire hydrologic system.

The NRC report goes on to state that other substantial improvements in hydrologic science are also necessary. Some of the areas identified are: the conjunctive use of surface and groundwater; water and chemical transport through the vadose zone; improved understanding of the interaction of physical, chemical, and biological components; and improvement in statistical and probabilistic modeling techniques.

2.2.2. Overriding Concepts and Constraints of Model Development

The NRC report concluded its summary of the 11 contributed papers with three very important findings. They are important concepts or constraints that must be considered in any natural resource model development.

(1) Neither empirical–optimization models nor increased causal understanding alone will enable hydrologists to respond with necessary effectiveness to current societal demands—a combination of the two is required. Most hydrologic problems occur at intermediate scales that require a blending of the two types of models.

(2) Problems of hydrologic scale impact not only scientific alternatives and research direction but also the management of water–resource systems. Explanations and predictions that must be made, in responding to the emerging problems previously described, require identification of the time and space dimensions along which they must hold if they are to be satisfactory.

Previous discussion emphasizes the need to concentrate on basic or fundamental research which is at the microscale. Frequently it is assumed that linking microscale solutions to form a causal chain will facilitate examination of problems at the macroscale. Unfortunately it seldom happens that way; at some scale or characteristic dimension, mechanistic dimension breaks down, and we must resort to unverified causal and statistical or probabilistic components.

The watershed is frequently taken as a convenient unit in hydrologic analyses and we represent the fundamental processes as fluxes connecting sets of storage reservoirs. Because of differences in scale, it is unreasonable to assume that we can use the same set of expressions and parameters for watersheds of all sizes. We need some measure of hydrologic scale that can help discriminate among alternative forms of analysis and aid in optimal development of causal tools at all watershed scales.

(3) The institutional constraints that impact the project must be recognized. Following are some examples: (a) Some institutions collect data whereas others cannot collect it but need it—this leads to problems of fragmented and incomplete data bases; (b) watersheds and political boundaries rarely coincide; and (c) some institutions deal primarily with water quality and others primarily with water quantity.

2.2.3. Scientific Issues that Must be Addressed

The NRC report described some of the highest priority scientific issues that need to be addressed. These were prepared in the form of questions. A few years ago (AGU, 1983) research indicated that various federal, state, and private research groups were starting to direct efforts toward answering some of these questions, yet much remains to be done.

(1) Can the concept of scale—temporal and spatial—be included in hydrology so that some of the processes can unambiguously and efficiently be studied and managed at one level and some at another?

(2) How can chemistry and biology be coupled with our understanding of the physics of water movement? How can this understanding be applied to the management of groundwater and surface water quality?

(3) What is the impact on the food web—and, ultimately, on human life—of toxic substances in water? Do the long–time horizons, explicit in the migration of toxic substances, impose the need for new methods of risk assessment?

(4) Given various theoretical models, what advances and techniques are needed to measure and define statistically stable estimates of the requisite parameters? Are existing data sets adequate for both causal and statistical usage? What information needs to be collected to meet the requirements of both theoretical and applied hydrology?

(5) Can models of sequentially dependent processes be linked satisfactorily to form a causal chain of understanding leading to optimal control and the predictive power necessary to address the relevant societal issues? Can existing data bases be manageably integrated to formulate and calibrate such chains of models and their parameters?

(6) What is the proper role of statistics in developing and assessing models? How far should we press for predictive accuracy and precision? When is a model "good enough" for scientific understanding or engineering applications? At what scale should we strive for understanding hydrologic phenomena, and at what scale is empiricism adequate?

(7) What algorithms should be used to design protective works and establish nonstructural defenses against hydrologic extremes? Instead of concentrating on attempts to define a flood of a given return period, we need to evaluate the socioeconomic consequences of protection against events two times, five times, ... etc. the maximum flood of record and the cost of failure of a structure designed for such large events considering the design life of the structure.

The scientific issues described above should be seriously considered at the time that a philosophy of research to guide the development of a research watershed is developed. The NRC report expands on these seven questions. Even though the report is several years old and progress has been made in responding to the above questions, the 11 papers that make up the body of the report very aptly describe the current state of knowledge in hydrology and how that knowledge base interacts with management and planning of our water resources. As such they should be extremely helpful in structuring data bases that will have maximum utility in water resource model development.

2.2.4. Empirical vs. Causal Models

Klemes (1982) presents an excellent discussion of empirical and causal models used in hydro-logic evaluation (see also Haan et al., 1982; Dooge, 1973; Chapman and Dunin, 1975). Klemes discusses the advantages and disadvantages of each type of model. Empirical models are generally based on observations of change in one quantity that corresponds to a change in another quantity. They are of necessity simple and lead to limited knowledge and under-standing that describe what happens but do not derive the outcome from the dynamic mechanisms governing the process.

The strengths of empirical models, according to Klemes, are: (1) the possibility of develop-ing them without much understanding of the modeled phenomenon, (2) simplicity which is achieved by short–circuiting complex causal chains, and (3) potential for making collected data usable without much delay and hence a high cost effectiveness. The weaknesses of such models are: (1) attempts to improve the model, without additional data, lead to over–fitting or concentration on unimportant aspects; (2) the models have no justification outside the range of underlying data sets—extrapolation involves risk of large errors; (3) uncertainty about model structure leads to false assumptions about cause and effect; (4) arbitrariness in the form of the model can lead to adoption of a basically incorrect approach; and (5) the availabil-ity of input/output data and lack of understanding tend to lead to a "let–the–data–speak–for themselves" philosophy that can be misleading or untenable.

With these rather minor strengths and significant weaknesses one wonders why empirical models appear to work. Klemes suggests that they work because: (1) They are essentially an interpolation formula; (2) they represent only small, well understood components; (3) they are untestable; (4) decisions are insensitive to the embedded model; (5) they are irrelevant to results obtained; (6) by chance, the model may be actually descriptive of the physical system; (7) reputation is based on superficial appearance; (8) they work for the wrong reasons; (9) they have not been adequately tested and are considered good by default; and (10) models that do not work are not published.

Klemes describes the advantages of causal models as: (1) The ability to derive the behavior of a hydrologic process for a given state of the system from dynamic mechanisms without recourse to calibration by empirical fitting—thus, they can be used to predict hydrologic response under conditions that did not exist during process–recorded history; (2) the inherent ability to assess the effects of environmental changes on water resources; (3) the potential to point out ways to efficient short cuts and thereby to better empirical models; and (4) a system-ization of research that can provide a rational methodology for hydrologic model use.

Klemes describes the disadvantages of causal models as the facts that they require large volumes of data to develop and extensive basic research programs to test their concepts. It is also easy for a hydrologist, because of the lack of knowledge in related fields (geology, biology, chemistry and climatology) and extensive data requirements, to make short cuts that lead to empiricism. Thus a model can degenerate to one that is a somewhat structured empirical model.

In the last few years the USDA – Agricultural Research Service (ARS) has made significant progress in the development of causal models. See DeCoursey (1985), Knisel (1980), Rovey et al. (1977), Saxton and Bluhm (1982), Shaffer and Larson (1987), Smith (1981), Smith and Ferreira (1988), DeCoursey and Rojas (1990), Wight (1987), and Williams et al. (1983). These models vary considerably in the degree of causality and empiricism imbedded within them. However, the degree of empiricism is generally a matter of temporal and spatial scale. What is one person's causal description is macroscopic and empirical in another's view. Are causal models a necessary condition for formulating working hypotheses about macroscopic systems? Perhaps not. This is not an argument against understanding components of a causal chain, but rather suggesting that advances are made by an orderly interplay between theory and careful observation.

Time and space do not permit a review of the status of recent model development in various subject matter areas in hydrology; however, the 11 essays in the NRC Report and three recent AGU publications, U.S. National Report on Hydrology (AGU, 1979, 1983, and 1987), provide excellent reviews. Collectively they provide coverage of the entire field of hydrology, including sedimentation and water quality. Proceedings of the International Symposium on Rainfall–Runoff Modeling held May 18–21, 1981 at Mississippi State University, Mississippi State, Mississippi, provide a wealth of information. The proceedings appear as four separate hardback Water Resources Publications (Singh, 1982a–d). Another very useful publication is an American Society of Agricultural Engineers monograph on Hydrologic Modeling of Small Watersheds (Haan et al., 1982). The proceedings of the International Symposium on Water Quality Modeling of Agricultural Non–Point Sources (DeCoursey, 1990) provide an excellent review of nonpoint source water quality models and related subjects.

2.3. Coordination of Model Development and Field Data Collection

In the previous section of this chapter issues involved in the selection of mathematical models were discussed. Emphasis was placed on the development of causal models oriented toward water resources decision makers' needs. Of equal importance is the problem of model testing or validation, i.e. data collection. Many of the contributions in the NRC report stress the need for coordination of field data collection and model development.

In the last few years, field experiments have improved our understanding of the mechanics and the spatial variability of runoff processes (e.g. Lane et al., 1978), and have stimulated the development of causal models of hydrologic processes. The causal models have in turn provoked better field research by providing unified interpretation of data from different environments and suggesting critical measurements that must be made. It is particularly important that this synergism between model development and field work continue. As we continue the development of causal models, field work is needed to verify that the model is an adequate description of the processes and that parameter values are correct and realistic.

Coordination in the design of field experiments and model development will also minimize the aimless collection of field data that has been so obvious in the past. Too many field experiments have been set up without any thought given to how the data were to be specifically used. The experiments were designed to observe the hydrologic response when a change was made in some input or variable that describes the state of the hydrologic system. Frequently no thought was given as to why the change in that input or state variable provoked the hydrologic response. As such, the data were of limited value and of almost no value in a predictive sense because of a lack of transferability of the data. Dunne (1982) stresses the need to strengthen this coordinative effort:

"... the most vigorous and sophisticated current developments in hydrology are due to the efforts of researchers concerned with physically based mathematical models. However, this expanding frontier will be hollow unless it is matched by equally sophisticated field experiments to discover unexpected hydrologic phenomena, to develop new concepts about familiar processes, and to guide the development of mathematical models based on sound physical insights into field conditions."

2.4. Model Structure and Field Data Collection

The need for continual causal model development has been emphasized because this is where answers to the most pressing problems are to be found. However, I have also stressed the point that our applications models—those models developed for decision making—will very likely be a combination of causal and empirical models. Consequently, continued development of empirical models can be expected. Also, they provide a means of bridging the scale problems previously described. However, we can expect the empirical models to be developed with the aid of causal models.

The causal—mechanistic—descriptions of hydrologic processes require more field data for development, testing, and validation than any other model type. Thus, I will confine my comments to their data requirements. Previous references to several causal models, AGU and ASAE publications, proceedings of symposia, and the 11 NRC report contributions cite papers that describe excellent examples of causal models of hydrologic processes. Rather than try to describe the variety of data requirements needed for different models, I will show by a few examples what is needed.

The Opus model (Smith and Ferreira, 1988) was developed to predict the hydrologic response of a field–size area. The model assumes that a field is made up of a series of overland flow areas and channels. It is an elementary model in that it assumes uniform soils and climatic inputs for each overland flow plane or element, and it simulates the movement of water, sediment, and chemicals in and from the area. It is a good example of a causal type model. Following are some of the model's features, showing inputs and how various components of the fully dynamic model are simulated.

(1) Input data –
- rainfall – defined by breakpoint, i.e. short time–interval intensities
- solar energy – total daily values
- temperature – average daily values
- relative humidity – average daily values pan evaporation – daily values (used as an alternative in ET calculations)
- wind – total daily miles
- soils – physical and chemical properties by layer
- topography – length and slope of overland flow elements, profile and cross section shape of concentrated flow areas, roughness, aspect
- land use – crops, management, and chemical application rates
- system – values for the above variables, and state parameters

(2) Infiltration – Smith–Parlange or Green and Ampt equations applied to a layered soils description – parameters obtained from soils data (soil moisture release curves)

(3) Surface flow – kinematic solution of flow equations – requires description of flow cross section and resistance parameters

(4) Erosion
- rainfall and surface flow detachment
- adjusts to flow history and management changes
- transport capacity uses alternate expressions

(5) Snow melt
- uses inputs of degree day plus heat from soil
- runoff from snow pack simulated
- frozen soil calculated from max/min temperature

(6) Pond through–flow – complete routing through ponded areas

(7) Soil temperature – solves heat diffusion equation and accounts for water convection

(8) Soil evaporation – linked to soil flux limits

(9) Plant growth – mechanistic model with heat, water and nutrient stress. Driven by radiation and degree days

(10)Nutrient accounting – all components of nutrient and carbon cycles throughout soil profile, accounted for by mechanistic algorithms

(11)Pesticide accounting – uses both equilibrium and kinematic adsorption/desorption coefficients—accounted for by mechanistic approach throughout soil profile

(12)Management practices supported – (feedback between all hydrologic components maintained) – standing to fallen residue conversion; double cropping, variable rotation; simulation of perennials; efficient specifications for effect of cultivation on soil porosity, fertilization, irrigation, tile drainage, manure application, and grazing.

It does not take much imagination to appreciate the magnitude of the inputs required to use a model such as Opus. However, material is being prepared to aid the user by identifying the less critical parameters and providing guidelines for parameter selection. It will be several years before this development is completed, even though the basic structure and computer coding are complete. Much testing remains to be done.

This example was presented to illustrate the tremendous data collection facilities that are required to develop and fully test such models. In the case of the above model, the development and testing took place over several years. Data facilities at several locations were used to develop and test individual components. Cost of a complete data collection installation at one location would be prohibitive.

Freeze (1982) describes a distributed hillslope model which focuses on water movement on a simple rectangular catchment. His model demonstrates the significance and effects of spatial variability of soil properties on hydrologic response. Spatial variability was accomplished by laying a grid over a map of the surface of the hillslope and randomly assigning soil characteristics to each element of the grid. The assigned values were representative of the range and correlation structure of values normally found in a field. See Figure 1 for an example of one distribution from the set developed. The model, with its assigned soil parameters, was then used in a Monte Carlo experiment with synthetically generated rainfall characteristics. Output shows the storm–to–storm variability that is frequently observed in a field: areas with and without surface runoff caused by high or low infiltration rates (Hortonian conditions), and areas where a high water table rises to the surface (Dunne mechanism) creating ponded conditions. See parts b, c, and d in Figure 1. The model also shows how different parts of the hillslope respond to storms of different duration and intensity. To demonstrate the significance of the soils properties on the hydrographs, Freeze used the same series of storms and the same underlying probability distribution of soils parameters, but he generated different assignments for each of the grid elements. He analyzed the output hydrographs, tabulated the peak rates, and plotted them on lognormal probability paper; see Figure 2. The study showed: (1) The hydrographs produced were very similar in peak, duration, lag, shape and volume to field catchments with similar soils properties; (2) the probability curves of peak flow rate show extremely different characteristics—at the narrowest point (50 percent probability), the umbrella curve are nearly an order of magnitude apart; see Figure 3; and (3) the value of accurate specific soils information in estimating hydrologic response.

Discussion of the Opus and Freeze models is not intended to illustrate an example of an applications model or to suggest that all fields in a research watershed be instrumented to the extent necessary to define all their characteristics, or sampled well enough to define all the spatial characteristics. It is intended to illustrate the use of detailed hydrologic models in the design of an experiment, i.e., (1) studying a research site; (2) selecting appropriate parameters; (3) evaluating the accuracy and precision required of variables, characteristics, and parameters; (4) studying the significance of spatial and temporal representativeness; and

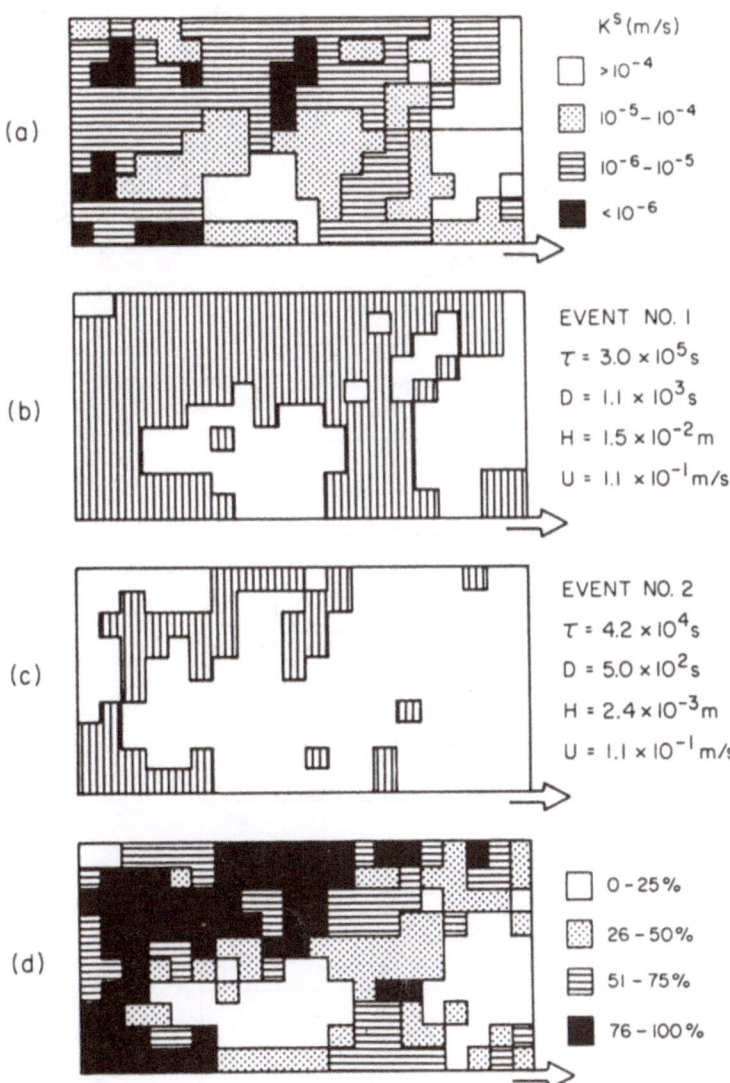

Figure 1. Hydraulic conductivity and sources of surface runoff from one realization using Freeze's distributed hillslope model. (a) Hydraulic conductivity distribution, (b) source areas for overland flow for event 1, (c) source areas for overland flow for event 2, and (d) source area summary for the 100–event Monte Carlo simulation showing frequency of contribution (Freeze, 1982).

602

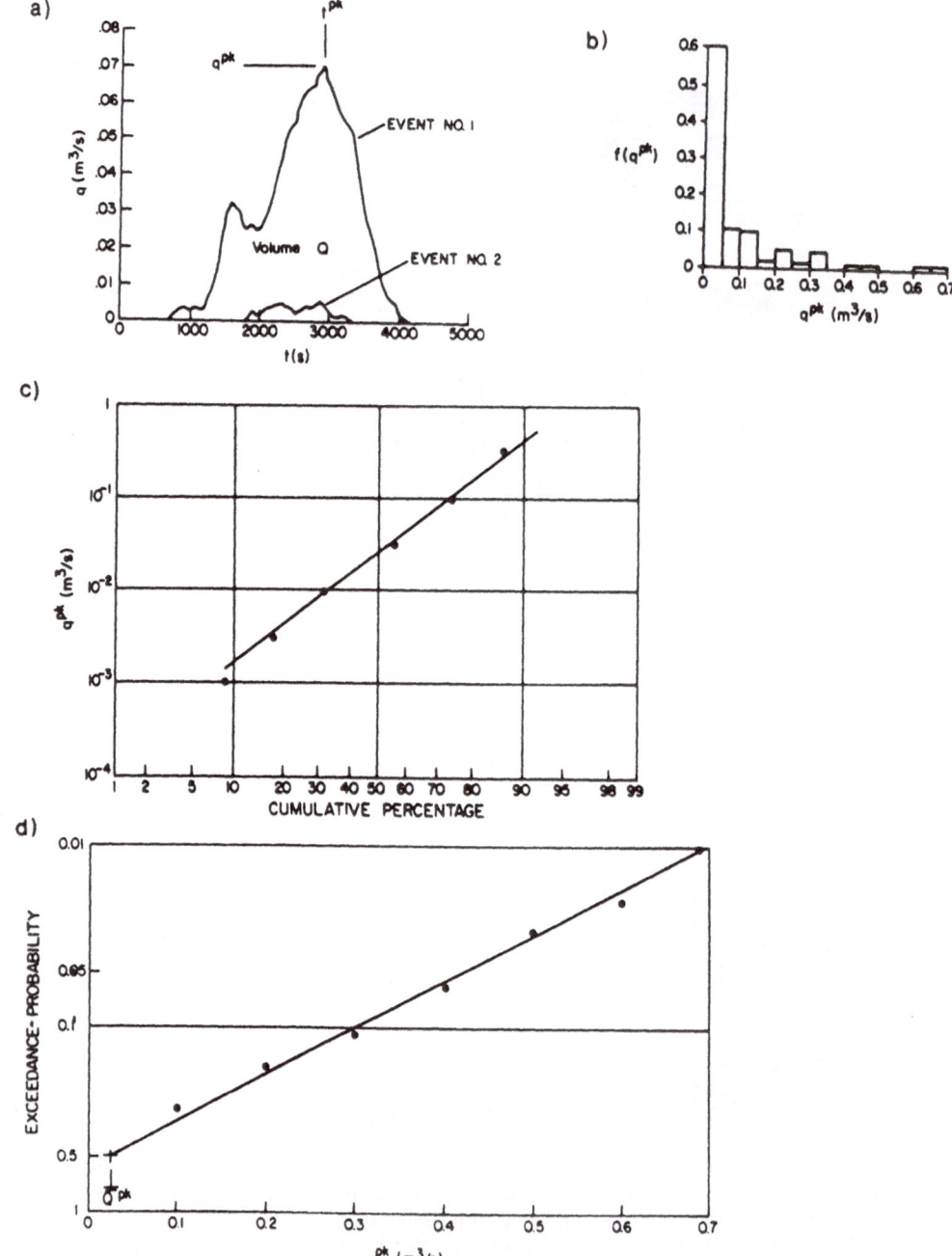

Figure 2. Representative hydrographs and frequency distribution of events from one realization of soil properties distribution, (a) output hydrographs for events 1 and 2, (b) frequency distribution of peak rates, q^{pk}, (c) lognormal probability plot of peak values, and (d) exceedance probability plot of peak values (Freeze, 1982).

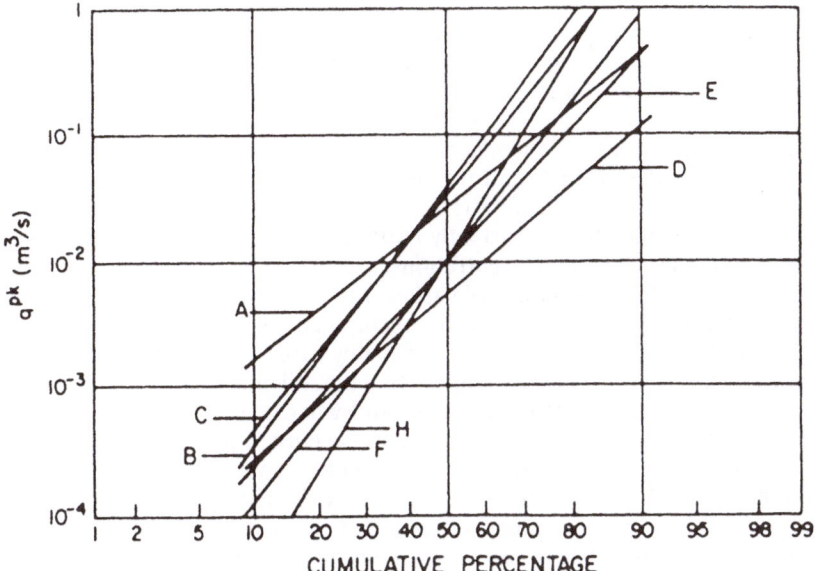

Figure 3. Log normal probability plots of frequency distribution for q^{pk} for several different realizations (A–H) of soil properties selected from the same probability distribution of soil properties. The same series of storms were used to develop the exceedance probability plots for each of the realizations (Freeze, 1982).

(5) illustrating time/space tradeoff in the value of additional data. Each of these factors will be discussed in the following sections using information from these examples.

On September 21–23, 1982 an International Symposium on Hydrological Research Basins and their Use in Water Resources Planning was held in Berne, Switzerland. The proceedings of that Symposium appeared as three separate volumes (Vischer and Emmenegger, 1982). The material presented in these proceedings is divided into: Data Acquisition and Management, Water and Heat Balance and Their Components, Rainfall–Runoff Processes, Nutrient and Sediment Budget and Meltwater Runoff, and Application and Transfer of Results of Catchment Research to Water Resources Planning and Management. Collectively, these papers present a good summary of watershed research throughout the world. Even though there are some excellent papers in the proceedings, the bulk of the papers presented showed that much of the data collection supported empirical model development rather than causal models, thus indicating a lack of conceptual model development to aid in data collection. The following sections are intended to show why such development is needed.

2.4.1. Studying a Research Site

The following hypothetical situation will be used to illustrate the use of a causal hydrologic model in studying a research site. Assume that the research catchment is about one hectare in size planted in a row crop. It is being used to study the erosion characteristics of the soil as an aid in development of an applications model that reflects different management conditions. The catchment has a mild slope to the outlet. The soil is assumed to be fairly uniform both in areal extent and in depth, i.e. no impervious or restricting layers. Assume that characteristic soil moisture release curves are available. Further assume that initial instrumentation consists of: a flow measuring station with a pumping sampler; a climatological station to measure rainfall, relative humidity, temperature, total incoming solar energy, and an evaporation pan; three tubes which provide access for neutron measurements of the soil moisture conditions. The Opus model, previously described, will be used as a guide for the field research.

The following example shows how the model can be used in a synergistic way with the watershed data to both study the site and improve the model. Suppose that some data have been collected and attempts have been made to simulate the watershed response using the model. Results show that the model is consistently under predicting the volume of surface runoff and over predicting the sediment concentrations. Assuming there is no error in measurements, the error in estimating the volume of surface runoff could be caused by one or a combination of several things. The soils map may be accurate, but even within a field of uniform soil type, the saturated hydraulic conductivity probably varies over two to three orders of magnitude (Freeze, 1982). Thus, it is highly unlikely that the soil moisture release curves of the particular soil are representative of the average (and what is average, how does one develop an average curve, can one even develop an average curve for such nonlinear systems?). With this kind of variability different parts of the field are going to respond differently to the same rainfall, and the "average" release curve will be different for each storm (Smith and Hebbert, 1979). Additional measurements of saturated conductivity will show whether the error is in this parameter or somewhere else. It will not be likely that all error in prediction can be eliminated by a redefinition of the saturated conductivity, but one may find that over the range of observed storms that "on-the-average," the new values are much closer.

Another source of error in predicting the volume of surface flow could be the model's estimate of initial soil moisture. If this were the problem the soil moisture record should have indicated it; the error in the model could be one of several things: (1) the model's crop evapotranspiration, (ET), rates could be in error; (2) the problems cited previously concerning the soil characteristics could have influenced the model's handling of unsaturated flow during interstorm periods; (3) the effects of tillage on soil evaporation and soil water redistribution are not adequately modeled; and (4) the plant model may not provide the correct distribution of roots which could affect soil moisture simulation and especially the vertical soil moisture distribution. Each of these problems could be investigated with additional instrumentation or sampling. For example, the use of net radiometers and additional measurement of wind, and air and wet bulb temperature should enable better estimation of transpiration and soil evaporation rates. Results of these additional studies could show that some component of crop growth or ET in the model was in error or that parameter values in the algorithms describing the process were incorrectly obtained or set. The source of error may also have been a poor simulation of the effect of tillage on soil evaporation. We could go on and on speculating on the causes of problems such as those mentioned above, but that is not the point to be made. This discussion is intended to show how a causal model, such as described, can be used to improve our understanding of the mechanisms taking place in a research catchment and how data from the catchment in turn aids in improving model performance. To obtain the optimal benefit from a research catchment, a causal description of the hydrologic processes involved must accompany the watershed development.

2.4.2. Parameter Selection

The previous example can be used to show the importance of appropriate parameter selection and evaluation. In the example, the model over predicted the sediment concentrations even though it under predicted the runoff volume. Just as in the previous example, this may be caused by several factors: (1) the effect of tillage operations and subsequent reconsolidation as defined in the model is in error; (2) the estimate of tillage parameter values used in the model is in error; (3) parameters describing raindrop detachment, overland flow detachment, transport capacity, particle size characteristics, or the effects of canopy cover are either not being properly evaluated or the phenomenon is not adequately described in the model; (4) description of the water as shallow overland vs. concentrated flow in rills/channels and subsequent estimates of shear and its distribution between plants, soils, microtopography, channels, etc., are in error; and (5) the prototype data were in error because of the type of instrumentation used. Solution to this problem is more difficult.

Perhaps one of the best ways to approach the problem would be to perform sensitivity analyses on each of the parameters to find how much change in a parameter value would be needed to bring prediction in line with observation. In some cases it may not be possible, especially if the error is caused by more than one problem, but the procedure should enable the researcher to narrow down the list of possible causes. If error in prediction is caused by a fundamental problem in the detachment or entrainment components, it will be necessary to carry out laboratory or rainfall simulator studies to look at individual subprocesses such as raindrop or rill–flow detachment.

In this case assume the problem is created by an improper assumption as to the effect of canopy cover on detachment by rainfall. Every plant has a characteristic drip pattern and energy level associated with the drops. Corn, for example, will funnel much of the water to the stalk where it runs down; only the ends of the leaves, beyond the break point, will drip. Some plants such as cotton will form a drip line somewhat inside the perimeter of the plant. If the leaves of a plant are close to the ground, drops from the leaves will have very little energy. If canopy cover is relatively sparse it may have very little impact on natural rainfall erosion characteristics.

It is obvious that the model user must understand the role that the parameter plays and how to evaluate it. In this case, if the model did not have provision to modify the effect of raindrop impact for the effect of canopy cover, it would not have been possible to account for the effect of that component. If the component were missing, then error in model response would have been incorporated into some other parameter thus biasing parameter values, possibly making them unreasonable.

2.4.3. Accuracy and Precision of Variables, Characteristics and Parameters

In situations similar to those described above, it is possible to determine the accuracy that the researcher should use in describing the variables, characteristics, and parameters by a sensitivity analysis (Vol. 1, Chapter 6 by Lane and Ferriera in Knisel, 1980; Coleman and DeCoursey, 1976; Saxton, 1975). If the model is completely analytical and continuous in the first derivative, then a partial derivative of the expression to be evaluated with respect to the variable or parameter in the equation will show the sensitivity of that variable or parameter. If the model is not completely analytical, then the same type information can be obtained by finite difference. If the model is relatively insensitive to one or more of the variables or parameters, the researcher should consider fixing the parameters to the mean value or at least investigate its performance. It may be possible to reduce the frequency of measurement of a particular variable or stop data collection completely.

It is possible, by expressing sensitivity in terms relative to the magnitude of the phenomena in question (evapotranspiration rate, for example), to compare parameter sensitivity to alternative expressions. A plot of the relative sensitivity throughout the year will show how it changes with season depending upon the magnitude of the hydrologic process in question. See Figure 4 for an example of seasonal changes for several typical expressions. A sensitivity analysis can also provide a measure of the overall variance associated with measurement error. This is extremely useful in comparing alternate expressions and studying the effect of instrument error variance. See Coleman and DeCoursey (1976) for more information.

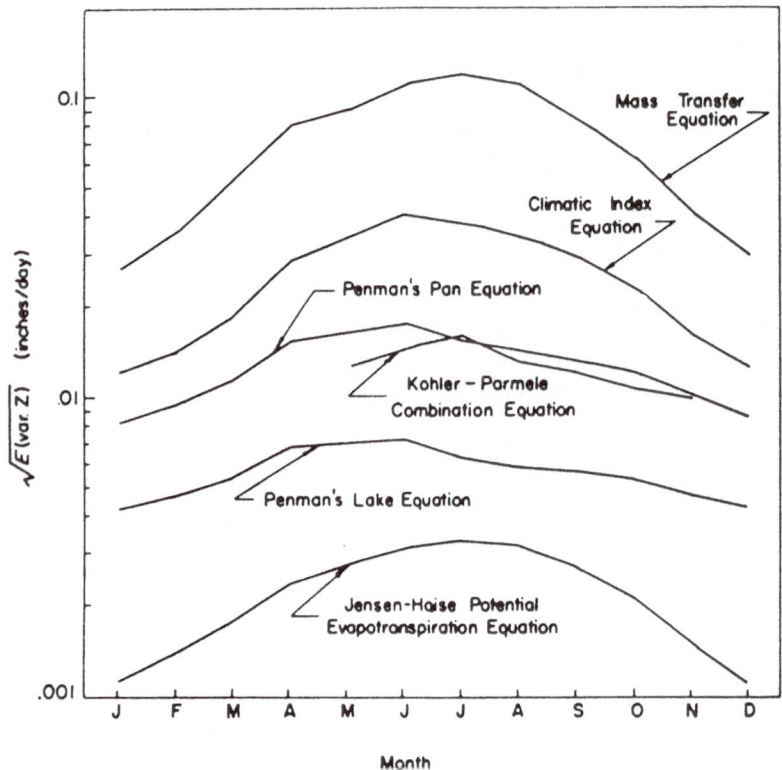

Figure 4. Expected standard deviation $[E(\text{Var } Z)]^{1/2}$ of error, for six evaporation and evapotranspiration models, by monthly averages (Coleman and DeCoursey, <u>Water Resources Research</u>, 12(5), pg. 878, 1976, copyright by the American Geophysical Union).

2.4.4. Spatial and Temporal Representativeness

The degree to which a model should be spatially and temporally representative of the real world situation depends upon its ultimate use. A basin–scale model, intended for preliminary design of water resources systems, cannot be expected to represent spatial detail or temporal variability in much detail. However, spatial and temporal effects should be built into the model's response by providing the user with realistic probability statements of performance. This will require that recommendations for values of input parameters be presented in the form of probability distributions that adequately represent the effect of spatial and temporal variability. The development of such models, which will probably be empirical in nature, should evolve from causal models so that they have a mechanistic structure that will enhance their application to a wide variety of situations. The Freeze model, previously described, is obviously too complex, in its present form, for application–type situations. Scientists and engineers from the U.S.D.A.–A.R.S. are now starting to refine a very comprehensive Small Watershed Model (SWAM) (Alonso and DeCoursey, 1985) and develop it into an applications model for small catchments. The first step in that process will be the use of

SWAM as a simulation tool to study the best way of spatially and temporally integrating the wide variety of soils, land uses, and topographic characteristics that influence the hydrologic response of "large" watershed basins. The resultant model will have many empirical components, but it will have a mechanistic structure and fundamental basis.

Models such as the spatially variable hillslope model of Freeze, previously discussed, and the work of Smith and Hebbert (1979) are extremely helpful in providing examples of the influence of spatial variability on hydrologic response. By using models such as these in a simulation mode, using Monte Carlo experiments or analytically generated inputs (Eagleson, 1978), it is possible to determine the hydrologic response in probabilistic terms. For example, Eagleson uses a derived distribution of rainfall probability (amount and time) as the driving mechanism for a simplified causal model that includes infiltration, exfiltration, transpiration, percolation to groundwater, and capillary rise to derive probability distributions of various independent water balance elements: surface runoff, evapotranspiration, and groundwater.

The effects of spatial variability on hydrologic response, see Freeze (1982) and Smith and Hebbert (1979), are dramatic enough that their effects should be incorporated into decision makers' models, probably in the form of probability statements of response. This suggests that additional work similar to that of Freeze, Eagleson, and Smith and Hebbert is needed. It will, of course, require a significant commitment in both time and resources to develop. Almost all input values to the models previously described, i.e. slope, channel roughness, flow cross-sectional shape, extent of plant canopy cover, climatic parameters, and soils exist in the form of probability distributions—in both time and space for some parameters. Analyses, based on observed measures of these distributions, are needed to determine parameter significance. This will require a significant commitment to field–data collection and sampling.

2.4.5. Model Evaluation and Time–Space Trade–Off

"Clearly, models and data are twinned components of the modeling procedure. Model assessment implies a concomitant assessment of data availability and worth. Certain models may have no predictive value unless certain data are available. Certain data may have great value as input to one model and no value at all as input to another" (Freeze, 1982).

This quotation introduces the problem of the worth of data, and leads to the concept of time–space tradeoffs and comparison of models. Assume that two models, capable of predicting hydrologic response, are available. For ease in presenting the concept, assume that both are event-type models. One of the models is a regression model that relates surface runoff to rainfall intensity and amount, antecedent soil moisture, extent of canopy cover, and infiltration rate which is assumed to be a function of tillage and time of the year. The other model is a simple conceptual model in which the catchment is divided into three regions that drain either into the channel or cascade across another to reach the channel. Infiltration is expressed by the Green and Ampt equation with parameters that differ for each of the three areas and also vary as a function of tillage. ET is calculated by the Penman equation and seepage beyond the root zone is expressed as a function of soil moisture tension. The period of record will be divided into two parts: a calibration period, in which parameter values will be determined, and a validation period. The following evaluation will be performed on the validation period only. The error or residual variance, F^2, of predicted and observed values is calculated for both models (see Freeze, 1982) as:

$$F^2 = \sum (q' - q)^2 \qquad (1)$$

in which q and q' are the observed and computed volumes of surface runoff. The initial variance, F_o^2, of the volumes of surface runoff is defined as:

$$F_2^o = \sum (q - \bar{q})^2 \tag{2}$$

in which $\bar{q}$ is the mean of the observed q's. The efficiency of the model is defined in terms of the proportion of the initial variance accounted for by the model as:

$$R^2 = \frac{F_o^2 - F^2}{f_o^2} \tag{3}$$

The efficiency, as defined by Equation (3), will be a minimum when the model is explaining much of the initial variance. The efficiency term, R^2, can be used to compare performance of the two models. Also, it is possible to use the same concept to compare models of individual events. For example, hydrographs are compared (using the R^2 criteria), by letting the q's in Equations (1–3) be the consecutive observations of the hydrograph.

It is possible to use the efficiency term to study stability of a model. In this case the efficiency, R^2, is evaluated for both the calibration and the validation periods. In most cases the efficiency of the validation period is not as good as that of the calibration period. However, if the efficiency of the validation period is considerably less than that of the calibration period, then the model may be unstable.

In models used to predict the response of a system that contains unexplained or stochastic components, there are two ways that a modeler can improve efficiency by minimizing R^2: one is by increasing the number of events, i.e., extending the length of the data set; the other is by increasing the spatial accuracy of the model. This is best illustrated by considering a mechanistic model that described the watershed as composed of subareas. Freeze (1982) and Smith and Hebbert (1979) show how accuracy of prediction can be improved by increasing the density of measurement points in the catchment. Thus, instead of increasing the period of observation, we may be able to improve efficiency more by dividing the catchment into more subareas. The cost of obtaining a better index of the soil variability (i.e. conductivity and related data) may be considerably less than maintaining the record for a longer period of time. This is one example of a space–time tradeoff. The concept of a space–time tradeoff can also be applied to a network of gages where overall performance of a model can be improved by introducing more gages or applying it to more watershed sites rather than waiting for a longer period of record.

2.4.6. Model Optimization

This is not a general discussion of optimization, but a description of a type of optimization that should be of benefit to hydrologists in the next few years.

Most researchers, working on hydrologic models, concentrate their efforts to improve model performance by analyzing the response of one quantity: surface runoff, evapotranspiration rate, infiltration rate, or groundwater flow, etc. Frequently, performance of the entire model is based on how well the model predicts surface runoff volumes, peaks, etc. Larger, more complex models are composed of submodels with outputs, or fluxes, that have been independently evaluated or optimized. Sensitivity analyses and testing of complete models rarely consider an optimal evaluation of how all the submodels interact. Overall performance of the model may be improved by an optimization of the entire model. Fiering and Kuczera (1982) describe a robust (strong or more powerful) method of optimization that considers five concurrent elements: surface runoff, surface detention, plant storage, vadose zone storage, and groundwater storage. Continuity equations are defined for each of the four storage elements. Fluxes into and out of each of the storages, values of state variables, and constraints on all parameters are identified. Even though the continuity equations are deterministic, a random disturbance term is added to respond to spatial heterogeneity, noise in measurement processes, and misspecification of state variables and fluxes. Parameters of the four storage continuity equations, the runoff, and the random term are obtained by constrained least squares or one of several other optimization techniques. The principal advantage of the method is that parameter values are constrained by observations of runoff and the four main storage compartments in the model, rather than only one compartment. Because some parameter values influence more than one storage compartment, this method of optimization in effect considerably increases the number of observations that are available to fit the data. The result is a set of parameters that are less likely to be unique to the data set. It is also possible to improve the robustness of the model by imposing tighter constraints on parameters with values that are known to lie within specified ranges. This approach to model optimization is now under development, but should have considerable value as more causal models are developed in the next few years.

3. Conclusions

The development of research watersheds must be accompanied by an analytical description of the hydrologic processes being investigated. The description could be as simple as an empirical regression equation or as complex as a detailed casual model. However, it should have its roots in the needs of the decision maker. The research may be fundamental in character, but it must respond to some gap in knowledge needed to improve the decision maker's ability to make decisions based on adequate information.

In the past decade it has become obvious that human influence on the hydrologic cycle has become so significant that we must critically look at both the temporal and spatial scale of our hydrologic modeling activity. Extensive use of the hydrosphere for storage, detoxification, waste assimilation, energy production, agricultural development, and urbanization have shifted emphasis in hydrologic research to the need to study long–term and global effects. The health and well–being of much of the world's population is affected; thus, there is a need to develop new and improved procedures that require increasingly precise and accurate predictive statements.

Response to these problems requires that we concentrate on the development of causal models. Obviously, the decision maker, and those providing the information upon which decisions are to be based, cannot afford to work with complex, data–intensive models. The tools are going to have to be a combination of empirical and causal models. However, the empirical models must have a realistic structure that will enable them to be used in the wide variety of locations and situations with which the decision maker is forced to deal. These arguments stress the need for continued emphasis on causal model development.

The highest priority needs in the development of these models are considered to be: (1) improved ability to deal effectively with problems of temporal and spatial scale, (2) better understanding of the coupling of chemistry and biology, (3) better understanding of the effect of toxic substances on the food chain, (4) improved techniques to provide statistically stable parameter estimates, (5) continued emphasis in development of causal chains of hydrologic understanding, (6) development of more responsive statistical tools and methods of treating confidence intervals, and (7) development of improved algorithms to deal with hydrologic extremes. These high–priority needs should permeate our thinking when we are developing analytical models to guide our watershed research programs.

Because of the large quantity of literature describing recent hydrologic model development, this chapter concentrates on the synergism between model development and field data collection. This was illustrated by using very brief descriptions of hypothetical models. The descriptions also illustrate the impact of model structure on data collection. Causal–type models, designed to improve understanding of the physical processes, require considerably more field data and site description than do conventional empirical models; information frequently needed may include soil conductivity, soil roughness, tillage, canopy cover, solar energy, wind, relative humidity, soil particle characteristics, etc. As both field research and model development progress, there is an obvious need for additional information; however, it may be possible to reduce the collection of data in one area to support increased effort in another.

Model development that parallels field data collection aids in studying a research site by providing insight into the physical processes. Errors in predicting watershed response can frequently be traced to a lack of field definition, missinterpretation of parameter values, and/ or an erroneous description of the physical processes. Thus, the interaction between modeling and field data collection is very important to both efforts and selection of appropriate field instrumentation.

Hydrologic models also provide valuable assistance in assessing required instrumentation accuracy. On the other hand, if sensitivity of the instrumentation is known, variance in model performance can be obtained. A sensitivity analysis of the model with respect to variables and parameters provides the information needed for this assessment.

Spatial and temporal variability may be treated by performing hydrologic analyses using a short–time interval and segmenting the watershed site into "homogeneous" areas. This approach is not entirely satisfactory because in many cases it does not adequately represent the true effects of spatial variability. The few models that have been developed to demonstrate the significant impact that spatial variability can have on hydrologic response require too much information to be of use to the decision maker. Continued effort to place probability estimates on parameter values is needed to properly assess risk.

Another benefit of concurrent model and watershed development is the assessment of alternative model performance. By calculating residual or error variance in the prediction of an observed record, and thus the efficiency of model performance, it is possible to compare alternative models and assess the value of additional data. Given this information, it is possible to determine the cost of this data and thus assess its worth. The efficiency of model performance can also be used to study the significance of alternative ways of improvement, i.e. increasing the length of the observation period or increasing the number of observational points within the watershed. The use of additional spatial information instead of extending the record is a space for time tradeoff that can be extremely helpful to hydrologic model and watershed development.

A discussion of hydrologic model optimization is also presented. The method described consists of using all the major water storage reservoirs in the hydrologic cycle: the vadose zone, groundwater, surface detention, and plant interception/storage in addition to runoff to determine optimal parameter values. The method takes advantage of the fact that parameter values are constrained by all five components; thus, observations provide an effectively larger data set. The net effect should be a more robust model and more accurate simulation of the physical processes.

4. References

Alonso, C. V., and D. G. DeCoursey: 1985, 'Small Watershed Model,' *Proceedings of Natural Resources Modeling Symposium*, Pingree Park, Colorado, Oct. 16–21, 1983, USDA, ARS–30, Cons. Res. Report, pp. 40–46.

American Geophysical Union: 1987, *'Contributions in Hydrology,'* U.S. National Report 1983–1986, Nineteenth General Assembly, International Union of Geodesy and Geophysics, Vancouver, British Columbia, Canada, pp. 101–178.

American Geophysical Union: 1983, *'Contributions in Hydrology,'* U.S. National Report 1979–1982 to the Eighteenth General Assembly of the International Union of Geodesy and Geophysics, Hamburg, Federal Republic of Germany, pp. 697–776.

American Geophysical Union: 1979, *'Papers in Hydrology,'* U.S. National Report 1975–1978 to the Seventeenth General Assembly of the International Union of Geodesy and Geophysics, Canberra, Australia, pp. 1165–1351.

Chapman, T. G., and F. X. Dunin (eds.): 1975, *Prediction in Catchment Hydrology*, Australian Academy of Science, Griffin Press, 482 pp.

Coleman, G., and D. G. DeCoursey: 1976, 'Sensitivity and Model Variance Analysis Applied to Some Evaporation and Evapotranspiration Models,' *Water Resources Res.* **12**(5), 873–879.

DeCoursey, D. G. (ed.): 1990, *Proceedings International Symposium on Water Quality Modeling of Agricultural Non–Point Sources, Parts I and II*, Logan, UT, USDA, Ag. Res. Serv. ARS–81, 881 pp.

DeCoursey, D. G. (ed.): 1985, *Proceedings of Natural Resources Modeling Symposium*, Pingree Park, Colorado, Oct. 16–21, 1983, USDA, ARS Cons. Res. Report, 496 pp.

DeCoursey, D. G., and R. W. Rojas: 1990, 'RZWQM–A Model for Simulating the Movement of Water and Solutes in the Root Zone,' in D. G. DeCoursey (ed.), *Proceedings International Symposium on Water Quality Modeling of Agricultural Non–Point Sources, Part II*, June 19–23, 1988, Logan, UT, USDA, Ag. Res. Serv. ARS–81, pp. 813–822.

Dooge, J. C. I.: 1973, *'Linear Theory of Hydrologic Systems,'* USDA, Agricultural Research Service Technical Bulletin No. 1468, 327 pp.

Dunne, T.: 1982, 'Models of Runoff Processes and Their Significance,' *Scientific Basis of Water–Resources Management*, Studies in Geophysics by the National Research Council, National Academy Press, Washington, D.C., pp. 17–30.

Eagleson, P. S.: 1978, 'Climate, Soil and Vegetation, 1. Introduction to Water Balance Dynamics,' *Water Resources Research* **14**(5), 705–776.

Fiering, M. B., and G. Kuczera: 1982, 'Robust Estimators in Hydrology,' *Scientific Basis of Water–Resources Management*, Studies in Geophysics, by the National Research Council, National Academy Press, Washington, D.C., pp. 85–94.

France, J., and J. H. M. Thornley: 1984, *Mathematical Models in Agriculture*, Butterworths, 335 pp.

Freeze, R. A.: 1982, 'The Influence of Hillslope Hydrological Processes on the Stochastic Properties of Streamflow,' in V. P. Singh (ed.), *Statistical Analysis of Rainfall and Runoff*, Water Resources Publications (P.O. Box 2841, Littleton, CO 80161–2841, U.S.A.), pp. 155–172.

Haan, C. T., H. P. Johnson, and D. L. Brakensiek (eds): 1982, *'Hydrologic Modeling of Small Watersheds,'* American Society of Agricultural Engineers, Monograph No. 5, Amer. Soc. of Agr. Engrs., St. Joseph, Michigan, 533 pp.

Hewlett, J. D.: 1981, *'Models in Land Use Hydrology: The Need for Closing the Gap between Theory and Practice,'* Keynote address at Workshop on the Effect of Rural Land Use and Catchment Management on Water Resources, Dept. of Water Affairs, Forestry and Environmental Conservation, Pretoria, South Africa, 13 pp.

Klemes, V.: 1982, 'Empirical and Causal Models in Hydrology,' *Scientific Basis of Water–Resources Management*, Studies in Geophysics, by the National Research Council, National Academy Press, Washington, D.C., pp. 95–104.

Knisel, W. G. (ed.): 1980, *'CREAMS: A Field–Scale Model for Chemicals, Runoff, and Erosion for Agricultural Management Systems,'* U.S. Department of Agriculture, Conservation Report No. 26, 640 pp., illus.

Lane, L. J., M. H. Diskin, D. E. Wallace, and R. M. Dixon: 1978, 'Partial Area Response on Small Semiarid Watersheds,' *Water Resources Bulletin* **14**(5), 1143–1158.

Matalas, N. C., J. M. Landwehr, and M. G. Wolman: 1982, 'Prediction in Water Management,' *Scientific Basis of Water–Resources Management*, Studies in Geophysics, by the National Research Council, National Academy Press, Washington, D.C., pp. 118–127.

National Research Council: 1982, *Scientific Basis of Water–Resource Management*, Studies in Geophysics, National Academy Press, Washington, D.C., 127 pp.

Office of Technology Assessment: 1982, *'Use of Models for Water Resources Management, Planning, and Policy,'* Summary, Six Chapters and Appendix, U.S. Government Printing Office, Washington, D.C.

Rovey, E. W., D. A. Woolhiser, and R. E. Smith: 1977, *'A Distributed Kinematic Model of Upland Watersheds,'* Hydrology Paper No. 93, Colorado State University, Fort Collins, CO, 52 p.

Saxton, K. E.: 1975, 'Sensitivity Analyses of the Combination Vapotranspiration Equation,' *Agr. Meteorol.* **15**, 343–353.

Saxton, K. E., and G. C. Bluhm: 1982, 'Regional Prediction of Crop Water Stress by Soil Water Budgets and Climatic Demand,' *Trans. ASAE* **25**(1), 105–111, 115.

Shaffer, M. J., and W. E. Larson (eds.): 1987, '*NTRM, A Soil–Crop Simulation Model for Nitrogen, Tillage, and Crop–Residue Management,'* USDA, ARS, Cons. Res. Report 34–1, 103 pp.

Singh, V. P. (ed.): 1982a, 'Statistical Analysis of Rainfall and Runoff,' *Proc. Int. Symp. on Rainfall–Runoff Modeling,* Mississippi State, Mississippi, Water Resources Publications, 700 pp.

Singh, V. P. (ed.): 1982b, 'Modeling Components of Hydrologic Cycle,' *Proc. Int. Symp. on Rainfall–Runoff Modeling,* Mississippi State, Mississippi, Water Resources Publications, 590 pp.

Singh, V. P. (ed.): 1982c, 'Rainfall–Runoff Relationship,' *Proc. Int. Symp. on Rainfall–Runoff Modeling,* Mississippi State, Mississippi, Water Resources Publications, 582 pp.

Singh, V. P. (ed.): 1982d, 'Applied Modeling in Catchment Hydrology,' *Proc. Int. Symp. on Rainfall–Runoff Modeling,* Mississippi State, Mississippi Water Resources Publications, 563 pp.

Smith, R. E.: 1981, 'A Kinematic Model for Surface Mine Sediment Yield,' *Trans. ASAE* **24**(6), 1509–1514.

Smith, R. E., and V. A. Ferriera: 1988, '*Opus, An Advanced Simulation Model for Non–Point Source Pollution Transport at the Field Scale,'* Draft Documentation.

Smith, R. E., and R. H. B. Hebbert: 1979, 'A Monte Carlo Analysis of the Hydrologic Effects of Spatial Variability of Infiltration,' *Water Resources Research* **15**(2), 419–429.

Vischer, D., and C. Emmenegger (Organizers): 1982, *International Symposium on Hydrologic Research Basins and Their Use in Water Resources Planning,* Landeshydrologie–Postfack 2742, Ch.–3001, Berne, Switzerland, 3 Vol.

Williams, J. R., P. T. Dyke, and C. A. Jones: 1983, 'EPIC – A Model for Assessing the Effects of Erosion and Soil Productivity,' *Proc. Third International Conference on State–of–the–Art in Ecological Modeling,* Colorado State University, Fort Collins, May 24–28, 1982, 20 pp.

Wight, J. R. (ed.): 1987, '*SPUR – Simulation of Production and Utilization of Rangelands: Documentation User Guide,'* USDA–ARS Publication No. ARS–63, 367 pgs.

Chapter 29

Numerical Simulations and Field Experiments of Unsaturated Flow and Transport: The Roles of Hysteresis and State-Dependent Anisotropy

James T. McCord
Sandia National Laboratory
Waste Management Systems Division
P.O. Box 5800
Albuquerque, New Mexico 87185–5800 U.S.A.

and

Daniel B. Stephens and **John L. Wilson**
New Mexico Institute of Mining and Technology
Department of Geoscience, Research and
  Development Division
Socorro, New Mexico 87801 U.S.A.

Abstract. This chapter describes a series of detailed soil–water tracer experiments conducted in a natural landscape. It also discusses three different approaches taken to numerically model the observed flow and transport. These field experiments provide the first conclusive evidence for the existence of a variable, state–dependent anisotropy in the hydraulic conductivity of an unsaturated, layered medium in a natural field setting. This concept has been previously postulated in a number of independent theoretical and experimental investigations. While previous studies have identified layered heterogeneity as the principal cause of state–dependent anisotropy, we develop a simple scenario to show how hysteresis in the soil–moisture characteristics (θ–ψ relationship) can contribute to the anisotropic behavior of a texturally homogeneous, porous media profile under transient, unsaturated conditions. Three approaches at modeling the observed flow and transport are implemented and compared. The results of the computer simulations indicate that even relatively uniform media (such as the sands from our field site) exhibit variable anisotropy, and that local (laboratory scale) hysteresis makes only minor contributions to macroscopic anisotropic flow behavior.

1. Introduction

1.1. Anisotropy in Hydraulic Conductivity

Research in the past decade indicates that at field scales, spatial variability in the hydraulic properties of a porous medium exerts a large influence on flow and transport through that medium (Bakr et al., 1978; Gelhar and Axness, 1983). With respect to variably saturated flow, several recent studies have provided important contributions to our understanding of the nature of anisotropy in hydraulic conductivity. Using a variety of approaches ranging from analytical mathematical models (Zaslavsky and Sinai, 1981a,b,c; Mualem, 1984; Yeh et al.,

D. S. Bowles and P. E. O'Connell (eds.), Recent Advances in the Modeling of Hydrologic Systems, 613–640.
© 1991 *Kluwer Academic Publishers.*

1985b,c; Mantoglou and Gelhar, 1987) to computer (Bear et al., 1987), laboratory (Stephens and Heermann, 1988), and field (McCord and Stephens, 1987) experiments, a number of different researchers arrived at the same general conclusion: anisotropy in hydraulic conductivity of a porous media can vary significantly as the saturation of the media varies. Work by Lynn Gelhar and his associates (Yeh et al., 1985b; Mantoglou and Gelhar, 1987) indicates that anisotropy can depend on several other factors in addition to saturation. These include the magnitude and direction of the hydraulic gradient and a variety of parameters which characterize the variability and spatial correlative structure of the hydraulic properties for a heterogeneous media. None of these studies, however, provide conclusive evidence for this phenomena in a natural field setting.

In developing predictive mathematical models for flow through unsaturated soils, no one has considered the possibility of a variable anisotropy in hydraulic conductivity. To solve the governing partial differential equation for unsaturated flow with nonuniform initial and/or boundary conditions and/or for irregularly shaped spatial domains, one typically employs a finite difference or finite element numerical approach. Bear and Braester (1987) point out that in the past, many (if not all) researchers who numerically modeled variably saturated flow, represented the effective hydraulic conductivity of an anisotropic porous medium as a product of the saturated hydraulic conductivity tensor, k_{ij}, and the relative hydraulic conductivity, K_r, which is a function of saturation, S. The same relative conductivity function, $K_r(S)$, is assumed to apply to all components of the effective hydraulic conductivity, $K_{ij}(S)$; i.e. the magnitude of the anisotropy remains constant throughout the entire range of saturations. Based on the theoretical and experimental studies mentioned in the first paragraph, we join Bear and Braester (1987) in questioning the validity of this commonly used approach of handling anisotropy in variably saturated flow models.

1.2. Objectives and Approach

This chapter will address two important problems: (1) field validation of the concept of variable macroscopic anisotropy in the hydraulic conductivity of heterogeneous media, and (2) how one might account for variable macroscopic anisotropy in a computer code which models flow and transport in unsaturated media.

With insufficient direct field evidence for state–dependent anisotropy, we decided to conduct a series of tracer experiments in a natural field setting with simultaneous hydraulic head monitoring. This chapter briefly describes the experiments, which show large horizontal downslope flows occurring in the presence of a near–vertical downward hydraulic gradient field. We believe these experiments substantially validate, or at least provide strong evidence for, the concept of state–dependent anisotropy.

In addition to the layered heterogeneity issues analyzed by Mualem (1984), Yeh et al. (1985c) and Mantoglou and Gelhar (1987), there is another factor which contributes to the anisotropic behavior observed in the tracer experiments. The term given this other contributing factor is "moisture–dependent heterogeneity in hydraulic conductivity." We will show how this phenomena can cause texturally homogeneous materials to behave as hydraulically anisotropic media when viewed at field scales, and how hysteresis in the soil moisture characteristics (water content–capillary pressure relationships) tends to magnify this phenomena.

To model the flow and transport observed at the field site, we use three distinct approaches. For the first two approaches, we employ the commonly used techniques of assuming a constant anisotropy at all levels of saturation. These approaches will differ, however, in the consideration of hysteresis in the soil–moisture characteristics as measured in the laboratory. For the first approach the soils will be considered nonhysteretic, whereas in the second, hysteresis will be accounted for. Cyclical wetting and drying events at the soil surface cause the development of saturation–dependent, hydraulically conductive heterogeneity in a porous media profile. This heterogeneity, which is magnified by hysteresis, should cause the occurrence of horizontal downslope components to the unsaturated flow, a phenomenon which was observed at the field site. In the third approach, local hysteresis is neglected. Instead, an algorithm was included in the model which allows the anisotropy in the principal directions of the effective conductivity tensor to vary as the state (capillary pressure) of the system varies. To our knowledge, the phenomenon has not been implemented previously in any computer model of variably saturated flow through porous media.

Through this structured modeling approach, we can evaluate the relative importance of hysteresis and variable anisotropy as factors causing the lateral flow behavior observed at the field site. In addition, we address the practical issue of which approach is best suited to modeling "real world" problems.

2. Methods and Materials

2.1. Field Experiments

2.1.1. Site Description

The site comprises an area of about 900 m² situated in the Sevilleta National Wildlife Refuge about 25 km north of Socorro, New Mexico. The site lies on a small sand dune on the south side of the Rio Salado, an ephemeral, braided tributary to the Rio Grande. A generalized location map is shown on Figure 1. The area receives about 20 cm of precipitation per year, and the gross annual potential lake evaporation is about 178 cm.

The surficial lithology consists of fine-to-medium uniform sands to a depth of 6.5 meters, the maximum depth which was sampled. Trenches excavated in the sands revealed no

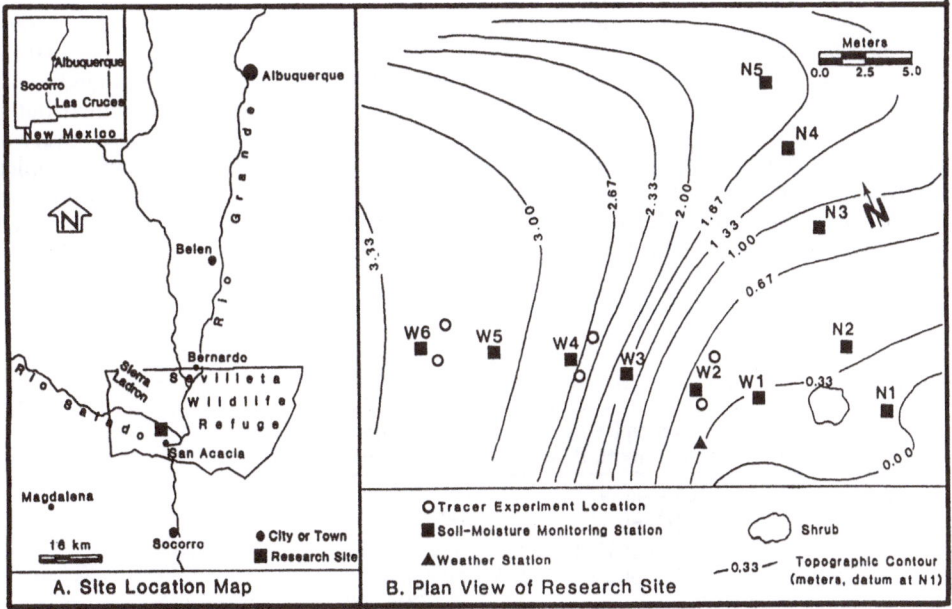

Figure 1. Location map and plan view of field experiment site.

apparent fine–grained layers, although we could discern medium–to–large–scale eolian cross–stratifications. The foreset beds were on the order of 1 to 3 millimeters thick and were oriented approximately parallel to the ground surface slope.

Figure 1 also presents a topographic map of the ground surface at the study location. Included with the topographic map is a layout of the subsurface monitoring equipment.

As depicted in Figure 1 the instrumentation array included 11 soil–water monitoring stations and a fully automated weather station. Each soil–water monitoring station consisted of: (1) an aluminum tube buried to a depth of 3 meters to provide access for a neutron probe to measure moisture content, and (2) tensiometers installed at 30 cm increments down to a 244 cm depth to permit hydraulic head measurements. The soil–water stations were monitored on a weekly basis. Gypsum electrical resistance blocks were placed at 10, 20, and 30 cm depths at most stations on the west transect (W2 through W6) to provide hydraulic head estimates in the near–surface sands which often dry to tensions which exceed the operation range of tensiometers.

After all of the instrumentation was in place, we conducted a two–month long baseline monitoring program to test the monitoring array. Once it was established that all of the equipment was operating properly, the experiment commenced with the tracer tests. The investigation reported here is based on a series of tracer experiments conducted along the east–west trending instrumentation transect.

2.1.2. Tracer Experiments

Three tracer experiments were performed on the west transect at three different topographic positions: (1) on a relatively level position near the top of the slope (station W6), (2) on a sloping position near the middle of the transect (between stations W3 and W4), and (3) on a level position at the base of the slope (station W2). The location of each experiment is shown in the plan view presented in Figure 1.

The tracer experiments were conducted as follows: At each selected position, a 30 cm long, narrow (~3 cm) trench was excavated perpendicular to the ground surface slope to a 40 cm depth. A small amount (50 grams) of a dry bromide salt ($CaBr_2$) was then tremmied to the bottom of the excavation between two parallel metal plates. The bromide salt was distributed as evenly as possible along the bottom of the trench so that there essentially was a source of tracer that was 30 cm long. Following precipitation, soils from each location were sampled along transects perpendicular and parallel to the line source. The soil samples were then analyzed for bromide concentration using an ion specific electrode. The sequence of precipitation and sampling events for the tracer experiments is summarized in Table 1.

Table 1. Sequence of events for 1986 tracer experiments.

Date Tracer Emplaced	Precipitation mm depth/date	Date Sampled
Sept. 3	9.0/Sept. 8	–
	0.8/Sept. 10	–
	8.8/Sept.13–15	–
	7.6/Sept. 23	–
	–	Sept. 25
	20.4/Oct. 5–6	–
	48.6/Oct. 10–12	Oct. 16

Figures 2, 3, and 4 present the tracer test results. The figures present profile views of solute plume contours, as well as plan views of solute contours derived from integrating the concentrations over the depth sampled. By integrating the plumes in the plan view over the area covered, one can obtain an estimate of bromide mass contained within each plume. Normalizing this mass with respect to mass originally implanted gives an indication of the amount of tracer recovered. Tracer recovery percentages are presented in Table 2. Also shown in the profile view of Figures 2, 3, and 4 are contours of average total head measured during the tracer experiment. Average values are presented rather than specific values for specific dates because the head remained relatively constant throughout the experiments.

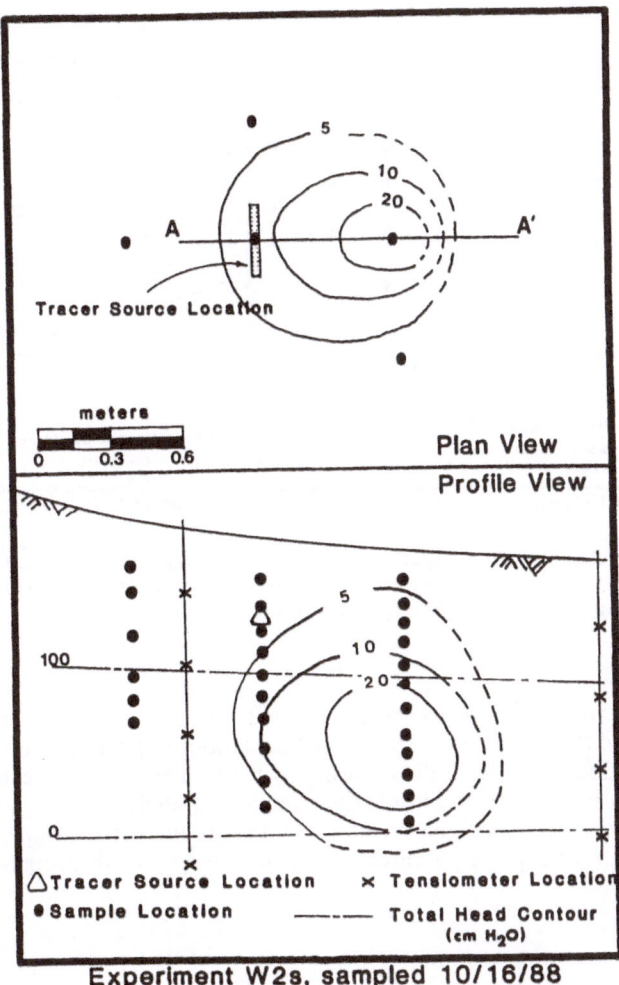

Figure 2. Plan and profile views of solute plume from experiment where tracer was planted at a 40 cm depth just south of station W2. Tracer concentrations are given in (moles/kg-soil) and total head in (cm-water). Profile view taken along axis AA' shown in the plan view.

618

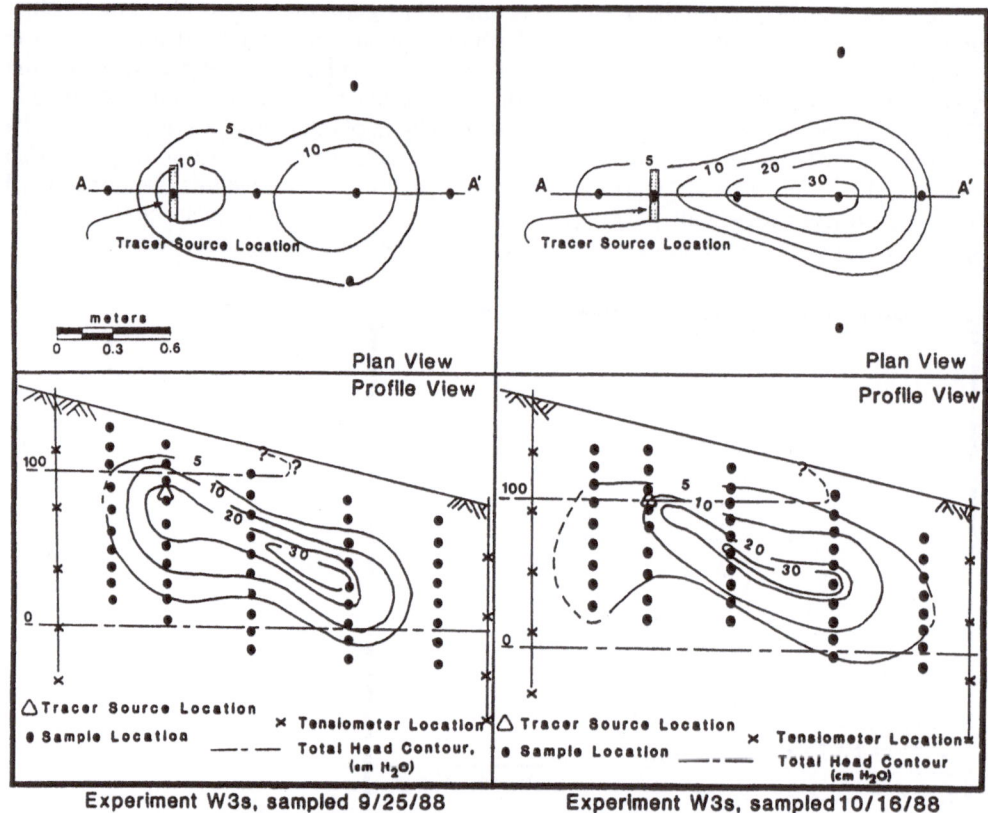

Figure 3. Plan and profile views of solute plume from experiment where tracer was planted at a 40 cm depth just south of station W3. Tracer concentrations are given in (moles/kg–soil) and total head in (cm–water). Profile view taken along axis AA′ shown in the plan view.

By using the tracer plume orientation in conjunction with the head measurements, one can estimate the effective anisotropy of the soil profile. Diffusion calculations indicate that advective transport dominates over diffusive transport in these experiments. Therefore, one can use the plumes to infer components of the soil water flux in the direction parallel to the ground surface slope, s, and the direction normal to the land surface, n. If the ground surface slopes at an angle, β, and if n and s represent the principal directions of anisotropy, the directional fluxes can be computed as:

$$q_s = K_s\left(\frac{\partial \psi}{\partial s} + \sin\beta\right) \tag{1}$$

$$q_n = K_n\left(\frac{\partial \psi}{\partial n} + \cos\beta\right) \tag{2}$$

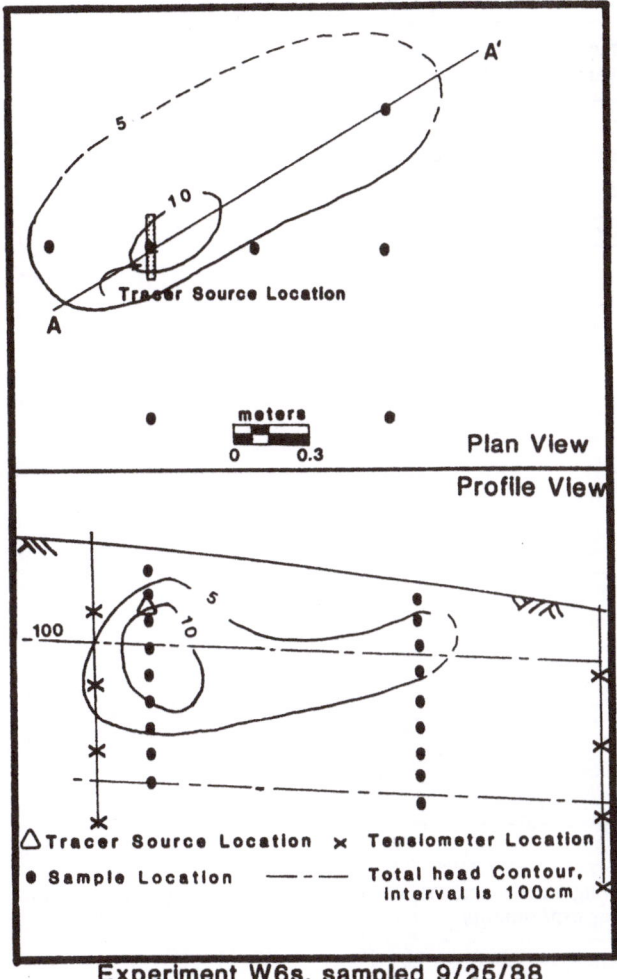

Plan View

Profile View

△ Tracer Source Location   × Tensiometer Location

● Sample Location   —— – —— Total head Contour,
Interval is 100cm

Experiment W6s, sampled 9/25/88

Figure 4. Plan and profile views of solute plume from experiment
where tracer was planted at a 40 cm depth just south of station W6.
Tracer concentrations are given in (moles/kg–soil) and total head
in (cm–water). Profile view taken along axis AA' shown in the plan
view.

where q_s is the soil–water flux in the direction parallel to the slope, q_n is the soil–water flux in the direction normal to the slope, and K_i is the hydraulic conductivity in the i direction. The situation is depicted schematically in Figure 5. One can use Equations (2) and (3) coupled with the measured hydraulic heads to compute the effective anisotropy expressed as K_s/K_n. Depending on the plume selected, q_s/q_n varies between 0.7 and 4.5. Owing to the fairly uniform, near–unitary downward hydraulic gradient observed below a 30 cm depth, the estimated effective anisotropy ranges from 5.1 to 16.9 at a fairly constant pressure head of 50 cm of water. The average effective anisotropy for all of the experiments is 10.7. Table 2 presents the tracer data used to estimate effective anisotropy.

Table 2. Summary of tracer recovery and anisotropy estimates from tracer experiments.

Plume ID	Sampling Date	% Tracer Recovered	Ground Surf. Slope, β	q_s/q_n	Anisotropy Ratio
W6South	09/25/86	92	6°	1.0	9.5
W3South	09/25/86	110	15°	4.5	16.9
W3South	10/16/86	140	15°	3.0	11.3
W2South	10/16/86	100	8°	0.7	5.1

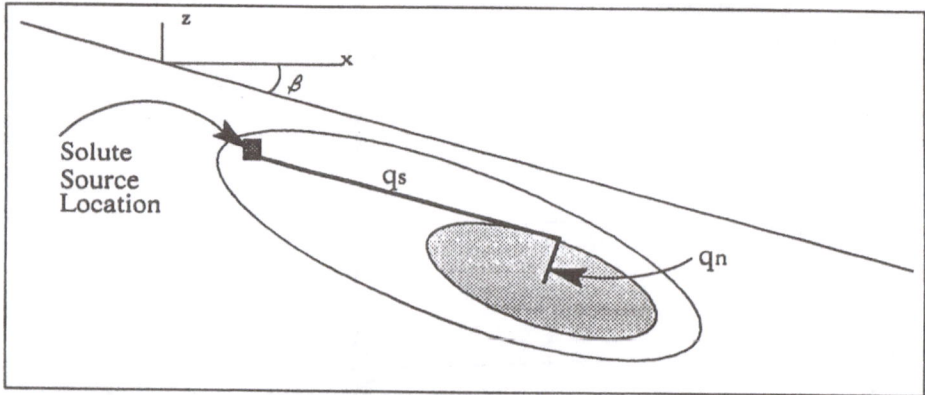

Figure 5. Schematic diagram showing how macroscopic anisotropy is estimated using plume geometries observed in the field tracer experiments.

2.1.3. Soil Characterization

As mentioned in the Site Description section, a trench excavated at the site revealed no obvious fine–grained horizons of lower permeability. However, since visual appearances may be misleading, we performed a detailed grainsize analysis at several locations. A plot of the D_{10} and the D_{60} grain sizes versus depth for one such location (station W1) is presented in Figure 6.

For comparison to the anisotropy estimates obtained from tracer plume geometry, anisotropy was also calculated using alternative, independent techniques for the entire range of moisture contents.

Two different methods were employed to estimate the anisotropy ratio of the soils at 100 percent saturation. The first method consisted of obtaining large–scale (7.6 cm diameter x 25 cm length) samples oriented both parallel and perpendicular to the dune stratifications and performing constant head permeameter experiments on these samples. The anisotropy ratio was computed simply by dividing the measured saturated conductivity of the samples

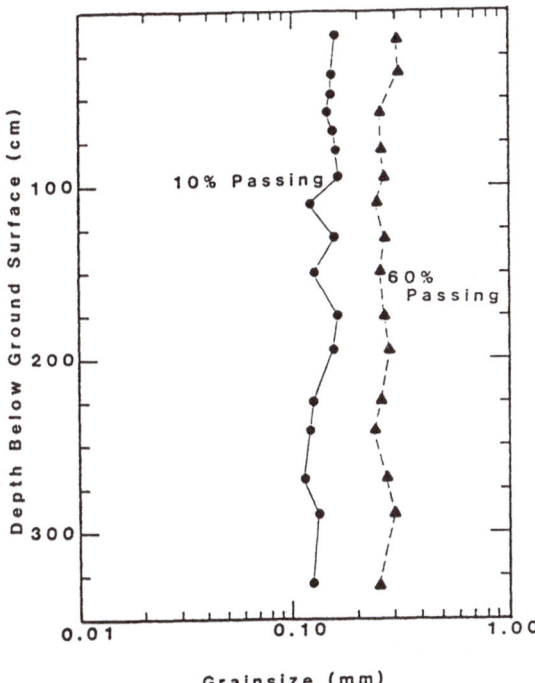

Figure 6. Depth versus grainsize (d10 and d60) plot for one location from the field tracer experiment site. This is typical of the results obtained from several test holes at the site.

oriented parallel to the bedding by that of the samples oriented perpendicular to the bedding. This method predicted an anisotropy ratio at saturation equal to 1.54. Table 3 presents the field data obtained from these large–scale permeameter analyses. The other method used to estimate anisotropy at complete saturation is based on the fact that the equivalent saturated conductivity of a layered soil in the direction parallel to stratification can be calculated by simply determining the arithmetic mean of the individual conductivities for each layer. Consequently, the equivalent effective conductivity in the direction perpendicular to stratification can be calculated as the harmonic mean. If we have a large enough sample size such that we can determine the probability density function of the saturated conductivity, $f(k_s)$, then the arithmetic mean can be defined as:

$$K_A = \int_{k_l}^{k_u} k_s \, f(k_s) \, dk_s \tag{3}$$

the harmonic mean can be defined as:

$$K_H = \left[\int_{k_l}^{k_u} \frac{1}{k_s} \, f(k_s) \, dk_s \right]^{-1} \tag{4}$$

Table 3. Summary of results from large scale saturated permeameter analyses.

Sample ID	Orientation	Ksat (cm/sec)
W2V@30	Perpendicular	1.40×10^{-2}
W2V@75	Perpendicular	1.64×10^{-2}
W3V@30	Perpendicular	1.45×10^{-2}
W3V@75	Perpendicular	1.44×10^{-2}
W3V@60	Perpendicular	1.35×10^{-2}
W3V@75	Perpendicular	2.04×10^{-2}
W3V2@60	Perpendicular	1.49×10^{-2}
W3V2@75	Perpendicular	1.62×10^{-2}
W2W@75	Parallel	1.37×10^{-2}
W3W@30	Parallel	2.19×10^{-2}
W3W@30	Parallel	2.66×10^{-2}
W3W2@30	Parallel	2.28×10^{-2}
W3W4@30	Parallel	2.28×10^{-2}
W2S@30	Parallel	2.23×10^{-2}
W3S@75	Parallel	2.26×10^{-2}
W3S2@30	Parallel	1.92×10^{-2}
W3S4@30	Parallel	1.78×10^{-2}

with k_l representing the lower limit and k_u the upper limit of conductivity. The anisotropy ratio can be calculated as K_A/K_H. To apply this approach, constant head permeameter experiments were conducted on 82 samples of the dune sand. Then both the actual conductivities and the natural log of the conductivities were plotted on Gaussian probability paper. These plots indicated that the random conductivity field is best described by a lognormal distribution. The mean and variance estimates of the $ln(k_s)$ field are 3.53 and 0.82, respectively (k_s in units of cm/sec). Substituting this lognormal distribution into Equations (3) and (4), an anisotropy ratio at full saturation of 1.51 can be obtained.

For moisture contents less than 100 percent saturation, two other approaches were followed to estimate anisotropy. Both methods were developed based on analyses of layered heterogeneous media. The first approach, that of Mualem (1984), extends Equations (3) and (4) to unsaturated conditions, and the second approach will make use of the stochastic result of Yeh et al. (1985b).

To use Mualem's (1984) method, an assumption is made that the unsaturated conductivity of each stratification in a layered soil can be characterized as a function of pressure head, ψ, and k_s, and that k_s is a random variable. Depending on the relative conductivity function selected, there may be any number of other parameters, referred to here as α_i for generality, which affect the character of the unsaturated conductivity function:

$$K(\psi) = g(\psi, k_s, \alpha_i) \tag{5}$$

Substituting this function into Equations (3) and (4) and employing the definition of the anisotropy ratio as K_A/K_H, one can then derive the following expression for anisotropy as a function of pressure head:

$$U(\psi) = \frac{\left[\int_{k_1}^{k_u} g(\psi, k_s, \alpha_i)\, f(k_s)\, dk_s\right]}{\left[\int_{k_1}^{k_u} \frac{f(k_s)}{g(\psi, k_s, \alpha_i)}\, dk_s\right]^{-1}} \tag{6}$$

Random α can be accounted for by also making each α_i some function of k_s. To determine this anisotropy function for the sands from the field site, we numerically evaluated the above integral (Equation 6) subject to the following conditions: (1) we used the probability density function for k_s which was discussed earlier; (2) we estimated the relative conductivity curve from moisture retention data for 12 soil samples using a nonlinear, least squares fit to van Genuchten's (1980) conductivity model; (3) this relative conductivity curve was fit to an exponential conductivity function, $K(\psi) = k_s e^{-\alpha\psi}$; (4) functional relationships between k_s and α were determined by linear regression. The resulting anisotropy function was found to be quite sensitive to the choice of the $\alpha(k_s)$ functions. Figure 7 shows the $U(\psi)$ relationship obtained for both power curve, semi–log, and linear fits for α versus k_s.

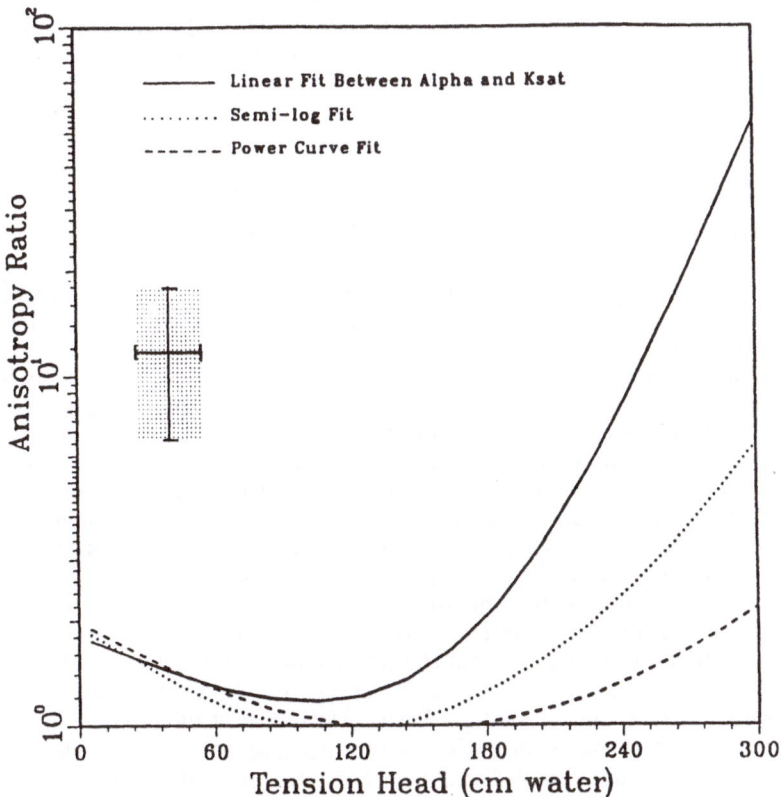

Figure 7. Anisotropy as a function of pressure head for Sevilleta dune sands calculated by the method of Mualem (1984) for three different functional relationships between alpha and Ksat. Shaded area shows mean in situ pressure heads and estimated anisotropies (+ /- one standard deviation) from field tracer experiments.

As discussed in the Introduction, Yeh et al. (1985a,b) used spectral stochastic techniques to analyze steady, unsaturated flow in porous media. In their study, Yeh et al. (1985a) used an exponential conductivity function:

$$K(k_s, a, \psi) = k_s e^{-a\psi} \tag{7}$$

where both k_s and α can be considered random fields. Among the results they derived, is the following expression (Yeh et al., 1985b) for the anisotropy ratio for layered soils with a mean unit downward hydraulic gradient, and uncorrelated $ln(k_s)$ and α fields:

$$\frac{K_{11}}{K_{12}} = \exp\left(\frac{\sigma_f^2 + \sigma_a^2 \, H^2}{1 + \lambda A \cos \beta}\right) \tag{8}$$

where σ_f^2 is the variance of the log conductivity (lnk_s) random field, σ_α^2 is the variance of the slope of the lnK versus ψ relationship, H is the mean value of ψ, λ is the correlation length in the direction perpendicular to stratification, A is the mean slope of lnK versus ψ, and β is the angle between the soil layer stratification and the horizontal plane. Estimating σ_f^2, σ_α^2, H, and A from field and laboratory data, K_{11}/K_{12} was calculated for $\beta = 0$ and various values of λ. These results are plotted in Figure 8 along with the results using Mualem's approach for linear fit between α and k_s.

2.2. Numerical Modeling of Flow and Transport

2.2.1. Modeling Considerations – Anisotropy of Texturally Homogeneous Media

Although the heterogeneity issues discussed above certainly must play a significant role in contributing to the lateral flow evidenced in the tracer tests, another contributing factor which would occur even in a truly homogeneous profile needs to be addressed.

In a medium which is homogeneous in both texture and saturated hydraulic conductivity, a macroscopic anisotropic behavior can be induced by moisture–dependent heterogeneities in hydraulic conductivity. It is well known that fluid flow paths deflect as they pass from a medium of one conductivity into another of a different conductivity. Stochastic, climatic inputs at the soil surface will lead to moisture content heterogeneities (and consequent hydraulic conductivity) in any soil profile, even one which may be considered homogeneous. The development of a moist, highly conductive layer parallel to the ground surface of a hill-slope would lead to horizontal downslope flow components if the hydraulic gradient field were oriented nearly vertical downwards. Figure 9 schematically shows such a situation.

Hysteresis in the soil moisture characteristics (moisture content–capillary pressure relationship) measured at the laboratory–scale would tend to magnify the anisotropic behavior of homogeneous soils at the field–scale. Figure 10 shows the hysteretic soil moisture characteristics for the Sevilleta sands from the field site. Hysteresis would prolong the effect of moisture–induced heterogeneities discussed above because this process slows the redistribution of infiltrated water. Simulation studies comparing one–dimensional infiltration and redistribution in hysteretic and nonhysteretic soils (Rubin, 1967; Bresler et al., 1969) show that hysteretic media will develop a wetted zone which is less diffuse and has a higher peak water content than the wetted zone which would develop in comparable nonhysteretic media under the same boundary conditions. This higher peak water content will result in a larger contrast in hydraulic conductivity and consequently a greater deflection of flow paths.

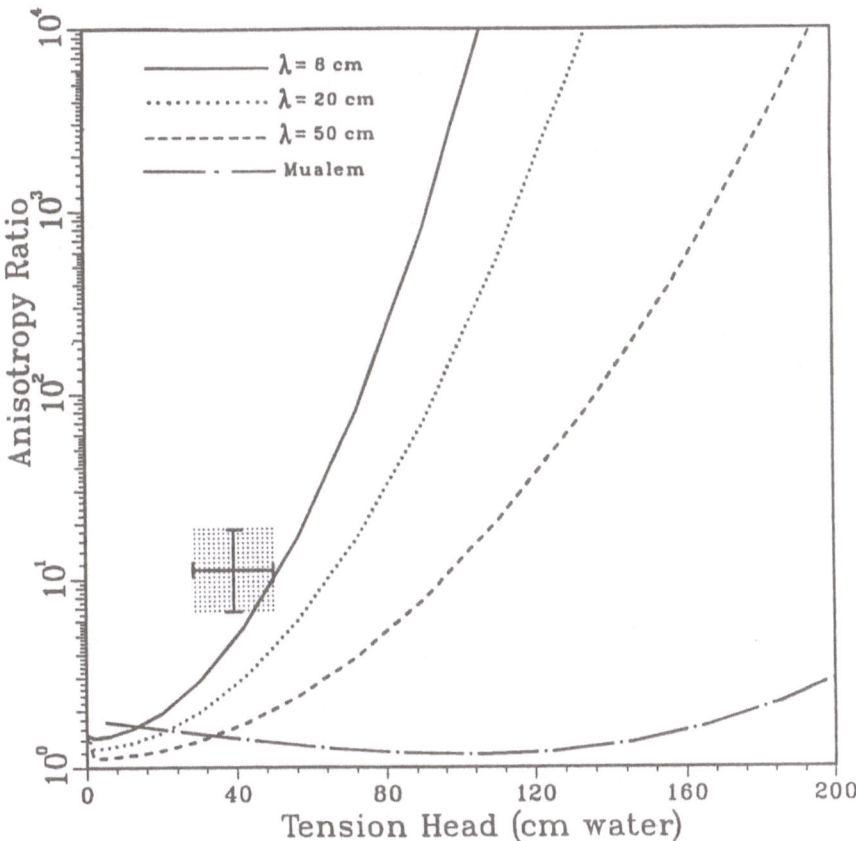

Figure 8. Anisotropy as a function of pressure head for Sevilleta dune sands calculated using the stochastic estimator of Yeh et al. (1985b) for three different correlation lengths. Shown for comparison is the result for Mualem's (1984) method with linear alpha–Ksat relationship. Shaded area shows mean in situ pressure heads and estimated anisotropies (+/– one standard deviation) from field tracer experiments.

2.2.2. Mathematical Formulation

In predictive modeling of flow through variably saturated media, one typically uses the Richards equation:

$$\nabla \cdot \left[\underline{\underline{K}}(\psi) \nabla (\psi + z) \right] = C(\psi) \frac{\partial \psi}{\partial t} \tag{9}$$

where $\underline{\underline{K}}(\psi)$ is the unsaturated hydraulic conductivity tensor, ψ is the capillary pressure, z is the vertical coordinate (increasing downward), $C(\psi)$ is a storage term, and t is time. The

626

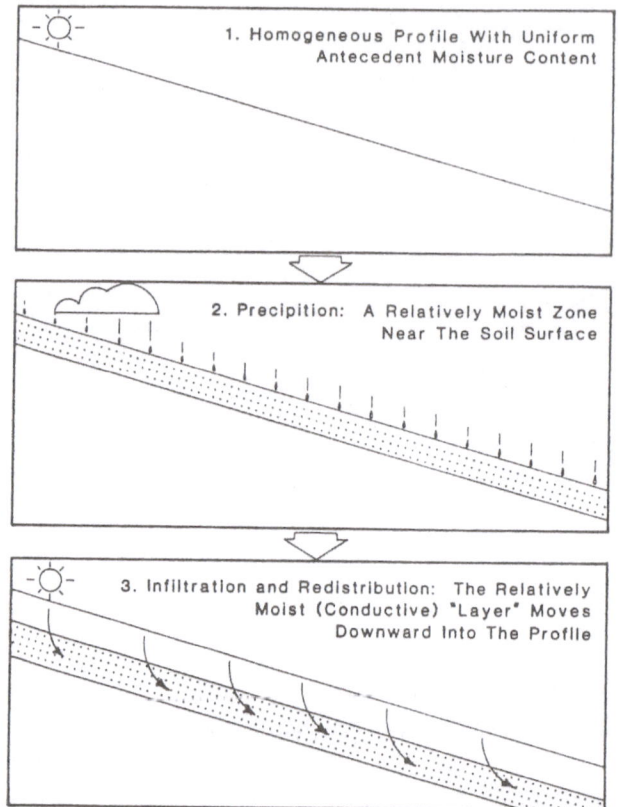

Figure 9. Schematic diagram showing how transient climatic inputs at the ground surface can lead to moisture–dependent, unsaturated hydraulically conductive heterogeneities in texturally uniform soils.

hydraulic parameters K and C are local properties which depend on the spatial coordinate. The Richards equation is a special case simplification of the more general equations for iso-thermal, multi–phase flow; which have been found to adequately approximate observed flow behavior for soils at intermediate saturations. For the case of two–dimensional flow with the (x,z) cartesian computational coordinates aligned with the principal directions of anisotropy and horizontally stratified media, Equation (9) can be rewritten:

$$\frac{\partial}{\partial x}\left[K_{xx}(\psi)\frac{\partial \psi}{\partial x}\right] + \frac{\partial}{\partial z}\left[K_{zz}(\psi)\frac{\partial}{\partial z}(\psi + z)\right] = C(\psi)\frac{\partial \psi}{\partial t} \tag{10}$$

The commonly employed approach of handling the directional unsaturated conductivities is to consider them as the product of the directional conductivity at saturation and a relative

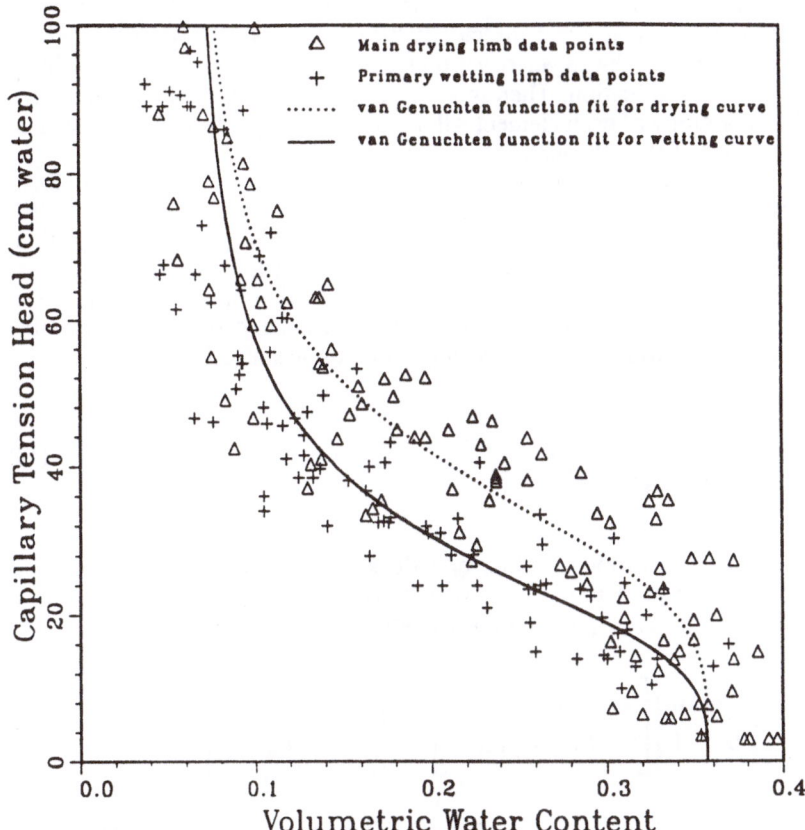

Figure 10. Hysteretic soil moisture characteristic curves obtained from hanging column analysis of 12 samples of Sevilleta dune sands.

conductivity function. The same relative conductivity is assumed to apply in all directions resulting in the following formulation:

$$\frac{\partial}{\partial x}\left[k_{xx}K_r(\psi)\frac{\partial \psi}{\partial x}\right] + \frac{\partial}{\partial z}\left[k_{zz}K_r(\psi)\frac{\partial}{\partial z}(\psi + z)\right] = C(\psi)\frac{\partial \psi}{\partial t} \tag{11}$$

where k_{ii} represents the saturated hydraulic conductivity in direction i. We see that this commonly used formulation leads to an anisotropy ratio ($K_{xx}(\psi)/K_{zz}(\psi)$) which is constant at all saturations.

Based on field and laboratory analyses, along with conclusions drawn in other theoretical and experimental work, it is apparent that the hydraulic anisotropy of a porous medium might vary significantly as the degree of saturation changes. Therefore, the commonly applied mathematical models (11) should be amended to allow for consideration of a variable, state–dependent anisotropy. The stochastic researchers (Yeh et al., 1985b; Mantoglou and Gelhar, 1987) suggest that anisotropy may depend on a number of parameters which characterize the media's spatial variability and at least two important variables: the mean capillary pressure and the angle of stratification relative to the hydraulic gradient (see

Equation 8). A first order sensitivity analysis (McCord, 1988) of Equation (8) indicates that for media with a mean hydraulic gradient directed vertically downward and with relatively gently dipping strata (dip $< 30°$), the anisotropy is much more sensitive to the mean capillary pressure than the angle of stratification. Therefore, in our implementation of variable anisotropy, the hydraulic anisotropy of each element will be made a simple function of the state (centroidal pressure head) of that element.

In the paper by Bear et al. (1987), it is suggested that the orientation of the principal directions of permeability tensor may rotate as the saturation varies. Figure 5b in Bear et al. (1987) show the major axis at one level of saturation becomes the minor axis at another saturation. Since the results of our tracer experiments offer no indication of such a flip–flop in principal axes and for the sake of simplicity we will not consider the rotation of axes issue.

As a first cut to this important modeling concern, rather than solving the standard variably saturated flow equation (Equation 11), the following equation is suggested:

$$\frac{\partial}{\partial x_i}\left[U_{ij}(\psi)k_{ij}K_r'(\psi)\frac{\partial(\psi + z)}{\partial x_j}\right] = C(\psi)\frac{\partial\psi}{\partial t} \tag{12}$$

where $U(\psi)$ is the variable anisotropy function, k_{ij} is the saturated conductivity tensor, and $K_r'(\psi)$ is a relative permeability function. For a two dimensional system, if the computational coordinates coincide with the principle directions of anisotropy, we define:

$$U_{ij}(\psi)k_{ij}K_r'(\psi) = \begin{bmatrix} A(\psi) & 0 \\ 0 & 1 \end{bmatrix}\begin{bmatrix} k_{11}K_r'(\psi) & 0 \\ 0 & k_{22}K_r'(\psi) \end{bmatrix} = \begin{bmatrix} K_{11}(\psi) & 0 \\ 0 & K_{22}(\psi) \end{bmatrix} \tag{13}$$

Therefore, $K_{11}(\psi) = k_{11}U(\psi)K_r'(\psi)$ and $K_{22}(\psi) = k_{22}K_r'(\psi)$. If the geometric mean of the directional conductivities for the variable anisotropy formulation (Equation 13) equals the geometric mean of the directional conductivities for the commonly employed formulation (Equation 11), then:

$$K_r'(\psi) = \frac{K_r(\psi)}{\sqrt{U(\psi)}} \tag{14}$$

This formulation (14) was found to be reasonable for the range of pressure heads simulated (McCord et al., 1991). The initial and boundary conditions which the variably saturated flow equation (Equation 12) is subject to are:

$$\psi(x_i, 0) = \psi_0(x_i)$$

$$\psi(x_i, t) = \Psi(x_i) \text{ on } B1$$

$$\frac{\partial\psi(x_i, t)}{\partial n_i} = \Delta_n \text{ on } B2a \tag{15}$$

and

$$V_i n_i(t) = V_n \quad \text{on } B2b$$

where ψ_0 is the initial head value, $B1$ is the portion of the flow boundary where ψ is prescribed as Ψ, n_i is the outward normal unit vector, ∇_n is the outward normal prescribed head gradient, V_i is the Darcy velocity vector, V_n is the outward normal prescribed velocity, and $B2$ is the portion of the flow boundary where the outward normal head gradient or Darcy velocity is specified.

The governing equation for two dimensional transport of a conservative solute species in a variably saturated soil is:

$$\frac{\partial}{\partial x_i}\left(D_{ij}\frac{\partial c}{\partial x_j}\right) - V_i\frac{\partial c}{\partial x_i} = \phi S_w\frac{\partial c}{\partial t} \tag{16}$$

where D_{ij} is the apparent hydrodynamic dispersion tensor, c is the solute concentration in the soil water, V_i is the Darcy velocity, ϕ is porosity of the media, and S_w is the water saturation. Darcy velocities required as input for the transport analysis are obtained from solution of the flow problem (Equations 12 and 15). The initial and boundary conditions associated with Equation 16 are:

$$c(x_i, 0) = C_0(x_i)$$

$$c(x_i, t) = \bar{C}(x_i) \quad \text{on } B1'$$

$$D_{ij}\frac{\partial c}{\partial x_j}n_i = q_c^D \quad \text{on } B2' \tag{17}$$

and

$$D_{ij}\frac{\partial c}{\partial x_j}n_i - V_i n_i c = q_c^T \quad \text{on } B3'$$

where $C_0(x_i)$ is the initial concentration field, $B1'$ is the portion of the boundary where c is prescribed as C, $B2'$ is the portion of the boundary where the dispersive mass flux is prescribed as q_c^D, and $B3'$ is the boundary through which the total mass flux is prescribed as q_c^T.

2.2.3. Numerical Modeling Approach

To model the flow and transport observed at our field site, three independent approaches were employed. In our first attempt at modeling the flow behavior observed at our field site, the Richards equation was solved, neglecting hysteresis and employing the commonly used approach of considering hydraulic anisotropy to be constant. For one set of simulations the soils are defined to be isotropic (anisotropy ratio is 1:1) and for another set, based on our soil characterization efforts, a constant anisotropy ratio of 1.5:1 is specified. In the second approach, the flow equation is once again solved, using a constant anisotropy ratio, but this time capillary hysteresis is accounted for. As discussed previously, consideration of hysteresis will enhance the development of moisture–dependent, hydraulically conductive heterogeneities in the porous media profile following infiltration and subsequent redistribution. And finally, in the third approach hysteresis is neglected, but an algorithm is included which allows anisotropy in the hydraulic conductivity of an element to vary as the capillary pressure in that element varies. Through this procedure, an attempt was made to identify the relative importance of the primary contributors to the lateral flow behavior observed in the field experiments.

All subsequent modeling utilized a finite element flow and transport code, VAMD2 (developed by Peter Huyakorn of HydroGeoLogic, Inc.), which was modified to allow for hysteresis and state–dependent anisotropy. Details on numerical techniques employed to solve the flow and transport equations can be found in papers by Huyakorn and Nilkuha (1979), Huyakorn et al. (1984), and the textbook by Huyakorn and Pinder (1983). The code required modification to allow for consideration of hysteresis and variable anisotropy. To implement hysteresis we added an algorithm which accounts for linear scanning curves (Hanks et al., 1969; Jaynes, 1984). Consideration of state–dependent anisotropy is achieved using the stochastic estimator developed by Yeh et al. (1985b), Equation (8) in this chapter.

In all simulations, the same finite element mesh (composed of rectangular elements) was used to discretize the spatial domain. By simply changing the Wx and Wy values in the finite element discretization, modeling a hillslope of any desired spatial dimension using the same mesh is possible. Figure 11a schematically depicts the finite element mesh of the soil profile modeled. Areas in the figure identified as Soil I are specified as regions of horizontally stratified media. Soil II represents media with a principal direction oriented parallel to the ground surface, since media stratifications were observed to roughly parallel the ground surface slope. Soil properties used are those determined for the Sevilleta sands from the field site.

For the flow problem, the boundary conditions applied to the system in all simulations were no flow on the sides, free drainage (specified unit gradient) at the base, and specified flux (precipitation/evaporation) at the ground surface. Figure 11b summarizes system initial and boundary conditions, including the temporal distribution of the surface flux rate. For the hysteretic simulations as an additional initial condition, all nodes are specified to be drying along the main drying limb of the hysteretic $\theta-\psi$ curve. In essence, the response of a hillslope profile (composed of Sevilleta dune sands) to a 5 cm precipitation event is simulated, followed by a period of evaporation and redistribution.

In all proceeding solute transport simulations, an antecedent solute concentration is specified to be zero everywhere except at one spot at a depth of about 30 cm near the top of the slope, which has a specified initial concentration equal to 1.0.

Two different spatial domains are modeled using the finite element discretization depicted in Figure 11a. One set of simulations models a fairly large region while the other set of simulations models flow and transport in a relatively small region. This is achieved by simply changing the nodal spacings depicted in Figure 11a. Since the primary objective of this modeling exercise is to determine the causes of the unsaturated, lateral flow behavior, and not necessarily to accurately model the dispersive transport mechanisms, particular attention was given to the general transport behavior exemplified by movement of the center of mass of the solute plume. To address a secondary objective, the impact of nodal spacing on the resulting solution of the nonlinear flow problem, simulations were performed with a fairly coarsely discretized spatial domain. This coarse grid allowed a significant amount of numerical dispersion to creep in on the subsequent modeling of the transport equation (Equation 16). However, comparing results from a coarsely gridded domain to results from a finely gridded domain allowed a demonstration that lateral flows induced by moisture–dependent, permeable heterogeneities will be most pronounced when nodal separation distances are relatively small (so that moisture content and consequent permeability contrasts are not attenuated significantly by the numerical averaging of element properties). Therefore for one group of simulations, Δx_4 through Δx_{17} (see Figure 11a) were specified the relatively small value of 12.5 cm, and Δy_{10} through Δy_{24} were assigned a value of 6.7 cm; the other Δx's and Δy's were scaled accordingly. The finite element mesh obtained from these assigned nodal spacings will henceforth be referred to as the fine grid. The depth and breadth of the domain modeled by the fine grid are 270 cm and 300 cm, respectively. For another set of simulations Δx_4 through

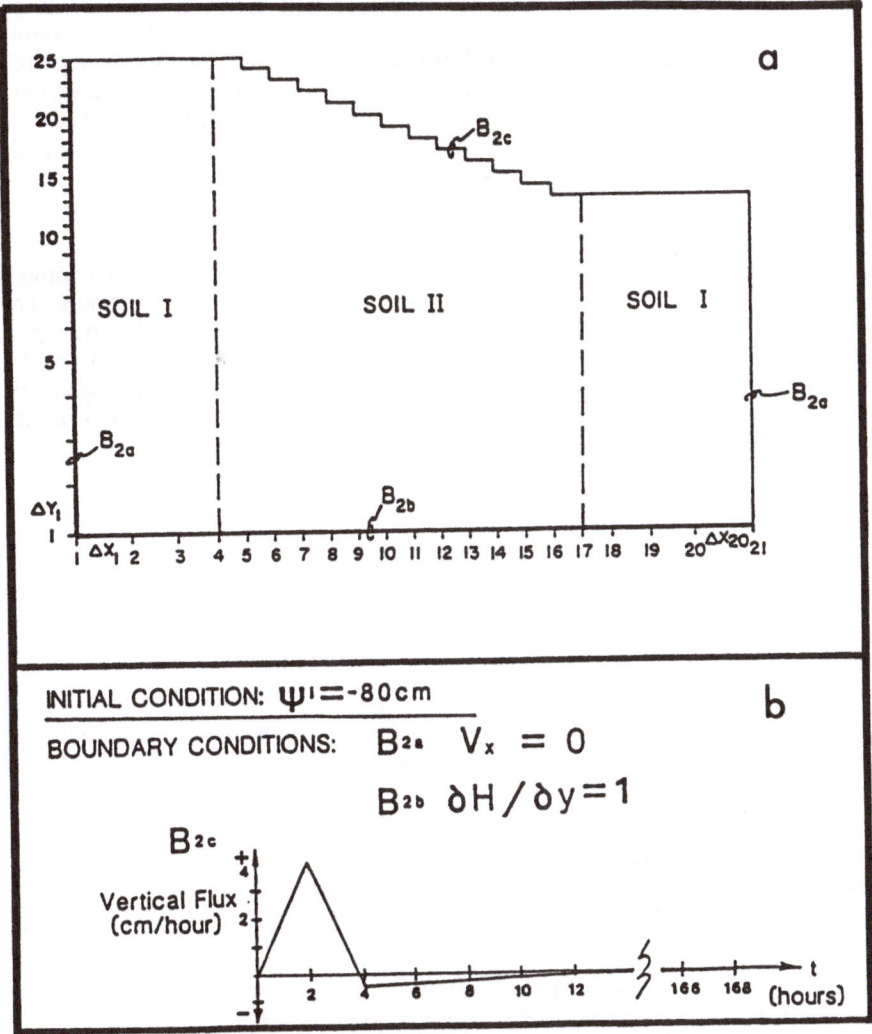

Figure 11. (a) Spatial domain, boundary conditions, and porous media modeled in the numerical simulations. Soil I and II have exactly the same properties, except Soil I is specified to be horizontally stratified while Soil II dips at an angle parallel to the ground surface slope. (b) Summary of initial and boundary conditions used in flow modeling.

Δx_{17} were specified to be 50 cm and Δy_{10} through Δy_{24} to be 20 cm. The resulting mesh simulates a profile 800 cm in depth and 1200 cm in length and will henceforth be referred to as the coarse grid. By simulating domains of two different spatial dimensions using the same finite element mesh, conclusions concerning the scale of the problem which can be addressed using a given level of computer (CPU) time can be drawn.

The three modeling approaches used are not implemented for both gridding options. A comparative analysis of the impact of nonhysteretic versus hysteretic media is performed only for the fine grid. The importance of nodal spacing in modeling field–scale behavior was

632

investigated by comparing simulation results for hysteretic media of constant anisotropy from the fine grid to results for the same media conditions from the coarse grid. The variable anisotropy approach was implemented only with the coarsely gridded domain. In a related document, McCord (1988) presents a technique for objectively determining an appropriate nodal spacing to use when modeling the transport equation so that numerical dispersion is minimized. Table 4 summarizes the porous media properties and grid conditions for the six simulations discussed.

2.2.4. Modeling Results

Figure 12 shows the velocity vector field along with the water content and total head contours for simulation Case II (fine grid, hysteretic, isotropic media; see Table 4) at $t = 3.5$ hours. The velocity vector field from each time–step is subsequently used as input for the solute transport simulations. The simulated solute plume contours at $t = 10$ hours for Cases III and IV, presented in Figure 13, allow one to compare solute transport in hysteretic and nonhysteretic media. Figure 14 provides a comparison of the fine grid versus coarse grid transport results,

Table 4. Summary of gridding options and porous media properties used in the numerical simulations.

4a. Simulation Case Summary

Case I.D.	Spatial Discretization	Media Characteristics
I	Fine Grid	Nonhysteretic and Isotropic
II	Fine Grid	Hysteretic and Isotropic
III	Fine Grid	Nonhysteretic, Const. Anisotropy = 1.5
IV	Fine Grid	Hysteretic, Const. Anisotropy = 1.5
V	Coarse Grid	Hysteretic, Const. Anisotropy = 1.5
VI	Coarse Grid	Nonhysteretic, Variable Anisotropy

4b. Summary of Porous Media Properties

∇ Flow Problem

k_{11} (minor axis of k_s tensor) for all simulations	= 36 cm/hour
Effective porosity	= 0.35
Residual water saturation	= 0.18
Variable anisotropy parameters $\sigma^2 f$	= 0.82
$\sigma^2 \alpha$	= 0.002
λ	= 5 cm
A	= 0.13 cm^{-1}
$\cos \beta$	= 1.0

∇ Transport Problem

Longitudinal dispersivity in hydrodynamic dispersion	= 8.00 cm
Transverse dispersivity in hydrodynamic dispersion	= 1.00 cm

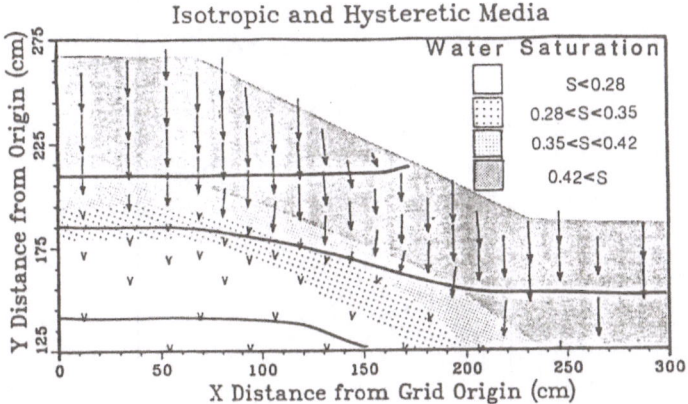

Figure 12. Velocity vector, total head, and water saturation fields for simulation Case II at t = 3.5 hours. Total head contour interval is 50 cm–water.

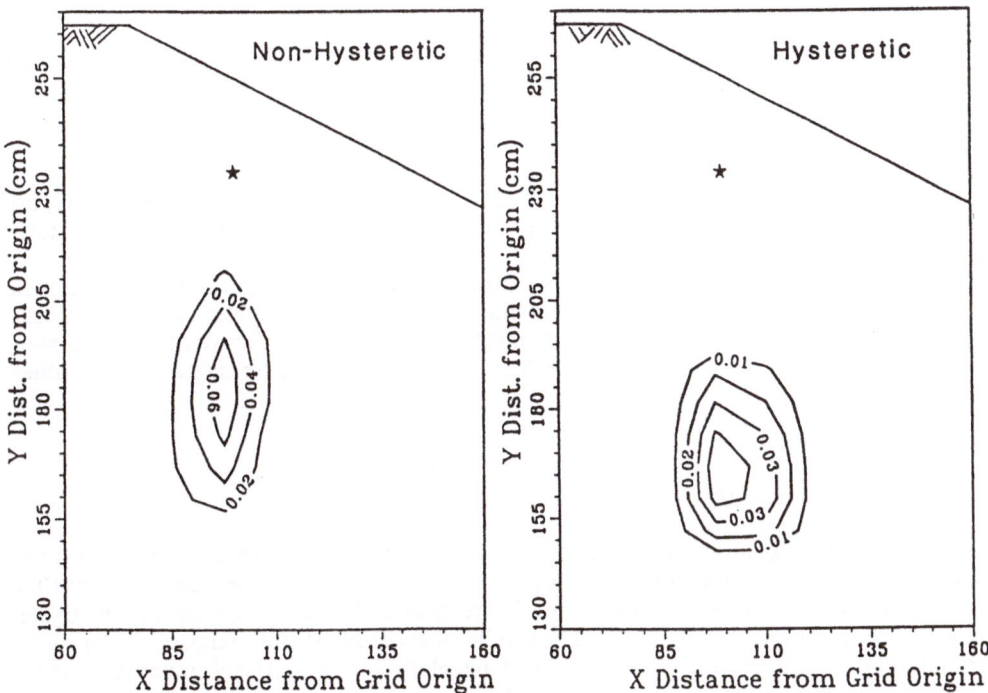

Figure 13. Simulated solute plumes at t = 12 hours for both nonhysteretic and hysteretic media with constant anisotropy ratio = 1.5:1 (simulation Cases III and IV). Stars represent solute source locations.

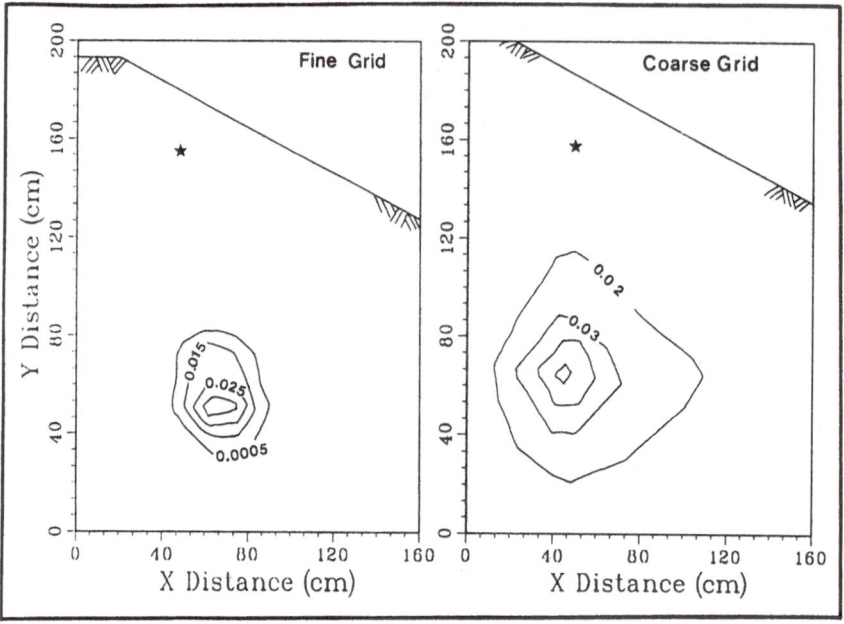

Figure 14. Simulated solute plumes at t = 20.6 hours for hysteretic media with constant anisotropy ratio = 1.5:1 for both fine and coarse gridding options (simulation Cases IV and V). Stars represent solute source locations.

given the same porous media properties (Case IV and V) at t = 72 hours. And finally, Figure 15 presents the solute plume development at various stages in time for the variable, state-dependent anisotropy simulation (Case VI).

One can estimate effective, macroscopic anisotropies of the media profile by applying the same simple Darcian considerations to the simulated plumes that were employed to estimate macroscopic anisotropies from the field tracer experiments (see Figure 5). Table 5 summarizes the results of such anisotropy estimates for each simulation case at various times.

3. Discussion of Results

3.1. Field and Laboratory Analyses

Determination of the bromide concentration in the soil water of the samples from the tracer site allows one to obtain the smooth solute plumes depicted in the profile views of Figures 3, 4, and 5. The relatively high percent of bromide recovered (Table 2) suggests that the figures reasonably depict the actual plumes. The profile views demonstrate a large downslope horizontal component to the soil–water flow, while the relatively flat total head contours indicate a relatively steady, near vertical downward hydraulic gradient below a 30 cm depth. The non-colinearity of the inferred flux and gradient vectors lead to the conclusion that the dune sand behaves as an anisotropic medium at the macroscopic scale. Simple Darcian calculations applied to the plume geometries indicate that the degree of anisotropy ranges from 5.1 to 16.9. However, two independent experimental procedures indicate that the anisotropy at complete

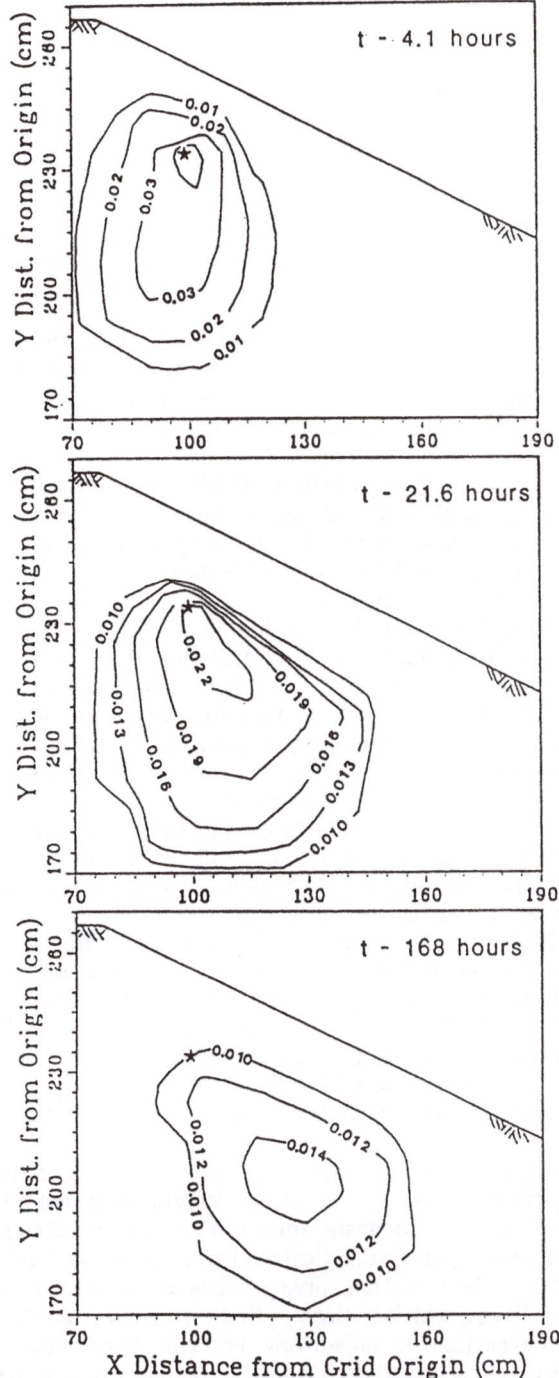

Figure 15. Solute plume development at various times for simulation of flow and transport in nonhysteretic, variably ansiotropic media (simulation Case VI). Stars represent solute source locations.

Table 5. Macroscopic anisotropy estimates at various times for selected numerical simulation cases.

Time Since Rainfall Began (hours)	Anisotropy Estimate for Simulation Cases			
	Case 3	Case 4	Case 5	Case 6
10	0.90	1.40	————	1.45
72	————	1.78	1.15	3.77
168	————	————	————	6.70

saturation is about 1.5. Clearly, the effective macroscopic anisotropy of the dune sand increases significantly upon desaturation. This field evidence substantially validates the qualitative result of Mualem (1984), Yeh et al. (1985b), Bear et al. (1987), and others who used analytical and experimental techniques to predict that anisotropy depends (at least in part) on the degree of saturation.

This result once again calls into question deterministic methods which are commonly employed in predictive numerical modeling of fluid movement in unsaturated media. Bear and Braester (1987) recommend that the typical approach of applying the same anisotropy ratio at all levels of saturation should not be used when modeling flow through anisotropic porous media. The grain size analyses (Figure 6) and the saturated conductivity measurements of the sands from our field site may lead one to assume that they closely approximate a homogeneous, isotropic medium. In this light, the results of our tracer experiments may lead one to question the prevalence (or perhaps, even the existence) of naturally deposited media which are hydraulically isotropic at all moisture contents. Therefore, following the lead of Bear and Braester (1987), we feel that the commonly used parameterization techniques for modeling variably saturated flow are inaccurate in most cases. Consequently, one may ask "what is the best estimator of the variable anisotropy function"?

One possible method for calculating the anisotropy of a layered medium is that proposed by Mualem (1984) (Equation 7). We applied this equation to hydraulic conductivity data for the dune sand, and the results are presented in Figure 7. Although anisotropy is predicted to increase greatly with decreasing pressure head, the anisotropy calculated by Mualem's method at in situ tensions (–30 to –70) is much less than that inferred from the tracer experiments at the field site. Although Mualem's (1984) method accounts for randomness in saturated conductivity, k_s, there are probably many other factors which influence hydraulic anisotropy. For example, to make the integrals (Equation 7) involved tractable, any variability in α can only occur through functional relationships to k_s. However, it is likely that α is better described by statistical cross–correlations to k_s, or perhaps α may vary statistically independent from k_s. In addition Mualem's approach fails to account for any spatial correlation structure in the parameters.

As previously shown, stochastic methods which account for spatial correlation structure can also be used to characterize hydraulic anisotropy in variably saturated, layered media (Yeh et al., 1985b). As depicted in Figure 8, the stochastic approach predicts that anisotropy increases much more rapidly than Mualem's (1984) statistical method would predict, particularly for smaller correlation lengths (λ). The $\lambda = 8$ cm curve predicts an anisotropy ratio at the in situ pressure head ($-70 < \psi < -30$ cm) which is close to that estimated from the field tracer data. It is quite obvious, however, that the anisotropy ratio calculated is extremely sensitive to the correlation length. A previous investigation into the spatial variability in saturated conductivity of the dune sands at the site (Leavitt, 1986) suggests that correlation lengths are on the order of the length of the sample rings (5 cm). Although the anisotropy predicted using the stochastic result of Yeh et al. (1985) with $\lambda = 8$ cm is within the range of values estimated from the tracer experiments, the difficulty of estimating some of the required input parameters based on field and laboratory testing leads to considerable uncertainty in applying this result (Equation 8) to the soils from this particular site.

As discussed, in addition to the impacts of heterogeneities, hysteresis in the soil moisture characteristic would tend to enhance a macroscopic anisotropic behavior, even in relatively homogeneous, uniform media as schematically presented in Figure 9. Figure 10 demonstrates that the dune sands from this field site are highly hysteretic which would be expected, considering the granular nature of the soils.

3.2. Numerical Simulations

Figure 12, which presents the velocity vector, moisture content, and total head fields for simulation Case II at time = 3.5 hours, demonstrates that the direction of the velocity vectors do diverge slightly from the direction of the total head gradient, particularly beneath the hillslope in the moistened zone behind the wetting front. This supports the assertion that moisture-dependent permeable heterogeneities can cause even homogeneous, uniform soils to behave anisotropically when viewed at the field–scale. Note that near the wetting front, where capillary gradients dominate, velocity vectors have a strong component into (normal to) the hillslope. This type of behavior is not unexpected and was observed in all simulations.

Figures 13 through 15 show simulated solute plumes for a variety of gridding and material property options. Perhaps the most obvious (and most important) conclusion which can be drawn is that the simulation which accounts for state–dependent anisotropy does a much better job at predicting transport behavior similar to that which we observed in the field experiments.

Although the visual presentation of the plumes in Figures 13 through 15 dramatically highlights the impact of considering variable anisotropy, quantitative analyses using the locations of the plume centroids provide a means of evaluating the subtle differences evidenced in the other plumes. Recalling the discussion on spatial discretization options presented in section 2.2.3., only plume centroids are used in this comparative analysis due to the numerical dispersion induced in the simulations of the coarsely gridded (large scale) domain.

Table 5 presents macroscopic media anisotropies estimated from simulated plume geometries, using the procedure discussed in section 2.1.2 and Figure 5. Comparison of simulation Cases III and IV at $t = 10$ hours suggest that hysteresis does slightly enhance macroscopic anisotropic behavior, with the macroscopic anisotropy of the hysteretic media about 50 percent higher than that for the nonhysteretic media. Comparison of Cases IV and V clearly demonstrates that macroscopic anisotropy arising from moistured–dependent permeable heterogeneities can be modeled only by using a finely gridded spatial discretization, with the fine grid anisotropy (at $t = 72$ hours) more than 50 percent higher than that for the coarsely gridded system. An important result (which has implications concerning the macroscopic anisotropies estimated from the field tracer experiments) is the fact that all of the macroscopic anisotropies at short times ($t = 10$ hours) are less than the local anisotropy specified as a material property (local anisotropy was specified to be 1.5). This should be expected, since this simple anisotropy estimation procedure assumes unit downward hydraulic gradient. Whereas in reality, at short times capillary gradients will dominate and there will be a significant flow component normal to the ground surface slope. Therefore, macroscopic anisotropies estimated from the field tracer experiments using this procedure may be slightly on the low side. And finally, Table 5 clearly shows that only for Case VI (the state–dependent anisotropy simulation) do the estimated macroscopic anisotropies come close to the values measured in this field experiment.

3.3. Practical Implications

The first practical implication of these results is that state–dependent anisotropy, a media property which has been proposed in recent theoretical studies, is indeed a real and significant phenomenon in some (if not all) natural landscapes at field–scales. Given the potential for orders of magnitude variation in hydraulic anisotropy, this can have important consequences on the understanding of any number of processes, i.e., from groundwater recharge and watershed hydrology to erosion and gully formation, and from hazardous waste containment by burial to stability of natural and engineered slopes.

A second practical implication lies in mathematical modeling of variable, saturated flow and transport. If variable anisotropy is an actuality, then widely used quantitative modeling approaches are deficient in this respect. Analytical solutions for unsaturated flow problems may be even more difficult to obtain and numerical flow models may require modifications to allow for consideration of a state-dependent anisotropy.

These results suggest the need for additional research on this topic. It is imperative that more large–scale field and/or laboratory experiments be designed and performed to determine if this phenomenon is important for a wide variety of soil types. The methods used to characterize the hydraulic properties of porous media should also be examined, and how local (laboratory scale) measurements should be incorporated into large–scale models.

4. Conclusions

Detailed soil–water tracer experiments conducted on a sand dune indicate that significant horizontal downslope flow components develop despite the presence of a near vertical downward, hydraulic gradient. This suggests that at in situ pressure heads (–30 to –70 cm), the dune sands behave as a highly anisotropic medium. Field and laboratory permeameter analyses show that the sands are nearly isotropic at complete saturation. These field and laboratory measurements provide strong supporting evidence for the theoretical (Mualem, 1984; Yeh et al., 1985b,c) and experimental (Bear et al., 1987; Stephens and Heermann, 1988) postulations that anisotropy in hydraulic conductivity of unsaturated, heterogeneous, porous media varies as the state (capillary pressure, saturation, etc.) of the media varies. Simple arguments developed herein show that moisture content heterogeneities in a texturally uniform, homogeneous medium can cause the medium to behave anisotropically with respect to fluid flow. Hysteresis in the soil moisture characteristics should enhance the macroscopic anisotropy of a texturally homogeneous medium. Three different approaches were taken to modeling the flow and transport observed at the field site: constant anisotropy and no hysteresis, constant anisotropy with hysteresis, and state dependent anisotropy without hysteresis. Results of the numerical simulations demonstrate that hysteresis enhanced, moisture–dependent, permeable heterogeneities do contribute slightly to the macroscopic anisotropy evidenced at the field site. These contributions are most pronounced when the spatial domain is finely discretized. However, only the simulation which accounted for state-dependent, variable anisotropy predicted downslope flow and solute transport to the same degree which was observed in the field experiments. These significant lateral flows were modeled using relatively large nodal spacings which suggests that the variable anisotropy approach is well suited to modeling field–scale problems.

In summary, the results of this study indicate that state–dependent anisotropy in hydraulic conductivity and, to a much smaller extent, hysteresis in the soil–moisture characteristics are important phenomena which can have dramatic implications on our understanding and mathematical modeling of field–scale flows in variably saturated media.

5. Acknowledgments

We wish to acknowledge the United States Geological Survey, U.S. Department of the Interior (Matching Grant No. 1-4-23660) and the New Mexico Water Resources Research Institute for their financial support of the work performed. We would like to thank the U.S. Fish and Wildlife Service and the Nature Conservancy for providing us with access to the Sevilleta National Wildlife Refuge site for our field experiments. The first author also wishes to recognize valuable discussions with Dr. T.-C. Jim Yeh of the University of Arizona, in addition to dialogue with Drs. Allan Gutjahr and Bob Bretz of New Mexico Tech, all of whom have contributed to the ideas expressed in this chapter.

6. References

Bakr, A. A., L. W. Gelhar, A. L. Gutjahr, and R. J. MacMillan: 1978, 'Stochastic Analysis of Spatial Variability in Subsurface Flows; 1. Comparison of One- and Three-dimensional Flows,' *Water Resour. Research* **14**(2), 233–271.

Bear, J., C. Braester, and P. Menier: 1987, 'Effective and Relative Permeabilities of Anisotropic Porous Media,' *Transport in Porous Media* **2**, 301–316.

Bear, J., and C. Braester: 1987, Comment on 'A Three-dimensional Finite Element Model for Simulating Water Flow in Variably Saturated Porous Media,' by P. S. Huyakorn et al., *Water Resour. Res.* **23**(8), 1705–1706.

Bresler, E., W. D. Kemper, and R. J. Hanks: 1969, 'Infiltration, Redistribution, and Subsequent Evaporation of Water From Soil as Affected by Wetting and Hysteresis,' *SSSA Proc.* **33**, 832–840.

Gelhar, L. W., and C. L. Axness: 1983, 'Three-dimensional Stochastic Analysis of Macrodispersion in Aquifers,' *Water Resour. Res.* **19**(1), 161–180.

Hanks, R. J., A. Klute, and E. Bresler: 1969, 'A Numeric Method for Estimating Infiltration Redistribution, Drainage, and Evaporation of Water From Soil,' *Water Resour. Res.* **5**(5), 1064–1069.

Huyakorn, P. S., S. D. Thomas, and B. M. Thompson: 1984, 'Techniques for Making Finite Elements Competitive in Modeling Flow in Variably Saturated Porous Media,' *Water Resour. Res.* **20**(8), 1099–1115.

Huyakorn, P. S., and G. F. Pinder: 1983, *Computational Methods in Subsurface Flow*, Academic Press, 473 pp.

Huyakorn, P. S., and K. Nilkuha: 1979, 'Solution of Transient Transport Equation Using an Upstream Finite Element Scheme,' *Applied Mathematical Modeling* **3**, 7–17.

Jaynes, D. B.: 1984, 'A Comparison of Soil-Water Hysteresis Models,' *Journal of Hydrology* **75**, 287–299.

Leavitt, M.: 1986, 'Spatial Variability of Fluvial and Eolian Soils at the Sevilleta NWR,' Unpublished Master Sci. Indep. Study, NMIMT, Socorro, NM.

Mantoglou, A., and L. W. Gelhar: 1987, 'Effective Hydraulic Conductivities of Transient Unsaturated Flow in Stratified Soils,' *WRR* **23**(1), 57–67.

McCord, J. T.: 1988, 'Hysteresis and State-Dependent Anisotropy in Variably Saturated Flow: Field, Laboratory, and Numerical Modeling Investigation,' Ph.D. Dissertation, New Mexico Inst. of Mining and Technology, Socorro, NM, 354 pp.

McCord, J. T., and D. B. Stephens: 1987, 'Lateral Moisture Movement on Sandy Hillslope in the Apparent Absence of an Impeding Layer,' *Journal of Hydrological Processes* **1**(3), 225–228.

McCord, J. T., D. B. Stephens, and J. L. Wilson: 1991, 'The Importance of Hysteresis and State-dependent Anisotropy in Modeling Hillslope Hydrologic Processes,' Accepted for publication in *Water Resour. Res.*

Mualem, Y.: 1984, 'Anisotropy of Unsaturated Soils,' *SSSA Journal* **48**, 505–509.

Rubin, J.: 1967, 'Numerical Method for Analyzing Hysteresis Affected Post-Infiltration Redistribution of Soil Moisture,' *SSSA Proc.* **31**, 13–20.

Stephens, D. B., and S. Heermann: 1988, 'Dependence of Anisotropy on Saturation in a Stratified Sand,' *Water Resour. Res.* **24**(5), 770–778.

van Genuchten, M. T.: 1980, 'A Closed-form Equation for Predicted the Hydraulic Conductivity of Unsaturated Soils,' *SSSA Journ.* **44**, 892–898.

Yeh, T.-C. J., L. W. Gelhar, and A. L. Gutjahr: 1985a, 'Stochastic Analysis of Unsaturated Flow in Heterogeneous Soils 1. Statistically Isotropic Media,' *Water Resour. Res.* **21**, 447–456.

Yeh, T.-C. J., L. W. Gelhar, and A. L. Gutjahr: 1985b, 'Stochastic Analysis of Unsaturated Flow in Heterogeneous Soils 2. Statistically Anisotropic Media With Variable α,' *Water Resour. Res.* **21**, 457–464.

Yeh, T.-C. J., L. W. Gelhar, and A. L. Gutjahr: 1985c, 'Stochastic Analysis of Unsaturated Flow in Heterogeneous Soils 3. Observations and Applications,' *Water Resour. Res.* **21**, 465–471.

Zaslavsky, D., and G. Sinai: 1981a, 'Surface Hydrology: I. Explanation of Phenomena,' Proc. Paper 15958, *Journ. of Hydraulics Div.*, ASCE, **107**(HY1), 1–16.

Zaslavsky, D., and G. Sinai: 1981b, 'Surface Hydrology: III. Causes of Lateral Flow,' Proc. Paper 15960, *Journ. of Hydraulics Div.*, ASCE, **107**(HY1), 37–52.

Zaslavsky, D., and G. Sinai: 1981c, 'Surface Hydrology: IV. Flow in Sloping, Layered Soil,' Proc. Paper 15961, *Journ. of Hydraulics Div.*, ASCE, **107**(HY1), 53–64.

Appendix – Other Presentations Given at the NATO ASI

For more information on these presentations, the reader should contact the authors at the addresses given below.

Effects of Spatial Variability of Soil Type and Land Use on the Estimation of Runoff Production

Baldaccaro Bacchi, Marco Mancini, and **Renzo Rosso**
Institute of Hydraulics, Politecnico di Milano, 32 Leonardo da Vinci, 20133–I Milan, Italy

Abstract. Distributed and lumped approaches to the estimation of runoff production are compared relying on SCS–CN model calibration from non–parametric data on soil type and land use for elementary areas of a basin. The analyzed basin, located in the South of Italy, has an extension of 412 sq km and has been discretized in square cells of 200 x 200 m. Fluctuations and mutual relationship between cell Curve Numbers are analyzed in space.

A relevant aspect of spatial distribution is concerned with autocorrelation in space, which has shown a high degree of persistence at different spatial lag that highlight the dependence of absorption parameter from the "history" of a particular basin. Therefore, one could observe that it is quite arbitrary to assume a synthetic field of absorption parameter for a basin area, which is hypothesized to be spatial independent, or to have a preassigned correlation structure, as proposed by some recent simulation studies (respectively, Matias et al., 1988; Wood et al., 1987).

The difference between lumped and distributed surface runoff volume estimates is explained by the effects of changing the scale of the CN parameter, from the most detailed grid discretization (200 x 200 m) to the global approach, where the mean CN value is assumed to be representative of the whole basin. The changing scale effects on the CN field, obtained using a moving averaged process, have been investigated and highlight: a) variance and mutual relationship of CN parameter on the basin surface, b) surface runoff production and its statistics, and c) the descending variability of surface runoff produced by a combined effect of increasing storm event and changing the CN scale.

A moderate bias is found to be provided by lumped estimates of the expected value of runoff production at the basin scale, such bias rapidly decreasing with increasing storm depth. Nevertheless, mean square error between lumped and distributed estimates of runoff production may still involve a large variability of surface runoff in space as it is produced by the local response of contributing areas. The same pattern is also observed assuming that the variability of the CN field estimates could be described with a Beta distribution. In this case the probability density function of the runoff production is analytically derived and used to: a) fit the CN frequency curve values, and b) evaluate the mean surface runoff at different scale of GIS detail.

Adequate Derivation of Flood Frequencies in Karst River Basins

Ertugrul Benzeden
Dokuz Eylul University, Faculty of Engineering and Arch., Bornova, Izmir, Turkey

Abstract. The frequency distribution models with physically meaningful lower bounds for the annual peak flow sequences of 21 streamgaging stations in seven karstic river basins of Turkey are investigated. It is taken into account that goodness of fit is a necessary but not a sufficient condition for acceptance of a certain model and it should be supported by hydrologic justification, if possible.

643

D. S. Bowles and P. E. O'Connell (eds.), Recent Advances in the Modeling of Hydrologic Systems, 643–660.
© 1991 *Kluwer Academic Publishers.*

644

Furthermore, annual peak flows observed at seven representative streamgaging stations are separated into their surface and base flow components by introducing a graphical procedure based on annual hydrographs of daily flows. Then, frequency analyses with eight theoretical distribution models are carried out on the surface and base flow components in order to infer how the base flow affects sample statistics and frequency distributions.

Computational results revealed that: (a) Karst media and the base flow components, consequently, considerably reduce the coefficient of variation, however, their impacts on the frequency distribution model is not so evident, (b) A physical meaning can be attributed to the lower bounds when they are compared with the average and/or minimum karst spring estimates. Hence, a three parameter distribution model may have a physical justification and should be preferred in karst river basins, especially if its goodness of fit is also comparable, (c) The base flow components may be assumed almost independent of the surface flow components and of the recurrence intervals, and (d) It seems quite possible to draw more reliable information and to derive valuable conclusions about floods in karst river basins if the flow components can be separated more precisely and if they are treated on the basis of a bivariate distribution of two independent variables. Any reasonable solution to this problem will accelerate the regional analysis studies in karst river basins.

Hydrological Modeling of Acidified Canadian Watersheds

A. G. Bobba and **D. C. L. Lam**
Environment Canada, National Water Research Institute, Burlington, Ontario L7R 4A6, Canada

Abstract. Hydrology plays an important role in the acidification of watersheds in North America by airborne pollutants such as the wet and dry deposition of sulfate and nitrate. The pathways of anions through the soil layers, the contact times with soil matrix, the snow accumulation and melting sequences, and the groundwater flows are all important hydrological factors that influence the acidification processes. A geochemical model including organic acids has been coupled with a hydrological model and the combined model has been called Turkey Mersey Watershed Acidification Model (TMWAM). The model has been applied to different acidification watersheds in Canada.

As an example, Turkey lakes results are presented on the calibration and confirmation of the model using the data from these watersheds. Particular emphasis is placed on the hydrological factors and their relationships to the varying degrees of acidification observed at different stations. Preliminary results of linking the hydrological and hydrogeochemical models are discussed in the context of model application as part of the national acidification assessment efforts in Canada.

Operational Flood Forecasting on the River Meuse, Belgium

F. P. De Troch, P. A. Troch, and **D. Van Erdeghem**
State University Ghent, Laboratory of Hydrology, Coupure Links 653, B9000 Ghent, Belgium

Abstract. The paper describes the first phase of the development of a real-time flood forecasting model by the Laboratory of Hydrology of the State University Ghent, in cooperation with the Service for Hydrological Studies of the Ministry of Public Works, Belgium. This forecasting model can be considered as a non-structural (software) measure enhancing structural (hardware) measures in combatting high economic damages caused by flooding of the river Meuse and its tributaries.

The river catchment is divided in different subcatchments and, in the described model, these subcatchments are represented by different moduli. Each modulus consists of a mathematical model simulating the dynamic relationship between the hydrological (input/output) variables for the considered subcatchment (section 2). Input data are collected by a data monitoring system operating in real-time and in teletransmission. The different moduli are linked by an hydraulic flood routing model based on the governing equations of unsteady water flow in the river channels (section 3).

In the present stage of development, the model is able to forecast discharges and water levels at Liège with a lead time of 8–10 hours. Future developments, including extension of the forecasting horizon and real-time management of the weirs, are indicated.

Unit Hydrograph Revisited Through Differencing and Deconvolution on Multi–Event Data Set: The FDTF Approach

D. Duband[1], I. Nalbantis[2], Ch. Obled[2], J. Y. Rodriguez[2], and P. Tourasse[1]
[1]Division Technique Regionale (EDF), 37 rue Diderot, BP 41, 38040 Grenoble, France
[2]Institut de Mécanique de Grenoble, BP 53 X, 38041 Grenoble Cedex, France

Abstract. A new approach is proposed and illustrated for the identification of rainfall–runoff relationships on small mountainous watersheds. The underlying model is basically the same as in the original Unit Hydrograph (U.H.) method, i.e., the input rainfall, called the Raw Precipitation (RP), is first split into losses and an Excess Precipitation (ER), efficient for rapid runoff. This ER is then routed linearly through a UH or Transfer function (TF) to provide a catchment outflow (Q).

Some uncertainty in this classical UH method comes from the a priori choice of a model for the loss function. Based either on theoretical or physical considerations from laboratory or plot experiments, it cannot be calibrated separately from the TF, thus allowing uncontrolled interactions between one part of the model and the other.

One major goal of our new approach is to provide, simultaneously, an estimate of the TF or UH, together with an identified EPs series, from a multi–event set of floods. Any loss model may then be calibrated separately by fitting appropriate relationships between observed RPs and identified EPs series.

Evaluating the Performance of Rainfall–Runoff Models at Extreme Events

Thomas A. Fontaine
Oak Ridge National Laboratory, Environmental Sciences Division, P.O. Box 2008, Oak Ridge, Tennessee 37831-6038 U.S.A.

Abstract. Probable Maximum Flood (PMF) estimates are usually developed using a rainfall–runoff model (or "runoff model") in a deterministic manner with no evaluation of error or uncertainty. Runoff models have also been used in recent research of methods for predicting the exceedance probabilities of floods up to the PMF. In either application a critical but unknown issue is the accuracy of the simulated extreme flood event. This issue is being investigated by applying several runoff models to well-documented extreme events. Differences between the simulated and observed flood hydrographs can be caused by errors in the model structure, calibration of parameters, input data (e.g., precipitation, potential evapotranspiration, antecedent conditions, and soil characteristics), and discharge measurements of the observed event. With careful selection of the runoff models and flood events, and careful calibration and verification of the models, it is possible to measure both the overall error in the simulated flood and the relative contribution from each potential source of error. This information could be used to improve existing runoff models or to develop new models specifically designed for simulating extreme floods. The results will also include an evaluation of the existing data collection networks that provide precipitation, evaporation, discharge, and other data crucial for accurate modeling of extreme floods.

Hydrologic Simulation of Forested Catchments using the BROOK–Model

Felix Forster and **Hans M. Keller**
Swiss Federal Institute of Forestry Research, CH-8903 Birmensdorf, Switzerland

Abstract. BROOK is a descriptive, physically–based water yield model for small areas. It was designed to study changes in streamflow from forests that are likely to occur because of changes in cover type caused by forest management. The model operates with daily intervals using precipitation and temperature as input.

BROOK tries to include each important hydrologic process and to describe it with a physically realistic equation. Parameters were defined as physically measurable properties of a basin. The model does not allow for the simulation of instantaneous flood peaks. A modified version of the model allows subdivision of the catchment in subcatchments. Each subcatchment has its own parameter set.

First, the test results of two different catchments (in terms of forest cover and soil conditions) are discussed. Second, the results of simulations under contrasting conditions (fully forested and clearcut) using identical weather conditions in both catchments are shown.

This simulation experiment shows that a forest stand has a direct link to streamflow behavior, particularly where deep-rooted soils with ample storage capacity prevail. Impervious, shallow soils with little soil water storage capacity show less differences between fully forested and clearcut conditions.

Methods of Artificial Intelligence for the Control of Urban Sewer Systems

L. Fuchs and **A. Neumann**
Institut für Wasserwirtschaft, University of Hannover, Appelstra 9A, D-3000, Hannover 1, Federal Republic of Germany

Abstract. To reduce combined sewer overflows, to equalize the runoff to treatment facilities, and to reduce energy costs, on–line and real–time control of a sewer system is a possible solution. To find out the best control strategy, methods of artificial intelligence can be used besides numerical optimization. The general working algorithm of a learning production system is described and explained by simple examples. For the sewer network of the city of Bremen/West Germany with a catchment area of about 1000 ha, 3500 inlet areas, detention basins, and pumping stations, a learning production system was developed and tested in a simulated on–line mode. The control strategies of the learning production system for some events were compared with those of numerical optimization and existing control rules. It is shown that the learning production system leads to control strategies that are comparable to those derived from numerical optimization, and they cause similar energy costs or overflow volumes. The advantages of the learning production system lie in high flexibility, transparent control decisions, and fast computation.

Surface Runoff Modeling Through Integration of Remotely Sensed Data in a Geographic Information System

Mark S. Gregory
Oklahoma State University, 107 Thatcher Hall, Stillwater, Oklahoma 74078 U.S.A.

Abstract. A methodology is presented for modeling surface runoff by integrating remotely sensed data with ancillary data in a geographic information system. The model is based upon characteristics of previously developed lumped models, now implemented in a more distributed manner within a GIS framework.

Three basic types or layers of information were integrated within the GIS for model development. Hydrologically significant land cover types were produced by analysis of Landsat Thematic Mapper digital satellite data. Soil mapping units with associated attributes were digitized from Soil Conservation Service Soil Surveys. Digital elevation data was incorporated into the GIS to derive slope angle and slope aspect (direction) information. All data were gathered at the 30 meter resolution of the TM digital satellite data.

Key to model development was the utilization of the digital elevation data to automatically derive catchment boundaries and flow networks. The resulting 'connection' matrix of surface flow directions or paths allowed computation for any location in the study area of the up-stream drainage patterns and area and the distance between the selected location and any or all other locations. The distance measures were combined with the land cover information and empirical formulas to estimate runoff velocity and produce travel times within the matrix. Runoff volumes and travel times were then summed across the connection matrix to produce a hydrograph.

The result is a model, which for a given rainfall event, is capable of estimating a runoff hydrograph for any selected location within the defined study area. Model results compared favorably with observed hydrographs from several precipitation events on a 16,000 hectare watershed in south-central Oklahoma, U.S.A. Model inputs are minimal and can easily be updated with remote sensing and GIS techniques. The model is highly automated to allow easy use by planners and managers concerned with runoff estimates.

Transfer of Hydrologic Information in Karstic and Non-Karstic River Basins

Nilgun Harmancioglu
1407 Sokak No. 4/17, Alsancak, Izmir 35220, Turkey

Abstract. Rivers, fed by karstified limestones, have basins characterized by a significant degree of heterogeneity with regard to their geology, topography, climate, and hydrology. As a result of the diversity of areal and temporal distributions of such characteristics, hydrologic information transfer between runoff-runoff and precipitation-runoff processes reflects specific characteristics in karstic basins. Thus, this phenomenon needs to be investigated by more advanced methods than the classical ones in such basins where karstic springs make a major contribution to runoff.

This study uses two types of measures to find how much information can be transferred between points along a river and how much information regression analyses actually transfer. The concept of "transferable information" is defined on the basis of information theory to represent an upper limit for "transferred information" measured by regression. Both measures are analyzed for information transfer between runoff–runoff and precipitation–runoff processes in karstic and non–karstic basins. The results show within–the–year variation of transferred and transferable information to reflect how each basin responds to seasonally varying inputs. Effects of the structural properties of runoff and precipitation processes are also accounted for in information transfer, together with other factors such as the selection of a particular regression function, or that of a particular probability distribution function for the variables involved.

The above–mentioned methods are demonstrated in two case studies in Turkey, one with river points fed by karstic contributions and the other with negligible karstic springs effluents. The measures of transferable and transferred information serve here to identify the characteristics of runoff yields from karstified and non–karstic basins. The results of such investigations may also be used to guide future attempts at modeling runoff in the selected basins since the basic properties of the recharge systems are recognized and input/output relations are evaluated.

Modeling Snowmelt Runoff in an Arctic Watershed

Larry D. Hinzman and **Douglas L. Kane**
Water Research Center/Institute of Northern Engineering, University of Alaska, Fairbanks, Alaska 99775–1760 U.S.A.

Abstract. The Swedish HBV model has been successfully adapted to model runoff during snowmelt periods in subarctic Alaska and, more recently, in the Alaskan Arctic. The soil moisture routine of the model is varied to account for lower infiltration rates and storage capacities in frozen soil. Arctic basins are characterized by snow accumulation throughout the winter and a brief, yet intense snowmelt runoff event in late May or early June. The spring melt is usually the dominant hydrologic event of the year.

The test basin was Imnavait Creek, a zero–order watershed on the North Slope of Alaska that has sustained flow following snowmelt until freeze–up. The model was calibrated and tested with meteorological and streamflow data collected in 1985, 1986, 1987, and 1988. Permafrost, which underlies the entire basin, severely limits infiltration and soil storage and thus the role of the subsurface system is less significant than that of a more temperate region. The active layer thickness ranges from 25 to 60 cm at the end of the thaw season, however during the spring melt, the downslope subsurface flow occurs within the top 10 cm of a highly organic soil. Although the organic soil is quite desiccated at the end of winter, only about 1.5 cm of water is stored in the soil before water begins to runoff.

Portuguese Experience with NWSRFS

João Nuno Hipólito
Instituto Superior Técnico, Dep. Engenharia Civil, Av. Rovisco Pais, 1000 Lisboa, Portugal

Abstract. A succinct review of the evolution of the deterministic mathematical modeling of the physical processes that can be integrated into the hydrologic cycle is presented. The various types of models that can simulate the runoff portion of the referred cycle are classified and briefly connected to the logistics and peripheric scientific and technologic domains whose consideration is now mandatory.

The NWSRFS4 model, used by the United States National Weather Service River Forecast Centers, and its application to four portuguese river basins are described.

For these applications it was necessary to build computer subsystems for data management, parameter optimization, sensitivity analysis, and graphical representation which are also described.

Modifications were introduced in the modeling processes to overcome some anomalous results. The modified processes include those referring to surface retention, infiltration intensity control, distribution of the infiltrated water, percolation to the water table, and distribution of the percolated water.

The new system, comprehending the modified model and the referred subsystems, was named NWSIST, National Weather Service – Instituto Superior Técnico, and has been applied to the same four portuguese river basins.

In terms of the objective function used, the obtained results did not display significant differences, as compared to the ones produced by the original version. Nevertheless, they are a better fit to the daily observed base flow, they result from more realistic values of the subsurface water volume capacities, they benefit from a better physically–based description of the implicated phenomena, and they do not show anomalous results, facts which significantly increase their reliability.

A Comparison Between Lumped and Semi–Distributed Approaches to Modeling Isolated Flood Events

D. A. Hughes
Hydrological Research Unit, Rhodes University, Grahamstown, 6140, South Africa

Abstract. One topic within the field of deterministic flood modeling relatively neglected in the hydrological literature is the level of model sophistication that is needed to satisfy the requirements of different model applications. This paper reports on a study of two isolated event models, both of which can be operated in lumped or semi–distributed format.

The results of modeling 174 storm events from 16 medium sized catchments in the USA and South Africa are compared for the two versions of each model. The lumped models perform as well as the semi–distributed when the rainfall input is relatively spatially uniform. The semi–distributed models only show demonstrable improvements for storms with spatially variable rainfall if the rainfall input is adequately defined by the available data. The simpler of the two models reproduces observed hydrographs as well or better than the more complex one, but only at the expense of consistency of parameter values for a single catchment. Some of the parameter values of both models are broadly interpretable with respect to known catchment physiographic characteristics.

Further studies are in progress to investigate, inter alia, the development of more precise relationships between parameter values and characteristics of the catchments or the storm inputs in order to further assess the usefulness of the models in applied situations.

Use of Remotely Sensed Data in a Distributed Hydrologic Model

Geoff Kite
Hydrometeorological Research Division, Canada Climate Centre, National Hydrology Research Centre, 11 Innovation Blvd., Saskatoon, Saskatchewan, S7N 3H5, Canada

Abstract. A distributed hydrologic model using data from satellite as well as conventional ground-based meteorological and hydrological data is used on a microcomputer. Daily visible and near infrared data from a NOAA satellite are used, both from digital tape and from photograph. The data are combined in a database written specifically as a 'front-end' for hydrologic modeling. A hydrological model, developed by INRS-Eau, is applied to two basins in western Canada, the Kootenay and the Souris.

The study's objectives are to investigate the utility of satellite data in a hydrometeorological model in different physiographic and climatic areas, to use the model as a test-bed for different physically-based components, and to develop a land–phase component for global circulation models.

It was found that NOAA data provide a good source of distributed snow cover information for both watersheds which improves the simulation.

A Design of a Sediment Trap to Study Soil Erosion

A. Laguna and J. V. Giráldez
Depts. of Applied Physics and Agronomy, University of Córdoba, Apdo. 3048, 14080 Córdoba, Spain

Abstract. Soil erosion, a great menace for agriculture and environment, is not well known, especially in many countries where its effects are sometimes dramatic. Soil erosion may be measured by either recording soil losses or collecting sediments downslope. Most of the sediment traps are designed without any rational basis, based exclusively on previous field experience. The use of erosion models can supply a basis for design.

There are several erosion models based on fundamental formulations of the conservation laws, like the Saint Venant equations for water transport and the mass conservation equations for the sediment. The simplification of the water equations leading to the kinematic wave approach and the reduction to the rill and interrill erosion and deposition to some potential relations with rainfall rate and water depths like the Singh and Regl (1983) proposal affords a straight way to calculate sediment loads and net erosion rates.

The output of the model are the flow and sediment rates in function of time and space. The total sediment mass collected by the trap can be obtained by integration of the sediment flux density in the time domain for a fixed position.

Spatially Distributed Modeling, Conceptual Approach Coupling Surface Water and Groundwater

E. LeDoux, J. Deschenes, G. Girard, and J. P. Villeneuve
Ecole des Mines de Paris, Centre d'Informatique Géologique, 35, rue Saint-Honoré, 77305 Fontainebleau, France

Abstract. This paper discusses the joint modeling of surface and groundwater flows by presenting and describing the MC model. The purpose of this deterministic physically based model is to simulate the behavior of available water resources for one or more watersheds. The model integrates surface flow, streamflow, flow in the nonsaturated zone, groundwater flow, and the interactions between rivers and water tables. Its formulation and its structure, especially its nested square meshes of variable sizes, confer flexibility to the model; this facilitates adaptation to variable modeling scales and to a wide range of geological, geographical, and climatological conditions. An application of the MC model on the Caramy watershed (France) is presented.

Integration of Geographic Information System and Modeling for Evaluation of Pollution from Agricultural Nonpoint Sources

M. T. Lee
Illinois State Water Survey, 2204 Griffith Dr., Champaign, Illinois 61820 U.S.A.

Abstract. State–of–the–art appraisals of water quality impacts induced by agricultural nonpoint sources are difficult to make for many reasons: lack of an existing research data base, the diffusive characteristics of nonpoint source pollutants, and the technical difficulties in relating instream water quality to topographic features, soil characteristics, land use, land cover, soil moisture conditions, and management practices in the watershed.

To remedy these deficiencies, an experimental program known as the Rural Clean Water Program (RCWP) was established in the United States in 1980. Highland Silver Lake Watershed near Highland, Illinois, was one of five sites selected for comprehensive monitoring and evaluation nationwide. For this project, which began in 1981, three basic tasks were performed: field monitoring, application of a geographic information system (GIS), and hydrologic modeling. The field monitoring provided the baseline measured data for model verification. The GIS, which can describe the spatial characteristics of a hydrologic system, was used to provide inputs and to display outputs for hydrologic modeling. The Agricultural Nonpoint Source Pollution Model (AGNPS) was used. The results of this effort and suggestions for future research integrating the GIS and hydrologic modeling are presented.

Procedure for Distributed Snowpack Estimation

C. H. Leu[1], D. S. Bowles[2], R. W. Gunderson[3], and J. P. Riley[4]
[1]Department of Civil Engineering, National Chung–Hsing University, Taichung, Taiwan, Republic of China
[2]Utah Water Research Laboratory, Utah State University, Logan, Utah 84322–8200 U.S.A.
[3]Department of Electrical Engineering, Utah State University, Logan, Utah 84322–4120 U.S.A.
[4]Department of Civil and Environmental Engineering, Utah State University, Logan, Utah 84322–4110 U.S.A.

Abstract. Instantaneously distributed Landsat Thematic Mapper (TM) data and Digital Elevation Model (DEM) data are used with a Fuzzy C–Varieties (FCV) classification algorithms to identify representative prototypical points for snowpack modeling. The snowpack models at prototypical points use instantaneous point meteorological input data. Estimates of snowpack properties from the snowpack models are continuous time point estimates. The snowpack accumulation and ablation simulation models use a Kalman filtering technique for updating the continuous estimates of snowpack state variables (snowpack water equivalent, snowpack depth, one–third depth snowpack temperature, two–thirds depth snowpack temperature, surface albedo, and free water content) with instantaneous point snowpack variable measurements. Interpolation formulae based on the FCV classification algorithms are then used to obtain spatially distributed snowpack variable estimates which are also continuous in time. The spatially distributed snowpack variable estimates could then be input to a hydrology model in order to obtain spatially integrated estimates of streamflow and also temporarily integrated estimates of runoff volumes.

The results of a preliminary evaluation of the spatially distributed snowpack estimation procedure are described. The evaluation is based on an application to the Dry Lake Watershed in northern Utah.

Application of an Integrated Finite Difference Model to Land Use Problems

Lee H. MacDonald
Department of Earth Resources, Colorado State University, Fort Collins, Colorado 80523 U.S.A.

Abstract. Hydrologic concerns are central to several land use problems in the United States. An integrated finite difference groundwater model (Trust) is being applied on a trial basis to two quite different issues—wetland delineation and snowmelt runoff.

The Trust model combines the integral formulation of the basic saturated and unsaturated groundwater flow equation with the finite difference method to estimate hydraulic gradients (Narasimhan et al., 1986). Required input data includes: (1) establishment of a geometric mesh which reflects differences in materials; (2) specification of K_s, $K(\Psi)$, and the moisture retention curve $S(\Psi)$; (3) explicit source and sink terms to account for precipitation, evapotranspiration, and regional groundwater flow; and (4) initial conditions for each node. As a physically–based model there are no empirical parameters to be estimated by a calibration procedure. Soil moisture retention curves were determined from minimally–disturbed soil cores, while the saturated and unsaturated hydraulic conductivities were either measured in the field or calculated using empirical formulas.

Two potential wetland areas in California were evaluated by a series of transient and steady–state simulations. The purpose of the simulations was to determine whether the federally–mandated hydrologic criteria for wetlands would be satisfied in an average year. One–dimensional steady–state simulations were used to predict soil moisture for each node in the unsaturated zone under typical mid–winter conditions. These provided the initial conditions for transient simulations where specified amounts of precipitation were applied, and the probable depth and duration of saturated conditions was determined for each site.

Limited field data indicates that the steady–state simulation was able to accurately predict soil moisture content in the unsaturated zone. This suggests that the predictions regarding the development of saturated conditions from storms of varying magnitude and intensity would also be accurate, but there was no field data to verify the transient simulations. The results were relatively insensitive to the estimated rate of evapotranspiration, the depth of the rooting zone, and the distance between nodes. The dominant factor in determining the likelihood and extent of saturation was the pre–storm height of the water table.

The second application of the Trust model has been to assess the hydrologic effects of different forest management practices on the volume and timing of snowmelt runoff. This is an increasingly important water supply question in much of the western United States and Canada. The two–dimensional mesh used for these simulations represents a 130 m long and 30 m thick hillslope segment containing six distinct layers. The use of Trust for this problem has been hampered by the need to expand the capacity of the model and uncertainty over the boundary conditions. Results to date suggest that small changes in parameters which are typically ignored—such as the compressibility of the soil—can have a significant effect on the rate and timing of outflow.

In conclusion, simulations on several sites indicate that physically–based models such as Trust can facilitate land use management decisions by: (1) providing quantitative results in situations where interacting nonlinearities preclude simpler solutions; (2) extrapolating results from known situations to other sites or different conditions; (3) helping identify the most important physical parameters in the system under study; and (4) facilitating an understanding of the system being modeled. One limitation to the use of a physically–based model is the amount of input data required, and this often restricts their use to research or particularly contentious management issues.

The extent to which the estimated soil and rock properties reflect physical reality will limit the accuracy of the simulations. Macropores, entrapped air, spatial variability, and soil compressibility are some of the factors which are difficult to incorporate into a physical model. In some cases creating a more complex model (e.g., more nodes and materials) can hinder understanding by confounding the factors which cause a particular result. An evaluation of the relative importance of the physical properties and boundary conditions is essential to understanding the limitations of a physically–based model.

Storm Runoff Investigations Related to Effects of Urbanization in Urban Watersheds

U. S. Panu

Civil Engineering Dept., Lakehead University, Thunder Bay, Ontario, P7B 5EI, Canada

Abstract. A study of urban runoff in the St. John's area was conducted as part of comprehensive investigations of the effects of urbanization on water resources in the Waterford River Basin, Newfoundland. The main objective of this study, which included both field and analytical investigations, was to develop principles of hydrological design of urban drainage in Newfoundland. In field investigations, rainfall, runoff and its composition were monitored in a 13 ha residential subdivision in the St. John's area. The collected data were used to evaluate three runoff computational procedures – the rational method, the Illinois Urban Drainage Simulator (ILLUDAS), and the Storm Water Management Model (SWMM). Each of these procedures produced good results within their operational limits. The rational method somewhat underestimated the observed or extrapolated runoff peak flows. Satisfactory results were obtained by applying this method in conjunction with locally–derived composite runoff coefficients and inlet times, and a partly–contributing catchment area. Both runoff hydrograph models, the ILLUDAS and SWMM, reproduced the observed runoff hydrographs fairly well. The modeling results were particularly good for intermediate storms during which practically all runoff was generated on impervious elements. Only for a few very severe storms was it necessary to improve the modeling results by accounting for antecedent moisture conditions. Observed characteristics of runoff quality indicated pollution arising from urban activities. Such indications were particularly strong for zinc and lead whose concentrations exceeded the safe guidelines for aquatic life. Recommendations for addressing runoff problems in future urban developments were given. These included various runoff management strategies and technical guidelines for drainage design.

Application of the MC–Model (Girard & Al. 1981) to the Hydrologic Modeling of the Fecht Catchment, Vosges (France)

Jean–Louis Perrin, Bruno Ambroise, and **Joel Humbert**

Centre d'Etudes et de Recherches Eco–Géographiques, Universite Louis Pasteur, 3 rue de l'Argonne, F–67083 Strasbourg Cedex, France

Abstract. The MC–Model (Girard, LeDoux, Villeneuve: 1981) is a spatially distributed and conceptual hydrologic model. The basin is discretized with nested square meshes. Precipitation and potential evapotranspiration values, drainage network, and physiographic informations must be defined for each mesh.

The computation is accomplished for each mesh by a four storages' model. The first one simulates the water budget and the others, the water transfer. Usually the ten parameters which control the simulation are calibrated.

The Fecht study catchment (211 km2, 279–1363 m.a.s.l.) is an essentially granitic and grauwackous basin with coarse–textured soils. The climate is a transitional one between oceanic to continental with big climatological gradients. We disposed of a dense network of climatological measurements. The streamflow are measured on the outlets of three nested basins. A map of hydrodynamic units have been realized from 500 samples which have been taken in the whole basin.

Three meshes sizes have been used for the discretization of the Fecht catchment: 1000, 500 and 250 m side respectively. The knowledge of the inter units variabilities of the soil hydric properties permits to associate the two parameters which control the budget storage with physically–based measurements: soil water retention.

This study is the first test to simulate outlets flows and water balance from climatological daily data for two civil years. In spite of some climatological problems (precipitation and evapotranspiration gradients still not well known), the hydrological balance and the comparison between observed and calculated streamflows are good. These results confirm the validity of the catchment discretization into homogeneous units and the validity of the hypothesis which have allowed us to associate parameters with soil water retention measurements.

Hydrologic Model of Water Conservation Structures in SE Spain

José Roldán[1], José L. Ayuso[2], Juan V. Giráldez[1], and José G. López[3]
[1]University of Córdoba, Department of Agronomy, ETSIA, Apdo 3048, 14080 Córdoba, Spain
[2]University of Córdoba, Department of Rural Engineering, ETSIA, Apdo 3048, 14080 Córdoba, Spain
[3]IARA, Delegation of Almeria, Hermanos Machado s/n, 04071 Almería, Spain

Abstract. Over the centuries farmers have developed several structures for water conservation in southeastern Spain, a region whose scarce rainfall (around 200 mm during the last forty years) and the irregularity with which it happens makes agricultural practice extremely hard. However, a large area devoted to dry farming, takes profit of the scarce water resources through harvesting techniques originated at least since the bronze civilization.

The main water harvesting systems, that still remain, use an earth dyke to divert the flood water to terraces. This structure is locally known as "boquera." In every terrace, incoming floodwater joins the upper terrace excess water.

A simple model is proposed for evaluating the water harvesting system. The inclusion of an infiltration function in a series of linear storage reservoirs is interesting because of the linkage of the terraces of the describing harvesting system. The different components (storage, precipitation, runoff and infiltration) which made the model cell (a terrace) can be related by a water balance where evaporation is not considered. Storage is assumed to be linearly related to runoff and infiltration is expressed as the Horton function in this model.

The computed hydrographs allow an estimation of the opportunity time of infiltration in each terrace. Results indicate that the larger the event duration the larger the opportunity time is. On the contrary, the magnitude of the incoming flow rate into the terrace does not suppose any appreciable difference in the accumulated infiltration in the soil.

The implication of the results is that the most important purpose of the system could be the trap of the sediment to yield a good soil. Farmers are more cautious about flood damage than eager to get a larger water volume.

Piave River Daily Discharge Model

Giancarlo Rossi
ENEL – CRIS, Servizio Idrologico, Corso del Popolo 245, 30172 Venezia Mestre, Italy

Abstract. The "Vallata Model", which actually is going to be operating by the Hydrological Service of the Italian National Electricity Board (ENEL), provides a runoff forecasting system in order to optimize the hydropower management policy of the Piave River Upper Basin, which has the following features:

River	Piave
Reservoir	Comelico
Gauging station	Ponte della Lasta
Basin surface	357 km2
min. elevation	855 m s.l.m.m.
max. elevation	3092 m s.l.m.m.

The Vallata Model includes: a) a meteorological data gauging network with data transmission, b) a short term runoff forecasting, c) a daily scale runoff forecasting, and d) a daily scale snow cover hydro producibility assessment.

The two last operations are performed by the "Piave River Daily Discharge Model" which is composed by a snowcover evolution model based on the degree–day joining a stochastic model of "lumped integral" type for the runoff simulation (the well known C.L.S.); input and output data are managed by the following data bases:

a) meteorological data input,

b) physiographic and parameter data input, and

c) hydrological data input/output.

The meteo data base is updated with the data gathered by the network and, if available, with forecasting produced by meteorological models; the parameter data base is updated when necessary by calibration, and hydrological one is updated both with direct measurements, and with remote sensing information.

The model simulation of a winter period starts at the time of the last available snowcover survey and continues step by step until the date of the next survey; since the thaw period the simulation continues until the next winter season.

The exposed results concern a routing over the three year period ranging from 1985 to 1987. The comparison between computed and measured snowdepths shows the good performance of the model, mainly during thaw periods, and likewise computed and measured runoff show a good agreement, whose evaluation may be quantified by the explained variance of the series, which assumes the value 0.85.

Background to, Structure and Applications of 'ACRU' – A Multipurpose Multilevel Distributed Agrohydrological Modeling System

Roland E. Schulze
Department of Agricultural Engineering, University of Natal, Pietermaritzburg, 3200, Republic of South Africa

Abstract. A wide range of agrohydrological decisions have frequently to be made under widely differing 1st and 3rd world conditions of expertise, data availability and adequacy of model input information, e.g. decisions on catchment water yield, sizing of dams and their rates of siltation, influences of proliferation of small dams on downstream water resources, impacts of land use practices such as afforestation, irrigation water demand or the estimation of crop yield. The ACRU modeling system has been developed and provides multiple decision–making options for varying levels of data availability for a range of typical agrohydrological problems.

ACRU is a multi–layer soil moisture budgeting model structured to be highly sensitive to land use changes. It uses readily available daily climatic input data with certain more cyclic variables input as typical monthly values and reduced internally by Fourier analysis to daily values. Driven by a user-friendly, interactive input menu, many routines have alternative hierarchical (usually YES–NO) pathways with default options depending on the level of sophistication of available input, e.g. on potential evaporation or interception loss estimates, soil water retention characteristics, peak discharge calculations, reservoir storage: area relationships, etc. In a data time series the model handles changes in land cover and management (be they abrupt e.g. clearfelling of a forest, construction of a reservoir; or gradual e.g. growth of a forest or an urban area in time) by way of dynamic information files. ACRU furthermore has options for simulations at a point location or as a lumped or distributed catchment, where (in the last case) a "menubuilder" facilitates coping in simple, practical steps with the problem of inputting complex, time consuming soil land cover information in distributed mode.

Case studies are presented by the model's application, e.g. to the problem of the impact of upstream afforestation on inflows to a proposed irrigation project, impact of upstream reservoir storage on the hydrology of a wetland and the implications of deficit irrigation on the crop yields and profitability of an irrigation scheme.

Use of Synthetic Flow Series in Long Range Planning of Multiple Reservoir Systems

Mahmut Sert
Tübítak – Marmara Research Institute, Operations Research Division, Gebze, Kocaeli, Turkey

Abstract. In the long range planning of multi–reservoir and multiple purpose systems, optimization and simulation techniques may be combined to obtain solutions which account for the stochastic nature of the hydrological events and produce optimal benefits under certain risks. For a chosen set of dimensions of the system (dam heights, tunnel diameters, installed powers), optimal operating policies are determined by dynamic programming. In this way, firm power is maximized using observed dry period flow data, giving the "minimum storage levels" for the reservoirs, and, the total energy produced by the system is maximized on the basis of monthly mean flows, giving the "optimal storage levels" for the reservoirs in the system. Then, these policies are used in a simulation model together with the synthetic flow series to determine the risks on the firm power and the total energy produced over the economic life of the system. The same procedure can be repeated for different sets of dimensions of the system, and the dimensions and corresponding operation policies producing maximum net benefits can be chosen for application. (Here, the objective function for optimization is based on the energy benefit of the system and the water releases from each reservoir for purposes like irrigation, navigation, and pollution control are taken as constraints).

Synthetic flow series used in the simulation model are obtained by modeling the available flow data on the basis of the Thomas–Fiering model (1,2,3) which may be considered as a kind of Markov chain (a first order, linear, auto–regressive model) with parameters adjusted for the yearly period of the hydrological events. Since the probability distribution of the monthly flows is closer to a log–normal distribution than a normal distribution, the model is based on the logarithms of the monthly flows. Usually the synthetic series are generated at a number of different hydrological gauging stations and cross-correlation amongst these must be provided.

In an application for a system of five reservoirs (4), simulation model outputs for long range planning provide percentage firm power deficit and percentage failure of firm power realization as well as the total average power realized for each parametrical value of the firm power within each 50 year period of operation. It has been found that the frequency distribution of these quantities for 15x50 years period were near to normal distribution, thus 95% confidence interval has been taken as 2 x "standard deviation." The variations of percentage firm power deficit and percentage failure of firm power realization display increasing trends with the increasing parametrical values of firm power while the variation of the total average power with firm power is not very significant (within a range where firm power does not go very much beyond the average power).

A New Stochastic Rainfall Field Model

S. M. S. Shah and **P. E. O'Connell**
University of Newcastle upon Tyne, Department of Civil Engineering, Claremont Road, Newcastle upon Tyne, NE1 7RU, United Kingdom

Abstract. The relationship between the spatial variability of rainfall and the behavior of hydrological systems is central to many forecasting and prediction problems in hydrology. Some efforts have been made to develop moving–storm stochastic rainfall field models to allow this relationship to be explored using distributed rainfall–runoff models but with only limited success, partly because the models have not been applied to specific catchments. The aim of this research has been to develop a stochastic rainfall field model which can be calibrated with data from a limited number of recording raingauges, as a prelude to exploring the relationship between the spatial variability of rainfall and catchment response using a distributed model.

The stochastic rainfall field model uses as a starting point the fractionally differenced ARIMA (1,d,0) time series model to generate a discrete series of values at equidistant points along a line in space; the Turning Bands Method (TBM) is then employed to transform the Gaussian line process into a multidimensional spatial Gaussian process with a specified space–time correlation structure. As short duration (e.g. hourly rainfall data) exhibit non–Gaussian and non–stationary behavior, a non–stationary structure has been given to the generated fields by applying a scaling function to transform the multidimensional process to have the required non–Gaussian and non–stationary structure. Two different scaling functions have been proposed for this purpose. A factor has also been included to take account of the effect of topography on rainfall.

The model has been applied to hourly rainfall data from three raingauges located within the upland Wye catchment in mid Wales (area 10.55km^2) which has a predominantly frontal rainfall regime. Two schemes have been used to generate storms over the basin:

(i) an unconditional scheme where the generated storms reproduce various statistics of the historic storms (mean, variance, serial and cross–correlation structure); and

(ii) a conditional scheme in which the historical storm profiles are reproduced exactly at the measured points, and the generated values at all other points are conditioned on the measured values.

The model is to be extended to cope with rainfall regimes dominated by convective activity, and effort will also be devoted to improving the current empirical approach to model fitting.

MIKE II – A Micro Computer Based Modeling System for Rivers and Channels

Børge Storm
Danish Hydraulic Institute, Agern Allé 5, Dk–2970, Hørsholm, Denmark

Abstract. The use of numerical hydraulic models for dynamic analysis of open channel flow is a widely accepted and trusted technique. Existing models have generally been developed to be theoretically sound, but are often not polished to the extent where it offers a user–friendly tool for design, management, and operation of river basins and channel networks.

The demonstration describes the application of a commercial software package, MIKE II, for modeling of flows, water levels, and transport in any river or channel system.

MIKE II is suitable for a wide range of application comprising, e.g., flood forecasting and reservoir operation, and simulation of flood control measures, sedimentation studies, water quality and environmental and ecological studies in rivers, reservoirs, and estuaries. The modeling system includes a flexible interactive menu system for data handling, execution and graphic presentations.

KARMEN – System for Hydrological Simulation and Forecasting

Svein Taksdal

Hydrological Division, Norwegian Water Resources and Energy Administration, P.O. Box 5091, Majorstua, N–0301 OSLO 3, Norway

Abstract. Hydrological models have been used operationally in Norway since 1976, when the first forecasting model was implemented in a short term planning system for the Sira–Kvina powerplant. At present, forecasting models are used in approximately 20 catchments, operated by major power plants, river regulation boards and the Norwegian Water Resources and Energy Administration.

The model system was developed at the Hydrological division, NVE. The dynamical part was mainly based on the Swedish HBV model, whose main components are: a non–linear soil moisture submodel and a piecewise linear storage system. This part of the model has three state variables and eight model parameters, plus optional routing parameters.

The snow accumulation/melt submodel is more elaborated. Distributed snow pack and liquid water content, snow covered area, and short wave radiation reflectivity are state variables and are simulated in up to 10 altitude intervals. Precipitation measurement losses may be compensated for by wind observations.

Snow melt is mainly calculated by temperature index, but radiation, convection, and condensation components may be simulated separately from temperature and precipitation observations.

The model system is programmed in standard Fortran 77, with GPGS graphics, and is available in mainframe and MS-DOS versions.

Through a joint effort by the Norwegian Hydrodynamical Laboratory, the Electricity Research Institute, and the Norwegian Water Resources and Energy Administration, the model system is now to be updated and interfaced to other simulation systems and standard data systems used by the Norwegian hydropower industry. The main objective is to facilitate implementation and promote use of hydrological forecasting techniques. This project is supported and funded by the Water System Management Association of Norway.

Water Level Management on Navigable Waterways

R. Verhoeven[1], P. Verdonck[1], and J. Heynderickx[2]
[1]Hydraulics Laboratory, State University of Ghent, Sint – Pietersnieuwstraat 41, B-9000 Ghent, Belgium
[2]Ministry of Public Works, Belgium

Abstract. The realization of a flood control program on a complete waterway network requires a permanent knowledge of the discharge. Because the installation of a network for direct field survey of the discharge is very expensive, the discharge is mostly deduced from water–level gauging.

The discharge variation as a function of the time can be calculated from the stage variation in the few cases where a particular relationship between flowrate, Q, and water surface elevation, h, exists. Therefore the existence of a fixed check point is required, which is difficult to realize on navigable rivers.

If Q and h are not directly related, the problem can be solved by defining the discharge as a function of two different water–level recordings along a river reach [$Q = f(h_1, h_2)$].

Determinating a rating curve $Q = f(h_1, h_2)$ from field data is a complex and time–consuming procedure, since too much data are required. To meet this difficulty, the program "DISCHARGE" has been developed which is especially interesting in the following aspects: 1) it allows the river discharge to be calculated from the geometrical characteristics, the friction parameters and the upstream, and downstream water level; and 2) it creates the possibility to determine the friction, starting from a known discharge. All calculations performed are based on the water surface profiles theory.

Starting from a restricted number of measurements (Q, h_1, h_2), the program first determines the friction characteristics of the river reach and then creates new couples (Q, h_1, h_2). The latter combined with in situ measurements are used in a curve fitting procedure which determines the parameters of a rating curve.

The method can be applied to each river reach. With a minimum of data, a reliable rating curve is quickly delivered. Once known, the formula can be used for various purposes as it is especially suited to calculations with minicomputers and even small microprocessors.

Once the rating curves for a number of river reaches are known and stored in a small computer, a centralized discharge control of a complete waterway network can be brought about by connecting the different limnimeters to a teletransmission system.

Dangerous situations or sudden changes in the discharge can immediately be registered by the system so that action can be taken within a very limited space of time.

The method presented above, has been developed in cooperation with the hydrological Department of Belgian Ministry of Public Works. In the near future, rating curves for different important navigable rivers will be ready for practical application.

Estimation of Baseflow Recession Constants

Richard M. Vogel and **Charles N. Kroll**
Department of Civil Engineering, Tufts University, Medford, Massachusetts 02155 U.S.A.

Abstract. The recession portion of a streamflow hydrograph is shown to be an autoregressive process, where the autoregressive parameters are nonlinear functions of the hydrograph recession constants. Estimates of the baseflow recession constant are obtained using thousands of hydrograph recessions available at 23 sites in Massachusetts. Procedures are developed for estimating a baseflow recession constant for a watershed, using a large ensemble of individual hydrograph recessions. These estimates are compared with estimates derived from traditional baseflow separation procedures. Evidence of the physical and statistical significance of estimates of the baseflow recession constant is provided by the development of regional regression models which relate low–flow statistics to three independent variables: drainage area, mean basin elevation, and the baseflow recession constant. As anticipated, when approximately unbiased estimates of the baseflow recession constant are obtained, those estimates explain a significant portion of the regional variability of low flows.

Recent Mesoscale Experiments Concerning Watershed Evaporation

Wilfried Brutsaert
Cornell University, School of Civil and Environmental Engineering, Hollister Hall, Ithaca, New York 14853–3501 U.S.A.

Boundary Elements in Ground Water Flow Modeling

J. G. Ganoulis
Division of Hydraulics & Environmental Engineering, Aristotle University of Thessaloniki, 54006 Thessaloniki, Greece

A Dynamic Model of Gully Growth

Anne C. Marchington
Department of Geography, University of Bristol, University Road, Bristol B58 155, United Kingdom

On the Role of the Groundwater Component in Bulk Water Balances of Seepage Lakes

H. O. Pfannkuch
University of Minnesota, Department of Geology and Geophysics, 108, Pillsburg Hall, Minneapolis, Minnesota 55455 U.S.A.

Portugal. Overall Analysis of its Hydrological Characteristics. State-of-the-Art of Hydrological System Modeling

A. C. Quintela
Civil Engineering Department, Instituto Superior Técnico (I.S.T.), Av. Rovisco Pais, 1096 Lisboa Codex, Portugal

Simulation of the Model SHELL

Rui Rodrigues
Laboratorio Nacional de Engenharia Civil, DH–NHHF, Av. do Brasil, 101, 1799 Lisboa Codex, Portugal

The ARNO Real Time Flood Forecasting Model

Ezio Todini
E. Todini & Partners, S. r. 1., Via Zanardi, 16, 40131 Bologna, Italy

Index